90 0714833 4

ADVANCED DAIRY CHEMISTRY

ADVANCED DAIRY CHEMISTRY—1 PROTEINS

3rd Edition

PART B

Edited by

P. F. FOX and **P. L. H. McSWEENEY**

University College
Cork, Ireland

KLUWER ACADEMIC / PLENUM PUBLISHERS
New York/Boston/Dordrecht/London/Moscow

ISBN 0-306-47271-6

233 Spring Street, New York, New York 10013

http://www.wkap.com

10 9 8 7 6 5 4 3 2 1

A C.I.P. record for this book is available from the Library of Congress

Printed in the United States of America

CONTRIBUTORS

Francesco Addeo
Dipartimento di Scienza degli Alimenti, Facolta di Agraria, Università degli Studi di Napoli Federico II, Portici, Italy
Istituto di Scienze dell'Alimentazione del C.N.R. Via Roma 52 A/C, 83100 Avellino, Italy

Keith Brew
Department of Biomedical Sciences, Florida Atlantic University, 777 Glades Road, PO Box 3091, Boca Raton, Florida 33431-0991, USA

Eleanor M. Brown
Eastern Regional Research Center, US Department of Agriculture, Agricultural Research Service, Wyndmoor, Pennsylvania 19038, USA

A. J. Carr
Fonterra Research Centre, Private Bag 11 029, Palmerston North, New Zealand

L. K. Creamer
Fonterra Research Centre, Private Bag 11 029, Palmerston North, New Zealand

A. Corsetti
Dipartimento di Scienze degli Alimenti, Universita degli Studi di Perugia, via S. Costanzo, 06126 Perugia, Italy

C. G. de Kruif
NIZO Food Research, PO Box 20, 6710BA Ede, The Netherlands

Eric Dickinson
Procter Department of Food Science, University of Leeds, Leeds LS2 9JT, UK

Didier Dupont
INRA-SRTAL, Department of Immunochemistry, BP 89, 39801 Poligny Cedex, France

M. P. Ennis
Department of Food Science, Food Technology and Nutrition, University College, Cork, Ireland

Nana Y. Farkye
Dairy Products Technology Center, Dairy Science Department, California Polytechnic State University, San Luis Obispo, California 93407, USA

Harold M. Farrell, Jr.
Eastern Regional Research Center, US Department of Agriculture, Agricultural Research Service, Wyndmoor, Pennsylvania 19038, USA

Pasquale Ferranti
Dipartimento di Scienza degli Alimenti, Facolta di Agraria, Università degli Studi di Napoli Federico II, Portici, Italy
Istituto di Scienze dell'Alimentazione del C.N.R. Via Roma 52 A/C, 83100 Avellino, Italy

C. M. Fleming
Department of Food Science, Food Technology and Nutrition, University College, Cork, Ireland

R. J. FitzGerald
Life Sciences Department, University of Limerick, Castletroy, Limerick, Ireland

P. F. Fox
Department of Food Science, Food Technology and Nutrition, University College, Cork, Ireland

M. Gobbetti
Dipartimento di Protezione delle Piante e Microbiologia Applicata, Universita degli Studi di Bari, Via G. Amendola 165/a, 70126 Bari, Italy

H. Douglas Goff
Department of Food Science, University of Guelph, Guelph, Ontario N1G 2W1, Canada

F. Grosclaude
Laboratoire de Génétique Biochimique et de Cytogénétique, INRA, 78352 Jouy-en-Josas Cedex, France

T. P. Guinee
Dairy Products Research Centre, Teagasc, Moorepark, Fermoy, Co. Cork, Ireland

Leif Hambræus
Karolinska Institute, Department of Bioscience, Unit for Preventive Nutrition, SE 141 57, Huddinge, Sweden

Palatasa Havea
Institute of Food, Nutrition and Human Health, Massey University, Private Bag 11 222, Palmerston North, New Zealand

Peter D. Hoagland
Eastern Regional Research Center, US Department of Agriculture, Agricultural Research Service, Wyndmoor, Pennsylvania 19038, USA

C. Holt
Hannah Research Institute, Ayr, KA6 5HL, Scotland, UK

D. S. Horne
Charis Food Research, Hannah Research Park, Ayr, KA6 5HL, Scotland, UK

Walter L. Hurley
Department of Animal Sciences, University of Illinois, 1207 W. Gregory Dr., Urbana, IL 61801

D. B. Hyslop
Department of Food Science, University of Wisconsin-Madison, 1605 Linden Drive, Madison, WI 53706

A. L. Kelly
Department of Food Science, Food Technology and Nutrition, University College, Cork, Ireland

Shuichi Kaminogawa
Department of Applied Biological Chemistry, The University of Tokyo, 1-1-1 Yayoi, Bunkyo-ku, Tokyo 113-8657, Japan

Marie-France Laporte
Dairy Research Centre STELA, Université Laval, Pavillon Paul Comtois, room 0308, Ste-Foy, Québec, Canada G1K 7P4

A. J. R. Law
Charis Food Research, Hannah Research Park, Ayr, KA6 5HL, Scotland, UK

J. Leaver
Hannah Research Institute, Ayr, KA6 5HL, Scotland, UK

Joëlle Léonil
Institut National de la Recherche Agronomique, Laboratoire de Recherches en Technologie Laitière, 65, rue de Saint-Brieuc, 35042 Rennes Cedex, France

Christine Leroux
Laboratoire de Genetique biochimique et de Cytogenetique, Institut National de la Recherche Agronomique, Domaine de Vilvert, 78352 Jouy-en-Josas Cedex, France

Bo Lonnerdal
Department of Nutrition, University of California, Davis, California 95616, USA

John A. Lucey
Department of Food Science, University of Wisconsin-Madison, 1605 Linden Drive, Madison, Wisconsin 53706, USA

Edyth L. Malin
Eastern Regional Research Center, US Department of Agriculture, Agricultural Research Service, Wyndmoor, Pennsylvania 19038, USA

Patrice Martin
Laboratoire de Genetique Biochimique et de Cytogenetique, Institut National de la Recherche Agronomique, Domaine de Vilvert, 78352 Jouy-en-Josas Cedex, France

P. L. H. McSweeney
Department of Food Science, Food Technology and Nutrition, University College, Cork, Ireland

H. Meisel
Bundesanstalt fur Milchforchung, Insitut fur Chemie und Technologie der Milch, PO Box 6069, D-24121, Kiel, Germany

D. M. Mulvihill
Department of Food Science, Food Technology and Nutrition, University College, Cork, Ireland

K. F. Ng-Kwai-Hang

Faculty of Agricultural and Environmental Sciences, Department of Animal Science, McGill University, MacDonald Campus, 21,111 Lakeshore, Ste-Anne-de-Bellevue, Québec, Canada H9X 3V9

Johannes A. Nieuwenhuijse
Friesland Consumer Products Europe, Research and Development Leeuwarden, PO Box 226, 8901 MA Leeuwarden, Netherlands

J. E. O'Connell
Department of Food Science, Food Technology and Nutrition, University College, Cork, Ireland

Gunilla Olivecrona
Department of Medical Biosciences, University of Umeå, S-90187 Umeå, Sweden

Thomas Olivecrona
Department of Medical Biosciences, University of Umeå, S-90187 Umeå, Sweden

M. Ollivier-Bonsquet
Laboratoire de Biologie Cellulaire et Moleculaire, INRA, 78352 Jouy-en-Josas Cedex, France

Paul Paquin
Dairy Research Centre STELA, Université Laval, Pavillon Paul Comtois, room 1401, Ste-Foy, Québec, Canada G1K 7P4

Kenneth M. Pruitt
University of Alabama at Birmingham, 701 S. 19th St., Room 134, Lyons-Harrison Research Building, Birmingham, Alabama 35294-0007

Shakeel-Ur-Rehman
Dairy Products Technology Center, Dairy Science Department, California Polytechnic State University, San Luis Obispo, California 93407

Lindsay Sawyer
Structural Biochemistry Group, The University of Edinburgh, Swann Building, King's Buildings, Mayfield Road, Edinburgh EH9 3JR, Scotland, UK

D. B. Shennan
Hannah Research Institute, Ayr, KA6 5HL, Scotland

Harjinder Singh
Institute of Food, Nutrition and Human Health, Massey University, Private Bag 11 222, Palmerston North, New Zealand

C. R. Southward
Fonterra Research Centre, Private Bag 11 029, Palmerston North, New Zealand

L. Stepaniak
Department of Food Science, Agricultural University, N-1430 Aas, Norway

Harold E. Swaisgood
Department of Food Science, Southeast Dairy Foods Research Center, North Carolina State University, Raleigh, North Carolina 27695-7624, USA

Mamoru Totsuka
Department of Applied Biological Chemistry, The University of Tokyo, 1-1-1 Yayoi, Bunkyo-ku, Tokyo 113-8657, Japan

Louise Tremblay
Dairy Research Centre STELA, Université Laval, Pavillon Paul Comtois, room 1316, Ste-Foy, Québec, Canada G1K 7P4

Martinus A. J. S. van Boekel
Department of Food Science, Wageningen Agricultural University, PO Box 8129, 6700 EV, Wageningen, Netherlands

Senen Vilaro
Department de Biologia Cellular, University of Barcelona, 080288 Barcelona, Spain

J. L. Vilotte
Laboratoire de Genetique Biochimique et de Cytogenetique and Laboratoire de Biologie Cellulaire et Moleculaire, INRA, 78352 Jouy-en-Josas Cedex, France

C. B. A. Whitelaw
Division of Molecular Biology, Roslin Institute (Edinburgh), Roslin, Midlothian EH25 9PS, UK

PREFACE TO THE THIRD EDITION

Advanced Dairy Chemistry—1: Proteins is the first volume of the third edition of the series on advanced topics in Dairy Chemistry, which started in 1982 with the publication of *Developments in Dairy Chemistry*. This series of volumes is intended to be a coordinated and authoritative treatise on Dairy Chemistry. In the decade since the second edition of this volume was published (1992), there have been considerable advances in the study of milk proteins, which are reflected in changes to this book.

All topics included in the second edition are retained in the current edition, which has been updated and considerably expanded from 18 to 29 chapters. Owing to its size, the book is divided into two parts; Part A (Chapters 1–11) describes the more basic aspects of milk proteins while Part B (Chapters 12–29) reviews the more applied aspects. Chapter 1, a new chapter, presents an overview of the milk protein system, especially from an historical viewpoint. Chapters 2–5, 7–9, 15, and 16 are revisions of chapters in the second edition and cover analytical aspects, chemical and physiochemical properties, biosynthesis and genetic polymorphism of the principal milk proteins. Non-bovine caseins are reviewed in Chapter 6. Biological properties of milk proteins, which were covered in three chapters in the second edition, are now expanded to five chapters; a separate chapter, Chapter 10, is devoted to lactoferrin and Chapter 11, on indigenous enzymes in milk, has been restructured and expanded. Nutritional aspects, allergenicity of milk proteins, and bioactive peptides are discussed in Chapters 12, 13, and 14, respectively. Because of significant developments in the area in the last decade, Chapter 17 on genetic engineering of milk proteins has been included. Various aspects of the stability of milk proteins are covered in Chapter 18 (enzymatic coagulation), Chapter 19 (heat-induced coagulation), Chapter 20 (age gelation of sterilized milk), Chapter 21 (ethanol stability), and Chapter 22 (acid coagulation, a new chapter).

The book includes four chapters on the scientific aspects of protein-rich dairy products (milk powders, Chapter 23; ice cream, Chapter 24; cheese, Chapter 25; functional milk proteins, Chapter 26) and three chapters on technologically important properties of milk proteins (surface properties, Chapter 27; thermal denaturation aggregation, Chapter 28; hydration and viscosity, Chapter 29).

Like its predecessors, this book is intended for academics, researchers at universities and industry, and senior students; each chapter is referenced extensively.

We wish to thank sincerely the 60 contributors to the 29 chapters of this volume, whose cooperation made our task as editors a pleasure. The generous assistance of Ms. Anne Cahalane is gratefully acknowledged.

P. F. Fox
P. L. H. McSweeney
University College
Cork, Ireland

CONTENTS

12

NUTRITIONAL ASPECTS OF MILK PROTEINS

L. Hambræus and B. Lönnerdal

12.1 Introduction

The composition of milk obtained from various species shows great variations (Table 12.1). One of the major characteristics of the nutritional situation during the neonatal period in mammals is the fact that the offspring has to rely upon milk as their single source of nutrients. Thus, it is logical to assume, on a teleological basis, that the composition of the milk of different species represents the optimum composition of nutrients required during the neonatal period of that species. Consequently, it has rightly or wrongly been assumed that the composition of the milk of any mammalian species is the best indicator of the nutrient requirements of the offspring during the neonatal period.

It has been estimated that out of more than 4000 species of mammals, less than 200 types of milk have been analysed with respect to their content of macro-nutrients, and data from only 50 species are considered to be reliable (Oftedal, 1980; Widdowson, 1984). Several authors have anticipated that there is a relationship between the rate of growth of the offspring and nutrient density, i.e., the amount of essential nutrients in relation to its energy content, e.g., g protein per 10 MJ or 1000 kcal (Jenness, 1974; Blanc, 1981). Björnhag *et al.* (1979) have stressed the need to correct the relative growth rate in relation to the influence of size on growth rate. There also seems to exist some similarities in the composition of milk within the same zoological group of animals. Milks of rodents and carnivores show a high protein content and moderately large amounts of fat and carbohydrate. Those species with long intervals between feeding their offspring have milk rich in fat and energy. The milks of ruminants have relatively low levels of

Advanced Dairy Chemistry Volume 1: Proteins, 3rd edn.
Edited by P.F. Fox and P.L.H. McSweeney, Kluwer Academic/Plenum Publishers, 2003.

TABLE 12.1
Composition of milks obtained from different mammals and its relation to growth rate of the offspring

Species	*Concentration in milk (g per 100 ml)*				*Growth rate*
	Protein	*Fat*	*Lactose*	*kcal*	*(days to double birthweight)*
Primates					
Human	0.9	3.8	7.0	66	120–180
Ruminants					
Cow	3.4	3.7	4.8	66	30
Goat	2.9	4.5	4.1	69	6
Camel	3.7	4.2	4.1	69	
Reindeer	10.3	16.9	2.8	205	10
Rodents					
Rat	8.1	8.8	3.8	127	2
Perissodactyla					
Horse	1.9	1.3	6.9	47	13
Carnivores					
Cat	10.6	10.8	3.7	154	4
Marine carnivores					
Seal	10.2	49.4	0.1	486	5
Lagomorpha					
Rabbit	10.3	15.2	1.8	185	4
Cetacea					
Blue whale	10.9	42.3	1.3	430	10
Marsupials					
Kangaroo	4–10	1–16	5–0	45–184	

From Jenness (1974), Hambræus (1976), Björnhag *et al.* (1979), Oftedal (1980) and Widdowson (1984).

protein and fat and a high carbohydrate content. The length of the lactation period and the time allotted for the meal may also have an impact. It is said that dolphin babies must reach the surface every 30 s and consequently can suckle for only a few seconds. Thus, the milks of the dolphin and the blue whale are extremely rich in fat (30–40%), with a protein content of 7–10%. In some species, the composition of milk changes drastically from the start to the end of lactation. This is especially marked for marsupials. Atkinson *et al.* (1980, 1981) reported that the composition of milk from pre-term human mothers differed significantly from that of women who gave birth to full-term infants.

The nutritional significance of milk proteins in the human diet can be divided essentially into two areas: a *macro-nutrient* aspect, e.g., the role of milk proteins as a source of dietary protein, and a *micro-nutrient* or specific physiological aspect in which the specific physiological functions of certain protein fractions are taken into consideration.

The macro-nutrient aspect comprises, principally, the content of amino acids, especially the essential amino acids, and nitrogen in milk proteins in order to meet the requirements for protein synthesis for anabolic processes, as sources of material for growth and repair of tissues during convalescence. In this perspective, the nutritional role must be based on protein bioavailability as well as its relation to the specific protein needs of the consumers. This calls not only for accurate methods for estimating the nutritional value of milk protein, but also scientific support for data regarding protein requirements.

The specific physiological function of certain milk proteins can be exemplified by their potential role in the defence against infections, as well as in the transport of certain nutrients (e.g., trace elements or vitamins) or for certain metabolic implications (e.g., the potential hypercholesterolemic effect of casein) or endocrine functions (e.g., epidermal growth factor and casomorphins). From a strictly nutritional point of view, only their roles in the transport of certain nutrients, as well as in the metabolism of other metabolites of nutritional interest, should be considered in addition to the macro-nutrient aspects. Their role in the defence mechanisms against infections is, however, not essentially a nutritional problem although there is a close relationship between malnutrition and infection.

A strict nutritional approach to the role of milk proteins versus the specific physiological role of certain milk proteins will, however, lead to a slight confusion in our evaluation of the nutritional role of milk proteins in the diet. As will be commented upon later in this review, this is especially obvious when the true protein content of human milk is discussed in relation to that which is nutritionally available as a source of amino acids (van Woelderen, 1987).

12.2 The role of dairy products in the human diet

In the industrialised world, dairy products constitute an essential part of the staple diet, especially in northern Europe and North America. In these areas, milk products contribute around 30% of the total dietary protein intake and about 65% of the intake of animal protein. The increased interest in osteoporosis as a dominant public health problem has also accentuated interest in the role of dairy products in the diet as 75% of the calcium intake in Northern Europe and North America comes from dairy products.

The type of dairy products in the diet differ; in some countries, liquid milk products, in the form of standardised consumer's milk differing in fat content, dominate. In other countries, dairy products are consumed mainly in the form of cheese or fermented products, e.g., yoghurt.

In most low-income countries, however, milk and dairy products constitute only a minor component of the diet, e.g., 0.4 g protein per person per day compared to about 25 g protein per person per day in Sweden and the United States (FAOSTAT, 1999). Furthermore, they are usually accessible only to privileged people. As it is almost impossible to transport fresh milk without refrigeration, fermented milk products and cheese are dominant in low-income countries. Recent developments using the lacto-peroxidase system for preservation purposes (IDF, 1988) may increase the availability of fluid milk products in tropical areas also.

In India and Pakistan, 6–12 g protein in the diet are derived from dairy products (FAOSTAT, 1999). As a matter of fact, India ranks third in the world as a milk producer after the United States and Russia and the supply of whole milk per caput increased from 22 kg in 1966 to 42 kg in 1997. Interestingly, the role of dairy products decreased in affluent societies during the same 30 year period, as the consumption of whole milk decreased from 173 kg to 117 kg per person per year in the USA and from 118 to 91 kg per person per year in Sweden (FAOSTAT, 1999).

Another important factor is that dairy products are usually consumed by those who are in special need of a nutritionally well-balanced diet, i.e., vulnerable groups, including infants and pre-school children. Various milk products, e.g., fresh milk, fermented products, as well as cheese, all represent very valuable components of the human diet. Cheese manufacture also represents a method of preserving the nutritionally high-value proteins in milk, which itself cannot be stored readily.

Milk represents one of the most complete single food items in the human diet. However, after weaning, milk and dairy products are used rarely as a single food item but in combinations with other food components. This means that their nutritional role in the diet can be viewed as a nutritional supplement in the diet for humans at all ages from the weaning period up to old age. From this perspective, milk and dairy products should be viewed essentially as sources of macro-nutrients compared to food items in general like other foods of animal origin, such as meat, fish and eggs. Thus, most interest regarding the nutritional significance of dairy products in the diet should be devoted to the nutrients in which dairy products are especially rich, or where dairy products may play a significant role as nutritional supplements in a conventional diet.

Dairy products can also be used as a component in the diet of young children, e.g., in formulae and gruel. The specific characteristics of milk composition which might be related to the metabolic capacity and nutritional requirements of the offspring lead to the conclusion that bovine milk is intended for calves and human milk for human infants. It is obvious that it is possible to formulate high quality breast milk substitutes to be used when the mother, for one reason or another, is unable or unwilling to breast-feed her child. However, this has lead to a conflict between formula feeding and breast-feeding, which led to an international code for marketing breast milk substitutes (WHO, 1981).

The role of dairy products in the diet of small children should be based on its role as an alternative constituent in the daily food supply to other food items with high nutritional value, e.g., eggs, fish or meat. Bovine milk as such is not especially suitable to infants and small children.

The growing interest in functional foods has also increased the interest in milk products containing certain components and their potential role in milk products for specific physiological effects in the body. The use of milk proteins in non-dairy products might then be based on the exploitation of the functional characteristics of certain milk proteins. There is reason to believe that the specific physiological or metabolic roles of certain milk proteins e.g., lactoferrin, α-lactalbumin, or minor proteins or biologically active peptides, (e.g., casein phosphopeptides and epidermal growth factor), are valid also in the diet of adults.

Another perspective on the specificity of dairy products is their role as probiotics, based on their role as carriers of lactobacilli and bifidobacteria.

12.3 Composition of milk proteins

Milk from all species analysed to date contains protein although the concentration varies from less than 1% to more than 10%. The protein content of human milk is the lowest at 0.8 g per 100 ml, while milk samples obtained from cat, rabbit, reindeer, seal and blue whale contain more than 10%, and are reported to be the highest (Table 12.1). There are also very pronounced qualitative and quantitative variations in the occurrence of various types of protein in the milk from different species.

Milk proteins have been studied intensively by protein biochemists since 1838, when the Dutch chemist, Mulder, devised one of the first acid precipitation methods for the separation of bovine casein, fibrin and egg white. Interestingly, in the same year he introduced the concept of protein. This term was suggested by the Swedish chemist, Berzelius, in a letter to Mulder dated July 10, 1838 (Vickery, 1950).

The proteins of bovine milk have, for obvious reasons, been studied more widely than those of the milk of any other species, human included. Originally, casein was assumed to be the only milk-specific protein while most of the less abundant milk proteins were thought to be derived from blood. Later, extensive studies have shown that many of the non-casein proteins in milk are also synthesised in the mammary gland and must be considered as milk-specific (Kon and Cowie, 1961; McKenzie, 1970, 1971; Chapter 1).

Studies on the composition and the specific nutritional and physiological functions of bovine milk proteins have been extensive during the last decades (see Galesloot and Tinbergen, 1985; Barth and Schlimme, 1989). Increased interest has also been devoted to the composition of human milk and its specific nutritional and metabolic implications in relation to cows' milk (Blanc, 1981; Hambræus *et al.*, 1977; Hambræus, 1984; Lönnerdal, 1985). The nomenclature of milk proteins is, however, still essentially based on that of the proteins of bovine milk (Whitney *et al.*, 1976).

Milk contains a heterogeneous mixture of proteins. Consequently, a variety of methods have been used to separate and characterise the various milk proteins since acid precipitation of casein was introduced at the beginning of the 19th century. Isolation and characterisation of milk proteins is complicated by their considerable tendency to associate and form complexes which makes the definition of their specific physiological, nutritional and immunological characteristics difficult. (For more specific data regarding the chemical characteristics of the various protein fractions, the reader is referred to separate chapters in this book.)

As dairy products are essentially based on bovine milk, most nutritional studies on milk proteins have focussed on those of bovine milk. In countries where nutritional problems may be more serious than in the industrialised world, the nutritional implications of differences in milk protein composition may be accentuated. In this respect, it should be remembered that in less developed countries, milks from other species are consumed, e.g., buffalo, camel, sheep and goat, making the specific characteristics of their milk proteins worth considering. The principal nutritional problems are, however, the same, i.e., the potential hazards and benefits of consuming non-species-specific milk proteins.

This review will deal essentially with the nutritional aspects of the differences between bovine and human milk proteins, as most data are available regarding the composition of bovine milk.

As indicated in Table 12.2, there are four major differences between bovine and human milk:

1. Total concentration of proteins.
2. The ratio of protein-N to non-protein N
3. The ratio of casein to non-casein N
4. The profile of individual proteins in the casein and whey protein fractions.

Unfortunately, we still do not fully understand the physiological implications, neither nutritionally nor metabolically, of these differences. This, of course, limits our complete understanding of the specific nutritional role of milk proteins.

12.4 Protein requirement in humans

The macro-nutrient aspect, or nutritive value, of milk protein has two dimensions: *quantitative and qualitative*. The quantitative aspects are related

TABLE 12.2
Protein composition of mature human and bovine milk

Component	*Human milk*	*Bovine milk*
Total nitrogen ($g.L^{-1}$)	1.8	5.3
Non-protein nitrogen ($g.L^{-1}$)	0.4	0.3
Casein ($g.L^{-1}$)	**0.7**	**25**
% thereof as		
α_{s1}- + α_{s2}-casein	–	54
β-casein	> 85	33
κ-casein	< 15	13
Whey proteins ($g.L^{-1}$)	**5**	**6**
% thereof as		
α-lactalbumin	39	19
β-lactoglobulin	–	49
lactoferrin	26	trace
Immunoglobulins (total)		11
Secretory IgA	13	trace
IgG_1	<2	
IgG_2	<2	
Lysozyme	3	trace
Serum albumin	13	5
Miscellaneous	2	16

(source: Renner, 1983; Lönnerdal and Adkins, 1999).

to the total protein content in relation to energy (e.g., energy percent, E%), or to weight. The qualitative aspect is related to its amino acid composition and the bioavailability of these amino acids. The content of essential amino acids, i.e., those that cannot be synthesised in the body and consequently must be supplied through the diet, is of special concern. This should then be related to the protein and amino acid requirement of humans.

The protein requirements of humans can be divided into two categories: (1) *growth* requirement and (2) *maintenance* requirement. The first refers particularly to the situation in infants and children but also to pregnant and lactating women, and to adults during convalescence, physical training or body-building, when there is increased protein synthesis. The maintenance requirement, on the other hand, refers to the need to cover the daily turnover which in adults is estimated to be about 400 g of body protein per day, of which < 25% is derived from the diet.

Interestingly, diets in the industrialised, as well as low-income, countries have a protein content corresponding to 10–12 E%, regardless of whether they are based on a vegetarian or a mixed animal-vegetable diet. There are obvious reasons to conclude that this concentration of protein will guarantee a safe intake of protein by adults as long as the energy needs are met. Bovine milk, which contains about 20 E% protein and, especially, low-fat milk containing 0.5% fat (36 E% protein), represent good protein sources in the diet. However, it is questioned whether there really is a need for protein-rich food items, e.g., milk products with >30 E% protein, under normal conditions, as there is little if any problem with protein intake in adults when energy needs are met by a conventional diet. However, in pre-school children, food items containing high-value dietary proteins are still considered to be of interest. In this context, it is interesting to note that human milk contains only 5 E% protein and still meets the protein requirements during the intensive post-natal growth period.

With respect to the qualitative aspects of dietary protein intake, it is still an open question whether there is a qualitative difference between maintenance and growth requirement with regard to the need for essential amino acids. When formulating the 1985 edition of protein requirements, the FAO/WHO/UNU expert consultation followed the same principle as the 1973 FAO/WHO committee by which they established two fundamental reference points: (i) the maintenance protein requirement of the young child and (ii) the requirement of the young male adult. The 1985 FAO/WHO/UNU consultation concluded that few normal diets provide insufficient amounts of essential amino acids for most groups, with the exception of infants and pre-school children.

The amino acid scoring pattern for pre-school children is based on a limited number of amino acid balance studies performed by Torun *et al.* (1981) and Pineda *et al.* (1981) at INCAP, Guatemala, and which, unfortunately, have still not been published in peer-reviewed journals. It has been questioned whether the subjects studied were representative of normal healthy pre-school children as they were recovering from malnutrition. The studies on adult males were performed by Young *et al.* (1989). There is still some controversy regarding the optimal amino acid requirements of the adult (Millward, 1997; Young *et al.*, 1989; Young and El-Khoury, 1996).

12.5 What is meant by the nutritive value of a protein?

In calculations of the nutritive value of dietary proteins, attention should be given not only to the amino acid composition but also to the digestibility of the proteins. In the 1985 edition of the FAO/WHO/UNU protein recommendations, the expert consultation stressed that, in a mixed diet, the availability of dietary protein for all age groups can be affected significantly by digestibility. This is considered to be the most important factor for the nutritional value of dietary proteins for adults (FAO/WHO/UNU, 1985). Interestingly, this was almost neglected in earlier conclusions of expert committees on the protein requirement of humans.

True digestibility is defined as the per cent of protein or nitrogen absorbed from a given protein (nitrogen) intake:

$$\frac{100 \times (\text{nitrogen intake} - \text{faecal nitrogen})}{\text{nitrogen intake}}$$

The relatively low relevance of amino acid composition for the nutritive value of dietary proteins for adults means that the superiority of milk proteins, owing to their relatively high content of essential amino acids, is less valid for adults than for children. However, in comparison with vegetable proteins, the high digestibility of milk proteins is still a nutritional advantage.

Bovine milk protein, as well as egg protein, has often been used as references for evaluation of the nutritive value of food proteins (FAO, 1970; FAO/WHO, 1973). This may be due essentially to the fact that they seem to be the only proteins originally intended to function as a single source of nutrients for the offspring during a nutritionally critical period in life. On the other hand, it is obvious that there is quite a difference in the protein composition and thereby also the amino acid composition, not only between milk and egg, but also between milks from various species, as stated earlier. Consequently, from the nutritional point of view, it seems more logical to use human milk composition as a reference for the optimum protein and amino acid composition, at least during the neonatal period. This is also accepted for infants in the latest edition of protein requirements (FAO/WHO/UNU, 1985).

As stated by the expert committee (FAO/WHO/UNU, 1985), above the age of 12 years, most normal diets provide sufficient amounts of essential amino acids when a mixed diet is consumed.

As a consequence of the increased interest with respect to digestibility in the evaluation of protein quality, the concept of protein digestibility-corrected amino acid score (PDCAAS) was proposed by the FAO/WHO Expert Consultation on Protein Quality Evaluation (1991). This committee concluded that protein quality can be assessed by comparing the content of the first limiting essential amino acid of the test protein with the content of the same amino acid in a reference pattern of essential amino acids and by correcting this ratio by a factor for true faecal digestibility of the test protein as measured in a rat assay. Values higher than 1 are adjusted to 1 as concentrations of essential amino acids above those in the reference pattern are considered to provide no additional nutritional value.

Although this concept has its advantages, it also has the following limitations:

- The accuracy of the reference amino acid score pattern (see above),
- The validity of the true faecal digestibility correction based on rat experiments,

- The role of high-value animal proteins, e.g., milk, as a supplement in diets in order to balance the amino acid composition of unbalanced proteins (e.g., wheat and collagen) is not considered.

12.6 Estimation of the nutritional value of food proteins

From a biochemical point of view, protein is defined essentially as polypeptides containing more than about 50 amino acids and/or a molecular weight above 6000 Da. From a nutritional viewpoint, however, the primary function of dietary proteins is as a source of amino acids for protein synthesis needed for growth and maintenance. Consequently, the size of the protein molecule is of less importance, as small peptides are relevant also. In addition, some proteins may have a potential and specific physiological role, which in turn might have an impact on nutritional status; this might be considered as a secondary role.

Metabolic balance studies in humans, including measures of growth and/or other metabolic indicators, e.g., nitrogen balance and plasma amino acid pattern, represent the most accurate assessment of protein quality. However, for obvious economic and ethical reasons, such methods cannot be used for routine analysis of protein quality. Consequently, assessment of the nutritive value of food proteins has traditionally involved a variety of assay techniques, i.e., from biochemical analysis to metabolic balance studies in human subjects (Table 12.3).

12.6.1 Biochemical evaluations

Biochemical analysis involves determination of the total *protein content* as well as determination of *specific protein components. Amino acid analysis* after acid hydrolysis or enzymatic digestion of the proteins *in vitro* will also give some information on the nutritive value of various protein fractions.

(a) Analysis of total protein

There is no absolute method for the determination of total protein content in a complex mixture of proteins in solutions containing peptides, amino acids and other nitrogen-containing components, e.g., milk. In food tables, the protein content of various food items is almost always based on indirect calculations derived from nitrogen analyses by the Kjeldahl method. The Kjeldahl method is an adequate reference method when the correct digestion temperature and catalyst are chosen. The nitrogen value derived from the analysis is then multiplied by a conversion factor, usually 6.25, since most proteins have a nitrogen content of ~16%. However, the conversion factor is dependent on:

TABLE 12.3
Methods for evaluation of the nutritive value of protein

In vitro methods	
Biochemical methods:	nitrogen analysis amino acid analysis
Microbiological methods:	amino acid analysis digestion by proteolytic microorganisms *Streptococcus zymogenes* *Tetrahymena pyriformis*
Enzymatic methods	enzyme preparations proteolytic microorganisms
In vivo methods	
Biological tests with experimental animals:	screening methods (e.g., PER)*
	nitrogen balance studies (e.g., NPU, BV, PDCAAS)* indirect nitrogen analysis toxicological studies
Human tests (in adults and infants):	acceptability tests nitrogen balance studies metabolic studies using stable isotopes

*Abbreviations: PER = Protein Efficiency Ratio; NPU = Net protein utilisation; BV = Biological Value; PDCAAS = Protein Digestibility-Corrected Amino Acid Score.

- the ratio between non-protein nitrogen and protein nitrogen
- the amino acid composition, as the nitrogen content of various amino acids differ (Table 12.4).

Consequently, in the interest of standardisation, the protein content of food items calculated using a conversion factor of 6.25 should be referred to as *crude protein*, as the true conversion factor varies between different proteins, as illustrated in Table 12.5. The problem with using crude protein is exemplified by human milk.

The protein content of human milk, calculated on the basis of total nitrogen determinations using a conversion factor of 6.38, has been reported to be 1.1–1.2 g per 100 ml (FAO, 1970; Fomon, 1974). However, in human milk as much as 25% of the nitrogen is present as non-protein nitrogen, while only 5% of the nitrogen in bovine milk is non-protein. Interestingly, human and bovine milk contain almost the same absolute concentration of non-protein nitrogen (40–50 mg per 100 ml) (Table 12.6). However, the total amount of nitrogen is much lower in human milk, being only about 200 mg per 100 ml *versus* 540 mg per 100 ml in bovine milk. Consequently, the value of crude protein in human milk based on a conversion factor of

TABLE 12.4
Nitrogen content of amino acids

	Amino acid	*Nitrogen content (%)*
Less than 16%:	Tyrosine	7.7
	Phenylalanine	8.5
	Methionine	9.4
	Glutamic acid	9.5
	Aspartic acid	10.5
	Hydroxyproline	10.7
	Isoleucine	10.7
	Leucine	10.7
	Cyst(e)ine	10.7
	Threonine	11.8
	Valine	12.0
	Proline	12.2
	Serine	13.3
	Tryptophan	13.7
	Alanine	15.7
More than 16%:	Asparagine	18.7
	Glycine	18.7
	Glutamine	19.2
	Lysine	19.2
	Histidine	27.1
	Arginine	32.2

TABLE 12.5
Nitrogen conversion factors

Wheat (whole grain)	5.83
Maize	6.25
Rice	5.95
Potato	6.25
Bean (*Phaseolus vulgaris*)	6.25
Soybean	5.71
Sunflower	5.30
Beef	6.25
Egg (white)	6.25
Cow's milk	6.38
Human milk	5.18*

*Corrected for non-protein nitrogen content (Lönnerdal *et al.*, 1976a).

6.25 or 6.38 represents an over-estimation of true protein by about 20%. Analysis of total nitrogen minus non-protein nitrogen, as well as calculations based on amino acid analysis of true protein, have shown that the protein content is only 0.8–0.9 g per 100 ml in milk from well-nourished, as

TABLE 12.6
Nutritionally available protein in human milk (g per 100 ml)

Based on Kjeldahl analysis of total N	
(a) using 6.38 as conversion factor	1.1–1.2
(b) using 5.18[a] as conversion factor	0.8–0.9
Based on true protein content	
(a) total protein[b]	0.8–0.9
(b) with correction for nutritionally not available protein[c]	0.5–0.7

[a]Assuming that 25% of total N is derived from non-protein nitrogen (Hambræus *et al.*, 1984).
[b]Based on the sum of different major protein fractions analysed by specific methods.
[c]Assuming that physiologically active proteins remain intact throughout and are not digested in the gastro-intestinal system.

well as from malnourished, lactating women (Lindblad and Rahimtoola, 1974; Lönnerdal *et al.*, 1976a,b; Hambræus *et al.*, 1976, 1978), representing a conversion factor of 5.18 (Table 12.6). Such a low protein content in human milk was reported by Macy *et al.* (1931), but their data seem to have been neglected for many years.

Since the Kjeldahl method for the estimation of total nitrogen is not as simple and rapid as needed when large numbers of samples are to be analysed for protein content, other analytical approaches have been assessed, e.g., dye-binding methods and UV and infrared spectrophotometric methods. However, results obtained by different methods are not entirely comparable.

(b) Amino acid analysis

Amino acid analysis of milk samples is usually performed by ion exchange chromatography using amino acid analysers or by gas liquid chromatography (GLC) or high performance liquid chromatography (HPLC). Since these analyses do not differentiate between free and protein-bound amino acids, they are of limited practical nutritional significance as they give no information regarding their bioavailability. However, of greater practical significance is the fact that some amino acids are destroyed during analysis, e.g., tryptophan, or may react to form unavailable components, e.g., lysine, due to the Maillard reaction during heat treatment in the dairy industry, e.g., roller drying, or during food preparation (Abrahamsson *et al.*, 1977).

The amino acid content of a food protein is often expressed in relation to that of a reference protein, using the term 'chemical score', first introduced by Mitchell and Block (1946). The amino acid that occurs in the lowest concentration in relation to the amount in the reference protein is called the 'first limiting amino acid' and the percentage represents the 'chemical score' or 'amino acid score'. However, one serious limitation of all amino acid reference patterns used to date is the fact that sulphur-containing amino acids, methionine and cysteine/cystine, are added together, as are the two aromatic amino acids, tyrosine and phenylalanine. This has some nutritional implications, especially when dietary proteins for use during early infancy are evaluated as sources of amino acids. Another disadvantage is the fact that the score value is influenced only by the content of the limiting amino acid. It gives no information whatsoever regarding any changes in the content of other essential amino acids as long as there is no change in the first limiting amino acid. Thus, any destruction of lysine in milk products as a result of the Maillard reaction will not affect the chemical score value until lysine has become the first limiting amino acid. Finally, the usefulness of amino acid score as an index of the nutritive value of a protein is of course completely dependent on the selection of a relevant reference protein.

International groups have used various amino acid scoring systems throughout the years. This is illustrated in Table 12.7 where the egg protein reference pattern recommended by FAO (1965), and the provisional amino acid score patterns recommended by FAO (1957, 1973, 1985), respectively, are shown in relation to the essential amino acid content of casein and skim milk protein. It shows that with the egg protein pattern as a reference used by

TABLE 12.7

Essential amino acid content of skim milk powder (SMP), casein, whey proteins and human milk, as well as reference patterns [Egg reference (FAO, 1965); Provisional amino acid scoring pattern (FAO, 1973); FAO/WHO/UNU, 1985); FAO/WHO 1991]

	Ile	*Leu*	*Lys*	*Met + Cys*	*Phe + Tyr*	*Thr*	*Tyr*	*Val*	*Total essential amino acids*
SMP	52	97	71	34	96	41	14	63	468
Casein	54	95	81	32	111	47	16	75	511
Whey protein	76	118	113	52	70	84	24	72	609
Human milk	49	91	65	37	76	44	na	52	414
Reference patterns									
1965[1]	66	88	64	55	100	51	16	73	513
1973[2]	40	70	55	35	60	40	10	50	360
1985[3]									
adult	13	19	16	17	19	9	5	13	111
pre-school	28	66	58	25	63	34	11	35	320
infant	46	93	66	42	72	43	17	55	434
1989[4]	38	65	70	27	65	35	10	40	350

[1]Egg reference (FAO, 1965).
[2]Provisional amino acid scoring pattern (FAO, 1973).
[3]FAO/WHO/UNU (1985).
[4]Recommended for all ages except infants (Young *et al.*, 1989).

FAO (1965), the sulphur-containing amino acids are the limiting ones, giving a chemical score of 62 for skim milk protein and 58 for casein, while the scores are 81 and 76, respectively, with the FAO (1985) reference pattern.

The FAO/WHO Expert Consultation (1991) recommended that the amino acid composition of human milk should be the basis of the scoring pattern to evaluate protein quality in foods for infants under 1 year of age. For all other age groups, the FAO/WHO/UNU (1985) reference pattern for children of pre-school age (Table 12.7) should be used.

(c) Biological evaluations

The use of amino acid score is limited also by the fact that it does not take into consideration poor bioavailability or digestibility. Thus, biochemical analyses are often supplemented by the use of some biological methods in order to establish whether the amino acids analysed are available for digestion and absorption in the gastrointestinal tract and may promote growth or nitrogen balance. Various biological methods can be used (Table 12.3), including *in vitro* techniques using enzymatic digestion or microbiological assays using *Streptococcus zymogenes* or *Tetrahymena pyriformis*. However, most biological methods for assessing protein quality are based on some type of feeding study with experimental animals (usually rats) in metabolic cages. Since we are usually interested in the nutritive value of proteins for human consumption, it should always be remembered that each species has its own requirements for protein and amino acids. Consequently, the final proof of the nutritional value of a protein in the human diet can, consequently, be based only on metabolic studies on humans.

The protein efficiency ratio (PER) has for a long time been used as a predictor of clinical tests. However, it is known that this index overestimates the value of some animal proteins for human growth and underestimates that of some vegetable proteins. In their review, the joint FAO/WHO (1991) Expert Consultation recommended, as criteria for assessing the suitability of protein quality, a combination of amino acid score and protein digestibility in order to provide results which are consistent with those obtained from clinical studies designed to assess protein quality. The methods should be applicable to the entire range of foods used in human diets and collaborative studies should demonstrate excellent repeatability within a laboratory and reproducibility between laboratories. This led to the recommendation for adopting the protein digestibility-corrected amino acid score (PDCAAS) method as an official method at the international level (FAO/WHO, 1991). The PDCAAS is calculated by multiplying the lowest amino acid ratio by true protein digestibility. PDCAAS values above 1.00 would be considered as 1.00 or 100%. This has led to some controversy where those with an interest in dairy products argue that the potential supplementary value of high quality animal proteins is not appreciated fully (IDF, 1998).

12.7 Composition of milk protein

The milk protein composition and content, from a qualitative point of view, differ in various dairy products (e.g., fresh or fermented milks, hard and soft cheeses) as a result of the technology used in the manufacture of dairy products, which influence the protein nutritive value of the product. Thus, the specific nutritional quality of different dairy products is dependent on the specific nutritional characteristics of various milk proteins.

Essentially as a consequence of the technology used in the dairy industry, milk proteins are usually subdivided into two major groups: the caseins and the whey proteins. Maximum precipitation of bovine casein is obtained at pH 4.6 and this is also the most commonly used definition of casein (see Chapters 1, 2 and 3). However, the conditions for maximum precipitation of casein vary for milks from different species, e.g., pH 4.0 is reported to be optimal for precipitation of casein from rat milk while human casein is very difficult to precipitate (Mellander and Folsch, 1972; Lönnerdal and Forsum, 1985) unless a combination of low pH (4.3) and calcium addition (60 mM) is used (Kunz and Lönnerdal, 1989). Furthermore, there is often co-precipitation of whey proteins or non-casein proteins with the caseins. In the manufacture of some cheeses, this is considered desirable in order to increase the recovery of proteins from milk. From a nutritional point of view, this is also an advantage since whey proteins have a higher nutritional quality than the caseins (Forsum, 1973).

12.7.1 Caseins

The caseins comprise a group of proteins with unique molecular structures which aggregate into larger complexes or micelles (see Chapters 1, 4 and 5). Casein micelles contain calcium and inorganic phosphate and represent some of the few naturally occurring phosphoproteins. Overall, caseins are milk-specific proteins, but the types of casein show inter-species variation. Some of the inter-species similarities may have nutritional implications, e.g., a high content of ester-bound phosphate, high proline content, low content of sulphur-containing amino acids, especially of cysteine, and low solubility at pH 4 to 5. The caseins constitute the dominant protein fraction in bovine milk, in which they represent about 80% of the total protein. In human milk, however, casein constitutes only 20–40% of total protein (Lönnerdal and Forsum, 1985; Kunz and Lönnerdal, 1989) (Table 12.2). Caseins are not quantified easily due to their heterogeneity and micellar composition. They are, however, classically determined by acid precipitation at pH 4.6, which is the isoelectric point of casein. Since the properties of caseins differ in milks from various species, this also influences their isoelectric points. Human casein does not have the same distinct isoelectric point as bovine casein and some whey protein components also precipitate between pH 4 and 5 (Lönnerdal and Forsum, 1985; Kunz and Lönnerdal, 1989). Furthermore, certain whey proteins co-precipitate with casein at this pH. Consequently, acid precipitation over-estimates the casein content of human milk (Lönnerdal and Forsum, 1985). However, addition of calcium makes the precipitation of casein at pH 4.3 more specific (Kunz and Lönnerdal, 1989). By means of isoelectric precipitation and ultracentrifugal methods, as well as indirect quantitation of casein by analysis of total protein and subtracting the values for the major whey proteins, determined by immunochemical methods, casein has been shown to constitute less than 30% of the total protein in human milk (Hambræus *et al.*, 1978).

α_s-Caseins are the dominant casein in ruminant milks, but it is present at a very low concentration in human milk (Kunz and Lönnerdal, 1989), in which β-casein is the dominant form (Jenness and Sloan, 1970). The γ-caseins represent C-terminal segments of β-casein from which an acidic phosphopeptide is missing; γ-caseins are, consequently, no longer referred to as a separate casein group. κ-Casein is unique since it remains soluble in calcium solution under conditions that precipitate all other caseins and it can consequently also occur in the whey fraction. It stabilises the other caseins in the presence of calcium and plays a key role in the formation of calcium-containing casein micelles. κ-Casein is heavily glycosylated, with the carbohydrate content of human κ-casein being higher than that of bovine κ-casein (Yamauchi *et al.*, 1984). When chymosin or pepsin acts on κ-casein, a glycomacropeptide is formed which is suggested to be part of the so-called "bifidus factor" (György *et al.*, 1953) which plays a significant role during infancy as it promotes the growth of and colonisation of the gut by *Bifidobacterium* spp. and inhibits gut colonisation by *Escherichia coli* (Bezkorovainy *et al.*, 1979). This initiated the development and production of special infant formulae containing bifidus factors.

It has been suggested that the primary function of the caseins is to provide nutrition to the newborn. This is due not only to their role as a source of amino acids but also to their role as a source of calcium and inorganic phosphate. Calcium-containing casein micelles make the content of calcium and inorganic phosphate in milk far higher than would be expected from their physico-chemical solubility in milk.

Curd formation may also be of physiological interest for the function of the digestive tract and may enable gastric enzymes to start protein digestion in the newborn, but such functions are difficult to evaluate. There are, however, physico-chemical differences between human and bovine caseins resulting in different curd-forming characteristics in the stomach. The casein micelles in human milk are smaller than those in bovine milk. Casein is very heat-resistant but is precipitated at low pH values, resulting in curd formation; it is also precipitated by enzymes, particularly chymosin, which are secreted by the stomach. The coagulum from human milk is soft and flocculent while that from bovine milk is resistant and firmer, the physiological significance of which is still not fully understood but it is believed that the digestibility of bovine casein in infants is lower than that of human casein.

Being very specific phosphoproteins, caseins probably also play a role in the phosphorus balance in adults. From a nutritional point of view, this means that casein not only functions as a protein source but also as a source of calcium and phosphorus. A milk-free diet will consequently also have an impact on the calcium intake of a population at all ages. For example, in Sweden, 75% of the dietary calcium intake is derived from dairy products.

From a nutritional point of view it is also of interest to comment on the uniqueness of the amino acid composition of the caseins. The physiological role of casein in the human diet is still not well known. It is consequently somewhat difficult to establish whether other dietary proteins can substitute for casein nutritionally. This is illustrated in Table 12.8. It can be seen that the very low cysteine content of the caseins results in a methionine/cysteine ratio that is 2–3 times higher than that of other animal proteins and more than 7 times higher than that of human milk. As a matter of fact, human milk seems to be the only animal protein source that has a methionine/cysteine ratio below or close to 1.0, which is probably due to its high content

TABLE 12.8
Amino acid content in casein and whey proteins and other food proteins (mg amino acid per g total nitrogen)

	Human Milk	*Bovine Casein*	*Bovine Whey protein*	*Egg*	*Beef*	*Wheat*	*Soybean*
Isoleucine	254	345	476	393	301	204	284
Leucine	471	607	736	551	507	417	486
Lysine	337	518	704	436	556	179	399
Methionine	78	178	151	210	169	94	79
Cysteine	114	23	174	152	80	159	83
Phenylalanine	171	334	224	358	275	282	309
Tyrosine	223	371	214	260	225	187	196
Threonine	228	297	527	320	287	183	241
Tryptophan		103	147	93	70	68	80
Valine	296	430	449	428	313	276	300
Arginine	171	239	175	381	395	288	452
Histidine	114	186	144	151	213	163	158
Alanine	166	196	341	370	365	226	266
Aspartic acid	451	455	766	601	562	308	731
Glutamic acid	1000	1406	1231	796	955	1866	1169
Glycine	98	126	126	207	304	245	261
Proline	513	738	450	260	236	621	343
Serine	228	385	374	478	252	287	320

Ref FAO (1970), Forsum *et al.* (1973).

of whey proteins and concomitant low casein content. This is of considerable metabolic and nutritional interest since it is known that the premature infant lacks the enzymatic capacity to convert methionine to cysteine which makes cysteine or cystine an essential amino acid during early life (Sturman *et al.*, 1970). The lack of cysteine in casein is consequently a nutritional disadvantage. Interestingly, cysteine in milk is derived almost exclusively from the whey proteins, β-lactoglobulin and α-lactalbumin. Therefore, the use of casein-rich infant formulae warrants further consideration from a nutritional and metabolic point of view.

Casein has been used as a reference protein for many years in the evaluation of the nutritive value of proteins, especially in biological assays in rats. It is, however, well known that casein is by no means an optimal protein from a nutritional point of view since several animal proteins (e.g., egg protein, whey proteins), as well as a few vegetable proteins (e.g., soy protein, rapeseed proteins), show a higher nutritive value (FAO, 1970). As the sulphur-containing amino acids are limiting, casein is sometimes supplemented with methionine when used as a reference protein.

β-*Casomorphins*

β-Casomorphins are sequences in the caseins which are believed to be released during digestion and have been shown to exert an opioid, morphine-like activity (Brantl and Teschemacher, 1982; see Chapter 14). Furthermore, they have been shown to stimulate insulin release postprandially, but they do not seem to have a hypercholesterolemic effect (Pfeuffer and Earth, 1988).

12.7.2 Whey proteins

Whey is a highly nutritious by-product of cheese and casein manufacture and contains about 0.6% protein. The whey proteins represent the non-casein proteins as well as fragments of the caseins which remain soluble when the caseins have been precipitated enzymatically by chymosin or isoelectrically by acid precipitation at their isoelectric point, pH 4.6. Whey contains a very heterogeneous group of proteins (Woodhouse and Lönnerdal, 1988) which share very few common characteristics except that of being soluble under conditions that precipitate casein. Milk from carnivorous mammals contains a considerable amount of milk-specific non-casein proteins while milk from other species, especially rodents, contain much less. One of the characteristics of the amino acid pattern of the whey protein fraction is the relatively low content of aromatic amino acids, especially phenylalanine and tyrosine.

One important function of the whey proteins is their contribution to the nutritive value of milk proteins. The nutritive value of casein, which does not seem to be optimal from the human point of view, is inferior to that of the whey proteins. It has also been postulated that whey proteins in human milk counterbalance casein in order to balance the supply of essential amino acids for the newborn. The high nutritive value of whey proteins is due mainly to their high content of essential amino acids (Forsum and Hambræus, 1974). This is illustrated in Tables 12.7 and 12.8 which show the essential amino acid content of whey protein concentrate (WPC), as well as of wheat, soybean, egg and beef. It can be seen that WPC has a surplus of all essential amino acids. Of special interest is the high content of isoleucine, leucine, threonine and tryptophan. Wheat is low in lysine and threonine, while maize is low in lysine and tryptophan. Rice, on the other hand, has a relatively well-balanced amino acid composition although the contents of lysine and threonine are fairly low.

Some of the whey proteins seem to have distinct physiological and biochemical roles, e.g., lactoferrin strongly binds iron, α-lactalbumin is a constituent of lactose synthase and lysozyme is an enzyme that destroys the bacterial cell wall, to mention a few of the most widely discussed whey proteins. With respect to β-lactoglobulin, which is the dominant whey protein in bovine milk, we still lack unequivocal information regarding a specific physiological role.

(a) β-Lactoglobulin

β-Lactoglobulin is the dominant whey protein in bovine milk. It is relatively heat-labile and occurs in the milk of all ruminants studied but has been reported to be absent from milk of the camel and guinea pig (Bell and McKenzie, 1964) as well as human milk, although minor amounts have been reported to occur in human milk (Conti *et al.*, 1980). Although the physiological role of β-lactoglobulin is still not fully understood, it has been speculated that it plays a regulatory role in phosphorus metabolism in the mammary gland (Farrell and Thompson, 1971) and retinol binding (see Chapter 7). Some of the changes in the properties of milk that occur on heating are due to denaturation and aggregation of denatured β-lactoglobulin with other proteins such as the formation of a disulphide bond between β-lactoglobulin and κ-casein. It has been postulated that β-lactoglobulin may, in part, be a cause of intolerance to cows' milk protein observed in formula-fed infants.

(b) α-Lactalbumin

α-Lactalbumin is a well-defined whey protein, second only to β-lactoglobulin in bovine whey and one of the three dominant whey proteins in human milk (Blanc, 1981). The reader should be aware of the fact that there is still some confusion as to nomenclature. The term 'α-lactalbumin' is not equivalent to 'lactalbumin', which was used earlier as a common name for the total whey protein fraction and still is used occasionally in the nutritional or paediatric literature. α-Lactalbumin occurs in the milk of all mammals although it is present at very low levels in milks containing little or no lactose, e.g., milk from the northern fur seal and sea lion. α-Lactalbumin has a stable configuration between pH 5.4 and 9 and it is also a relatively heat-stable whey protein. It is part of the enzyme lactose synthase (EC 2.4.1.22) and the biosynthesis of lactose seems to be regulated by the production of α-lactalbumin. Consequently, there appears to be a relationship between the content of lactose and that of α-lactalbumin when milk samples obtained from various mammals are compared. There is, however no direct correlation between the concentrations of lactose and α-lactalbumin concentrations of human milk according to transversal and longitudinal studies, which showed that the lactose concentration increases whilst α-lactalbumin content decreases throughout lactation (Lönnerdal *et al.*, 1976a,b). Although the primary role of α-lactalbumin seems to be its function as regulator in lactose biosynthesis, it is obvious that it also plays a significant nutritional role as it has a high nutritive value (Forsum, 1973). Its amino acid composition seems to be optimal for the requirements of the human infant and it has a high digestibility.

α-Lactalbumin has a metal ion-binding site, which normally binds calcium (Lönnerdal and Glazier, 1985). Therefore, it has been suggested to play a key role in the biosynthesis of casein in the mammary gland (Hiraoka *et al.*, 1980).

(c) Iron-binding proteins

Milk contains two kinds of iron-binding proteins: transferrin, which seems to be almost identical to transferrin in blood, and lactoferrin, which is closely related to transferrin but is characteristic of milk and exocrine glands (see Chapter 10). All mammals seem to be able to produce these two iron-binding proteins although they do not occur to the same extent in all milks (Table 12.9). Lactoferrin has been identified, but at a much lower concentration than in human milk, in milk from cow, goat, buffalo, horse,

TABLE 12.9
Concentration (μg/ml) of iron and iron-binding proteins in milk from various species

Concentration μg/ml	*Iron*	*Lactoferrin*	*Transferrin*
0.3–0.6	Human, cow, goat		
1.4–2.4	Sow		
2–4	Rabbit		
3–14	Rat, dog		
< 50		Rat, rabbit, dog	Dog, human
20–200		Cow, goat, sow	Sow, cow, goat, mare
200–2000		Guinea-pig, mouse, mare	Guinea-pig, mouse
> 2000		Human	Rabbit, rat

pig, mouse and guinea-pig. It seems to be absent from the milk of the rabbit, rat and dog, in which transferrin is the dominant iron-binding protein.

The potential physiological role of lactoferrin was first described as a bacteriostatic one since lactoferrin is present in human milk in a highly unsaturated form. Therefore, lactoferrin can compete effectively for iron with the siderophores of iron-requiring bacteria. Consequently, lactoferrin was considered to be at least partly responsible for the defence against gastrointestinal infections during the neonatal period (Bullen *et al.*, 1972).

However, of more nutritional interest, is a second possible function of lactoferrin, i.e., to facilitate iron absorption. It is well-known that, although the iron content in human milk is very low, the bioavailability of the iron is high (Saarinen *et al.*, 1977). It has been suggested that this is due to lactoferrin which may function as an iron transporter. Further evidence for such a role has been gained in recent years with the demonstration that a lactoferrin receptor exists in the brush-border membrane of the small intestine in several species (Cox *et al.*, 1979; Mazurier *et al.*, 1985; Davidson and Lönnerdal, 1988). Thus, it has been suggested that receptor-mediated uptake of lactoferrin-iron is responsible for the reported high bioavailability of iron in breast milk. Lactoferrin in human milk is very unsaturated with respect to iron (3–5% of its total iron-binding capacity; Fransson and Lönnerdal, 1980); therefore, it can compete effectively with bacterial siderophores, making iron unavailable to bacteria. Only a minor fraction of the iron in human milk is actually bound to lactoferrin whilst about one-third is bound to the fat fraction (fat globule membrane) and another one-third to low-molecular weight components (Fransson and Lönnerdal, 1980). Saturation of lactoferrin has been found to reverse its bacteriostatic effect in nature (Reiter, 1978). However, it is possible that iron present in other compartments may be released and transferred to lactoferrin during digestion. Interestingly, iron-saturated lactoferrin is more resistant to proteolysis than the apo-protein (Brock *et al.*, 1976).

The specific role of lactoferrin, both with respect to the defence against gastrointestinal infections and as a promoter of iron transport in the intestinal tract, depends on the fact that it is not digested in the gastrointestinal tract. However, this also means that it is nutritionally unavailable as a source of amino acids.

(d) Immunoglobulins

The whey protein fraction also contains several immunoglobulins, which represent a very heterogeneous group of proteins. Immunoglobulin A (IgA) seems to be most specific for milk and occurs at the highest levels in certain species, i.e., human milk and the milk from other primates, while it is a minor component in milk from ruminants where the less milk-specific immunoglobulins, IgG and IgM, dominate. Immunoglobulin A occurs primarily as secretory IgA, which is synthesised in the mammary gland by binding two molecules of IgA with two other components, a joining chain (J-chain) and secretory component (see Chapter 9). Secretory IgA is resistant to proteolysis in the gastrointestinal tract, which is probably necessary for it to exert its physiological action as part of the defence mechanism against infections during the neonatal period. Intact IgA has been found in significant amounts in the stools of breast-fed infants (Davidson and Lönnerdal, 1987).

The immunoglobulins are of little interest from a strictly nutritional point of view, since their role is not primarily as nutrients. Whether or not they have a double role, both intact in the defence mechanism against gastrointestinal infections and in more or less digested form as sources of amino acids, still has to be elucidated. This has a practical implication in evaluation of the nutritional value of those milks that contain immunoglobulins.

(e) Enzymes

Milk contains a number of enzymes; about 60 different enzymes have been detected in bovine, human and some other milks (see Chapter 11). Some of them may, at least in part, resist digestion and be involved in the defence mechanism against bacteria in the gastrointestinal tract (e.g., lysozyme, lactoperoxidase) and are therefore of limited interest from strictly nutritional viewpoint. Others (e.g., lipase, α-amylase and proteases) may be of nutritional significance, as they can affect the absorption of other nutrients. Lysozyme (EC 3.2.1.17) is present in especially high amounts in human milk and is suggested to have a bactericidal function in the gastrointestinal tract since it destroys the cell wall structure of Gram-positive bacteria. The digestion of lipids is enhanced in breast-fed infants by a bile salt-stimulated lipase (see Chapter 11). It has been postulated that α-amylase, which occurs in human and bovine milks, survives passage through the stomach and consequently helps in the digestion of starch and other oligosaccharides. The role of proteases and protease inhibitors in milk remains unknown although the nutritional significance seems to be limited. It is possible, however, that the relatively high concentration of α_1-antitrypsin in human milk may help some bioactive proteins to survive digestion (Davidsson and Lönnerdal, 1987).

(f) Vitamin-binding proteins

Two vitamin-binding proteins that occur in milks are of special nutritional interest: folate-binding protein (Ford *et al.*, 1969) and cobalamin-binding protein (Sandberg *et al.*, 1981). Both occur at low concentrations but there is some evidence that they facilitate the uptake of the corresponding vitamin from milk in the intestinal tract as they survive low pH and some of the proteolytic enzymes (Andersen and von der Lippe, 1979; Salter and Mowlem, 1983). It is also possible that they aid in the accumulation of these vitamins in milk during secretion from the mammary gland.

(g) Hormone-binding proteins

Several growth-promoting factors have been reported to occur in milk, e.g., epidermal growth factor, insulin, insulin-like growth factor and transforming growth factor (Koldovsky, 1989). They occur at very low concentrations and since they essentially act as hormones, with very little nutritional value, they are beyond the scope of this review.

(h) Proteins of the milk fat globule membrane

A membrane formed from the epithelial cells in the mammary gland (Freudenstein *et al.*, 1979) surrounds the fat globules in milk. The membrane contains a number of enzymes, among them xanthine oxidase and alkaline phosphatase. We still know little about the specific physiological and nutritional roles of these proteins. However, they may be of nutritional significance since they may influence the bioavailability of minerals such as iron, zinc and manganese, which occur in significant amounts in the protein fraction of the milk fat globule membrane. This might explain the findings of trace elements, e.g., iron, zinc and manganese, in human milk fat. It has already been mentioned that a substantial amount of the iron in milk is found in the lipid fractions where it occurs in the fat globule membrane proteins (Fransson and Lönnerdal, 1980).

12.8 Nutritional *versus* physiological roles of milk proteins

Some of the proteins in human milk are considered to have a specific physiological role, e.g., secretory IgA, lactoferrin and lysozyme. However, this necessitates that they can withstand hydrolysis in the gastrointestinal tract. As these proteins exert their physiological role only if they are not digested and absorbed in the gastrointestinal tract, they should not be included in calculations of the nutritionally available proteins in milk. Consequently, there is obviously a conflict between a nutritional approach to the role of milk proteins based on the total protein content and that of a physiological approach to their role. The practical implication of this is illustrated in the case of human milk. A determination of protein based only on analysis of nitrogen indicates a total protein content of 1.1–1.2 g per 100 ml using the conventional correlation factors, 6.25 or 6.38. If compensated for the high non-protein nitrogen content, the protein content is 0.8–0.9 g per 100 ml. However, since secretory IgA and lactoferrin constitute a considerable amount of the whey protein fraction in human milk, the content of nutritionally available protein might be as low as 0.5–0.6 g per 100 ml (Hambræus *et al.*, 1984) (Table 12.7) if they are not digested at all in the gastrointestinal tract. If so, human milk might have the lowest protein content of all milks and the protein fraction constitutes only 4% or even less of the total energy content. This illustrates that the macro-nutrient aspect of milk proteins must take any specific physiological role of certain protein fractions and their bioavailability into consideration. Evaluating the nutritive value of milk protein without compensation for proteins of specific physiological importance which are not digested in the gastrointestinal tract consequently might be greatly misleading (Hambræus *et al.*, 1984; van Woelderen, 1987). The implications of this from the perspective on protein requirement during the neonatal period remains to be evaluated. So far, most data regarding the protein requirement of neonates are based on the estimation of the crude protein content in human milk, i.e., using a conversion factor of 6.38 or 6.25. There has been no correction either for the high non-protein nitrogen content, or for the fact that physiologically active protein fractions might not be nutritionally available.

12.9 Can the nutritional value of milk proteins be damaged?

Most milk products are consumed after being subjected to some form of heat treatment. This might lead to a number of chemical reactions, which may interfere with their nutritional quality. In the case of milk proteins, all industrial heat treatments will lead to some denaturation of whey proteins and all proteins can react with each other as well as with other milk constituents, even if the heat treatment is mild. One of the essential amino acids, lysine, which is often the limiting one in most diets of vegetable origin, can participate in two types of reactions, which reduce its bioavailability. Firstly, the ε-amino group of lysyl residues can react with aldehydes from carbohydrates in the so-called Maillard reaction (Eriksson *et al.*, 1981). This chemical reaction, which is complicated and not fully understood in detail, can cause nutritional problems as lysine is no longer available for absorption and utilisation (Abrahamsson *et al.*, 1977). As trypsin can no longer cleave the peptide bond between the modified lysyl residue and the next amino acid, protein digestibility is impaired and abnormal peptides are formed which, it has been suggested, are toxic or harmful.

Lysine can also form lysinoalanine, which occurs when casein is heated in dilute alkali leading to the elimination of phosphate from phosphoserine with the formation of dehydroalanine which reacts with the ε-group of lysine (Bohak, 1964). Processes that include heating in dilute alkali, e.g., the manufacture of sodium caseinate and the preparation of textured proteins, may lead to the formation of lysinoalanine. Since this peptide improves the physical properties of textured casein products, it may be desirable from this perspective, but the nutritional value is reduced concomitantly since lysine will be less available. It has also been suggested that lysinoalanine may be potentially toxic, but this has still not been verified completely.

Treatment of milk proteins, especially casein, with proteolytic enzymes in order to produce soluble peptides which may be more readily absorbed or may have reduced antigenic characteristics, may give rise to a bitter taste, probably caused principally by hydrophobic peptides from β-casein (Clegg *et al.*, 1974).

12.10 Role of milk proteins in the human diet

12.10.1 How much milk is needed?

The role of milk products for the nutritional status of the population varies. In some industrialised countries, i.e., in northern Europe and the USA, milk products constitute about 30% of the protein intake and 75% of the calcium intake. As protein deficiency is not a public health problem in these countries, neither qualitatively nor quantitatively, there is no minimal quantity of milk needed to cover the protein needs. The amount of dairy products needed in the diet is therefore essentially determined by their role as a source of calcium, since there are few other natural calcium sources in the Western diet. This leads to a recommended intake for adults of slightly less than 500 ml milk, or its equivalent, per person per day.

12.10.2 Supplementary effect

As illustrated in Table 12.6, the amounts of essential amino acids in milk protein are close to what is expected to be the optimal amino acid requirement and composition for the humans. Consequently, the best application of milk proteins seems to be as a supplement for cheaper vegetable proteins, thus taking advantage of its abundance of essential amino acids which enhance the nutritive value of a mixed diet as it nutritionally supplements inferior proteins (Table 12.8). The enhancing effect is most obvious when small amounts of milk are used to supplement cereals. This is also of considerable international importance with respect to global food scarcity since by far the larger part of the world's population relies on cereals as their staple food. The new approach to amino acid scoring by the FAO/WHO/UNU (1985, 1991) Expert Consultations, however, reduces the nutritional impact of milk protein supplementation of diets.

Whey may be a serious pollution problem for the dairy industry. To date, the whey proteins have not been used to any great extent in the human diet in ways that recognise their high nutritive value; when they are used, it is mainly for their functional properties in beverages, creams, gels, etc. There has been an increased interest in the use of whey proteins in nutrition supplements for athletes because of their high digestibility and high content of branched-chain amino acids (Forsum, 1973; Batterman, 1986). Some studies also indicate that whey protein hydrolysates have a significantly higher biological value than the intact protein (Boza *et al.*, 1994) and, consequently, they are used in some nutrition supplements for athletes.

12.10.3 Infant formulae

Formulae intended for use as a substitute for breast milk are traditionally based on skim milk and the nutrient composition is modified to correspond to the infant's requirements. It was early found, however, that, in contrast to bovine milk, breast milk has a much higher whey protein:casein ratio (Figure 12.1). Therefore, when WPC became commercially available, it was added to skim milk to alter the whey protein:casein ratio and make it more

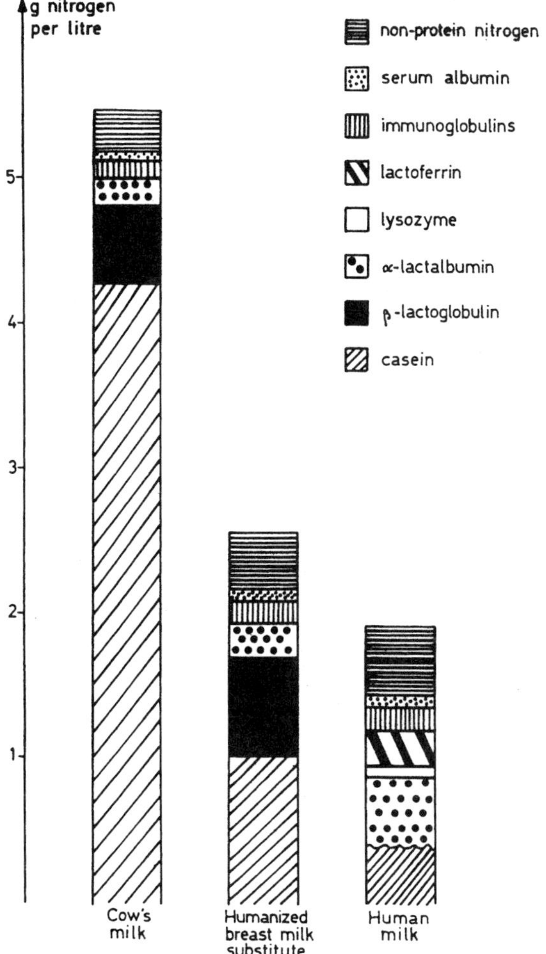

Figure 12.1 Protein composition of bovine milk, humanised infant formula (adapted formula) and human milk. Nitrogen derived from the various proteins and non-protein nitrogen are given as g of nitrogen/L (from Hambræus *et al.*, 1977).

similar to the 60/40 ratio in breast milk. As a consequence, two major types of bovine milk-based formulae are available: casein-predominant (80% casein) and whey-predominant (40% casein). More recently, other milk protein fractions have become available for inclusion in infant formulae. For example, enriched α-lactalbumin, the amino acid composition of which improves the overall amino acid profile of formula (see below), has been proposed as a potentially valuable component of formulae. Other components include lactoferrin, lactoperoxidase and immunoglobulins.

Clinical studies on infants fed various milk formulae with the normal concentration of protein (~15–18 g/L) frequently show high plasma concentrations of some amino acids and of blood urea nitrogen (BUN) (Lönnerdal *et al.*, 1990a,b). Since these levels are considered unnecessarily high and potentially stressful to the developing liver and kidneys, two approaches have been taken in an attempt to change this: to alter the amino acid composition of the protein in infant formulae, or to reduce the protein concentration in the formulae. The first option was discussed above, and is currently being tested in clinical studies. Lowering the protein concentration in formulae has been attempted, but while a reduction in the concentration of some of the amino acids, which were very high in the blood of infants fed the formula in question, was achieved, the concentration of other amino acids were found to be too low (Räihä *et al.*, 1986a,b). In particular, plasma tryptophan was found to be lower than in breast-fed infants in some studies on formula containing 13 g protein/L. Supplementation with tryptophan has been attempted (Hanning *et al.*, 1992; Steinberg *et al.*, 1992), but even if plasma concentrations were elevated, the long-term consequences of feeding free tryptophan to infants have not been examined.

A common mistake when evaluating infant formulae is to estimate protein content by conventional methods. Some whey-predominant formulae contain whey protein fractions that contain substantial quantities of urea and other low molecular weight compounds (Donovan and Lönnerdal, 1989). Since protein concentration is based on nitrogen × 6.25 (or 6.38), a significant level of urea will result in an over-estimation of the "true" protein content. Further, a high concentration of dietary urea in the formula can artificially increase the infant's BUN level by being absorbed and excreted, but not utilised (Donovan *et al.*, 1990). High BUN levels are otherwise indicative of excessive protein intake and amino acid breakdown. Therefore, it is important to characterise the protein and urea content in some detail when performing clinical and field studies, so that the relative importance of each component can be assessed in some detail.

Milk formulae are, to various degrees, exposed to heat treatment, which affects protein digestion and amino acid utilisation. Liquid formulae are usually heat-treated by so-called retort sterilisation, which exposes the product to a high temperature (110–115°C) for a relatively long period of time (1–2 h total). Powdered formulae are manufactured conventionally by spray-drying which exposes the protein to a lower temperature and for shorter time than the retort process. More recently, ultra-high-temperature (UHT) treatment has become increasingly popular; at a temperature of ~130°C for 6–8 s, bacteria and spores are killed, but the effect on protein quality is less severe than by roller drying or spray-drying. Retort sterilisation in particular, but also spray-drying, have been shown to reduce available lysine in formulae and also to reduce protein digestibility (Rudloff and Lönnerdal, 1992),

Milk proteins have been the traditional basis for special formulae for allergic infants and children. First, casein was hydrolysed to small peptides to eliminate any epitopes that may trigger an allergic reaction. However, casein hydrolysate formulae do not taste very pleasant and, more recently, whey protein hydrolysates have also been used for so-called "hypoallergenic formulae", in an attempt to improve the taste of the product. These "hypoallergenic" formulae had some commercial success since it was believed that they could prevent milk allergy. There is very little support for this, and since some sensitive infants reacted against the product as the degree of hydrolysis was not enough to avoid potential anaphylactic shock, the term "hypoallergenic" is now avoided and the term "hypoantigenic" is used instead. The merits of such "hypoantigenic" products warrant further investigation.

While regular milk formulae were traditionally intended for infants up to 1 year of age, a new type of formula, so-called "follow-on" formula, was introduced with the notion that this product should be more appropriate for older infants. "Follow-on" formula is often based on skim milk and contains a higher concentration of protein than formulae for newborn infants. The nutritional need for such products has been questioned by paediatric experts, as the protein requirement decreases, not increases, with age.

12.11 MILK PROTEINS IN CLINICAL DIETETICS

The benefits of dairy products are due to their high content of nutritionally valuable proteins as well as of calcium and a number of vitamins. Negative factors in a milk-dominated diet are the low levels of fibre and iron, which often lead to constipation and iron deficiency anaemia, respectively. Another disadvantage is the milk fat, which constitutes a much-discussed and dominating source of saturated fatty acids in many industrialised countries.

Milk proteins are considered still by some to be the nutritionally best-valued food proteins due to their content of essential amino acids and a high digestibility. Dairy products are, therefore, often used in the treatment of malnutrition in children in various health programmes in developing, as well as developed, countries; the use of milk and dairy products in school-feeding programmes in many countries is one such example.

Dairy products are used in specific preparations for clinical dietetics as a major ingredient in formulae due to their high nutritive value and high acceptability. Often, milk and milk products are accepted well metabolically and easy to take in small quantities. They are, therefore, suitable for use in the formulation of well-balanced small meals. In the case of advanced gastrointestinal dysfunction, however, the lactose content must be reduced as such diseases may include a secondary lactase deficiency.

The increased interest in enteral feeding of patients with nutritional problems pre- and post-operatively, as an alternative or supplement to parenteral nutrition, has also increased the interest for high-quality, easily digested proteins.

Dairy products have also been used for a long time in the diets of patients with peptic ulcers. This was motivated by their high content of protein and phosphate, which gave them a good buffering capacity. However, modern treatment with antacid preparations has reduced the need for diets with a high buffering capacity. The ingestion of large quantities of milk and dairy products by these patients often leads to a high intake of saturated fat and an increased risk for coronary heart disease.

The use of partly hydrolysed milk protein products for patients with reduced digestive capacity, e.g., patients with cystic fibrosis, has also been tested but is hampered by the formation of bitter-tasting products.

The low content of phenylalanine and tyrosine in whey proteins has made them suitable as a protein source in formulae used in the dietetic treatment of patients with disturbed metabolic capacity for these amino acids, i.e., phenylketonuria, tyrosinemia and hyperphenylalaninemia. When using such a formula, it was possible to obtain a more positive nitrogen balance and a more pronounced nitrogen retention than with a conventional humanised breast-milk substitute, which illustrated the high nutritional value of the whey proteins (Hambræus *et al.*, 1974).

Fermented milk products, e.g., products with *Lactobacillus acidophilus*, have been fed to geriatric patients with good results. The nutritive value of milk protein is not disturbed but the effect on the intestinal flora reduces the risk of constipation and may also have a slight antibiotic effect on the intestinal microflora (Nahaisi, 1986).

12.12 Milk for high-energy consumers

High energy consumers such as athletes, especially those involved in weight lifting and body-building, but also those engaged in endurance sports, have been using milk products as a source of high-value dietary protein supplements (Di Pasquale, 1997). There are some reports in which the investigators attempted to verify a positive effect of high protein intake on their performance. However, there is still a lot to be learned about any potential benefit of a high protein intake as well as whether or not there is an increased demand for protein by athletes.

Protein does not contribute to muscular energy to any great extent according to studies on macro-nutrient utilisation and urea excretion in the urine (Forslund *et al.*, 1998, 1999). During endurance exercise, reduced protein synthesis (Millward *et al.*, 1982) and increased protein degradation (Lemon and Nagle, 1981; Young and Torun, 1981) have been observed. The contribution of protein to energy expenditure may also be related to the intensity and length of exercise (Lemon, 1985). Recently, a number of 24 h metabolic studies have been performed to investigate the energy-protein interaction during moderate endurance exercise on a bicycle ergometer (45% of VO_2 max) at a normal and high protein intake based on skim milk protein (El-Khoury *et al.*, 1997; Forslund *et al.*, 1998, 1999). The results indicated that although amino acid oxidation is increased during muscle activity, this is compensated for post-exercise, leading to no change in protein balance from a 24 h perspective. There was no difference with respect to a normal and a high protein intake and the losses in sweat were small even on a high protein diet. Thus, there was no increased protein requirement in athletes at energy balance; neither did a high protein intake lead to a positive nitrogen balance as it resulted in increased amino acid oxidation. However, protein intake post-exercise may have a positive effect on the rate of restoration of muscle protein in a short-term perspective.

Exercise-induced hypertrophy of the muscles depends on increased protein synthesis, which may, in theory, result in an increased protein requirement. Excessive protein intake may, however, affect performance by causing reduced carbohydrate intake and stressing the liver and kidneys. Thus, a carbohydrate-rich diet, constituting 60–70% of the energy intake, and a protein intake corresponding to 2 g per kg per day has been recommended for athletes. In order to reduce any waste of protein due to a low quality, animal protein sources are usually recommended due to their high digestibility and nutritionally complete amino acid composition. On the basis of these characteristics, milk protein should be a favourable source of dietary protein. Furthermore, it has been argued that the high protein bioavailability and high content of branched-chain amino acids in milk proteins have an ergogenic effect (Dragan *et al.*, 1985; Di Pasquale, 1997).

The high energy consumption by athletes, however, indirectly leads to a concomitant high protein intake when they consume a conventional mixed diet. Thus, an athlete with a daily energy turnover of 21 000 kJ (5000 kcal) on a conventional diet containing protein at 12% of energy will consume 150 g of protein. This is more than twice their protein requirement (0.8 g kg^{-1} body weight for a 75 kg man) and consequently will meet any increased demand of extra protein demand due to muscle hypertrophy or compensate for an increased turnover of muscle protein. In addition, most athletes consume a diet which has more than 12% of energy from protein. Furthermore, the higher daily intake of protein reduces the need for high-value protein, i.e., content of essential amino acids, since there is an efficient re-utilisation of amino acids, especially of essential amino acids, in the body. Thus, there seems to be very limited scientific evidence so far that milk protein is especially needed by athletes for nutritional reasons. However, most dairy products are well tolerated and could be consumed as convenient foods. Especially, high-energy consumers and athletes in intensive training have problems to meet their energy needs during training periods and need easily digested, highly nutritious meals. This makes dairy products especially well suited for athletes.

12.13 Can milk proteins be hazardous?

12.13.1 Milk intolerance

Intolerance to milk, essentially cows' milk as this is most commonly used in the diet, has been reported and will of course interfere with the nutritional value of dairy products. The intolerance could be due to various mechanisms and thereby be related to different constituents in the milk. Perhaps the most common syndrome is due to *lactose intolerance*, but quite a number of syndromes are also due to milk protein components and should be commented on here, as they interfere with the nutritional value of milk protein.

One of the practical problems with food intolerance related to milk and dairy products is the fact that milk or milk components are often used in combination with other food items in processed food. The use of such products in clinical dietetics may thus lead to serious complications in dietetic treatment of patients with milk protein intolerance.

Food intolerance can include various forms of adverse reactions from psychological reasons, which often are referred to as *food aversion*, to allergic reactions due to an abnormal immunological reaction. In addition, adverse reactions could be due to a lack of certain enzymes, e.g., in lactose intolerance, or to a pharmacological effect of certain compounds, irritation of mucous membranes, leading to gastritis or diarrhoea. Increased gas production and colic may also occur due to fermentation of unabsorbed food residues. Often, it can be difficult to differentiate between adverse effects of milk proteins due to food intolerance, food allergy or food aversion. In most cases, the clinical diagnosis of food intolerance can be established only if the symptoms and signs disappear when an elimination diet is administered.

Damage to the small intestinal mucosa has been reported as a result of intolerance to cows' milk protein in children, which can lead to malnutrition and irritable bowel syndrome has been experienced in some adults and is considered to be related to milk protein intolerance. However, the incidence and prevalence of milk protein intolerance is still not known.

12.13.2 Milk protein allergy

Hypersensitivity reactions to cows' milk are confined to milk proteins, as antigenic reactions do not occur to fat or carbohydrate components. Consequently, all different milk protein fractions, i.e., any of the four caseins as well as various whey proteins, may give rise to an allergenic reaction. Interestingly, it has been reported that the capacity to induce an immune response can be increased markedly by the Maillard reaction (Eriksson, 1984). Digestion of milk proteins may also give rise to new and quite distinct patterns of immune response.

It has been shown that early introduction to cows' milk-based formula can lead to allergic reactions against milk proteins as well as other food proteins. All proteins occurring in cows' milk are potential allergens. Hydrolysis of milk proteins has been used to reduce the antigenicity (Vandenplas *et al.*, 1995), but partial digestion of milk proteins may not result necessarily in a hypoallergic milk product (Schmidt *et al.*, 1995). Small amounts of cows' milk protein may lead to an increased risk of anaphylactic sensitisation of infants when compared to a full cows' milk protein diet. Since there are reasons to believe that the problem in milk allergy is more likely to be due to the response by the infant than to the food itself, the concept of hypoallergenic milks as a result of partly digested or reduced cows' milk protein content is not considered scientifically sound (Ferguson and Watret, 1988). Pasteurisation does not seem to reduce the potential allergenic effect of milk proteins in sensitised individuals but intensive heat treatment and extensively hydrolysed casein or whey products do. However, such treatment may destroy the nutritional value of other essential components in the milk, e.g., vitamins. In the evaluation of the nutritional significance of the use of hypoallergenic milk formulae, the immunogenicity of the products must be related to the nutritional quality.

Atopic eczema has been reported to occur as a result of intolerance to cows' milk protein in children and has been considered to occur also in exclusively breast-fed infants due to carry-over of cows' milk protein antigen in the mothers' milk. However, intolerance to cows' milk protein does not seem to be an important component in adult atopic eczema (Barnetson *et al.*, 1981).

Intolerance to cows' milk protein, leading to diarrhoea and failure to thrive, has been reported in a number of studies since the 1960s but these cases must be differentiated diagnostically from celiac disease as well as lactose intolerance (Kuitunen *et al.*, 1975). Rectal bleeding, resulting in loose mucousy and bloody stools, has also been described in infants under the age of one year as a result of intolerance to cows' milk protein. Recently, increased interest has been focused on the appearance of infantile colic in breast-fed infants as a result of exposure to antigens of cows' milk protein that had to be eliminated from the mother's diet. Lothe *et al.* (1982) and Lothe and Lindberg (1989) found that cows' milk whey proteins can elicit symptoms of infantile colic in colicky infants and that infants with infantile colic had significantly higher macromolecular absorption. Their studies indicated that the gut mucosa is affected in infantile colic which could be due to damage to mucosa or immature intestinal function (Jakobsson and Lindberg, 1978; Lothe *et al.*, 1982; Lothe and Lindberg, 1989).

12.13.3 Are milk proteins atherogenic?

In the nutritional debate on the relationship of diet and health there has been an ongoing discussion regarding the potential risks of homogenised milk (Oster, 1971). This has been related to the role of xanthine oxidase in the milk fat globule membrane as homogenisation might increase the effect of xanthine oxidase which would then oxidise plasmalogens and increase the risk of heart infarcts. However, this has been much criticised and is not generally accepted (Oster and Ross, 1982; Clifford *et al.*, 1983).

Of special nutritional concern is the discussion of the role of casein as a promotor of atherosclerosis (Terpstra *et al.*, 1983). As early as the 1920s, rabbits fed casein were shown to develop arterial lesions. In the 1950s, increased levels of serum cholesterol were reported in rats fed casein-based diets. However, in strictly controlled human studies, casein was not found to increase serum cholesterol (Sacks *et al.*, 1983). Although such effects have been shown to occur in experimental animals, usually rabbits and rats, no clear differential effect of casein and soy protein has been demonstrated on plasma cholesterol in humans (West and Beynen, 1985). Thus, there seems to be a species-dependent effect of dietary casein on cholesterol metabolism.

The hypothesis regarding the hypercholesterolemic effect of casein has been divided in three groups:

- an effect of the amino acid composition of the dietary protein on the endocrine response;
- an increased intestinal absorption of cholesterol, and
- reduced bile acid synthesis.

It has been suggested that casein-induced hypercholesterolemia is caused primarily by the effect of casein on the enterohepatic circulation of steroids. The hypercholesterolemic effect of dietary casein depends on a low intestinal phosphatase activity in combination with a high ratio of glycine-conjugated/taurine-conjugated bile acids. However, the intestinal phosphatase activity in humans seems to counteract any casein-induced hypercholesterolemia. Of nutritional interest is the fact that dietary calcium phosphate may have a favourable effect on cholesterol metabolism. Thus, as dairy products are the main sources of calcium phosphate as well as of casein, more research is needed to understand better the potentially hazardous effect of casein in causing atherosclerosis.

12.13.4 Milk protein intake and diabetes mellitus

Exposure to cow milk-based formulae at early age in infancy has been reported to be associated with an increased risk of insulin-dependent diabetes mellitus (Virtanen *et al.*, 1996). However, studies on the possible pathogenic mechanisms linking cows' milk and insulin dependent diabetes mellitus are contradictary. It has been postulated that the finding is explained by oral immunisation to bovine insulin present in cows' milk (Vaarala *et al.*, 1998, 1999). The risk is suggested to be essentially a result of feeding cows' milk as a trigger of immunity to insulin and that this is of special concern for those with genetic predisposition for Type I diabetes, e.g., siblings of children with diabetes. Long-term multi-center studies are being performed currently to investigate these hypotheses.

12.14 Conclusion

Milk represents one of the most complete single foods and the different composition of milks from various species might be an indicator by Nature of the specific nutrient requirements of that species. The use of various types of milk in the human diet without doubt takes advantage of their content of essential nutrients and dairy products function as high-valued supplements to many diets in different regions of the world. To date, most interest has been devoted to the macro-nutrient aspects of the role of milk proteins in the human diet. There is, however, a conflict between the strict nutritional aspects, which essentially take into account the milk proteins as a source of amino acids for protein synthesis, and the more specific physiological roles of certain proteins in the milk, which exert well-defined metabolic effects. There are also a number of potentially very active proteins and peptides which may have significant physiological roles during the neonatal period and which may also have effects in adults. There are, consequently, still many questions to be answered regarding the physiological implications of the various characteristics of milk proteins and their relation to the nutritional and metabolic effects in humans. More studies are needed and it is possible that various specific milk protein components could be used in the future for specific nutritional and pharmacological purposes. Thus, by taking advantage of advanced dairy technology, whey proteins may be utilised in human nutrition to a considerably greater extent than at present, thereby increasing their economic value.

References

Abrahamsson, L., Bengtsson, O., Hambræus, L. and Holm, H. (1977) Protein quality of milk-cereal based foods for infants and children in relation to processing methods and composition of the products. *J. Food Technol.*, **14**, 429–40.

Andersen, K.J. and von der Lippe, G. (1979) The effect of proteolytic enzymes on the vitamin B_{12}-binding proteins of human gastric juice and saliva. *Scand. J. Gastroent.*, **14**, 833–8.

Atkinson, S.A., Anderson, G.H. and Bryan, M.H. (1980) Human milk: comparison of the nitrogen composition in milk from mothers of premature and full-term infants. *Am. J. Clin. Nutr.*, **33**, 811–5.

Atkinson, S.A., Anderson, G.H. and Bryan, M.H. (1981) Energy and macro-nutrient content of human milk during early lactation for mothers giving birth prematurely and at term. *Am. J. Clin. Nutr.*, **34**, 258–65.

Barnetson, R.StC., Merrett, T.G. and Ferguson, A. (1981) Studies on hyperimmunoglobulinaemia E in atopic diseases with particular reference to food allergens. *Clin. Exper. Immunol.*, **46**, 54–60.

Battermann, W. (1986) Whey protein for athletes. *Deutsch. Milchwirtsch.*, **37**, 1010–2.

Barth, C.A. and Schlimme, E. (1989) *Milk Proteins-Nutritional, Clinical, Functional and Technological Aspects*, Springer Verlag, New York.

Bell, K. and McKenzie, H.A. (1964) β-Lactoglobulins. *Nature*, **204**, 1275–9.

Bezkorovainy, A., Grohlich, D. and Nichols, J.H. (1979) Isolation of a glycopolypeptide fraction with *Lactobacillus bifidus* subspecies *pennsylvanicus* growth-promoting activity from whole human milk casein. *Am. J. Clin. Nutr.*, **32**, 1428–32.

Björnhag, G., Knutsson, P.-G. and Sperber, I. (1979) Postnatal growth and milk composition. *Swedish J. Agric. Res.*, **9**, 65–74.

Blanc, B. (1981) Biochemical aspects of human milk-comparison with bovine milk, *World Rev. Nutr. Diet.*, **36**, 1–89.

Bohak, Z. (1964) N-Epsilon-(DL-2-amino-2-carboxyethyl)-L-lysine, a new amino acid formed on alkaline treatment of proteins. *J. Biol. Chem.*, **239**, 2878.

Boza, J.J., Jimenez, J., Martinez, O., Suarez, M.D. and Gil, A. (1994) Nutritional value and antigenicity of two milk protein hydrolysates in rats and guinea pigs. *J. Nutr.*, **124**, 1978–86.

Brantl, V. and Teschemacher, H. (1982) Opiartig wirkende Stoffe in Milch und Milchprodukten. 2. Isolierung, Strukturaufklärung und biologische Wirkung der β-casomorphine. *Milchwissenschaft*, **37**, 641–4.

Brock, J.R., Arzabe, F., Lampreave, F. and Pineiro, A. (1976) The effect of trypsin on bovine transferrin and lactoferrin. *Biochim. Biophys. Acta*, **446**, 214–25.

Bullen, J.J., Rogers, J.J. and Leigh, L. (1972) Iron-binding proteins in milk and resistance to *Escherichia coli* infection in infants. *Brit. Med. J.*, **1** (792), 69–75.

Clegg, K.M., Lim, C.L. and Manson, W. (1974) The structure of a bitter peptide derived from casein by digestion with papain. *J. Dairy Res.*, **41**, 283–7.

Clifford, A.J., Ho, C.Y. and Swenerton, H. (1983) Homogenized bovine milk xanthine oxidase: a critique of the hypothesis relating to plasmalogen depletion and cardiovascular disease. *Am. J. Clin. Nutr.*, **38**, 327–32.

Conti, A., Liberatori, J. and Napolitano, L. (1980) Isolation and preliminary physico-chemical characterization of human β-lactoglobulin. *Milchwissenschaft*, **35**, 65–8.

Cox, T.M., Mazurier, J., Spik, G., Montreuil, J. and Peters, T.J. (1979) Iron binding proteins and influx of iron across the duodenal brush border. Evidence for specific lactoferrin receptors in the human intestine. *Biochim. Biophys. Acta*, **588**, 120–8.

Davidson, L. and Lönnerdal, B. (1987) Persistence of human milk proteins in the breast-fed infant. *Acta Paediatr. Scand.*, **76**, 733–40.
Davidson, L. and Lönnerdal, B. (1988) Specific binding of lactoferrin to brush border membrane: ontogeny and effect of glycan chain. *Am. J. Physiol.*, **254**, G580–5.
Di Pasquale, M.G. (1997) *Amino Acids and Proteins for the Athlete. The Anabolic Edge*, CRC Press, Boca Raton, FL.
Donovan, S. and Lönnerdal, B. (1989) Non-protein nitrogen and true protein in infant formulas. *Acta Paediatr. Scand.*, **78**, 497–504.
Donovan, S., Lönnerdal, B. and Atkinson, S.A. (1990) Bioavailability of urea nitrogen for the low birthweight infant. *Acta Paediatr. Scand.*, **79**, 899–905.
Dragan, G.I., Vasiliu, A. and Georgescu, E. (1985) Effects of refit on elite weightlifters. *J. Sports Med. Phys. Fitness*, **25**, 246–50.
El-Khoury, A.E., Forslund, A., Olsson, R., Branth, S., Sjödin, A., Andersson, A., Atkinson, A., Selvaraj, A., Hambræus, L. and Young, V.R. (1997) Moderate exercise at energy balance does not affect 24-h leucine oxidation or nitrogen retention in healthy men. *Am. J. Physiol.*, **273** (*Endocrinol. Metab. 36*), E394–407.
Eriksson, C. (ed.) (1981) *Maillard Reactions in Food: Chemical, Physiological and Technological Aspects*, Pergamon Press, Oxford.
FAO (1957) *Protein Requirements*. FAO Nutritional Studies, No. 16. FAO, Rome.
FAO (1970) *Amino Acid Content of Foods*. FAO Nutritional Studies, No. 24. FAO, Rome. FAOSTAT Database 1999. Internet: http://www.fao.org.
FAO/WHO (1965) *Protein Requirements*. WHO Technical Report Series, No. 301. WHO, Geneva.
FAO/WHO (1973) *Protein and Energy Requirements*. WHO Technical Report Series, No. 522. WHO, Geneva.
FAO/WHO (1991) *Protein Quality Evaluation*. Report of a Joint FAO/WHO Expert Consultation FAO Nutrition paper 51, Rome.
FAO/WHO/UNU Expert Consultation (1985) *Energy and Protein Requirements*. WHO Technical Report Series, No. 724. World Health Organization, Geneva.
Farrell, H.M., Jr. and Thompson, M.P. (1971) Biological significance of milk protein polymorphism. *J. Dairy Sci.*, **54**, 1219–28.
Ferguson, A. and Watret, K.C. (1988) Cows' milk intolerance. *Nutr. Rep. Rev.*, **1**, 1–22.
Fomon, S. (1974) *Infant Nutrition*, 2nd edn, WB Saunders, Philadelphia.
Ford, J.E., Salter, D.N. and Scott, K.J. (1969) The folate-binding protein in milk. *J. Dairy Res.*, **36**, 435–46.
Forslund, A., Hambræus, L., Olsson, R., El-Khoury, A.E., Yu, Y.-M. and Young, V.R. (1998) The 24-h whole body leucine and urea kinetics at normal and high protein intake with exercise in healthy adults. *Am. J. Physiol.*, 275 (*Endocrinol. Metab.* 38), E310–20.
Forslund, A., El-Khoury, A.E., Olsson, R., Sjödin, A., Hambræus, L. and Young, V.R. (1999) Effect of protein intake and physical activity on 24-h pattern and rate of macro-nutrient utilization. *Am. J. Physiol.*, **276** (*Endocrinol. Metabol. 39*), E964–76.
Forsum, E. (1973) Nutritional evaluation of whey protein concentrates and their fractions. *J. Dairy Sci.*, **57**, 665–70.
Forsum, E. and Hambræus, L. (1974) Utilization of whey proteins in human nutrition, in *Joint IUPAC/IUFoST Symposium-The Contribution of Chemistry to Food Supplies*, (I. Morton and D.N. Rhodes eds.) Butterworths, London, pp. 375–83.

Forsum, E., Hambræus, L. and Siddiqi, I.H. (1973) Fortification of wheat by whey protein concentrate, dried skim milk, fish protein concentrate and lysine monohydrochloride. *Nutr. Rep. Int.*, **8**, 39–48.

Fransson, G.B. and Lönnerdal, B. (1980) Iron in human milk. *J. Pediatr.*, **96**, 380–4.

Freudenstein, C., Keenan, T.W., Eigel, W.N., Sasaki, M., Stadler, J. and Franke, W.W. (1979) Preparation and characterization of the inner coat material associated with fat globule membranes from bovine and human milk. *Exp. Cell Res.*, **118**, 227–94.

Galesloot, T.E. and Tinbergen, B.J., eds. (1985) *Milk Proteins* '84. *Proc. Intern. Congr. Milk Proteins*, Agricultural Publishing and Documentation, Wageningen.

György, P., Norris, R.F. and Rose, C.S. (1953) Bifidus factor. I. A variant of *Lactobacillus bifidus* requiring a special growth factor. *Arch. Biochem. Biophys.*, **48**, 193–201.

Hambræus, L. (1984) Human milk composition. *Nutr. Abstr. Rev.*, **54**, 219–36.

Hambræus, L., Forsum, E. and Lönnerdal, B. (1977) Nutritional aspects of breast milk versus cows' milk formula. In *Food and Immunology*, *Symposia of the Swedish Nutrition Foundation XIII*, (L. Hambræus, L.Å. Hanson and H. McFarlane eds.) Almqvist and Wiksell, Uppsala, pp. 116–24.

Hambræus, L., Hardell, L.I., Forsum, E. and Lorentsson, R. (1974) Use of a formula based on a whey protein concentrate in the feeding of an infant with hyperphenylalaninemia. *Nutr. Metabol.*, **17**, 84–90.

Hambræus, L., Lönnerdal, B., Forsum, E. and Gebre-Medhin, M. (1978) Nitrogen and protein components of human milk. *Acta Paediatr. Scand.*, **67**, 561–5.

Hambræus, L., Fransson, G.-B. and Lönnerdal, B. (1984) Nutritional availability of breast milk protein. *Lancet* **ii**, 167–8.

Hanning, R.M., Paes, B. and Atkinson, S.A. (1992) Protein metabolism and growth of term infants in response to a reduced-protein, 40:60 whey:casein formula with added tryptophan. *Am. J. Clin. Nutr.*, **56**, 1004–11.

Hiraoka, Y., Segawa, T., Kuwajima, K., Sugai, S. and Murai, N. (1980) α-Lactalbumin: a calcium metalloprotein. *Biochem. Biophys. Res. Commun.*, **95**, 1098–104.

International Dairy Federation (IDF) (1988) Code of practice for preservation of raw milk by the lactoperoxide system, Bulletin 234, International Dairy Federation, Brussels.

International Dairy Federation (IDF) (1998) Nutrition properties of milk proteins. F-Doc 286, International Dairy Federation, Brussels.

Jakobsson, I. and Lindberg, T. (1978) Cows' milk as a cause of infantile colic in breast-fed infants. *Lancet*, **ii**, 437–40.

Jenness, R. and Sloan, R.E. (1970) The composition of milks of various species: a review. *Dairy Sci. Abstr.*, **32**, 599–612.

Koldovsky, O. (1989) Search for role of milk-borne biologically active peptides for the suckling. *J. Nutr.*, **119**, 1543–51.

Kon, S.K. and Cowie, A.T. (1961) *Milk-The Mammary Gland and Its Secretion*, Vols. 1 and 2. Academic Press, New York.

Kuitunen, P., Visakorpi, J.K., Savilahti, E. and Pelkonen, P. (1975) Malabsorption syndrome with cows' milk intolerance. Clinical findings and course in 54 cases. *Arch. Dis. Childhood*, **50**, 351–6.

Kunz, C. and Lönnerdal, B. (1989) Human milk proteins: separation of whey proteins and their analysis by polyacrylamide gel electrophoresis, fast protein liquid chromatography (FPLC) gel filtration, and anion-exchange chromatography. *Am. J. Clin. Nutr.*, **49**, 464–70.

Lemon, P.W.R. (1985) Metabolic aspects of proteins for top sportsmen. In *Milk Proteins '84, Proc. Int. Congr. Milk Proteins*, (T.E. Galesloot and B.J. Tinbergen eds.) Agricultural Publishing and Documentation, Wageningen, pp. 88–98.

Lemon, P.W.R. and Nagle, F.J. (1981) Effects of exercise on protein and amino acid metabolism. *Med. Sci. Sports Exer.*, **13**, 141–43.

Lindblad, B.S. and Rahimtoola, R.J. (1974) A pilot study of the quality of human milk in a lower socio-economic group in Karachi, Pakistan. *Acta Paediatr. Scand.*, **63**, 125–8.

Lönnerdal, B. (1985) Biochemistry and physiological function of human milk proteins. *Amer. J. Clin. Nutr.*, **42**, 1299–317.

Lönnerdal, B. and Adkins, Y. (1999) Developmental changes in breast milk protein composition during lactation. In: *Development of the Gastrointestinal Tract*, (I.R. Sanderson and B.C. Walker eds.) Decker Inc., Hamilton, Ontario, Canada, pp. 227–44.

Lönnerdal, B. and Chen, C.L. (1990a) Effects of formula protein level and ratio on infant growth, plasma amino acids and serum trace elements. I. Cows' milk formula. *Acta Paediatr. Scand.*, **79**, 257–65.

Lönnerdal, B. and Chen, C.L. (1990b) Effects of formula protein level and ratio on infant growth, plasma amino acids and serum trace elements. II. Follow-up formula. *Acta Paediatr. Scand.*, **79**, 266–73.

Lönnerdal, B. and Forsum, E. (1985) Casein content of human milk. *Am. J. Clin. Nutr.*, **41**, 113–20.

Lönnerdal, B. and Glazier, C. (1985) Calcium binding by α-lactalbumin in human milk. *J. Nutr.*, **115**, 1209–16.

Lönnerdal, B., Forsum, E. and Hambræus, L. (1976a) The protein content of human milk. I. A transversal study of Swedish normal material. *Nutr. Rep. Int.*, **13**, 125–34.

Lönnerdal, B., Forsum, E. and Hambræus, L. (1976b) A longitudinal study of the protein, nitrogen and lactose contents of human milk from Swedish well-nourished mothers. *Am. J. Clin. Nutr.*, **29**, 1127–33.

Lothe, L. and Lindberg, T. (1989) Cows' milk whey protein elicits symptoms of infantile colic in colicky formula-fed infants: A double-blind cross-over study. *Pediatrics*, **83**, 262–6.

Lothe, L., Lindberg T. and Jakobsson, I. (1982) Cows' milk formula as a cause of infantile colic: a double blind study. *Pediatrics*, **70**, 7–10.

Macy, J.G., Nims, B., Brown, M. and Hunscher, H.A. (1931) Human milk studies. VII. Chemical analysis of milk representative of the entire first and last halves of the nursing period. *Am. J. Dis. Child.*, **42**, 569–89.

Mazurier, J., Montreuil, J. and Spik, G. (1985) Visualization of lactoferrin brush-border receptors by ligand-blotting. *Biochim. Biophys. Acta*, **821**, 453–60.

McKenzie, H.A. (1970, 1971) *Milk Proteins, Chemistry and Molecular Biology*, Vol. 1 and 2, Academic Press, New York.

Mellander, O. and Folsch, G. (1971) Enzyme resistance and metal binding of phosphorylated peptides. In *Protein and Amino Acid Functions*, (E.J. Bigwood ed.) Pergamon Press, Oxford, pp. 569–79.

Millward, D.J., Davies, C.T.M., Halliday, D., Wolman, S.L., Matthews, D. and Rennie, M. (1982) Effect of exercise on protein metabolism in humans as explored with stable isotopes. *Fed. Proc.*, **41**, 2686–91.

Millward, D.I. (1997) Human amino acid requirements. *J. Nutr.*, **124**, 1842–6.

Mitchell, H.H. and Block, R.J. (1946) Some relationships between the amino acid contents of proteins and their nutritive values for the rat. *J. Biol. Chem.*, **163**, 599–630.

Nahaisi, M.H. (1986) *Lactobacillus acidophilus*: therapeutic properties, products and enumeration. *Develop. Food Microbiol.*, **2**, 153–78.

Oftedal, O. (1980) Milk composition and formula selection for hand-rearing young mammals. In *The Nutrition of Captive Wild Animals*, (E.R. Maschgan M.E. Allen and L.E. Fischer eds.) First Annual Dr Scholl Nutrition Conference, pp. 67–83.

Oster, K.A. (1971) Plasmalogen diseases: a new concept of the etiology of the atherosclerotic process. *Am. J. Clin. Res.*, **2**, 30–5.

Oster, K.A. and Ross, D.J. (1982) Flawed objectives to the bovine milk xanthine oxidase theory. *J. Am. Coll. Nutr.*, **1**, 420.

Pfeuffer, M. and Barth, C.A. (1988) Influences of casomorphines on plasma lipid levels and lipid secretion rates. In *Milk Proteins, Nutritional, Clinical, Functional and Technological Aspects*, (C.A. Barth and E. Schlimme eds.) Springer Verlag, New York, pp. 291–2.

Pineda, O., Torun, B., Viteri, F.E. and Arroyave, G. (1981) Protein quality in relation to estimates of essential amino acid requirements. In *Protein Quality in Humans*, (C.E. Bodwell, J.S. Adkins and D.T. Hopkins eds.) AVI publishing Co. Inc., Westport, Connecticut, pp. 29–42.

Räihä, N.C.R., Minoli, I. and Moro, G. (1986a) Milk protein intake in the term infant: I. Metabolic responses and effects on growth. *Acta Paediatr. Scand.*, **75**, 881–6.

Räihä, N.C.R., Minoli, I., Moro, G. and Bremer, H. (1986b) Milk protein intake in the term infant: II. Effects on plasma amino acid concentrations. *Acta Pediatr. Scand.*, **75**, 887–92.

Reiter, B. (1978) Review of the Progress of Dairy Science: antimicrobial systems in milk. *J. Dairy Res.*, **45**, 131.

Renner, E. (1983) *Milk and Dairy Products in Human Nutrition*, Verlag, Munich.

Rudloff, S. and Lönnerdal, B. (1992) Solubility and digestibility of milk proteins in infant formulas exposed to different heat treatment. *J. Pediatr. Gastroenterol. Nutr.*, **15**, 25–33.

Saarinen, U.M., Siimes, M.A. and Dallman, P.R. (1977) Iron absorption in infants. High bioavailability of breast milk iron as indicated by extrinsic tag method of iron absorption and by the concentration of serum ferritin. *J. Pediatr.*, **91**, 36–9.

Sacks, F.M., Breslow, J.L., Wood, P.O. and Kass, E.H. (1983) Lack of an effect of dairy protein (casein) and soy protein on plasma cholesterol of strict vegetarians. An experiment and a critical review. *J. Lipid Res.*, **24**, 1012–20.

Salter, D.N. and Mowlem, A. (1983) Neonatal role of milk folate-binding proteins. Studies on the course of digestions of goat's milk folate binder in the 6-d child. *Brit. J. Nutr.*, **50**, 589–96.

Sandberg, D.P., Begley, J.A. and Hall, C.A. (1981) The content, binding and forms of vitamin B_{12} in milk. *Am. J. Clin. Nutr.*, **34**, 1717–24.

Schmidt, D.G., Maijer, R.J., Slangen, C.J. and van Beresteijn, E.C. (1995) Raising the pH of the pepsin-catalysed hydrolysis of bovine whey proteins increases the antigenicity of the hydrolysates. *Clin. Exp. Allergy*, **25**, 1007–17.

Steinberg, L.A., O'Connell, N.C., Hatch, T.F., Picciano, M.F. and Birch, L.L. (1992) Tryptophan intake influences infants sleep latency. *J. Nutr.*, **122**, 1781–91.

Sturman, J.A., Gaull, G.E. and Räihä, N.C.R. (1970) Absence of cystathionase in human fetal liver: Is cystine essential? *Science*, **169**, 74–5.

Terpstra, A.H.H., Hermus, R.J.J. and West, C.E. (1983) The role of dietary protein in cholesterol metabolism. *World Rev. Nutr. Diet.*, **42**, 1–55.

Torun, B., Pineda, O., Viteri, F.E. and Arroyave, G. (1981) Use of amino acid composition data to predict protein nutritive value for children with special reference to new estimates of their essential amino acid requirements. In *Protein*

Quality in Humans, (C.E. Bodwell, J.S. Adkins and D.T. Hopkins eds.) AVI Publishing Co. Inc., Westport, Connecticut, pp. 374–93.

Vaarala, O., Paronen, J., Otonkoski, T. and Åkerblom, H.K. (1998) Cow milk feeding induces antibodies in children – a link between cow milk and insulin-dependent diabetes mellitus? *Scand. J. Immunol.*, **15**, 131–5.

Vaarala, O., Knip, M., Paronen, J., Hämälainen, A.M., Muona, P., Väätäinen, M., Ilonen, J., Simell, O. and Åkerblom, H.K. (1999) Cows' milk formula feeding induces primary immunization to insulin in infants at genetic risk for type 1 diabetes. *Diabetes*, **48**, 1389–94.

Vandenplas, Y., Hauser, B., Van der Borre, C. and Clybouw, C. (1995) The long-term effect of a partial whey hydrolysate formula on the prophylaxis of atopic disease. *Eur. J. Pediatr.*, **154**, 488–94.

Vickery, H.B. (1950) The origin of the word protein. *Yale J. Biol. Med.* May 22.

Virtanen, S.M., Hyppönen, E., Läärä, E., Vähäsalo, P., Kulmala, P., Savola, K., Räsänen, L., Aro, A., Knip, M. and Åkerblom, H.K. (1996) Diabetes and cows' milk. *Lancet*, **15**, 1656–7.

West, C.E. and Beynen, A.C. (1985) Milk proteins in contrast to plant proteins: effects on plasma cholesterol. In *Milk Proteins* '84. *Proc. Int. Congr. Milk Protein*, (T.E. Galeshot and B.J. Tinbergen eds.) Agricultural Publishing and Documentation, Wageningen, pp. 80–7.

Whitney, R.McL., Brunner, J.R., Ebner, K.E., Farrell, H.M., Jr., Josephson, R.V., Morr, C.V. and Swaisgood, H.E. (1976) Nomenclature of the proteins of cows' milk: Fourth revision. *J. Dairy Sci.*, **59**, 795–815.

Widdowson, E.M. (1984) Lactation and feeding patterns of different species. In *Health Hazards of Milk*, (D.L.J. Freed ed.) Bailliére Tindall Eastbourne, England, pp. 85–90.

van Woelderen, B.F. (1987) Changing insight into milk proteins: some implications. *Nutr. Abstr. Rev.* (*Ser. A*)., **57**, 129–34.

Woodhouse, L.R. and Lönnerdal, B. (1988) Quantitation of the major whey proteins in human milk, and the development of a technique to isolate minor whey proteins. *Nutr. Res.*, **8**, 853–64.

WHO (1981) *International Code of Marketing Breast-milk Substitute*, World Health Organization, Geneva.

Yamauchi, K., Azuma, N., Kobayashi, H. and Kaminogawa, S. (1984) Isolation and properties of human κ-casein. *Agric. Biol. Chem.*, **48**, 771–6.

Young, V.R. and Torun, B. (1981) Physical activity: impact on protein and amino acid metabolism and implications for nutritional requirements. In *Nutrition in Health and Disease and International Development*, (A.E. Harper and G.K. Davis eds.) A.R. Liss Inc., New York, pp. 57–85.

Young, V.R., Bier, D.M. and Pellett, P.L. (1989) A theoretical basis for increasing current estimates of the amino acid requirements in adult man, with experimental support. *Am. J. Clin. Nutr.*, **50**, 80–92.

Young, V.R. and El-Khoury, A.E. (1996) Human amino acid requirements: a re-evaluation. *Food Nutr. Bull.*, **17**, 191–203.

13

ALLERGENICITY OF MILK PROTEINS

S. Kaminogawa and M. Totsuka

13.1 Introduction

The proportion of the population with allergies caused by food has increased recently in developed nations. It is unfortunate, indeed, when the food that we eat makes us ill, because food is indispensable for life. The Ministry for Health and Welfare of Japan reported that 12.6% of Japanese children aged six years or less have food allergies (Iikura *et al.*, 1999). There are many kinds of food allergens, the substances that cause food allergies. Cow's milk protein is one of the most common food allergens in many developed countries (Eigenman and Sampson, 1997; Chiaramonte *et al.*, 1999).

In this chapter, we describe the characteristics of allergy to cow's milk and its mechanism of onset, and discuss the allergenicity, the ability to trigger an allergic reaction, of various cow's milk proteins and the sites of the molecules recognized by the immune system. We also describe the results of recent studies on the suppression of allergic reactions by introducing changes in the antigenic sites of an allergen through amino acid substitution.

13.2 Characteristics of milk allergy

The onset of cow's milk allergy is generally during infancy. Varying estimates of the incidence from 1.8 to 7.5% of infants have been reported (Høst, 1997). The variation seems to be attributable to differences in diagnostic criteria, study design and the population being studied. The incidence in developed countries with strict diagnostic criteria can be estimated as

Advanced Dairy Chemistry Volume 1: Proteins, 3rd edn.
Edited by P.F. Fox and P.L.H. McSweeney, Kluwer Academic/Plenum Publishers, 2003.

2–5% (for reviews see Bahna and Heiner, 1980; Savilahti, 1981; Savilahti and Kuitunen, 1992; Businco and Bellanti, 1993; Høst, 1994, 1997; Høst *et al.*, 1995; Wal, 1998).

One of the reasons for the increased proportion of infants with allergy to cow's milk seems to be the decrease in the number of mothers who breast-feed their infants for a prolonged period (Savilahti, 1981). The safest food for infants, with few exceptions, is breast milk. Cow's milk formulae which have components similar in kind and proportion to those of breast milk, are the usual replacement for breast milk. Studies on prevalence have shown that 0.5% of exclusively breast-fed infants, 7% of bottle-fed babies and 1.5 to 2.0% of all children have allergy to milk (Bahna and Heiner, 1980). During infancy, a small part of milk proteins that are allergenic may be absorbed through the intestinal wall without being digested because the intestinal tract is immature. This immaturity is considered to be one important cause of the onset of food allergy in infancy, although the degree of permeability of the intestine is not directly related to the risk of developing food allergy in general. The other main factor seems to be the insufficient functioning of the immune system of infants. A fully functioning immune system would suppress the onset of food allergies (see Section 13.3 below).

Atopic dermatitis and gastrointestinal symptoms such as diarrhea, common in patients with food allergies, are seen in many patients with milk allergies. Reactions in the respiratory tract and anaphylactic reactions are also seen (Businco and Bellanti, 1993). Spontaneous resolution of the symptoms with age in most patients is a striking characteristic of milk allergy (Høst, 1994; Høst *et al.*, 1995). The reasons for this remain unclear.

13.3 Mechanism of onset of milk allergy

When food allergens, such as those in milk, enter the body, they react in complex ways with the organs of the gastrointestinal tract. First, allergens are degraded in the stomach and intestine into substances of low molecular weight, such as peptides and amino acids. Substances of sufficiently low molecular weight are not recognized by the immune system and do not cause allergic reactions. Although at present there is no definite molecular weight below which peptides are considered not to be allergenic, it has been suggested that, in general, peptides with a molecular weight <1800 Da are not allergenic (Businco *et al.*, 1999). When allergens that are intact or insufficiently degraded are absorbed *via* the intestinal tract, they may be recognized by the immune system and consequently cause an allergic reaction.

However, if an allergy developed every time an allergen appeared in the blood, there would be a great many more people with allergies. The immune system operating in the intestinal tract has a special response that usually suppresses immune responses to orally-ingested proteins. This phenomenon is called oral tolerance (Thompson and Staines, 1990; Weiner, 1994; Weiner *et al.*, 1994; Kaminogawa, 1996; Kaminogawa *et al.*, 1999) which can be observed experimentally in mice first given a protein orally and then given the same protein parenterally. In such mice, immune responses, such as proliferation of T cells and the production of antibodies to the protein, are less pronounced than those in mice not first given the protein.

Food allergy develops when food allergens are absorbed *via* the intestinal tract intact, without being degraded, provided that various mechanisms of the intestinal tract immune system (primarily, oral tolerance), which serve to suppress hypersensitivity, do not function, and genetic factors predispose the person to allergy. Except for the involvement of the intestinal tract immune system, the basic mechanism of onset in food allergy seems to be the same as that in the case of other allergies. There are four kinds (types I to IV) of hypersensitivity reactions. The overwhelming majority of cases of milk allergy involve type I hypersensitivity reactions (Bruijnzeel-Koomen *et al.,* 1995), the general mechanism of which will be described below.

The sequence of events occurring in type I allergy is illustrated in Figure 13.1. Antigen-presenting cells, T cells, B cells and mast cells, all participate in the onset of type I reactions. First, allergens are internalized by antigen-presenting cells and are then degraded by proteinases. Of the peptides thus produced, the ones that bind to major histocompatibility complex (MHC) class II molecules are presented on the cell surface.

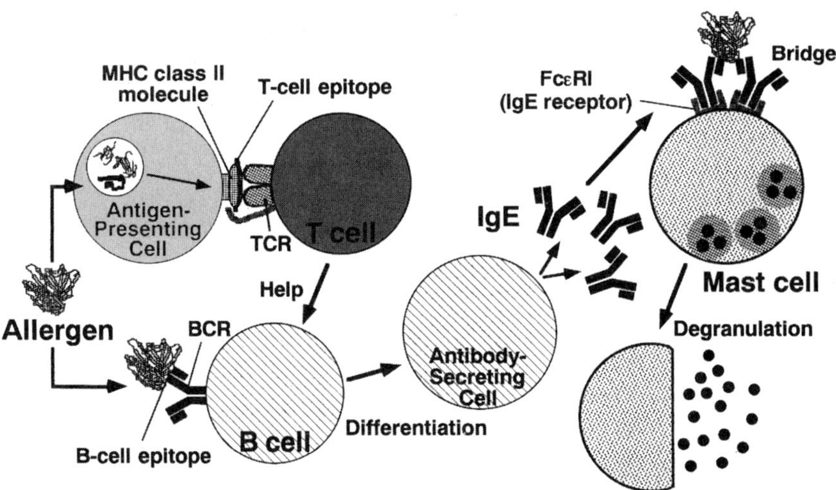

Figure 13.1 Schematic representation of the onset of Type I allergy (Reproduced from Kaminogawa, 1996). BCR: B-cell receptor, IgE: immunoglobulin E, MHC: major histocompatibility complex, TCR: T-cell receptor.

Complexes of an antigenic peptide with a MHC class II molecule are recognized *via* T-cell antigen receptors (TCR) on the surface of T cells. On recognizing such complexes, T cells are activated. The sites on the antigen that bind to MHC molecules and that are recognized by TCR are called T-cell epitopes or T-cell antigenic determinants. T cells are classified as $CD4^+$ (helper) T cells or $CD8^+$ (cytotoxic) T cells according to differences in their cell-surface antigens and differences in their functions. $CD4^+$ T cells are classified as type 1 or type 2 helper T cells (Th1 and Th2 cells) depending on the cytokines they produce (Abbas *et al.*, 1996). The helper T cells involved in the onset of type I allergy are Th2 cells, which produce cytokines such as interleukins -4, 5, 6, 10 and 13.

B cells bind to an intact antigen *via* B-cell antigen receptors (BCR) on their cell surface. The sites on the antigen recognized by BCR are called the B-cell epitopes or B-cell antigenic determinants. Antibodies produced by B cells bind to these same sites. T-cell and B-cell epitopes are usually in different parts of the antigen molecule. B cells degrade antigens bound to BCR into peptides and present them on the cell surface as complexes with MHC class II molecules. When activated T cells interact with B cells, presenting peptides that are recognized by the TCR, they activate the B cells to differentiate into antibody-producing cells. Some of the activated B cells differentiate into cells that produce immunoglobulin E (IgE) which binds to IgE receptors (FcεRI) on the surface of mast cells and, when allergens bind in bridge fashion to allergen-specific IgE bound to FcεRI, chemical mediators which cause inflammation, such as leukotriene and histamine, are released by the mast cells. These substances cause allergic symptoms.

A distinctive feature of the type I allergic process is that the antibodies involved are of the IgE class. Like IgG, IgM and IgA, IgE is produced during normal immune responses, but in very small amounts. Allergic patients, however, often have high levels of IgE.

13.4 Allergenicity of milk proteins

13.4.1 Methods for identifying milk allergens

In almost all cases, the cause of milk allergy is a protein. Bovine milk contains ~3.5% protein, of which 80% is casein, and the remainder is whey proteins. Any of the more than 20 types of protein may cause allergy, but their degree of allergenicity differs.

Typically, four methods are used to identify allergens: 1) challenge tests to provoke symptoms by oral administration of an allergen, 2) skin reaction tests, 3) IgE-binding tests, and 4) *in vitro* tests for histamine release and lymphocyte proliferation. In the first method, i.e., oral administration, an allergen is given by mouth to a person suspected of having a food allergy and the patient is examined later to see if a skin reaction or gastrointestinal symptoms have developed. For skin reaction tests, various methods are used. In the most common method, an allergen is injected under the skin of a patient suspected of having an allergy and the injection site is examined later for a visible reaction. If the injected substance is the allergen causing the patient's allergy, the area of erythema will be larger than that in a person without allergy. IgE-binding tests examine the binding specificity of IgE in the blood of a patient with type I allergy. The allergen is identified by finding which allergen binds specifically to the IgE of the patient. A radioallergosorbent test (RAST) or enzyme-linked immunosorbent assay (ELISA) is used. In the *in vitro* tests for histamine release or lymphocyte proliferation, leukocytes from the patient are incubated with a possible allergen and the amount of histamine released or the proliferative response, respectively, is measured.

13.4.2 Identification of the principal milk allergens

In the 1960s, Goldman *et al.* (1963a,b) studied which components of milk are allergens. These authors performed symptom-provoking challenge tests by orally administering allergens to patients allergic to milk, and found that 62% of the patients tested reacted to β-lactoglobulin, 60% reacted to casein and 53% reacted to α-lactalbumin (Goldman *et al.*, 1963a). Kuitunen *et al.* (1975) performed symptom-provoking challenge tests by orally administering several milk protein fractions separately to eight patients with gastrointestinal symptoms of milk allergy and found that the protein which most frequently provoked an allergic response was β-lactoglobulin. Freier *et al.* (1969) reported similar findings. β-Lactoglobulin has the highest allergenicity, probably because this protein is stable even at a low pH, making it resistant to digestion in the stomach and because it is not present in human breast milk, making it a foreign substance to human infants.

The results of testing to find which proteins are the principal allergens depend on the method used. When Goldman *et al.* (1963b) used skin reaction tests, 36% of the patients tested reacted to casein, 27% to α-lactalbumin, 24% to bovine serum albumin and 23% to β-lactoglobulin. The results of histamine-release and lymphocyte-proliferation tests also showed that casein is strongly allergenic (Baudon *et al.*, 1987). In a study in which immunoblotting was used to identify the milk allergens that bind with IgE in 80 milk-allergic patients (Docena *et al.*, 1996), casein was identified as an allergen for all 80 patients, and β-lactoglobulin (10 patients) and α-lactalbumin (5 patients) were next in frequency. The striking finding in that study was that the patients for whom casein was an allergen greatly outnumbered patients for whom the other proteins were allergens. Bernard *et al.* (1998) investigated which of the four types of casein (α_{s1}-, α_{s2}-, β- and κ-casein) had the greatest allergenicity. Using an ELISA to examine the molecules that bound with IgE in the sera of 58 milk-allergic patients, they found that IgE binds mostly to α_{s1}-casein and β-casein (in that order) and in much smaller amounts to α_{s2}- and κ-casein.

Thus, the allergens that cause milk allergy depend on the individual and different assay methods give different results. Sometimes, multiple allergens together trigger allergic symptoms. In addition, there is no unequivocal relationship between the allergen in a given case and what symptoms develop. However, results of various studies show that β-lactoglobulin and α_{s1}-casein are the principal allergens in cow's milk.

13.5 Allergenicity of heat-processed milk proteins and hydrolysed infant formulae

All of the allergens described above occur naturally in milk. However, many of the dairy products we ingest have been subjected to processes such as heat sterilization and evaporation. At this stage, the proteins in the milk are in new forms and differ from the naturally occurring proteins. For infants with cow's milk allergy, formulae based on milk protein hydrolysates have been developed. In this section, we will describe the allergenicity of heat-processed milk proteins and hydrolysed infant formulae (HF).

13.5.1 Heat-processed milk proteins

Whey proteins such as β-lactoglobulin (Papiz *et al.*, 1986; Brownlow *et al.*, 1997) and α-lactalbumin (Stuart *et al.*, 1986; Pike *et al.*, 1996) have a folded tertiary structure. This conformation can be disrupted by heat. Such denaturation affects their allergenicity (Lee, 1992).

As explained in Section 13.3, allergens have T-cell epitopes which are regions that bind to MHC class II molecules and TCR, and B-cell epitopes which bind with antibodies, including IgE. T-cell epitopes are regions where binding occurs after the allergen has been denatured and degraded to peptides. Thus, T-cell epitopes are affected little by heat. There are two kinds of B-cell epitopes: those that depend on tertiary structure (conformational, or discontinuous, epitopes) and those that depend on only the primary structure (sequential, or continuous, epitopes) (Barlow *et al.*, 1986). When IgE that can recognize a tertiary structure encounters an allergen, the tertiary structure of which was destroyed by heat, this IgE cannot bind and thus the allergenicicty of the antigen has been reduced. In this way, denaturation can reduce the allergenicity of allergens that have a globular tertiary structure. Further, denaturation seems to reduce allergenicity by increasing digestibility in the gastrointestinal tract (Enomoto *et al.*, 1993;

del Val *et al.*, 1999). However, caseins do not have a fixed tertiary structure (Kumosinski *et al.*, 1991; Chapter 4), and so undergo little, if any, structural change when they are heated. Thus, it should be noted that normal processing procedures such as pasteurization, spray drying and evaporation do not eliminate, but only partly reduce, the allergenicity of milk proteins.

When heated, milk proteins can interact with lactose *via* the Maillard reaction. The tertiary structure of the protein conjugated with lactose sometimes changes as a consequence of this reaction. The complexes formed during the Maillard reaction sometimes are more allergenic than the native protein. When skin reactions to β-lactoglobulin isolated from pasteurized milk were tested, its allergenicity was found to be about 100 times higher than that of unheated β-lactoglobulin (Bleumink and Young, 1968). β-Lactoglobulin prepared from the heated milk contained lactose bound to it. Substances that cause milk allergy also were found in a non-dialyzable macromolecular fraction prepared from a crude lactose preparation (Spies, 1971). This fraction appeared to be produced by the Maillard reaction between milk proteins and lactose, because it was composed of proteins and sugars and because the starting material for the preparation of the lactose was the whey fraction of milk (Kaminogawa *et al.*, 1984).

13.5.2 Hydrolysed infant formulae

Hydrolysed formulae have been developed for the purpose of reducing the allergenicity of cow's milk proteins, and have been used for feeding infants with cow's milk allergy. Categorized on the basis of the protein source, bovine casein- and bovine whey-based hydrolysed formulae are available on the market (Wahn *et al.*, 1992; Therattil and Bahna, 1999). Further, according to the degree of hydrolysis, hydrolysed formulae can be defined as partially hydrolysed or extensively hydrolysed. The molecular weight of the peptides is usually <1500 Da in the extensively hydrolysed formulae (eHF), whereas partially hydrolysed formulae (pHF) contain a significant number of peptides with a molecular weight >15000 Da (Businco *et al.*, 1999).

Casein HF have been used for more than 40 years and there have been few reports of adverse reactions to these formulae. The safety of a casein HF have been confirmed in 20 infants with cow's milk allergy (Sampson *et al.*, 1991). It has been shown that whey-based eHF can be introduced safely in most cases, whereas whey-based pHF can trigger symptoms in about 50% of children with cow's milk allergy (Oldaeus *et al.*, 1991; Ragno *et al.*, 1993; Isolauri *et al.*, 1995). This difference is likely due to a striking difference between the two kinds of formulae in terms of the number of peptides with a high molecular weight and intact milk proteins (Makinen and Sorva, 1993; Buts *et al.*, 1994; Chiancone *et al.*, 1995). However, even in eHF, small amounts of intact milk proteins may still be present (Makinen and

Sorva, 1993; Buts *et al.*, 1994; Chiancone *et al.*, 1995) and this may explain the reactions to eHF described in highly sensitive children (Businco *et al.*, 1989; Saylor and Bahna, 1991; Ragno *et al.*, 1993). These observations demonstrate that none of the 'hypoallergenic' formulae are 'non-allergenic' (Businco *et al.*, 1993).

Hydrolysis of milk proteins, however, can generate molecules recognized by the immune system as molecules different from those with the structure before such change and can give rise to "new antigens" (Spies *et al.*, 1972). New peptides with antigenicity different from that of β-lactoglobulin could be produced by digestion with pepsin (Spies *et al.*, 1972; Haddad *et al.*, 1979). As will be explained below, in the case of α_{s1}-casein, highly hydrophobic regions, unlikely to be on the surface in the native molecule, become the primary binding sites for human IgE (Spuergin *et al.*, 1996). Food proteins differ from other allergens in that they are degraded and digested in the gastrointestinal tract, and those with structural changes after being digested can be responsible for an allergic reaction.

13.6 Antigenic and allergenic structure of milk allergens

The ability of an antigen to evoke an immune response is called its immunogenicity, and its ability to react with already induced antibodies and activated T cells is called its antigenicity. Allergic reactions are one kind of immune response, but they differ in many ways from general immune responses to antigens. For example, in type I allergy, IgE is the only kind of antibody involved in the allergic response, unlike in the general immune response, in which other classes, such as IgG and IgM, are involved. Therefore, the immunogenicity and antigenicity of a particular protein, which need not involve an IgE response, can differ from its allergenicity, for which an IgE response is necessary.

However, for studies on the allergenicity of milk proteins, it is important to study their immunogenicity and antigenicity, the basis for their allergenicity. The immunogenicity and antigenicity of milk proteins may differ between humans and experimental animals such as mice, but information obtained using experimental animals provides important information for both clinical studies and basic experiments about possible strategies for allergy suppression by manipulation of milk allergens. The identification of epitopes recognized in humans (both persons with and without milk allergy) could suggest targets for such manipulation.

Accordingly, in this section we will describe the T- and B-cell epitopes of the principal milk allergens, α_{s1}-casein and β-lactoglobulin, as determined in animal experiments. We will discuss these proteins also in terms of the epitopes recognized by IgE antibodies in the sera of allergic patients and the epitopes recognized by T cells in peripheral and cord blood.

13.6.1 α_{s1}-Casein

(a) Antigenic structure of α_{s1}-casein

α_{s1}-Casein is the principal component of the micelles which are dispersed as a colloid in milk. Bovine α_{s1}-casein consists of 199 amino acids and accounts for 36% (w/w) of the total casein (see Chapters 1 and 3).

The locations of the epitopes of α_{s1}-casein recognized by T cells in mice were identified by means of a T-cell proliferation assay (Enomoto *et al.*, 1990; Shon *et al.*, 1991). In those experiments, 13 synthetic overlapping peptides (19–20 amino acid residues), encompassing the entire sequence of the protein, were tested for the ability to stimulate the proliferation of lymph node cells from three different strains of mice (BALB/c, C3H/He and C57BL/6), each possessing a distinct type of MHC (H-2^d, H-2^k and H-2^b, respectively). The rationale for the use of three different strains of mice was that with different MHC molecules, the repertoire of peptides that bind to the molecules is different; consequently, the regions of the molecule that evoke T-cell responses also differ. As shown in Figure 13.2, it is clear that only limited regions of α_{s1}-casein were recognized by the T cells and the profiles of the T-cell epitopes differed among the three murine strains. The number and location of T-cell epitopes were dependent on the strain of mice examined. Responses to some epitopes were strong (dominant) while those to others were weak (subdominant). The immunodominance of T-cell epitopes is dependent mainly on the MHC haplotype of the mouse and may be explained by the reactivity of the peptide with MHC molecules, the availability of a specific T-cell repertoire and the antigen presentation pattern.

The locations of B-cell epitopes on the α_{s1}-casein molecule were examined also in studies using the same three strains of mice mentioned above

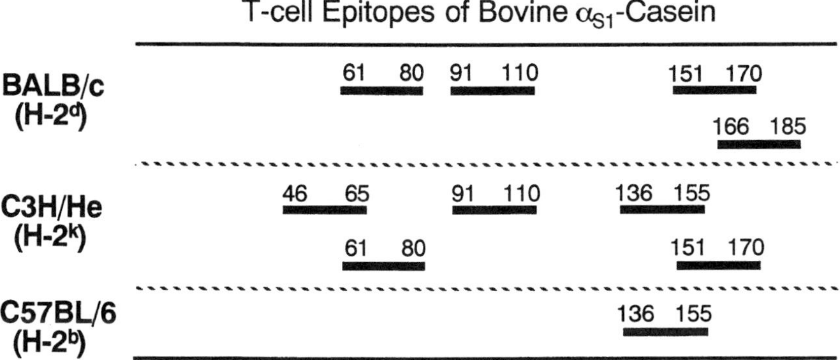

Figure 13.2 T-cell epitopes of bovine α_{s1}-casein recognized in three strains of mice (Reproduced from Kaminogawa, 1996).

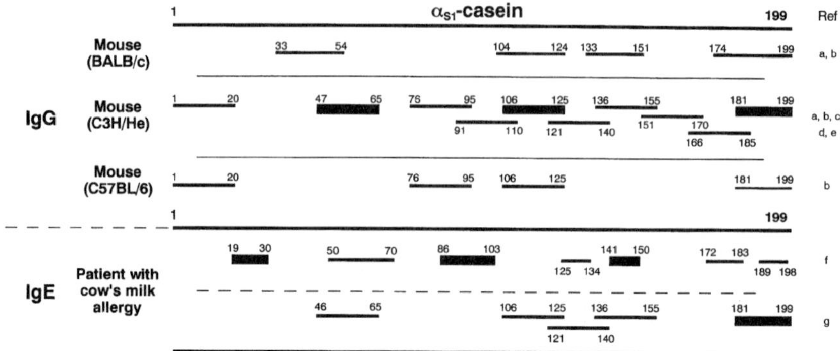

Figure 13.3 B-cell epitopes of bovine α_{s1}-casein. Thick lines indicate dominant epitopes recognized by antibodies in the sera examined. [a]Ametani *et al.* (1987); [b]Ametani *et al.* (1988); [c]Enomoto *et al.* (1990); [d]Shon *et al.* (1991); [e]Kim *et al.* (1993); [f]Spuergin *et al.* (1996); [g]Nakajima-Adachi *et al.* (1998).

(Figure 13.3) (Ametani *et al.*, 1987,1988; Enomoto *et al.*, 1990; Shon *et al.*, 1991; Kim *et al.*, 1993). Mice were immunized intraperitoneally with α_{s1}-casein, and the extent of binding of each peptide to IgG from these mice was measured by ELISA. Only a few regions of the protein molecule were recognized and the response varied depending on the strain of mice tested.

These results obtained through parenteral administration of α_{s1}-casein to mice were compared with those obtained through oral administration (Kim *et al.*, 1993). A feed for mice containing whole casein as the sole protein source was prepared and supplied *ad libitum* to C3H/He mice. On day 20 and thereafter, IgG antibodies specific for α_{s1}-casein were found in the sera of the mice. The antibodies in the sera reacted strongly with residues 46–65 and moderately with residues 166–185, but bound poorly to the other regions. These results differed from those obtained through intraperitoneal immunization of C3H/He mice, which showed three dominant B-cell epitopes, residues 46–65, 106–125 and 181–199 and seven subdominant epitopes, residues 1–20, 76–95, 91–110, 121–140, 136–155, 151–170 and 166–185. The difference probably occurred because, when orally administered, allergens are degraded by digestive enzymes and oral tolerance appears.

The onset of food allergy is suppressed by oral tolerance (Thompson and Staines, 1990; Weiner, 1994; Weiner *et al.*, 1994; Kaminogawa, 1996; Kaminogawa *et al.*, 1999), an important function of the gastrointestinal-tract immune system. When oral tolerance is established for a food allergen, T-cell response to each of the epitopes on the allergen may be suppressed in equal efficiency, or alternatively there may be a difference among the epitopes in terms of their ability to allow the response to that site to be suppressed. If the latter is true, it is possible that allergy is induced by a particular epitope(s) which can be suppressed only with difficulty. To clarify this point, the possibility that there are differences in the epitopes involved in T-cell and B-cell responses comparing mice given α_{s1}-casein orally and mice immunized with α_{s1}-casein was examined (Hachimura *et al.*, 1994). The results showed that in the case of certain epitopes, oral tolerance develops slightly. Such epitopes may be responsible for causing allergy.

(b) Regions of α_{s1}-casein recognized by IgE from milk-allergic patients
Spuergin *et al.* (1996) used a PEPSCAN method with 188 12-mer peptides, together encompassing the entire α_{s1}-casein sequence of 199 amino acid residues, to identify epitopes recognized by IgE and IgG in the sera of 15 milk-allergic and 10 healthy persons. With PEPSCAN, epitopes were searched for with a series of peptides designed with a window 8-mer to 15-mer long that shifts in one- or two-residue steps (Geysen *et al.*, 1984; Geysen *et al.*, 1987; Maeji *et al.*, 1990). The entire sequence of a given protein was searched in this way. Such a series of peptides is often synthesized on a small scale using chemically activated polypropylene pins. Spuergin *et al.* (1996) found that the three peptide regions, 19–30, 93–98 and 141–150, were recognized by IgE in the sera from all 15 allergic subjects. Nonpolar and aromatic amino acid residues are abundant in these regions, where antibodies cannot bind unless α_{s1}-casein has been degraded or disturbed in its secondary structure. IgE in the serum from individual subjects reacted with the above-mentioned three regions and elsewhere, at residues 50–70, 125–134, 172–183 and 189–198. The sera from healthy subjects showed a similar but weaker reaction pattern. Subsequently, the effects of these peptides on the binding of IgE from the sera of the subjects allergic to α_{s1}-casein were examined. A peptide with the sequence of residues 86–103 inhibited the binding of IgE from the sera of all subjects allergic to α_{s1}-casein, as did a peptide with the sequence of residues 141–150 for all allergic subjects except one. The site at residues 86–103 seemed to be the most important epitope, because inhibition was greatest there.

Nakajima-Adachi *et al.* (1998) studied the regions of α_{s1}-casein recognized by IgE antibodies in sera of nine pediatric patients with cow's milk allergy who had higher serum titers of anti-α_{s1}-casein IgE antibodies than the reference range. The sites recognized by IgE antibodies in the sera of these patients were analyzed by ELISA, using 13 peptides synthesized on the basis of part of the amino acid sequence of α_{S1}-casein. Despite the patients being completely unrelated as far as was known, the anti-α_{s1}-casein antibodies of all of the patients were found to bind to the same site, residues 181–199, at the C-terminus. The difference in the dominant IgE epitopes identified in comparison with those found by Spuergin *et al.* (1996) may be due to regional and racial differences. Nakajima-Adachi *et al.* (1998) next examined whether the epitopes of anti-α_{s1}-casein IgG_4 from these nine subjects were shared, as seen in the case of the IgE epitopes. Studies on the biosynthesis of immunoglobulins have shown that some B cells that produce IgE antibodies do so only after first producing IgG_4 antibodies. The sites recognized by the I_gG_4 antibodies were found to be scattered throughout the α_{s1}-casein molecule, unlike those recognized by IgE antibodies, and many recognition sites were found in the N- and C-terminal regions, in particular. Some of the sites recognized by IgG_4 were the same as those recognized by IgE antibodies. In contrast, Spuergin *et al.* (1996) found no essential difference in specificity between IgE and IgG.

The IgE binding sites on a given allergen tend to be the same for different patients, as was found for β-lactoglobulin in cases of milk allergy (Ball *et al.*, 1994; Heinzmann *et al.*, 1999) and for some allergens that are inhaled (van't Hof *et al.*, 1991; van Milligen *et al.*, 1994; Ong *et al.*, 1995; Kobayashi *et al.*, 1996). Therefore, special B cells producing IgE antibodies that recognize only a particular site or sites may contribute to the onset of allergies involving IgE antibodies.

(c) Epitopes of T-Cell lines derived from peripheral blood of allergic patients reacting to α_{s1}-casein

Using peripheral blood from four patients allergic to cow's milk, Nakajima *et al.* (1996) established 26 T-cell lines that recognized α_{s1}-casein specifically. Twenty-three T-cell lines were examined to determine their surface marker expression, and on this basis they could be classified into three groups: those consisting of predominantly $CD4^+CD8^-$ T cells (10 lines), those consisting of predominantly $CD4^-CD8^+$T cells (3 lines), and those consisting of both $CD4^+CD8^-$ and $CD4^-CD8^+$ T cells (10 lines). The most unexpected and important result was that the T-cells lines consisting of $CD8^+$ T cells formed almost half of all the T-cell lines established, considering that most $CD8^+$ T cells recognize fragments of proteins synthesized in cells presenting them, such as viral proteins, but not fragments of exogenous proteins, such as allergens. The $CD8^+$ subset of T cells is considered to include regulatory T cells that can suppress immune responses and $CD8^+$ T cells often produce much interferon-γ, which suppresses Th2-type responses. The induction of $CD8^+$ T cells specific for α_{s1}-casein, found with considerable frequency in cases of milk allergy, may account for the frequent resolution of milk allergy with age (Høst, 1994; Høst *et al.*, 1995).

Next, the sites of α_{s1}-casein molecules that are recognized by the T-cell lines were studied (Nakajima-Adachi *et al.*, 1998). All seven cell lines that reacted to at least one α_{s1}-casein peptide were $CD4^+CD8^-$ cells, which produced much more interleukin-4 than interferon-γ, and thus showed Th2-type responses. All of the T-cell lines recognized different epitopes of α_{s1}-casein. However, when the sequences of the peptides recognized by the T cells were examined, sequence motifs were found, each consisting of two

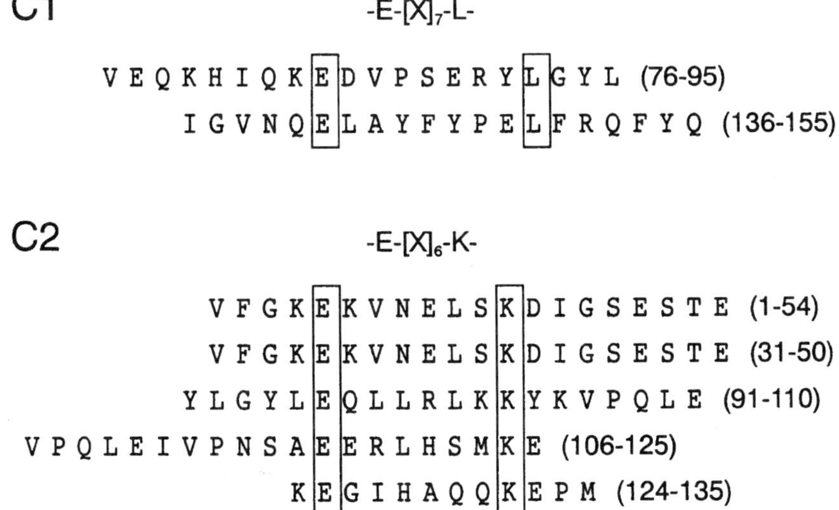

Figure 13.4 Common structures found in peptides recognized by α_{s1}-casein-specific T-cell lines established from patients (C1 and C2) with cow's milk allergy (Reproduced from Nakajima-Adachi *et al.*, 1998).

conserved amino acid residues separated by several intervening residues, common to the epitope peptides recognized by T-cell lines established from individual patients. That is, two of the cell lines established from patient C1 recognized sequences containing a motif consisting of a glutamic acid residue and a leucine residue separated by seven intervening residues (-E-$[X]_7$-L-) and five of the cell lines established from patient C2 recognized sequences with a motif consisting of a glutamic acid residue and a lysine residue separated by six intervening residues (-E-$[X]_6$-K-) (Figure 13.4). The relationship between these shared structures and the structures of MHC molecules and the TCR on the T cells, both of which bind to the epitope peptides, is not known. The existence of T cells that recognize specific regions on a given allergen suggests that these T cells would be important in the production of IgE antibodies binding to specific regions on the allergen, as, mentioned above.

13.6.2 β-Lactoglobulin

(a) Antigenic structure of β-lactoglobulin

β-Lactoglobulin is a globular protein consisting of 162 amino acid residues and with a molecular weight of *ca.* 18,000 Da. Its primary (Braunitzer *et al.*,

1973; Alexander *et al.*, 1989) and tertiary structures (Papiz *et al.*, 1986; Brownlow *et al.*, 1997) are known (see Chapter 7). The authors' group prepared monoclonal antibodies that bind specifically to various sites on β-lactoglobulin and identified their binding sites (Kaminogawa *et al.*, 1987, 1989). Further, we showed that monoclonal antibodies can be used to analyze the process by which the rigid structure of β-lactoglobulin becomes denatured (Kaminogawa *et al.*, 1989; Hattori *et al.*, 1993; Katakura *et al.*, 1994). Recently, Venien *et al.* (1997) and Négroni *et al.* (1998) prepared several dozen distinct monoclonal antibodies against bovine β-lactoglobulin, classified them into several groups, and identified their binding sites on the antigen. These antibodies were prepared for use in examining the structural changes in β-lactoglobulin caused by heat and the structure of the protein in relation to allergic reactions.

Takahashi *et al.* (1990) injected β-lactoglobulin, together with an adjuvant, intraperitoneally into BALB/c mice and used the sera obtained to identify B-cell epitopes. The results showed such epitopes at residues 21–40, 41–60, 102–124 and 149–162 (Figure 13.5). The first three of these four B-cell epitopes are at random-coil areas between the β-sheets. A similar pattern for the location of epitopes has been observed in other globular proteins (Benjamin *et al.*, 1984; Barlow *et al.*, 1986). These results confirmed some of the observations reported by Otani and Hosono (1987).

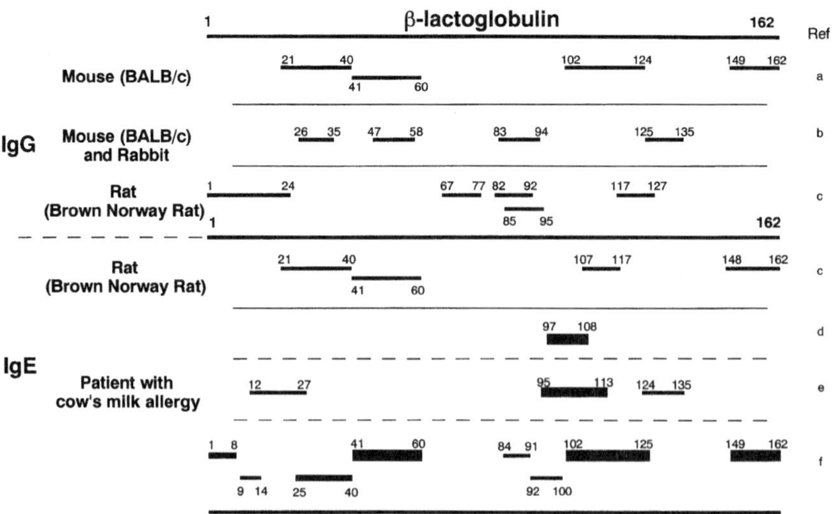

Figure 13.5 B-cell epitopes of bovine β-lactoglobulin. Thick lines indicate dominant epitopes recognized by antibodies in the sera examined. [a]Takahashi *et al.* (1990); [b]Williams *et al.* (1998); [c]Miller *et al.* (1999); [d]Ball *et al.* (1994); [e]Heinzmann *et al.* (1999); [f]Sélo *et al.* (1999).

Mizumachi *et al.* (1990) immunized rabbits, goats and mice with β-lactoglobulin and used many synthetic peptides to identify the epitopes recognized by IgG antibodies in sera of the animals. Six epitopes were identified; the one at residues 149–162 was especially important because antibody binding was lost almost completely when His_{161} was replaced.

Williams *et al.* (1998) used PEPSCAN and phage-display techniques to study the binding sites on β-lactoglobulin recognized by IgG antibodies in sera from rabbits and BALB/c mice immunized with this protein. In the phage-display method, 9-mer peptides with random sequences were displayed on the surface of the coat protein of a filamentous phage, phage particles were selected on the basis of their binding to antibodies and then epitopes were searched for by analysis of the inserted sequences. Using the two epitope-mapping techniques, they found that residues 26–35, 47–58, 83–94 and 125–135 were recognized by antibodies from both rabbits and mice. All of the regions identified by Takahashi *et al.* (1990) as B-cell epitopes in BALB/c mice, including regions 102–124 and 149–162, were shown to contain epitopes recognized in at least one of BALB/c mice tested by Williams *et al.* (1998), thus agreeing with the earlier results.

Miller *et al.* (1999) identified IgE- and IgG- epitopes of β-lactoglobulin recognized in Brown Norway rat, which has been shown to produce IgE to food proteins, by ELISA using overlapping synthetic peptides and tryptic peptides. They found that fragments 21–40, 41–60, 107–117 and 148–162 are epitopes that are recognized predominantly by IgE, whereas fragments 1–24, 67–77, 82–92, 85–95 and 117–127 appeared to be more selective for IgG antibody recognition.

T-cell epitopes on β-lactoglobulin recognized in three strains of mice (C57BL/6, BALB/c, and C3H/He) with different MHC molecules were identified by the authors' group using a PEPSCAN method (Tsuji *et al.*, 1993; Totsuka *et al.*, 1997a). We prepared 148 15-mer peptides, which together covered the entire sequence of bovine β-lactoglobulin and identified the regions that triggered proliferation of lymph-node cells from mice immunized with β-lactoglobulin (Totsuka *et al.*, 1997a). Figure 13.6 shows the positions of these epitopes on β-lactoglobulin. The core regions of the immunodominant epitopes were residues 122–130 in C57BL/6 mice, residues 140–148 in C3H/He mice, and residues 67–75, 71–79 and 80–88 in BALB/c mice.

Thus far, we have discussed the identification of T- and B-cell epitopes of α_{s1}-casein and β-lactoglobulin. The immune system seems to select T- and B-cell epitopes independently, but they interact. To clarify this point, we used peptides derived from α_{s1}-casein or β-lactoglobulin that can elicit antibodies to show that the relative positions of T-cell and B-cell epitopes partly govern the efficiency of the interactions between T and B cells in antibody production (Shon *et al.*, 1991; Sakurai *et al.*, 1993).

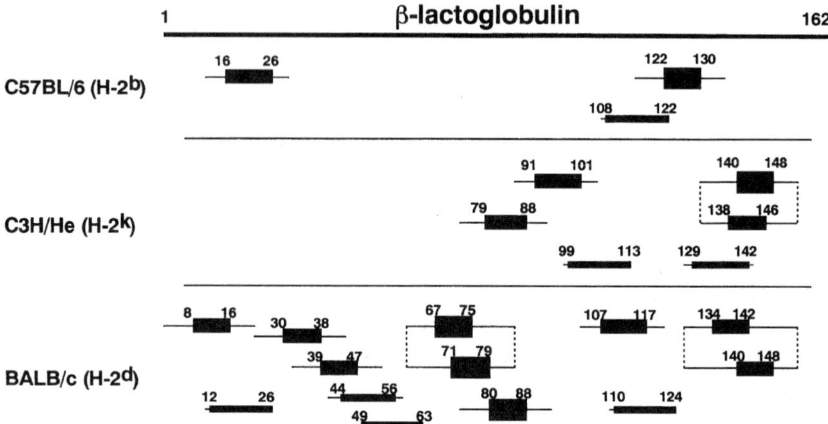

Figure 13.6 T-cell epitopes of bovine β-lactoglobulin recognized in three strains of mice (Reproduced from Totsuka *et al.*, 1997a). The location of each T-cell epitope is indicated as the combination of a region identified as containing T-cell epitope(s) (thin line) and a putative core sequence (solid box). The thick, intermediate and thin boxes show the core sequences of dominant, subdominant, and minor epitopes, respectively. The numbers at both ends of each core sequence show the position of the first and the last residues of the core sequence. Regions determined to contain two determinants are indicated as two separate lines, which are bonded with vertical broken lines, each with a box indicating the core sequence.

(b) Epitopes on β-lactoglobulin recognized by IgE from patient allergic to milk

Adams *et al.* (1991) used the hydrophilicity of regions of the amino acid sequence of β-lactoglobulin as the basis for computer prediction of epitopes. On the basis of the results obtained, they synthesized a peptide corresponding to residues 124–134. Examining the serum of a patient allergic to β-lactoglobulin, they found that this peptide inhibited 60% of the total binding of IgE to β-lactoglobulin.

Ball *et al.* (1994) used 12-mer peptides that overlapped to identify sequential epitopes recognized by IgE in the sera of allergic patients. IgE in the sera of 14 of the 16 patients bound to residues 97–108. Similar results were reported by Heinzmann *et al.* (1999). Using the PEPSCAN method with 10-mer peptides, they identified epitopes recognized by IgE in sera from 14 patients with cow's milk allergy. One region of β-lactoglobulin corresponding to residues 95–113 showed binding to IgE in the sera of all 14 patients. They also found that regions 12–27 and 124–135 were recognized by IgE in serum from individual patients.

Sélo *et al.* (1998) used peptides generated from β-lactoglobulin by cyanogen bromide degradation to identify the binding sites recognized by IgE in sera from 19 allergic patients. Their results showed that almost all of the peptides obtained (fragments 8–24, 25–107, 108–145 and 146–162) bound with IgE. They also found that antibodies in almost all of the sera recognized more than one peptide. They concluded that human IgE reactions to β-lactoglobulin seem to be heterogeneous and variable. Further, using tryptic peptides from β-lactoglobulin, they found the major IgE epitopes in fragments 41–60, 102–125 and 149–162, which were recognized by 92, 97 and 89%, respectively, of the sera from 46 milk-allergic patients (Sélo *et al.*, 1999). Fragments 1–8 and 25–40 were recognized by 58 and 72% of the sera, and fragments 9–14, 84–91 and 92–100 were recognized by more than 40% of the sera.

Taken together, although many IgE epitopes have been identified, a major epitope of β-lactoglobulin recognized by IgE in sera from milk allergic patients appears to exist in region 95–125. Takahashi *et al.* (1990) and Miller *et al.* (1999) identified a B-cell epitope in this region in mice and rats. Williams *et al.* (1998) also found a B-cell epitope corresponding to this region recognized in rabbits and mice.

(c) Epitopes recognized by β-lactoglobulin-specific human T cells

Piastra *et al.* (1994) examined the response to β-lactoglobulin by cord-blood mononuclear cells (MNCs) from neonates at high risk for milk allergy (at least one parent had atopy) and control-group neonates (neither parent had atopy). In primary cultures, cord-blood MNCs from seven of the 13 high-risk neonates reacted to β-lactoglobulin, but MNCs from the controls did not react. When cord-blood MNCs from 10 non-responders were cultured with β-lactoglobulin for eight days ('*in vitro* priming'), cell proliferation occurred in seven of the cultures. They used peptides corresponding to residues 7–23, 24–106, 107–144 and 145–161 from degraded β-lactoglobulin for epitope mapping of primed cord-blood MNCs, and found no difference in the epitope specificities of the two groups, i.e., none of the peptides was recognized by MNCs from only one of the groups. At least 40% of the cultures examined reacted with the fragment 145–161, and almost all of the cultures recognized more than one peptide.

13.7 Antigenic Microstructure of Milk Allergens and Suppression of Immune Allergic Reactions by its Alteration

The basic treatment for cow's milk allergy is complete avoidance of cow's milk and cow's milk product (Businco *et al.*, 1999), and at present there is no sure treatment for food allergies. Recently, however, it has become possible, at least in experimental systems, to suppress immune response safely and effectively on the basis of information about the molecular mechanisms of immune responses. We will describe how it is possible to suppress immune responses, particularly by altering the structure of antigens, and how this method may be used in the treatment and prevention of allergy.

13.7.1 Analysis of the antigen microstructure of β-lactoglobulin

The initial step involved in an allergic reaction is the interaction of antigen-presenting cells with T cells. Since T cells are central in the regulation of immune responses, an allergic reaction can be suppressed by suppression of the T-cell response. T cells are activated when TCRs bind complexes of MHC class II molecules and antigen-derived peptides, therefore, the interactions among these three components are particularly important in regulating immune responses. In peptides containing T-cell epitopes, there are residues with side chains important in binding with MHC class II molecules and residues with side chains important in binding with TCR. To manipulate the immune response to an allergenic protein by modifying the T-cell response to the protein through use of a mutant form of the protein or related peptides with amino acid substitution, it is necessary first to clarify the microstructure of the T-cell epitopes of the protein.

As previously mentioned, we identified the T-cell epitopes of β-lactoglobulin recognized in three mouse strains with different MHC (Totsuka *et al.*, 1997a) (Figure 13.6). The distribution map of the epitopes was simplest in the case of C57BL/6 mice, so we thought it would be easiest to examine the effects of amino acid substitution on the immune response in this strain. We first identified the MHC-binding residues within the region of residues 122–130, the core region of the epitope which elicits the strongest T-cell response in C57BL/6 mice. Side chains of residues Pro_{126} and Val_{128} were shown to be important in binding with I-A^b, a murine MHC class II molecule. Next, we identified the residues important for binding with TCR within a peptide corresponding to residues 119–133 of β-lactoglobulin (f119–133). We established G1.19, a murine $CD4^+$ T-cell clone that specifically recognizes f119–133 of β-lactoglobulin (Totsuka *et al.*, 1997b), and examined its proliferative response to the analogues of (f119–133) with a single amino acid substitution. With this T-cell clone, all residues from 122 to 130, except those important for binding with I-A^b, had some involvement in TCR binding. These results are summarized in Figure 13.7.

13.7.2 Modification of the immunogenicity of β-lactoglobulin by amino acid substitution

On the basis of the results of our detailed analysis of the immunodominant T-cell epitope of β-lactoglobulin described above, we designed an experiment to show if the immunogenicity of β-lactoglobulin can be reduced by

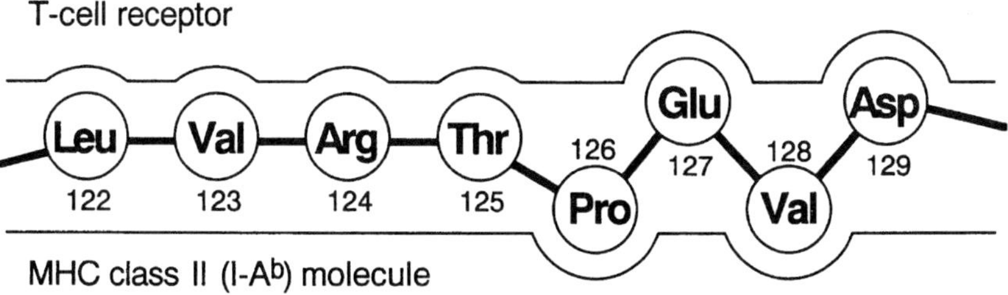

Figure 13.7 T-cell receptor- and I-A^b-contact residues of a T-cell epitope in region 119–133 of bovine β-lactoglobulin (Reproduced from Kaminogawa, 1996).

amino acid substitution. Using a yeast expression system (Totsuka *et al.*, 1990; Katakura *et al.*, 1994), we prepared two β-lactoglobulin variants, one with Pro_{126}, which is important for binding with MHC class II molecules, replaced by alanine (P126A) and one with Asp_{129}, which binds with TCR, replaced by alanine (D129A), and examined their ability to elicit antibodies. The replacement of Pro_{126} by alanine had been shown to reduce greatly the binding capacity of f119–133 to I-A^b molecules. When native β-lactoglobulin or the D129A mutant protein was used for immunization, a strong and specific response was observed. However, when the P126A mutant protein was used, almost no specific antibodies were found, showing greatly reduced immunogenicity (unpublished observation). These results show that the immune response to the entire allergen molecule can be reduced markedly by the replacement of a single amino acid residue that normally binds with MHC class II molecules in a single immunodominant T-cell epitope.

13.7.3 Suppression of immune responses by analogues of a peptide derived from β-lactoglobulin

T cells are activated when they recognize, *via* TCR, complexes of MHC molecules and antigenic peptides. The signals transmitted to T cells *via* these three-molecule complexes had been thought to be transmitted by an all-or-nothing system, that is, by a binary switch that controls whether T cells are on and off. However, recently reported findings have disproved this view. Some, but not all, of various cell responses (e.g., proliferation and cytokine production) that occur after T-cell activation are elicited by stimulation using single amino acid substituted variants of an antigenic peptide (Evavold and Allen, 1991; Sloan-Lancaster and Allen, 1996). These variant epitope peptides are called partial agonists. TCR antagonism has been observed also: suppression of the T-cell proliferative response and other responses to the native antigenic peptide when at the same time a variant of this peptide is present (De Magistris *et al.*, 1992; Sette *et al.*, 1994). Together, these two kinds of variants are called 'altered peptide ligands' (APL) (Sloan-Lancaster and Alien, 1996). These results suggest that the three-molecule complexes do not convey simple 'on' or 'off' signals, and that immune reactions can be controlled artificially by the use of APL.

Since TCR antagonists can control specifically and exclusively the T-cell responses to a specific antigen, they may be a safe and effective method for treatment and/or prevention of diseases caused by deleterious immune responses such as autoimmune diseases and allergy. Studies done with murine models of autoimmune diseases have shown that the onset of the diseases can be suppressed by parenteral administration of APLs with TCR-antagonist activity (Franco *et al.*, 1994; Karin *et al.*, 1994; Kuchroo *et al.*, 1994; Singh *et al.*, 1994; Nicholson *et al.*, 1995, 1997; Brocke *et al.*, 1996). All of the models employed are those of T-cell-mediated autoimmune diseases and antigen-specific inhibition of immune responses might be achieved also in treatment of antibody-mediated immune diseases, such as type I hypersensitivity. Therefore, we examined the efficacy of TCR antagonists derived from β-lactoglobulin in the inhibition of antigen-specific antibody production *in vivo*.

Using analogues of f119–133 containing the immunodominant T-cell epitope of β-lactoglobulin in C57BL/6 mice, we identified those with TCR-antagonist activity that suppresses the proliferative response of T-cell clone G1.19, which specifically recognizes this region (Totsuka *et al.*, 1997b; Toda *et al.*, 2000). The analogue peptides D129A and D129S with Asp_{129} replaced by alanine or serine, respectively, were identified as TCR antagonists, because they suppressed the response of G1.19 to the antigen. D129A and D129S also suppressed antigen-specific proliferative responses of two of three other T-cell clones specific for f119–133. We found that D129A mutant protein with Asp_{129} replaced by alanine, also suppressed T-cell responses, as did D129A peptide (Totsuka *et al.*, 1997b). These results show that TCR antagonism can occur with protein variants as well as with peptide variants.

Next, we studied the effects of the TCR-antagonist peptides on immune responses *in vivo*, especially the antigen-specific antibody response (Toda *et al.*, 2000). When we administered f119–133 together with D129S peptide to mice, mouse-derived lymph-node T cells proliferated less than without D129S, and the serum titer of anti-f119–133 antibodies including IgE was also much less. These results show that both T-cell and antibody-producing responses can be suppressed by TCR-antagonist peptides *in vivo*. Further studies are needed to ascertain whether immune responses to intact β-lactoglobulin molecules can be suppressed by TCR-antagonist peptides, but these results suggest that TCR antagonists can suppress IgE antibody production and thus allergic reactions.

13.7.4 Increased production of interferon-γ by CD8$^+$ T cells stimulated with analogues of a peptide derived from α_{s1}-casein

As described above, many CD8$^+$ T-cell lines were established from the peripheral blood of milk-allergic patients reacting to α_{s1}-casein (Nakajima-Adachi *et al.*, 1998). Probably, the induction of CD8$^+$ T cells against this milk allergen is related to the recovery from milk allergy with age, as seen in the case of many affected persons. CD8$^+$ T cells usually produce much interferon-γ, which suppresses Th2-type responses of CD4$^+$ T cells (Abbas *et al.*, 1996). This cytokine suppresses the production of IgE antibodies, which contribute to type I allergy. Therefore, we thought that if we could identify antigen analogues derived from α_{s1}-casein that increase the production of interferon-γ by CD8$^+$ T cells, it might be possible to suppress the allergic response effectively and promote resolution of the allergy.

We immunized C57BL/6 mice with α_{s1}-casein subcutaneously and, using lymph-node cells from these mice, we established CD8$^+$ T-cell clones specific for α_{s1}-casein (Hisatsune *et al.*, 1995; Nishijima *et al.*, 1997). All of the clones responded specifically to a peptide corresponding to residues 142–149 of α_{s1}-casein (f 142–149). We used these clones to identify peptides that increased interferon-γ production (Kohyama *et al.*, 1998; Totsuka *et al.*, 1998).

Peptide L142I, which is a variant of f142–149 with residue Leu_{142} replaced by isoleucine, increased interferon-γ production in all three CD8$^+$ T-cell clones studied (Totsuka *et al.*, 1998). Peptide L142I increased interferon-γ production also in polyclonal CD8$^+$ T cells derived from mice immunized with p142–149. CD8$^+$ T cells from mice immunized with L142I produced more interferon-γ after stimulation with f142–149 *in vitro* than did CD8$^+$ T cells from mice immunized with f142–149. We found that L142I binds more strongly, forming more stable complexes, with H-2 K^b molecules, a kind of murine MHC class I molecule, than does f142–149. This is an important factor and it seems to be involved in the effect of L142I in increasing interferon-γ production.

13.8 Concluding remarks

In this chapter, we have discussed the allergenicity of milk proteins from the viewpoints of immunology, allergology, and protein chemistry, in particular. In the past, research on questions related to milk allergy was done mainly by pediatricians, but now biochemists and cytobiologists are also participating in such studies, and much progress has been made. However, important problems remain unsolved: for example, why are there so many cases of milk allergy, why α_{s1}-casein and β-lactoglobulin readily become allergens, and whether some of these molecules have a structure which increases their allergenicity. We also described efforts to suppress milk allergy by alteration of the structure of allergens. There has been some success in animal studies examining these problems, but they have not been studied in humans. These are important problems that must be solved in the future. To that end, we urge researchers from a wide range of fields to cooperate in studies of milk allergy and allergens.

References

Abbas, A.K., Murphy, K.M. and Sher, A. (1996) Functional diversity of helper T lymphocytes. *Nature*, **383**, 787–93.

Adams, S.L., Bamett, D., Walsh, B.J., Pearce, R.J., Hill, D.J. and Howden, M.E. (1991) Human IgE-binding synthetic peptides of bovine β-lactoglobulin and α-lactalbumin. *In vitro* cross-reactivity of the allergens. *Immunol. Cell Biol.*, **69**, 191–7.

Alexander, L.J., Hayes, G., Pearse, M.J., Beattie, C.W., Stewart, A.F., Willis, I.M. and Mackinlay, A.G. (1989) Complete sequence of the bovine β-lactoglobulin cDNA. *Nucl. Acids Res.*, **17**, 6739.

Ametani, A., Kaminogawa, S., Shimizu, M. and Yamauchi, K. (1987) Rapid screening of antigenically reactive fragments of α_{s1}-casein using HPLC and ELISA. *J. Biochem.*, **102**, 421–5.

Ametani, A., Kim, S.M., Kaminogawa, S. and Yamauchi, K. (1988) Antibody response of three different strains of mice to α_{s1}-casein analyzed by using proteolytic and synthetic peptides. *Biochem. Biophys. Res. Commun.*, **154**, 876–82.

Bahna, S.L. and Heiner, D.C. (1980) *Allergies to Milk*, Grune and Stratton, New York.

Ball, G., Shelton, M.J., Walsh, B.J., Hill, D.J., Hosking, C.S. and Howden, M.E. (1994) A major continuous allergenic epitope of bovine β-lactoglobulin recognized by human IgE binding. *Clin. Exp. Allergy*, **24**, 758–64.

Barlow, D.J., Edwards, M.S. and Thornton, J.M. (1986) Continuous and discontinuous protein antigenic determinants. *Nature*, **322**, 747–8.

Baudon, J.J., Mougenot, J.F. and Didry, J.R. (1987) Lymphoblastic stimulation test with food proteins in digestive intolerance to cow's milk and in infant diarrheas. *J. Pediatr. Gastroenterol. Nutr.*, **6**, 244–51.

Benjamin, D.C., Berzofsky, J.A., East, I.J., Gurd, F.R.N., Hannum, C., Leach, S.J., Margoliash, E., Michael, J.G., Miller, A., Prager, E.M., Reichlin, M., Sercarz, E.E., Smith-Gill, S.J., Todd, P.E. and Wilson A.C. (1984) The antigenic structure of proteins: a reappraisal. *Annu. Rev. Immunol.*, **2**, 67–101.

Bernard, H., Créminon, C., Yvon, M. and Wal, J.-M. (1998) Specificity of the human IgE response to the different purified caseins in allergy to cow's milk proteins. *Int. Arch. Allergy Immunol.*, **115**, 235–44.

Bleumink, E. and Young, E. (1968) Identification of the atopic allergen in cow's milk. *Int. Arch. Allergy Appl. Immunol.*, **34**, 521–43.

Braunitzer, G., Chen, R., Schrank, B. and Stangl, A. (1973) The sequence of β-lactoglobulin. *Hoppe Seyler's Z. Physiol. Chem.*, **354**, 867–78.

Brocke, S., Gijbels, K., Allegretta, M., Ferber, I., Piercy, C., Blankenstein, T., Martin, R., Utz, U., Karin, N., Mitchell, D., Veromaa, T., Waisman, A., Gaur, A., Conlon, P., Ling, N., Fairchild, P.J., Wraith, D.C., O'Garra, A., Fathman,

C.G. and Steinman, L. (1996) Treatment of experimental encephalomyelitis with a peptide analogue of myelin basic protein. *Nature*, **379**, 343–6.

Brownlow, S., Morais, C.J., Cooper, R., Flower, D.R., Yewdall, S.J., Polikarpov, I., North, A.C. and Sawyer, L. (1997) Bovine β-lactoglobulin at 1.8 Å resolution-still an enigmatic lipocalin. *Structure*, **5**, 481–95.

Bruijnzeel-Koomen, C., Ortolani, C., Aas, K., Bindslev-Jensen, C., Bjorksten, B., Moneret-Vautrin, D. and Wuthrich, B. (1995) Adverse reactions to food. European Academy of Allergology and Clinical Immunology Subcommittee. *Allergy*, **50**, 623–35.

Businco, L., Cantani, A., Longhi, M.A. and Giampietro, P.G. (1989) Anaphylactic reactions to a cow's milk whey protein hydrolysate (Alfa-Re, Nestle) in infants with cow's milk allergy. *Ann. Allergy*, **62**, 333–5.

Businco, L. and Bellanti, J. (1993) Food allergy in childhood. Hypersensitivity to cows' milk allergens. *Clin. Exp. Allergy*, **23**, 481–3.

Businco, L., Dreborg, S., Einarsson, R., Giampietro, P.G., Høst, A., Keller, K.M., Strobel, S., Wahn, U., Björkstén, B., Kjellman, M.N. Sampson, H. and Zeiger, R. (1993) Hydrolysed cow's milk formulae. Allergenicity and use in treatment and prevention. An ESPACI position paper. *Pediatr. Allergy Immunol.*, **4**, 101–11.

Businco, L., Bruno, G. and Giampietro, P.G. (1999) Prevention and management of food allergy. *Acta Paediatr. Suppl.*, **430**, 104–9.

Buts, J.P., Cadranel, S., Deprettere, A., Scaillon, M., Sokal, E., Van Caillie-Bertrand, M., Vandenplas, Y. and Vanwinckel, M. (1994) Hypo-allergenic formulae: what's in a name? *Eur. J. Pediatr.*, **153**, 390–2.

Chiancone, E., Gattoni, M., Giampietro, P.G., Ragno, V. and Businco, L. (1995) Detection of undegraded β-lactoglobulins and evaluation of the molecular weight of peptides in hydrolysate cow's milk formulae. *J. Investig. Allergol. Clin. Immunol.*, **5**, 228–31.

Chiaramonte, L.T., Bassett, C.W. and Altman, D.R. (1999) Food Allergy Perception and Reality, in *Food Hypersensitivity and Adverse Reactions. A Practical Guide for Diagnosis and Management*, (M. Fried and B. Kettelhut eds.) Marcel Dekker, Inc., New York, pp. 155–64.

De Magistris, M., Alexander, J., Coggeshall, M., Altman, A., Gaeta, F.C., Grey, H.M. and Sette, A. (1992) Antigen analog-major histocompatibility complexes act as antagonists of the T cell receptor. *Cell*, **68**, 625–34.

del Val, G., Yee, B.C., Lozano, R.M., Buchanan, B.B., Ermel, R.W., Lee, Y.-M. and Frick, O.L. (1999) Thioredoxin treatment increase digestibility and lowers allergenicity of milk. *J. Allergy Clin. Immunol.*, **103**, 690–7.

Docena, G.H., Fernandez, R., Chirdo, F.G. and Fossati, C.A (1996) Identification of casein as the major allergenic and antigenic protein of cow's milk. *Allergy*, **51**, 412–6.

Eigenman, P.A. and Sampson, H.A. (1997) Adverse reactions to food, in *Allergy*, 2nd edn., (A.P. Kaplan ed.) W.B. Sauders Company, Philadelphia, pp. 542–65.

Enomoto, A., Shon, D.H., Aoki, Y., Yamauchi, K. and Kaminogawa, S. (1990) Antibodies raised against peptide fragments of bovine α_{s1}-casein cross-react with the intact protein only when the peptides contain both B and T cell determinants. *Mol. Immunol.*, **27**, 581–6.

Enomoto, A., Konishi, M., Hachimura, S. and Kaminogawa, S. (1993) Milk whey protein fed as a constituent of the diet induced both oral tolerance and a systemic humoral response, while heat-denatured whey protein induced only oral tolerance. *Clin. Immunol. Immunopathol.*, **66**, 136–42.

Evavold, B.D. and Allen, P.M. (1991) Separation of IL-4 production from Th cell proliferation by an altered T cell receptor ligand. *Science*, **252**, 1308–10.

Franco, A., Southwood, S., Arrhenius, T., Kuchroo, V.K., Grey, H.M., Sette, A. and Ishioka, G.Y. (1994) T cell receptor antagonist peptides are highly effective inhibitors of experimental allergic encephalomyelitis. *Eur. J. Immunol.*, **24**, 940–6.

Freier, S., Kletter, B., Gery, I., Lebenthal, E. and Geifman, M. (1969) Intolerance to milk protein. *J. Pediatr.*, **75**, 623–31.

Geysen, H.M., Meloen, R.H. and Barteling, S.J. (1984) Use of peptide synthesis to probe viral antigens for epitopes to a resolution of a single amino acid. *Proc. Natl. Acad. Sci. USA.*, **81**, 3998–4002.

Geysen, H.M., Rodda, S.J., Mason, T.J., Tribbick, G. and Schoofs, P.G. (1987) Strategies for epitope analysis using peptide synthesis. *J. Immunol. Meth.*, **102**, 259–74.

Goldman, A.S., Anderson, J.D.W., Sellers, W.A., Saperstein, S., Kniker, W.T. and Halpern, S.R. (1963a) Milk allergy I. Oral challenge with milk and isolated milk proteins in allergic children. *Pediatrics*, **32**, 425–43.

Goldman, A.S., Sellers, W.A., Halpern, S.R., Anderson, D.W., Furlow, T.E. and Johnson, J.C.H. (1963b) Milk allergy II. Skin testing of allergic and normal children with purified milk proteins. *Pediatrics*, **32**, 572–9.

Hachimura, S., Fujikawa, Y., Enomoto, A., Kim, S.M., Ametani, A. and Kaminogawa, S. (1994) Differential inhibition of T and B cell responses to individual antigenic determinants in orally tolerized mice. *Int. Immunol.*, **6**, 1791–7.

Haddad, Z.H., Kalra, V. and Verma, S. (1979) IgE antibodies to peptic and peptic-tryptic digests of β-lactoglobulin: significance in food hypersensitivity. *Ann. Allergy*, **42**, 368–71.

Hattori, M., Ametani, A., Katakura, Y., Shimizu, M. and Kaminogawa, S. (1993) Unfolding/refolding studies on bovine β-lactoglobulin with monoclonal antibodies as probes. Does a renatured protein completely refold? *J. Biol. Chem.*, **268**, 22414–9.

Heinzmann, A., Blattmann, S., Spuergin, P., Forster, J. and Deichmann, K.A. (1999) The recognition pattern of sequential B cell epitopes of β-lactoglobulin does not vary with the clinical manifestations of cow's milk allergy. *Int. Arch. Allergy Immunol.*, **120**, 280–6.

Hisatsune, T., Nishijima, K., Kohyama, M., Kato, H. and Kaminogawa, S. (1995) $CD8^+$ T cells specific to the exogenous antigen. Mode of antigen recognition and possible implication in immunosuppression. *J. Immunol.*, **154**, 88–96.

Høst, A. (1994) Cow's milk protein allergy and intolerance in infancy: some clinical, epidemiological and immunological aspects. *Pediatr. Allergy Immunol.*, **5**(Suppl.), 5–36.

Høst, A. (1997) Cow's milk allergy. *J. R. Soc. Med.*, **90**(Suppl. 30), 34–9.

Høst, A., Jacobsen, H.P., Halken, S. and Holmenlund, D. (1995) The natural history of cow's milk protein allergy/intolerance. *Eur. J. Clin. Nutr.*, **49**(Suppl. 1), S13–8.

Iikura, Y., Imai, Y., Imai, T., Akasawa, A., Fujita, K., Hoshiyama, K., Nakura, H., Kohno, Y., Koike, K., Okudaira, H. and Iwasaki, E. (1999) Frequency of immediate-type food allergy in children in Japan. *Int. Arch. Allergy Immunol.*, **118**, 251–2.

Isolauri, E., Sutas, Y., Makinen, K.S., Oja, S.S., Isosomppi, R. and Turjanmaa, K. (1995) Efficacy and safety of hydrolyzed cow milk and amino acid-derived formulas in infants with cow milk allergy. *J. Pediatr.*, **127**, 550–7.

Kaminogawa, S. (1996) Food allergy, oral tolerance and immunomodulation—Their molecular and cellular mechanisms. *Biosci. Biotech. Biochem.*, **60**, 1749–56.

Kaminogawa, S., Kumagai, K., Yamauchi, K., Iwasaki, E., Mukoyama, T. and Baba, M. (1984) Allergic skin reactivity and chemical-properties of allergens in 2 grades of lactose. *J. Food Sci.*, **49**, 529–35.

Kaminogawa, S., Hattori, M., Ando, O., Kurisaki, K. and Yamauchi, K. (1987) Preparation of monoclonal-antibody against bovine β-lactoglobulin and its unique binding-affinity. *Agric. Biol Chem.*, **51**, 797–802.

Kaminogawa, S., Shimizu, M., Ametani, A., Hattori, M., Ando, O., Hachimura, S., Nakamura, Y., Totsuka, M. and Yamauchi, K. (1989) Monoclonal antibodies as probes for monitoring the denaturation process of bovine β-lactoglobulin. *Biochim. Biophys. Acta*, **998**, 50–6.

Kaminogawa, S., Hachimura, S., Nakajima-Adachi, H. and Totsuka, M. (1999) Food allergens and mucosal immune systems with special reference to recognition of food allergens by gut-associated lymphoid tissue. *Allergol. Int.*, **48**, 15–23.

Karin, N., Mitchell, D.J., Brocke, S., Ling, N. and Steinman, L. (1994) Reversal of experimental autoimmune encephalomyelitis by a soluble peptide variant of a myelin basic protein epitope: T cell receptor antagonism and reduction of interferon gamma and tumor necrosis factor alpha production. *J. Exp. Med.*, **180**, 2227–37.

Katakura, Y., Totsuka, M., Ametani, A. and Kaminogawa, S. (1994) Tryptophan-19 of β-lactoglobulin, the only residue completely conserved in the lipocalin superfamily, is not essential for binding retinol, but relevant to stabilizing bound retinol and maintaining its structure. *Biochim. Biophys. Acta*, **1207**, 58–67.

Kim, S.M., Enomoto, A., Hachimura, S., Yamauchi, K. and Kaminogawa, S. (1993) Serum antibody response elicited by a casein diet is directed to only limited determinants of α_{s1}-casein. *Int. Arch. Allergy Immunol.*, **101**, 260–5.

Kobayashi, I., Sakiyama, Y., Tame, A., Kobayashi, K. and Matsumoto, S. (1996) IgE and IgG_4 antibodies from patients with mite allergy recognize different epitopes of *Dermatophagoides pteronyssinus* group II antigen (Der p 2). *J. Allergy Clin. Immunol.*, **97**, 638–45.

Kohyama, M., Kakehi, M., Totsuka, M., Hachimura, S., Hisatsune, T. and Kaminogawa, S. (1998) Selective induction of $CD8^+$ T cell functions by single substituted analogs of an antigenic peptide: distinct signals for IL-10 production. *FEBS Lett.*, **423**, 138–42.

Kuchroo, V.K., Greer, J.M., Kaul, D., Ishioka, G., Franco, A., Sette, A., Sobel, R.A. and Lees, M.B. (1994) A single TCR antagonist peptide inhibits experimental allergic encephalomyelitis mediated by a diverse T cell repertoire. *J. Immunol.*, **153**, 3326–36.

Kuitunen, P., Visakorpi, J.K., Savilahti, E. and Pelkonen, P. (1975) Malabsorption syndrome with cow's milk intolerance. Clinical findings and course in 54 cases. *Arch. Dis. Child.*, **50**, 351–6.

Kumosinski, T.F., Brown, E.M. and Farrel, H.M., Jr. (1991) Three-dimensional molecular modeling of bovine α_{s1}-casein. *J. Dairy Sci.*, **74**, 2889–95.

Lee, Y.H. (1992) Food-processing approaches to altering allergenic potential of milk-based formula. *J. Pediatr.*, **121**, S47–50.

Maeji, N.J., Bray, A.M. and Geysen, H.M. (1990) Multi-pin peptide synthesis strategy for T cell determinant analysis. *J. Immunol. Methods*, **134**, 23–33.

Makinen, K.S. and Sorva, R. (1993) Bovine β-lactoglobulin levels in hydrolysed protein formulas for infant feeding. *Clin. Exp. Allergy*, **23**, 287–91.

Miller, K., Meredith, C., Sélo, I. and Wal, J.M. (1999) Allergy to bovine β-lactoglobulin: specificity of immunoglobulin E generated in the Brown Norway rat to tryptic and synthetic peptides. *Clin. Exp. Allergy*, **29**, 1696–704.

Mizumachi, K., Kurisaki, J. and Tsuji, N.M. (1990) Modification of antigenic structure in the C-terminal region of bovine β-lactoglobulin. *Proc XXIII Int. Dairy Congr.* (Montreal)., 284.

Nakajima, H., Hachimura, S., Nishiwaki, S., Katsuki, T., Shimojo, N., Ametani, A., Kohno, Y. and Kaminogawa, S. (1996) Establishment and characterization of α_{s1}-casein-specific T-cell lines from patients allergic to cow's milk: unexpected higher frequency of $CD8^+$ T-cell lines. *J. Allergy Clin. Immunol.*, **97**, 1342–9.

Nakajima-Adachi, H., Hachimura, S., Ise, W., Honma, K., Nishiwaki, S., Hirota, M., Shimojo, N., Katsuki, T., Ametani, A., Kohno, Y. and Kaminogawa, S. (1998) Determinant analysis of IgE and IgG_4 antibodies and T cells specific for bovine α_{s1}-casein from the same patients allergic to cow's milk: existence of α_{s1}-casein-specific B cells and T cells characteristic in cow's-milk allergy. *J. Allergy Clin. Immunol.*, **101**, 660–71.

Négroni, L., Bernard, H., Clement, G., Chatel, J.M., Brune, P., Frobert, Y., Wal, J.-M. and Grassi, J. (1998) Two-site enzyme immunometric assays for determination of naive and denatured β-lactoblobulin. *J. Immunol. Meths.*, **220**, 25–37.

Nicholson, L.B., Greer, J.M., Sobel, R.A., Lees, M.B. and Kuchroo, V.K. (1995) An altered peptide ligand mediates immune deviation and prevents autoimmune encephalomyelitis. *Immunity*, **3**, 397–405.

Nicholson, L.B., Murtaza, A., Hafler, B.P., Sette, A. and Kuchroo, V.K. (1997) AT cell receptor antagonist peptide induces T cells that mediate bystander suppression and prevent autoimmune encephalomyelitis induced with multiple myelin antigens. *Proc. Natl. Acad. Sci. USA.*, **94**, 9279–84.

Nishijima, K., Kohyama, M., Hisatsune, T., Kato, H., Kakehi, M. and Kaminogawa, S. (1997) Enhancing effect of interleukin-4 on the secretion of interferon-gamma by α_{s1}-casein-specific $CD8^+$ T cells. *Biosci. Biotech. Biochem.*, **61**, 1156–62.

Oldaeus, G., Bjorksten, B., Einarsson, R. and Kjellman, N.I. (1991) Antigenicigy and allergenicity of cow's milk hydrolysates intended for infant feeding. *Paediatr. Allergy Immunol.*, **4**, 156–60.

Ong, E.K., Knox, R.B. and Singh, M.B. (1995) Mapping of the antigenic and allergenic epitopes of Lol pVB using gene fragmentation. *Mol. Immunol.*, **32**, 295–302.

Otani, H. and Hosono, A. (1987) Antigenic reactive regions of *S*-carboxymethylated β-lactoglobulin. *Agric. Biol. Chem.*, **51**, 531–6.

Papiz, M.Z., Sawyer, L., Eliopoulos, E.E., North, A.C., Findlay, J.B., Sivaprasadarao, R., Jones, T.A., Newcomer, M.E. and Kraulis, P.J. (1986) The structure of β-lactoglobulin and its similarity to plasma retinol-binding protein. *Nature*, **324**, 383–5.

Piastre, M., Stabile, A., Fioravanti, G., Castagnola, M., Pani, G. and Ria, F. (1994) Cord blood mononuclear cell responsiveness to β-lactoglobulin: T-cell activity in 'atopy-prone' and 'non-atopy-prone' newborns. *Int. Arch. Allergy Immunol.*, **104**, 358–65.

Pike, A.C., Brew, K. and Acharya, K.R. (1996) Crystal structures of guinea-pig, goat and bovine α-lactalbumin highlight the enhanced conformational flexibility of regions that are significant for its action in lactose synthase. *Structure*, **4**, 691–703.

Ragno, V., Giampietro, P.O., Bruno, G. and Businco, L. (1993) Allergenicity of milk protein hydrolysate formulae in children with cow's milk allergy. *Eur. J. Pediatr.*, **152**, 760–2.

Sakurai, T., Ametani, A., Nakamura, Y., Shimizu, M., Idota, T. and Kaminogawa, S. (1993) Cryptic B cell determinant in a short peptide: T cells do not induce antibody response of B cells when their determinants entirely overlap each other. *Int. Immunol.*, **5**, 793–800.

Sampson, H.A., Bemhisel-Broadbent, J., Yang, E. and Scanlon, S.M. (1991) Safety of casein hydrolysate formula in children with cow milk allergy. *J. Pediatr.*, **118**, 520–5.

Savilahti, E. (1981) Cow's milk allergy. *Allergy*, **36**, 73–88.

Savilahti, E. and Kuitunen, M. (1992) Allergenicity of cow milk proteins. *J. Pediatr.*, **121**, S12–20.

Saylor, J.D. and Bahna, S.L. (1991) Anaphylaxis to casein hydrolysate formula. *J. Pediatr.*, **118**, 71–4.

Sélo, I., Négroni, L., Créminon, C. Yvon, M., Peltre, G. and Wal, J.-M. (1998) Allergy to bovine β-lactoglobulin: specificity of human IgE using cyanogen bromide-derived peptides. *Int. Arch. Allergy Immunol.*, **117**, 20–8.

Sélo, L., Clement, G., Bernard, H., Chatel, J., Creminon, C., Peltre, G. and Wal, J.-M. (1999) Allergy to bovine β-lactoglobulin: specificity of human IgE to tryptic peptides. *Clin. Exp. Allergy*, **29**, 1055–63.

Sette, A., Alexander, J., Ruppert, J., Snoke, K., Franco, A., Ishioka, G. and Grey, H.M. (1994) Antigen analogs/MHC complexes as specific T cell receptor antagonists. *Annu. Rev. Immunol.*, **12**, 413–31.

Shon, D.H., Enomoto, A., Yamauchi, K. and Kaminogawa, S. (1991) Antibodies raised against peptide fragments of bovine α_{s1}-casein cross-react with the native protein, but recognize sites distinct from the determinants on the protein. *Eur. J. Immunol.*, **21**, 1475–80.

Singh, D.P., Kikuchi, T., Singh, V.K. and Shinohara, T. (1994) A single amino acid substitution in core residues of S-antigen prevents experimental autoimmune uveitis. *J. Immunol.*, **152**, 4699–705.

Sloan-Lancaster, J. and Allen, P.M. (1996) Altered peptide ligand-induced partial T cell activation: Molecular mechanisms and role in T cell biology. *Annu. Rev. Immunol.*, **14**, 1–27.

Spies, J.R. (1971) New antigens in lactose. *Proc. Soc. Exp. Biol. Med.*, **137**, 211–4.

Spies, J.R., Stevan, M.A. and Stein, W.J. (1972) The chemistry of allergens. XXI. Eight new antigens generated by successive pepsin hydrolyses of bovine β-lactoglobulin. *J. Allergy Clin. Immunol.*, **50**, 82–91.

Spuergin, P., Mueller, H., Walter, M., Schiltz, E. and Forster, J. (1996) Allergenic epitopes of bovine α_{s1}-casein recognized by human IgE and IgG. *Allergy*, **51**, 306–12.

Stuart, D.I., Acharya, K.R., Walker, N.P., Smith, S.G., Lewis, M. and Phillips, D.C. (1986) α-Lactalbumin possesses a novel calcium binding loop. *Nature*, **324**, 84–7.

Takahashi, T., Yamauchi, K. and Kaminogawa, S. (1990) Comparison between the antigenicity of native and unfolded β-lactoglobulin. *Agric. Biol. Chem.*, **54**, 691–7.

Therattil, J.M. and Bahna, S.L. (1999) Hypoallergenic formulas, in *Food Hypersensitivity and Adverse Reactions. A Practical Guide for Diagnosis and Management*, (M. Frieri and B. Kettelhut eds.) Marcel Dekker, Inc., New York, pp. 477–83.

Thompson, H.S. and Staines, N.A. (1990) Could specific oral tolerance be a therapy for autoimmune disease? *Immunol. Today*, **11**, 396–9.

Toda, M., Totsuka, M., Furukawa, S., Yokota, K., Yoshioka, T., Ametani, A. and Kaminogawa, S. (2000) Downregulation of antigen-specific antibody production by T-cell receptor antagonist peptides *in vivo*. *Eur. J. Immunol.*, **30**, 403–14.

Totsuka, M., Katakura, Y., Shimizu, M., Kumagai, I, Miura, K. and Kaminogawa, S. (1990) Expression and secretion of bovine β-lactoglobulin in *Saccharomyces cerevisiae*. *Agric. Biol. Chem.*, **54**, 3111–6.

Totsuka, M., Ametani, A. and Kaminogawa, S. (1997a) Fine mapping of T-cell determinants of bovine β-lactoglobulin. *Cytotechnology*, **25**, 101–13.

Totsuka, M., Furukawa, S., Sato, E., Ametani, and Kaminogawa, S. (1997b) Antigen-specific inhibition of $CD4^+$ T-cell responses to β-lactoglobulin by its single amino acid-substituted mutant form through T-cell receptor antagonism. *Cytotechnology*, **25**, 115–26.

Totsuka, M., Kakehi, M., Kohyama, M., Hachimura, S., Hisatsune, T. and Kaminogawa, S. (1998) Enhancement of antigen-specific IFN-γ production from $CD8^+$ T cells by a single amino acid-substituted peptide derived from bovine α_{s1}-casein. *Clin. Immunol. Immunopathol.*, **88**, 277–86.

Tsuji, N.M., Kurisaki, J., Mizumachi, K. and Kaminogawa, S. (1993) Localization of T-cell determinants on bovine β-lactoglobulin. *Immunol. Lett.*, **37**, 215–21.

van Milligen, F.J., van't Hof, W., van den Berg, M. and Aalberse, R.C. (1994) IgE epitope on the cat (*Felix domesticus*) major allergen Fel d I: a study with overlapping synthetic peptides. *J. Allergy Clin. Immunol.*, **93**, 34–43.

van't Hof, W., Driedijk, P.C., van den Berg, M., Beck-Sickinger, A.G., Jung, G. and Aalberse, R.C. (1991) Epitope mapping of the *Dermatophagoides pteronyssinus* house dust mite major allergen Der p II using overlapping synthetic peptides. *Mol. Immunol.*, **28**, 1225–31.

Venien, A., Levieux, D., Astier, C, Briand, L., Chobert, J.M. and Haertle, T. (1997) Production and epitopic characterization of monoclonal antibodies against bovine β-lactoglobulin. *J. Dairy Sci.*, **80**, 1977–87.

Wahn, U., Wahl, R. and Rugo, E. (1992) Comparison of the residual allergenic activity of six different hydrolyzed protein formulas. *J. Pedeatr.*, **121**, S80–4.

Wal, J.-M. (1998) Cow's milk allergens. *Allergy*, **53**, 1013–22.

Weiner, H.L. (1994) Oral tolerance. *Proc. Natl. Acad. Sci. USA.*, **91**, 10762–5.

Weiner, H.L., Friedman, A., Miller, A., Khoury, S.J., Al-Sabbagh, A., Santos, L., Sayegh, M., Nussenblatt, R.B., Trentham, D.E. and Hafler, D.A. (1994) Oral tolerance: immunologic mechanisms and treatment of animal and human organ-specific autoimmune diseases by oral administration of autoantigens. *Annu. Rev. Immunol.*, **12**, 809–37.

Williams, S.C., Badley, R.A, Davis, P.J., Puijk, W.C. and Meloen, R.H. (1998) Identification of epitopes within β-lactoglobulin recognised by polyclonal antibodies using phage display and PEPSCAN. *J. Immunol. Meth.*, **213**, 1–17.

14

MILK PROTEIN HYDROLYSATES AND BIOACTIVE PEPTIDES

R.J. FitzGerald and H. Meisel

14.1 Introduction

The basic function of milk proteins has long been thought to be that of providing nitrogen and essential amino acids for young mammals (Hambræus, 1992). In addition, intact milk proteins have a range of biological activities, e.g., immunoglobulins have an immunoprotective effect, lactoferrin displays antibacterial activity while low concentrations of growth factors and hormones, mainly present in colostrum, appear to play a significant role in post-natal development (Schanbacher *et al.*, 1998). Milk proteins also contain a large range of bioactive peptide sequences which are encrypted within their primary structures (Table 14.1, Meisel, 1998). These include opioid agonist and antagonist peptides, potential hypotensive peptides which inhibit angiotensin-I-converting enzyme (ACE), mineral binding, immunomodulatory, antibacterial and antithrombotic peptides (FitzGerald, 1998; Meisel, 1998; Schanbacher *et al.*, 1998; Takano, 1998; Tomé and Debabbi, 1998; Meisel and Bockelmann, 1999).

14.2 Structure and biological activity

Biologically-active peptides have been characterised in milk proteins following

- their *in vitro* digestion with both specific and crude proteinase and exopeptidase preparations
- their *in vivo* digestion with gastrointestinal proteinases/peptidases
- the action of bacterial proteinase and peptidase activities during the generation of fermented milk products.

Advanced Dairy Chemistry Volume 1: Proteins, 3rd edn.
Edited by P.F. Fox and P.L.H. McSweeney, Kluwer Academic/Plenum Publishers, 2003.

TABLE 14.1
Summary of bioactive peptides derived from milk proteins

Bioactive peptide	*Protein precursor*	*Bioactivity*
Casomorphins	α_{s1}- and β-Caseins	Opioid agonist
α-Lactorphin	α-Lactalbumin	Opioid agonist
β-Lactorphin	β-Lactoglobulin	Opioid agonist
Lactoferroxins	Lactoferrin	Opioid antagonist
Casoxins	κ-Casein	Opioid antagonist
Casokinins	α_{s1}- and β-Caseins	ACE-inhibitory
Lactokinins	α-Lactalbumin, β-lactoglobulin and serum albumin	ACE-inhibitory
Immunopeptides	α_{s1}-, β- and κ-Caseins	Immunomodulatory
Lactoferricin	Lactoferrin	Antimicrobial
Casocidin	α_{s2}-Casein	Antimicrobial
Isracidin	α_{s1}-Casein	Antimicrobial
Casoplatelins	κ-Casein	Antithrombotic
Phosphopeptides	α_{s1}-, α_{s2}-, β- and κ-Caseins	Mineral binding

Source: Author. ■ ■ ■

Furthermore, the structure and biological activity of various peptides corresponding to specific sequences of milk proteins has been determined and confirmed by peptide synthesis studies.

14.2.1 Opioid peptides

Typical endogenous opioid peptides are derived from enkephalins, endorphins and dynorphins, and act as opioid receptor ligands. The physiological effect of these peptides depends on receptor type, e.g., μ receptors are linked to the control of intestinal motility and emotional behaviour, δ receptors control emotional behaviour and κ receptors are linked with analgesia and satiety (Höllt, 1983). Milk protein-derived opioid peptides can display either agonist or antagonistic activity. Exogenous casein-derived opioid peptides are known as casomorphins and/or casoxins, whereas whey protein-derived opioid peptides are known as lactorphins (Table 14.2). Opioid peptides, i.e., opioid receptor (μ-, δ- or κ-type) ligands with agonistic activity, originate from different milk proteins and exert naloxone-inhibitable opioid activities in both receptor studies and during bioassays (Brantl *et al.*, 1981; Loukas *et al.*, 1983; Chiba and Yoshikawa, 1986). The atypical opioid peptides derived from milk proteins have N-terminal sequences different from that of the 'typical' endogenous opioid peptides (Teschemacher *et al.*, 1997). The major exogenous opioid peptides, β-casomorphins, are fragments of β-casein (f 60–70) and have been characterized as μ-type ligands (Meisel, 1986; Teschemacher and Brantl, 1994). β-Casomorphins were also found in

TABLE 14.2
Opioid peptides derived from bovine milk proteins

Opioid peptide	*Peptide fragment/ analogue*		*Primary structure (one letter code)*	IC_{50}^{a} *(μmol/l)*	*Enzymatic hydrolysis*	*Peptide synthesis*	*Reference*
Opioid agonists							
β-casomorphin-11	β-casein	(f 60–70)	YPFPGPIPNSL	10	+[b]	–	Meisel (1986)
β-casomorphin-7		(f 60–66)	YPFPGPI	14	+	+	Brantl *et al.* (1981) Teschemacher *et al.* (1997)
β-casomorphin-6		(f 60–65)	YPFPGP	3.2	–	+	Brantl *et al.* (1981) Haileselassie *et al.* (1999)
β-casomorphin-5		(f 60–64)	YPFPG	1.1	+	+	Brantl *et al.* (1981)
β-casomorphin-4		(f 60–63)	YPFP	2.7	–	+	Brantl *et al.* (1981)
morphiceptin		(f 60–63)	YPFP.NH_2	3	+	+	Chang *et al.* (1981) Chang *et al.* (1985) Yoshikawa *et al.* (1986)
α_{s1}-casein exorphin	α_{s1}-casein	(f 90–96)	RYLGYLE	1.2	+	+	Loukas *et al.* (1983)
α_{s1}-casein exorphin		(f 90–95)	RYLGYL	12	+	+	Loukas *et al.* (1983)
α_{s1}-casein exorphin		(f 91–96)	YLGYLE	45	–	+	Loukas *et al.* (1983)
α-lactorphin	α-lactalbumin	(f 50–53)	YGLF	67	–	+	Chiba and Yoshikawa (1986) Antila *et al.* (1991)
β-lactorphin	β-lactoglobulin	(f 102–105)	YLFF	38	–	+	Chiba and Yoshikawa (1986) Antila *et al.* (1991)
serorphin	serum albumin	(f 399–404)	YGFNA	85	+		Tani *et al.* (1994)
Opioid antagonists							
casoxin A	κ-casein	(f 35–42)	YPSYGLNY	400	+	+	Chiba *et al.* (1989) Yoshikawa *et al.* (1994)
casoxin B		(f 58–61)	YPYY		–	+	Chiba *et al.* (1989) Yoshikawa *et al.* (1994)
casoxin C		(f 25–34)	YIPIQYVLSR	50	+	+	Chiba *et al.* (1989) Yoshikawa *et al.* (1994)

[a]Peptide concentration required to inhibit [3H]-ligand binding by 50%.
[b]*in vivo* hydrolysis product isolated from intestinal chyme (duodenum) of Goettingen mini pig.
Source: Adapted from Meisel and FitzGerald (2000).

analogous positions in sheep, water buffalo and human β-casein (see Fiat and Jollès, 1989, for review).

Opioid antagonists have been found in bovine and human κ-caseins and in human α_{s1}-casein (Chiba *et al.*, 1994; Yoshikawa *et al.*, 1994). Various casoxins, which are opioid receptor ligands of the μ-type, have been isolated as C-terminally methoxylated peptides. Chemically-modified casoxins were more active than the non-methoxylated fragments (Chiba and Yoshikawa, 1986).

14.2.2 Peptide inhibitors of angiotensin-I-converting enzyme

Angiotensin-I-converting enzyme (ACE; peptidyldipeptide hydrolase, EC 3.4.25.1) has been classically associated with the renin-angiotensin system which regulates peripheral blood pressure. The enzyme is a zinc-dependent carboxydipeptidase. The enzyme can increase blood pressure by converting angiotensin I to the potent vasoconstrictor, angiotensin II. ACE can also catalyse the degradation of bradykinin, a vasodilatory peptide, and enkephalins. It is evident, therefore, that inhibition of ACE may exert an anti-hypertensive effect as a consequence of a decrease in angiotensin II and a concomitant increase in bradykinin activity. Indeed, potent synthetic inhibitors of ACE, such as Captopril®, are used extensively in the treatment of hypertension in humans (Wyvratt and Patchet, 1985). Several other endogenous peptides, such as β-endorphin, substance P and ACTH also act as competitive substrates and inhibitors of ACE (Koide *et al.*, 1980). ACE is, therefore, a multifunctional enzyme since its inhibition can lead to anti-hypertensive, immuno-stimulating and neuroactivatory effects.

Casokinin sequences [casein-derived peptide inhibitors of ACE (Meisel and Schlimme, 1994)] have been found in α_{s1}-, β- and κ-caseins (Table 14.3), and lactokinins (FitzGerald and Meisel, 1999) in α-lactalbumin, β-lactoglobulin and bovine serum albumin (Table 14.4). Two strategies have generally been used in the identification and characterisation of such peptides, i.e., isolation of inhibitory peptides from *in vitro* enzymatic digests of milk proteins and chemical synthesis of peptides or peptide analogues having similar structures to those known to inhibit ACE. Potent casokinins include α_{s1}-CN (f 25–27), β-CN (f 74–76), β-CN (f 169–174) and κ-CN (f 108–110) which have ACE IC_{50} values of 2, 5, 5 and 5 μmol/l, respectively (see Table 14.3). The most potent major whey protein-derived lactokinin reported to date is β-Lg (f 142–148) which has an ACE IC_{50} value of 43 μmol/l. The peptide corresponding to (f 208–218) of serum albumin, a minor whey protein, has a reported ACE IC_{50} value of 3 μmol/l (see Table 14.4). The structure-activity relationship for ACE-inhibitory peptides is not yet fully understood (FitzGerald and Meisel, 1999). However, it appears that binding to ACE is influenced significantly by the C-terminal tripeptide region of the substrate. It appears that peptides containing

TABLE 14.3
Bovine casein-derived angiotensin–I-converting enzyme (ACE) inhibitory peptides[a]

Peptide fragment/analogue		*Primary structure (one letter code)*	IC_{50}^{b} *(μ mol/l)*	*Enzymatic hydrolysis*	*Peptide synthesis*	*References*
α_{s1}-casein	(f 23–34)	FFVAPFPEVFGK	77	+	–	Maruyama and Suzuki (1982) Maruyama *et al.* (1985)
	(f 23–37)	FFVAP	6	+	–	Maruyama *et al.* (1985)
	(f 24–27)	FVAP	10	–	+	Maruyama *et al.* (1987a)
	(f 25–27)	VAP	2	–	+	Maruyama *et al.* (1987a)
	(f 27–30)	PFPE	>1000	–	+	Maruyama *et al.* (1987a)
	(f 28–34)	FPEVFGK	140	+	–	Maruyama *et al.* (1987a)
	(f 32–34)	FGK	160	–	+	Maruyama *et al.* (1987a)
	(f 104–109)	YKVPQL	22	+	–	Maeno *et al.* (1996)
	(f 142–147)	LAYFYP	65	+	–	Philanto–Leppälä *et al.* (1998)
	(f 143–148)	AYFYPE	106[c]	+	–	Yamamoto *et al.* (1994)
	(f 157–164)	DAYPSGAW	98	+	–	Philanto–Leppälä *et al.* (1998)
	(f 194–199)	TTMPLW	16	+	–	Maruyama *et al.* (1987b)
	(f 197–199)	PLW	36	–	+	Maruyama *et al.* (1987b)
	(f 198–199)	LW	50	–	+	Maruyama *et al.* (1987b)
β-casein	(f 57–64)	SLVLPVPE	39	+	–	Yamamoto *et al.* (1994)
	(f 60–66)	YPFPGPIP	500	–	–	Meisel and Schlimme (1994)
	(f 74–76)	IPP	5	–	–	Nakamura *et al.* (1995)
	(f 84–86)	VPP	9	–	–	Nakamura *et al.* (1995)
	(f 108–113)	EMPFPK	423[c]	–	–	Pihlanto–Leppälä *et al.* (1998)
	(f 169–174)	KVLPVP	5	–	–	Maeno *et al.* (1996)
	(f 169–175)	KVLPVPQ	1000	–	–	Maeno *et al.* (1996)
	(f 177–179)	AVP	340	–	–	Maruyama *et al.* (1987a)
	(f 177–181)	AVPYP	80	–	–	Maruyama *et al.* (1987a)
	(f 177–183)	AVPYPQR	15	–	–	Maruyama *et al.* (1987a)
	(f 179–181)	PYP	220	–	–	Maruyama *et al.* (1987a)

TABLE 14.3 (Continued)

Peptide fragment/analogue		*Primary structure (one letter code)*	IC^{b}_{50} *(μ mol/l)*	*Enzymatic hydrolysis*	*Peptide synthesis*	*References*
	(f 181 –183)	PQR	> 400	–	–	Maruyama *et al.* (1987a)
	(f 193–198)	YQQPVL	280	–	–	Pihlanto–Leppaia *et al.* (1998)
	(f 193–202)	YQQPVLGPVR	300	–	–	Meisel and Schlimme (1994)
κ–casein	(f 25–34)	YIPIQYVLSR	nd	–	–	Chiba and Yoshikawa (1991)
	(f 35–41)	YPSYGLNY	nd	–	–	Chiba and Yoshikawa (1991)
	(f 58–59)[d]	YP	720	–	–	Yamamoto *et al.* (1999)
	(f 108–110)	IPP	5	–	–	Nakamura *et al.* (1995)

[a]Details of other casein–derived peptides/or related peptides which inhibit ACE are available within the references used to generate this Table.
[b]Peptide concentration required to inhibit ACE by 50%.
[c]IC_{50} value given in μg/ml.
[d]This sequence also in α_{s1}-casein (f 146–147) and (f 159–160), and in β-casein (f 114–115); nd – not determined.
Source: FitzGerald and Meisel (2000).

TABLE 14.4
Whey protein - derived angiotensin-I-converting enzyme (ACE) inhibitory peptides

Peptide fragment/analogue		*Primary structure (one letter code)*	IC_{50}^{a} (μ *mol/l*)	*Enzymatic hydrolysis*	*Peptide synthesis*	*Reference*
α-Lactalbumin	(f 50–51)	YG	1523	–	+	Mullally *et al.* (1996)
	(f 50–53)	YGLF	733	+	+	Mullally *et al.* (1996)
	(f 52–53)	LF	349	–	+	Mullally *et al.* (1996)
	(f 105–110)	LAHKAL	621	+	–	Pihlanto–Leppälä *et al.* (1998)
β-Lactoglobulin	(f 9–14)	GLDIQK	580	+	–	Pihlanto–Leppälä *et al.* (1998)
	(f 15–20)	VAGTWY	1682	+	–	Pihlanto–Leppälä *et al.* (1998)
	(f 102–103)	YL	122	–	+	Mullally *et al.* (1996)
	(f 102–105)	YLLF	172	+	+	Mullally *et al.* (1996)
	(f 104–105)	LF	349	–	+	Mullally *et al.* (1996)
	(f 142–148)	ALPMHIR	43	+	+	Mullally *et al.* (1997a)
	(f 146–148)	HIR	953	–	+	Mullally *et al.* (1997a)
	(f 146–149)	HIRL	1153	–	+	Mullally *et al.* (1996)
	(f 147–148)	IR	695	–	+	Mullally *et al.* (1996)
	(f 148–149)	RL	2439	–	+	Mullally *et al.* (1996)
Bovine serum albumin	(f 208–216)	ALKAWSVAR	3	–	+	Chiba and Yoshikawa (1991)

[a]Peptide concentration required to inhibit ACE by 50%.
Source: FitzGerald and Meisel (2000).

hydrophobic amino acid residues at the C-terminal region display highest inhibitory potency (Cheung *et al.*, 1980; Ondetti and Cushman, 1982). Many potent casokinins and lactokinins contain a C-terminal proline, lysine or arginine residue (Tables 14.3 and 14.4). Structure-activity data suggest that the presence of amino acids having a positively charged side group contributes significantly to ACE inhibitory potency (Ondetti and Cushman, 1982; Meisel, 1993).

14.2.3 Mineral binding peptides

All the caseins contain phosphate which is covalently bound *via* monoester linkages to seryl residues. The extent of phosphorylation depends on the individual casein and is influenced by genetic polymorphism (Chapter 16). Highly phosphorylated regions in the caseins display a common motif, i.e., a sequence of three phosphoseryl residues followed by two glutamic acid residues (West, 1986; Reynolds, 1994). The excellent bioavailability of minerals such as calcium from dairy products has been attributed, in part, to the presence of caseinophosphopeptides (CPPs) which, due to their highly polar acidic domains, can bind and solubilise minerals (Kitts and Yuan, 1992; FitzGerald, 1998). The apparent calcium-binding constant (K_{app}) for tryptic CPPs corresponding to β-CN (f 1–25)4P, α_{s1}-CN (f 43–58)2P and α_{s1}-CN (f 59–79)5P was 629, 328 and 841 L/mol, respectively (Meisel, 1997).

14.2.4 Antimicrobial peptides

Antimicrobial peptides have been derived from the whey protein, lactoferrin (Tomita *et al.*, 1991). Lactoferrin is an iron-binding glycoprotein present in most biological fluids of mammals, including milk. The existence of an antimicrobial sequence, named lactoferricin (f 17–41) near the N-terminus of lactoferrin, in a region distinct from its iron-binding sites, has been reported (Bellamy *et al.*, 1992). This peptide region contains one intramolecular disulphide bond. Lactoferricin displayed antimicrobial activity against various Gram-positive and -negative bacteria, yeasts and filamentous fungi (Bellamy *et al.*, 1992). The antimicrobial activity of lactoferricin and synthetic analogs seems to be correlated with the net positive charge of the peptides. These cationic peptides kill sensitive microorganisms by increasing cell membrane permeability (Bellamy *et al.*, 1993). Lactoferricins are reported to have antibacterial activity against enterotoxigenic *Escherichia coli* and *Listeria monocytogenes* (Dionysius and Milne, 1998). The antibacterial activity of lactoferricin-derived peptides against clinical isolates of *E. coli* O157:H7 has been reported also (Shin *et al.*, 1998). The fragment, α_{s2}-CN (f 165–203), casocidin-I, which contains a high proportion of basic amino acid residues (10 of 39) was found to inhibit the growth of *E. coli* and *Staphylococcus carnosus* (Zucht *et al*., 1995). An antibacterial peptide, isracidin, α_{s1}-CN (f 1–23), was significantly effective against lethal infection of mice by *Staphylococcus aureus* (Lahov and Regelson, 1996).

14.2.5 Antithrombotic peptides

Antithrombotic peptides, casoplatelins, are derived from the C-terminal part (caseinoglycomacropeptide) of bovine κ-casein. These peptides inhibit the aggregation of ADP-activated platelets and the binding of human fibrinogen Y-chain to platelet surface fibrinogen receptors (Jollès *et al*., 1986; Bouhallab *et al*., 1992). The main antithrombotic peptides of κ-casein correspond to κ-CN (f 106–116) and smaller fragments thereof, i.e., (f 106–112), (f 112–116) and (f 113–116). The smaller peptides, (f 106–112) and (f 113–116), are less potent inhibitors of platelet aggregation and do not inhibit fibrinogen binding (Fiat and Jollès, 1989).

14.2.6 Immunomodulatory peptides

Casein-derived immunopeptides, including α_{s1}-CN (f 194–199) and β-CN (f 63–68), (f 193–202) and (f 191–193) stimulate phagocytosis of red blood cells of sheep by murine peritoneal macrophages and exert a protective effect against *Klebsiella pneumoniae* infection in mice after intravenous administration of peptides (Migliore-Samour *et al*., 1989). The C-terminal sequence of β-casein, β-CN (f 193–209), which contains β-casokinin-10, (f 193–202), induced a significant proliferative response in rat lymphocytes (Coste *et al*., 1992). Depending on peptide concentration, β-CN (f 60–66), i.e., β-casokinin-10 and β-casomorphin-7, showed either a suppressive or a stimulatory effect on lymphocyte proliferation. β-CN (f 60–66) inhibited the proliferation of human colonic lamina propria lymphocytes where the anti-proliferative effect was reversed by the opiate receptor antagonist, naloxone (Elitsur and Luk, 1991). It has been shown that Tyr-Gly [α-La (f 18–19)] and (f 50–51), and [κ-CN (f 38–39)] as well as Tyr-Gly-Gly [α-La (f 18–20)] significantly enhanced the proliferation of human peripheral blood lymphocytes *in vitro* (Kayser and Meisel, 1996). Casein hydrolysates obtained on incubation with gastrointestinal proteinases and with *Lactobacillus GG*-derived enzymes displayed lymphocyte stimulatory and suppressive effects (Sutas *et al*., 1996; Rokka *et al*., 1997). Phosphopeptides from α_{s1}-CN (f 59–79) and β-CN (f 1–25) had mitogenic activity and enhanced immunoglobulin production in mouse spleen cells (Hata *et al*., 1998). β-Casomorphins and α_{s1}-casein exorphins are reported to inhibit the proliferation of human prostate cancer cell lines by a mechanism partly linked to opioid receptors (Kampa *et al*., 1997).

It is interesting to note that many milk-derived peptides reveal multifunctional properties, i.e., specific peptide sequences with two or more different bioactivities. For example, β-CN (f 60–66) displays opioid, ACE-inhibitory and immunomodulatory activity. Furthermore, some regions in the primary structure of the caseins contain overlapping peptide sequences which exert different biological functions. These regions have been considered as 'strategic zones' which are partially protected from proteolytic breakdown (Fiat and Jollès, 1989; Meisel, 1998; Schanbacher *et al.*, 1998).

14.3 Generation of bioactive peptides

The manifestation of specific bioactivities requires pre-digestion of intact milk proteins. This can occur during *in vivo* gastrointestinal transit and/or during food processing. It has been shown that relatively large amounts of bioactive peptides could potentially be produced during proteolysis of individual caseins and whey proteins (Meisel, 1998).

14.3.1 Production *in vivo*

Several reports definitively show that bioactive peptides are produced *in vivo* in mammals following intake of milk proteins. β-Casomorphins were found in the intestinal contents of humans fed bovine casein (Svedberg *et al.*, 1985). The opioid peptide, β-casomorphin-11 (β-casein (f 60–70)) has been found in the intestinal contents of minipigs (Meisel, 1986; Meisel and Frister, 1989). β-Casomorphin precursors have been reported in the intestinal contents of pre-ruminant calves following milk ingestion (Scanff *et al.*, 1992). Crude (unidentified) CPPs as well as specific CPPs sequences, i.e., α_{s1}-CN (f 66–74), have been found in the intestinal chyme and stools of mammals (minipigs and rats) during studies involving ingestion of whole casein or purified β-casein (Naito *et al.*, 1972; Sato *et al.*, 1983; Meisel and Frister, 1988; Kasai *et al.*, 1992). Chabance *et al.* (1998) detected CPPs in the stomach and duodenum of adult humans after ingestion of milk or yoghurt. Meisel *et al.* (2001) recently reported, for the first time, the presence of CPPs in the distal small intestine (ileum) of adult humans following ingestion of either milk and CPP meals. Lactoferricin has been isolated from the gastrointestinal contents of rats fed a diet containing lactoferrin (Tomita *et al.*, 1994).

14.3.2 Production *in vitro*

Hydrolysate products obtained following the *in vitro* digestion of milk proteins are used in a wide range of commercial product outlets, e.g., reduced/hypoallergenic infant formulae, enteral/sports and parenteral nutrition (FitzGerald *et al.*, 1996). β-Casomorphins, fragments of β-CN (f 60–70), have been released from casein by the action of purified gastrointestinal proteinases and an amide derivative of β-casomorphin-4 was identified in a commercial casein hydrolysate (Brantl *et al.*, 1981; Chang *et al.*, 1981; Teschemacher and Brantl, 1994; Teschemacher *et al.*, 1997). α_{s1}-Casein exorphins, fragments of α_{s1}-CN (f 90–96), were observed in peptic digests of casein (Ziodrou *et al.*, 1979; Loukas *et al.*, 1983; Pihlanto-Lepällä *et al.*, 1994). Opioid antagonists, κ-casoxins, were identified in peptic and tryptic digest of κ-casein (Yoshikawa *et al.*, 1986; Chiba and Yoshikawa, 1989). Peptic digests of α-La were shown to contain α-lactorphin while pepsin plus trypsin digestion of β-Lg resulted in the appearance of β-lactorphin (Antila *et al.*, 1991). Both purified and crude proteinase preparations have been used in the *in vitro* release of CPPs from whole and individual caseins (Manson and Annan, 1971; Berrocal *et al.*, 1989; Brulé *et al.*, 1989; Juillerat *et al.*, 1989; Reynolds, 1992; Adamson and Reynolds, 1995; Goepfert and Meisel, 1996; McDonagh and FitzGerald, 1998). Significant differences have been observed in the calcium-binding and solubilising abilities of CPPs generated by proteinases from mammalian, bacterial or plant sources. These differences were, in part, attributed to the specificity of the enzymatic activities used to hydrolyse the starting casein substrate (McDonagh and FitzGerald, 1998).

Tryptic hydrolysates of casein contain ACE inhibitors (Maruyama and Suzuki, 1982; Maruyama *et al.*, 1985, 1987b; Karaki *et al.*, 1990). Peptides corresponding to tryptic digests of α-lactalbumin and β-lactoglobulin (i.e., α- and β-lactorphin) were shown to have ACE inhibitory activity (Mullally *et al.*, 1996). Proteinase K digests of whey protein also possessed ACE inhibitory activity (Abubakar *et al.*, 1998). Significant differences were also observed in the ACE inhibitory potency of whey protein hydrolysates generated with different gastrointestinal proteinase preparations (Mullally *et al.*, 1997b; Pihlanto-Leppälä *et al.*, 1998). A detailed knowledge of the activities (proteinase and exopeptidase) present in commercially available proteinase preparations may provide some insight on those activities which favour the production of specific bioactivities during the *in vitro* hydrolysis of milk proteins (Mullally *et al.*, 1994; Smyth and FitzGerald, 1998).

Tryptic hydrolysis of the C-terminal glycomacropeptide region of κ-casein yielded antithrombotic peptides (Jollès *et al.*, 1986; Fiat and Jollès, 1989). Tryptic hydrolysis of bovine and human casein yielded peptides having immunomodulatory activity (Jollès *et al.*, 1981; Migliore-Samour *et al.*, 1989; Fiat *et al.*, 1993). Tryptic digestion of α_{s1}-casein yielded peptides which displayed mitogenic activity and which stimulated immunoglobulin production in mouse spleen cells (Hata *et al.*, 1998, 1999). Peptic digests of lactoferrin yielded peptides having antimicrobial activity (Tomita *et al.*, 1991; Bellamy *et al.*, 1992). Hydrolysis of α-La and β-Lg with gastrointestinal proteinases yielded hydrolysate products which, at relatively high concentrations, inhibit the growth of an indicator *E. coli* strain (Pihlanto-Leppälä *et al.*, 1999).

14.3.3 Production by bacteria

Bioactive peptides may be produced by starter and non-starter bacteria used in the production of fermented dairy products. The bacterial cell wall-associated proteinase and intracellular peptidases, which are released following cell lysis, lead to the production of various small peptides. CPPs have been found, for instance, in cheese (Roudot-Algaron *et al.*, 1994; Singh *et al.*, 1997). Incubation of β-casein with the PI-type proteinase from *Lactococcus lactis* yielded more than 100 different peptides. β-CN (f 60–68), which is part of β-casomorphin-11, and of β-CN (f 190–193), which is an immunopeptide, were observed in the hydrolysate (Juillard *et al.*, 1995). Several β-casein-derived casokinins, including β-CN (f 169–175), were produced on incubating casein with the cell wall-associated proteinase from *Lactobacillus helveticus* CP790 (Yamamoto *et al.*, 1994; Maeno *et al.*, 1996). Fermentation of acidified milk with a culture containing *Lactobacillus helveticus* CP790 and *Saccharomyces cerevisiae* yielded hypotensive β-CN (f 74–76) and (f 84–86) (Yamamoto *et al.*, 1994). Secondary proteolysis during cheese ripening also led to the formation of various ACE inhibitory peptides (Meisel *et al.*, 1997). Varying amounts of β-casomorphin immunoreactive material were detected in bovine milk samples incubated with caseolytic bacterial species such as *Pseudomonas aeruginosa* and *Bacillus cereus* (Hamel *et al.*, 1985). The presence of intracellular general and proline-specific exopeptidase activities in lactic acid starter bacteria (Bouchier *et al.*, 1999) would suggest that the occurrence of biologically active peptides in fermented milks should be rare (Meisel and Bockelmann, 1999). However, it has been demonstrated that the enzymatic degradation of β-casomorphins by cell lysates of *Lactococcus lactis* subsp. *cremoris* was highly influenced by the environmental conditions (Muehlenkamp and Warthesen, 1996). β-Casomorphins were more resistant to degradation at the high salt concentration, relatively low pH and incubation temperature used during cheese ripening.

14.4 Enrichment of bioactive peptides

Several strategies have been developed to enrich/purify specific biologically active peptides from milk protein hydrolysates. Selective precipitation with hydrophilic solvents (e.g., ethanol) after mineral salt-induced (e.g., calcium salt) aggregation has been used to produce CPPs (Reeves and Latour, 1958; Adamson and Reynolds, 1995; McDonagh and FitzGerald, 1998). Several chromatographic procedures using ion-exchange matrices have been described for the enrichment of CPPs from casein hydrolysates (Berrocal *et al.*, 1989; Juillerat *et al.*, 1989; Kunst, 1990; Koide *et al.*, 1991; Lihme *et al.*, 1994; Park *et al.*, 1998; Ellegård *et al.*, 1999). Calcium-aggregated CPPs have been separated from non-phosphorylated peptides by ultrafiltration (Brulé *et al.*, 1989; Reynolds, 1992). Ultrafiltration through defined nominal molecular mass cut-off membranes has also shown potential for the enrichment of ACE inhibitory peptides from whey protein hydrolysates (Mullally *et al.*, 1997b; Pihlanto-Leppälä *et al.*, 1998) and for opioid peptides from whey protein hydrolysates (Pihlanto-Leppälä *et al.*, 1996). Ultrafiltration membrane reactor systems have been investigated for the continuous production of permeates enriched in CPPs or casomorphin precursors (Righetti *et al.*, 1997), and antithrombotic peptides (Bouhallab *et al.*, 1992). Ion-exchange membrane systems have been used recently for the *in situ* production of cationic antibacterial peptides from lactoferrin (Recio and Visser, 1999). Hen egg yolk antibodies (IgY) have been raised against β-CN (f 193–202), an ACE inhibitory peptide (Meisel, 1994). However, the potential of an immunoaffinity approach for the enrichment of milk protein-derived bioactive peptides does not appear to have been investigated extensively. β-Casomorphins have also been produced in bacteria as fusion protein complexes following genetic manipulation (Carnie *et al.*, 1989).

14.5 Physiological significance

Bioactive peptides must reach their target site(s) in order to exert a specific physiological response. Furthermore, orally ingested peptides and/or bioactive peptides produced by intestinal proteinases may need to survive gastrointestinal proteinases/peptidases, luminal intestinal peptidases and serum peptidase activities in order to mediate a physiological effect (Figure 14.1).

The physiological effects of several milk protein-derived bioactive peptides have been studied in mammalian systems. Opioid peptides derived from milk proteins appear to have physiological significance in mammals (due to the liberation of casomorphin in the mammary gland) and in neonates, and they appear to participate in the control of gastrointestinal functions in adults (Teschemacher *et al.*, 1997). Opioid receptors are located in the nervous, endocrine and immune systems as well as in the intestinal tract of mammals and can interact with their endogenous ligands as well as with exogenous opioids and opioid antagonists (Teschemacher and Brantl, 1994). Orally administered opioid peptides derived from milk proteins are able to modulate absorption processes in the gut. The enhancement of net water and electrolyte absorption by β-casomorphins in the small and large intestine is a major component of their antidiarrhoeal action (Daniel *et al.*,

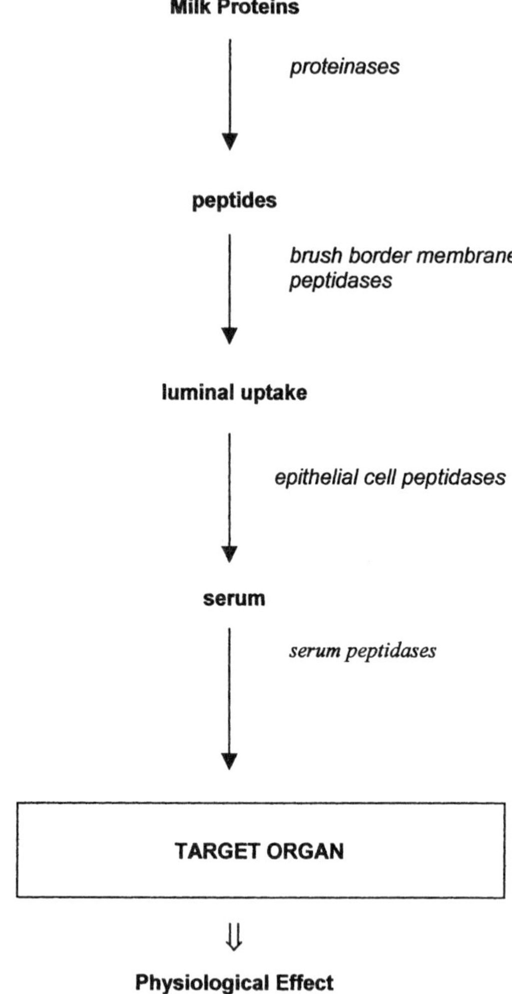

Figure 14.1 Route from intact protein to target organ. Source: author ■ ■ ■.

1990a,b). This effect could be mediated *via* subepithelial opioid receptors or specific luminal binding sites at the brush border membrane (Tomé *et al.*, 1987; Brandsch *et al.*, 1994). β-Casomorphins, β-CN (f 60–64), (f 60–66) and (f 60–70), are claimed to be degraded rapidly once they enter the blood stream. However, the presence of β-casomorphin-7 immunoreactive material has been demonstrated in the plasma of newborn calves after their first milk intake (Umbach *et al.*, 1985). This material revealed a greater molar mass than β-casomorphin-7, β-CN (f 60–66), and thus has been considered as a β-casomorphin precursor. Such pre-casomorphins could reach any potential site of action in the organism to elicit effects after liberation of the protected active sequence from the precursor molecule. Opioid casein fragments have not been detected in the plasma of adult mammals (Umbach *et al.*, 1985; Teschemacher *et al.*, 1986). Thus, only the neonatal intestine appears to be permeable to (pre-)casomorphins. A variety of opioid activities have been described in other test systems which may be of pharmacological but not of physiological significance (see Teschemacher *et al.*, 1997, for review).

Inhibition of the multifunctional enzyme, ACE, which is located in different tissues (e.g., plasma, lung, kidney, heart, skeletal muscle, pancreas, brain, mammary arteries, testes, uterus, intestine) may influence different regulatory systems (Ondetti and Cushman, 1982; Bruneval *et al.*, 1986; Johnston, 1992; Meisel, 1993). ACE has been associated classically with the renin-angiotensin system regulating peripheral blood pressure where ACE-inhibition results in an anti-hypertensive effect. When ACE-inhibitory peptides were administered orally to rats, blood pressure was reduced in a dose-dependent manner (Suetsuna and Osajima, 1989). It is known that small peptides, such as di- and tri-peptides, are absorbed easily in the intestine without being degraded by digestive enzymes. Accordingly, small ACE-inhibitory peptides can pass directly across the intestine to reach peripheral target sites (see Yamamoto, 1997, for review). In addition, ACE-inhibitory peptides having Pro residues, especially at the C-terminus, are highly resistant to degradation by digestive enzymes. For example, the dipeptide Tyr-Pro, (Table 14.3), mediates significant hypotensive effects in spontaneously hypertensive rats (Yamamoto *et al.*, 1999). Recently, a placebo-controlled study showed that the daily intake of 95 ml of a fermented sour milk drink product (Calpis Co., Japan) resulted in a significant decrease in blood pressure in hypertensive human subjects (Hata *et al.*, 1996; Takano, 1998). The sour milk drink, which was prepared by fermenting reconstituted skim milk powder with a starter culture of *Lactobacillus helveticus* CP790 and *Saccharomyces cerevisiae*, contained two ACE-inhibitory tripeptides, Val-Pro-Pro (β-CN (f 84–86)) and Ile-Pro-Pro (β-CN (f 74–76)) in addition to κ-CN (f 108–110), which were not present in the placebo drink (Nakamura *et al.*, 1995). It is worth noting that the ingested dose of these peptides was in the range of only 1.2 to 1.6 mg per day.

The formation of soluble complexes with minerals and the resistance to proteolysis of CPPs in the intestinal lumen is a prerequisite for their potential function as mineral carriers (FitzGerald, 1998). Under normal physiological conditions, the vitamin D-independent paracellular transport of calcium plays a major role in calcium absorption from the distal small intestine (see Kitts and Yuan, 1992, for review). Hence, CPPs might exert an influence on the absorption of calcium or other minerals and trace elements in the intestine. A recent human feeding trial showed an increase in calcium and zinc absorption following CPP incorporation into a rice-based infant food (Hansen *et al.*, 1997). Furthermore, it has been shown that calcium-binding CPPs can have an anticariogenic effect in that they inhibit caries lesions through recalcification of the dental enamel (see Reynolds, 1987).

Bactericidal peptides may assist in protecting against a microbial challenge, especially in the neonatal intestinal tract, and thus support the non-immune defence of the gut (Schanbacher *et al.*, 1997). It is not yet clear how the antibacterial effects of food-derived peptides are of physiological importance. However, an 11-residue fragment of lactoferricin was shown to have a lower hemolytic activity while maintaining its antimicrobial activity (Kang *et al.*, 1996). This peptide is a likely target for the future development of peptide antibiotics as therapeutic agents having low toxicity.

Data on the physiological significance of food-derived peptides with anti-thrombotic activity are not yet available. However, the presence of case-inomacropeptides in the plasma of 5 day-old newborn infants following ingestion of human milk and bovine milk-based formula has been demonstrated (Chabance *et al.*, 1995).

The immunopotentiatory peptides, Tyr-Gly (α-La (f 18–19), (f 50–51) and κ-CN (f 38–39)) and Tyr-Gly-Gly (α-La (f 18–20)), were found to be the active components in a dialysed leucocytes extract from normal donors that was used in a large multi-center trial to inhibit the development of infections in patients with pre-AIDS (Hadden, 1991). It has been suggested that casein-derived peptides are implicated in stimulation of the newborn's immune system (Migliore-Samour *et al.*, 1989). These immunostimulating activities may also have a direct effect on the resistance to bacterial and viral infections in human adults.

14.6 Conclusion

Physiologically, functional food ingredients can be defined as foods, or part of naturally occurring food(s), which on consumption can contribute to reducing the risk of developing specific diseases. Nutraceuticals or medicinal foods, on the other hand, may be considered as bioactive substances, the beneficial health effects of which relate to the prevention and treatment of specific diseases which normally require the use of pharmacologically active compounds (Meisel, 1998). While numerous studies have demonstrated significant potential *in vitro* and *in vivo* physiological effects of biologically active peptides, much work is still required before their widespread application as functional food ingredients and nutraceuticals. Detailed clinical trials involving double-blind placebo-controlled studies are necessary to validate the efficacy of milk protein-derived bioactive peptides. Current information from human studies suggests potential functional food ingredient and nutraceutical applications for milk protein-derived peptides as hypotensive (Hata *et al.*, 1996), immunomodulatory (Hadden, 1991) and mineral carrying (Reynolds, 1987; Hansen *et al.*, 1997) agents. Further human and cell culture studies are required to confirm and expand these initial findings in addition to demonstrating *in vivo* physiological effects of other milk protein-derived bioactive peptides (Meisel, 1998).

Acknowledgement

Prof. E. Schlimme is gratefully acknowledged for continued support and helpful discussions.

References

Abubakar, A., Saito, T., Kitazawa, H., Kawai, Y. and Itoh, T. (1998) Structural analysis of new antihypertensive peptides derived from cheese whey protein by proteinase K digestion. *J. Dairy Sci.*, **81**, 3131–8.

Adamson, N.J. and Reynolds, E.G. (1995) Characterisation of tryptic casein phosphopeptides prepared under industrially relevant conditions. *Biotech. Bioeng.*, **45**, 196–204.

Antila, P., Paakkari, I., Järvinen, A., Mattila, M.J., Laukkanen, M., Pihlanto-Leppälä, A., Mänstsälä, P. and Hellman, J. (1991) Opioid peptides derived from *in-vitro* proteolysis of bovine whey proteins. *Int. Dairy J.*, **1**, 215–29.

Bellamy, W., Takase, M., Yamauchi, K., Kawase, K., Shimamura, S. and Tomita, M. (1992) Identification of the bactericidal domain of lactoferrin. *Biochim. Biophys. Acta*, **1121,** 130–6.

Bellamy, W., Wakabayashi, H., Takase, M., Kawase, K., Shimamura, S. and Tomota, M. (1993) Role of cell-binding in the antibacterial mechanism of lactoferricin B. *J. Appl. Bacteriol.*, **75**, 478–84.

Berrocal, R., Chanton, S., Juillerat, M.A., Pavillard, B., Scherz, J.-C. and Jost, R. (1989) Tryptic phosphopeptides from whole casein: II. Physicochemical properties related to the solubilisation of calcium. *J. Dairy Res.*, **56**, 335–41.

Bouchier, P., FitzGerald, R.J. and O'Cuinn, G. (1999) Hydrolysis of α_{s1}- and β-casein-derived peptides with a broad specificity aminopeptidase and with proline specific aminopeptidases from *Lactococcus lactis spp. cremoris* AM2. *FEBS Lett.*, **445**, 321–4.

Bouhallab, S., Mollé, D. and Léonil, J. (1992) Tryptic hydrolysis of caseinomacropeptide in a membrane reactor, preparation of bioactive peptides. *Biotechnol. Lett.*, **14**, 805–10.

Brandsch, M., Brust, P., Neubert, K. and Ermisch, A. (1994) β-Casomorphins - chemical signals of intestinal transport systems, in, *β-Casomorphins and Related Peptides: Recent Developments,* (V. Brantl and H. Teschemacher eds.) VCH, Weinheim, pp. 207–19.

Brantl, V., Teschemacher, H., Bläsig, J., Henschen, A. and Lottspeich, F. (1981) Opioid activities of β-casomorphins. *Life Sci.*, **28**, 1903–9.

Brulé, G., Roger, L., Fauquant, J. and Piot, M. (1989) Casein phosphopeptide composition. *US Patent*, 4,816,398.

Bruneval, P., Hinglais, N. and Alhenc-Gelas, F. (1986) Angiotensin I converting enzyme in human intestine and kidney. Ultrastructural immunohistochemical localization. *Histochem.*, **86**, 73–80.

Carnie, J., Minter, S., Oliver, S., Perra, F. and Metzlaff, M. (1989) Nutritional compositions containing β-casomorphins. UK Patent Application GB 2214810 A.

Chabance, B., Jollès, P., Izquierdo, C., Mazoyer, E., Francoual, C., Drouet, L. and Fiat, A.-M. (1995) Characterisation of an antithrombotic peptide from κ-casein in newborn plasma after milk ingestion. *Br. J. Nutr.*, **73**, 583–90.

Chabance, B., Marteau, P., Rambaud, J.C., Migliore-Samour, D., Boynard, M., Perrotin, P., Guillet, R., Jollès, P. and Fiat, A.-M. (1998) Casein peptide release and passage to the blood in humans during digestion of milk or yoghurt. *Biochimie*, **80**, 155–65.

Chang, K.J., Killian, A., Hazum, E. and Cuatrecasas, P. (1981) Morphiceptin: A potent and specific agonist for morphine (μ) receptors. *Science*, **212**, 75–7.

Chang, K.J., Fu Su, Y., Brent, D.A. and Chang, J.-K. (1985) Isolation of a specific μ-opiate receptor peptide, morphiceptin, from an enzymatic digest of milk proteins. *J. Biol. Chem.*, **260**, 9706–12.

Cheung, H.-S., Feng-Lai, W., Ondetti, M.A., Sabo, E.F. and Cushman, D.W. (1980) Binding of peptide substrates and inhibitors of angiotensin-converting-enyzme. *J. Biol. Chem.*, **255**, 401–7.

Chiba, H. and Yoshikawa, M. (1986) Biologically functional peptides from food proteins: New opioid peptides from milk proteins, in *Protein Tailoring for Food and Medical Uses*, (R.E. Feeney and J.R. Whitaker eds.) Marcel Dekker Inc., New York, pp. 123–53.

Chiba, H., Tani, F. and Yoshikawa, M. (1989) Opioid antagonist peptides derived from κ-casein. *J. Dairy Res.*, **56**, 363–6.

Chiba, H. and Yoshikawa, M. (1991) Bioactive peptides derived from food proteins. *Kagaku to Seibutsu.*, **29**, 454–8.

Coste, M., Rochet, V., Léonil, J., Mollé, D., Bouhallab, B. and Tomé, D. (1992) Identification of the C- terminal peptides of bovine β-casein that enhance proliferation of rat lymphocytes. *Immunol. Lett.*, **33**, 41–6.

Daniel, H., Vohwinkel, M. and Rehner, G. (1990a) Effect of casein and β-casomorphins on gastrointestinal motility in rats. *J. Nutr.*, **120**, 252–7.

Daniel, H., Wessendorf, A., Vohwinkel, M. and Brantl, V. (1990b) Effect of D-$Ala^{2,4}Tyr^{5}$-β-casomorphin-5-amide on gastrointestinal functions, in β-*Casomorphins and Related Peptides*, (F. Nyberg and V. Brantl eds.) Fyris-Tryck AB, Uppsala, pp. 95–104.

Dionysius, D.A. and Milne, J.M. (1998) Antibacterial peptides of bovine lactoferrin: purification and characterizaton. *J. Dairy Sci.*, **80**, 667–74.

Elitsur, Y. and Luk, G.D. (1991) β-Casomorphin (BCM) and human colonic lamina propria lymphocyte proliferation. *Clin. Exper. Immunol.*, **85**, 493–97.

Ellegård, K.H., Gammelgård-Larsen, C., Sørensen, E.S. and Fedosov, S. (1999) Process scale chromatographic isolation, characterisation and identification of tryptic bioactive casein phosphopeptides. *Int. Dairy. J.*, **9**, 639–52.

Fiat, A.-M. and Jollès, P. (1989) Caseins of various origins and biologically active casein peptides and oligosaccharides: structural and physiological aspects. *Mol. Cell. Biochem.*, **87**, 5–30.

Fiat, A.-M., Migliore-Samour, D., Jollès, P., Drouet, L., Collier, C. and Caen, J. (1993) Biologically active peptides from milk proteins with emphasis on two examples concerning antithrombotic and immunomodulating activities. *J. Dairy Res.*, **76**, 301–10.

FitzGerald, R.J. (1998) Potential uses of caseinophosphopeptides. *Int. Dairy J.*, **8**, 451–7.
FitzGerald, R.J. and Meisel, H. (1999) Lactokinins: Whey protein-derived ACE inhibitory peptides. *Nahrung.*, **43**, 165–7.
FitzGerald, R.J. and Meisel, H. (2000) Milk protein – derived peptide inhibitors of angiotensin-I-converting enzyme. *Br. J. Nutr.*, **84** (S1), 33–7.
FitzGerald, R.J., Smyth, M., McDonagh, D. and Slattery H. (1996) Applications of milk protein hydrolysates, in *Proceedings 3rd Food Ingredients Symposium*, (M.K. Keogh ed.) Teagasc, Dublin, pp. 76–86.
Goepfert, A. and Meisel, H. (1996) Semi-preparative isolation of phosphopeptides derived from casein and dephosphorylation of casein phosphopeptides. *Nahrung.*, **40**, 245–8.
Hadden, J.W. (1991) Immunotherapy of human immunodeficiency virus infection. *Trends Pharm. Sci.*, **12**, 107–11.
Haileselassie, S.S., Lee, B.H. and Gibbs B.F. (1999) Purification and identification of potentially bioactive peptides from enzyme-modified cheese. *J. Dairy Sci.*, **82**, 1612–7.
Hambræus, L. (1992) Nutritional aspects of milk proteins, in *Advanced Dairy Chemistry–1. Proteins*, 2nd edn., (P.F. Fox ed.) Elsevier Applied Science, London, pp. 57–490.
Hamel, U., Kielwein, G. and Teschemacher, H. (1985) β-Casomorphin immunoreactive materials in cow's milk incubated with various bacterial species. *J. Dairy Res.*, **52**, 139–48.
Hansen, M., Sandstöm, B., Jensen, M. and Sörensen, S.S. (1997) Casein phosphopeptides improve zinc and calcium absorption from rice-based but not from whole grain infant cereal. *J. Pediatr. Gastroenterol. Nutr.*, **24,** 56–62.
Hata, Y., Yamamoto, M., Ohni, M., Nakajima, K., Nakamura, Y. and Takano, T. (1996) A placebo-controlled study of the effect of sour milk on blood pressure in hypertensive subjects. *Am. J. Clin, Nutr.*, **64**, 767–71.
Hata, I., Higashiyama, S. and Otani, H. (1998) Identification of a phosphopeptide in bovine α_{s1}-casein digest as a factor influencing proliferation and immunoglobulin production in lymphocyte cultures. *J. Dairy Res.*, **65**, 569–79.
Hata, I., Ueda, J. and Otani, H. (1999) Immunostimulatory action of commercially available casein phosphopeptide preparations, CPP-III, in cell cultures. *Milchwisswenchaft*, **1**, 3–7.
Höllt, V. (1983) Multiple endogenous opioid peptides. *Trends Neurosci.*, **6**, 24–6.
Johnston, C.I. (1992) Renin-angiotensin system: a dual tissue and hormonal system for cardiovascular control. *J. Hyperten.*, **10**, S13–26.
Jollès, P., Parker, F., Floc'h, F., Migliore, D.. Alliel, P., Zerial, A. and Werner, G.G. (1981) Immunostimulating substances from human casein. *J. Pharmacol.*, **3**, 363–9.
Jollès, P., Lévy-Toledano, S., Fiat, A.-M., Soria, C., Gillessen, D., Thomaidis, A., Dunn, F.W. and Caen, J.B. (1986) Analogy between fibrinogen and casein. *Eur. J. Biochem.*, **158**, 379–84.
Juillard, V., Laan, H., Kunji, E.R.S., Jeronimus-Stratingh, C.M., Bruins, A.P. and Konings, W.N. (1995) The extracellular PI-type proteinase of *Lactococcus lactis* hydrolyzes β-casein into more than one hundred different oligopeptides. *J. Bacteriol.*, **177**, 3472–8.
Juillerat, M.A., Baechler, R., Berrocal, R., Chanton, S., Scherz, J.-C. and Jost, R. (1989) Tryptic phosphopeptides from whole casein I; Preparation and analysis by FPLC. *J. Dairy Res.*, **56**, 603–11.
Kampa, M., Loukas, S., Hatzoglou, A., Daminaki, A., Martin, P.M. and Castanas, E. (1997) Opioid alkaloids and casomorphin peptides decrease the proliferation of

prostatic cancer cell lines (LNCaP, PC3 and DU145) through partial interaction with opioid receptors. *Eur. J. Pharmacol.*, **335**, 255–65.

Kang, J.H., Lee, M.K., Kim, K.L. and Hahm, K.S. (1996) Structure biological activity relationship of 11-residue highly basic peptide segment of bovine lactoferrin. *Int. J. Pep. Prot. Res.*, **48**, 357–63.

Karaki, H., Doi, K., Sugano, S., Uchiya, H., Sugai, R., Murakami, U. and Takemoto, S. (1990) Antihypertensive effect of tryptic hydrolysate of milk casein in spontaneously hypertensive rats. *Comp. Biochem. Physiol.*, **96**, 367–71.

Kasai, T., Honda, T. and Kiriyama, S. (1992) Caseinophosphopeptides (CPP) in faeces of rats fed a casein diet. *Biosci. Biotechnol. Biochem.*, **56**, 1150–1.

Kayser, H. and Meisel, H. (1996) Stimulation of human peripheral blood lymphocytes by bioactive peptides derived from bovine milk proteins. *FEBS Lett.*, **383**, 18–20.

Kitts, D.D. and Yuan, Y.V. (1992) Caseinophosphopeptides and calcium bioavailability. *Trends Food Sci. Technol.*, **3**, 31–5.

Koide, H., Ito, K., Miyamoto, M. and Nishino, H. (1980) Effect of long-term blockade of angiotensin-converting enzyme with captopril (SQ 14 22S) on haemodynamics and circulating blood volume in SHR. *Hypertension*, **2**, 229–303.

Koide, K., Itoyama, K., Fukushima, T., Miyazawa, F. and Kuwata, T. (1991) Method for separation and concentration of phosphopeptides. European Patent Application EP, 0 443 718 A2.

Kunst, A. (1990) Process to isolate phosphopeptides. European Patent Application EP 0 476 199 Al.

Lahov, E. and Regelson, W. (1996) Antibacterial and immunostimulating casein-derived substances from milk: casesidin, isracidin peptides. *Food Chem. Toxicol.*, **34**, 131–45.

Lihme, A.O.F., Aagesen, M.I., Gamalgard-Larsen, C. and Ellegard, K.H. (1994) Method for isolating biomolecules by ion exchange. World Patent WO 94/06822.

Loukas, S., Varoucha, D., Zioudrou, C., Streaty, R.A. and Klee, W.A. (1983) Opioid activities and structures of α-casein-dervied exorphins. *Biochem.*, **22**, 4567–73.

Maeno, M., Yamamoto, N. and Takano T. (1996) Identification of antihypertensive peptides from casein hydrolysate produced by a proteinase from *Lactobacillus helveticus* CP790. *J. Dairy Sci.*, **73**, 1316–21.

Manson, W. and Annan, W.D. (1971) The structure of a phosphopeptide derived from β-casein. *Arch. Biochem. Biophys.*, **145**, 16–26.

Maruyama, S. and Suzuki, H. (1982) A peptide inhibitor of angiotensin-I-converting enzyme in the tryptic hydrolysate of casein. *Agric. Biol. Chem.*, **46**, 1393–94.

Maruyama, S., Nakagomi, K., Tomizuka, N. and Suzuki, H. (1985) Angiotensin I-converting enzyme inhibitor derived from an enzymatic hydrolysate of casein. II. Isolation and bradykinin-potentiating activity on the uterus and the ileum of rats. *Agric. Biol. Chem.*, **49**, 1405–9.

Maruyama, S., Mitachi, H., Tanaka, H., Tomizuka, N. and Suzuki, H. (1987a) Studies on the active site and antihypertensive activity of angiotensin I-converting enzyme inhibitors derived from casern. *Agric. Biol. Chem.*, **51**, 1581–6.

Maruyama S., Mitachi, H., Awaya, J., Kurono, M., Tomizika, N. and Suzuki H. (1987b) Angiotensin I converting enzyme inhibitory activity of the C-terminal hexapeptide of α_{s1}-casein. *Agric. Biol. Chem.*, **51**, 2557–61.

McDonagh, D. and FitzGerald, R.J. (1998) Production of caseinophosphopeptides (CPPs) from sodium caseinate using a range of commercial protease preparations. *Int. Dairy J.*, **8**, 39–45.

Meisel, H. (1986) Chemical characterisation and opioid activity of an exorphin isolated from *in vivo* digests of casein. *FEBS Lett.*, **196**, 223–7.

Meisel, H. (1993) Casokinins as inhibitors of angiotensin-I-converting enzyme, in *New Perspectives in Infant Nutrition*, (G. Sawatski and B. Renner eds.) Thieme, Stuttgart, pp. 153–9.

Meisel, H. and Frister, H. (1988) Chemical characterization of a caseinophosphopeptide isolated from *in vivo* digests of a casein diet. *Biol. Chem. Hoppe-Seyler*, **369**, 1275–9.

Meisel, H. and Frister, H. (1989) Chemical characterization of bioactive peptides from *in vivo* digests of casein. *J. Dairy Res.*, **56**, 343–9.

Meisel, H. (1994) Antibodies from egg yolk of immunised hens against a bioactive caseinophosphopeptide (β-casokinin-10). *Biol. Chem. Hoppe-Seyler*, **375**, 401–5.

Meisel, H. and Schlimme, E. (1994) Inhibitors of angiotensin-converting-enzyme derived from bovine casein (casokinins), in, *β-Casomorphins and Related Peptides: Recent developments*, (V. Brantl and H. Teschemacher eds.) VCH, Weinheim, pp. 27–33.

Meisel, H. (1997) Biochemical properties of bioactive peptides derived from milk proteins: potential nutraceuticals for food and pharmacological applications. *Liv. Prod. Sci.*, **50**, 125–38.

Meisel, H., Goepfert, A. and Günther, S. (1997) Occurrence of ACE inhibitory peptides in milk products. *Milchwissenschaft*, **52**, 307–11.

Meisel, H. (1998) Overview of milk protein-derived peptides. *Int. Dairy J.*, **8**, 363–73.

Meisel, H. and Bockelmann, W. (1999) Bioactive peptides encrypted in milk proteins: proteolytic activation and trophofunctional properties. *Ant. van Leeuwen.*, **76**, 207–15.

Meisel, H. and FitzGerald, R.J. (2000) Opioid peptides encrypted in intact milk protein sequences. *Br. J. Nutr.*, **84**(S1), 27–31.

Meisel, H., Bernard, H., Fairweather-Tait, S., FitzGerald, R.J., Hartmann, R., Lane, C.N., McDonagh, D., Teucher, B. and Wal, J.-M. (2001) Nutraceutical and functional food ingredients for food and pharmaceutical applications. *Br. J. Nutr.* **85**, 635 (1 page).

Migliore-Samour, D., FlocTi, F. and Jollès, P. (1989) Biologically active casein peptides implicated in immunomodulation. *J. Dairy* Res., **56**, 357–62.

Muehlenkamp, M.R. and Warthesen, J.J. (1996) β-Casomorphins: analysis in cheese and susceptibility to proteolytic enzymes from *Lactococcus lactis* ssp. *cremoris*. *J. Dairy Sci.*, **79**, 20–6.

Mullally, M.M., O'Callaghan, D.M., FitzGerald, R.J., Donnelly, W.J. and Dalton, J.P. (1994) Proteolytic and peptidolytic activities in commercial pancreatic proteinase preparations and their relationship to some whey protein hydrolysate characterisitics. *J. Agric. Food Chem.*, **42**, 2973–81.

Mullally, M.M., Meisel, H. and FitzGerald, R.J. (1996) Synthetic peptides corresponding to α-lactalbumin and β-lactoglobulin sequences with angiotensin-I-converting enzyme inhibitory activity. *Biol. Chem. Hoppe-Seyler*, **377**, 259–60.

Mullally, M.M., Meisel, H. and FitzGerald, R.J. (1997a) Identification of a novel angiotensin-I-converting enzyme inhibitory peptide corresponding to a tryptic digest of bovine β-lactoglobulin. *FEBS Lett.*, **402**, 99–101.

Mullally, M.M., Meisel, H. and FitzGerald, R.J. (1997b) Angiotensin-I-converting enzyme inhibitory activities of gastric and pancreatic proteinase digests of whey proteins. *Int. Dairy J.*, **7**, 299–303.

Naito, H., Kawakami, A. and Inamura, T. (1972) *In vivo* formation of phosphopeptide with calcium-binding property in the small intestinal tract of the rat fed on casein. *Agric. Biol. Chem.*, **36**, 409–15.

Nakamura, Y., Yamamoto, N., Sakai, K., Okubo, A., Yamazaki, S. and Takano, T. (1995) Purification and characterization of angiotensin I-converting enzyme inhibitors from a sour milk. *J. Dairy Sci.*, **78**, 777–83.

Ondetti, M.A. and Cushman, D.W. (1982) Enzymes of the renin-angiotensin system and their inhibitors. *Ann. Rev. Biochem.*, **51**, 283–308.

Park, O., Swaisgood, H.E. and Alien, J.C. (1998) Calcium binding of phosphopeptides derived from hydrolysis of α_{s1}-casein or β-casein using immobilised enzymes. *J. Dairy Sci.*, **81**, 2850–57.

Pihlanto-Leppällä, A., Antila, P., Mäntsälä, P. and Hellman, J. (1994) Opioid peptides produced by *in vitro* proteolysis of bovine caseins. *Int. Dairy J.*, **4**, 291–301.

Pihlanto-Leppälä, A., Koskinen, P., Paakkari, I., Tupasela, T. and Korhonen, H. (1996) Opioid whey protein peptides obtained by membrane filtration, *Bulletin* **311**, International Dairy Federation, Brussels, pp. 36–8.

Pihlanto-Leppälä, A., Rokka, T. and Korhonen, H. (1998) Angiotensin I converting enzyme inhibitory peptides derived from bovine milk proteins. *Int. Dairy J.*, **8**, 325–31.

Pihlanto-Leppälä, A., Marnila, P., Rokka, T., Korhonen, H. and Karp, M. (1999) The effect of α-lactalbumin and β-lactoglobulin hydrolysates on the metabolic activity of *Escherichia coli* JM103. *J. Appl. Microbiol.*, **87**, 540–5.

Recio, I. and Visser, S. (1999) Two ion-exchange methods for the isolation of antibacterial peptides from lactoferrin – *in situ* enzymatic hydrolysis on an ion-exchange membrane. *J. Chromatogr.*, **831**, 191–201.

Reeves, R.E. and Latour, N.G. (1958) Calcium phosphate sequestering phosphopeptide from casein. *Science*, **128**, 472. (1 page)

Reynolds, E. (1987) Phosphopeptides. *PCT Int.* Patent Application WO 87/07615 A1.

Reynolds, E.C. (1992) Production of phosphopeptides from casein. World Patent Application WO 92/18526.

Reynolds, E.C. (1994) Anticariogenic Casein Phosphopeptides. *Proc. 24th Int. Dairy Congr.*, Brief Communications Melbourne, Australia, p. 24(abstr).

Righetti, P.G., Nembri, F., Bossi, A. and Mortarino, M. (1997) Continuous enzymatic hydrolysis of beta-casein and isoelectric collection of some of the biologically active peptides in an electric field. *Biotechnol. Prog.*, **13**, 258–64.

Rokka, T., Syvaoja, E.-L., Tuominen, J. and Korhonen, H. (1997) Release of bioactive peptides by enzymatic proteolysis of *Lactobacillus GG* fermented UHT-milk. *Milchwissenschaft*, **52**, 675–8.

Roudot-Algaron, F., Le Bars, D., Kerhoas, L., Einhorn, J. and Gripon, J.C. (1994) Phosphopeptides from Comte cheese: Nature and origin. *J. Food Sci.*, **59**, 544–7, 60.

Sato, R., Noguchi, T. and Naito, H. (1983) The necessity for the phosphate portion of casein molecules to enhance Ca absorption from the small intestine. *Agric. Biol. Chem.*, **47**, 2415–7.

Scanff, P., Yvon, M., Thirouin, S. and Pelissier, J.-P. (1992) Characterisation and kinetics of gastric emptying of peptides derived from milk proteins in the preruminant calf. *J. Dairy Res.*, **59**, 437–47.

Schanbacher, F.L., Talhouk, R.S. and Murray, F.A. (1997) Biology and origin of bioactive peptides in milk. *Liv. Prod. Sci.*, **50**, 105–23.

Schanbacher, F.L., Talhouk, R.S., Murray, F.A., Gherman, L.I. and Willett, L.B. (1998) Milk-borne bioactive peptides. *Int. Dairy J.*, **8**, 393–403.

Shin, K., Yamauchi, K., Teraguchi, S., Hayasawa, H., Tomita, M., Otsuka, Y. and Yamazaki, S. (1998) Antibacterial activity of bovine lactoferrin and its peptides against enterohaemorragic *E. coli* O157:H7. *Lett. Appl. Microbiol.*, **26**, 407–11.

Singh, T.K., Fox, P.F. and Healy, A. (1997) Isolation and identification of further peptides in the diafiltration retentate of the water soluble fraction of Cheddar cheese. *J. Dairy Res.*, **64**, 433–43.

Smyth, M. and FitzGerald, R.J. (1998) Relationship between some characteristics of WPC hydrolysates, and the enzyme complement in commercially available proteinase preparations. *Int. Dairy J.*, **8**, 819–27.

Suetsuna, K. and Osajima, K. (1989) Blood pressure reduction and vasodilatory effects *in vivo* of, peptides originating from sardine muscle (in Japanese). *J. Jap. Soc. Nutr. Food Sci.*, **42**, 47–54.

Sutas, Y., Soppi, E., Korhonen, H., Syvaoja, E.-L., Saxelin, M., Rokka, T. and Isolauri, E. (1996) Suppression of lymphocyte proliferation *in vitro* by bovine caseins hydrolysed with *Lactobacillus GG*-derived enzymes. *J. Allergy Clin. Immunol.*, **98**, 216–24.

Svedberg, J., de Haas, J., Leimenstoll, G., Paul, F. and Teschemacher, H. (1985) Demonstration of β-casomorphin immunoreactive materials in *in vitro* digests of bovine milk and in small intestine contents after bovine milk ingestion in adult humans. *Peptides*, **6**, 825–30.

Takano, T. (1998) Milk derived peptides and hypertension reduction. *Int. Dairy J.*, **8**, 375–81.

Tani, F., Shiota, A., Chiba, H. and Yoshikawa, M. (1994) Serorphin, an opioid peptide derived from bovine serum albumin, in, *β-Casomorphins and Related Peptides: Recent Developments,* (V. Brantl and H. Teschemacher eds.) VCH, Weinheim, pp. 49–53.

Teschemacher, H., Umbach, M., Hamel, U., Praetorius, K., Ahnert-Hilger, G., Brantl, V., Lottspeich, F. and Henschen, A. (1986) No evidence for the presence of β-casomorphins in human plasma after ingestion of cows' milk or milk products. *J. Dairy Res.*, **53**, 135–8.

Teschemacher, H. and Brantl, V. (1994) Milk protein derived atypical opioid peptides and related compounds with opioid antagonist activity, in, *β-Casomorphins and Related Peptides: Recent Developments*, (V. Brantl and H. Teschemacher eds.) VCH, Weinheim, pp. 3–17.

Teschemacher, H., Koch, G. and Brantl, V. (1997) Milk protein-derived opioid receptor ligands. *Biopoly.*, **43**, 99–117.

Tomé, D. and Debabbi, H. (1998) Physiological effects of milk protein components. *Int. Dairy J.*, **8**, 383–92.

Tomé, D., Dumontier, A.M., Hautefeuille, M. and Desjeux, J.F. (1987) Opiate activity and transepithelial passage of intact β-casomorphins in rabbit ileum. *Am. J. Physiol.*, **253**, G737–44.

Tomita, M., Bellamy, W., Takase, M., Yamauchi, K., Wakabayashi, H. and Kawase, K. (1991) Potent antibacterial peptides generated by pepsin digestion of bovine lactoferrin. *J. Dairy Sci.*, **74**, 4137–42.

Tomita, M., Takase, M., Bellamy, W. and Shimämura, S. (1994) A review: the active peptide of lactoferrin. *Acta Pediatr. Jpn.*, **36**, 585–91.

Umbach, M., Teschemacher, H., Praetorius, K., Hirschhauser, R., and Bostedt, H. (1985) Demonstration of a β-casomorphin immunoreactive material in the plasma of newborn calves after milk intake. *Reg. Pep.*, **12**, 223–30.

West, D.W. (1986) Structure and function of the phosphorylated residues of casein. *J. Dairy Res.*, **53**, 333–52.

Wyvratt, M.J. and Patchet, A.A. (1985) Recent developments in the design of angiotensin-converting enzyme inhibitors. *Med. Res. Rev.*, **5**, 485–531.

Yamamoto, N., Akino, A. and Takano, T. (1994) Antihypertensive effects of peptides derived from casein by an extracellular proteinase from *Lactobacillus helveticus* CP790. *J. Dairy Sci.*, **77**, 917–22.

Yamamoto, N. (1997) Antihypertensive peptides derived from food proteins. *Biopoly.*, **43**, 129–34.

Yamamoto, N., Maeno, M. and Takano, T. (1999) Purification and characterisation of an antihypertensive peptide from yoghurt-like product fermented by *Lactobacillus helveticus* CPN4. *J. Dairy Sci.*, **82**, 1388–93.

Yoshikawa, M., Tani, F., Yoshimura, T. and Chiba, H. (1986) Opioid peptides from milk proteins. *Agric. Biol. Chem.*, **50**, 2419–21.

Yoshikawa, M., Tani, F., Shiota, H., Usui, H., Kurahashi, K. and Chiba, H. (1994) Casoxin D, an opioid antagonist/ileum-contracting/vasorelaxing peptide derived from human α_{s1}-casein, in, β-*Casomorphins and Related Peptides: Recent Developments*, (V. Brantl and H. Teschemacher eds.) VCH, Weinheim, pp. 43–8.

Ziodrou, C., Streaty, R.A. and Klee, W.A. (1979) Opioid peptides derived from food proteins. *J. Biol. Chem.*, **254**, 2446–9.

Zucht, H.D., Raida, M., Andermann, K., Mägert, H.-J. and Forssman, W.G. (1995) Casocidin-I: a casein-α_{s2} derived peptide exhibits antibacterial activity. *FEBS Lett.*, **372**, 185–8.

15

BIOSYNTHESIS OF MILK PROTEINS

J.L. Vilotte, C.B.A. Whitelaw, M. Ollivier-Bousquet and D.B. Shennan

15.1 Introduction

During lactation, mammary epithelial cells secrete large quantities of milk proteins. More than 90% of these proteins are derived from the transcription of a few tissue-specific genes, expression of which is under a complex multi-hormonal regulation that involves both transcriptional and post-transcriptional mechanisms. Furthermore, to fulfil its bioreactor activity, the mammary gland needs an optimal supply of amino acids as well as efficient translation and transport systems during lactation.

The previous editions of this chapter in Fox (1982, 1992) described in detail the hormonal regulation of milk protein gene expression, their mRNA and gene structures, their co- and post-translational modifications, and the transport and the secretion of milk proteins. The aim of this revised version will be to briefly summarize our knowledge on the structure of the milk protein genes to put into context the rapid growth of information on the regulatory elements involved in controlling the expression of these genes. This has revolved around the identification of STAT5-mediated transduction of the prolactin signal pathway. We will also focus on the amino acid supply to the mammary gland, and on the intracellular routing and sorting of milk proteins in mammary cells. However, two important topics will not be covered. The first, the widespread presence of casein variants will be covered by Chapter 16, while the practical applications of these studies for the dairy field, notably in terms of modification of the milk protein composition by transgenesis will be described in Chapter 17.

Advanced Dairy Chemistry Volume 1: Proteins, 3rd edn.
Edited by P.F. Fox and P.L.H. McSweeney, Kluwer Academic/Plenum Publishers, 2003.

15.2 Milk protein gene structure

The major milk-protein genes have been sequenced in several species. Overall, the mosaic structure of these genes has been well conserved during evolution and observed inter-species differences in the length of their transcription unit can often be attributed to the occurrence of repetitive DNA within some introns, mainly artiodactyl retroposons. Similarly, deletions or insertions of amino acids in caseins from different species or variants appear to occur mainly by exon skipping. Details on the incidence of casein pre-RNA splicing on protein structure are given in Chapter 16. These genes share the canonical structure of tissue-specific eukaryotic genes with a cap site, numerous donor, branch point and acceptor sites for splicing, and polyadenylation signals. All sequenced milk protein genes contain within their proximal 5'-flanking sequence a TATA box and often a CAAT box. Beyond these basic similarities, these genes differ substantially in their genomic organisation.

15.2.1 The casein-encoding genes

(a) General organization of the gene cluster

Classical genetic studies on the transmission of bovine casein variants have demonstrated that the four casein genes are closely linked and a general organization of the gene cluster, based on estimated genetic distances between the four casein loci, was deduced (reviewed by Grosclaude, 1979). Since then, structural analysis at the DNA level on the gene cluster in the murine, bovine and human species has confirmed and refined the protein data (Figure 15.1), Tomlinson *et al.*, 1996; Fujiwara *et al.*, 1997; George *et al.*, 1997; Rijnkels *et al.*, 1997a,b,c). The overall size of the casein locus varies between species from around 250 kb in murine and bovine species to 350 kb in humans, but the order and orientation of the genes within the cluster are conserved (Figure 15.1). In mice, the γ- and δ-casein genes, which are linked within 60 kb, encode an α_{s2}-casein-like protein and sequence analysis suggests that the ancestral α_{s2}-casein gene duplicated at the time of radiation between rodents and artiodactyla (George *et al.*, 1997; Rijnkels *et al.*, 1997a). A similar duplication of the α_{s2}-casein gene is also suspected to have occurred in rabbits (Dawson *et al.*, 1993). The human casein locus is the largest identified so far, spanning about 350 kb (Rijnkels *et al.*, 1997c; Fujiwara *et al.*, 1997). Interestingly, this longer size appears to result from an insertion event that occurred between the β-casein and the α_{s2}-casein genes. It is thought that this may represent the insertion of yet another gene into the locus, but that its expression profile may not be restricted to the mammary gland (Rijnkels, pers. comm.). The significance of this observation remains to be determined.

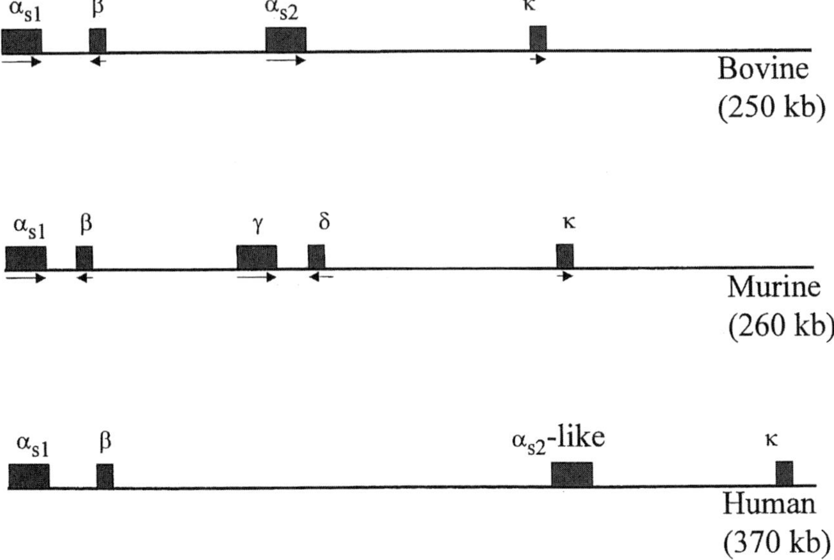

Figure 15.1 Overall organization of the casein locus in various species. Boxes represent the transcription unit of the genes. When available, orientation of the gene transcription unit is indicated by an arrow. Origins of the data are mentioned in the text.

The casein locus has been localized to chromosome 6 in the cow, sheep and goat (Threadgill and Womack, 1990; Ansari *et al.*, 1992; Hayes *et al.*, 1992, 1993; Gallagher *et al.*, 1994); chromosome 5 in mice (Geissler *et al.*, 1988); chromosome 12 in rabbits (Gellin *et al.*, 1985); chromosome 4 in humans and chromosome 3 in chimpanzees (McConkey *et al.*, 1996). Gene knock-out experiments, where an individual casein has been disrupted, e.g., β-casein (Kumar *et al.*, 1994), did not prevent expression of the remaining caseins within the locus. Thus, although the casein genes are clustered spatially to a single chromosomal locus they are expressed independently of each other.

(b) Individual gene structures

Internal homologies within α_{s1}- and α_{s2}-casein proteins have suggested that the cognate genes evolved through intragenic duplications, a result confirmed at the DNA level by the observed duplication of intron-exon-intron stretches (Figure 15.2). The ubiquity of the major phosphorylation site and the striking homology of signal peptides of α_{s1}-, α_{s2}- and β-casein proteins, indicated a possible common origin for these calcium-sensitive casein-encoding genes. This hypothesis was further substantiated by the identification of common sequence motifs in the proximal 5'-flanking region and the similar structural organization of the first four exons (reviewed in

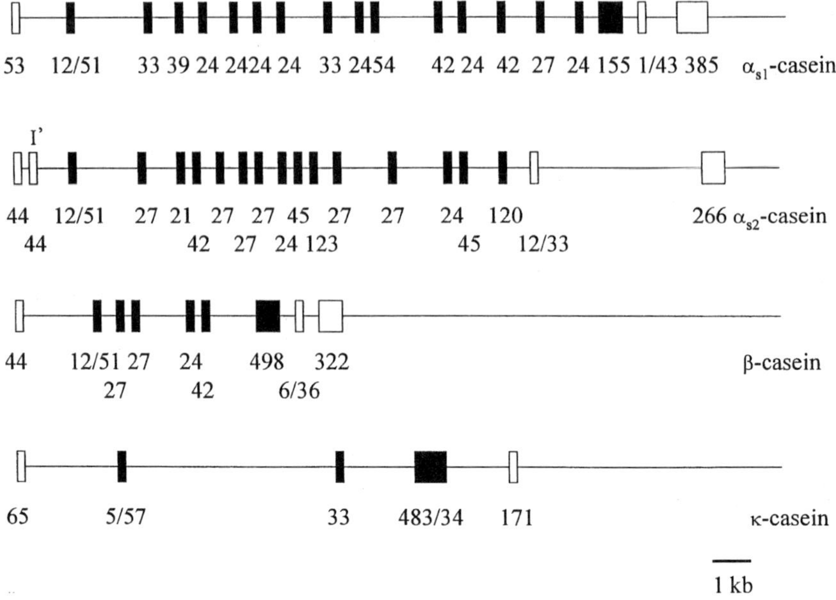

Figure 15.2 Organization of the bovine casein-encoding genes. Exons are represented by boxes. White boxes: untranslated regions. Black boxes: coding frame. Exons are not to scale and their sizes are indicated as base-pairs. Two numbers are indicated below exons that comprise both untranslated and coding sequences. Exon I' from the α_{s2}-casein gene corresponds to a partially skipped exon in the sheep species. Origins of the data are mentioned in the text.

Mercier and Vilotte, 1993). Despite its location within the casein cluster, the κ-casein gene appears to have no evolutionary relationship with the calcium-sensitive casein-encoding genes.

The structure of the α_{s1}-casein gene has been partially analyzed in rat (Yu-Lee *et al.*, 1986), and fully in cow (Koczan *et al.*, 1991), goat (Leroux *et al.*, 1992) and rabbit (Jolivet *et al.*, 1992). The transcription unit of the gene is split into 19 exons and spans around 17.5 kb in ruminants and 16 kb in rabbits (Figure 15.2). Within the reading frame, like that of the all calcium-sensitive casein genes, no coding triplet is disrupted by a splice junction.

The structure of the α_{s2}-casein gene has, so far, been described only in the cow (Groenen *et al.*, 1993). The transcription unit of the gene is split into 18 exons and spans 18.5 kb (Figure 15.2). The first intron of this gene contains a non-coding exon (exon I' in Figure 15.2) that is known to be retained in 4% of the ovine mRNA (Boisnard *et al.*, 1991). In this spcies, exon VI is also partially skipped. Sequence comparisons suggest that this gene is more closely related to the β-casein gene than it is to the α_{s1}-casein gene (Groenen *et al.*, 1993).

The structure of the β-casein gene is known in mouse (Yoshimura and Oka, 1989), cow (Bonsing *et al.*, 1988), rat (Jones *et al.*, 1986), rabbit (Thépot *et al.*, 1991), goat (Roberts *et al.*, 1992), human (Hansson *et al.*, 1994) and sheep (Provot *et al.*, 1995). Its transcription unit is composed of 9 exons and spans between 8 and 10 kb according to differences between species in the length of intronic sequences (Figure 15.2). Exon 3 is skipped in humans, leading to the deletion of 9 amino acids in the mature protein.

Characterization of the κ-casein gene has been reported in cow (Alexander *et al.*, 1988; Kapelinskaia *et al.*, 1989), human (Edlund *et al.*, 1996) and rabbit (Baranyi *et al.*, 1996). The transcription unit of this gene comprises 5 exons, and its length varies from 7.5 kb in rabbit to 12.5 kb in cow (Figure 15.2). Again, this observed difference in size is due to the length of the fourth intron. The fourth exon contains most of the coding sequence. This gene is unrelated to the calcium-sensitive casein gene family, but instead is evolutionarily related to fibrinogens (Jollès *et al.*, 1978). Indeed a 24-bp sequence located at the 5' end of exon IV of the gene was found to be similar with the end of exon II of the γ-fibrinogen gene, suggesting that it represents an exon from the ancestral gene (Alexander *et al.*, 1998).

15.2.2 The major whey protein-encoding genes

(a) The β-lactoglobulin-encoding gene and pseudogenes

Structure of the β-lactoglobulin-encoding gene has been reported in sheep (Ali and Clark, 1988; Harris *et al.*, 1988), cow (Alexander *et al.*, 1993), goat (Folch *et al.*, 1994), tammar wallaby (Collet and Joseph, 1995) and horse (Masel *et al.*, 1998, unpublished; EMBL: AF107201). The structure of this gene, with seven exons and a transcription unit length of around 4.8 kb, was conserved during evolution (Figure 15.3). In dogs and horses, two functional genes are present in the genome. These genes are closely linked and encode β-lactoglobulin type I and II (Halliday *et al.*, 1990, 1993; Lear *et al.*, 1998). In felines, a third functional gene is present (Halliday *et al.*, 1990; Pena *et al.*, 1999).

In ruminants, a β-lactoglobulin pseudogene has been identified (Passey and Mackinlay, 1995; Folch *et al.*, 1996), while in both cow and goat, a gene conversion event has occurred. The bovine and caprine pseudogene has been sequenced. It contains seven exons, and the ancestral protein that it encodes for is related to the monomeric β-lactoglobulin II protein. The bovine pseudogene has been mapped to a cosmid that contains the true bovine β-lactoglobulin gene. It is located 14 kb 5' from the functional gene, in a similar orientation, i.e., in an head-to-tail arrangement (Passey and Mackinley, 1995). This differs from the duplication events that are thought to have occurred in the casein locus where duplicated casein genes are found in a tail-to-tail arrangement.

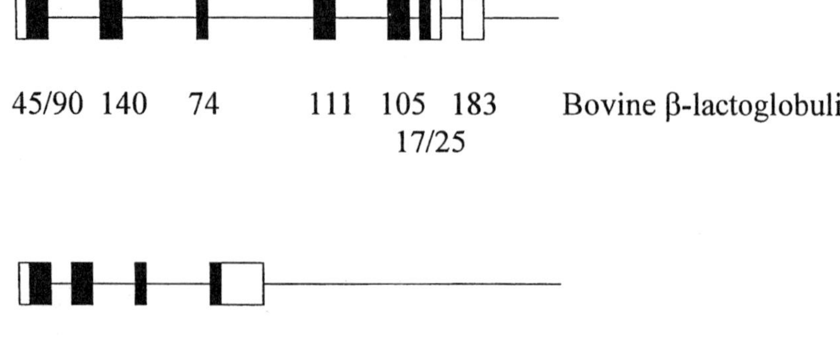

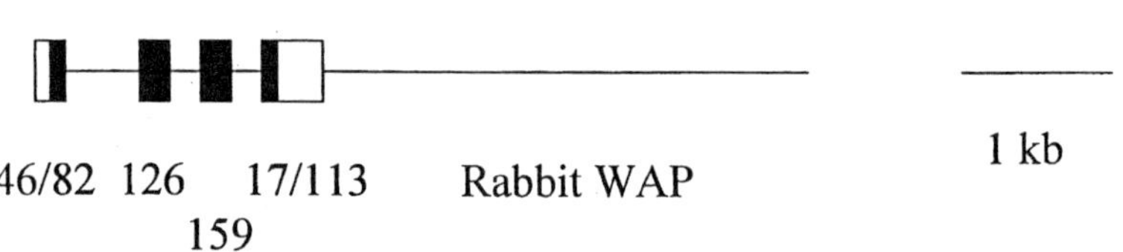

Figure 15.3 Organization of the major whey protein-encoding genes. Exons are represented by boxes. White boxes: untranslated regions. Black boxes: coding frame. Exons are not to scale and their sizes are indicated as base-pairs. Two numbers are indicated below exons that comprise both untranslated and coding sequences. Origins of the data are mentioned in the text (WAP: whey acidic protein).

β-Lactoglobulin belongs to the lipocalin protein family (Flower, 1996; Chapter 7 for review). Comparison of the structure of various lipocalin genes with that of β-lactoglobulin gene has revealed striking similarities, confirming further their evolutionary relationship (see Ali and Clark, 1988; Mercier and Vilotte, 1993 for review). The β-lactoglobulin gene(s) and pseudogene were assigned to chromosome 11 in cow and goat, 3 in sheep (Hayes and Petit, 1993; Folch *et al.*, 1996) and 28 in horse (Lear *et al.*, 1998).

(b) The α-lactalbumin gene and pseudogenes

The α-lactalbumin gene has been sequenced in seven species: rat (Quasba and Safaya, 1984), cow (Vilotte *et al.*, 1987), human (Hall *et al.*, 1987), guinea pig (Laird *et al.*, 1988), goat (Vilotte *et al.*, 1991), mouse (Vilotte and Soulier, 1992) and tammar wallaby (Collet and Joseph, 1995). In eutherians, the gene is composed of four exons and its transcription unit is about 2 kb in length (Figure 15.3). It shares the same structural organization with the lysozyme gene, corroborating the hypothesis of a common ancestor (Quasba and Safaya, 1984). The structure of the tammar wallaby α-lactalbumin gene appears different, with the occurrence of a putative 5' untranslated first exon that has yet to be identified (Collet and Joseph, 1995). In ruminants, the occurrence of related sequences has been reported (Soulier *et al.*, 1989; Vilotte *et al.*, 1991, 1993). Since two of these sequences show about 80% similarity with the transcription unit of the gene downstream of exon II and apparently no similarity upstream from it, they probably represent pseudogenes. Pulse-field gel electrophoresis and *in situ* hybridization have indicated that all α-lactalbumin-related sequences are closely linked (Hayes *et al.*, 1993; Gallagher *et al.*, 1993), and analysis of goat BAC clones suggested that they are located 3' to the functional gene (quoted in Stinnakre *et al.*, 1999). The α-lactalbumin-encoding gene (and related sequences) has been assigned to human chromosome 12 (Davies *et al.*, 1987), sheep chromosome 3, bovine and goat chromosomes 5 (Hayes *et al.*, 1993) and pig chromosome 5 (Rohrer *et al.*, 1997).

(c) The whey acidic protein (WAP)-encoding gene

The WAP gene has been sequenced in rat, mice (Campbell *et al.*, 1984) and rabbit (Thepot *et al.*, 1990). The WAP gene was thought to be present only in rodents, but homologous sequences have been identified recently in the pig. The 2 kb transcription unit is composed of four exons (Figure 15.3) and as such it is the smallest of all milk protein genes. The structural organization of the transcription unit is conserved in the pig WAP gene (Rival, pers. comm.). The functional role for WAP in milk has yet to be defined, although it bears a similarity to a family of protease inhibitors. Both WAP and protease inhibitors of the Kunitz family are characterized by highly conserved cysteine residues. These cysteine residues are present in two domains, which in the WAP gene are encoded by exons II and III. The murine WAP gene has been assigned to chromosome 11 (Gupta *et al.*, 1982).

15.3 Milk protein-encoding gene expression and regulation

15.3.1 Tissue-specificity and developmental regulation

The major milk protein genes are defined as mammary-specific and developmentally-regulated expressed genes. As such, they represent markers of mammary differentiation. The amount of milk protein mRNA in mammary epithelial cells increases steadily from mid-pregnancy to lactation, but at different rates according to the different genes (Harris *et al.*, 1990; Robinson *et al.*, 1995). It reflects an increase in the transcription rate of these genes as well as a stabilization of the transcripts (Guyette *et al.*, 1979). The overall observed abundance of a specific mRNA is also dependent on the level of the gene expression per cell and the number of epithelial cells that express it. The observed asynchrony of mammary epithelial cell maturation during pregnancy (Robinson *et al.*, 1995) is paralleled by heterogeneous expression during lactation of the major milk protein genes in sheep and cattle (Molenaar *et al.*, 1992). This heterogeneous pattern of expression is not seen in lactating mouse mammary glands (Dobie, *et al.*, 1996). Nevertheless, a short closure of lactating murine mammary gland resulted in local perturbation of milk protein gene expression, leading to the appearance of a mosaic pattern (Faerman *et al.*, 1995). These results strongly suggest that the regulation in mammary epithelial cells of the major milk protein gene expression is indeed under a complex regulation that could involve both a graded and a binary mechanism.

Over the last few years, the restricted expression in the mammary gland of the milk protein genes has been questioned. Expression of α-lactalbumin in the rat epididymis was reported (Qasba *et al.*, 1983) and denied (Moore *et al.*, 1990; Tang, 1993). This murine α-lactalbumin gene was also reported to be expressed, alongside the β-casein gene, in the sebaceous glands during lactation (Maschio *et al.*, 1991), but this observation has not been confirmed (Persuy *et al.*, 1992; Vilotte *et al.*, 1992). Recently, RT-PCR experiments have suggested that the rat α-lactalbumin gene is also expressed at low levels in the brain of some lactating animals (Fujiwara *et al.*, 1999) and casein mRNA expression has been observed in cytotoxic T-lymphocyte-derived cell lines (Grusby *et al.*, 1990). Similarly, the occurrence of casein-like immuno-reactive substances in diverse organs, including the thymus, has been described recently in the rat (Onoda and Inano, 1997). As yet, the significance of these findings is not apparent; however, it is curious that many of the tissues in which ectopic expression has been observed also contain some of the transcription factors required for mammary expression of the milk protein genes, e.g., in T-cells, interleukin 2 activates STAT5 (Gilmour *et al.*, 1995).

15.3.2 Hormonal regulation and identification of *cis*-regulatory elements

Expression of the major milk protein-encoding genes is under a complex multi-hormonal regulation, resulting from the interplay of steroid and polypeptide hormones. In addition, local growth factors, cell-cell and cell-substratum interactions are also involved. Since several reviews have already focused on this topic (Forsyth, 1983, 1986; Rosen *et al.*, 1986, 1988; Lin and Bissell, 1993; Mercier and Vilotte, 1993; Groner *et al.*, 1994), including the previous editions of this chapter (in Fox, 1982, 1992), we will give only a brief survey.

Schematically, lactogenic hormones, such as insulin, prolactin and glucocorticoids, activate the transcription of the major milk protein genes whereas other hormones, such as progesterone, inhibit this activation in the early stages of pregnancy to favour cell proliferation over cell differentiation. It should be noted that in many systems it is difficult to separate direct induction of milk protein transcription from indirect differentiation-related events. As already mentioned, transcription activation of the major milk protein genes is not concomitant during pregnancy, perhaps due to the presence of various hormonal micro-environments and/or different responses to this environment displayed by different milk protein genes. For example, expression of calcium-sensitive casein genes, which are activated at mid-pregnancy, relies on prolactin and is increased by the synergetic action of glucocorticoids, while the transcriptional activity of the WAP promoter, a late expressed gene, is reciprocally regulated by these two hormones. In addition, whereas the glucocorticoid induction of the WAP gene expression is rapid, its action on the β-casein gene occurs only with a significant time-lag and requires *de novo* protein synthesis. Expression of the β-lactoglobulin gene, although displaying a temporal expression profile similar to the caseins, appears to be less dependent on lactogenic hormones than the calcium-sensitive casein genes. Finally, the α-lactalbumin gene although displaying a temporal expression profile similar to WAP, is induced by prolactin in the presence of low concentrations of glucocorticoids whereas high concentrations of glucocorticoids inhibit its expression, at least in eutherians (Funder, 1989). Induction of the α-lactalbumin gene in the absence of prolactin could be observed in murine pregnant mammary explants in the presence of insulin and cortisol (Warner *et al.*, 1993), while in marsupials, α-lactalbumin gene expression depends only on prolactin (Collet *et al.*, 1990). The mechanistic rationale for these intriguing differences has not been identified.

Regulation of mammary gene expression is also controlled by the epithelial cell basement membrane. For example, laminin can induce expression of α-lactalbumin and α_{s1}-casein by 160 fold (Aggeler *et al.*, 1988; Blum *et al.*, 1989). The differences observed between milk protein genes with regard to hormonal-induction are also evident in their varying requirement for a basement membrane. Transcriptional control of the β-casein promoter appears less dependent on the three dimensional structure of the mammary epithelial cells than does the WAP promoter, although both of them are sensitive to extracellular matrix (ECM) components (Lin *et al.*, 1995). At least for the β-lactoglobulin gene, regulation of expression by the ECM occurs through activation of STAT5 (Streuli *et al.*, 1995) and this may occur through an ECM-dependent modulation of protein-tyrosine phosphatase activity (Edwards *et al.*, 1998). Cell sub-stratum components as well as glucocorticoids can, at least for the casein genes, also act at the post-transcriptional level (Eisenstein and Rosen, 1988).

Beside these differences, milk protein genes share specific hormonal responses. For example, in at least the majority of species, progesterone inhibits expression of milk protein genes. It is still unclear, however, as to exactly how progesterone acts, although possible mechanisms include binding to its own receptor and/or competing with glucocorticoids and by local and/or systemic mechanisms. For example, it has been reported recently that progesterone might repress expression of the long-form of the prolactin receptor mRNA in the mammary gland (Mizoguchi *et al.*, 1997). This suggests the intriguing possibility that progesterone may interfer with prolactin signaling.

Transfection and transgenic studies have revealed that the promoters of most of the major milk protein-encoding genes are responsive to lactogenic hormones and target mammary-specific expression of reporter genes in transgenic animals. However, these promoters cannot on their own sustain full expression (compare Webster *et al.*, 1995 with Whitelaw *et al.*, 1992). Indeed, intragenic sequences of some of the major milk protein-encoding genes were shown to be able to contribute to their hormonal regulation (Lee *et al.*, 1989) and important regulatory elements have been identified within the introns (Kang *et al.*, 1998) and/or in the 3'UTR and flanking regions (Dale *et al.*, 1992). Furthermore, and as already mentioned, milk protein genes are also regulated at a post-transcriptional level.

Several approaches have allowed the identification of *cis*-regulatory elements within milk protein-encoding genes. Sequence comparison of a particular gene between several species or from different milk protein genes has led to the identification of DNA motifs that were conserved and thus suspected to be involved in the control of the gene transcription. A classical example is the high conservation of sequence elements in the proximal 5'-flanking region (−200/+1) of the calcium-sensitive casein genes (Rosen, 1987). These early identified elements were subsequently found to be recognized by regulatory nuclear factors. Consensus sequences recognized by effectors known to be involved in lactogenesis were also identified within the major milk protein gene promoter sequences by computer searches. But the true evidence of the presence of a *cis*-regulatory element within a DNA fragment came from transfection experiments in cell cultures, transgenic studies, DNAse I protection, footprinting and/or gel shift assays. Site-directed mutagenesis of the identified binding-sites and observations either in cell culture or in transgenics of the consequences on the promoter transcriptional regulation were performed sometimes to further define the functional role of these elements.

15.3.3 Transcriptional control of milk protein genes

Binding sites for several transcription factors have been identified within the promoters of most of the major milk protein encoding genes, such as binding sites for OCT-1, NF-1, C/EBP, STAT5, GR, Ets-1 and YY1 (see

Rosen *et al.*, 1996, for review). Other DNA elements have been shown to interact with yet unidentified effectors, such as the negative regulatory elements of the WAP promoter (Kolb *et al.*, 1994) and of the β-casein promoter (Lee and Oka, 1992; Altiok and Groner, 1993, 1994). Most of these sequences appear to be clustered within short DNA fragments of several hundred base-pairs in length, that encompass both positive and negative regulatory elements. Such composite response elements have been identified in the proximal 5'-flanking regions of the β-lactoglobulin gene (region −406/+1; Watson *et al.*, 1991), of the calcium-sensitive casein genes (region −200/+1; Rosen *et al.*, 1996 for review), in more distal regions of the bovine (BCE-1 element: region −1613/−1562; Schmidhauser *et al.*, 1992; Myers *et al.*, 1998) and human (region −4700/−4550; Winklehner-Jennewein *et al.*, 1998) β-casein genes, of the rabbit α_{s1}-casein gene (region −3442/−3118; Pierre *et al.*, 1994) and of the rat WAP gene (region −949/−720; Li *et al.*, 1994; Raught *et al.*, 1994). Thus, it appears that the transcriptional regulation of the major milk protein genes is under a combinatorial control with the binding of multiprotein complexes that can either repress or activate gene expression (see Wolberger, 1998, for review).

Differences in the composition of these composite regulatory elements may explain the observed hormonal differences in the transcriptional developmental regulation between the various major milk protein genes. In the WAP gene, for example, an Ets-1 binding site located at −110 appears to be important for the stage-specific transcriptional activation of the gene but not for its stable expression during lactation (McKnight *et al.*, 1995). Activation of promoters by transcription factors can be mediated by the relief of the binding of transcription repressors through both the competitive binding of these activators and the hormonal regulation of the expression or activation of these factors (Schmitt-Ney *et al.*, 1991). Thus, some negative binding factors appear to be present in the mammary gland only during pregnancy but not during lactation (Lee and Oka, 1992). Some of them mediate the inhibitory action of progesterone. Similarly, expression of C/EBPβ protein isoforms, that are essential both for mammogenesis and lactogenesis (Robinson *et al.*, 1998; Seagroves *et al.*, 1998), are regulated during pregnancy and lactation. The ratio between LIP, a dominant negative transcriptional repressor, and LAP, which is an activator of transcription, is high during pregnancy and decreases during lactation due, in part, to the inhibition of LIP expression by the glucocorticoids (Raught *et al.*, 1995). Expression of C/EBPα is also increased during lactation, possibly following stimulation by some extra-cellular matrix components (Raught *et al.*, 1995). The action of another factor, the transcription factor YY1, which represses expression of the β-casein promoter by binding to it at position −120/−110, is counteracted by its replacement from its binding site by the prolactin-activated STAT5 protein, which in turn positively regulates the promoter activity (Meier and Groner, 1994). A more surprising example of binding regulation of negative regulatory factors is given by the two proteins that bind the upper strand of the β-casein promoter at position −221/−170 during pregnancy and involution (Altiok and Groner, 1994). During lactation, a molecule inhibits the binding of these factors to the gene promoter, and it is suspected that this molecule could be the β-casein mRNA itself which possesses high-affinity binding sites for the two proteins in its 5' UTR (Altiok and Groner, 1994).

Much has been discovered about how milk protein genes are regulated and the involvement of many transcription factors described. Notwithstanding all this information, the identification of STAT5 as the end point of prolactin signalling in the mammary gland heralded a new era in our understanding of mammary gene regulation. All the more so, since the STAT proteins are central to all cytokine responses (Heim, 1999). Much of this work has been pioneered by studies involving mammary genes (Watson *et al.*, 1991; Schmitt-Ney *et al.*, 1991).

15.3.4 Prolactin signal transduction

In the mammary gland, prolactin induction results in the expression of the milk protein genes. This occurs through a rapid but transient signalling transduction pathway (Heim, 1999). First, prolactin-induced dimerisation of its receptor induces *trans*-phosphorylation of the kinase which is associated constitutively with the cytoplasmic domain of the receptor. The kinase is called JAK2 (for janus kinase 2). The activated JAK2 phosphorylates the receptor, creating a docking site for STAT5 through its SH2 domain. Subsequent phosphorylation and dimerisation of STAT5 result in its translocation to the nucleus where it binds to GAS (γ-interferon activation sequences) elements in target genes, e.g., β-lactoglobulin. Many other proteins are associated with this pathway, e.g., MAPK, resulting in a logistical problem for the cell to manage if it is to maintain tight regulation of the signal. It does this by balancing the activation signal with the generation of factors which inhibit the signal (Starr and Hilton, 1999).

STAT5 is an important positive transcription factor in the transcriptional regulation of milk protein genes. It is involved in the transduction of the prolactin signal. Functional STAT5-binding sites have been identified in the promoter region of almost all major milk protein-encoding genes, with the possible exception of the κ-casein-encoding gene (Adachi *et al.*, 1996). Within calcium-sensitive casein and β-lactoglobulin composite elements, the occurrence of multiple STAT5 binding sites is observed. These STAT5-binding sites were shown to be essential to confer prolactin transcriptional stimulation to the linked promoter (Schmitt-Ney *et al.*, 1991; Demmer *et al.*, 1995; Jolivet *et al.*, 1996; Soulier *et al.*, 1999). Furthermore, several experiments suggest that STAT5 effects are limited to the modulation of expression level and are not involved in determining the tissue-specificity of expression. For example, mutation of one or several of the STAT5-binding sites within the β-lactoglobulin promoter did not affect the tissue-specific expression of this gene in transgenic mice (Burdon *et al.*, 1994).

The STAT5 transcription factor actually consists of two proteins, STAT5a and STAT5b. These highly similar proteins are encoded by two genes, with differences in their binding affinities due to a single amino acid substitution (Boucheron *et al.*, 1998). The role of these proteins has been studied using gene knock-out approaches, with STAT5a emerging as the major factor required for expression of milk protein genes. STAT5a-deficient mice exhibit defective mammary gland development (Liu *et al.*, 1997). As one might expect, given the different response of the various milk protein genes to hormones, STAT5a-knock-out mice express the milk protein genes to different levels. Essentially normal levels of β-casein and α-lactalbumin mRNA levels are detected, with only WAP mRNA levels showing a reduction (Liu *et al.*, 1997). Indeed, there is no correlation between the expression of the STAT5a and the β-casein genes (Kazansky *et al.*, 1995). Studies with STAT5b knock-out mice indicate that this protein is associated with growth hormone effects (Udy *et al.*, 1997).

Taking these observations together generates the hypothesis that STAT5 may facilitate the interaction of other transcription factors to allow the transcriptional activation of the promoter. Indeed, functional interaction between STAT5 and the glucocorticoid receptor was reported recently and this molecular complex cooperates in the induction of the β-casein promoter, independently of the DNA-binding function of the glucocorticoid receptor (Stöcklin *et al.*, 1996; Lechner *et al.*, 1997; Stoecklin *et al.*, 1997).

Altogether, the better understanding of the nature and structural arrangement of the *cis*-regulatory elements involved in the control of the transcription of the major milk protein genes has given an insight into the differential developmental regulation of their expression. However, many questions remain. Why is expression of WAP in preference to β-casein affected in the STAT5a-null mice given that the reciprocal requirement for prolactin is observed? If it is not STAT5, and it appears not to be, what confers mammary-specificity to milk protein gene expression? How do the various transcription factors involved interact with each other to generate a production transcription complex? How does this transcription complex interact with the underlying nucleosomes?

15.3.5 Milk protein gene chromatin domains

It is now generally accepted that chromatin plays a central role in the regulation of gene expression. For transcription factors to bind to DNA, a reorganization of the chromatin structure is often needed. Furthermore, activation of a specific locus, while other linked genes remain silent, requires the formation of independent functional domains. To facilitate this, some gene loci are associated with dominantly-acting transcriptional regulators or locus control regions (LCR). In addition, the transcriptional apparatus and higher-order organisation of genes probably requires association with the nuclear matrix.

To date, very little is known about milk protein gene chromatin, although speculations abound. Indirect, although compelling, evidence that chromatin reorganisation is important for expression of milk protein genes comes from studies on the ovine β-lactoglobulin gene. Concomitant to increases in β-lactoglobulin gene expression during pregnancy, reflecting activation of STAT5, changes in DNase l hypersentisitivity over the β-lactoglobulin gene occur (Whitelaw and Webster, 1998). This implies a reorganisation of the chromatin in response to the interaction of transcription factors. Furthermore, this assay has detected species-specific differences in this reorganisation (Pena *et al.*, 1998): these differences are important clues to understand the different levels of expression of a same gene between species or between alleles. Similar studies imply that chromatin reorganisation events occur in response to induction of expression for the rat WAP (Li and Rosen, 1994). However, the clearest evidence yet describing a role for chromatin in the regulation of milk protein gene expression comes from analysis of the β-casein BCE-1 element. This element is not activated in transient transfections, where true chromatin is not formed, and is responsive to the acetylation state of the histones (Myers *et al.*, 1998).

The transcription domains of the major milk protein genes are poorly defined. In general, they are as yet defined only through functional analyses. Transgenic studies have revealed that while the α_{s1}- and the β-casein genes from various species can be well expressed in mice, only weak expression of the α_{s2}- and the κ-casein transgenes was observed. It is thus hypothesized that a mammary LCR might control the expression of the casein locus, and that this putative element is located close to the α_{s1}- and the β-casein loci. As yet, no direct evidence for the existance of such an element has been provided. Furthermore, if the human casein cluster does indeed carry an additional non-mammary gene, its expression can not be affected by the putative casein-LCR.

A consequence of chromatin gene domains is that boundaries to these domains exist. There are several documented examples (see Geyer, 1997, for review) but again, as yet, none has been conclusively identified for a milk protein gene. However, by gene comparison, some predications are possible. The boundaries of the α-lactalbumin gene are not known, but those of the related lysozyme gene are well defined (Phi-Van and Strätling, 1988) and located several kb 5' and 3' from the transcription unit of the gene. It is likely that the α-lactalbumin gene contains similar elements, probably in a similar location. This hypothesis is supported by the observed position-independent and copy number- related expression of a human 250 kb YAC- and of a 160 kb goat BAC-α-lactalbumin transgenes (Fujiwara *et al.*, 1997; Stinnakre *et al.*, 1999).

The transcription domain of the β-lactoglobulin gene might be the best studied within milk protein genes so far. Several DNase l-hypersensitive sites, which reflect its expression status, have been located within the promoter, intronic and 3'-flanking regions (Whitelaw and Webster, 1998). These sites spatially reflect the limit of the chromatin domain as defined by nuclease sensitivity. The 5' limit of this domain resides very close to the promoter. In addition, the proximal 3'-flanking β-lactoglobulin sequences can interact with the nuclear matrix *in vitro*. This suggests that the β-lactoglobulin gene resides in a very small chromatin domain (Whitelaw, unpublished results). Sequences which interact with the nuclear matrix, usually AT-rich in nature, may be involved in regulating chromatin structure, thereby facilitating gene expression. Although several studies have addressed the ability of such sequences to enhance milk protein gene expression, a clear picture has not yet emerged (e.g., Attal *et al.*, 1995; McKnight *et al.*, 1996).

15.3.6 Milk protein mRNAs

The structure of milk protein mRNAs has been investigated in numerous species (see Mercier and Vilotte, 1993, for review). In lactation, these RNAs account for up to 60 to 80% of the total RNA present in mammary epithelial cells. The calcium-sensitive casein mRNAs are characterized by a better interspecies conservation of the untranslated regions and of the sequence-encoding the signal peptide compared to the mature protein coding frame. Since part of the regulation of milk protein gene regulation is exerted at a post-transcriptional level (e.g., Aggeler *et al.*, 1988; Blum *et al.*, 1989; Golden and Rillema, 1995), conservation of the untranslated region (UTR) might reflect their importance for mRNA processing. Indeed, sequences located in the 3' UTR are known to interact with proteins to either stabilize or degrade the RNA, and/or alter mRNA translation efficiency.

As already mentioned, and as illustrated in Chapters 3 and 16 of this issue, total or partial exon-skipping during the splicing of the pre-mRNA is responsible for the differences observed between casein variants and between homologous proteins from different species. In this section, we will describe only variations which affect the overall level of gene expression. Deletion or transition of a single nucleotide within coding exons of a caprine α_{s1}-casein allele and of two β-casein alleles create nonsense codons (Leroux *et al.*, 1992; Persuy *et al.*, 1996; Rando *et al.*, 1996). These alleles are characterized by a much lower level of mRNAs compared to other alleles and, for some of them, with multiple exon skipping events. This phenomenon has been observed for several other genes where nonsense codons have been found to be associated with mRNA decay and/or exon skipping (see Valentine, 1998; Hentze and Kulozik, 1999, for reviews). A goat and a bovine α_{s1}-casein allele, also associated with a reduced amount of mRNA and milk protein, were found to contain a truncated inserted LINE element in the 3' UTR (Perez *et al.*, 1994; Rando *et al.*, 1998). Insertion of these elements is supposed to reduce the stability of this allele mRNA.

15.3.7 Regulation of translation: from RNA to protein

Although we now know in some detail the players involved in the regulation of milk protein gene transcription, our understanding of what happens to the resulting transcripts once they have passed into the cytoplasm is still vague. The basic mammalian RNA processing events are well known (Siomi and Dreyfus, 1997) but investigation of RNA regulatory mechanisms is overdue. It appears that experiments *in vitro* can be informative of *in vivo* events (Donofrio *et al.*, 1996). In addition, initial studies suggest that it may be at the level of translation control where insulin and prolactin interact to synergistically enhance milk protein levels (Baruch *et al.*, 1998). Furthermore, since this aspect of gene control can have major effects on final protein levels, an increased understanding of how milk protein RNAs are processed may well have application in mammary biotechnology.

15.4 Intracellular transport and co- and post-translational processing of milk proteins

As described above (Section 15.3.6), the structure of milk protein-encoding genes, the control of their expression and the primary structure of the main milk proteins are now partly elucidated. How the mammary epithelial cell (MEC) is able to process any polypeptide along the secretory pathway in order to release a wide variety of milk proteins with specific physico-chemical and structural characteristics (soluble proteins and caseins aggregated in micelles) is an emerging question. In this section, the cellular secretory machinery of the lactating MEC will be described briefly. Data about the processing of milk proteins in relation to their intracellular transport and a brief survey of the hormonal regulation of milk protein secretion will be reported.

15.4.1 Morphological organisation of mammary epithelial cells

The mammary epithelium consists of a single layer of cuboidal cells, separated from the interstitial space by a basement membrane. This epithelium is organised into acini, the lumen of which are connected by ductules to the teat. Passage from the interstitial space to the lumen of the acini is restricted during lactation by tight junctionnal complexes, located at the luminal borders of the cells. This barrier partly controls the permeability between the blood compartment and the milk. During lactation, MECs are highly polarised and display features typical of secretory cells (Pitelka and Hamamoto, 1983). Their cytoplasm contains numerous parallel lamellar cisternae of endoplasmic reticulum (ER) and an extended Golgi apparatus in a peri- and supra-nuclear region. This latter organelle is composed of stacked sacculae more or less distended, sometimes containing filamentous products, and is surrounded by an extensive network of tubulo-vesicular structures. On the *cis* face of the Golgi apparatus, which faces the ER, these tubulo-vesicular structures form the ER-Golgi intermediate compartment (ERGIC). On the *trans* face of the Golgi, the *trans* Golgi network (TGN) is also surrounded by numerous vesicles, some of them coated with the typical spike structure characteristic of clathrin. The TGN faces a region rich in secretory vesicles. These vesicles contain filamentous structures and casein micelles suspended in an electron-lucent fluid. These compartments are schematically described and illustrated in Figures 15.4 and 15.5.

15.4.2 Intracellular transport and processing of milk proteins

Transport of newly-synthesised proteins proceeds in MECs according to the general pathway described in other exocrine glands (Palade, 1975). Following a 3 min pulse labelling with a radioactive amino acid, flow of newly-synthesised proteins is detectable in the ER 5 min after the beginning of the pulse, concentrated in the Golgi region at 15 min, accumulated in secretory vesicles located in an apical position at 45 min and predominantly located in the lumen of the acini at 60 min (Seddiki *et al.*, 1991).

(a) Transport in the endoplasmic reticulum

Entry into the ER requires a visa, the signal peptide. Most proteins targeted to the ER and subsequently carried through the secretory pathway have cleavable amino terminal targeting sequences, called signal sequences. These sequences are recognised directly by receptors on the ER (Schatz and Dobberstein, 1996). Translocation of the nascent polypeptide from the cytoplasmic side to the *trans* side of the ER membrane involves association between ribosomes, from which the signal sequence emerges, and a *trans* membrane channel-forming heteromeric protein complex. The alignment of the peptide exit site with the channel, maintenance of the open conformation of the channel and recruitment of the ER luminal chaperones, which trap or actively pull the incoming precursor, are now well described in yeasts and mammalian cells (Pilon and Scheckman, 1999).

A significant evolutionary conservation of the primary structures of the signal peptides (interspecies homology $\geq 87\%$) of the calcium-sensitive

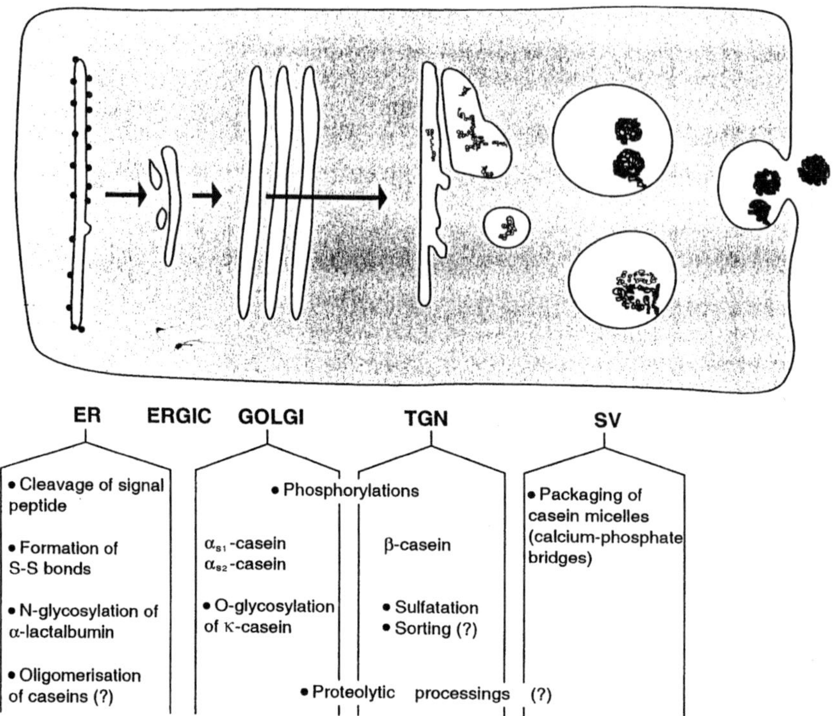

Figure 15.4 Schematic representation of intracellular compartments of the mammary epithelial cell in which milk proteins are processed. The polypeptide chains undergo various co-translational modifications during their translocation into the lumen of the endoplasmic reticulum (ER). Proteins are carried through an ER-Golgi intermediate compartment (ERGIC) to the *cis* face of the Golgi apparatus. Proteins undergo various post-translational modifications along the different compartments of the Golgi apparatus. Caseins are packaged into micelles within the secretory vesicles (SV).

caseins, α_{s1}-, α_{s2}- and β-casein, has been observed, suggesting that an optimal and definite conformation of the signal peptide is required for efficient translocation of casein polypeptide chains into the ER (previous edition of this chapter in Fox, 1992).

Once translocated into the lumen of the ER, proteins are in contact with an oxidizing environment. This environment promotes the formation of disulfide bonds in some peptides that are rich in cysteine residues, by a process catalysed by protein disulfide isomerase (PDI) (Freedman, 1984). The folding of nascent polypeptides is additionally assisted by proteins with chaperone activities which include BiP/GRP78, Erp77, calnexine and calreticulin (Pelham, 1989). These ER-resident proteins bind to misfolded, newly-synthesised proteins and retain them in the ER until they are completely folded or redirect them to a degradative pathway. A correct folding

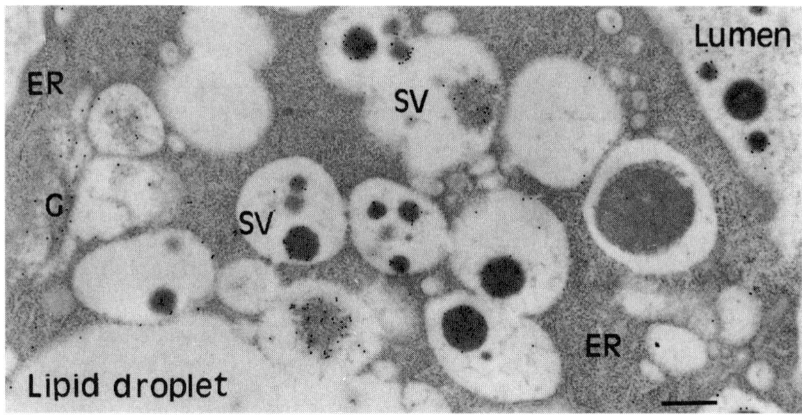

Figure 15.5 Immuno gold electron microscopy localisation of caseins in the mammary epithelial cell of the lactating mouse. The tissue was fixed with 2% paraformaldehyde in 0.1 M sodium cacodylate buffer, dehydrated with ethanol and embedded in Unicryl. Sections were immunolabelled with an antiserum against mouse caseins followed by a gold-conjugated secondary antibody (as described in Devinoy *et al.*, 1995). Label is present in the endoplasmic reticulum (ER), on Golgi saccules (G) and in various types of secretory vesicles (SV). Filamentous structures, contained in dilated saccules and in secretory vesicles, condensated aggregates and typical casein micelles are labelled with gold particles. Bar, 0.5 μm.

of these proteins is a prerequisite for their transport into the Golgi complex and thus it appears that folding acts as a "quality control" (Hammond and Hellenius, 1995).

In MECs, the role of ER-resident proteins on the folding and the retention of milk proteins is not well defined. However, BiP, GRP, calreticulin and PDI have been detected in rat and goat lactating MECs (Ghosal *et al.*, 1994; Chanat *et al.*, 1999). Interestingly, both calreticulin and PDI have been detected not only in the ER but also in the cytosol from homogenates of rat mammary gland (Ghosal *et al.*, 1994). These authors suggested that these proteins could be also involved in the formation of lipid droplets, raising questions about a possible link between the enzymes implied in protein and in lipid synthesis.

Elucidating the primary structure of α-lactalbumin and β-lactoglobulin has revealed the presence of four and two disulfide bridges in the peptide chain, respectively (Brew *et al.*, 1970; Papiz *et al.*, 1986). WAP contains numerous cysteine residues which are able to form multiple intramolecular disulfide bonds (Hennighausen *et al.*, 1982; Devinoy *et al.*, 1988). Moreover, WAP consists of two subunits linked by at least one disulfide bridge (Baranyi *et al.*, 1995). A role of PDI in the formation of disulfide bonds during the transport of these milk proteins in the ER can be suggested. Lastly, *N*-glycosylation of α-lactalbumin polypeptide chains has been detected in the mammary microsomal membranes (Mercier, 1981).

It was believed that the forward transport out of the ER does not require a specific signal (Wieland *et al.*, 1987). However, whether or not cargo proteins contain sorting information to move to the Golgi apparatus is still controversial (Bremsel *et al.*, 1999). Oligomerisation of protein subunits also acts as a quality control. Thus, the assembly of newly synthesised subunits in the ER for oligomeric proteins such as thyroglobulin and IgM or for membranous proteins (hemaglutinin of influenza virus) is necessary for their transport to the pre-Golgi compartment (Copeland *et al.*, 1986; Kim and Arven, 1991).

The question of quality control associated with the forward transport of caseins out of the ER has been addressed. The occurrence of an extensive polymorphism of the goat α_{s1}-casein gene has make it possible to study intracellular transport of caseins in MEC from goats naturally deficient for a single casein (Chanat *et al.*, 1999). In the absence of expression of α_{s1}-casein, other caseins (α_{s2}, β and κ) accumulate in the distended ER and the export of these caseins from the ER is slowed down. This stage of transport appears to be dependent on the relative proportion of α_{s1}-casein in the ER. This supports the notion that α_{s1}-casein can interact with the other caseins in the ER and that the formation of a primary complex between the four caseins is required for an efficient transport from the ER to the Golgi.

(b) Transport through the Golgi apparatus and trans Golgi network

The export of proteins out of the ER relies on the production of small vesicles or on the transformation of the membranes occurring at the ER exit site. An active recruitment and concentration of soluble cargo at the exit sites could also occur. This concentration mechanism is likely to involve positive sorting signals (Nishimura and Balch, 1997). Recent data suggest that the tubulo vesicular network functions to collect cargo emerging from the ER and to form a transitional structure, ERGIC, in the delivery of cargoes to the Golgi stacks. Cytoplasmic proteins (COPII) are involved in the machinery responsible for vesicle budding (Springer *et al.*, 1999). Intra-Golgi trafficking has been believed for a long time to be the result of a vesicle shuttle mechanism. However, an alternative model is now favoured (Glick and Malhotra, 1998). In the cisternal maturation model, fusion of ERGIC clusters generate a new *cis*-Golgi cisterna and this cisterna progresses through the stack to the *trans*-face of the Golgi. A retrograde transport of Golgi-resident proteins by COPI-coated vesicles assures the maturation of saccules and maintains the enzymatic composition specific of each compartment of the Golgi apparatus. In lactating MECs, the morphological observation that electron-dense structures are detectable in the Golgi stacks but are excluded from the Golgi-associated vesicles is in agreement with this cisternal maturation model (Clermont *et al.*, 1993).

Enzymes which reside in the Golgi stacks and in the TGN carry out functions which include glycosylation, phosphorylation, sulphation and proteolytic processings. In MECs, glycosyltransferases are involved in the

O-glycosylation of κ-casein and also act jointly with α-lactalbumin to promote synthesis of lactose from glucose and UDP-galactose. Galactosyltransferase activity has been detected within the Golgi apparatus but also in the secretory vesicles and in the milk (Witsell *et al.*, 1990; R. Boisgard, personal communication). Phosphorylation of caseins is a key event for calcium binding and micelle formation. Kinases, located within the Golgi apparatus, phosphorylate the caseins (Bingham and Farrell, 1981; West and Clegg, 1984). Disorganisation of the Golgi apparatus, after treatment of MECs with brefeldin A (BFA), interrupts the transport of the proteins between the Golgi stacks and the TGN (Pauloin *et al.*, 1997). In mouse MECs, this treatment inhibits the phosphorylation of β- and γ-caseins but not of α_{s1}- or α_{s2}-casein (Turner *et al.*, 1993). In lactating rabbit mammary tissue, *in vitro* metabolic labelling followed by analysis of the conversion of newly synthesised forms of caseins to their mature forms revealed that the maturation of β-casein occurred in a more distal compartment than that of the α_{s1}-casein (Boisgard *et al.*, to be published). It can be concluded that the phosphorylation of the α_s-caseins occurs in the Golgi cisternae whereas that of the β- and γ-caseins (in mouse MECs) and of the β-casein (in rabbit MECs) occur in TGN. The sulphation of the proteins takes place in the TGN. There is no evidence of sulphation of the major milk proteins but the sulphation of proteoglycans occurs in the TGN from rabbit MEC (Boisgard R., personal communication).

Proteolytic processing of a variety of precursor proteins by a family of cellular endoproteases has been demonstrated to occur within the secretory pathway (Steiner *et al.*, 1992). Although no direct evidence of cleavage of milk proteins by these enzymes has been shown, the presence of furin, a serine protease of this family, has been detected in the lactating rabbit MEC (Pauloin *et al.*, personal communication). This endoprotease, known to be localized in the TGN of many cells, is relatively abundant in clathrin-coated vesicles from lactating rabbit MEC. Because in this species, the quantity of clathrin coated vesicles is very high in the Golgi region, it can be suggested that furin could have an important role.

(c) Transport in secretory vesicles

As mentioned (in Section 15.4.2 (*b*)), the caseins are phosphorylated in the Golgi apparatus. The formation of micelles, by binding of calcium ions to casein ester phosphate groups, depends on the presence of a high Ca^{2+} concentration in the terminal Golgi apparatus (Neville and Waiters, 1983). The exact site where micelle formation begins is not clear. Morphological observations reveal filamentous proteins in dilated Golgi saccules. Two types of secretory vesicles are visible, one type being filled with chains of small spherical particles and filamentous structures, the other type containing, in addition to these structures, closely-packed casein micelles, with the characteristic honeycombed texture. In addition, aggregates intermediate between these two types of structures are visible (Figure 15.5). By which processes caseins mature to give dense micellar structures remains to be elucidated.

Submicelles and micelles are present in MECs from β-casein-deficient transgenic mice (Kumar *et al.*, 1994). In MECs and milk from α_{s1}-casein-deficient goats, micelles are also present. Thus, the presence of the four caseins is not a prerequisite for the formation of micelles. However, the size of micelles is variable, depending on the species and κ-casein is a modulator of the size of the micelles (Gutienez-Adam *et al.*, 1996). In most species, the number of micelles per vesicle during the lactation ranges from one micelle to more than ten. At the beginning of mammary gland involution, large vesicles containing a great number of micelles accumulate in the cell. It is notable that the modification of the number of micelles occurs in transgenic mice deficient for the α-lactalbumin gene. MECs from transgenic animals do not synthesise lactose and the diameter of secretory vesicles is reduced and contain only one micelle (Stinnakre *et al.*, 1994). The question whether there is a link between lactose synthesis, micelle formation and the generation of secretory vesicles should be addressed.

Lactoferrin in MECs from lactating mice (Neville *et al.*, 1998), WAP in MEC from lactating rabbit (Ollivier-Bousquet, unpublished observation) and recombinant human and bovine GH expressed in transgenic mice (Devinoy *et al.*, 1995) are detectable in the luminal part of secretory vesicles containing casein micelles, confirming that whey proteins and aggregated caseins are carried in the same vesicles during the last stage of exocytosis.

15.4.3 Hormonal regulation of milk protein secretion

As discussed above (Section 15.4.2(*b*)), milk proteins are processed during their journey along the secretory pathway. Whether modifications of transport kinetics might have repercussions on the processing is still unclear. Nevertheless, hormonal regulation of the intracellular transport of milk protein is now established. There is no substantial storage of newly synthesised proteins inside MECs (Devinoy *et al.*, 1995; Pauloin *et al.*, 1997). In spite of that, hormones such as oxytocin and prolactin are able, *in vitro*, to exert a direct stimulatory effect on protein secretion in lactating MEC of rabbits and rodents (see Ollivier-Bousquet, 1993, 1997): oxytocin increases the transport of newly-synthesised proteins from the ER to the Golgi region and to the secretory vesicles. Prolactin increases the release of newly-synthesised proteins by an effect on the last stages of exocytosis.

15.5 Amino acid transport by the mammary gland

Amino acids, extracted from interstitial fluid, are the major source of amino-nitrogen for milk protein synthesis; therefore, a knowledge of mammary tissue amino acid transport mechanisms and their regulation is important if we are to fully understand the process of milk protein secretion. Such knowledge will help those wishing to manipulate milk protein content *via* dietary means. The uptake of amino acids across the basolateral membranes of mammary secretory cells, the major point of entry, is accomplished by an array of distinct transport mechanisms. The amino acid transporters differ from one another with respect to kinetics, substrate specificity and ion dependency, however, it is evident that they operate in a co-ordinated fashion to supply amino-nitrogen to support milk protein synthesis.

The study of mammary epithelial amino acid transport is hampered by the relatively complex anatomy of the mammary gland. Nevertheless, the use of mammary tissue explants and the perfused mammary gland has enabled the transport of amino acids by mammary tissue to be characterised. Mammary explants are easy to prepare and have the advantage that the cellular architecture remains intact. Furthermore, the preparation of mammary explants does not require digestive agents: enzymes such as collagenase could ultimately alter the properties of membrane transport proteins. Although mammary tissue explants isolated from lactating animals are comprised of more than one cell type, it can be assumed that the vast majority of the surface area of explants is that of the basolateral membrane of the secretory cells. Therefore, it is reasonable to assume that mammary tissue explants can be used to give a measure of amino acid transport across the blood-facing pole of the mammary epithelium. The large tissue extracellular space associated with mammary explants does, however, place limitations on the design of experiments. The perfused mammary gland allows the transport of amino acids across the blood-facing aspect of the mammary epithelium to be measured under near physiological conditions; perfusates can be delivered to the gland with a flow and pressure profile similar to that found *in vivo*. The perfused mammary gland used in combination with a rapid, paired-tracer dilution technique allows the transport of amino acids to be measured over very short time-courses (Mepham *et al.*, 1985; Calvert and Shennan, 1996).

15.5.1 Mammary tissue amino acid transport systems

The identification of individual amino acid transport systems is difficult on account of the fact that a single amino acid may be able to utilize several transport systems (see Barker and Ellory, 1990). Furthermore, the difficulty in characterising amino acid transport is compounded by the lack of specific inhibitors of amino acid transporters. In spite of these drawbacks, significant progress has been made towards identifying the amino acid transport systems that are present in mammary tissue (Shennan *et al.*, 1997).

Mammary tissue amino acid transport mechanisms fall into two categories: Na^+-dependent and Na^+-independent systems.

(a) Na^+-dependent transport mechanisms
Lactating mammary cells are able to concentrate free amino acids (particularly the non-essential ones) with respect to plasma, suggesting that there must be an input of free energy (Shennan *et al.*, 1997). It is apparent that several amino acid transport systems utilize the electrochemical Na^+-gradient to drive the 'uphill' movement of amino acids into mammary cells. The Na^+-dependent mechanisms which have been identified in mammary tissue include systems A, ASC, X^-_{AG} and β.

System A prefers short-chain neutral amino acids as substrates and is characterised by its tolerance of N-methylated amino acids such as N-methylaminoisobutyrate (MeAIB). Indeed, Na^+-dependent amino acid uptake inhibited by MeAIB is usually taken as a measure of transport *via* system A. It appears that mouse, rat and bovine mammary tissue possesses system A activity (Neville *et al.*, 1980; Baumrucker 1985; Verma and Kansal, 1993; Shennan and McNeillie 1994a). In contrast, no evidence for the presence of system A at the blood-facing aspect of the guinea-pig mammary gland could be found (Mepham *et al.*, 1985). The limited kinetic data available suggest that system A in lactating mammary tissue operates with relatively low affinity: the K_M of methionine and α-aminoisobutyric acid uptake *via* system A in mouse mammary tissue is 0.47 mM and 2.0 mM, respectively (Neville *et al.*, 1980; Verma and Kansal, 1993). It is notable that the precise substrate specificity of system A in mammary tissue remains to be defined. Several lines of evidence suggest that system A is regulated by milk stasis, starvation and by the stage of lactation (Neville *et al.*, 1980; Shennan and McNeillie, 1994c; Verma and Kansal, 1995). In this connection, there is the strong possibility that system A activity is regulated by prolactin: treating animals with bromocryptine, a drug which inhibits prolactin secretion from the pituitary gland, markedly reduces the arterio-venous concentration differences of amino acids which are potential substrates of system A (Vina *et al.*, 1981).

Mammary tissue from several species (e.g., bovine, mouse, guinea pig) has been shown to posses system ASC (Baumrucker, 1985; Mepham *et al.*, 1985; Verma and Kansal, 1993). It has been established in other tissues that this mechanism co-transports neutral amino acids such as alanine, threonine and cysteine with the Na^+. System ASC in mammary tissue, like system A, operates with relatively low affinity: the K_M of methionine uptake by lactating mouse mammary tissue *via* system ASC is 0.46 mM (Verma and Kansal, 1993). At present, there appears to be a paucity of information on the regulation of mammary tissue amino acid transport *via* system ASC except for the finding that starvation gives rise to a large increase in system ASC activity in mouse mammary tissue (Verma and Kansal, 1995).

The transport of anionic amino acids by lactating mammary tissue has been studied extensively (Millar *et al.*, 1996, 1997). The predominant, if not the only, pathway for L-glutamate and L-aspartate transport is a high affinity (K_M = 18 μM for L-glutamate), Na^+-dependent system analogous to system X^-_{AG} (Kanai *et al.*, 1994). The Na^+-dependent anionic amino acid carrier is very selective for anionic amino acids: it does not interact with neutral or cationic amino acids (Millar *et al.*, 1996, 1997). An unusual feature of system X^-_{AG} is its ability to discriminate between the optical isomers of glutamate but not those of aspartate. The high affinity anionic amino acid carrier is able to act as an exchange system as well as a co-transport mechanism, suggesting that the transport of L-glutamate will affect the intracellular concentration of L-aspartate (and *vice versa*) (Millar *et al.*, 1997). Several high affinity anionic amino acid carriers which have identity with system X^-_{AG} have been cloned and characterised (e.g. see Kanai and Hediger, 1992; Pines *et al.*, 1992; Storck *et al.*, 1992). Two of these clones, namely GLAST and GLT-1, have been identified in rat mammary tissue (Martinez-Lopez *et al.*, 1998). However, the exact contribution of these carriers to the transport of anionic amino acids across the basolateral membranes of mammary secretory cells is unknown.

Taurine, a non-protein amino acid, is taken up by lactating rat mammary tissue *via* a high affinity ($K_M \sim 43$ μM), Na^+-dependent transport system analogous to system β (Shennan and McNeillie, 1994b). It appears that the rat mammary (Na^+ + taurine) co-transport system also requires Cl^- for maximal activity. System β has a narrow substrate specificity: only β amino acids such as taurine and β-alanine interact with the transporter. The lactating gerbil mammary gland expresses a high affinity (Na^+ + taurine) co-transport system which, unlike the rat mammary taurine transporter, is not dependent upon Cl^- (Shennan, 1995).

(b) Na^+-independent transport mechanisms

System L has been identified in mouse, rat, guinea pig and bovine mammary tissue (Neville *et al.*, 1980; Mepham *et al.*, 1985; Verma and Kansal, 1993; Shennan and McNeillie, 1994). System L is a Na^+-independent transport mechanism which has a wide substrate specificity. Indeed, system L may be the most important transport system for the uptake of essential neutral amino acids by the lactating mammary gland. Na^+-independent amino acid transport sensitive to 2-aminobicyclo-heptane-2-carboxylic acid (BCH) is taken as a measure of system L activity. It is generally accepted that system L can act as an amino acid exchange mechanism. Accordingly, methionine uptake by mouse mammary gland *via* system L can be *trans*-accelerated by intracellular amino acids. However, amino acid efflux from rat mammary tissue, *via* system L, is not *trans*-stimulated by extracellular amino acids, suggesting that the transporter operates with asymmetric kinetics which could favour the retention of substrates within the gland.

The transport of cationic amino acids by lactating mammary tissue is a Na^+-independent process (Baumrucker, 1984; Shennan *et al.*, 1994). The pathway for lysine and arginine uptake by bovine mammary tissue is not affected by replacing extracellular Na^+ and appears to be relatively specific for cationic amino acids. On this basis, Baumrucker (1984) concluded that cationic amino acids are transported *via* system y^+. Similarly, the uptake of cationic amino acids by lactating rat mammary tissue is not dependent on Na^+ but it is evident that neutral amino acids may share a pathway with lysine and arginine for transport (Shennan *et al.*, 1994; Calvert and Shennan, 1996). Thus, certain neutral amino acids, leucine and glutamine in particular, inhibit and stimulate, respectively, lysine uptake by and efflux from rat mammary tissue. The exact nature of the pathway which accepts both cationic and neutral amino acids as substrates remains to be determined.

System T, a mechanism specific for aromatic amino acids, has been described in lactating mouse mammary tissue (Kansal and Kansal, 1996). Thus, the moiety of tyrosine uptake by mouse mammary tissue which is not dependent upon extracellular Na^+ and is not sensitive to BCH has been ascribed to system T. This mechanism operates with low affinity; the K_M for tyrosine transport is approximately 15 mM.

(c) Volume-activated amino acid transport

To survive, cells have to regulate their volume within relatively narrow limits (see Hoffmann and Simonsen, 1989, for a review). Cell membranes are very permeable to water which means that cell volume, otherwise termed the cellular hydration state, will be determined by the osmolality of the extracellular fluid and by the intracellular content of osmotically active solutes. The cellular hydration state can be changed as a consequence of anisosmotic conditions, cellular accumulation of solutes or oxidative metabolism. Cells are able to regulate their volume following swelling or shrinking. Cell volume regulation involves the *trans*-membrane movement of solutes together with osmotically obliged water (Hoffmann and Simonsen, 1989).

A knowledge of volume-regulatory processes in mammary tissue is of particular importance given that cell volume changes markedly affect the rate of milk protein synthesis (Millar *et al.*, 1997). It has been established that cell-swelling activates the transport of amino acids in mammary tissue (Shennan *et al.*, 1994, 1996). Thus, cell-swelling, induced by a hyposmotic shock, increases the efflux of amino acids such as taurine and glycine *via* a pathway which has the characteristics of a channel rather than a carrier. The swelling-induced amino acid efflux pathway appears to be situated in the blood-facing aspect of the mammary epithelium (Calvert and Shennan, 1998). The volume-activated efflux of taurine is dependent upon the extent of cell swelling and can be inhibited by a number of pharmacological agents such as niflumic acid (Shennan *et al.*, 1996). There is the strong possibility that volume-activated amino acid efflux may participate in mammary cell volume regulation given that amino acids are concentrated within mammary cells with respect to plasma.

15.5.2 Transport and metabolism of peptides

Although free amino acids are the major source of amino-nitrogen for milk protein synthesis, it is apparent that the uptake of several essential amino acids is less than their output in milk, suggesting that other circulating forms of amino acids, such as peptides, may be available for casein production (e.g., see Backwell *et al.*, 1996). In this connection, it has been demonstrated that the goat mammary gland is able to use intravenously administered peptides for milk protein synthesis (Backwell *et al.*, 1994, 1996). The *in vivo* studies were unable to show whether the mammary gland transported peptides intact or whether the peptides were hydrolysed extracellularly, followed by uptake of the liberated amino acids. It has been shown, albeit indirectly, that mouse mammary tissue does not significantly hydrolyse peptides extracellularly but is able to transport peptides intact (Wang *et al.*, 1996). On the other hand, studies using the rat mammary gland as a model suggest that mammary tissue is able both to transport and hydrolyse dipeptides (Shennan *et al.*, 1998, 1999). It is evident that the perfused lactating rat mammary is able to transport dipeptides which are resistant to hydrolysis (i.e., D-Phe-L-Gln; D-Phe-L-Glu); the nature of the pathway remains to be identified precisely. However, it is apparent that the capacity of the rat mammary gland to transport peptides intact across the basolateral pole of the epithelium is relatively limited (Shennan *et al.*, 1998). Rat mammary tissue is able to extensively hydrolyse peptides extracellularly: this has been demonstrated using both direct and indirect experimental approaches (Shennan *et al.*, 1998, 1999). Anionic dipeptides presented to rat mammary tissue can *trans*-stimulate D-aspartate efflux *via* the high affinity anionic amino acid carrier. If it is accepted that dipeptides do not interact directly with the amino acid carrier, then it can be assumed that anionic amino acids, produced as a consequence of extracellular hydrolysis, act to stimulate D-aspartate efflux. In support of the notion that mammary tissue is capable of hydrolysing peptides extracellularly is the finding that rat mammary tissue can hydrolyse a variety of aminoacyl-*p*-nitroanilides (peptide analogues). Quantitatively, it appears, at least in the rat, that hydrolysis of peptides followed by uptake of the free amino acids may be more important than the transport of peptides.

The identity of the peptidases involved in the extracellular hydrolysis of peptides by mammary tissue has not been established. However, γ-glutamyltranspeptidase appears to be involved given that glutathione can stimulate D-aspartate efflux from rat mammary tissue in a fashion sensitive to acivicin (Shennan *et al.*, 1998). The hydrolysis of glutathione may be an important route for providing the mammary gland with cysteine.

References

Adachi, T., Ahn, J.Y., Yamamoto, K., Aoki, N., Nakamura, R. and Matsuda, T. (1996) Characterization of the bovine kappa-casein gene promoter. *Biosci. Biotech. Biochem.*, **60**, 1937–40.

Aggeler, J., Park, C.S. and Bissell, M.J. (1988) Regulation of milk protein and basement membrane gene expression: the influence of the extracellular matrix. *J. Dairy Sci.*, **71**, 2830–42.

Alexander, L.J., Hayes, G., Bawden, W., Stewart, A.F. and Mackinlay, A.G. (1993) Complete nucleotide sequence of the bovine beta-lactoglobulin gene. *Animal Biotech.*, **4**, 1–10.

Alexander, L.J., Stewart, A.F., MacKinlay, A.G., Kapelinskaya, T.V., Tkach, T.M. and Gorodetsky, S.I. (1988) Isolation and characterization of the bovine κ-casein gene. *Eur. J. Biochem.*, **178**, 395–401.

Ali, S. and Clark, J. (1988) Characterization of the gene encoding ovine beta-lactoglobulin. Similarity to the genes for retinol binding protein and other secretory proteins. *J. Mol. Biol.*, **199**, 415–26.

Altiok, S. and Groner, B. (1993) Interaction of two sequence-specific single-stranded DNA-binding proteins with an essential region of the beta-casein promoter is regulated by lactogenic hormones. *Mol. Cell. Biol.*, **13**, 7303–10.

Altiok, S. and Groner, B. (1994) Beta-casein mRNA sequesters a single stranded nucleic acid-binding protein which negatively regulates the beta-casein gene promoter. *Mol. Cell. Biol.*, **14**, 6004–12.

Ansari, H.A., Broad, T.E., Burkin, D.J., Cambridge, L.M., Carpenter, M.A., Davidson, B.P., Jones, C.A., Lewis, P.E., Maher, D.W., Malcom, A.A., McNabb, L.R., Phua, S.H. and Wood, N.J. (1992) Regional assignment of five genes onto sheep chromosomes. *Tenth European Colloquium on the Cytogenetics of Domestic Animals, Utrecht*, The Netherlands, p. 27 (one page).

Attal, J., Cajero-Juarez, M., Petitclerc, D., Theron, M.C., Stinnakre, M.G., Bearzotti, M., Kann, G. and Houdebine, L.M. (1995) The effect of matrix attached regions (MAR) and specialized chromatin structure (SCS) on the expression of gene constructs in cultured cells and transgenic mice. *Mol. Biol. Rep.*, **22**, 37–46.

Backwell, F.R.C., Bequette, B.J., Wilson, D., Calder, A.G., Metcalf, J.A., Wray-Cahen, D., McRae, J.C., Beever, D.E. and Lobley, G.E. (1994) Utilization of peptides by the caprine mammary gland for protein synthesis. *Am. J. Physiol.*, **267**, R1–6.

Backwell, F.R.C., Bequette, B.J., Wilson, D., Metcalf, J.A., Franklin, M.F., Beever, D.E., Lobley, G.E. and MacRae, J.C. (1996) Evidence for the utilization of peptides for milk protein synthesis in the lactating dairy goat *in vivo*. *Am. J. Physiol.*, **271**, R955–60.

Baranyi, M., Aszodi, A., Devinoy, E., Fontaine, M.L., Houdebine, L.M. and Bosze, Z. (1996) Structure of the rabbit kappa-casein encoding gene: expression of the cloned gene in the mammary gland of transgenic mice. *Gene*, **174**, 27–34.

Baranyi, M., Brignon, G., Anglade, P. and Ribadeau-Dumas, B. (1995) New data on the proteins of rabbit (*Oryctolagus cuniculus*) milk. *Comp. Biochem. Physiol. B. Biochem. Mol. Biol.*, **111**, 407–15.

Barker, G.A. and Ellory, J.C. (1990) The identification of neutral amino acid transport systems. *Exp. Physiol.*, **75**, 3–26.

Baruch, A., Shani, M. and Barash, I. (1998) Insulin and prolactin synergize to induce translation of human serum albumin in the mammary gland of transgenic mice. *Transgenic Res.*, **7**, 15–27.

Baumrucker, C.R. (1984) Cationic amino acid transport by bovine mammary tissue. *J. Dairy Sci.*, **67**, 2500–6.

Bingham, E.W. and Farrell, H.M., Jr (1974) Casein kinase from the Golgi apparatus of lactating mammary gland. *J. Biol. Chem.*, **249**, 3647–51.

Blum, J.L., Zeigler, M.E. and Wicha, M.S. (1989) Regulation of mammary differentiation by the extracellular matrix. *Environ. Health Perspect.*, **80**, 71–83.

Boisgard, R. and Chanat, E. (2000) Phospholipase D-dependent and -independent mechanisms are involved in milk protein secretion in rabbit mammary epithelial cells. *Biochim. Biophys. Acta,* **1458**, 1–16.

Boisnard, M., Hue, D., Bouniol, C., Mercier, J.C. and Gaye, P. (1991) Multiple mRNA species code for two non-allelic forms of ovine alpha-s2-casein. *Eur. J. Biochem.*, **201**, 633–41.

Bonsing, J., Ring, J.M., Stewart, A.F. and MacKinlay, A.G. (1988) Complete nucleotide sequence of the bovine beta-casein gene. *Austr. J. Biol. Sci.*, **41**, 527–37.

Boucheron, C., Dumon, S., Santos, S.C., Moriggl, R., Hennighausen, L., Gisselbrecht, S. and Gouilleux, F. (1998) A single amino acid in the DNA binding regions of Stat5 A and Stat5B confers distinct binding specificities. *J. Biol. Chem.*, **273**, 33936–41.

Bremser, M., Nickel, W., Schweikert, M., Ravazzola, M., Amherdt, M., Hughes, C.A., Söllner, T.H., Rothman, J.E. and Wieland, F.T. (1999) Coupling of coat assembly and vesicle budding to packaging of putative cargo receptors. *Cell*, **96**, 495–506.

Brew, K., Castellino, F.J., Vanaman, T.C. and Hill, R.L. (1970) The complete amino acid sequence of bovine alpha-lactalbumin. *J. Biol. Chem.*, **245**, 4570–82.

Burden, T.G., Maitland, K.A., Clark, A.J., Wallace, R. and Watson, C.J. (1994) Regulation of the sheep beta-lactoglobulin gene by lactogenic hormones is mediated by a transcription factor that binds an interferon gamma activation site-related element. *Mol. Endo.*, **8**, 1528–36.

Calvert, D.T. and Shennan, D.B. (1996) Evidence for an interaction between cationic and neutral amino acids at the blood-facing aspect of the lactating rat mammary epithelium. *J. Dairy Res.*, **63**, 25–33.

Calvert, D.T. and Shennan, D.B. (1998) Volume-activated taurine efflux from the *in situ* perfused lactating rat mammary gland. *Acta Physiol. Scand.*, **162**, 97–105.

Campbell, S.M., Rosen, J.M., Henninghausen, L.G., Strech-Jurk, U. and Sippel, A.E. (1984) Comparison of the whey acidic protein genes of the rat and mouse. *Nucleic Acids Res.*, **12**, 8685–97.

Chanat, E., Martin, P. and Ollivier-Bousquet, M. (1999) α_{s1}-Casein is required for the efficient transport of β- and κ-casein from the endoplasmic reticulum to the Golgi apparatus of mammary epithelial cells. *J. Cell Sci.*, **112**, 3399–412.

Clermont, Y., Xia, L., Rambourg, A., Turner, J.D. and Hermo, L. (1993) Transport of casein submicelles and formation of secretion granules in the Golgi apparatus of epithelial cells of the lactating mammary gland of the rat. *Anat. Rec.*, **235**, 363–73.

Collet, C. and Joseph, R. (1995) Exon organization and sequence of the genes encoding alpha-lactalbumin and beta-lactoglobulin from the Tammar Wallaby (Macropodidae, Marsupialia). *Biochem. Genet.*, **33**, 61–72.

Collet, C., Joseph, R. and Nicholas, K. (1990) Cloning, cDNA analysis and prolactin-dependent expression of a marsupial alpha-lactalbumin. *Reprod. Fertil. Dev.*, **2**, 693–701.

Copeland, C.S., Doms, R.W., Bolzau, E.M., Webster, R.G. and Helenius, A. (1986) Assembly of influenza hemagglutinin trimers and its role in intracellular transport *J. Cell. Biol.*, **103**, 1179–91.

Dale, T.C., Krnack, M.J., Schmidhauser, C., Yang, C.L.Q., Bissell, M.J. and Rosen, J.M. (1992) High-level expression of the rat whey acidic protein gene is mediated by elements in the promoter and 3' untranslated region. *Mol. Cell. Biol.*, **12**, 905–14.

Davies, M.S., West, L.F., Davis, M.B., Povey, S. and Craig, R.K. (1987) The gene for human alpha-lactalbumin is assigned to chromosome 12ql3. *Ann. Hum. Genet.*, **51**, 183–8.

Dawson, S.P., Wilde, C.J., Tighe, P.J. and Mayer, R.J. (1993) Characterization of two novel casein transcripts in rabbit mammary gland. *Biochem. J.*, **15**, 777–84.

Demmer, I., Burdon, T.G., Djiane, J., Watson, C.J. and Clark, A.J. (1995) The proximal milk protein binding factor binding site is required for the prolactin responsiveness of the sheep beta-lactoglobulin promoter in Chinese hamster ovary cells. *Mol. Cell. Endo.*, **107**, 113–21.

Devinoy, E., Stinnakre, M.G., Lavialle, F., Thepot, D. and Ollivier-Bousquet, M. (1995) Intracellular routing and release of caseins and growth hormone produced into milk from transgenic mice. *Exp. Cell. Res.*, **221**, 272–80.

Dobie, K.W., Lee, M., Fantes, J.A., Graham, E., Clark, A.J., Springbett, A., Lathe, R. and McClenaghan, M. (1996) Variegated transgene expression in mouse mammary gland is determined by the transgene integration locus. *Proc. Natl. Acad. Sci. USA*, **93**, 6659–64.

Donofrio, G., Bignetti, E., Clark, A.J. and Whitelaw, C.B. (1996) Comparible processing of beta-lactoglobulin pre-mRNA in cell culture and transgenic mouse models. *Mol. Gen. Genet.*, **25**, 465–9.

Edlund, A., Johansson, T., Leidvik, B. and Hansson, L. (1996) Structure of the human kappa-casein gene. *Gene*, **174**, 65–9.

Edwards, G.M., Wilford, F.H., Liu, X., Hennighausen, L., Djiane, J. and Streuli, C.H. (1998) Regulation of mammary differentiation by extracellular matrix involves protein-tyrosine phosphatases. *J. Biol. Chem.*, **17**, 9495–500.

Eisenstein, R.S. and Rosen, J.M. (1988) Both cell substratum regulation and hormonal regulation of milk protein gene expression are exerted primarily at the posttranscriptional level. *Mol. Cell. Biol.*, **8**, 3183–90.

Faerman, A., Barash, I., Puzis, R., Nathan, M., Hurwitz, D.R. and Shani, M. (1995) Dramatic hereogeneity of transgene expression in the mammary gland of lactating mice: a model system to study the synthetic activity of mammary epithelial cells. *J. Histochem. Cytochem.*, **43**, 461–70.

Flower, D.R. (1996) The lipocalin protein family: structure and function. *Biochem. J.*, **318**, 1–14.

Folch, J.M., Coll, A. and Sanchez, A. (1994) Complete sequence of the caprine beta-lactoglobulin gene. *J. Dairy Sci.*, **77**, 3493–7.

Folch, J.M., Coll, A., Hayes, H.C. and Sanchez, A. (1996) Characterization of a caprine beta-lactoglobulin pseudogene, identification and localization by *in situ* hybridization in goat, sheep and cow. *Gene,* **24**, 87–91.

Forsyth, I.A. (1983) The endocrinology of lactation. In *Biochemistry of Lactation*, (T.B. Mepham ed.) pp. 309–49, Elsevier, Amsterdam.

Forsyth, I.A. (1986) Variation among species in the endocrine control of mammary growth and function: the roles of prolactin, growth hormone, and the placental lactogen. *J. Dairy Sci.*, **69**, 886–903.

Fox, P.F. (ed.) (1982) *Developments in Dairy Chemistry-1-Proteins*, Applied Science Publishers, London.

Fox, P.F. (ed.) (1992) *Developments in Dairy Chemistry-1-Proteins*, Elsevier Applied Science Publishers Ltd, London.

Freedman, R.B. (1984) Native disulfide bond formation in protein biosynthesis: evidence for the role of protein disulfide isomerase. *Trends Biochem. Sci.*, **9**, 438–41.

Fujiwara, Y., Miwa, M., Nogami, M., Okumura, K., Nobori, T., Suzuki, T. and Ueda, M. (1997) Genomic organization and chromosomal localization of the human casern gene family. *Hum. Genet.*, **99**, 368–73.

Fujiwara, Y., Miwa, M., Takahashi, R., Kodaira, K., Hirabayashi, M., Suzuki, T. and Ueda, M. (1999) High-level expressing YAC vector for transgenic animal bioreactors. *Mol. Reprod. Dev.*, **52**, 14–20.

Funder, J.W. (1989) Hormonal regulation of gene expression. *Biochem. Soc. Symp.*, **55**, 105–14.

Gallagher, D.S., Schelling, C.P., Groenen, M.M.A. and Womack, J.E. (1994) Confirmation that the casein gene cluster resides on cattle chromosome 6. *Mammalian Genome*, **5**, 524.

Gallagher, D.S., Treadgill, D.W., Ryan, A.M., Womack, J.R. and Irwin, D.M. (1993) Physical mapping of the lysozyme gene family in cattle. *Mammalian Genome*, **4**, 368–73.

Geissler, E.N., Cheng, S.V., Gusella, J.F. and Housman, D.E. (1988) Genetic analysis of the dominant white-spotting (W) region on mouse chromosome 5: identification of clones DNA markers near W. *Proc. Natl. Acad. Sci.*, USA, **85**, 9635–9.

Gellin, J., Echard, G., Yerle, M., Dalens, M., Chevalet, C. and Gillois, M. (1985) Localization of the alpha and beta casein genes to the q24 region of chromosome 12 in the rabbit (*Oryctolagus cuniculus L.*) by *in situ* hybridization. *Cytogenet. Cell Genet.*, **39**, 220–3.

George, S., Clark, A.J. and Archibald, A.L. (1997) Physical mapping of the murine casein locus reveals the gene order as alpha-beta-gamma-epsilon-kappa. *DNA Cell. Biol.*, **16**, 477–84.

Geyer, P.K. (1997) The role of insulator elements in defining domains of gene expression. *Current Opinion Genet. Develop.*, **7**, 242–8.

Ghosal, D., Shappell, N.W. and Keenan, T.W. (1994) Endoplasmic reticulum lumenal proteins of rat mammary gland. Potential involvement in lipid droplet assembly during lactation. *Biochim. Biophys. Acta*, **1200**, 175–81.

Gilmour, K.C., Pine, R. and Reich, N. (1995) Interleukin 2 activates STAT5 transcription factor (mammary gland factor) and specific gene expression in T lymphocytes. *Proc. Natl. Acad. Sci. USA*, **92**, 10772–6.

Click, B.S. and Malhotra, Y. (1998) The curious status of the Golgi apparatus. *Cell*, **95**, 883–9.

Golden, K.L. and Rillema, J.A. (1995) Effects of prolactin on galactosyl transferase and alpha-lactalbumin mRNA accumulation in mouse mammary gland explants. *Proc. Soc. Exp. Biol. Med.*, **209**, 392–6.

Groenen, M.A.M., Dijknof, R.J.M., Verstege, A.J.M. and van der Poel, J.J. (1993) The complete sequence of the gene encoding bovine alpha-s2-casein. *Gene*, **123**, 187–93.

Groner, B., Altiok, S. and Meier, V. (1994) Hormonal regulation of transcription factor activity in mammary epithelial cells. *Mol. Cell Endocrinol.*, **100**, 109–14.

Grosclaude, F. (1979) Polymorphism of milk proteins: some biochemical and genetical aspects. In *Proc. 16th Int. Conf. Anim. Blood Grps Biochem. Polymorphisms*, (International Society for Animal Blood Group Research, ed.) pp. 54–92, International Society for Animal Blood Group Research, Leningrad.

Grusby, M.J., Mitchell, S.C., Nabavi, M. and Glimcher, L.H. (1990) Casein expression in cytotoxic T lymphocytes. *Proc. Natl. Acad. Sci. USA*, **87**, 6897–901.

Gupta, P., Rosen, J.M., D'Eustachio, P. and Ruddle, F.H. (1982) Localization of the casein gene family to a single mouse chromosome. *J. Cell Biol.*, **93**, 199–204.

Gutierrez-Adan, A., Maga, E.A., Meade, H., Shoemaker, C.F., Medrano, J.F., Anderson, G.B. and Murray, J.D. (1996) Alterations of the physical characteristics of milk from transgenic mice producing bovine kappa-casein. *J. Dairy Sci.*, **79**, 791–9.

Guyette, W.A., Matusik, R.J. and Rosen, J.M. (1979) Prolactin-mediated transcriptional and post-transcriptional control of casein gene expression. *Cell*, **17**, 1013–23.

Hall, L., Emery, D.C., Davies, M.S., Parker, D. and Craig, R.K. (1987) Organization and sequence of the human alpha-lactalbumin gene. *Biochem. J.*, **242**, 735–42.

Halliday, J.A., Bell, K., McKenzie, H.A. and Shaw, D.C. (1990) Feline whey proteins: identification, isolation and initial characterization of alpha-lactalbumin, beta-lactoglobulin and lysozyme. *Comp. Biochem. Physiol.*, **95**, 773–9.

Hammond, C. and Helenius, A. (1995) Quality control in the secretory pathway. *Curr. Opin. Cell. Biol.*, **1**, 523–9.

Hansson, L., Edlund, A., Johansson, T., Hernell, O., Strömqvist, M., Lindquist, S., Lönnerdal, B. and Bergström, S. (1994) Structure of the human beta-casein encoding gene. *Gene*, **139**, 193–9.

Harris, S., Ali, S., Anderson, S., Archibald, A.L. and Clark, A.J. (1988) Complete nucleotide sequence of the genomic ovine beta-lactoglobulin gene. *Nucleic Acids Res.*, **16**, 10379–80.

Harris, S., McClenaghan, M., Simons, J.P., Ali, S. and Clark, A.J. (1990) Gene expression in the mammary gland. *J. Reprod. Fert.*, **88**, 707–15.

Hayes, H., Petit, E., Bouniol, C. and Popescu, P. (1993) Localization of the alpha-S2-casein gene (CASAS2) to the homeologous cattle, sheep and goat chromosomes 4 by *in situ* hybridization. *Cytogent. Cell. Genet.*, **64**, 282–5.

Hayes, H., Petit, E., Lemieux, N. and Dutrillaux, B. (1992) Chromosomal localization of the ovine beta-casein gene by non-isotopic *in situ* hybridization and R-banding. *Cytogent. Cell. Genet.*, **61**, 286–8.

Hayes, H., Popescu, P. and Dutrillaux, B. (1993) Comparative gene mapping of lactoperoxidase, retinoblastoma and alpha-lactalbumin genes in cattle, sheep and goats. *Mammalian Genome*, **4**, 593–7.

Hayes, H.C. and Petit, E.J. (1993) Mapping of the beta-lactoglobulin gene and of an immunoglobulin M heavy chain-like sequence to homoeologous cattle sheep, and goat chromosomes. *Mammalian Genome*, **4**, 207–10.

Heim, M.H. (1999) The Jak-STAT pathway: cytokine signalling from the receptor to the nucleus. *J. Recept. Signal Transduct. Res.*, **19**, 75–120.

Hennighausen, L.G. and Sippel, A.E. (1982) Mouse whey acidic protein is a novel member of the family of four-disulfide corel proteins. *Nucleic Acids Res.*, **10**, 2677–84.

Hentze, M. and Kulozik, A.E. (1999) A perfect message: RNA surveillance and nonsense-mediated decay. *Cell*, **96**, 307–10.

Hoffmann, E.K. and Simonsen, L.O. (1989) Membrane mechanisms in volume and pH regulation in vertebrate cells. *Physiol. Rev.*, **69**, 315–82.

Hurtley, S.M. and Helenius, A. (1989) Protein oligomerization in the endoplasmic reticulum. *An. Rev. Cell. Biol.*, **5**, 277–307.

Jolivet, G., Devinoy, E., Fontaine, M.L. and Houdebine, L.M. (1992) Structure of the gene encoding rabbit alpha sl-casein. *Gene*, **113**, 257–62.

Jolivet, G., L'Hotte, C., Pierre, S., Tourkine, N. and Houdebine, L.M. (1996) The MGF/STAT5 binding site is necessary in the distal enhancer for high prolactin induction of transfected rabbit alpha-sl-casein-CAT gene transcription. *FEBS Lett.*, **8**, 257–62.

Jollès, J., Fiat, A.M., Schoentgen, F., Alais, C. and Jollès, P. (1974) The amino acid sequence of sheep kappa-A-casein. II. Sequence studies concerning the kappa-A-caseinoglycopeptide and establishment of the complete primary structure of the protein. *Biochim. Biophys. Acta*, **365**, 335–42.

Jones, W.K., Yu-Lee, L.Y., Clift, S.M., Brown, T.L. and Rosen, J.M. (1986) The rat casein multigene family. Fine structure and evolution of the beta-casein gene. *J. Biol. Chem.*, **260**, 7042–50.

Kanai, Y. and Hediger, M.A. (1992) Primary structure and functional characterization of a high-affinity glutamate transporter. *Nature*, **360**, 467–71.

Kanai, Y., Smith, C.R. and Hediger, M.A. (1994) A new family of neurotransmitter transporters: the high affinity glutamate transporters. *FASEB J.*, **8**, 1450–9.

Kang, Y.K., Lee, C.S., Chung, A.S. and Lee, K.K. (1998) Prolactin-inducible enhancer activity of the first intron of the bovine beta-casein gene. *Mol. Cells*, **30**, 259–65.

Kansal, R. and Kansal, V.K. (1996) Discrimination of transport systems of L-tyrosine in mouse mammary gland: Characterisation of system T. *Ind. J. Exp. Biol.*, **34**, 750–7.

Kapelinskaia, T.V., Tkach, T.M., Smirnov, I.K. and Gorodetskii, S.I. (1989) The *Bos taurus* casein genes. Isolation and characterization of the kappa-casein gene. *Genetika*, **25**, 15–23.

Kazansky, A.V., Raught, B., Lindsey, S.M., Wang, Y.F. and Rosen, J.M. (1995) Regulation of mammary gland factor/Stat5 during mammary gland development. *Mol. Endo.*, **9**, 1598–609.

Kim, P.S. and Arvan, P. (1991) Folding and assembly of newly synthesized thyroglobulin occurs in a pre-Golgi compartment. *J. Biol. Chem.*, **266**, 12412–8.

Koczan, D., Hobom, G. and Seyfert, H.M. (1991) Genomic organization of the bovine α_{s1}-casein gene. *Nucleic Acids Res.*, **19**, 5591–6.

Kolb, A.F., Günzburg, W.H., Albang, R., Brem, G., Erfle, V. and Sahnons, B. (1994) Negative regulatory element in the mammary specific whey acidic protein promoter. *J. Cell. Biochem.*, **56**, 245–61.

Kumar, S., Clarke, A.R., Hooper, M.L., Horne, D.S., Law, A.J., Leaver, J., Springbett, A., Stevenson, E. and Simons, J.P. (1994) Milk composition and lactation of beta-casein deficient mice. *Proc. Natl. Acad. Sci. USA*, **91**, 6138–42.

Laird, J.E., Jack, L., Hall, L., Boulton, A., Parker, D. and Craig, R.K. (1988) Structure and expression of the guinea-pig alpha-lactalbumin gene. *Biochem. J.*, **254**, 85–94.

Lear, T.L., Brandon, R., Masel, A., Bell, K. and Bailey, E. (1998) Molecular cloning and chromosomal localization of horse alpha-l-antitrypsin (AAT), beta-lactoglobulin 1 and 2 (BLG, BLG2), lactotransferrin (LTF) and transferrin (TF). *Animal Genet.*, **29**, 43.

Lechner, J., Welte, T., Tomasi, J.K., Bruno, P., Cairns, C., Gustafsson, J.A. and Doppler, W. (1997) Promoter-dependent synergy between glucocorticoid receptor and Stat5 in the activation of beta-casein gene transcription. *J. Biol. Chem.*, **272**, 20954–60.

Lee, C.S. and Oka, T. (1992) A pregnancy-specific mammary nuclear factor involved in the repression of the mouse beta-casein gene transcription by progesterone. *J. Biol. Chem.*, **267**, 5797–801.

Lee, K.F., Atiee, S.H., Henning, S.J. and Rosen, J.M. (1989) Relative contribution of promoter and intragenic sequences in the hormonal regulation of rat beta-casein transgenes. *Endocrinology*, **3**, 447–53.

Leroux, C., Mazure, N. and Martin, P. (1992) Mutations away from splice site recognition sequences might cis-modulate alternative splicing of goat α_{S1}-casein transcripts. Structural organization of the relevant gene. *J. Biol. Chem.*, **267**, 6147–57.

Li, S. and Rosen, J.M. (1994) Distal regulatory elements required for rat whey acidic protein gene expression in transgenic mice. *J. Biol. Chem.*, **269**, 14235–43.

Li, S. and Rosen, J.M. (1994) Glucocorticoid regulation of rat whey acidic protein gene expression involves hormone-induced alterations of chromatin structure in the distal promoter region. *Mol. Endocrinol.*, **8**, 1328–35.

Li, S. and Rosen, J.M. (1995) CTF/NF1 and mammary gland factor (Stat5) play a critical role in regulating rat whey acidic protein gene expression in transgenic mice. *Mol Cell. Biol.*, **15**, 2063–70.

Lin, C.Q. and Bissell, M.J. (1993) Multi-faceted regulation of cell differentiation by extracellular matrix *FASEB J.*, **7**, 737–43.

Lin, C.Q., Dempsey, P.J., Coffey, R.J. and Bissell, M.J. (1995) Extracellular matrix regulates whey acidic protein gene expression by suppression of TGF-alpha in mouse mammary epithelial cells: studies in culture and in transgenic mice. *J. Cell Biol.*, **129**, 1115–26.

Liu, X., Robinson, G.R., Wagner, K.U., Garrett, L., Wynshaw-Boris, A. and Hennighausen, L. (1997) Stat5a is mandatory for adult mammary gland development and lactogenesis. *Genes Develop.*, **11**, 179–86.

Martinez-Lopez, I., Garcia, C., Barber, T., Vina, J.R. and Miralles, V.J. (1999) The L-glutamate transporters GLAST (EAAT1) and GLT-1(EAAT2): expression and regulation in the rat lactating mammary gland. *Mol. Mem. Biol.*, **15**, 237–42.

Maschio, A., Brickell, P.M., Kioussis, D., Mellor, A.L., Katz, D. and Craig, R.K. (1991) Transgenic mice carrying the guinea-pig alpha-lactalbumin gene transcribe milk protein genes in their sebaceous glands during lactation. *Biochem. J.*, **275**, 459–67.

McConkey, E.H., Menon, R., Williams, G., Baker, E. and Sutherland, G.R. (1996) Assignment of the gene for beta-casein (CSN2) to 4q13 -> q21 in humans and 3p13 -> p12 in chimpanzees. *Cytogenet. Cell Genet.*, **72**, 60–2.

McKnight, R.A., Spencer, M., Dirtmer, J., Brady, J.N., Wall, R.J. and Hennighausen, L. (1995) An Ets site in the whey acidic protein gene promoter mediates transcriptional activation in the mammary gland of pregnant mice but is dispensable during lactation. *Mol. Endocrinol.*, **9**, 717–24.

McKnight, R.A., Spencer, M., Wall, R.J. and Hennighausen, L. (1996) Severe position effects imposed on a 1 kb mouse whey acidic protein gene promoter are overcome by heterogeneous matrix attachment regions. *Mol. Reprod. Dev.*, **44**, 179–84.

Meier, V.S. and Kroner, B. (1994) The nuclear factor YY1 participates in repression of the beta-casein gene promoter in mammary epithelial cells and is counteracted by mammary gland factor during lactogenic hormone induction. *Mol. Cell. Biol.*, **14**, 128–37.

Mepham, T.B., Overthrow, J.I. and Short, A.H. (1985) Epithelial cell entry and exit competition amongst amino acids in the perfused isloated lactating mammary gland of guinea pig. In *Carrier Mediated Transport of Solutes from Blood to Tissue*, (D.L. Yudilevich and G.E. Mann eds.) Longman, London, pp. 369–72.

Mercier, J.C. (1981) Phosphorylation of caseins, present evidence for an amino acid triplet code posttranslationally recognized by specific kinases. *Biochimie*, **63**, 1–17.

Mercier, J.C. and Vilotte, J.L. (1993) Structure and function of milk protein genes. *J. Dairy Sci.*, **76**, 3079–98.

Mercier, J.C., Vilotte, J.L. and Provost, C. (1990) Structure and function of milk protein genes. In *Genome, Analysis in Domestic Animals*, (H. Geldermann and F. Ellendorf eds.) Verlag VCH, Weinheim, pp. 233–58.

Millar, I.D., Barber, M.C., Lomax, M.A., Travers, M.T. and Shennan, D.B. (1997) Mammary protein synthesis is acutely regulated by the cellular hydration state. *Biochem. Biophys. Res., Commun.*, **230**, 351–5.

Millar, I.D., Carvert, D.T., Lomax, M.A. and Shennan, D.B. (1996) The mechanism of L-glutamate transport by lactating rat mammary tissue. *Biochim. Biophys. Acta*, **1282**, 200–6.

Millar, I.D., Calvert, D.T., Lomax, M.A. and Shennan, D.B. (1997) Substrate specificity of the mammary tissue anionic amino acid carrier operating in the cotransport and exchange modes. *Biochim. Biophys. Acta*, **1326**, 92–102.

Mizoguchi, Y., Kim, J.Y., Enami, J. and Sakai, S. (1997) The regulation of the prolactin receptor gene expression in the mammary gland of early pregnant mouse. *Endo. J.*, **44**, 53–8.

Molenaar, A.J., Davis, S.R. and Wilkins, R.J. (1992) Expression of alpha-lactalbumin, alpha-s1-casein and lactoferrin genes is heterogeneous in sheep and cattle mammary tissue. *J. Histochem. Cytochem.*, **40**, 611–8.

Moore, A., Hall, L. and Hamilton, D.W. (1990) An 18-kDa androgen-regulated protein that modifies galactosyltransferase activity is synthesized by the rat caput epididymidis, but has no structural similarity to rat milk alpha-lactalbumin. *Biol. Reprod.*, **43**, 497–506.

Myers, C.A., Schmidhauser, C., Mellentin-Michelotti, J., Fragoso, G., Roskelley, C.D., Casperson, G., Mossi, R., Pujuguet, P., Hager, G. and Bissell, M.J. (1998) Characterization of BCE-1, a transcriptional enhancer regulated by prolactin and extracellular matrix and modulated by the state of histone acetylation. *Mol. Cell. Biol.*, **18**, 2184–95.

Neville, M.C., Chatfield, K., Hansen, L., Lewis, A., Monks, J., Nuijens, J., Ollivier-Bousquet, M., Schanbacher, F., Sawicki, V. and Zhang, P. (1998) Lactoferrin secretion into mouse milk. Development of secretory activity, the localization of lactoferrin in the secretory pathway, and interactions of lactoferrin with milk iron. *Adv. Exp. Med. Biol.*, **443**, 141–53.

Neville, M.C., Lobitz, C.J., Ripoll, E.A. and Tinney, C. (1980) The sites of α-aminoisobutyric acid uptake in normal mammary gland and ascites tumor cells. *J. Biol. Chem.*, **255**, 7311–6.

Nishimura, N. and Balch, W.E. (1997) A di-acidic signal required for selective export from the endoplasmic reticulum. *Science*, **277**, 556–8.

Ollivier-Bousquet, M. (1993) Secrétion des caséines: régulation hormonale. In *Biologie de la Lactation*, INSERM/INRA, (J. Martinet and L.M. Houdebine eds.) pp. 367–87.

Ollivier-Bousquet, M. (1997) Milk protein transfer in the mammary cell. *Flem. Vet. J.*, **66**, Suppl., 125–42.

Onoda, M. and Inano, H. (1997) Distribution of casein-like proteins in various organs of rat. *J. Histochem. Cytochem.*, **45**, 663–74.

Palade, G. (1975) Intracellular aspects of the process of protein synthesis. *Science* **189**, 347–58.

Papiz, M.Z., Sawyer, L., Eliopoulos, E.E., North, A.C.T., Findlay, J.B.C., Sivaprasadavao, R., Jones, T.A., Newcomer, M.E. and Kraulis, P.J. (1986) The structure of beta-lactoglobulin and its similarity to plasma retinol-binding protein. *Nature*, **324**, 383–5.

Passey, R.J. and McKinlay, A.G. (1995) Characterization of a second, apparently inactive, copy of the bovine beta-lactoglobulin gene. *Eur. J. Biochem.*, **233**, 736–43.

Pauloin, A., Delpal, S., Chanat, E., Lavialle, F., Aubourg, A. and Ollivier-Bousquet, M. (1997) Brefeldin A differently affects basal and prolactin-stimulated milk protein secretion in lactating rabbit mammary epithelial cells. *Eur. J. Cell. Biol.*, **72**, 324–36.

Pauloin, A., Tooze, S., Michelutti, I., Delpal, S. and Ollivier-Bousquet, M. (1999) The majority of clathrin coated vesicles from lactating rabbit mamary gland arises from the secretory pathway. *J. Cell Sci.*, **112**, 4089–100.

Pelham, H.R. (1989) Control of protein exit from the endoplasmic reticulum. *Anna. Rev. Cell. Biol.*, **5**, 1–23.

Pena, R.N., Folch, J.M., Sanchez, A. and Whitelaw, C.B.A. (1998) Chromatin structures of goat and sheep beta-lactoglobulin gene differ. *Biochem. Biophys. Res. Commun.*, **252**, 649–53.

Pena, R.N., Sanchez, A., Coll, A. and Folch, J.M. (1999) Isolation, sequencing and relative quantification by fluorescent-ratio PCR of feline beta-lactoglobulin I, II and III cDNAs. *Mamm. Genome*, **10**, 560–4.

Perez, M.J., Leroux, C., Bonastre, A.S. and Martin, P. (1994) Occurrence of a LINE sequence in the 3' UTR of the goat alpha-s1-casein E-encoding allele associated with reduced protein synthesis level. *Gene,* **147**, 179–87.

Persuy, M.A., Prinlz, C., Medrano, J.F. and Mercier, J.C. (1996) One mutation might be responsible for the absence of beta-casein in two breeds of goats. *Animal Genet.*, **27**, 96.

Persuy, M.A., Stinnakre, M.G., Printz, C., Mahe, M.F. and Mercier, J.C. (1992) High expression of the caprine beta-casein gene in transgenic mice *Eur. J. Biochem.*, **205**, 887–91.

Phi-Van, L. and Strätling, W.H. (1988) The matrix attachment regions of the chicken lysozyme gene co-map with the boundaries of the chromatin domain. *EMBO J.*, **7**, 655–64.

Pierre, S., Jolivet, G., Devinoy, E. and Houdebine, L.M. (1994) A combination of distal and proximal regions is required for efficient prolactin regulation of transfected rabbit alpha-α_{s1}-casein chloramphenicol acetyltransferase constructs. *Mol. Endocrinol.*, **8**, 1720–30.

Pilon, M. and Schekman, R. (1999) Protein translocation: how Hsp70 pulls it off. *Cell*, **97**, 679–82.

Pines, G., Danbolt, N.C., Bjoras, M., Zhang, Y., Bendahan, A., Eide, L., Koepsell, H., Storm-Mathisen, J., Seeberg, E. and Kanner, B.L. (1992) Cloning and expression of a rat brain L-glutamate transporter. *Nature*, **360**, 464–7.

Pitelka, D.R. and Hamamoto, S.T. (1983) Ultrastructure of the mammary secretory cell, in *Biochemistry of Lactation*, (T.B. Mepham ed.) Elsevier Science, Publishers B.V., pp. 29–70.

Provot, C., Persuy, M.A. and Mercier, J.C. (1995) Complete sequence of the ovine beta-casein-encoding gene and interspecies comparison. *Gene*, **154**, 259–63.

Qasba, P.K. and Safaya, S.K. (1984) Similarity of the nucleotide sequences of rat alpha-lactalbumin and chicken lysozyme genes. *Nature*, **608**, 377–80.

Qasba, P.K., Hewlett, I.K. and Byers, S. (1983) The presence of the milk protein alpha-lactalbumin and its mRNA in the rat epididymis. *Biochem. Biophys. Res. Commun.*, **30**, 306–12.

Rando, A., Di Gregorio, P., Ramunno, L., Mariani, P., Fiorella, A., Senese, C., Marletta, D. and Masina, P. (1998) Characterization of the CSNAG allele of the bovine alpha-sl-casein locus by the insertion of a relict of a long interspersed element. *J. Dairy Sci.*, **81**, 1735–42.

Rando, A., Pappalardo, M., Capuano, M., Di Gregorio, P. and Ramunno, L. (1996) Two mutations might be responsible for the absence of beta-casein in goat milk. *Animal. Genet.*, **27**, 31 (one page).

Raught, B., Liao, W.S.L. and Rosen, J.M. (1995) Developmentally and hormonally regulated CCAAT/enhancer-binding protein isoforms influence beta-casein gene expression. *Mol. Endo.*, **9**, 1223–32.

Rijnkels, M., Kooiman, P.M., de Boer, H.A. and Pieper, F.R. (1997b) Organization of the bovine casein gene locus. *Mammalian Genome*, **8**, 148–52.

Rijnkels, M., Meershoek, E., de Boer, H.A. and Pieper, F.R. (1997c) Physical map and localization of the human casein gene locus. *Mammalian Genome*, **8**, 285–6.

Rijnkels, M., Wheeler, D.A., de Boer, H.A. and Pieper, F.R. (1997a) Structure and expression of the mouse casein gene locus. *Mammalian Genome*, **8**, 9–15.

Roberts, B., DiTulho, P., Vitale, J., Hehir, K. and Gordon, K. (1992) Cloning of the goat beta-casein-encoding gene and expression in transgenic mice. *Gene*, **121**, 255–62.

Robinson, G.W., Johnson, P.F., Hennighausen, L. and Sterneck, E. (1998) The C/EBPbeta transcription factor regulates epithelial cell proliferation and differentiation in the mammary gland. *Genes Develop.*, **12**, 1907–16.

Robinson, G.W., McKnight, R.A., Smith, G.H. and Hennighausen, L. (1995) Mammary epithelial cells undergo secretory differentiation in cyclin virgins but require pregnancy for the establishment of terminal differentiation. *Development*, **121**, 2079–90.

Rohrer, G.A., Alexander, L.J. and Beattie, C.W. (1997) Mapping genes located on human chromosomes 2 and 12 to porcine chromosomes 15 and 5. *Anim. Genet.*, **28**, 448–50.

Rosen, J.M. (1987) Milk protein gene structure and expression. In *The Mammary Gland*, (M.C. Neville and C.W. Daniel eds.) Plenum Publishing Corporation, New York, pp. 301–22.

Rosen, J.M., Li, S., Raught, B. and Hadsell, D. (1996) The mammary gland as a bioreactor: factors regulating the efficient expression of milk protein-based transgenes. *Am. J. Clin. Nutr.*, **63**, 627–32.

Rosen, J.M., Mediona, D., Schlein, A.R., Eisenstein, R.S. and Yu-Lee, L.Y. (1988) Hormonal and cell-substratum regulation of casein gene expression at the posttranscriptional level. In *Steroid Hormone Action*, (G. Ringold ed.) Alan R. Liss. Inc., New York, pp. 269–78.

Rosen, J.M., Rodgers, J.R., Couch, C.H., Bisbee, C.A., David-Inouye, Y., Campbell, S.M. and Yu-Lee, L.Y. (1986) Multihormonal regulation of milk protein gene expression. *Ann. New York Acad. Sci.*, **478**, 63–76.

Rosen, J.M., Zahnow, C., Kazansky, A. and Raught, B. (1998) Composite response elements mediate hormonal and developmental regulation of milk protein gene expression. *Biochem. Soc. Symp.*, **63**, 101–13.

Schatz, G. and Dobberstein, B. (1996) Common principles of protein translocation across membranes. *Science*, **271**, 1519–26.

Schmidhausser, C., Casperson, G.F., Myers, C.A., Sanzo, K.T., Bolten, S. and Bissell, M.J. (1992) A novel transcriptional enhancer is involved in the prolactin- and extracellular matrix-dependent regulation of beta-casein gene expression. *Mol. Cell Biol.*, **3**, 699–709.

Schmitt-Ney, M., Doppler, W., Ball, R.K. and Groner, B. (1991) Beta-casein gene promoter activity is regulated by the hormone-mediated relief of transcriptional repression and a mammary-gland-specific nuclear factor. *Mol. Cell. Biol.*, **11**, 3745–55.

Seagroves, T.N., Krnacik, S., Raught, B., Gay, J., Burgess-Beusse, B., Darlington, G.J. and Rosen, J.M. (1998) C/EBPbeta, but not C/EBPalpha is essential for ductal morphogenesis, lobuloalveolar proliferation, and functional differentiation in the mouse mammary gland. *Genes Develop.*, **12**, 1917–28.

Seddiki, T. and Ollivier-Bousquet, M. (1991) Temperature dependence of prolactin endocytosis and casein exocytosis in epithelial mammary cells. *Eur. J. Cell. Biol.*, **55**, 60–70.

Shennan, D.B. (1995) Identification of a high affinity taurine transporter which is not dependent on chloride. *Bioscience Rep.*, **15**, 231–9.

Shennan, D.B. and McNeillie, S.A. (1994a) Characteristics of α-aminoisobutyric acid transport by the lactating rat mammary gland. *J. Dairy Res.*, **61**, 9–19.

Shennan, D.B. and McNeillie, S.A. (1994b) High affinity (Na^+ + Cl^-)-dependent transport by lactating mammary tissue. *J. Dairy Res.*, **61**, 335–43.

Shennan, D.B. and McNeillie, S.A. (1994c) Milk accumulation down regulates amino acid uptake *via* systems A and L by lactating mammary tissue. *Horm. Metab. Res.*, **26**, 611.

Shennan, D.B., Backwell, F.R.C. and Calvert, D.T. (1999) Metabolism of aminoacyl-*p*-nitroanilides by rat mammary tissue. *Biochim. Biophys. Acta*, **1427**, 227–35.

Shennan, D.B., Cliff, M.J. and Hawkins, P. (1996) Volume-sensitive taurine efflux from mammary tissue is not obliged to utilize volume-activated anion channels. *Bioscience Rep.*, **16**, 459–65.

Shennan, D.B., Millar, I.D. and Calvert, D.T. (1997) Mammary-tissue amino acid transport systems. *Proc. Nutr. Soc.*, **56**, 177–91.

Shennan, D.B., Calvert, D.T., Backwell, F.R.C. and Boyd, C.A.R. (1998) Peptide aminonitrogen transport by the lactating rat mammary gland. *Biochim. Biophys. Acta*, **1373**, 252–60.

Shennan, D.B., McNeillie, S.A. and Curran, D.E. (1994) The effect of a hyposmotic shock on amino acid efflux from lactating rat mammary tissue: stimulation of taurine and glycine efflux *via* a pathway distinct from anion exchange and volume-activated anion channels. *Exp. Physiol.*, **79**, 797–808.

Shennan, D.B., McNeillie, S.A., Jamieson, E.A. and Calvert, D.T. (1994) Lysine transport in lactating rat mammary tissue: evidence for an interaction between cationic and neutral amino acids. *Acta Physiol. Scand.*, **151**, 461–6.

Siomi, H. and Dreyfuss, G. (1997) RNA-binding proteins as regulators of gene expression. *Curr. Opin. Genet. Dev.*, **7**, 345–53.

Soulier, S., Lepourry, L., Stinnakre, M.G., Langley, B., L'Huillier, P.J., Paly, J., Djiane, J., Mercier, J.C. and Vilotte, J.L. (1999) Introduction of a proximal Stat5 site in the murine alpha-lactalbumin promoter induces prolactin dependency *in vitro* and improves expression frequency *in vivo*. *Transg. Res.*, **8**, 23–31.

Soulier, S., Mercier, J.C., Vilotte, J.L., Anderson, J., Clark, J. and Provot, C. (1989) Characterization of genomic clones homologous to the alpha-lactalbumin gene in the bovine and ovine species. *Gene*, **83**, 331–8.

Springer, S., Spang, A. and Schekman, R. (1999) A primer on vesicle budding. *Cell* **97**, 145–8.

Starr, R. and Hilton, D.J. (1999) Negative regulation of the JAK/STAT pathway. *Bioessays*, **21**, 47–52.

Steiner, D.F., Smeekens, P.S., Ohagi, S. and Chan, S.J. (1992) The new enzymology of precursor processing endoproteases. *J. Biol. Chem.*, **267**, 23435–8.

Stinnakre, M.G., Soulier, S., Schibler, L., Lepourry, L., Mercier, J.C. and Vilotte, J.L. (1999) Position-independent and copy-number-related expression of a goat bacterial artificial chromosome alpha-lactalbumin gene in transgenic mice. *Biochem. J.*, **339**, 33–6.

Stinnakre, M.G., Vilotte, J.L., Soulier, S. and Mercier, J.C. (1994) Creation and phenotypic analysis of alpha-lactalbumin-deficient mice. *Proc. Natl. Acad. Sci. USA*, **91**, 6544–8.

Stöcklin, E., Wissler, M., Gouilleux, F. and Groner, B. (1996) Functional interactions between Stat5 and the glucocorticoid receptor. *Nature*, **383**, 726–8.

Stöcklin, E., Wissler, M., Morrigl, R. and Groner, B. (1997) Specific DNA binding of Stat5, but not of glucocorticoid receptor, is required for their functional cooperation in the regulation of gene transcription. *Mol. Cell. Biol.*, **17**, 6708–16.

Storck, T., Schulte, S., Hofmann, K. and Stoffel, W. (1992) Structure, expression and functional analysis of a Na^{+}-dependent glutamate aspartate transporter from rat brain. *Proc. Natl. Acad. Sci. USA*, **89**, 10959–65.

Streuli, C.H., Edwards, G.M., Delcommenne, M., Whitelaw, C.B.A., Burdon, T.G., Schindler, C. and Watson, C.J. (1995) Stat5 as a target for regulation by extracellular matrix. *J. Biol. Chem.*, **270**, 21639–44.

Tang, Y. (1993) No alpha-lactalbumin-like activity detected in a low molecular mass protein fraction of rat epididymal extract. *Reprod. Fertil. Dev.*, **5**, 229–37.

Thépot, D., Fontaine, M.L., Houdebine, L.M. and Devinoy, E. (1990) Complete sequence of the rabbit WAP gene. *Nucleic Acids Res.*, **18**, 3641.

Thépot, D., Fontaine, M.L., Houdebine, L.M. and Devinoy, E. (1991) Structure of the gene encoding rabbit beta-casein. *Gene*, **97**, 301–6.

Threadgill, D.W. and Womack, J.E. (1990) Genomic analysis of the major bovine casein genes. *Nucleic Acids Res.*, **18**, 6935–42.

Tomlinson, A.M., Cox, R.D., Lehrach, H.R. and Dalrymple, M.A. (1996) Restriction map of two yeast artificial chromosomes spanning the murine casein locus. *Mammalian Genome*, **7**, 342–44.

Turner, M.D., Handel, S.E., Wilde, C.J. and Burgoyne, R.D. (1993) Differential effect of brefeldin A on phosphorylation of the caseins in lactating mouse mammary epithelial cells. *J. Cell. Sci.*, **106**, 1221–6.

Udy, G.B., Towers, R.P., Snell, R.G., Wilkins, R.J., Park, S.H., Ram, P.A., Waxman, D.J. and Davey, H.W. (1997) Requirement of STAT5b for sexual dimorphism of body growth rates and liver gene expression. *Proc. Natl. Acad. Sci. USA*, **94**, 7239–44.

Valentine, C.R. (1998) The association of nonsense codons with exon skipping. *Mutation Res.*, **411**, 87–117.

Verma, N. and Kansal V.K. (1995) Characterisation and starvation induced regulation of methionine uptake sites in mouse mammary gland. *Ind. J. Exp. Biol.*, **33**, 516–30.

Verma, N. and Kansal, V.K. (1993) Characterisation of the routes of methionine transport in mouse mammary glands. *Ind. J. Med. Res., [B]* **98**, 297–304.

Vilotte, J.L. and Soulier, S. (1992) Isolation and characterization of the mouse alpha-lactalbumin-encoding gene: interspecies comparison, tissue- and stage-specific expression. *Gene*, **119**, 287–92.

Vilotte, J.L., Soulier, S. and Mercier, J.C. (1993) Complete sequence of a bovine alpha-lactalbumin pseudogene: the region homologous to the gene is flanked by two directly repeated LINE sequences. *Genomics*, **16**, 529–32.

Vilotte, J.L., Soulier, S., Mercier, J.C., Gaye, P., Hue-Delahaie, D. and Furet, J.P. (1987) Complete nucleotide sequence of bovine alpha-lactalbumin gene. Comparison with its rat counterpart. *Biochimie*, **69**, 609–20.

Vilotte, J.L., Soulier, S., Printz, C. and Mercier, J.C. (1991) Sequence of the goat alpha-lactalbumin-encoding gene: comparison with the bovine gene and evidence of related sequences in the goat genome. *Gene*, **98**, 271–6.

Vina, J., Puertes, I.R., Saez, G.T. and Vina, J.R. (1981) Role of prolactin in amino acid uptake by the lactating mammary gland of the rats. *FEBS Lett.*, **126**, 250–2.

Warner, B., Janssens, P. and Nicholas, K. (1993) Prolactin-independent induction of alpha-lactalbumin gene expression in mammary gland explants from pregnant Balb/C mice. *Biochem. Biophys. Res. Commun.*, **194**, 987–91.

Watson, C.J., Gordon, K.E., Robertson, M. and Clark, A.J. (1991) Interaction of DNA-binding proteins with a milk protein gene promoter *in vitro*: identification of a mammary gland-specific factor. *Nucleic Acids Res.*, **19**, 6603–10.

Webster, J., Wallace, R.M., Clark, A.J. and Whitelaw, C.B.A (1995) Tissue-specific, temporally regulated expression mediated by the proximal ovine beta-lactoglobulin promoter in transgenic mice. *Cell. Mol. Biol. Res.*, **41**, 11–5.

West, D.W. (1981) Energy-dependent calcium sequestration activity in a Golgi apparatus fraction derived from lactating rat mammary glands. *Biochim. Biophys. Acta*, **673**, 374–86.

West, D.W. and Clegg, R.A. (1984) Casein kinase activity in rat mammary gland Golgi vesicles. Demonstration of latency and requirement for a transmembrane ATP carrier. *Biochem. J.*, **219**, 181–7.

Whitelaw, C.B.A. and Webster, J. (1998) Temporal profiles of appearance of DNase I hypersensitive sites associated with the ovine beta-lactoglobulin gene differ in sheep and transgenic mice. *Mol. Gen. Genet.*, **257**, 649–54.

Whitelaw, C.B.A., Harris, S., McClenaghan, M., Simons, J.P. and Clark, A.J. (1992) Position-independent expression of ovine beta-lactoglobulin in transgenic mice. *Biochem. J.*, **286**, 31–9.

Wieland, FT., Gleason, M.L., Serafini, T.A. and Rothman, J.E. (1987) The rate of bulk flow from the endoplasmic reticulum to the cell surface. *Cell*, **50**, 289–300.

Winklehner-Jennewein, P., Geymayer, S., Lechner, J., Welte, T., Hansson, L., Geley, S. and Doppler, W. (1998) A distal enhancer region in the human beta-casein gene mediates the response to prolactin and glucocorticoid hormones. *Gene*, **217**, 127–39.

Witsell, D.L., Casey, C.E. and Neville, M.C. (1990) Divalent cation activation of galactosyl transferase in native mammary Golgi vesicles. *J. Biol. Chem.*, **265**, 15731–7.

Wolberger, C. (1998) Combinatorial transcription factors. *Current Opinion Genet. Dev.*, **5**, 552–9.

Yoshimura, M. and Oka, T. (1989) Isolation and structural analysis of the mouse beta-casein gene. *Gene*, **78**, 267–75.

Yu-Lee, L.Y., Richter-Mann, L., Couch, C.H., Stewart, A.F., MacKinlay, A.G. and Rosen, J.M. (1986) Evolution of the casein multigene family conserved sequences in the 5' flanking and exon regions. *Nucleic Acids Res.*, **14**, 1883–902.

16

GENETIC POLYMORPHISM OF MILK PROTEINS

K.F. Ng-Kwai-Hang and F. Grosclaude

16.1 Introduction

Extensive studies on the qualitative and quantitative aspects of milk proteins in more than 100 mammalian species have demonstrated that the protein content varies from 1 to 20% between different species and within the same species of different genetic backgrounds under different environmental conditions. The milk of all species so far analysed contains an acid-precipitable fraction, commonly known as casein, and an acid-soluble fraction known as the whey protein or milk serum protein. Gel electrophoretic techniques have been used to reveal the identity of several types of caseins and whey proteins and to establish the presence of homologous proteins across several species. The discovery of two electrophoretically distinct forms of β-lactoglobulin by Aschaffenburg and Drewry (1955) initiated very active research in the field of genetic polymorphism of milk proteins in several countries around the world. Research activity from different teams has contributed significantly to knowledge on the biochemistry, molecular and population genetics, properties of milk proteins and their associations with production traits.

The purpose of this chapter is to present an overall picture of our present knowledge on the genetic polymorphisms of milk proteins in the following areas: methods of identification, number of variants already identified, differences in amino acid sequence, genetic basis, geographical and breed distribution, relationships with production traits, milk composition and milk quality. For obvious economic reasons, most of the research has been directed towards the bovine and, to a lesser extent, to the caprine species. Hence, most of the information given in this chapter pertains to milk of those two species.

Advanced Dairy Chemistry Volume 1: Proteins, 3rd edn.
Edited by P.F. Fox and P.L.H. McSweeney, Kluwer Academic/Plenum Publishers, 2003.

Since this chapter by the same authors (Ng-Kwai-Hang and Grosclaude, 1992) was published ten years ago, there has been an accumulation of research on various aspects of the genetic polymorphism of milk proteins. A survey of the literature since 1992 revealed in excess of 1000 publications on the topic. At a conference on milk protein variants organized by Forkingsparken in Hanko, Norway (Agricultural University of Norway, 1992), abstracts of the 21 presentations were made available to the participants. The first seminar on genetic polymorphism of milk proteins, organised by the International Dairy Federation, was held at Zurich, Switzerland in 1995 and had 32 communications which appeared as abstracts in IDF Bulletin No. 304. This was followed by a second seminar at Palmerston North, New Zealand, in 1997, at which the number of presentations increased to 71 and the full versions of those were published in a Special Issue of the International Dairy Federation (1997). There have been several comprehensive reviews on the subject (Lin *et al.*, 1992; Jakob and Puhan, 1992; Jakob, 1994; Ng-Kwai-Hang, 1998). The authors will attempt to update the general information since their last review (Ng-Kwai-Hang and Grosclaude, 1992) and suggest the reading of the above-mentioned documents for more in-depth details about specific topics.

16.2 Methods of detecting genetic polymorphism

16.2.1 Electophoretic techniques

The occurrence of genetic polymorphism among milk proteins was first reported by Aschaffenburg and Drewry (1955) who observed that when milk samples from individual cows were subjected to electrophoresis on filter paper in barbitone buffer, pH 8.6, and under 230 V for 16 h, these samples produced either one of two or a mixture of both electrophoretically distinct β-lactoglobulin bands which were denoted as β_1- and β_2-lactoglobulin in order of decreasing mobility. This nomenclature was later replaced by A and B, respectively, when it was discovered (Aschaffenburg and Drewry, 1957) that the synthesis of these two types of β-lactoglobulin was under genetic control, determined by two allelic autosomal genes. The discovery of Aschaffenburg and Drewry was followed by a very active search for the occurrence of genetic polymorphisms in the different caseins and whey proteins. Electrophoretic methods for the detection of polymorphs by far exceed any other method used because of their simplicity and ease of application for large numbers of samples. The original paper electrophoretic procedure has been replaced by several other electrophoretic methods which have been developed, modified or adapted in order to improve the resolving power for the purpose of phenotyping of different milk proteins from various species of animal.

The paper electrophoretic technique of Aschaffenburg and Drewry (1955) was applied by Blumberg and Tombs (1958) for the detection of genetic variants of α-lactalbumin in Nigerian White Fulani cattle. Paper electrophoresis of milk proteins was limited to the serum proteins because of the difficulties encountered in the separation of the native caseins which tend to associate and form micelles. However, by the inclusion of 6.0 M urea as dissociating agent in a citrate-phosphate buffer, pH 7.5, Aschaffenburg (1961) was able to overcome this limitation and demonstrated the presence of three polymorphs, labelled A, B and C in order of decreasing mobility, of β-casein in the milk of Jersey and Guernsey cows.

Thompson *et al.* (1962) introduced the use of urea starch gel electrophoresis at pH 8.6 to resolve three α_{s1}-casein variants, known as A, B and C, in Holstein, Brown Swiss and Ayrshire breeds. The resolution of β-lactoglobulin A, B and C was achieved by Bell (1962) through the use of a borate buffer, pH 8.5, without urea in a starch gel electrophoresis system. The method of Wake and Baldwin (1961) was modified (Grosclaude *et al.*, 1966a) by the inclusion of 2-mercaptoethanol in the gel for the separation of the A, B and D variants of β-lactoglobulin. The inclusion of small quantities of 2-mercaptoethanol in the sample buffer or electrophoretic gel contributed to the characterisation of two bands corresponding to κ-casein by three independent research groups (Neelin, 1964; Schmidt, 1964; Woychik, 1964). Genetic studies by Grosclaude *et al.* (1965) demonstrated that these electrophoretic bands represented two genetic variants of κ-casein. Starch gel electrophorphesis was used by Thompson *et al.* (1964) and Michalak (1967) to detect variants A, B and C of β-casein. Evidence of genetic polymorphism in α_{s2}-casein was first demonstrated by Grosclaude *et al.* (1976a) in a study involving milk from Nepalese yak, yak x zebu crosses, and bovine x yak crosses. Separation of three α_{s2}-casein variants was achieved by electrophoresis in starch gel slabs containing 2-mercaptoethanol and 7 M urea under alkaline conditions. Several methods have been applied for the phenotyping of genetic polymorphism of β-lactoglobulin, α_{s1}-casein, and β-casein under alkaline conditions in polyacrylamide gels (Aschaffenburg, 1964; Thompson *et al.*, 1964; Aschaffenburg and Thymann, 1965). The fastest single moving band of variant A of β-casein, which was not resolved under several alkaline conditions, was found later by Peterson and Kopfler (1966) to consist of three variants, now known as A^1, A^2 and A^3, which could be separated by electrophoresis at pH 2.8 in polyacrylamide gels containing 4.5 M urea. Resolution of these three distinct types of β-casein A was confirmed later by Grosclaude *et al.* (1966b) and Arave (1967) using starch gel electrophoresis at low pH. The use of polyacrylamide gels as the supporting electrophoretic medium instead of starch gels has some advantages in that the former are easier to prepare and manipulate because they can be formed without heat treatment and can be stained without slicing.

Other supporting media for electrophoresis include agar gels which were used by Aschaffenburg (1965) for the phenotyping of four polymorphs of β-lactoglobulin at pH 8.6 in veronal buffer. Bell and Stone (1978) obtained good resolution of β-lactoglobulin A and B by electrophoresis in cellulose acetate strips under alkaline conditions. The same support, in combination with tris-glycine buffer at pH 9.0 and containing 6.0 M urea in the presence of 2-mercaptoethanol, was used by Davoli (1981) for the phenotyping of variants of α_{s1}-casein, κ-casein and α-lactalbumin. Because β-lactoglobulin B had the same mobility as β-casein A, the author recommended separate runs for caseins and whey proteins. Cellulose acetate electrophoresis requires 4 h for separation and hence is much faster than the traditional starch gel electrophoresis, which needs 20–22 h.

The application of different zone electrophoretic techniques for the separation of the major milk proteins has contributed enormously to the study of genetic polymorphism in this particular group of proteins. Thompson (1970) presented a comprehensive review of different electrophoretic techniques that could be used for the phenotyping of milk proteins. It has to be recognised that the above listing of different methods used for identification of genetic variants of α_{s1}-casein, α_{s2}-casein, β-casein, κ-casein, β-lactoglobulin and α-lactalbumin is far from being complete. A search of the literature will reveal a bewildering array of electrophoretic procedures that have been modified for the phenotyping of the major milk proteins in the different breeds of cattle and in different species of mammal. A more complete and thorough survey of the different milk protein polymorphs that have been identified will be addressed in Section 16.3 of this chapter. For a new researcher in the field of milk protein polymorphism, the problem is not so much finding a suitable method for phenotyping but rather deciding which method to choose from the hundreds that have been published. In spite of this difficulty, for practical purposes, the choice can be narrowed down to either the use of starch gel electrophoresis, polyacrylamide gel electrophoresis or isoelectric focusing (see Section 16.2.2). The selection between these two gel supporting media depends on the availability of facilities because both types of electrophoresis have been developed along parallel lines and provide good resolution of the caseins when used in gel buffers containing urea and 2-mercaptoethanol at alkaline pH. Better separation of β-lactoglobulin and α-lactalbumin polymorphs is achieved in the absence of urea and 2-mercaptoethanol under alkaline conditions. Because alkaline gels cannot be used to resolve β-casein A into its components, A^1, A^2 and A^3, another electrophoretic run is required for their resolution. Three separate electrophoretic runs under different sets of conditions would be needed in order to establish the whole spectrum of genetic polymorphism for α_{s1}-casein, α_{s2}-casein, β-casein, κ-casein, β-lactoglobulin and α-lactalbumin.

More recently, capillary electrophoresis has been developed as an alternative technique for the analysis of milk proteins (Chen and Zang,

1992; de Jong *et al.*, 1993; Otte *et al.*, 1994; Fairise and Cayot, 1998). Recio *et al.* (1997a) reviewed the methods available for analysis of caseins, whey proteins, peptides and genetic variants by capillary electrophoresis. A procedure for the discrimination of variants A, B and C of β-lactoglobulin was described by Paterson *et al.* (1995). van Riel and Olieman (1995) were not able to separate the A and B variants of κ-casein using a hydrophobically-coated capillary in combination with 6 M urea in a citrate buffer. Genetic variants A and D of α_{s2}-casein, A^1, A^2, A^3, B and C of β-casein and C of α_{s1}-casein in bovine milk were characterised by capillary electrophoresis (Recio *et al.*, 1997b). This method was also used to resolve the genetic variants of caseins in caprine (Recio *et al.*, 1997c) and ovine (Recio *et al.*, 1997d) milk.

16.2.2 Isoelectric focusing techniques

Isoelectric focusing can be considered as a special type of gel electrophoresis whereby each protein migrates to a specific point across a pH gradient, i.e., its isoelectric point. Longer running times are required for isoelectric focusing than for polyacrylamide gel electrophoresis because the heating effect of the current interferes with the formation of a pH gradient in the gel and the mobility of the protein decreases as it approaches its isoelectric point. Whole casein, when subjected to electrophoresis across a pH gradient from 3.0 to 10.0 in 5% polyacrylamide containing urea, is resolved into as many as 20 bands. This separation method is capable of resolving proteins with isoelectric points differing by only 0.02 pH unit (Peterson, 1969). This technique could separate variants A^1, A^2, A^3, B and C of β-casein and Peterson (1969) claimed its usefulness for the separation of β-lactoglobulin and α-lactalbumin variants. Josephson (1972) reported the successful resolution of κ-casein A and B through the use of pH gradient electrophoresis of bovine milk caseins. Isoelectric focusing in the pH range 3.5 to 5.0 in 5% polyacrylamide gels containing 6 M urea was used by Pearce and Zadow (1978) to separate the A and B variants of β-lactoglobulin. A procedure for simultaneously phenotyping the four major caseins by isoelectric focusing was described by Addeo *et al.* (1983). Later, the same group (Di Luccia *et al.*, 1988) reported that κ-casein C had an isoelectric pH between the A and B variants. The use of an agarose gel containing 7 M urea as support medium in an isoelectric system with a pH gradient from 2.5 to 7.0 could also give good results for the phenotyping of α_{s1}-casein, β-casein, κ-casein and β-lactoglobulin (Bech and Munk, 1988). Seibert *et al.* (1985) described a rapid method for the simultaneous phenotyping of these four proteins by isoelectric focusing in ultra-thin (100 μm thick) 4.5% polyacrylamide gels containing 8 M urea and 2-mercaptoethanol. Using this method, Seibert *et al.* (1987) detected a new variant of κ-casein in the milk of German

Simmental cows, known at that time as variant D, which had the same mobility as the A variant in polyacrylamide or starch gel electrophoresis. The isoelectric point of variant D is 5.76 compared to 5.62 for A and 5.83 for B. Later, when it was found that variants D and C of κ-casein were the same, the former designation was dropped. Erhardt (1989a) reported on another newly discovered type of κ-casein, denoted as the E variant. Application of isoelectric focusing could also differentiate the existence of three β-lactoglobulin variants in sheep (Erhardt *et al.*, 1989).

The use of isoelectric focusing for the phenotyping of milk proteins can be considered as an alternative to starch gel or polyacrylamide gel electrophoresis. This method seems attractive due to the fact that a single run is adequate for phenotyping the seven known variants of β-casein. The application of isoelectric focusing procedures for the phenotyping of milk proteins has revealed the existence of additional genetic variants of α_{s1}-casein, α_{s2}-casein, κ-casein and β-lactoglobulin (Krause *et al.*, 1988; Erhardt 1989a, 1993a,b, 1996; Erhardt *et al.*, 1998) which were not detected by electrophoretic methods.

16.2.3 Column chromatographic techniques

During the past four decades, many chromatographic techniques have been developed and adapted for the study of milk proteins. Yaguchi and Rose (1971) prepared an extensive review covering the application of column chromatographic techniques for the separation and purification of milk proteins. The most commonly used stationary phase in those methods is based on the principle of either ion exchange or size exclusion chromatography. For the purpose of phenotyping milk proteins, size exclusion chromatography is of limited value because the differences in molecular mass between the several variants of a specific protein are not sufficient to result in differences in the elution profile. Although ion exchange chromatography may have some potential application as a method for detecting genetic polymorphism of milk proteins, its use would be impractical due to the fact that the process of equilibrating the column, doing the actual chromatographic separation and regenerating the column for the next sample may take several days. Another limitation pertains to the lack of discrimination between genetic variants. Nevertheless, it is interesting to note the possibility of separating the β-casein variant A from C in a DEAE-cellulose system containing urea (Thompson and Pepper, 1964) and by ion exchange chromatography on QAE-cellulose (Ng-Kwai-Hang and Pélissier, 1989). Chromatography on a DEAE-cellulose column using imidazole-HCl-urea at pH 7.0 (Thompson, 1966) could resolve variants A and B of κ-casein.

With the development of high performance liquid chromatography it became possible to use chromatographic columns of considerably smaller dimensions, for example, 50 mm long and 5 mm in diameter and hence significantly reduced the analysis time and also improved resolution of milk proteins. The HPLC system described by Pearce (1983) was able to resolve α-lactalbumin and variants A and B of β-lactoglobulin within 30 min. Similarly, Humphrey and Newsome (1984) and Andrews *et al.* (1985) reported on the resolution of β-lactoglobulin A and B by HPLC on an anion exchange column. Genetic variants A and B of κ-casein have been separated satisfactorily by HPLC (Dalgleish, 1986; Visser *et al.*, 1986; Guillou *et al.*, 1987; Ng-Kwai-Hang and Dong, 1994). The technique used by Guillou *et al.* (1987) could resolve neither variants B and C of α_{s1}-casein, nor variants A^1, A^2 and B of β-casein. Later, these three variants of β-casein were separated on a cation exchange resin (Hollar et al, 1991). Visser *et al.* (1991) described an HPLC method for the separation of some genetic variants of bovine caseins. In the process of fractionating bovine caseins by HPLC on a reversed phase column, Carles (1986) discovered a new variant of β-casein which is more hydrophobic than β-casein A^1 by having a leucine instead of a proline in the 114–169 region of β-casein. This newly-found variant is an example of the so-called 'silent variants' because the substitution of an amino acid in this case would not result in a change in the net electrical charge of the protein and hence such variants would not be resolved by electrophoresis or ion exchange chromatography. With this as background information, the search for other polymorphs of milk proteins by HPLC techniques opens up a new area of research as discussed in the next section.

16.2.4 HPLC techniques for silent variants

Until now, the detection of most genetic polymorphs of milk proteins has been achieved through various electrophoretic techniques and some of the identified variants have been ascertained by chromatographic methods. These methods are restricted to the resolution of proteins differing in net charge caused by the substitution of one amino acid for another. In theory, it has been estimated that mutations in proteins due to amino acid substitutions that do not lead to a change in the net charge on proteins should occur three times more frequently than those resulting in an alteration in net charge. The search for methods other than electrophoresis for the identification of further genetic variants of the major milk proteins should open up a new area in the biochemistry of milk proteins.

Considering that the total number of amino acid residues varies from 123 in α-lactalbumin to 209 in β-casein, changes in the net hydrophobic nature of the whole protein due to mutation caused by the substitution of one amino acid for another may be too small to detect by the reversed phase HPLC technique described by Carles (1986). Hydrolysis of the protein followed by peptide mapping, which has been used successfully to detect mutants in hemoglobin (Schroeder *et al.*, 1982), could be adapted for identification of the silent variants of milk proteins. Basically, this method involves digesting the purified protein of interest with trypsin. The resulting mixture of peptides is then subjected to HPLC on a reversed phase column and its elution profile compared with that of a standard protein of known primary structure. The eluent fraction corresponding to a peak with a different retention time from the standard elution profile is collected for further analysis. Often, mass spectrometric and amino acid composition analysis of the aberrant peptide is sufficient to determine the type and position of the substitution in the variant. In certain cases, amino acid sequencing is necessary if the aberrant peptide contains more than one residue of the amino acid involved in the mutation. Variants F (Visser *et al.*, 1995) and G (Dong and Ng-Kwai-Hang, 1998) of β-casein are examples of variants detected by methods other than electrophoresis. Both were identified as β-casein A^1 by electrophoresis and involved a proline to leucine substitution at positions 152 and 137, respectively. Should several genetic variants be discovered by application of this technique, then one will be faced with the problem of classification and nomenclature of the different milk proteins which so far has been based mainly on electrophoretic mobility. The feasibility of phenotyping milk proteins by peptide mapping will depend on the availability of equipment and facilities. This method is more time-consuming and considerably more involved than electrophoresis and hence would not be as practical as the latter method for large-scale routine screening of genetic polymorphism of milk proteins in animal populations.

16.2.5 Genetic variants at the DNA level

Because milk production is a sex-limited trait, determination of the genetic polymorphism in milk proteins has to await the initiation of lactation in the female of the species. The possibility of phenotyping female animals prior to lactation, at birth for example, and of phenotyping male animals for specific milk protein genes would have obvious profound advantages. With the development in molecular biology, several techniques applicable for the identification of variants of milk proteins at the DNA level are now available.

Changes in the amino acid sequence of different genetic variants are the result of nucleotide sequence changes in the codon which could be detected by several molecular genetic techniques. Levéziel *et al.* (1988) were the first to adapt Southern restriction fragment length polymorphism to differentiate the A from the B allele of κ-casein. This was followed by the use of other endonucleases (Rando *et al.*, 1988; Rogne *et al.*, 1989) to identify the two common variants of κ-casein. Techniques using the same principle have been applied for the genotyping of other major milk proteins (Rando *et al.*, 1990; Threadgill and Womack, 1990; Osta *et al.*, 1995a,b). The improvement in techniques for the amplification of gene segments by the polymerase chain reaction has facilitated the task of genotyping for milk proteins. The amplified portion of the gene which includes the nucleotide substitutions characteristic of the variant is digested with specific endonucleases and the digestion products are then resolved on an agarose gel. This procedure has been applied for the genotyping of κ-casein variants (Denicourt *et al.*, 1990; Zadworny *et al.*, 1990; Schlieben *et al.*, 1991; Schlee and Rottman, 1992; Nebola *et al.*, 1996; Prinzenberg *et al.*, 1996), β-lactoglobulin (Medrano and Aguilar-Cordova, 1990; Savva *et al.*, 1994; Chung *et al.*, 1994), α_{s1}-casein (Prinzenberg and Erhardt, 1994; Koczan *et al.*, 1993), α-lactalbumin (Schlee and Rottman, 1992; Mao, 1994) and β-casein (Schlee and Rottman, 1992; Velmala *et al.*, 1994). The single-strand conformation polymorphism technique described by Orita *et al.* (1989) has been adapted for the detection of the A, B, C and E variants of κ-casein (Barroso *et al.*, 1997, 1998). This method was used for the typing of A and B variants of β-lactoglobulin (Kaminski and Zabolewicz, 1997) and the D variant of α_{s1}-casein (Prinzenberg and Erhardt, 1997). Methods requiring analysis of restriction fragment length polymorphism rely on the recognition of mutation sites by the restriction enzymes. In cases where mutations do not lead to recognisable enzyme restriction sites, other procedures have been developed for the genotyping of milk proteins. The technique of amplification-created site has been applied to detect β-casein variants by Lien *et al.* (1992) and α_{s1}-casein A by Wilkins and Xie (1997b). Methods based on allele-specific polymerase chain reaction have been used to identify of β-casein B and C, β-lactoglobulin A and B (Schlee and Rottman, 1992), κ-casein A and B (Medrano and Aguilar-Cordova, 1990; Groot *et al.*, 1995) and α_{s1}-casein B and C (David and Deutch, 1992). Bidirectional allele-specific polymerase chain reaction (Damiani *et al.*, 1992) was used for the genotyping of β-casein.

In addition to alleles in the milk protein genes which result in genetic polymorphism of milk proteins, several other alleles in the non-coding regions of the genes have been identified. These variants at the DNA level have to be differentiated from those which are translated so as not to cause confusion in the nomenclature. The A and B alleles of the α-lactalbumin gene described by Voelker *et al.* (1997) are due to a nucleotide substitution in the 5′-flanking region of the gene and should not be confused with the already known A and B alleles at the protein level which are characterised by a Gln–Arg substitution at position 10. Hence, the "A and B" designation by Voelker *et al.* (1997) would not be appropriate. Polymorphism in the non-coding regions of the genes may have major influences on promoter sites, regulatory sites, binding with transcription factors, stability of m-RNA and overall gene expression. Geldermann *et al.* (1996) and Schild and Geldermann (1996) reported several variable sites located at approximately

1 kb in the 5′-flanking regions of the casein and β-lactoglobulin genes. The 5′-flanking regions of β-casein (Bleck *et al.*, 1996), κ-casein (Schild *et al.*, 1994; Kaminski, 1996), β-lactoglobulin (Wagner *et al.*, 1994) and α-lactalbumin (Osta *et al.*, 1995b; Voelker *et al.*, 1997) have also been shown to be highly polymorphic. Levéziel *et al.* (1994) have identified 6 alleles with different numbers of dinucleotide microsatellite repeats within the third intron of κ-casein and those are different from changes in nucleotide sequences.

The "two distinct gene mutations – one milk protein polymorphism" phenomenon reported by Wilkins and Xie (1997a) represents an interesting aspect in the field of genetic polymorphism of milk proteins. α_{s1}-Casein A has a deletion of 13 amino acids as a result of skipping exon 4 which may arise as a consequence of either a nucleotide deletion at position 4 or a nucleotide substitution at position 6 of intron 4. The insertion of 371 bp between nucleotides 58 and 59 in the 19th non-coding exon of α_{s1}-casein gives rise to the G variant which has an amino acid sequences similar to α_{s1}-casein B (Rando *et al.*, 1998). A similar situation is encountered in the case of caprine α_{s1}-casein E (see Section 16.5.1). This mutation at the DNA level does not alter the primary sequence of the protein, but has a dramatic effect on the expression of the gene. Variant G of α_{s1}-casein is associated with a reduction of approximately 40% in the level of α_{s1}-casein in the milk.

16.3 Occurrence of genetic polymorphism in bovine species

Most of the work on the genetic polymorphism of milk proteins has been performed on cattle. The different genetic variants of the major milk proteins so far identified are represented schematically in Figure 16.1. At present, seven genetic variants of α_{s1}-casein, designated as A, D, B, G, C, F and E in order of decreasing electrophoretic mobility in alkaline gels containing urea and 2-mercaptoethanol, have been reported. Variants A, B and C were discovered by Thompson *et al.* (1962) while variant D was first reported in French Flamande cattle (Grosclaude *et al.*, 1966a). Ten years elapsed before the discovery of the fifth variant, F, in Nepalese Yak's milk (Grosclaude *et al.*, 1976b). Evidence of a new variant of α_{s1}-casein with a slower mobility than the C variant was reported in milk from Bali cattle, *Bos javanicus* (Bell *et al.*, 1981c). Before assigning another letter to this particular variant, more information is required regarding its primary structure; this unique variant might be identical to either the D or E variant which have been reported previously. Variant F was found at a frequency of 0.009 among 375 German Black and White cattle (Erhardt, 1993b). From DNA studies, Rando *et al.* (1998) described variant G which is identical to

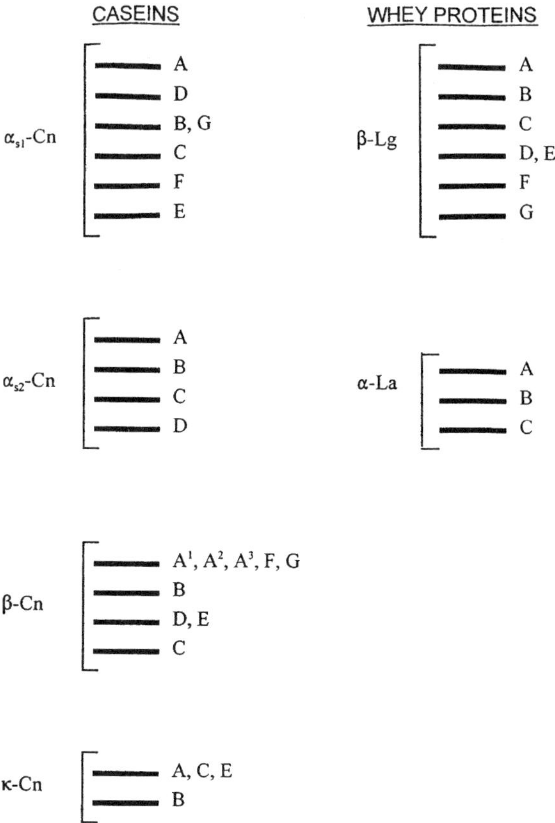

Figure 16.1 Schematic representation of zone electrophoresis at alkaline pH of major milk proteins to show the relative migrations of genetic variants. The positions shown for the different proteins are approximate and are only for illustration purposes. Migration of other variants described in the text (β-casein A^1, B^2, A^4, $A^3_{Mongolie}$; κ-casein F, G, H, I, J which are differentiated by isoelectric focussing; β-lactoglobulin Dr, H, I, J, W, X) are not shown.

the primary structure of α_{s1}-casein B. The latest disclosed variant is H, present in the African Kuri cattle in the Chad region (Mahé *et al.*, 1999).

The first report on genetic polymorphism of α_{s2}-casein was by Grosclaude *et al.* (1976a) who used the designation A for the already known variant common in European breeds and assigned the letter B to a newly discovered variant in cattle (*Bos taurus*) and Zebu (*Bos indicus*), and the letter C to another variant observed in yaks (*Bos grunniens*) from the Nepalese valley. Among the European breeds of *Bos taurus* in France, a fourth variant, D, was later identified in the Vosgienne and Montbéliarde breeds (Grosclaude *et al.*, 1979).

In search for genetic polymorphism of milk proteins in several dairy cattle breeds, Aschaffenburg (1961) discovered three variants, denoted as A, B and C, of β-casein. The A variant could be resolved into A^1, A^2 and A^3 by gel electrophoresis under acidic conditions (Peterson and Kopfler, 1966). Evidence for a rare variant, designated D, in the Deshis breed of India and the Borans of Kenya was reported by Aschaffenburg *et al.* (1968). A seventh variant, E, of β-casein was discovered by Voglino (1972) in the Piedmont breed of Italy. In alkaline gels containing urea, the decreasing order of electrophoretic mobility for the seven genetic variants of β-casein is: $A^1 = A^2 = A^3 > B > D, E > C$ (Kiddy, 1975). At least four additional variants of rare occurrence have been described. Abe *et al.* (1975) detected β-casein A^1, which had a significantly lower mobility than A^1, A^2 or A^3 at pH 1.7, among Japanese Brown cows. In the same year, the presence of variant B^2 in two cows of an unidentified breed was reported in New Zealand (Creamer and Richardson, 1975). A new variant of β-casein, provisionally denoted as A^4, was observed in Bali (Bell *et al.*, 1981a) and Korean (Han *et al.*, 1983, 1984) cattle. The identification of a new variant among Mongolian cattle led to the designation of β-casein $A^3_{Mongolie}$ (Grosclaude *et al.*, 1982). This variant has the same electrophoretic mobility as A^3 but different amino acid substitutions. Variants F (Visser *et al.*, 1995) and G (Dong and Ng-Kwai-Hang, 1998), which were not resolved from A^1 by electrophoretic procedures, could be characterised by peptide mapping followed by amino acid sequencing.

The inclusion of 2-mercaptoethanol in urea-containing gels disclosed the presence of genetic polymorphism in κ-casein (Grosclaude *et al.*, 1965). Each of the two variants, A and B, of κ-casein showed multiple bands when subjected to alkaline electrophoresis due to different degrees of glycosylation (Pujolle *et al.*, 1966; Doi *et al.*, 1979). κ-Casein has long been considered as diallelic, consisting of variants A and B. A third variant C has been described by Di Stasio and Merlin (1979) and two additional variants, D and E, were detected by Seibert *et al.* (1987) and Erhardt (1989a) using isoelectric focusing. Results of amino acid sequencing of variants C, D and E revealed that variants C and D were the same (Erhardt, G., pers. comm., 1989). Phenotyping by isoelectric focusing led to the discovery of variants F among Finnish Ayrshire (Ikonen *et al.*, 1996) and G among Pinzgauer cattle from Austria and Bavaria (Erhardt, 1996). However, the designations F and G had already been assigned (Sulimova *et al.*, 1993, 1996). To avoid any confusion, it is proposed that one of the F's and G's be renamed. While awaiting a change in nomenclature, the two types of F and G will be provisionally referred to as F^I (Ikonen), F^s (Sulimova), G^E (Erhardt) and G^s (Sulimova), respectively. The differences in the primary structure of the four variants will be discussed in Section 16.4.4. In the process of screening cattle from different European breeds and *Bos taurus* × *Bos indicus* crosses from a commercial herd in Namibia, Prinzenberg and Erhardt (1998) reported on the existence of variants H and I of κ-casein. Mahé *et al.* (1999) described the presence of variant J in the Baoulé cattle from Ivory Coast.

Genetic polymorphism of milk proteins was first recognised after the discovery of β-lactoglobulin A and B by Aschaffenburg and Drewry (1955). The presence of variant C (Bell, 1962) was later detected in the milk of Jersey cows. A fourth variant, D, of β-lactoglobulin was identified in the French Montbéliarde breed (Grosclaude *et al.*, 1966a). Another variant of β-lactoglobulin, designated Dr, unique to the Droughtmaster cattle, was described by Bell *et al.* (1970a). The Dr variant is identical to the A variant except for the presence of a covalently-bound carbohydrate moiety. This glycosylated form of β-lactoglobulin is not a true genetic variant but rather the product of post-translational modification. However, Bell *et al.* (1981a) indicated that β-lactoglobulin Dr differs from the A variant by the substitution of asparagine for aspartic acid at position 28, the site of attachment of the carbohydrate moiety. If this point mutation could be confirmed, then β-lactoglobulin Dr could be considered as a true genetic variant. Yak milk has been shown to contain only one type of β-lactoglobulin, D_{yak} (Grosclaude *et al.*, 1976a, 1982), which has the same electrophoretic mobility as bovine β-lactoglobulin D. To avoid the use of notation referring to species in milk protein nomenclature, Eigel *et al.* (1984) suggested that the variant in yak be labelled as E due to its order of discovery. There is strong evidence for the presence of two additional variants, F and G, among Banteng cattle (Bell *et al.*, 1981b). Variant H of β-lactoglobulin was characterised by Conti *et al.* (1988) using isoelectric focusing in immobilised pH gradients. Variant I has been identified in Polish Red (Erhardt *et al.*, 1998) and variant J in Hungarian Grey (Baranyi *et al.*, 1997) cows. Reference is made (Buchberger, 1995) to variant W found in the Murnau-Werdenfelser breed and to variant X found in Hungarian Grey cattle. It should be noted that variant W was found in the milk of only one cow and is not a polymorphism in the usual sense of the term.

Although Aschaffenburg and Drewry (1957) suspected the possible occurrence of genetic polymorphism in α-lactalbumin, they were unable to confirm it because all the individual milk samples analysed at that time contained only variant B. It was Blumberg and Tombs (1958) who discovered the presence of α-lactalbumin A among the Zebu cattle in Nigeria. A third variant of α-lactalbumin with slower mobility than α-lactalbumin B has been provisionally designated as C, pending amino acid sequencing (Bell *et al.*, 1981a). This latest variant of α-lactalbumin was identified in Bali cattle.

16.4 Molecular basis for genetic polymorphism in bovine species

Genetic polymorphism in the milk proteins is due to either substitutions of amino acids or deletion of a certain amino acid sequence along the peptide chain as a consequence of mutations causing changes in the sequence of base pairs of the DNA molecule which constitute the protein gene. Although various investigators have reported on the occurrence of polymorphs of the major milk proteins, distinction must be made between genetic polymorphism, which is due to a mutation resulting in a change in the amino acid sequence of the protein and polymorphism due to post-transcriptional modification such as different degrees of phosphorylation and/or glycosylation of the protein. For the 45 genetic variants identified in six types of bovine milk protein, mutations leading to the deletion of amino acids occur in only three cases (α_{s1}-casein A and H, α_{s2}-casein D) whereas the more frequent mutations leading to amino acid substitutions occur for the other genetic variants. Determination of the primary structure of a protein is a prerequisite for pinpointing the exact location where mutation has occurred and thus resulting in genetic polymorphism. The first milk protein to be totally sequenced was the relatively small α-lactalbumin which contains 123 amino acids (Brew *et al.*, 1970). The complete amino acid sequence of the second whey protein, β-lactoglobulin, with 162 amino acids, is now known (Braunitzer and Chen, 1972; Braunitzer *et al.*, 1972) after several attempts at partial sequencing (Frank and Braunitzer, 1967; McKenzie, 1971). Determination of the amino acid sequences of different variants in the family of caseins constitutes one of the major contributions to milk protein biochemistry by the research group at Jouy-en-Josas, France. A summary of the differences in amino acid sequences giving rise to genetic variants of the milk protein is presented in Table 16.1. Only the positions of the altered amino acids for a particular genetic variant are shown. For a more complete sequence of each of the milk proteins, the reader is referred to either the original publications, the last review on milk proteins nomenclature (Eigel *et al.*, 1984) or appropriate chapters in this book.

16.4.1 Bovine α_{s1}-casein

α_{s1}-Casein was resolved into two distinct electrophoretic bands which were characterised as α_{s1}-casein for the major component and α_{s0}-casein for the faster moving minor component. Determination of the primary sequence of the two components of α_{s1}-casein (Mercier *et al.*, 1971; Grosclaude *et al.*, 1973a) revealed identical amino acid sequences. α_{s0}-Casein has a faster electrophoretic mobility than α_{s1}-casein due to the presence of an additional phosphate group attached at the serine residue in position 41 (Manson *et al.*, 1976, 1977). Post-translational modification giving rise to a component with a different electrophoretic mobility could not be considered as genetic variation because the amino acid sequence specified by the gene is identical for both components. From the primary structure of α_{s1}-casein B, containing 199 amino acid residues, which was

Table 16.1
Positions and amino acid differences in genetic variants of milk proteins

Protein[a]	*Variant*	*Position and amino acid in the protein*									
		14–26			53	51–58	59	66	192		
α_{s1}-Cn	A	Deleted							Glu		
(199)	B								Glu		
	C				*Ala*		*Gln*	*SerP*	*Gly*		
	D				ThrP				Glu		
	E						Lys		Gly		
	F							Leu	Glu		
	G								Glu		
	H					Deleted			Glu		
		33	47			51–59		130			
α_{s2}-Cn	A	*Glu*	*Ala*					*Thr*			
(207)	B	complete sequence not yet determined									
	C		Gly	Thr				Ile			
	D					Deleted					
			18	35	36	37	67	106	122	137	152
β-Cn	A^1						His				
(209)	A^2		*SerP*	*SerP*	*Glu*	*Glu*	*Pro*	*His*	*Ser*	*Pro*	*Pro*
	A^3							Gln			
	B						His		Arg		
	C			Ser		Lys	His				
	D		Lys								
	E				Lys						
	F						His				Leu
	G						His			Leu	

Table 16.1
(Continued)

Protein[a]	*Variant*	*Position and amino acid in the protein*									
			10	97	104	135	136		148	155	
κ-Cn	A		*Arg*	*Arg*	*Ser*	*Thr*	*Thr*		*Asp*	*Ser*	
(169)	B						Ile		Ala		
	C			His							
	E									Gly	
	F^s								Val		
	F^I		His				Ile		Ala		
	G^s								Ala		
	G^E			Cys			Ile		Ala		
	H					Ile					
	I				Ala						
	J									Arg	
		45	50	59	64	70	78	108	118	126 129	158
β-Lg	A				Asp				Val		
(162)	B	Glu	*Pro*	*Gln*	*Gly*	*Lys*	*Ile*	*Glu*	*Ala*	*Pro Asp*	*Glu*
	C			His							
	D	Gln									
	E										Gly
	F		Ser							Tyr	Gly
	G						Met				Gly
	H				Asp	Asn			Val		
	I							Gly			
	J									Leu	
				10			?				
α-La	A			Gln							
(123)	B			*Arg*			*Asp*				
	C						Asn				

[a]Numbers in parentheses indicate the total number of amino acid residues in the protein.
Amino acids in the reference variants are in *Italics*.

established by Mercier *et al.* (1971) and Grosclaude *et al.* (1973a), the same group of researchers was able to deduce the sites of mutation giving rise to the genetic variants A, C, D and E of α_{s1}-casein (Grosclaude *et al.*, 1972, 1976b). Compared to variant B, there is a deletion of a sequence of 13 amino acids, residues 14–26, in variant A. Variants A, B, D, F, G and H differ from C and E by having glutamic acid instead of glycine at position 192. Variant D has threonine P instead of alanine at position 53 in variant C. The substitution of glutamine at position 59 of α_{s1}-casein C for lysine gives α_{s1}-casein E. Prizenberg *et al.* (1998) reported α_{s1}-casein F as a subtype of α_{s1}-casein B with a single amino acid substitution, serine P for leucine at position 66. Although variant G is different from B at the gene level (Rando *et al.*, 1998), their amino acid sequences are identical. Compared to α_{s1}-casein B, variant H lacks amino acids from 51 to 58 (Mahé *et al.*, 1999).

16.4.2 Bovine α_{s2}-casein

Gel electrophoretic separation of casein revealed a group of several fast-moving protein bands ahead of the β-casein zones which were labelled as: α_{s0}-, α_{s1}-, α_{s2}-, α_{s3}-, α_{s4}-, α_{s5}- and α_{s6}-casein in order of decreasing mobility (Annan and Manson, 1969; Hoagland *et al.*, 1971). The structures of the first two, α_{s0}- and α_{s1}-caseins, were described in the previous section. When Brignon *et al.* (1976, 1977) conclusively showed that the remaining α_s-caseins had the same amino acid sequence, but different degrees of phosphorylation, it was logically proposed to collectively denote this group as α_{s2}-casein. α_{s5}-Casein is a dimer consisting of α_{s2}-casein and α_{s4}-casein, linked by two disulphide bonds. The complete sequence of the 207 amino acids in α_{s2}-casein variant A was elucidated by Brignon *et al.* (1977). To the authors' knowledge, there is no report of the complete sequence of α_{s2}-casein B. Variant D of α_{s2}-casein is another example of a mutation in which there is a deletion of several amino acids in sequence. Here, nine amino acids are absent in one of the three following sequences: 50–58, 51–59 or 52–60 (Grosclaude *et al.*, 1978, 1979). It was later confirmed (Bouniol *et al.*, 1993) that the peptide deletion is located in the 51–59 position. The difference between the primary structures of variants A and C of α_{s2}-casein is due to the substitution of glutamic acid, alanine and threonine at positions 33, 47 and 130, for glycine, threonine and isoleucine, respectively (Mahé and Grosclaude, 1982). The substitution at position 130 affects the phosphorylation of serine residues at positions 129 and 131: in variant A, only serine 129 can be partially phosphorylated, whereas in variant C both serine 129 and serine 131 can be phosphorylated.

16.4.3 Bovine β-casein

Within β-casein, the A^2 variant was the first to be completely sequenced (Ribadeau Dumas *et al.*, 1972). The locations along this polypeptide chain of 209 amino acids where mutations have occurred to produce several genetic variants were later elucidated by (Grosclaude *et al.*, 1972, 1974a,b). Using the sequence of β-casein A^2 as a reference, variants A^1, B and C differ by having a histidine instead of a proline at position 67. Variant B differs from A^2 by having an arginine instead of a serine at position 122. β-Casein C also differs from β-casein A^2 by a glutamic acid to lysine substitution at position 37 and the absence of glutamic acid at position 37 leads to a non-phosphorylated serine at position 35 in variant C. Substitution of histidine in variant A^2 at position 106 for a glutamine leads to the rare variant A^3. Replacement of serine P at position 18 of A^2 by a lysine gives variant D of β-casein. Substitution of lysine for glutamic acid in variant A^2 at position 36 leads to variant E. The primary structures of variants A^1, B^2, A^4 and A^3_{Mongolie} are not yet known. Variants F and G of β-casein, which have electrophoretic mobilities similar to A^1, are different from the latter by a proline to leucine substitution at position 152 (Visser *et al.*, 1995) and 137 (Dong and Ng-Kwai-Hang, 1998), respectively.

16.4.4 Bovine κ-casein

The primary structures of two genetic variants, A and B, of κ-casein were established by Grosclaude *et al.* (1972b) and Mercier *et al.* (1973). κ-Casein A differs from the B variant by substitution of threonine for isoleucine at position 136 and of aspartic acid for alanine at position 148. Both amino acid substitutions occur in the macropeptide region, residues 106 to 169, resulting from hydrolysis by chymosin. The difference between κ-caseins A and B in Zebu involves similar amino acid substitutions at positions 136 and 148, and a third substitution at position 135, isoleucine (κ-A) or threonine (κ-B) (Grosclaude *et al.*, 1974a). Variants C and E of κ-casein were characterised by Miranda *et al.* (1993). Substitution of arginine in variant B for histidine at position 97 gives variant C. At position 155, variant A has serine while variant E has glycine. Compared to variant A, the F variant of Sulimova *et al.* (1996), F^s, has valine instead of aspartic acid at position 148. With alanine at position 148, the G variant of Sulimova *et al.* (1996), G^s is intermediate between the A and B variants. The replacement of arginine at positions 10 and 97 in κ-casein B by histidine and cysteine, respectively, gives rise to variants F of Ikonen *et al.* (1996), F^I, and G of Erhardt *et al.* (1996), G^E (Prinzenberg *et al.*, 1996). Because both variants F^I and G^E have isoleucine at position 136 and alanine at position 148, they could be considered as being derived from variant B. The use of the F^I, F^s, G^E and G^s notations is only temporary for the purpose of differentiating κ-casein variants in this chapter pending the adoption of a better and more appropriate nomenclature. The distinctive feature of variant B in Zebu cattle which was mentioned by Grosclaude *et al.* (1974a) corresponds to variant H and it differs from A by a threonine to isoleucine substitution at position 135 (Prinzenberg and Erhardt, 1998). From the DNA sequence of κ-casein I, it was deduced that this variant is different from A by the replacement of serine at position 104 by an alanine (Prinzenberg *et al.*, 1999). An exchange of serine at position 155 for arginine constitutes the difference between variants A and J (Mahé *et al.*, 1999).

16.4.5 Bovine β-lactoglobulin

To date, ten genetic variants of β-lactoglobulin, which contains 162 amino acids, are known. The primary structure of β-lactoglobulin A and B was determined by Braunitzer and Chen (1972), Braunitzer *et al.* (1972) and Kolde and Braunitzer (1983). The substitution of aspartic acid at position 64 and valine at position 118 by glycine and alanine, respectively, constitute the difference between variants A and B. Substitution of glutamine at position 59 of variant B by histidine gives variant C, while replacement of glutamic acid at position 45 of variant B by a glutamine results in variant D (Brignon and Ribadeau-Dumas, 1973). Variant E, which was discovered in yak's milk (Grosclaude *et al.*, 1976b), differs from variant B by having glycine instead of glutamic acid at position 158. The complete amino acid sequence of β-lactoglobulins F and G, which are restricted to Bali cows (Bell *et al.*, 1981b), have yet to be fully established. However, it has been suggested that variants F and G, like variant E, have glycine at position 158 replacing glutamic acid in variant B. In addition, β-lactoglobulin G has methionine at position 78 instead of isoleucine as in variant B. Variant F also differs from B by substitution of serine for proline at position 50 and tyrosine for aspartic acid either at position 129 or 130. The exact locations of amino acid substitutions in β-lactoglobulin F and G remain to be ascertained after the determination of the primary structures of these two variants. Variant H is probably different from A by a lysine to asparagine substitution at position 70. β-Lactoglobulin I and J, which were sequeneed by Godovac-Zimmermann *et al.* (1996), deviate from the B variant by the substitution of glutamic acid for glycine at position 108 and proline for leucine at position 126, respectively. The difference between variants B and W (found in only one cow) is due to the replacement of the isoleucine at position 56 by a serine.

16.4.6 Bovine α-lactalbumin

The complete amino acid sequences of α-lactalbumin A and B were determined by Brew *et al.* (1970). Variant A differs from B by a single substitution of glutamine for arginine at position 10, The exact amino acid substitutions that differentiate variant C, observed in *Bos javanicus*

(Bell *et al.*, 1981a), from A or B have not yet been established. It was tentatively proposed that variant C differs from B by an amide group, most probably by the substitution of asparagine for aspartic acid.

16.5 Occurrence and molecular basis of genetic polymorphism in species other than bovine

16.5.1 Caprine species

The polymorphism of milk proteins in the goat is highlighted by the remarkable example of α_{s1}-casein. Richardson and Creamer (1975) concluded from an analytical study of a group of caprine caseins that there was no protein corresponding to α_{s1}-casein. In contrast, Boulanger *et al.* (1984) not only identified goat α_{s1}-casein, but also showed that the electrophoretic polymorphism of α_{s1}-casein was significantly associated with its quantitative variations. Grosclaude *et al.* (1987) and Mahé and Grosclaude (1989) established the existence of at least seven alleles, associated with various amounts of α_{s1}-casein in the milk. While alleles α_{s1}-Cn^{A}, α_{s1}-Cn^{B} and α_{s1}-Cn^{C} each contributed approximately 3.6 g casein per litre of milk, α_{s1}-Cn^{E} contributed only 1.6 g, α_{s1}-Cn^{F} and α_{s1}-Cn^{D} 0.6 g, and α_{s1}-Cn^{O} appeared to be a null allele. Quantitative variations of α_{s1}-caseins were also observed in Italian breeds (Russo *et al.*, 1977, 1986; Ciafarone and Addeo, 1984; Addeo *et al.*, 1988; Ambrosoli *et al.*, 1988).

Although the D allele has been withdrawn, because it is probably identical to G, as many as 13 alleles are now recognized: A, B_1, B_2, B_3, B_4 (subdivisions of the former B), C, E, F, G, H, L, O_1 and O_2 (subdivisions of the former O, or "null" allele) (Grosclaude and Martin, 1997; Chianese *et al.*, 1997). Because variant B_1, which is associated with a high amount of α_{s1}-casein, shows the closest identity with the ovine and bovine sequences, this variant may be considered as the original type of the species. Also, compared to alleles A, B and C, which are associated with higher amounts of α_{s1}-casein, alleles E, F, G, O_1 and O_2 could be considered as defective mutants.

Variants F and G are characterized by deletions of 37 and 13 amino acids, respectively, which are the consequence of exon skipping caused by specific mutations of their genes (Brignon *et al.*, 1990; Leroux *et al.*, 1992; Martin and Leroux, 1994). Variant E has the same amino acid sequence as B_4. The lower level of α_{s1}-casein is attribuable to the insertion of a LINE sequence in the 3′UTR of the gene (Jànsa Pérez *et al.*, 1994). It is not possible to study the products of alleles α_{s1}-Cn^{O1} and α_{s1}-Cn^{O2} because they are associated with the absence of α_{s1}-casein. Allele α_{s1}-Cn^{O1} is characterized by a large deletion and α_{s1}-Cn^{O2} by a large insertion (C. Leroux, P. Martin and coworkers, in preparation). The amino acid substitutions specific to the other variants have all been identified (Figure 16.2).

In addition to the polymorphism of α_{s1}-casein, Boulanger *et al.* (1984) described a polymorphism of α_{s2}-casein, with two alleles, α_{s2}-Cn^A and α_{s2}-Cn^B, the former being predominant in the Alpine (0.85) and Saanen (0.87) breeds. Variant A was further subdivided into A and C, and the three variants were characterized. Variant B differs from A by having Lys instead of Glu at position 64. This substitution also induces the failure of phosphorylation of Ser_{62}. Variant C differs from A by the replacement of Lys by Ile at position 167 (Bouniol *et al.*, 1994).

In general, β-casein, κ-casein, β-lactoglobulin and α-lactalbumin have been considered to be monomorphic in the caprine species, although heterogeneous patterns of β-lactoglobulin (Boulanger, 1976) and α_{s1}-casein (Russo *et al.*, 1986) were observed in a few milk samples. Regarding β-casein, Dall'olio *et al.* (1989) found one milk sample from the Italian Garganica breed with no visible β-casein band. The existence of a null β-casein allele (β-Cn^O) was established by Mahé and Grosclaude (1993) in the Criollo breed of Guadeloupe. The same allele, characterized by the deletion of one nucleotide at the beginning of the seventh exon (Persuy *et al.*, 1996), was found at a frequency of 0.11 in the rare French Pyrénées population (Ricordeau *et al.*, 1999). Two additional variants of β-casein were also observed in the Criollo breed of Guadeloupe, one with slower and the other with faster electrophoretic mobility than the common type.

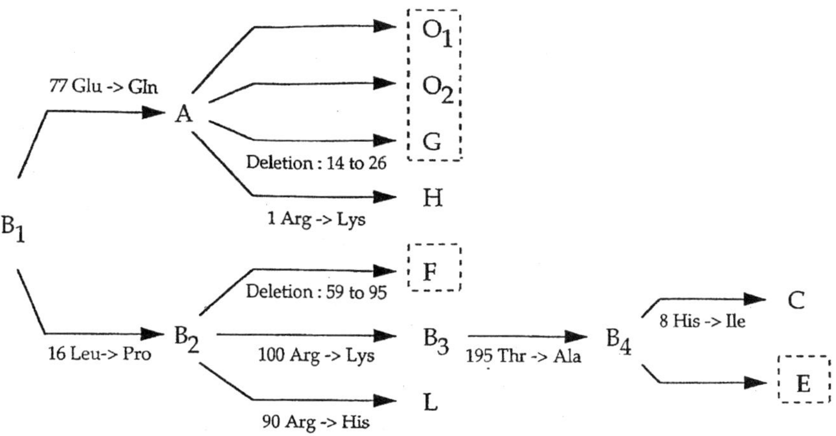

Figure 16.2 Proposed phylogeny of goat α_{s1}-Cn alleles and differences between the corresponding variants (Grosclaude and Martin, 1997; Chianese *et al.*, 1997). The defective alleles, E ("medium" type), F and G ("low"), O_1 and O_2 ("null") are in boxes. For the characteristics of alleles O_1, O_2 and E, see text.

16.5.2 Ovine species

The best documented polymorphism in the sheep is that of β-lactoglobulin, discovered by Bell and McKenzie (1964) and King (1969). Variants β-Lg A and β-Lg B differ by a substitution of Tyr in variant A for His in variant B at position 20 (Gaye *et al.*, 1986). Such a difference involving histidine versus a neutral amino acid explains why the polymorphism is seen by starch gel electrophoresis at pH 7.2 (King, 1969) but not at pH 8.6 (Chiofalo *et al.*, 1986). An additional variant, β-Lg C, was found by Erhardt *et al.* (1989), at a high frequency (0.175), in the German Merinoland breed. It differs from β-Lg A by a single substitution (Gln for Arg at position 148). Attempts to establish whether β-Lg^A or β-Lg^B is the original type in sheep by comparing the amino acid at position 20 in related species resulted in controversial conclusions. For example, with His at position 20, ovine β-Lg^B compares to β-lactoglobulin of the wild Sardinian mouflon, while with Tyr at the same position, ovine β-Lg A compares to caprine and bovine β-lactoglobulins (Godovac-Zimmermann *et al.*, 1987). In fact, Sardinian mouflon, which is now considered as a feral population (Poplin, 1979), does not represent the original wild type of sheep. In addition, the absence of β-Lg^A in this population is by no means certain. Because of its similarity with the goat and bovine proteins, and its higher frequency, β-Lg^A is probably the original form of ovine β-lactoglobulin.

Except for κ-casein, the other main proteins were also found, or suspected, to be polymorphic. In addition to the common α-lactalbumin type, α-La A, the rare α-La B variant was observed in three Italian breeds ($f = 0.004$–0.012) by Chiofalo and Micari (1987), and also reported in USSR (Stambekov *et al.*, 1977a,b) and Germany (Erhardt, 1989a). As early as 1966, King (1966) called attention to variations in the electrophoretic patterns of "α-casein", which is now known to consist of α_{s1}- and α_{s2}-casein fractions. To account for the observation in the Welsh Mountain, Clun Forest and Suffolk breeds of a pattern with two additional bands of "α-casein", King (1966) assumed the existence of a variant called 'Welsh'. Because of the complexity of the "α-casein" patterns, this variant, which was later shown to be α_{s1}-casein (Davoli *et al.*, 1990), proved to be the only clearly identifiable fraction. More recently, Chianese *et al.* (1993, 1996), using isoelectric focusing, identifed 3 variants of α_{s2}-casein (A, B, C) and 5 variants of α_{s1}-casein (A, B, C, D, E), but only 3 variants of α_{s1}-casein, A, C and D (the former "Welsh variant"), were characterized by Ferranti *et al.* (1995). Interestingly, their differences in charge are the consequence of changes in the degree of phosphorylation of the protein. Variant A differs from variant C, which is very likely the original type, by having Ser instead of Pro at position 13, which induces the addition of a phosphate group on Ser 12; variant D differs from A by the substitution of SerP by Asn at position 68, which causes, in addition, the loss of both phosphate groups on

Ser 64 and Ser 66. As regards β-caseins, heterogeneous patterns were described by King (1966), Micari *et al.*, (1986) and Chiofalo and Micari (1987).

16.5.3 Porcine species

The first descriptions of polymorphism of porcine milk proteins were made before those proteins had been identified and as a consequence, the designations used turned out to be inappropriate. Glasnak (1966, 1968a,b) reported on a polymorphism of the so-called β_1- and β_2-caseins, as well as of the 'X- and Z-proteins'. Likewise, Kraeling and Gerrits (1969) and Althen and Gerrits (1972) described the polymorphism of 'Whey$_1$' and 'Whey$_2$' protein, and Gerrits *et al.* (1969) studied a casein fraction called 'Cn$_3$'. In fact, 'β_1-casein' and 'Cn$_3$' were later identified as β-casein while 'Whey$_1$,' was shown to be β-lactoglobulin (Kalan *et al.*, 1971). 'Whey$_2$' and 'X' were apparently the same protein which according to Althen and Gerrits (1972) behaved electrophoretically like serum transferrin, although it was not this protein.

At present, the best characterised polymorphisms of porcine milk proteins are those of β-lactoglobulin. The A and B variants were identified at alkaline pH by Kraeling and Gerrits (1969) working at that time on 'Whey$_1$'. Variant A was predominant in Duroc pigs (0.95) and variant B in Yorkshire pigs (0.75). Because the A and B variants were shown to differ by an Ala (B)/Val (A) substitution, differences in conformation and/or ion binding resulting from compositional differences were considered to be responsible for the difference in electrophoretic mobility of the polymorphs (Kalan *et al.*, 1971). A third variant, C, was found by Bell *et al.* (1981d) who used SGE at pH 7.8. The overall frequency of allele β-Lg^C in the six breeds or crosses investigated was about 0.06. Interestingly, a quantitative variation was superimposed on this polymorphism with about two-thirds of the heterozygous animal having a much reduced amount of variant C. Variant C differs from variant A by two amino acid substitutions (Glu for Asp at position 9 and His for Gln at position 69). Erhardt and Senft (1987) and Erhardt (1989b), using acid PAGE, confirmed the existence of three β-lactoglobulin variants, A, B and C. Their identity with the previously described A, B and C, although not rigorously established, is probable. In particular, the C variant described by Erhardt (1989b) cannot be distinguished from A under alkaline conditions, a behaviour expected from the amino acid substitution (His/Gln) found by Bell *et al.* (1981d) for their C variant.

In the case of β-casein, the A and B shown by Erhardt and Senft (1987) and Erhardt (1989c) in three German breeds, are probably the same as the A and B variants described by Gerrits *et al.* (1969) and the B and C variants discovered by Glasnak (1966, 1968b). Variant A (B for Glasnak) appears to be predominant in all breeds investigated.

Using semi-discontinuous buffers, Bell *et al.* (1981c) discovered two variants of α-La, called A and B. Variant A was found to differ from variant B by a substitution of Arg for Lys. This particular difference explains why this polymorphism cannot be detected under the usual conditions. The polymorphism of α_{s2}-casein discovered by Erhardt (1989b) consists of a fast migrating band, in the presence or absence of a minor satellite. The present authors suggest that the latter band is the consequence of the incomplete saturation of a phosphorylation site in the protein, as in the case of bovine α_{s0}- and α_{s1}-caseins. The absence of this minor band would, thus, be due to a straightforward non-phosphorylation of the site that was induced by an amino acid substitution affecting the coding structure proposed by Mercier (1981). The designation α_{s2}-Cn^0, which suggests the existence of a null allele, would, thus, be inappropriate.

16.5.4 Equine species

The most clearly established polymorphism of equine milk protein is that of a minor whey protein, called protein of zone 1 by Bell *et al.* (1981a) and Whey$_1$ protein by Kingsbury and Gaunt (1977), which was later shown to be β-Lg II, one of the two β-Lgs of the species. The first group distinguished five variants (A, B, C, D, and E) while the second group showed four variants (A, B, C, and D). Using alkaline starch gel electrophoresis, Màcha and Novackova (1975) observed, in Czechoslovakian breeds, polymorphisms of "α-casein," β-casein, and unidentified "X3" protein, with in each case three alleles (A, B, and C). At acid pH, they also found two variants of β-Lg, called A and B. In contrast, Chiofalo *et al.* (1983) observed variations only in the patterns of one of the casein zones among Italian breeds.

16.5.5 Other species

Only fragmentary data are available for genetic polymorphism of milk proteins in other species. Among water buffalo (*Bubalus bubalis*), Mitra *et al.* (1998) reported a polymorphism (Thr/Ile) at position 135 of κ-casein of Egyptian buffaloes. Previous publications mentioned heterogeneity of different milk proteins, but without a characterization of these suggested polymorphisms.

Analysis of casein and whey protein fractions from Somalian camel's milk by Di Stasio *et al.* (1983) showed the occurrence of polymorphism in the β-Lg, but not in the other proteins. Two variants, A and B, of α-La were detected by Conti *et al.* (1985) in a Somalian camel's (*Camelus dromedarius*) milk. These two forms of α-La had a similar mobility in PAGE at pH 8.3, but differed in pI due to a difference in amino acid composition at the N terminal.

Voglino *et al.* (1975) reported two types of β-casein in human milk. They suggested that substitution of a charged amino acid in the protein resulted in the difference in electrophoretic mobility observed in acidic buffer. Casein samples isolated from milk collected from different populations of women (Turin, Sardinian, Kikuyu) showed polymorphism in the α- and β-casein systems (Voglino and Ponzone, 1972; Ponzone *et al.*, 1975; Ponzone and Voglino, 1976). Amino acid sequence analysis (Greenberg *et al.*, 1984) indicated the existence of genetic polymorphism involving uncharged residues in human β-casein.

Polymorphisms of milk proteins have also been identified in the milk of laboratory animals. For example, rat milk contains two variants of α-lactalbumin (McKenzie and Larson, 1978). Variants have been reported in the caseins and whey proteins of mouse milk (Piletz and Ganschow, 1981). Analysis of feline milk protein (Halliday *et al.*, 1990) revealed the presence of β-lactoglobulin-like proteins with complex electrophoretic patterns resulting from three distinct loci; two to five variants are expressed by each locus.

16.6 LINKAGE BETWEEN MILK PROTEIN GENES

One of the important features in milk protein genetics is the close linkage of the genes encoding the four caseins, a situation that was discovered and analysed in cattle. Grosclaude *et al.* (1964) presented the first evidence of a close linkage between loci α_{s1}-Cn and β-Cn, which was the first example of linkage ever found in cattle. Soon afterwards, having clarified the genetic determinism of the κ-caseins, they also concluded the linkage of locus κ-Cn with α_{s1}-Cn and β-Cn (Grosclaude *et al.*, 1965). Those results were confirmed by data of Larsen and Thymann (1966a). As regards α_{s2}-Cn, no genetic study was possible until polymorphism was found. This was achieved by Grosclaude *et al.* (1978) who showed that locus α_{s2}-Cn was also linked to the three other loci.

Table 16.2 recapitulates the data gathered by Grosclaude and coworkers and by Larsen and Thymann. As can be seen, no recombination was observed over 448 possibilities within the casein gene cluster, a result which demonstrates the close linkage between the members of this cluster.The tight genetic linkage of loci α_{s1}-Cn, β-Cn and κ-Cn was confirmed more recently by a single sperm typing experiment which revealed no recombinant among 293 meioses (Lien *et al.*, 1993). The casein gene cluster is located on chromosome 6; in contrast, the locus of α-lactalbumin, located on chromosome 5, and the locus of β-lactoglobulin, located on chromosome

TABLE 16.2
Number of informative dam-daughter pairs collected for analysing linkage between loci α_{s1}-casein, β-casein, α_{s2}-casein. No recombinant was found–a result indicating very close linkage

Linkage investigated	*Number of informative dam-daughter pairs*	*References*
α_{s1}-casein and β-casein	78	Grosclaude *et al.* (1964)
	45	Larsen and Thymann (1966a)
	25	Grosclaude (unpublished)
α_{s1}- or β-casein and κ-casein	108	Grosclaude *et al.* (1965a)
	143	Larsen and Thymann (1966a)
α_{s1}-, β-, κ-casein and α_{s2}-casein	49	Grosclaude *et al.* (1978)

11, are genetically independent (Eggen and Fries, 1995). In other species, linkage of casein genes was found whenever it was investigated, e.g., in the goat (Grosclaude *et al.*, 1987), sheep (Levéziel *et al.*, 1991) and pig (Archibald *et al.*, 1994).

The non-independent inheritance of casein alleles has observable consequences in the genotypic structure of populations. King *et al.* (1965) were the first to report the apparent absence of particular combinations of α_{s1}- and β-casein genotypes. In the Jersey breed, for example, the expected genotypes comprising α_{s1}-casein BC and β-casein B, as well as α_{s1}-casein C and β-casein AB or B, were missing. The authors postulated the following 'chromosomes', or 'combinations': α_{s1}-Cn^B, β-Cn^A; α_{s1} Cn^B, β-Cn^B and α_{s1}-Cn^C, β-Cn^A. Since the α_{s1}-Cn^B, β-Cn^A combination prevailed in all breeds examined, the other combinations were assumed to be the result of mutations in either α_{s1}-Cn or β-Cn. The combination α_{s1}-Cn^C, β-Cn^B was considered as a recombinant one and its apparent absence was interpreted as a consequence of a very close linkage between loci α_{s1}-Cn and β-Cn. Such an interpretation, dividing the combinations into an original type, mutant derivatives and recombinant types, was developed independently by Grosclaude *et al.* (1966b, 1973b, 1974a) and Grosclaude (1974) who positioned the additional variants as they were discovered.

In the early interpretations of King *et al.* (1965) and Grosclaude *et al.* (1966b), the combination α_{s1}-Cn^B, β-Cn^A, and later α_{s1}-CnB, β-Cn^{A2}, was considered as the original type because it prevailed in all western European breeds (*Bos taurus*). However, because the combination α_{s1}-Cn^C, β-Cn^{A2} strongly predominates in Zebu (*Bos indicus*) and Yak (*Bos grunniens*) populations (α_{s1}-Cn^B, β-Cn^{A2} being absent in the Yak), Grosclaude (1979) and Grosclaude *et al.* (1982) suggested that this combination is actually the original type. Their conclusion is substantiated by the fact that the amino

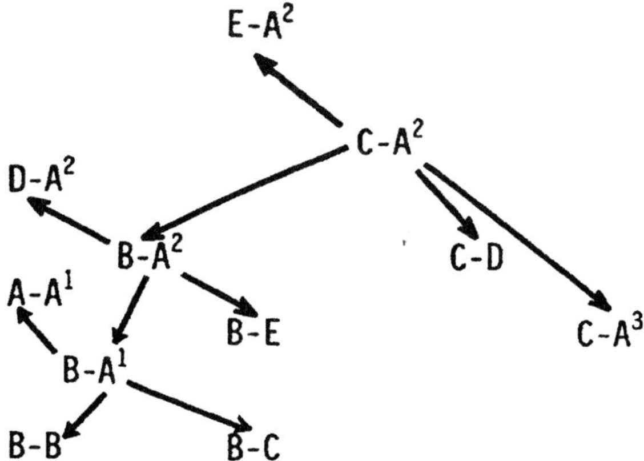

Figure 16.3 Phylogeny of the main alleles of the α_{s1}-casein, β-casein cluster. The α_{s1}-casein C-β-casein A^2 allele (C-A^2) in the figure may be considered as the original one from which the others have derived by successive mutations either at the α_{s1}-casein or at the β-casein locus.

acid residue found in the caprine as well as in the ovine α_{s1}-casein sequence at the position homologous to bovine 192 is Gly as in α_{s2}-Cn^C and not Glu, as in α_{s1}-Cn^B (Brignon *et al.*, 1989). Figure 16.3 shows the postulated relationships between the original α_{s1}-Cn^C, β-Cn^{A2} combination and its main mutant derivatives. The term 'haplotype', coined in human histocompatibility studies, will be used in place of 'combination' to designate a combination of alleles of closely linked genes.

The non-independent occurrence of casein genotypes in populations is an example of the phenomenon termed 'linkage disequilibrium'. Statistical tests may be applied to estimate the strength of linkage disequilibrium between alleles of two loci. In the case of bovine caseins, such tests, carried out in 32 breeds, were summarized by Graml *et al.* (1986b). On the whole, linkage disequilibrium between alleles of the α_{s1}-Cn and β-Cn loci is statistically significant in 18 of 32 breeds and, except in one case, always of the same sign. This means that the same haplotypes are either present or absent in nearly all breeds. This is not the case when alleles of either α_{s1}-Cn and κ-Cn, or β-Cn and κ-Cn, are being considered. However, linkage disequilibrium between alleles of β-Cn and κ-Cn is significant in 16 of 32 breeds; between alleles of α_{s1}-Cn and κ-Cn, it is significant in only 11 of 32 breeds. These results confirm the early suggestion by Grosclaude *et al.* (1965) that the strongest linkage disequilibrium occurs between alleles of the α_{s1}-Cn and β-Cn loci. Based on these observations, Grosclaude (1979) proposed a model in which the β-Cn and α_{s1}-Cn loci were very close to each other, in contrast to the κ-Cn locus which was further apart. Such a model is now confirmed (Ferretti *et al.*, 1990) and, moreover, it is known that the α_{s2}-Cn locus is situated between the β-Cn and κ-Cn loci. Because of such a close linkage between casein loci it is logical to consider, as units of inheritance, not only the alleles of the individual loci, but also the haplotypes (Grosclaude *et al.*, 1966b; Larsen and Thymann, 1966b). This is particularly true for the haplotypes of the α_{s1}-Cn, β-Cn pair of loci, which are so closely linked. Their occurrence and frequencies in the breeds will be discussed in the next section.

16.7 Frequency distribution of genetic variants in cattle species

Data available on the occurrence and frequencies of milk protein variants in the three main species of the genus *Bos*, *Bos taurus* (taurines), *Bos indicus* (zebus) and *Bos grunniens* (yaks), although numerous, are far from being complete. First, the proportion of populations investigated is higher for taurines than for zebus or yaks, but, even for taurines, important gaps exist. In Europe, for example, while Italian and French breeds have been investigated extensively, nothing is known about native Spanish and Portuguese cattle. Moreover, breeds in Great Britain, a country that has played an important role in the history of modern cattle, were only poorly investigated in their native area. Fortunately, those having an international distribution were studied in their outside settlements. Other reasons contributing to the heterogeneity of data include differences in the electrophoretic techniques used in the course of time or even contemporarily and also in the degree of attention given to particular polymorphisms. As discussed earlier, the last point is well exemplified by the polymorphism of α_{s2}-casein, which is certainly less restricted than the literature would indicate.

When considering data about the breed distribution of milk protein variants, it is of interest to know, for each locus, the relative appearance in time of different alleles, a process which can be represented by a phylogenetic tree. This implies the identification of the original variant, which can be based on two different criteria. First, the original variant should normally give the closest sequence alignment with the same protein of related species. Second, the corresponding allele may be expected to have, usually, the highest frequency over a large number of populations. Although this rule cannot be considered as absolute (Watterson and Guess, 1977), the availability of population data for three different species of the genus *Bos* makes this criterion fairly convincing in the particular case of milk protein polymorphism. Finally, when trying to infer conclusions from population data on protein polymorphism, attention should be paid to the risk of "cryptic heterogeneity". This does not refer to the well-known existence of 'silent variants', which cannot be distinguished by electrophoresis because differences in their structure have no effect on the net charge of the protein. It concerns the possibility that different amino acid substitutions, resulting from distinct mutations, have the same effect on the net charge of the protein, and thus give rise to variants indistinguishable by electrophoresis. Without a biochemical analysis, those variants would thus be considered as one and the same. As discussed in Section 16.3, variant β-CnA^3 provides an example of such heterogeneity. In the same way, the common type of β-lactoglobulin of the Yak and β-Lg D of European breeds are identical on electrophoresis although, compared to β-Lg B, they have different amino acid substitutions. Other similar situations may not have been discovered yet.

16.7.1 α-Lactalbumin

α-Lactalbumin was found to be polymorphic in all Zebu populations investigated so far, allele α-La^B being always predominant over α-La^A. As pointed out by Aschaffenburg (1968), the frequency of α-La^A appears to be higher in Indian (0.22 to 0.40) than in African (0.03 to 0.15) breeds. In Zebu, the most extensive survey was carried out in Madagascar where, in a sample of 778 cows, the frequency of α-La^A was 0.166, with no regional heterogeneity (Grosclaude *et al.*, 1974a). As far as *Bos taurus* is concerned, α-lactalbumin appears to be monomorphic in all northern European breeds, while allele α-La^A is present in African taurine populations (Mahé *et al.*, 1999) and in breeds of southern Europe (Figure 16.4). In this area, systematic investigations were carried out only in Italy where as many as 12 breeds were found to be polymorphic. Those breeds are essentially Podolic, or breeds known to have been crossed with Podolic cattle (Bettini and Masina, 1972; Mariani and Russo, 1977; Chianese *et al.*, 1988). This is also the case for the few polymorphic Romanian and Russian breeds. Verification of the biochemical uniqueness of variant α-La A found in different locations would be important. The variant of the north-eastern France Vosgienne breed, for example, is different from the Zebu variant (Grosclaude F and coworkers, in preparation). The frequency criterion unequivocally designates α-La B as the original type of the genus *Bos*, but curiously, when the amino acid residue at position 10 of α-lactalbumin variants are compared, variant B, with Arg, is identical to α-lactalbumin of the Italian water buffalo, *Bubalus arnee* (Addeo *et al.*, 1976) and of the Bali cattle *Bos* (*Bibos*) *javanicus* (Bell *et al.*, 1981a) while variant A, with Gln, is identical with goat and sheep proteins (McGillivray *et al.*, 1979; Gaye *et al.*, 1987). Whether this puzzling situation is only a consequence of chance remains to be established.

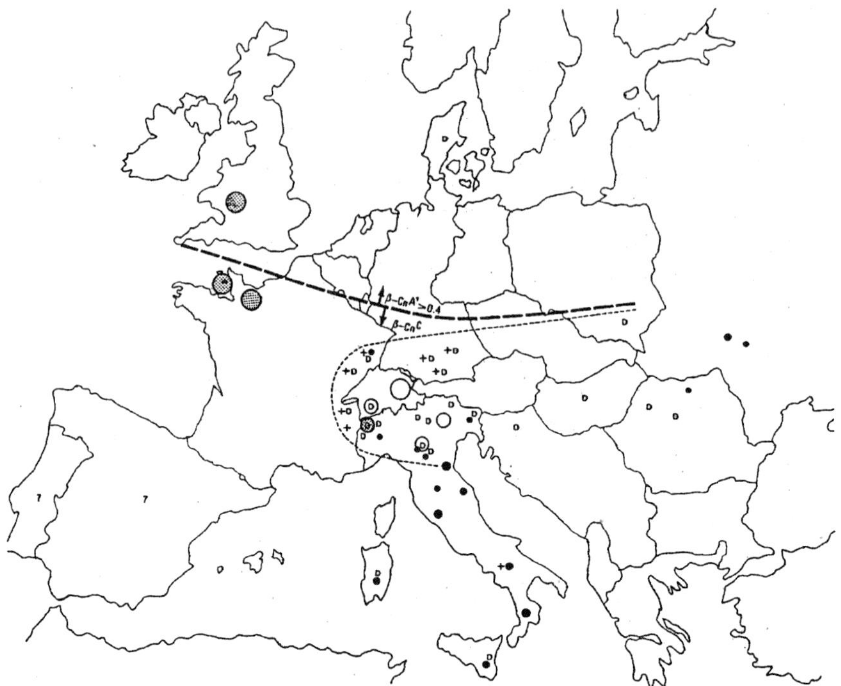

Figure 16.4 Schematic representation of the distribution of some milk protein variants in European cattle breeds. Heavy broken line: in the north, β-Cn A[1] tends to be the predominant β-casein variant and β-Cn C is essentially absent; in the south, β-Cn C is often present. Circles: breeds with a high frequency (about 0.3 or more) of β-Cn B: from south northwards. Reggiana, Valdotana and Castana, Rendena, Hérens, Brown Swiss, Normande, Jersey and Hereford; the breeds marked by a shaded circle also have a high frequency of κ-Cn B (haplotype α_{s1}-Cn^B, β-Cn^B); +, occurrence of a polymorphism of α_{s2}-Cn D in France and Germany, α_{s2}-Cn B in Italy) ●: occurrence of variant α-LaA (the 3 different sizes correspond to the following frequencies: less than 0.05, between 0.05 and 0.10, more than 0.10 and up to 0.23); D: occurrence of β-Lg D; the main extension area is surrounded by a light dotted line.

16.7.2 β-Lactoglobulin

The two common variants of β-lactoglobulin, β-Lg A and B, are probably universally distributed in bovine and zebu populations; β-Lg B is predominant in all zebu, as well as in about 75% of all taurine breeds so far investigated. Because this variant also gives the best alignment with β-lactoglobulins of other species (Godovac-Zimmermann and Braunitzer, 1987), it is clearly the original type for *Bos taurus* and *Bos indicus*. The Yak has its own β-lactoglobulin type, β-Lg E, initially termed β-Lg D_{yak}, which, interestingly, appears to be derived from β-Lg B (Grosclaude *et al.*, 1976b). Discovered by Bell (1962) in Australian Jersey, variant C is present in various settlements of that breed (Aschaffenburg, 1968; Bech and Kristiansen, 1990). Its occurrence in other populations remains uncertain because, as pointed out by Aschaffenburg (1968) a number of techniques used for β-lactoglobulin phenotyping fail to detect this variant. This is well exemplified by the case of the Normande, a breed having probable connections with the Jersey, where β-Lg C was identified only recently (Grosclaude and Mahé, unpublished) using isoelectric focusing instead of conventional techniques. On the other hand, the β-lactoglobulin C reported in a few central European breeds is probably β-Lg D. This is at least the case for the data of Seibert *et al.* (1985), as confirmed by Seibert (pers. comm., 1989). First observed in the Montbéliarde, a breed of French Jura (Grosclaude *et al.*, 1966a), β-Lg D is usually considered as a rare variant (Baker and Manwell, 1980). However, although its frequency never exceeds 0.04, this variant occurs (Figure 16.4) in most breeds of the Alps, Vosges and Jura mountains, like Simmental, Pinzgau, Brown Swiss, Tarentaise, Vosgienne (Russo and Mariani, 1978; Merlin and Di Stasio, 1982; Raimondi, 1982; Felenczak, 1983; Graml *et al.*, 1984a,b; Buchberger *et al.*, 1986; Grosclaude, 1988; Baranyi *et al.* 1997; Curik *et al.*, 1997). Its occurrence was also found outside of this area, e.g., in Danish Jersey (Larsen and Thymann, 1966a,b), in the Sicilian Modicana and in Sardinian cattle (Raimondi, 1982). This may be due to migrations and/or to cryptic heterogeneity. As a matter of fact, any new mutation of the common β-Lg^B allele that induces the loss of one charge in the protein would produce a D-type variant.

16.7.3 α_{s1}-Casein

The B and C variants are present in all *Bos taurus* and *Bos indicus* populations investigated so far, but, while the variant C is most frequent in all Zebu breeds, variant B was always found to be predominant in taurine breeds, including West African populations (Mahé *et al.*, 1999) except in two cases, the Jersey on the Isle of Jersey (Larsen *et al.*, 1974) and the Italian Calabrian (Bettini and Masina, 1972), a Podolic breed included by Baker and Manwell (1980) in the *primigenius* group. As a matter of fact, frequencies of allele α_{s1}-Cn^C higher than 0.25 are in the Channel Island area (Jersey, Guernsey, Normande) and in the *primigenius* group. As discussed above, α_{s1}-Cn^C is probably the original type in the genus *Bos*. Thus, while in yaks and zebus this original allele still prevails, in taurine breeds, including West African populations, the mutant allele, α_{s1}-Cn^B, predominates, almost tending to fixation in North European breeds. In contrast to B and C, the A and D variants, which are derived from B and are thus more recent, are far less common. Variant A is very rare, both in terms of breed distribution and of frequencies within breeds. Originally identified in the American Holstein (Thompson *et al.*, 1962), it was later found in three Red Danish cows (Larsen and Thymann, 1966a,b) and, more recently, in a few cows of the related Red German breed (Erhardt, 1993a). It is now established that the α_{s1}-Cn^{A} alleles found in Holstein and in Red German cattle are characterized by different mutations inducing the same exon 4 skipping resulting in the same protein variant (Wilkins and Xie, 1997a). Variant α_{s1}-CnA thus provides a good example of homoplasy. Variant D, first observed in the Flamande, a Red breed of Northern France (Grosclaude *et al.*, 1966a), is usually considered as a rare variant. However, although it occurs at a low frequency (0.002 to 0.08), it has now been identified in a total of 21 French, Dutch, Italian, Polish and Canadian breeds belonging to different phylogenetic groups (Corradini, 1969; Michalak, 1969; Russo and Mariani, 1978; Merlin and Di Stasio, 1982; Raimondi, 1982; Di Stasio and Dupont, 1983; Chianese *et al.*, 1988; Grosclaude, 1988). Variant E appears to be specific to the Yak (Grosclaude *et al.*, 1976a, 1982; Kawamoto *et al.*, 1992) while each of the very rare F, and the rare G and H variants, has been shown to date in only one breed (Erhardt, 1993a; Rando *et al.*, 1992; Mahé *et al.*, 1999).

16.7.4 α_{s2}-Casein

While the existence of genetic polymorphism of the three other caseins was established within a short period (1961–1965), the first example of polymorphism of the somewhat neglected α_{s2}-casein was discovered much later (Grosclaude *et al.*, 1976a) in Napalese yaks and crosses between yaks and local taurines or zebus. Besides the common type, α_{s2}-Cn A, two variants were suggested, α_{s2}-Cn B and α_{s2}-Cn C, the latter apparently specific to the Yak. Later, the same variants, α_{s2}-Cn B and C, were observed in Mongolian yaks with a frequency of 0.09 and 0.11, respectively (Grosclaude *et al.*, 1982). This confirmed that α_{s2}-casein was normally polymorphic in the Yak. In the meantime, the same group had discovered polymorphism of α_{s2}-casein in two breeds of eastern France, the Vosgienne and the Montbéliarde. Both shared an additional variant, α_{s2}-Cn D, with a frequency of only 0.01 in the Montbéliarde, but 0.09 in the Vosgienne (Grosclaude *et al.*, 1978). Later, this variant was also found in two French and five German alpine breeds: Abondance, Tarentaise, Gelbvieh, Vorderwälder, Hinterwälder, Fleckvieh and Braunvieh (Nuyts-Petit, 1991; Erhardt, 1993b). It thus emerges that variant α_{s2}-Cn D occurs in the same breed groups as β-Lg D (Figure 16.4). Because α_{s2}-Cn D is a result of exon skipping, it may, like α_{s1}-Cn A, be produced by different mutations. The rare occurrence of an α_{s2}-Cn D-like variant in the Finnish Ayrshire (Ikonen *et al.*, 1996) should thus be interpreted with caution. Chianese *et al.* (1988) reported on the occurrence of α_{s2}-Cn B at a low frequency (0.004) in the Italian Podolic breed. A search for this allele in other breeds, especially those of the podolic group, but also in Zebu, would certainly be worthwhile. One should also mention here the unexplained absence of α_{s2}-caseins, apparently balanced by more κ-casein, from the milk of 4 Red Danish cows in Poland (Michalak, 1969), which could possibly be due to the existence of a null α_{s2}-casein allele.

16.7.5 β-Casein

Subtyping of β-casein A into A^1, A^2 and A^3 by acid gel electrophoresis (Peterson and Kopfler, 1966) was performed in only part of the published breed investigations. Despite this, there is enough evidence to conclude that β-CnA^1 and β-CnA^2 are present in all *Bos taurus* and *Bos indicus* populations, while *Bos grunniens* may have only β-CnA^2 (Grosclaude *et al.*, 1982). Variant A^2 is predominant in all Zebu populations investigated so far and in the majority of taurine breeds, and may thus be considered as the original type of the genus *Bos*. Among the international breeds, the Guernsey has, by far, the highest frequency of allele β-Cn^{A2} (0.8 to 0.95 according to Rausch *et al.*, 1967; Kiddy *et al.*, 1968; Zikakis *et al.*, 1974). In the present state of knowledge on European cattle, the highest concentration of breeds with a high frequency of β-CnA^2 (more than 0.7) is found in the group of essentially beef breeds of central and southern France, including the Charolais, Limousin, Blonde d'Aquitaine, Salers, Aubrac and Ferrandaise (Grosclaude, 1988). On the other hand, variant β-CnA^1 occurs at a high frequency (0.4 to 0.7) in breeds of North European location or origin (Figure 16.4), like the Ayrshire (Kiddy *et al.*, 1968; Li and Gaunt, 1972; Lin *et ål.*, 1986), British Shorthorn (Grosclaude and Mahé, unpublished), Black and White Lowland breeds including the North-American Holstein (Kiddy *et al.*, 1968; Szmelik *et al.*, 1971; Li and Gaunt, 1972; McLean *et al.*, 1984; Ng-Kwai-Hang *et al.*, 1984; Bech and Kristiansen, 1990), European Red breeds such as Danish Red, Polish Red and the French Flamande (Szmelik *et al.*, 1971; Grosclaude, 1988; Bech and Kristiansen, 1990), but also in a few Italian breeds such as the Chianina and the Burlina (Merlin and Di Stasio, 1982; Raimondi, 1982) and in the Japanese Brown (Abe *et al.*, 1975; Komatsu and Abe, 1980). Interestingly, β-CnA^1 is also predominant in the West-African taurine populations (N'Dama, Baoulé, Lagune), which is in contrast with the neighbouring Zebu populations, where β-CnA^2 is the most frequent (Mahé *et al.*, 1999). Variant B, derived from A^1 and thus less ancient than A^2 and A^1, was found in nearly all *Bos taurus* and *Bos indicus* populations in the world. Considering only breed samples of more than 50 animals, the few exceptions were the Italian Calabrian (71 cows), the Japanese Brown (231) and the Madagascar Zebu (586) (Bettini and Masina, 1972; Grosclaude *et al.*, 1974a; Komatsu and Abe, 1980). The observation on Madagascar Zebu may be significant because of the fairly large sample size. In a majority of breeds, the frequency of allele β-Cn^B was less than 0.1. On the other hand, the highest frequencies were observed (Figure 16.4) in two distinct breed groups in Europe. The first group includes Brown Swiss and Brown Swiss-related European Brown breeds (0.2 to 0.4 according to Corradini, 1969; Horvath and Meszaros, 1972; Russo and Mariani, 1978; Raimondi, 1982; Samarieanu *et al.*, 1986; Graml *et al.*, 1984b; Grosclaude, 1988) as well as several rare Swiss and Italian breeds located close to the South of the Brown Swiss area, more particularly Valdotana Black-Pied (0.4), Hérens and Castana (0.3), Reggiano and Rendena (0.25) (Russo and Mariani, 1978; Merlin and Di Stasio, 1982; Raimondi, 1982; Bolla *et al.*, 1983; Di Stasio and Dupont, 1983; Pagnacco and Caroli, 1987). The second group located in north-western Europe, includes the various Jersey populations (approximately 0.3 according to Michalak, 1969; Szmelik *et al.*, 1971; Li and Gaunt, 1972; Larsen *et al.*, 1974; McLean *et al.*, 1984; Bech and Kristiansen, 1990), the French Normande (0.47 according to Grosclaude, 1974) which is the continental breed closest to the Jersey, and the Hereford (0.25 to 0.45) in rather small samples studied by Aschaffenburg (1968), Caldwell *et al.* (1971) and Komatsu and Abe (1980). Variant C, also derived from A^1, has not been observed in Zebu so far, and thus seems to be restricted to *Bos taurus*. Until the yet unexplored breeds are investigated, it may be considered as essentially absent from North European breeds. On the other hand, it is normally present in Mid- and Southern European populations (Figure 16.4). It was also found in Mongolian cattle (Grosclaude *et al.*, 1965), in Korean cattle (Han *et al.*, 1983) and in the Egyptian Baladi cattle (Graml *et al.*, 1986c). While its breed frequency is often less than 0.05, the highest values (0.10–0.23) were found, from east-westwards, in Polish Simmental (Michalak, 1969), in Hungarian Simmental and Spotted breeds (Horvath and Meszaros, 1972), in the Italian Chianina (Bettini and Maşina, 1972; Russo and Mariani, 1974), in the French Tarentaise or Italian Tarina (Raimondi, 1982; Grosclaude, 1988) and in the Bretonne Pie-Noire, in the extreme west of France (Grosclaude, 1988). Whenever present, variant A^3, derived from A^2, has a frequency less than 0.04. According to Baker and Manwell (1980), this variant is possibly indicative of some gene flow from Pied Lowlands into the Channel Island breeds. Indeed, this variant occurs in North European cattle (Ayrshire, Shorthorn, Holstein-Friesian) as well as in the Guernsey and its continental neighbour, the Normande (Grosclaude *et al.*, 1966b; Rausch *et al.*, 1967; Kiddy *et al.*, 1968; Li and Gaunt, 1972; Zikakis *et al.*, 1974; Hines *et al.*, 1977; Grosclaude and Mahé, unpublished). However, its occurrence was also reported in Polish Jersey (Szmelik *et al.*, 1971), German Simmental (Seibert *et al.*, 1987), Italian Burlina, Grigio Alpina, Piemontese and Rendena (Merlin and Di Stasio, 1982; Raimondi, 1982), as well as in Mongolian (Grosclaude *et al.*, 1982) and Korean native (Han *et al.*, 1984) cattle. In fact, variant A^3 provides an example of homoplasy because the A^3 variant of

Mongolian cattle was shown to be different from the A^3 variant characterised from a sample of Normande milk. Interestingly, the difference was suspected because A^3 was not in the same haplotype in the two populations (Grosclaude *et al.*, 1982). Variant D appears to be specific to *Bos indicus*, while each of the variants E, F and G has been shown to date in one breed (Voglino, 1972; Visser *et al.*, 1995; Dong and Ng-Kwai-Hang, 1998).

16.7.6 κ-Casein

The two common variants of κ-casein, A and B, have been found in all populations investigated. Variant A, which is predominant in a large majority of them, may thus be considered as the original type on the basis of the frequency criterion. On the other hand, sequence data (Mercier *et al.*, 1976) revealed a puzzling situation; while at position 136, variant A is identical to the κ-casein of other species (water buffalo, goat, sheep, pig, human), all having Thr, on the contrary, at position 148, it is variant B that is identical with the same proteins. Mercier *et al.* (1976) suggested that the original κ-casein type of the genus *Bos* was in fact an intermediate between A and B, having Thr at position 136 like A, and Ala at position 148 like B and thus migrating like B. Such a variant, termed κ-Cn G by Sulimova *et al.* (1996), was indeed found at a high frequency in yaks and European bisons. A possible explanation for the situation found in cattle would be to suggest that the ancestral form of type B was very early supplanted by a mutant of type A. Until further developments, and for greater simplicity, the present authors have considered variant A as the reference type for *Bos taurus* and *Bos indicus*. Curiously, as noted by Baker and Manwell (1980), the highest frequency of κ-Cn^A may be observed in two quite distant groups, e.g., *Bos indicus*, on one side, and on the other side, breeds of Northern Europe, the same also having high frequencies of α_{s1}-Cn^B and β-Cn^{A1} (Ayrshire, Shorthorn, Holstein-Friesian, Flamande, Danish Red, etc.). On the other hand, the highest values for κ-Cn^B are found in the various Jersey extensions [usually 0.7 or more, and up to 0.95 on the Isle of Jersey according to Larsen *et al.* (1974)] and in its continental neighbour, the Normande [0.66 in 1972, according to Grosclaude (1974); increased to 0.74, according to Balland (1987)], in the West-African taurine populations, Baoulé (0.64) and N'Dama (0.73) (Mahé *et al.*, 1999), but also in several breeds of the Italian podolic group (Calabrese, Maremmana, Marchigiana, Romagnola, Chianina and Modicana) which have a frequency close to 0.7 or higher (Russo and Mariani, 1978). Variants C and E, which are not uncommon, are apparently specific to different breed groups. Variant C was reported, like β-Lg D and α_{s2}-Cn D, in breeds of the Alpine area, like Italian Grey Alpine, Italian and German Brown, German, Hungarian and Croatian Simmental and French Tarentaise (Di Stasio and Merlin, 1979; Mariani, 1983; Nuyts-Petit, 1991;

Erhardt, 1993b; Baranyi *et al.*, 1997; Curik *et al.*, 1997), and also, unexpectedly, in the Kuri population of lake Chad (Mahé *et al.*, 1999). As for variant E, it was found in North-European Lowland and Red cattle, with a frequency as high as 0.307 in the Finnish Ayrshire (Erhardt, 1993b; Ikonen *et al.*, 1996). On the contrary, the very rare variants, F of Sulimova *et al.* (1996), F of Ikonen *et al.* (1996) and G of Erhardt (1996), may be restricted to Yakut cattle, Finnish Ayrshire and German Pinzgauer, respectively. Variant I seems to be restricted to *Bos indicus*, while H, first characterized by Grosclaude *et al.* (1974) in Madagascar zebus without being given a designation, and recently discovered in other Zebu populations as well as in Western-European breeds (Prinzenberg *et al.*, 1999), probably has a broad distribution. The distribution in African populations of variant J, found in West-African Baoulé and Fulani cattle (Mahé *et al.*, 1999), remains to be investigated more thoroughly.

16.7.7 Overall considerations for frequency distribution of alleles and haplotypes

According to Epstein (1971), all present *Bos taurus* and *Bos indicus* cattle have a common origin, in south-west Asia, where their early ancestors were selected from the same domestic cattle population derived from the wild auroch, *Bos primigenius*. However, on the basis of investigations on mitochondrial DNA, Loftus *et al.* (1994) concluded that the two subspecies had arisen through separate domestication events. In any case, the existence of sets of variants shared by *Bos taurus* and *Bos indicus* (α-La A and B, β-Lg A and B, α_{s1}-Cn B and C, β-Cn A^1, A^2 and possibly B, κ-Cn A and B) indicates that the corresponding mutations occurred before the separation of these two subspecies of the genus *Bos*. Note that Grosclaude *et al.* (1974a) ascertained that the amino acid substitutions between α_{s1}-Cn C and B, β-Cn A^1, A^2 and B as well as κ-Ca A and B, were the same in Zebu as in western humpless cattle. As discussed before, the original variants were α-La B, β-Lg B, α_{s1}-Cn C, α_{s2}-Cn A, β-Cn A^2 and possibly κ-Cn A. Interestingly, all those original variants are still predominant in *Bos indicus* while important shifts have occurred in *Bos taurus*, especially at the α_{s1}-Cn locus, but also at the β-Cn locus. Those shifts are best illustrated by taking into account the frequencies of haplotypes formed by the alleles of this pair of very closely linked loci (Grosclaude *et al.*, 1966b, 1973b). Table 16.3 gives the frequencies of such haplotypes in breeds representative of different phylogenetic groups. Clearly, Zebu has retained a high frequency of the original haplotype CA^2 (for α_{s1}-Cn^C, β-Cn^{A2}) while breeds of the Baltic and North Sea regions have nearly lost this haplotype, which has been replaced by BA^2, and especially BA^1; the latter reaches very high frequencies. Note that in this group, the rare α_{s1}-Cn^C allele tends to be associated with the rare β-Cn^{A3}. In contrast, several

Table 16.3
Frequency distribution of α_{s1}-Cn, β-Cn haplotypes for different breeds in various countries

Breeds and country	*α_{s1}-Cn, β-Cn haplotypes*									*References*
	AA^1	BA^1	BA^2	BB	BC	CA^1	CA^2	CA^3	$CI^{\text{-}}$	
Zebu Madagascar	–	0.027	0.139	–	–	0.072	0.762	–	–	Grosclaude *et al.* (1974)
Sudanese Fulani Burkina Faso	–	0.22	0.12	0.05	–	<0.01	0.58	–	0.02	Mahé *et al.* (1999)
N'Dama Ivory Coast and Senegal	–	0.61	0.24	0.03	–	–	0.12	–	–	Mahé *et al.* (1999)
Ayshire Canada	–	0.554	0.440	0.003	–	–	–	0.003	–	Lin *et al.* (1986)
Shorthorn Great Britain	–	0.671	0.321	0.004	–	–	–	0.004	–	Grosclaude and Mahé (1984 unpublished)
Holstein Canada	0.003	0.561	0.405	0.007	–	–	0.016	0.011	–	Ng-Kwai-Hang *et al.* (1984)
Red Danish Denmark	<0.01	0.71	0.23	0.06	–	–	<0.01	–	–	Bech and Kristiansen (1990)
Normande France	–	0.19	0.11	0.47	0.01	–	0.18	0.04	–	Grosclaude (1988)
Jersey USA	–	0.09	0.28	0.37	0.03	–	0.24	–	–	Li and Gaunt (1972)
Brown Swiss Italy	–	0.28	0.42	0.30	0.02	<0.01	0.07	–	–	Raimondi (1982)
Simmental France	–	0.26	0.55	0.03	0.06	0.02	0.08	–	–	Grosclaude (1988)
Charolaise France	–	0.10	0.68	0.13	0.01	–	0.08	–	–	Grosclaude (1988)
Podolica Italy	–	0.28	0.32	0.04	–	0.03	0.33	–	–	Ferrara *et al.* (1986)

North-Scandinavian rare breeds show rather high frequency (0.27 to 0.33) of allele α_{s1}-Cn^C (Lien *et al.*, 1999), undoubtly in haplotype CA^2 (not shown).

The taurine breed groups that have retained the highest frequencies of the original haplotype, CA^2, are the Channel Island and Podolic groups. Remarkably, as discussed in the Section 16.7.1, "α-lactalbumin," Podolic breeds are, like *Bos indicus*, also polymorphic at the α-La locus. This may be the result of introgression from *Bos indicus*, as suggested by Bettini and Masina (1972) for the Modicana breed in which about one-third of the animals have bifid processes to the last thoracic vertebrae, as usually observed in the Zebu. However, those observations are also consistent with the concept that both Zebu and Podolic cattle have retained, more than the other groups, genes of the common ancestor of all cattle populations.

It is noteworthy that the two taurine breed groups having a high frequency of haplotype BA^1 are found in quite distant parts of the world, the Baltic and North Sea regions, and West-Africa, but the main α_{s1}-Cn, β-Cn, κ-Cn haplotypes are not the same in the two groups, since BA^1 A predominates in northern Europe in contrast to BA^1 B in West-Africa. For the dairy industry, an interesting haplotype is BBB because it brings together, for cheese production, the positive effects of variant β-Cn B and κ-Cn B. As far as dairy breeds are concerned, this haplotype is frequent only in the Jersey and Normande breeds. Important additional work is needed to obtain a complete picture of the distribution of milk protein variants and haplotypes in cattle populations throughout the world.

16.8 Frequency distribution of genetic variants in other species

Data on the breed distribution of genetic variants of milk proteins in species other than cattle are far less complete. The best documented polymorphism is that of ovine β-lactoglobulin. Large breed differences were observed in the relative frequencies of alleles β-Lg^A and β-Lg^B, but, on the whole, β-Lg^A is predominant (Chiofalo and Micari, 1987). In addition, the C variant of sheep α_{s1}-casein, formerly called "Welsh", was found, at low frequencies, in breeds of quite different origins (King, 1966; Arave *et al.*, 1973; di Stasio, 1983; Barillet *et al.*, 1993).

At present, on account of its remarkable particularities, the most actively investigated polymorphism is that of caprine α_{s1}-casein. More than 30 breeds have been considered so far. Table 16.4 recapitulates the frequencies observed for the main alleles or groups of alleles ($B = B_1 + B_2 + B_3 + B_{4i}$; $O = O_1 + O_2$) in 16 of those breeds. Swiss breeds – Alpine, Saanen, Toggenburg– have a low frequency of the group of "strong" variants (<0.14). This has important consequences since the genes of these "international breeds" have been spread all over the world. Local breeds

TABLE 16.4
Estimated frequencies of α_{s1}-Cn alleles in samples of milk from 16 breeds or populations originating from 7 countries, according to Grosclaude *et al.* (1994), Jordana *et al.* (1996), Grosclaude and Martin (1997) and Enne *et al.* (1997)

Breed or population	*Country*	*Sample size*	*Alleles at the α_{s1}-Cn locus*					
			A	*B*	*C*	*E*	*F*	*O*
Alpine	France	213	0.14	0.05	0.01	0.34	0.41	0.05
	Switzerland	235	0.05	0.08	< 0.01	0.36	0.37	0.14
Saanen	France	159	0.07	0.06	–	0.41	0.43	0.03
	Switzerland	125	–	0.06	–	0.27	0.44	0.23
Toggenburg	Switzerland	218	< 0.01	0.02	< 0.01	0.12	0.69	0.17
Corse	France	106	0.06	0.13	–	0.14	0.59	0.08
Poitevine	France	209	0.05	0.35	< 0.01	0.45	0.14	–
Pyrénées	France	710	0.12	0.18	–	0.62	0.03	0.05
Malagueña	Spain	373	0.09	0.09	–	0.65	0.04	0.13
Payoya	Spain	111	0.05	0.19	–	0.76	–	–
Canaria	Spain	74	0.28	0.32	–	0.20	–	0.20
Sarda	Italy	115	0.11	0.60	–	0.07	0.14	0.08
Maltese	Italy	177	0.21	0.38	–	0.03	0.11	0.27
Balkan type	Albania	237	0.47	0.46	–	< 0.01	0.04	0.03
Damascus	Greece	74	0.87	0.03	–	0.10	–	–
Sahel	Chad	66	0.26	0.74	–	–	–	–

The very rare G allele is included in either O or F.

of continental France and Spain share a high frequency of the "medium" α_{s1}-Cn^{E} allele, followed, in terms of occurrence, by one of the "strong" alleles but in Mediterranean and Canarian breeds, the frequency of the null α_{s1}-Cn^{O} allele is fairly high (0.08 to 0.20). In the context of this chapter, the terms "strong, medium and null" refer to alleles associated with a high amount, a medium amount and the absence of α_{s1}-casein, respectively. Breeds of the Mediterranean area and of Africa have high frequencies of "strong" alleles. Breed differences also exist in the distribution of the subdivisions of allele α_{s1}-Cn^{B} (Grosclaude and coworkers, unpublished). Until now, in addition to the Creole population of Guadeloupe (Mahé and Grosclaude, 1993), the null β-Cn^{O} allele was observed, at quite significant frequencies (0.06–0.11), in the Italian Garganica and Maltese breeds, and in an Italian local goat population (Rammuno *et al.*, 1994) as well as is in the French Pyrénées population (Ricordeau *et al.*, 1999). This allele has probably a wider range of distribution.

16.9 Milk protein polymorphism as marker genes in bovine species

Several aspects emerging from the study of genetic polymorphism of milk proteins have been proposed to have practical applications for the purpose of improving the overall efficiency in different sectors of the dairy industry. Because the different polymorphic forms of the major milk proteins are controlled by autosomal genes which are inherited in accordance with Mendelian inheritance, selection of cows for a specific milk protein type is feasible, and it is possible to breed for desirable specific genetic variants. Consequently, researchers around the world have shown considerable interest in using milk protein genes as genetic markers for increasing milk production, improving production-related traits and altering milk composition. It must be emphasised that any effects a certain genetic variant of a specific milk protein would have on production traits are in the strictest sense statistical effects only, i.e, any observed differences are mere associations and in most cases little is known about the cause-and-effect relationship. In contrast, the effect of a genetic variant on the protein composition of milk may be considered as a more direct effect because mutation of the protein in question is located in the coding section of its gene and hence could alter its rate of biosynthesis.

16.9.1 Milk production

To give a summary of the extensive literature that deals with possible relationships between genetic polymorphism of milk proteins and milk production would be a rather difficult task because of differing trends and, at times, conflicting results that have been reported. In many cases, results from different studies are not comparable for various reasons, including population size, breeds of animal, frequency of genetic variants under consideration, methods of measuring production traits (test-day or entire lactation yield), and most importantly, the rigor of statistical analysis to adjust for other important factors contributing to milk production such as age of cow, season, stage of lactation, health status and effects of other genetic variants. Before any conclusions could be drawn concerning the association between milk protein polymorphism and milk yield, all the above-mentioned factors have to be considered. The relationships between genetic variants of β-lactoglobulin and milk production serve as an example of apparent confusion existing in the literature. Several studies (Brum *et al.*, 1968; Cerbulis and Farrell, 1975; McLean *et al.*, 1984; Ng-Kwai-Hang *et al.*, 1984, 1990a; Lin *et al.*, 1986; Gonyon *et al.*, 1987; Haenlein *et al.*, 1987) have reported no relationships between β-lactoglobulin variants and milk production, whereas others (Janicki, 1978; Yebroski and Komissarenko, 1982; Jairam and Nair, 1983) showed superior production for cows with phenotype BB. There are also reports of cows homozygous for β-lactoglobulin A are higher milk producers (Comberg *et al.*, 1964; Konevtzova, 1978; Atroshi *et al.*, 1982; Babukov *et al.*, 1982; Ng-Kwai-Hang *et al.*, 1986; Aleandri *et al.*, 1990; Bovenhuis *et al.*, 1992). The association of β-lactoglobulin AB with higher milk yield has also been reported (Pupkova, 1980; Curie *et al.*, 1993).

The effects of casein variants on milk production are also not well established and the information obtained so far is inconclusive and depends on the nature of the study. No significant associations were observed between milk yield and genetic variants of α_{s1}-, β- and κ-caseins (McLean *et al.*, 1984; Gonyon *et al.*, 1987; Haenlein *et al.*, 1987; Kim *et al.*, 1996, 1998). At the α_{s1}-casein locus, no associations were found between the different phenotypes and milk production (Ng-Kwai-Hang *et al.*, 1990a; Bovenhuis *et al.*, 1992). In contrast, the superiority of phenotype BB over AB or BC for milk yield has been reported (Pupkova, 1980; Yebrovski and Komissarenko, 1982; Ng-Kwai-Hang *et al.*, 1984, 1986, 1990a; Lin *et al.*, 1986; Aleandri *et al.*, 1990; Chung *et al.*, 1991; Winkelman and Wickham, 1996). The reports of Lin *et al.* (1986) and Aleandri *et al.* (1990) indicated no differences in milk yield among the variants of β-casein. High milk yield is usually associated with the A variant of β-casein (Babukov *et al.*, 1982; Yebrovski and Komissarenko, 1982; Matyukov, 1983; Ng-Kwai-Hang *et al.*, 1984, 1986; Lin *et al.*, 1986), more specifically A^2 and A^3 (Matyukov, 1983; Ng-Kwai-Hang *et al.*, 1986, 1990a; Chung *et al.*, 1991; Bovenhuis *et al.*, 1992; Kim *et al.*, 1996; Winkelman and Wickham, 1996). No definite conclusion could be drawn concerning the association between milk production and κ-casein variants. Some studies indicate no relationship

(McLean *et al.*, 1984; Ng-Kwai-Hang *et al.*, 1984, 1990a; Lin *et al.*, 1986), while others suggest that κ-casein AA (Gonyon *et al.*, 1987; Babukov *et al.*, 1982; Bovenhuis *et al.*, 1992) or κ-casein AB (Yebrovski and Komissarenko, 1982; Ng-Kwai-Hang *et al.*, 1986; Kim *et al.*, 1998), or κ-casein BB (Pupkova, 1980) is positively associated with milk production.

According to Famula and Medrano (1994), a direct analysis of cows with different genetic variants using an animal model would provide unbiased estimates of differences between the variants. They estimated that the contribution of milk protein genotypes to production traits is only minor. Because the casein genes are linked (see Section 16.6, "Linkage between milk protein genes"), it might be more appropriate to consider the haplotypes of α_{s1}-casein, β-casein and κ-casein in the analysis of data. No significant associations were found between milk yield and 10 haplotypes of caseins for Norwegian cattle (Lien *et al.*, 1995). Velmala *et al.* (1995) found that haplotype BA^2A was associated with higher milk yield than haplotype BA^1E. In Holstein and Jersey cattle, haplotype $BB/A^1A^2/AB$ was superior to $BB/A^1A^2/AA$ and $BB/A^1A^1/AB$ by 252 and 338 kg of milk, respectively, during the first lactation (Ojala *et al.*, 1997).

16.9.2 Milk composition

The absolute concentrations and relative proportions of various constituents present in milk will dictate the overall value of this farm produce in terms of economics, nutrition and technological properties. There are many reports that relate genetic polymorphism of milk proteins to milk content of fat and protein, and to a lesser extent of minerals and other organic ions that have a profound effect on the manufacturing properties of the milk. It appears that reports on the associations of genetic variants with different milk components is not as controversial as those with milk production.

(a) Milk fat

Although some studies (Cerbulis and Farrell, 1975; Graml *et al.*, 1985; Lin *et al.*, 1986; Gonyon *et al.*, 1987; Ng-Kwai-Hang *et al.*, 1990a; Kim *et al.*, 1996) indicate no significant correlation between genetic variants of β-lactoglobulin and the fat content of milk, others tend to show the contrary (Comberg *et al.*, 1964; Hoogendoorn *et al.*, 1969; Janicki, 1978; McLean *et al.*, 1984; Graml *et al.*, 1986a; Ng-Kwai-Hang *et al.*, 1984, 1986; Aleandri *et al.*, 1990; Bovenhuis *et al.*, 1992; Hill and Paterson, 1994). In most cases, β-lactoglobulin B, either in the homozygote or heterozygote form, AB, favours a higher fat percentage in the milk. For example, results of 25 000 test-day milk samples (Ng-Kwai-Hang *et al.*, 1986) from Holstein cows averaged 3.72 and 3.67% fat for β-lactoglobulin BB and AA, respectively. This difference of 0.05% in fat content may seem small but it is quite significant at the production level when one considers the volume of milk produced.

At the α_{s1}-casein loci, some reports (Zebrovskii *et al.*, 1977; Munro, 1978; Ng-Kwai-Hang *et al.*, 1986) show that a higher fat content is associated with the BC phenotype. Unfortunately, in all those studies, there was an absence of homozygous CC for α_{s1}-casein so that it was not possible to disclose whether variant C would be even more advantageous for fat content when it is present in the homozygous form. Other studies (Ng-Kwai-Hang *et al.*, 1984, 1990a; Gonyon *et al.*, 1987; Haenlein *et al.*, 1987) showed no relationships between α_{s1}-casein types and milk fat. Highest milk fat is associated with β-casein B (Babukov, 1978; McLean *et al.*, 1984; Ng-Kwai-Hang *et al.*, 1986; Bovenhuis *et al.*, 1992). For the different possible phenotypes of β-casein that were present, the fat content of the milk was in the following descending order according to genetic variants: A^1B, A^1A^1, A^1A^2, A^2A^2, A^1A^3, A^2A^3, A^2B (Ng-Kwai-Hang *et al.*, 1986). Higher fat content has also been correlated to κ-casein BB with 3.76% compared to 3.67% for κ-casein AA (Ng-Kwai-Hang *et al.*, 1986). In contrast, phenotype AA (Ng-Kwai-Hang *et al.*, 1990a; Chung *et al.*, 1991) and BC (Curic *et al.*, 1993) have been shown to be associated with more milk fat.

(b) Milk protein

The original discovery by Aschaffenburg and Drewry (1957) that β-lactoglobulin AA milk contains more protein than AB or BB milks has been confirmed by many researchers (Moustgaard *et al.*, 1960; Feagan *et al.*, 1972; Cerbulis and Farrell, 1975; Janicki, 1978; Mariani *et al.*, 1979b; McLean *et al.*, 1984; Ng-Kwai-Hang *et al.*, 1984, 1986, 1990a). There are also reports (Haenlein *et al.*, 1987; Hill, 1993; Hill and Paterson, 1994) indicating no difference in the protein content associated with variants A and B of β-lactoglobulin. The difference in protein content is due largely to altered rate of synthesis of β-lactoglobulin (Aschaffenburg and Drewry, 1957; Cerbulis and Farrell, 1975; McLean *et al.*, 1984; Ng-Kwai-Hang *et al.*, 1987). It was confirmed that the rate of synthesis of β-lactoglobulin was higher for the A allele in heterozygous AB German Simmental and German Brown cows (Graml, 1995) and Ayrshire, Jersey, Brown Swiss and Canadienne (Ng-Kwai-Hang and Kim, 1996). The higher β-lactoglobulin content associated with the AA phenotype will result in a higher whey protein fraction in milk and, if the rate of synthesis of casein is not altered to the same extent, there will be an alteration in the ratio of casein to total protein. This ratio, which is often known as 'casein number', is an important measure of casein- or cheese-yielding capacity of milk. Although a few reports indicate no influence of β-lactoglobulin polymorphism on total casein content (Cerbulis and Farrell, 1975; Mityukov, 1982), most work confirms that β-lactoglobulin AA is associated with less casein than the BB genotype (Mariani *et al.*, 1979a; Buchberger *et al.*, 1982; McLean *et al.*, 1984; Mariani, 1985; Ng-Kwai-Hang *et al.*, 1986, 1987; Rampilli *et al.*, 1988; van den Berg *et al.*, 1992; Hill and Paterson, 1994). The higher content of whey protein mainly due to β-lactoglobulin and the lower content of casein associated with the AA genotype contributes to the lower casein number observed with this phenotype (Schaar *et al.*, 1985; Ng-Kwai-Hang *et al.*, 1986; Reichardt *et al.*, 1995; Lodes *et al.*, 1997a). Studies on the effects β-lactoglobulin polymorphism on the concentrations of individual caseins and whey proteins (Aaltonen and Antila, 1987; Ng-Kwai-Hang *et al.*, 1987) indicated that β-lactoglobulin AA is associated with lower levels of α_s-casein, β-casein, κ-casein, α-lactalbumin, serum albumin and immunoglobulin, but a higher content of β-lactoglobulin. The α_s-casein notation is used to represent the sum of α_{s1}- and α_{s2}-casiens. From a study involving eight breeds of cattle, Komatsu *et al.* (1977) estimated that the mean content of β-lactoglobulin in milk is 2.58 for allele A compared to 1.70 mg/ml for allele B. Such a considerable difference in β-lactoglobulin content associated with the two types of this protein suggests that the structural gene for β-lactoglobulin is a major gene which determines the amount of this particular protein being synthesised. Genetic polymorphisms of α_{s1}-casein, β-casein and κ-casein are associated with either the amount of the corresponding casein, or with other casein or whey protein fractions so that there is an alteration in the total casein or protein content of milk. Some researchers (Gonyon *et al.*, 1987; Ng-Kwai-Hang *et al.*, 1990; Kim *et al.*, 1996, 1998) found no relationships between phenotypes of α_{s1}-casein and protein percentage in milk. Phenotype BC of α_{s1}-casein is associated with higher levels of α_{s1}-casein, total casein and total protein but lower β-lactoglobulin and whey protein than the BB or AB types (McLean *et al.*, 1984; Kroeker *et al.*, 1985; Ng-Kwai-Hang *et al.*, 1986, 1987; Lodes *et al.*, 1997a). Phenotype BC of α_{s1}-casein is also associated with higher amounts of α_{s1}-casein, β-casein and κ-casein and lower amounts of α-lactalbumin (Ng-Kwai-Hang *et al.*, 1987). Milk containing the newly discovered variant G of α_{s1}-casein (Rando *et al.*, 1998) contains less α_{s1}-casein, total casein and protein. This may be due to a decrease in the stability of the mRNA for α_{s1}-casein G. No differences were found in the milk protein content for the different phenotypes of β-casein (McLean *et al.*, 1984; Ng-Kwai-Hang *et al.*, 1990a; Aleandri *et al.*, 1990; Hartung *et al.*, 1993; Kim *et al.*, 1996, 1998). In contrast, Ng-Kwai-Hang *et al.* (1986), Gonyon *et al.* (1987), Bovenhuis *et al.* (1992) and Lodes *et al.* (1997a) reported that variant A^1 is associated with higher milk protein than variant A^2. For the polymorphic forms of β-casein, the A^1 and A^2 types are associated with a higher content of α_{s1}-casein and β-lactoglobulin, but lower β-casein, κ-casein and α-lactalbumin than the B type (Ng-Kwai-Hang *et al.*, 1987). The phenotype A^1B is associated with the highest levels of total protein (3.48%) and casein (2.78%) (Ng-Kwai-Hang *et al.*, 1986) whereas A^2A^2 is lowest in both components, 3.36 and 2.66%, respectively. Several reports have shown consistently that the B variant of κ-casein (Mariani *et al.*, 1976; McLean *et al.*, 1984; Kroeker *et al.*, 1985; Ng-Kwai-Hang *et al.*, 1986, 1987;

Rampilli *et al.*, 1988) is associated with a high amount of κ-casein in milk. In heterozygous κ-casein AB cows, the B allele is expressed more than the A (Van Eenennaam and Medrano, 1991). Variant B is also associated with high total casein (Ng-Kwai-Hang *et al.*, 1986; van den Berg *et al.*, 1992; Hartung *et al.*, 1993) and protein (Ng-Kwai-Hang *et al.*, 1984, 1986, 1990a; Bovenhuis *et al.*, 1992; Hartung *et al.*, 1993; Kim *et al.*, 1998), a relationship which has generated great interest within the cheese industry for the genetic polymorphism of κ-casein. κ-Casein BB is associated with higher levels of α_{s1}-casein, κ-casein, serum albumin and immunoglobulin but less β-casein, β-lactoglobulin and α-lactalbumin (Ng-Kwai-Hang *et al.*, 1987).

(c) Other milk components

Most of the research has concentrated on finding associations between protein polymorphisms and the two major constituents of milk, fat and protein. However, there are also fewer reports relating polymorphisms with minor constituents. Mariani *et al.* (1979a) showed that κ-casein A milk was associated with a higher citric acid content than κ-casein B milk, while Braunschweig and Puhan (1997) and Lodes *et al.* (1997b) reported no difference in citric acid content for the two variants. Mariani *et al.* (1984) reported a lower sialic acid content of total casein containing κ-casein A compared to κ-casein B. A higher concentration of calcium has been associated with κ-casein B (van den Berg, 1994; Castagnetti *et al.*, 1994). Hill and Paterson (1994) found no differences in the levels of minerals and lactose in β-lactoglobulin A and B milks. There were no significant differences in the sodium, calcium, potassium and magnesium contents of milk for the different phenotypes of α_{s1}-casein, β-casein, κ-casein and β-lactoglobulin, but less phosphate was associated with α_{s1}-casein CC (Lodes *et al.*, 1997b). Comparison of the composition of milk for different types of α-lactalbumin is rare because only one variant, B, is known to exist in the common dairy breeds. Research on Zebu cows showed that α-lactalbumin AB was associated with a higher potassium concentration in milk than the BB genotype (Perez-Beato and Ponce, 1985), but there were no significant differences in lactose and sodium for the two types of α-lactalbumin.

16.9.3 Dairy production-related traits

In addition to the search for associations between genetic polymorphs of milk proteins and milk production and milk composition, there have also been some studies aimed at finding relationships with production-related traits. The rationale is that milk production is dependent on such factors as growth rate of the heifer, reproductive performance, health status and productive life of the cow.

(a) Growth rate and body weight

The initial growth rate of calves from birth to production (Kriventsov, 1978; Singh *et al.*, 1981) was faster for those fed on β-lactoglobulin AA milk than those fed on β-lactoglobulin BB milk. One possible explanation for this observation is the fact that β-lactoglobulin AA milk has more protein, as discussed in Section 16.9.2(*b*), and hence is more nutritious. Other studies (Pokalov, 1975; Lin *et al.*, 1987) have shown that heifers with the gene for β-lactoglobulin A were heavier than those with the gene for β-lactoglobulin B. In this case, the difference observed is due to genetics and not nutrition because both groups of heifers were raised under the same conditions, including nutritional regime. Saama *et al.* (1992) reported no significant differences in the growth rate of Norwegian bulls between 90 and 360 days of age for the different variants of κ-casein and β-lactoglobulin. In a study of three Hereford cattle populations, Moody *et al.* (1996) concluded that the B variant of κ-casein was associated with lower body weight gain than the A variant. Cho *et al.* (1998) found that growth traits in Korean native cattle were affected by variants of κ-casein and not by those of β-lactoglobulin.

(b) Reproductive performance

The age of heifers at first calving is related to the variant of α-lactalbumin, β-lactoglobulin and κ-casein (Jairam and Nair, 1983; Ronda *et al.*, 1984). Heifers with BB α-lactalbumin and heterozygote AB for β-lactoglobulin or κ-casein are usually younger than heifers of other phenotypes at the start of the first lactation, and β-lactoglobulin AB heifers have a longer gestation period than either the AA or BB heifers (Lin *et al.*, 1987). β-Lactoglobulin AB heifers are younger at first conception and require fewer days from first service to conception. In a study involving more than 6000 cows and 700 sires, Hargrove *et al.* (1980) and Lin *et al.* (1987) found no significant differences between polymorphs of β-lactoglobulin, α_{s1}-casein, β-casein and κ-casein, and conception rate, average days open and the number of cows that conceived by the third service. From another study based on 4819 Holstein cows, Ng-Kwai-Hang *et al.* (1990b) concluded that there were no associations between genetic variants of the above four proteins and number of days to first breeding, days open and services per conception. Panicke *et al.* (1998) reported that there were no associations between the type of α_{s1}-casein and different measures of reproductive performance, but there were favourable effects of β-casein A^2A^2 and κ-casein AA on days open and number of services per conception in the German Black Pied cows.

(c) Mastitis

Mastitis is a disease caused by inflammation of the mammary gland and is of great economic importance to the dairy industry. Besides affecting the health of the cow, this condition drastically affects milk production and milk quality. From the literature, no definite conclusion can be drawn regarding the association of genetic polymorphism with the incidence of mastitis. In a study involving more than 2000 cows over two lactations (Jansen *et al.*, 1981), no difference was observed between the incidence of mastitis and polymorphs of β-lactoglobulin, β-casein and κ-casein. Similar conclusions were reached by Ng-Kwai-Hang, Monardes and Hayes (unpublished) who observed no association between lactational somatic cell count and genetic variants of α_{s1}-casein, β-casein, κ-casein and β-lactoglobulin for 2900 cows over three lactations. A lower incidence of mastitis was associated with β-lactoglobulin AB (Giesecke and Osterhoff, 1975; Stur *et al.*, 1976) but there was no relationship between the polymorphs of α_{s1}-casein, β-casein and κ-casein and the incidence of mastitis. While Leonhard-Kluz and Kaminska (1981) and Atroshi *et al.* (1982) reported that a higher resistance to mastitis was associated with β-lactoglobulin BB, Han *et al.* (1986) reported the opposite. With an increased infection of the mammary gland, it is known that the concentration of immunoglobulins in milk increases at the expense of β-lactoglobulin. Therefore, one would expect mastitis to be associated with protein variants which alter either the β-lactoglobulin or immunoglobulin content in milk. From other studies (for example, Ng-Kwai-Hang *et al.*, 1987), α_{s1}-casein BC, β-casein A^1B, κ-casein BB and β-lactoglobulin BB, which are associated with a high immunoglobulin concentration, could also be associated with a high incidence of mastitis. According to Jairam *et al.* (1994), casein types had no influence on the incidence of mastitis in Ongole, Jersey, Friesian and their crosses, but α-lactalbumin type had influence among the Friesian × Ongole crossbred and β-lactoglobulin type had influence among Jersey and Jersey × Ongole crossbred.

(d) Productive life

Phenotypes of α_{s1}-casein, β-casein and β-lactoglobulin were not associated with herd survival (Lin, 1995) whereas κ-casein AA, AB and BB had an effect with herd life of 55.1, 56.3 and 61.0 months, respectively. The higher frequency of κ-casein B observed in the seventh lactation suggested a more favourable survival rate for variant B. The breeding value for length of productive life of daughters of Austrian Simmental and Austrian Brown bulls were higher for κ-casein AA than for BB bulls by 15 and 97 days, respectively (Ortner *et al.*, 1995).

16.10 GENETIC POLYMORPHISM AND PROPERTIES OF BOVINE MILK

The properties of milk and dairy products are influenced to a large extent by the amounts and relative proportions of each of the milk constituents. The protein fraction, because of its diversity in concentration and type, contributes significantly in this regard. Polymorphism of a milk protein caused by deletion or substitution of amino acids is bound to alter the properties of this particular protein which depend on the linear array of amino acids endowing it with a unique primary structure. Differences in amino acid composition and sequence for genetic variants could partially explain changes in the properties of the molecules through a combination of a series of modification including net charge, hydrophobicity, degree of phosphorylation and glycosylation, all of which contribute to the behaviour of milk proteins and hence the overall manufacturing properties of the milk. Here, we encounter the direct effect of genetic polymorphism since there are modifications in protein structure due to mutations which lead to alterations in the overall properties of the protein.

16.10.1 Physico-chemical properties of genetic variants

A review of the relationships between genetic variants of the caseins and β-lactoglobulin, and physico-chemical properties was presented by Creamer and Harris (1997). X-Ray crystallographic studies (Bewley *et al.*, 1997) revealed that β-lactoglobulin A, B and C had a similar three-dimensional structure with an eight-stranded β-barrel. The substitution of Ala_{118} for Val (B → A) caused local internal readjustments which may affect the dimer interface and explain the differences in aggregation. The Gln_{59} to His substitution (B → C) resulted in a change of stabilising interactions on the surface, converting a hydrogen bond to an ion pair. Brittan *et al.* (1997) found that the free sulphydryl group of β-lactoglobulin was at position 121, irrespective of variant A, B or C.

Although there is a paucity of research that compares the physico-chemical properties of α_{s1}-casein polymorphs due to the very high frequency of the B variant and the lack of A or C variants in most of the dairy cattle population, the few observations that have been made are worth noting. The quality of cheese made from milk containing α_{s1}-casein A is inferior to that from milk with the variant B (Sadler *et al.*, 1968) and the former milk is more difficult to process (Thompson *et al.*, 1969a). The deletion of 13 amino acid residues in α_{s1}-casein A reduces the hydrophobic nature of the protein compared to α_{s1}-casein B (Creamer *et al.*, 1982), leading to a change in the rate of proteolysis during cheese ripening. Because of the lack of residues 14 to 26, variant A of α_{s1}-casein does not have the primary cleavage site of chymosin and a major cleavage site of plasmin. Coker *et al.* (1997) reported that α_{s1}-casein A is hydrolysed less rapidly by chymosin than variants B or C and that the three variants of α_{s1}-casein were degraded at the same rate by plasmin. Casein adhesives manufactured from casein containing α_{s1}-casein A are less viscous and have better adhesive strength than those containing either the B or C variants (Mountfort, 1984). This desirable attribute of

α_{s1}-casein A has prompted New Zealand researchers to establish a dairy herd consisting of cows homozygote for the very rare A for α_{s1}-casein. The rate of hydrolysis of β-casein by plasmin (Papoff *et al.*, 1995) was faster for the A^1 than the C variant. The same mixture, but different proportions, of peptides were obtained. Imafidon and Farkye (1994) reported that the effect of the genetic variant on chymosin hydrolysis of κ-casein depended on pH and temperature. At pH 6.0–6.7, no differences were observed between the A and B variants, whereas at pH 5.3–5.6, κ-casein B was hydrolysed more extensively than A. According to Smith *et al.* (1997), κ-casein C was less tightly bound to chymosin and was hydrolysed at a slower rate than variants A and B. There are differences in the relative susceptibilities of the 17 tryptic hydrolysis sites of β-lactoglobulin (Huang *et al.*, 1994), depending on the genetic variant – β-lactoglobulin B yielded larger peptides which were not present in β-lactoglobulin A hydrolysates. The influence of genetic variant on tryptic and chymotryptic hydrolysis of β-lactoglobulin is pressure-dependent (Willige and Fitzgerald, 1995). At atmospheric pressure, native β-lactoglobulin A is hydrolysed by the enzymes three times faster than the B variant, whereas at higher pressures (100 and 300 MPa), there were no differences in the rate of hydrolysis. The stability of casein micelles in the presence of calcium, lactate, citrate and phosphate is dependent on the genetic variants of α_{s1}-casein, β-casein and κ-casein present (El-Negoumy, 1974; Imafidon, 1990; Ng-Kwai-Hang and Imafidon, 1990). Micelles containing the combination BB-AB-AB of the three respective caseins were found to be more stable than micelles containing the combination BB-AA-BB. This difference in casein micelle stability may be due to differences in hydrophobicity between the polymorphs of β-casein (Eigel and Randolph, 1976). Anema and Creamer (1993) reported that solvation of micelles from milk containing α_{s1}-casein B and κ-casein A was lower than that from the A and B variants, respectively. α_{s1}-Casein B and κ-casein A were associated with lower κ-casein content in the milk and micelles, which is consistent with the lower solvation observed for two variants.

The B variant of β-casein is different from the A variant by having a lower water-binding capacity (Kirchmeir *et al.*, 1983). Differences in physico-chemical properties between variants A and B of κ-casein include higher heat stability (Kirchmeier *et al.*, 1983; McLean *et al.*, 1987) and smaller casein micelles (Morini *et al.*, 1975) in milk containing the B variant. Emulsifying properties measured in terms of interfacial area, protein load, particle size and creaming index were more favourable for α_{s1}-casein B, β-casein A^2 and κ-casein A (Gao *et al.*, 1996). Euston *et al.* (1997) reported that among β-casein variants, the A^1 variant had higher emulsifying activity than A^2 and among β-lactoglobulin, the trend of increasing emulsion stability was C > B > A. Another study (Bo and Pitotti, 1997) indicated that emulsion stability of β-lactoglobulin B was higher than that of A. Emulsion activity index was affected more by pH and salt concentration than by the genetic variant and the emulsion capacity of β-lactoglobulin B was less affected than A by the different pH and salt concentration. The influence of genetic variants of α_{s1}-casein, β-casein and κ-casein on surface-active properties have been reported by Gao *et al.* (1997). The heat stability of milk is affected by β-lactoglobulin variant (Feagan, 1979; Hillier *et al.*, 1979; McLean *et al.*, 1987; Imafidon, 1990; Nielsen *et al.*, 1996; Anema and McKenna, 1996; Allmere *et al.*, 1997; Holt *et al.*, 1998) as well as temperature, pH, protein concentration and method of measurement. Hillier *et al.* (1979) reported that at temperatures below 90°C, β-lactoglobulin A was more heat stable than β-lactoglobulin B, but at temperatures above 90°C, the situation was reversed. At pH 4 and 9, β-lactoglobulin A was less stable than β-lactoglobulin B over the entire range of temperature. Differential scanning calorimetric measurements in phosphate buffer at pH 6.8 (Imafidon, 1990; Ma *et al.*, 1990) indicated that β-lactoglobulin BB had a higher denaturation temperature than either the AB or AA phenotype. Manderson *et al.* (1997) and Lowe and Hill (1997) reported that β-lactoglobulin A unfolded and formed aggregates during heat treatment to give intermediates that were different from those obtained from variants B and C. The combination of β-lactoglobulin AA with κ-casein BB was more heat stable at 140°C than the BB/AA combination (Hill *et al.*, 1997a). During UHT treatment, milk containing the A variant of β-lactoglobulin fouled plate heat exchangers more rapidly than the B variant milk (Hill *et al.*, 1997b). Three genetic variants of β-lactoglobulin were tested for their ability to inhibit the hydrolysis of *p*-nitrophenyl phosphate by phosphoprotein phosphatase (Farrell and Thompson, 1971); the results indicated that the inhibitory effect of β-lactoglobulin was in the order A > B > C. The rate of reaction of β-lactoglobulin with chemically activated κ-casein or with 5′, 5′-dithiobis (2-nitrobenzoic acid) varied according to genetic variants: A > B > C (Thresher, 1997). Whey protein concentrates made from different β-lactoglobulin variants had different gelation properties (Foegeding *et al.*, 1997). At a high salt concentration (>100 mM NaCl), variant B formed a more rigid gel, while at a low salt concentration, the gels formed from phenotypes AA, AB, BB were of equal elasticity. These differences are related to structural changes in the β-lactoglobulin variants.

16.10.2 Coagulating properties of milk

The rate and extent of coagulation of milk by rennet and the characteristics of the resulting curd are very important factors which have to be considered in the process of cheesemaking. The first indication of effects of genetic polymorphism on coagulation properties (Sherbon *et al.*, 1967) was through the following observations: β-lactoglobulin B milk had a higher curd tension than milk containing the A variant; milk containing α_{s1}-casein BC was different from BB milk in terms of clotting time and consistency of the gel, and κ-casein AA milk took longer to coagulate and produced a softer gel than either AB or BB milk. To date, several laboratories have confirmed these original findings and most of the reports have concentrated on the effects of κ-casein polymorphism on rennet clotting time, rate of curd formation and curd firmness as determined by a formagraph. In general, most of the reports (Losi *et al.*, 1973, 1975; Mariani *et al.*, 1976; Tervala *et al.*, 1983, 1985; Schaar, 1984; Mariani and Leoni, 1985; Jacob and Puhan, 1986; Marziali and Ng-Kwai-Hang, 1986a; Aaltonen and Antila, 1987; Rampilli *et al.*, 1988; Horne and Muir, 1994; Kreuze *et al.*, 1996) are consistent in that milk containing κ-casein B is better than κ-casein A milk by having a shorter coagulation time, a faster rate of firming, and producing a firmer curd. This effect of κ-casein B may be due partially to higher levels of fat and casein in milk containing this variant. However, Marziali and Ng-Kwai-Hang (1986a) demonstrated the direct effect of κ-casein variant on coagulation properties after adjustments were made for the effects of milk composition. In terms of genetic polymorphism in other caseins, it appears that α_{s1}-casein BC is better than BB (Mariani *et al.*, 1988) and that β-casein BB is better than the AA type (Mariani *et al.*, 1986). Delacroix-Buchet and Marie (1994) compared milks containing the A with those of C variant of β-casein. They found that β-casein C was associated with casein micelles of larger diameters and lower calcium and thus exhibited non-typical behaviour during rennet coagulation, leading to softer curds, higher fat losses in whey and lower cheese yields. β-Casein A milk had better rheological properties (Castagnetti *et al.*, 1994) than β-casein B milk by having shorter clotting and gel firming times. Rennet coagulation time and gel firmness as measured by a gelograph were associated with variants C of α_{s1}-casein, B and C of β-casein, C and E of κ-casein, C, D and W of β-lactoglobulin (Lodes *et al.*, 1996).

Differences also exist in the coagulation behaviour between milks containing β-lactoglobulin A or β-lactoglobulin B (Marziali and Ng-Kwai-Hang, 1986a).

16.10.3 Cheesemaking

Several studies have demonstrated that the effects of genetic polymorphism of milk proteins on the coagulation properties are also reflected in cheese yield and cheese composition which is a determinant of cheese quality. In the manufacture of Parmesan cheese, Morini *et al.* (1979) obtained approximately 8% higher yield from κ-casein B than with κ-casein A milk. This yield differential in favour of the B variant was about 4% in the case of Cheddar-type cheese (Marziali and Ng-Kwai Hang, 1986b). However, no differences were observed in the yield of Svecia cheese (Schaar, 1986) between the two variants of κ-casein, possibly due to differences in cheesemaking processes. For the production of Tilsit cheese (Pabst, 1998), earlier cutting times, better rennetability and less cheese fines in whey were associated with κ-casein BB milk. Losses of fat and fines in the whey and moisture-adjusted yield of Edam cheese were influenced by genotype combinations of β-casein, κ-casein and β-lactoglobulin (Mayer *et al.*, 1997). Marie and Delacroix-Buchet (1994) compared the A and C variants of β-casein for making Beaufort cheese. The C variant cheese contained less fat, had an irregular structure, was firmer and less elastic than the A variant cheese. The differences in the hydrolysis of β-casein C, which is more frequent among the Tarentaise and Abondance breeds, were thought to be responsible for the differences in cheese characteristics and could be used as a marker for Beaufort cheese (Delacroix-Buchet and Marie, 1994). The higher yield for κ-casein B milk is also associated with higher levels of fat and protein in the cheese (Marziali and Ng-Kwai-Hang, 1986c). According to Morini *et al.* (1979), Parmesan cheese made from κ-casein BB milk is more desirable than that from κ-casein AA milk because of better organoleptic and physico-chemical properties. Cheese from κ-casein BB milk underwent greater syneresis during the first 24h, lost less water during ripening and contained more fat with a higher fat to protein ratio. Comparison of cheese yield from milks containing β-casein A^1A^1 or β-casein A^1A^2 revealed the superiority of the former (Marziali and Ng-Kwai-Hang, 1986b) due to greater retention of fat and protein in the cheese curd and lower losses of milk components in the whey (Marziali and Ng-Kwai-Hang, 1986c). Although no significant differences were observed in cheese yields for phenotypes AA, AB and BB of β-lactoglobulin, recovery of fat and protein in cheese was higher for the BB type due to lower losses in the whey (Marziali and Ng-Kwai-Hang, 1986b,c). Based on the results of a preliminary study, Graham *et al.* (1984) suggested that the combination of AB, BB, BB for κ-casein, β-casein and β-lactoglobulin, respectively, would be more favourable than AA, AA, AA in terms of casein content, coagulating properties and cheese yield.

There is a general consensus mat the cheesemaking properties of κ-casein B and β-lactoglobulin B milk are more favourable (Delacroix-Buchet *et al.*, 1993; Tong *et al.*, 1993; van den Berg *et al.*, 1994; Jakob, 1994b; Walsh *et al.*, 1995; Ng-Kwai-Hang, 1998).

16.11 Genetic polymorphism and properties of milk in other species

The influence of genetic polymorphism on the composition and properties of milk from species other than bovine, especially caprine and ovine, have also been studied extensively. With 11 genetic variants so far identified (Grosclaude and Martin, 1997), the caprine α_{s1}-casein locus is exceptionally polymorphic. Although there is no association between genetic variants of caprine α_{s1}-casein and milk yield (Grosclaude *et al.*, 1994), the composition and properties of the milk are affected by certain alleles. The alleles could be classified into four groups according to the amounts of α_{s1}-casein they are associated with as follows: (A, B_1, B_2, B_3, B_4, C); (E); (F, G); (O_1, O_2) with 3.5, 1.1, 0.45 and 0 mg/ml per allele, respectively. These values, which are slightly different from 3.6, 1.6, 0.6 and 0 mg/ml (Mahé and Grosclaude, 1989), represent a revision of the earlier estimates. Phenotype AA milk had the highest and FF the lowest concentrations of protein and fat. The lower calcium content and smaller diameter of casein micelles associated in the AA milk could explain the faster rate of firming and firmer gel associated with this type of caprine α_{s1}-casein. In the traditional cheesemaking of Pélardon des Cévennes-type cheese, there were differences of +7.4 and 14.8% in adjusted cheese yield when AA cheese milk was compared to EE and FF, respectively. The "goaty" flavour tended to be less pronounced in cheese made from AA milk.

The reports of Pirisi *et al.* (1997) and Chianese *et al.* (1997b) indicated that ovine α_{s1}-casein CC milk was associated with higher percentages of dry matter, fat, protein and casein, better coagulating properties and higher cheese yield when compared to AA and DD milk. The B variant of β-lactoglobulin is associated with a higher concentration of whey protein (Rampilli *et al.*, 1997; Di Stasio *et al.*, 1997), less casein and β-lactoglobulin, and inferior rennetability (Rampilli *et al.*, 1997).

16.12 Concluding remarks

The first report of genetic polymorphism in β-lactoglobulin approximately 47 years ago stimulated active research efforts in the areas of biochemistry and genetics of milk proteins. Initially, most research was focussed on developing methods for identification and on studying the mode of inheritance and frequency distribution of genetic variants of different milk proteins in various bovine breeds. Detection of genetic variants by electrophoresis has been possible because most of the mutants identified so far involve a change in the net charge of the protein. Additional variants have been identified by isoelectric focusing, HPLC techniques in combination with mass spectrometry and DNA sequencing. With the developments in molecular biology, methods are available for typing known genetic variants of the milk proteins at the DNA level and this is of practical importance because of the possibility of genotyping animals before initiation of lactation, as well as sires located in the artificial insemination centres. Polymorphisms detected in the non-coding regions of the milk protein genes might affect the rate of synthesis of milk proteins and not the properties of the specific proteins. Because milk protein genes are simply inherited in the Mendelian fashion, there is considerable interest in using them as markers for production and milk composition traits. Several studies have adapted genetic polymorphism of milk proteins as markers for the characterisation and differentiation of cattle population. From the literature, reports relating bovine milk protein variants with various aspects of dairy production are inconclusive and at times contradictory, mainly due to differences in the rigor of statistical analysis among the various studies. However, it has been shown consistently that the B variant of κ-casein and the A variant of β-lactoglobulin are associated with a higher protein content in the milk. The ratio of casein to whey protein is also altered by the variants of β-lactoglobulin and κ-casein present and this probably has a profound effect on the manufacturing properties of the milk. Studies on the genetic polymorphism of caprine α_{s1}-casein have shown convincing evidence that certain alleles of this protein are associated with higher concentrations of casein in milk. Before recommendations can be made regarding the selection for specific variants of milk proteins, it would be wise to establish that these variants are not associated with factors which are detrimental to the overall dairy industry. However, there is already selection for the B variant of bovine of κ-casein in Italy and for the "strong" variant of caprine α_{s1}-casein in France. Such a decision could be justifed by considering the overall advantages and inconveniences. Reports showing differences in the physico-chemical and coagulating properties of milk due to the presence of certain protein variants have to be confirmed by studies comparing isolated and pure forms of individual proteins without the interference of other milk components. Hence, this opens up a new area of research in the biochemistry of milk proteins.

References

Aaltonen, M.L. and Antila, V. (1987) Milk renneting properties and the genetic variants of proteins. *Milchwissenschaft*, **42**, 490–2.

Abe, T., Komatsu, M., Oishi, T. and Kageyama, A. (1975) Genetic polymorphism of milk proteins in Japanese and European cattle breeds in Japan. *Jap. J. Zootech. Sci.*, **46**, 591–9.

Addeo, F., Mercier, J.-C. and Ribadeau-Dumas, B. (1977) The caseins of buffalo milk. *J. Dairy Res.*, **44**, 455–68.

Addeo, F., Chianese, L., Di Luccia, A., Petrilli, P., Mauriello, R. and Anelli, G. (1983) Identification of bovine casein variants by gel isoelectric focusing. *Milchwissenschaft*, **38**, 586–8.

Addeo, F., Mauriello, R. and Di Luccia, A. (1988) A gel electrophoretic study of caprine casein. *J. Dairy Res.*, **55**, 413–21.

Agricultural University of Norway (1992) Conference on milk protein variants. Hanko, Norway, September 2–4, 1992.

Aleandri, R., Buttazzoni, L.G., Schneider, J.C., Caroli, A. and Davoli, R. (1990) The effects of milk protein polymorphisms on milk components cheese producing ability. *J. Dairy Sci.*, **73**, 241–55.

Allmere, T., Andren, A. and Bjorck, L. (1997) Interactions between different genetic variants of β-lactoglobulin and κ-casein A during heating of skim milk. *J. Agric. Food Chem.*, **45**, 1564–9.

Althen, T.G. and Gerrits, R.J. (1972) Polymorphism of $whey_2$ protein of sow's milk. *J. Dairy Sci.*, **55**, 331–3.

Ambrosoli, R., Di Stasio, L. and Mazzocco, P. (1988) Content of α_{s1}-casein and coagulation properties in goat's milk. *J. Dairy Sci.*, **71**, 24–8.

Andrews, A.T., Taylor, M.D. and Owen, A.J. (1985) Rapid analysis of bovine milk proteins by fast protein liquid chromatography. *J. Chrom.*, **348**, 177–85.

Anema, S.G. and Creamer, L.K. (1993) Effect of the A and B variants at both α_{s1}- and κ-casein on bovine casein micelle solvation and κ-casein content. *J. Dairy Res.*, **60**, 505–16.

Anema, S.G. and McKenna, A.B. (1996) Reaction kinetics of thermal denaturation of whey proteins in heated reconstituted whole milk. *J. Agric. Food Chem.*, **44**, 422–8.

Annan, W.D. and Manson, W. (1969) A fractionation of the α_s-casein complex of bovine milk. *J. Dairy Res.*, **36**, 259–68.

Arave, C.W. (1967) Procedure for simultaneous phenotyping of β-casein and β-lactoglobulin variants in cow's milk. *Dairy Sci.*, **50**, 1320–2.

Arave, C.W., Gillet, T.A., Price, D.A. and Matthews, D.H. (1973) Polymorphisms in casein of sheep milk. *J. Anim. Sci.*, **36**, 241–4.

Archibald, A., Couperwhite, S., Haley, H.S., Beattie, C.W. and Alexander, L.J. (1994) RFLP and linkage analysis of the porcine casein loci, CASAS1, CASAS2, CASB and CASK. *Anim. Genet.*, **25**, 349–51.

Aschaffenburg, R. (1961) Inherited casein variants in cow's milk. *Nature*, **192**, 431–2.

Aschaffenburg, R. (1964) Protein phenotyping by direct polyacrylamide gel electrophoresis of whole milk. *Biochim. Biophys. Acta*, **82**, 188–91.

Aschaffenburg, R. (1965) Variants of milk proteins and their pattern of inheritance. *J. Dairy Sci.*, **48**, 128–32.

Aschaffenburg, R. (1968) Genetic variants of milk proteins: their breed distribution. *J. Dairy Res.*, **35**, 447–60.

Aschaffenburg, R. and Drewry, J. (1955) Occurrence of different beta-lactoglobulins in cow's milk. *Nature*, **176**, 218–9.

Aschaffenburg, R. and Drewry, J. (1957) Genetics of the β-lactoglobulins of cow's milk. *Nature*, **180**, 376–8.

Aschaffenburg, R. and Thymann, M. (1965) Simultaneous phenotyping procedure for the principal proteins of cow's milk. *J. Dairy Sci.*, **48**, 1524–6.

Aschaffenburg, R., Sen A. and Thompson, M.P. (1968) Genetic variants of casein in Indian and African Zebu cattle. *Comp. Biochem. Physiol.*, **25**, 177–84.

Atroshi, F., Kangasniemi, R., Honkanon-Buzalski, T. and Sandholm, M. (1982) Beta-lactoglobulin phenotypes in Finnish Ayrshire and Friesian cattle with special reference to mastitis indicators. *Acta Vet. Scand.*, **23**, 135–43.

Babukov, A.V. (1978) Intrapopulation variability of the beta-casein locus in Black Pied cattle of the USSR and its effect on realisation of productive qualities. *Nau. Tru. Lenin. Sel'sko. Inst.*, **342**, 43–4.

Babukov, A.V., Mityutko, V.E. and Komissarenko, A.D. (1982) Use of genofond of polymorphic milk proteins for improvement of Black and White cattle. *Proc. XXI Intern. Dairy Congr. Moscow*, Vol. 1, Book 1, p. 39. (abstr.)

Baker, C.M.A. and Manwell, C. (1980) Chemical classification of cattle. 1. Breed groups. *Anim. Blood Grps. Biochem. Genet.*, **11**, 127–50.

Balland, O. (1987) Analyse et évolution du polymorphisme électrophorétique des lactoprotéines des races bovines Montbéliarde et Normande (Mimeographed report Département de Génétique animale, F-78350, Jouy-en-Josas).

Baranyi, M., Bosze, Z., Buchberger, J. and Krause, I. (1997) Genetic polymorphism of milk proteins in Hungarian cattle breeds and PCR amplification of β-lactoglobulin exon 5 to identify genetic variant J by RFLP. In *Milk Protein Polymorphism, Special Issue*, **9702**, International Dairy Federation, Brussels, Belgium, pp. 87–92.

Barillet, F., Mahé, M.F., Pellegrini, O., Grosclaude, F. and Bernard, S. (1993) Polymorphisme génétique des protéines du lait en race ovine de Lacaune. 5th *Int. Symp. On Machine Milking of Small Ruminants*, Budapest, Hungary, May 14–20.

Barroso, A., Dunner, S. and Canon, J. (1997) Use of a single-strand conformation polymorphism analysis to perform a simple genotyping of bovine κ-casein A and B variants. *J. Dairy Res.*, **64**, 535–40.

Barroso, A., Dunner, S. and Canon, J. (1998) Detection of bovine κ-casein variants A, B, C and E by means of polymerase chain reaction single strand conformation polymorphism (PCR-SSCP). *J. Anim. Sci.*, **76**, 1535–8.

Bech, A.M. and Kristiansen, K.R. (1990) Milk protein polymorphism in Danish dairy cattle and the influence of genetic variants on milk yield. *J. Dairy Res.*, **57**, 53–62.

Bech, A.M. and Munk, K.S. (1988) Studies on bovine milk protein polymorphism by electrofocusing in agarose gels containing 7M urea. *Milchwissenschaft*, **43**, 230–2.

Bell, J.W. and Stone W.K. (1978) Rapid separation of whey proteins by cellulose acetate electrophoresis. *J. Dairy Sci.*, **62**, 502–4.

Bell, K. (1962) One-dimension starch gel electrophoresis of bovine skim milk. *Nature*, **195**, 705–6.

Bell, K. and McKenzie, H.A. (1964) β-Lactoglobulins. *Nature*, **204**, 1275–9.

Bell, K., McKenzie, H.A., Murphy, W.H. and Shaw, D.C. (1970) β-Lactoglobulin Droughtmaster: a unique protein variant. *Biochim. Biophys. Acta*, **214**, 427–36.

Bell, K., McKenzie, H.A., Muller, U., Rogers, C. and Shaw, D.C. (1981a) Equine whey proteins. *Comp. Biochem. Physiol.*, **68B**, 225–36.

Bell, K., McKenzie, H.A. and Shaw, D.C. (1981b) Bovine β-lactoglobulin E, F and G of Bali (Banteng) cattle, *Bos* (*Bibos*) *javanicus*. *Aust. J. Biol. Sci.*, **34**, 133–47.

Bell, K., Hopper, K.E. and McKenzie, H.A. (1981c) Bovine α-lactalbumin C and α_{s1}-, β- and κ-caseins of Bali (Banteng) cattle, *Bos* (*Bibos*) *javanicus*. *Aust. J. Biol. Sci.*, **34**, 149–59.

Bell, K., McKenzie, H.A. and Shaw, D.C. (1981d) Porcine β-lactoglobulin A and C. Occurrence, isolation and chemical properties. *Mol. Cell. Biochem.*, **35**, 103–11.

Bell, K., McKenzie, H.A. and Shaw, D.C. (1981e) Porcine α-lactalbumin A and B. *Mol. Cell. Biochem.*, **35**, 113–9.

Bettini, T.M. and Masina, P. (1972) Proteine e polimorfismo proteico del latte vaccino. *Produz. Anim.*, **11**, 107–26.

Bewley, M.C., Qin, B.Y., Jameson, G.B., Sawyer, L. and Baker, E.N. (1997) Bovine β-lactoglobulin and its variants: a three-dimensional structural perspective. In *Milk Protein Polymorphism, Special Issue*, **9702**, International Dairy Federation, Brussels, Belgium, pp. 100–9.

Bleck, G.T., Conroy, J.C. and Wheeler, M.B. (1996) Polymorphisms in the bovine β-casein 5′-flanking region. *J. Dairy Sci.*, **79**, 347–9.

Blumberg, B.S. and Tombs, M.P. (1958) Possible polymorphism of bovine α-lactalbumin. *Nature*, **181**, 683–4.

Bo, A.D. and Pilotti, A. (1997) The effects of β-lactoglobulin genetic variants A and B on the functional properties of whey under different conditions. *Food Hydrocolloid*, **11**, 41–8.

Bolla, P., Caroli, A. and Ceriotti, G. (1983) Analisi genetica delle proteine del latte nella razza bovina Rendera. *Atti Soc. Ital. Sci. Vet.*, **37**, 432–6.

Boulanger, A. (1976) *Etude Biochimique et Génétique Des Proteines du Lait de Chévre* (Capra hircus), Thesis, University of Paris VII.

Boulanger, A., Grosclaude, F. and Mahé, M.F. (1984) Polymorphisme des caséines α_{s1} et α_{s2} de la chèvre (*Capra hircus*) *Génét. Sel Evol.*, **16**, 157–76.

Bouniol, C., Brignon, G., Mahé, M.F. and Printz, C. (1994) Biochemical and genetic analysis of variant C of caprine α_{s1}-casein (*Capra hircus*). *Anim. Genet.*, **25**, 173–7.

Bouniol, C., Printz, C. and Mercier, J.C. (1993) Bovine α_{s2}-casein D is generated by exon VIII skipping. *Gene*, **128**, 289–93.

Bovenhuis, H., van Arendonk, J.A.M. and Korver, S. (1992) Associations between milk protein polymorphisms and milk production traits. *J. Dairy Sci.*, **75**, 2549–59.

Braunitzer, G. and Chen, R. (1972) Die Spaltung des β-Lactoglobulins AB mit Bromcyan. *Hoppe-Seyler's Z. Physiol. Chem.*, **353**, 674–6.

Braunitzer, G., Chen, R., Schrank, B. and Strangi, A. (1972) Automatische Sequenzanalyse eines Proteins (β-Lactoglobulin AB). *Hoppe-Seyler's Z. Physiol. Chem.*, **353**, 832–4.

Braunschweig, M. and Puhan, Z. (1997) Relationship between κ-casein A and B variants versus citrate content of milk quantified by capillary electrophoresis. In *Milk Protein Polymorphism, Special Issue*, **9702**, International Dairy Federation, Brussels, Belgium, pp. 71–6.

Brew, K., Castellino, F.J., Vanaman, T.G. and Hill, R.L. (1970) The complete amino acid sequence of bovine α-lactalbumin. *J. Biol. Chem.*, **245**, 4570–82.

Brignon, G. and Ribadeau-Dumas, B. (1973) Localisation dans la chaine peptidique de la β-lactoglobuline bovine de la substitution Glu/Gln differenciant les variants génétique B et D. *FEBS Lett.*, **33**, 73–6.

Brignon, G., Ribadeau-Dumas, B. and Mercier, J.C. (1976) Premiers éléments de structure primaire des caséines α_{s2} bovines. *FEBS Lett.*, **71**, 111–6.

Brignon, G., Ribadeau-Dumas, B., Mercier, J.C. and Pélissier, J.P. (1977) Complete amino acid sequence of bovine α_{s2}-casein. *FEBS Lett.*, **76**, 274–9.

Brignon, G., Mahé, M.F., Grosclaude, F. and Ribadeau-Dumas, B. (1989) Sequence of caprine α_{s1}-casein and characterization of those of its genetic variants which are synthesized at a high level, α_{s1}-CnA, B and C. *Protein Seq. Data Anal.*, **2**, 181–8.

Brignon, G., Mahé, M.F., Ribadeau-Dumas, B., Mercier, J.C. and Grosclaude, F. (1990) Two of the three genetic variants of goat α_{s1}-casein which are synthesized at a reduced level have internal deletion possibly due to altered RNA splicing. *Eur. J. Biochem.*, **193**, 237–41.

Brittan, H., Mudford, J.C., Norris, G.E., Kitson, T.M. and Hill, J.P. (1997) Labelling the free sulfydryl group in β-lactoglobulin A, B and C. In *Milk Protein Polymorphism, Special Issue*, **9702**, International Dairy Federation, Brussels, Belgium, pp. 200–3.

Brum, F.W., Rausch, W.H., Mines, H. and Ludwick, T.M. (1968) Association between milk and blood polymorphism types and lactation traits of Holstein cattle. *J. Dairy Sci.*, **51**, 1031–8.

Buchberger, J. (1995) Genetic polymorphism of milk proteins: differences between breeds, *Bulletin 304*, International Dairy Federation Brussels, Belgium, pp. 5–6.

Buchberger, J., Kiermeier, F., Kirchmeier, O., Graml, R. and Pirchner, F. (1982) Effect of genetic variants of milk proteins on milk composition. *Proc. XXI Intern. Dairy Congr. Moscow*, Vol. 1., Book 1, pp. 40–1.

Buchberger, J., Graml, R. and Klostermeyer, H. (1986) Genfrequenzen der Milchproteine beim bayerischen Gelbvieh, beim Pinzgauer und beim Murnau-Werdenfelser Rind. *Bayer. Landwirtsch. Jahrb.*, **63**, 267–72.

Caldwell, J., Wesseli, D.F. and Cartwright, T.C. (1971) Occurrence of α_{s1}- and β-casein types in five breeds of beef cattle. *Anim. Sci.*, **32**, 601–4.

Carles, C. (1986) Fractionation of bovine caseins by reverse phase high performance liquid chromatography: identification of a genetic variant. *J. Dairy Res.*, **53**, 35–41.

Cerbulis, J. and Farrell, H.M. (1975) Composition of milks of dairy cattle. I. Protein, lactose and fat contents and distribution of protein fraction. *J. Dairy Sci.*, **58**, 817–27.

Chen, F.T.A. and Zang, J.H. (1992) Determination of milk proteins by capillary electrophoresis. *J. Assoc. Off. Anal. Chem. Int.*, **75**, 905–9.

Chianese, L., Di Luccia, A., Mauriello, R., Ferrara, L., Zehender, G. and Addeo, F. (1988) Polimorfismo biochimico delle proteine del latte in bovine di razza Podolica. *Zootec. Nutr. Anim.*, **14**, 189–97.

Chianese, L., Garro, G., Addeo, F., Lopez-Galvez, G. and Ramos, M. (1993) Discovery of an ovine α_{s2}-casein variants. *J. Dairy Res.*, **60**, 485–93.

Chianese, L., Garro, G., Mauriello, R., Laezza, P., Ferranti, P. and Addeo, F. (1996) Occurrence of five α_{s1}-casein variants in ovine milk. *J. Dairy Res.*, **63**, 49–59.

Chianese, L., Ferranti, P., Garro, G., Mauriello, R. and Addeo, F. (1997a) Occurrence of three novel α_{s1}-casein variants in goat milk. In *Milk Protein Polymorphism, Special Issue*, **9702**, International Dairy Federation, Brussels, Belgium, pp. 259–67.

Chianese, L., Mauriello, R., Ferranti, P., Tripaldi, C., Taibi, L. and Dell'Aguila, S. (1997b) Relationships between α_{s1}-casein variants and clotting capability of ovine milk. In *Milk Protein Polymorphism, Special Issue*, **9702**, International Dairy Federation, Brussels, Belgium, pp. 316–23.

Chiofalo, L. and Micari, P. (1987) Attuali conoscenze sulle varianti delle proteine del latte nelle popolazioni ovine allevate in Sicilia. Osservazioni sperimentali. *Sci. Tec. Latt. Casearia.*, **38**, 10–4.

Chiofalo, L., Micari, P. and Sturniolo, G. (1983) Polimorfismo delle proteine del latte in popolazioni cavalline allevate in Sicilia. *Zoot. Nutr. Anim.*, **9**, 311–8.

Chiofalo, L., Micari, P. and Gitmenia, A.M. (1986) Polimorfismo genetico del locus β-lattoglobulina nella razza ovina Comisana allevata in Sicilia. *Zootec. Nutr. Anim.*, **12**, 73–80.

Cho, B.W., Hwang, K-C., Lee, D.H., Lee, H.K., Jeon, G.J. and Han, J.Y. (1998) Effects of milk protein and growth hormone genotypes on economic traits for Hanwoo. *Korean J. Anim. Sci.*, **40**, 135–44.

Chung, E.R., Kim, D.K. and Han, S.K. (1991) Relationships between biochemical genetic markers and lactation traits Holstein dairy cattle. *Korean J. Dairy Sci.*, **13**, 240–53.

Chung, E.R., Kim, W.T. and Han, S.K. (1994) DNA typing of β-lactoglobulin locus using PCR-RFLP as a selection aid for genetic improvement of dairy cattle. *Korean J. Anim. Sci.*, **36**, 606–12.

Ciafarone, N. and Addeo, F. (1984) Composizione della caseina et proprieta del latte di capra. *I Vergaro*, **11**, 17–24.

Coker, C.J., Creamer, L.K., Burr, R.G. and Hill, J.P. (1997) The action of chymosin or plasmin on α_{s1}-casein A, B and C. In *Milk Protein Polymorphism, Special Issue*, **9702**, International Dairy Federation, Brussels, Belgium, pp. 175–81.

Comberg, G., Meyer, H. and Growing, M. (1964) Correlations between beta-lactoglobulin types in cattle and age at first calving, milk yield and fat contents. *Zuchtungskunde*, **36**, 248–55.

Conti, A., Godovac-Zimmermann, J., Napolitano, L. and Liberatori, J. (1985) Identification and characterization of two alpha-lactalbumin from Somali camel milk (*Camelus dromedarius*). *Milchwissenschaft*, **40**, 673–5.

Conti, A., Napolitano, L., Cantisani, A.M., Davoli, R. and Dall'Olio, S. (1988) Bovine β-lactoglobulin H: isolation by preparative isoelectric focusing in immobilized pH gradients and preliminary characterization. *J. Biochem. Biophys. Meth.*, **16**, 205–14.

Corradini, C. (1969) Distribution of the genetic variants of α_{s1}-, β- and κ-caseins in milk from Jersey cows in the Netherlands. *Neth. Milk Dairy J.*, **23**, 79–83.

Creamer, L.K. and Harris, D.P. (1997) Relationship between milk protein polymorphism and physico-chemical properties. In *Milk Protein Polymorphism, Special Issue*, **9702**, International Dairy Federation, Brussels, Belgium, pp. 110–23.

Creamer, L.K. and Richardson, B.C. (1975) A new genetic variant of beta-casein. *N.Z. J. Dairy Sci. Technol.*, **10**, 170–1.

Creamer, L.K., Zoerb, H.F., Olson, N.F. and Richardson, T. (1982) Surface hydrophobicity of α_{s1}-I, α_{s1}-casein A and B and its implication in cheese structure. *J. Dairy Sci.*, **65**, 902–6.

Curik, I., Kaps, M., Lukac-Havrarek, J. and Antunac, N. (1993) Association of milk protein genotypes with first-lactation traits in Croatian Simmentals. *Stocarstvo*, **47**, 15–22.

Curik, I., Havranek, J. and Samarzijà, D. (1997) Milk protein polymorphism and genetic structure of Croatian Simmental cattle. In *Milk Protein Polymorphism, Special Issue*, **9702**, International Dairy Federation, Brussels, Belgium, pp. 93–9.

Dalgleish, D. (1986) Analysis by fast protein liquid chromatography of variants of κ-casein and their relevance to micellar structure and renneting. *J. Dairy Res.*, **53**, 43–51.

Dall'olio, S., Davoli, R. and Russo, V. (1989) Una nuova variante di β-caseina caprina. *Sci. Tec. Latt.-Casearia.*, **40**, 2–8.

Damiani, G., Pilla, F., Leone, P. and Caccio, S. (1992) Direct sequencing and bidirectional allele specific polymerase chain reaction of bovine β-casein B variant. *Anim. Genet.*, **23**, 561–6.

David, V.A. and Deutch, A.H. (1992) Detection of bovine α_{s1}-casein genomic variants using the allele-specific polymerase chain reaction. *Anim. Genet.*, **23**, 425–9.

Davoli, R. (1981) Detection of genetic variants of milk proteins by electrophoresis on cellulose acetate. *Riv. Zoot. Vet.*, **9**, 96–100.

Davoli, R., Dall'olio, S. and Russo, V. (1990) Evidence supporting attribution of ovine Welsh variant to α_{s1}-casein. *Scien. Tecn. Latt.-Casear*, **41** (Suppl.), 327–33.

de Jong, N., Visser, S. and Olieman, C. (1993) Determination of milk proteins by capillary electrophoresis. *J. Chromatogr. A.*, **652**, 207–13.

Delacroix-Buchet, A., Lefier, D. and Nuyts-Petit, V. (1993) Polymorphism of κ-casein from three French breeds and its coagulability. *Lait*, **73**, 61–72.

Delacroix-Buchet, A. and Marie, C. (1994) Comparison of the A and C variants of β-casein in milk from Tarentaise cows used in Beaufort-type-cheese produc tion. I. Suitability for cheesemaking, and cheese yields. *Lait*, **74**, 343–60.

Denicourt, D., Sabour, M.P. and McAllister, A.J. (1990) Detection of bovine κ-casein genomic variants by the polymerase chain reaction method. *Anim. Genet.* **21**, 215–6.

Di Luccia, A., Addeo, F., Corradini, C. and Mariani, P. (1988) Bovine kappa-casein C: an electrophoretic study. *Sci. Tech. Latt-Cas.*, **39**, 413–22.

Di Stasio, L. (1983) New phenotypes of α_{s1}-casein in sheep. *Anim. Blood Grps. Biochem. Genet.*, **14**, 229–32.

Di Stasio, L. and Merlin, P. (1979) Polimorfismi biochimici del latte nella razza bovina Grigio Alpina. *Riv. Zootec. Vet.*, **2**, 64–7.

Di Stasio, L. and Dupont, E. (1983) La popolazione bovina 'castana': affinita genetiche con le razze Valdostana P.N. e Kerens. *Atti Soc. Ital. Sci. Vet.*, **37**, 505–7.

Di Stasio, L., Cristofori, F. and Sartore, G. (1983) Phenotypic variations in blood and milk of the Somali camel. *Anim. Blood Grps. Biochem. Genet.*, **14**, 225–8.

Di Stasio, L., Portolano, B., Todaro, M., Fiandra, P., Giaccone, P., Finocchiaro, R. and Alicata, M.L. (1997) Effect of ovine β-lactoglobulin phenotype on cheese yield and compostion. In *Milk Protein Polymorphism, Special Issue*, **9702**, International Dairy Federation, Brussels, Belgium, pp. 324–7.

Doi, H., Ibuki, F. and Kanamori, M. (1979) Heterogeneity of reduced bovine κ-casein. *J. Dairy Sci.*, **62**, 195–203.

Dong, C. and Ng-Kwai-Hang, K.F. (1998) Characterization of a non-electrophoretic genetic variant of β-casein by peptic mapping and mass spectrometric analysis. *Int. Dairy J.*, **8**, 967–72.

Eggen, A. and Fries, R. (1995) An integrated cytogenetic and meiotic map of the bovine genome. *Anim. Genet.*, **26**, 215–36.

Eigel, W.N., Butler, J.E., Ernstrom, C.A., Farrell, H.M., Jr., Harwalkar, V.R., Jenness, R. and Whitney, R.McL. (1984) Nomenclature of proteins of cow's milk. Fifth revision. *J. Dairy Sci.*, **67**, 1599–631.

Eigel, W.N. and Randolph, H.E. (1976) Comparison of calcium sensitivities of alpha s_1-B, beta-A2, and gamma-A2 casein and their stabilization by kappa-casein A. *J. Dairy Sci.*, **59**, 203–6.

El-Negoumy, A.M. (1974) Effect of polymorphism on casein stability in salt solutions of varying complexity before and after mixing. *J. Dairy Sci.*, **57**, 1170–6.

Enne, G., Feligini, M., Greppi, G.F., Iametti, S. and Pagani, S. (1997) Gene frequencies of caprine α_{s1}-casein polymorphism in dairy goat. In *Milk Protein Polymorphism, Special Issue*, **9702**, International Dairy Federation, Brussels, Belgium, pp. 275–9.

Epstein, H. (1971) *The Origin of the Domestic Animals of Africa*, Vol. 1. Africana Publishing Corporation, New York.

Erhardt, G. (1989a) κ-Kaseine in Raseine in Rindermilch-Nachweis eines weiteren Allels (κ-Cn^E) in rerschiedenen Rassen. *J. Anim. Breed. Genet.*, **106**, 225–31.

Erhardt, G. (1989b) Isolierung und Charakterisierung von Caseinfraktionen sowie deren genetische Varianten in Schweinemilch. *Milchwissenschaft*, **44**, 17–20.

Erhardt, G. (1989c) Isolierung und Charakterisierung der Molkenproteine sowie deren genetische Varianten in schweinemilch. *Milchwissenschaft*, **44**, 145–9.

Erhardt, G. (1993a) Allele frequencies of milk proteins in German cattle breeds and demonstration of α_{s2}-casein variants by isoelectric focusing. *Archiv. für Tierzucht.*, **36**, 145–52.

Erhardt, G. (1993b) A new α_{s1}-casein allele in bovine milk and its occurrence in different breeds. *Anim. Genet.*, **24**, 65–6.

Erhardt, G. (1996) Detection of a new κ-casein variant in milk of Pinzgauer cattle. *Anim. Genet.*, **27**, 105–7.

Erhardt, G. and Senft, B. (1987) Protein polymorphisms in porcine milk. *Proc. 20th Int. Conf. Anim. Blood Grps. Biochem. Polym. (Helsinki, 1986). Anim. Genet.*, **18** (supp. 1), 56–7.

Erhardt, G., Godovac-Zimmermann, J. and Conti, A. (1989) Isolation and complete primary sequence of a new ovine wild-type β-lactoglobulin C. *Biol. Chem. Hoppe-Seyler.*, **370**, 757–62.

Erhardt, G., Juszczak, J., Panicke, L. and Krick-Saleck, H. (1998) Genetic polymorphism of milk proteins in Polish red cattle: a new genetic variant of β-lactoglobulin. *J. Anim. Breed Genet.*, **115**, 63–71.

Euston, S.R., Hirst, R.L. and Hill, J.P. (1997) Emulsifying properties of milk protein genetic variants. In *Milk Protein Polymorphism, Special Issue*, **9702**, International Dairy Federation, Brussels, Belgium, pp. 395–411.

Fairise, J.F. and Cayot, P. (1998) New ultrarapid method for the separation of milk proteins by capillary electrophoresis. *J. Agric. Food Chem.*, **46**, 2628–33.

Famula, T.R. and Medrano, J.F. (1994) Estimation of genotype effects for milk proteins with animal and sire transmitting ability models. *J. Dairy Sci.*, **77**, 3153–62.

Farrell, H.M. and Thompson, M.P. (1971) Biological significance of milk protein polymorphism. *J. Dairy Sci.*, **54**, 1219–28.

Feagan, J.T. (1979) Factors affecting protein composition of milk and their significance to dairy processing. *Aust. J. Dairy Technol.*, **34**, 77–81.

Feagan, J.T., Bailey, L.A., Hehir, A.F., McLean, D.M. and Ellis, N.J.S. (1972) Coagulation of milk proteins. I. Effect of genetic variants of milk proteins on rennet coagulation and heat stability of normal milk. *Aust. J. Dairy Technol.*, **27**, 129–34.

Felenezak, A. (1983) 1. Genetic polymorphism of milk proteins. *Zesz. Nauk. Akad. Rol. Krak. Zootech.*, **22**, 175–91.

Ferranti, P., Malorni, A., Nitti, G., Laezza, P., Pizzano, R., Chianese, L. and Addeo, F. (1995) Primary structure of ovine α_{s1}-caseins: localization of phosphorylation sites and characterization of genetic variants A, C and D. *J. Dairy Res.*, **62**, 281–96.

Ferrara, L., Di Luccia, A., Astolfi, P., Mauriello, R. and Addeo, F. (1986) Polimorfismo delle proteine del latte in bovini di razza Podolica. II. Aspetti genetici. In *Convegno su l'allevamento del bovino Podolico nel Mezzogiorno d'Italia*, pp. 299–308.

Ferretti, L., Leone, P., Rognoni, G. and Sgaramella, V. (1990) Linkage of the four bovine casein genes as demonstrated by pulsed gene electrophoresis. *Proc. 4th World Congr. Genet. Appl. Livestock Prod.*, **XIII**, 75–8.

Foegeding, E.A., Lowe, R. and Hill, J.P. (1997) The properties of heat-induced whey protein concentrate gels produced from lactic whey containing only β-lactoglobulin A, only β-lactoglobulin B, or a mixture of β-lactoglobulin A and B. In *Milk Protein Polymorphism, Special Issue*, **9702**, International Dairy Federation, Brussels, Belgium, pp. 146–57.

Frank, G. and Braunitzer, G. (1967) Zur Primärstruktur der β-Lactoglobuline. *Hoppe-Seyler's Z. Physiol. Chem.*, **348**, 1691–2.

Gao, H., Britten, M. and Ng-Kwai-Hang, K.F. (1996) Emulsifying properties of the different genetic variants of α_{s1}-casein, β-casein and κ-casein, *J. Dairy Sci.*, **79**, Suppl. 1, 106.

Gao, H., Britten, M. and Ng-Kwai-Hang, K.F. (1997) The influence of genetic variants on surface-active properties of caseins. In *Milk Protein Polymorphism, Special Issue*, **9702**, International Dairy Federation, Brussels, Belgium, pp. 124–33.

Gaye, P., Hue-Delahaie, D., Mercier, J.C., Soulier, S., Vilotte, J.L. and Furet, J.P. (1986) Ovine β-lactoglobulin messenger RNA: nucleotide sequence and mRNA levels during fractional differentiation of the mammary gland. *Biochimie*, **68**, 1097–107.

Gaye, P., Hue-Delahaie, D., Mercier, J.C., Soulier, S., Vilotte, J.L. and Furet, J.P. (1987) Complete sequence of ovine α-lactalbumin mRNA. *Biochimie*, **69**, 601–8.

Geldermann. H., Gogol, J., Kock, M. and Tacea, G. (1996) DNA variants within the 5′-flanking region of bovine milk protein encoding genes. *J. Anim. Breed. Genet.*, **113**, 261–7.

Gerrits, R.J., Kraeling, R.R. and Kincaid, C.M. (1969) Polymorphism in a casein fraction of cow's milk. *Biochem. Genet.*, **3**, 355–8.

Giesecke, W.H. and Osterhoff, D.R. (1975) Milk protein phenotyping as a practicable genetic means of augmenting conventional methods of mastitis control and prevention. *Ann. Bull,* **85**, International Dairy Federation, Brussels, Belgium, pp. 223–33.

Glasnák, V. (1966) Protein polymorphism in cow's milk. Research note. In *Polymorphismes Biochimiques des Animaux*. Proc. Xth Eur. Conf. Anim. Blood Grps. Biochem. Polym. (Paris 1966), pp. 433–5.

Glasnák, V. (1968a) Polymorphism of β_1-caseins in sow's milk. *Folia Biol.*, **14**, 70–3.

Glasnák, V. (1968b) Inter- and intra-specific differences in milk proteins of cattle and swine. *Comp. Biochem. Physiol.*, **25**, 355–7.

Godovac-Zimmermann, J. and Braunitzer, G. (1987) Modern aspects of the primary structure and function of β-lactoglobulins. *Milchwissenschaft*, **42**, 294–7.

Godovac-Zimmermann, J., Shaw, D., Conti, A. and McKenzie, H. (1987) Identification and the primary structure of equine α-lactalbumin B and C (*Equus caballus*, Perissodactyla). *Biol. Chem. Hoppe-Seyler*, **368**, 427–33.

Godovac-Zimmermann, J., Krause, I., Baranyi, M., Fischer-Fruehholz, S., Juszczak, J., Erhardt, G., Buchberger, J. and Klostermeyer, H. (1996) Isolation and rapid sequence characterization of two novel bovine beta-lactoglobulin I and J. *J. Protein Chem.*, **15**, 743–50.

Gonyon, D.S., Mather, R.E., Hines, H.C., Haenlein, G.F.W., Arave, C.W. and Gaunt, S.N. (1987) Association of bovine blood and milk polymorphisms with lactation traits: Holsteins. *J. Dairy Sci.*, **70**, 2585–98.

Graham, E.R.B., McLean, D.M. and Zviedrans, P. (1984) The effect of milk protein genotypes on the cheesemaking properties of milk and on the yield of cheese. *4th Conf Aust. Ass. Anim. Breed. Genet.*, Adelaide, 4–6 June, pp. 136–7.

Graml, R. (1995) Dominance of alleles in heterozygous β-lactoglobulin genotypes of cattle. *Archiv. für Tierzucht.*, **38**, 613–9.

Graml, R., Buchberger, J., Kirchmeier, O., Kiermeier, F. and Pirchner, F. (1984a) Genfrequenschätzung bei Milchproteinen des bayerischen Fleckviehs. *Zuchtungskunde*, **56**, 73–87.

Graml, R., Buchberger, J., Klostermeyer, H. and Pirchner, F. (1984b) Untersuchungen über die Genfrequenzen der Caseine and β-lactoglobuline bei der Bayerisehen braunviehpopulation. *Zuchtungskunde*, **56**, 221–30.

Graml, R., Buchberger, J., Klostermeyer, H. and Pirchner, F. (1985) Pleiotrope Wirkungen von β-Lactoglobulin-und-Casein Genotypen auf Milchinhaltsstoffe der bayerischen Fleckviehs und Braunviehs. *Z. Tierzucht. Zuch. Biol.*, **102**, 353–70.

Graml, R., Buchberger, J., Klostermeyer, H. and Pirchner, F. (1986a) Pleiotrope Wirkungen von β-Lactoglobulin-und Caseine Genotypen auf Milchfett-und Milchproteinmengen der bayerischen Fleckviehs und Braunviehs. *Z. Tierzucht. Zuch. Biol.*, **103**, 33–45.

Graml, R., Buchberger, J. and Pirchner, F. (1986b) Genetic disequilibria between the α_{s1}-, β-, κ-casein and the β-lactoglobulin loci of the Bavarian Brown and Bavarian Simmental cattle. *Génét. Sél. Evol.*, **18**, 1–10.

Graml, R., Ohmayer, G., Pirchner, F., Erhardt, L., Buchberger, J. and Mostageer, A. (1986c) Biochemical polymorphism in Egyptian Baladi cattle and their relationship with other breeds. *Amm. Genet.*, **17**, 61–76.

Greenberg, R., Groves, M.L. and Dower, H.J. (1984) Human beta-casein. Amino acid sequence and identification of phosphorylation sites. *J. Biol.-Chem.*, **259**, 5132–8.

Groot, M.N.N., O'Hara, H., FritzGerald, R.J. and Ross, R.P. (1995) Evaluation of allele-specific primers for rapid identification of bovine κ-casein genotypes by the polymerase chain reaction. *Irish J. Agr. Food Res.*, **34**, 165–173.

Grosclaude, F. (1974) *Analyse Génétique et Biochimique du Polymorphisme Electrophoretique des Caséines* α_{s1}, *β et κ chez les Bovines* (Bos taurus) *et les Zebus* (Bos indicus), Thesis, University of Paris VII.

Grosclaude, F. (1979) Polymorphism of milk proteins: some biochemical and genetical aspects. *Proc. XVIth Int. Conf. Anim. Blood Grps. Biochem. Polym.*, Leningrad, 1978, Vol. 1, pp. 54–92.

Grosclaude, F. (1988) Le polymorphisme génétique des principales lactoprotéines bovines. Relations avec la quantité, la composition et les aptitudes fromagères du lait. *INRA Prod. Anim.*, **1**, 5–17.

Grosclaude, F., Garnier, J., Ribadeau-Dumas, B. and Jeunet, R. (1964) Etroite dépendance des loci controlant le polymorphisme des caséines α_s et β. *C.R. Hebd. Seanc. Acad. Sci. Paris*, **259**, 1569–71.

Grosclaude, F., Pujolle, J., Garnier, J. and Ribadeau-Dumas, B. (1965) Déterminisme genetique des caséines κ du lait de vache; étroite liaison du locus κ-Cn avec les loci α_{s1}-Cn et β-Cn. *C.R. Hebd. Seanc. Acad. Sci., Paris*, **261**, 5229–32.

Grosclaude, F., Pujolle, J. Garnier, J. and Ribadeau-Dumas, B. (1966a) Mise en évidence de deux variants supplémentaires des protéines du lait de vache: α_{s1}-Cn^D et Lg^D. *Ann. Biol. Anim. Biochim. Biophys.*, **6**, 215–22.

Grosclaude, F., Pujolle, J., Ribadeau-Dumas, B. and Garnier. J. (1966b) Analyse génétique du groupe de loci de structure synthetisant les caséines bovines. In *Polymorphismes Biochimiques des Animaux*, INRA, Paris, pp. 415–20.

Grosclaude, F., Mahé, M.F., Mercier, J.C. and Ribadeau-Dumas, B. (1970) Localisation, dans la partie NH_2-terminale de la caséine α_{s1} bovine, d'une délétion de 13 acides aminés différenciant le variant A des variants B et C. *FEBS Lett.*, **11**, 109–12.

Grosclaude, F., Mahé, M.F., Mercier, J.C. and Ribadeau-Dumas, B. (1972a) Caractérisation des variants génétiques des caseines α_{s1} et β bovines. *Eur. J. Biochem.*, **26**, 328–37.

Grosclaude, F., Mahé, M.F., Mercier, J.C. and Ribadeau-Dumas, B. (1972b) Localisation des substitutions d'acides aminsé differenciant les variants A et B de la caséine κ bovine. *Ann. Génét. Sél Anim.*, **4**, 515–21.

Grosclaude, F., Mahé, M.F. and Ribadeau-Dumas, B. (1973a) Structure primaire de la caséine α_{s1} et de la caséine β-bovine. *Eur. J. Biochem.*, **40**, 323–4.

Grosclaude, F., Mercier, J.C. and Ribadeau-Dumas, B. (1973b) Genetic aspects of cattle casein research. *Neth. Milk Dairy J.*, **27**, 328–40.

Grosclaude, F., Mahé, M.F. and Mercier, J.C. (1974a) Comparaison du polymorphisme génétique des lactoprotéines du Zébu et des bovins. *Ann. Génet. Sél. Anim.*, **6**, 305–29.

Grosclaude, F., Mahé, M.F. and Voglino, G.F. (1974b) Le variant βE et le code de phosphorylation des caséines bovines. *FEBS Lett.*, **45**, 3–5.

Grosclaude, F., Mahé M.F., Mercier, J.C., Bonnemaire, J. and Teissier, J.H. (1976a) Polymorphisme des lactoprotéines de bovines Népalais. II. Polymorphisme des caséines "α_{s1}-mineures"; le locus α_{s2}-Cn est-il lié aux loci α_{s1}-Cn, β-Cn et κ-Cn? *Ann. Génét. Sel. Anim.*, **8**, 481–91.

Grosclaude, F., Mahé, M.F., Mercier, J.C., Bonnemaire, J. and Teissier, J.H. (1976b) Polymorphisme des lactoproteines de bovines Népalais. I. Mise en évidence, chez le yak, et caractérisation biochimique de deux nouveaux variants: β-lactoglobuline D (yak) et caséine α_{s1}-E. *Ann. Génét. Sél. Anim.*, **8**, 461–79.

Grosclaude, F., Joudrier, P. and Mahé, M.F. (1978) Polymorphisme de la caséine α_{s2} bovine: etroite liaison du locus α_{s2}-Cn avec les loci α_{s1}-Cn, β-Cn et κ-Cn; mise en évidence d'une délétion dans le variant α_{s2}-CnD. *Ann. Génét. Sél. Anim.*, **10**, 313–27.

Grosclaude, F., Joudrier, P. and Mahé, M.F. (1979) A genetic and biochemical analysis of a polymorphism of bovine α_{s2}-casein. *J. Dairy Res.*, **46**, 211–13.

Grosclaude, F., Mahé, M.F. and Accolas, J.P. (1982) Note sur le polymorphisme génétique des lactoproteines de bovines de yaks Mongols. *Ann. Génét. Sél. Anim.*, **14**, 545–50.

Grosclaude, F., Mahé, M.F., Brignon, G., Di Stasio, L. and Jeunet, R. (1987) A Mendelian polymorphism underlying quantitative variations of goat α_{s1}-casein. *Génét. Sél Evol.*, **19**, 399–412.

Grosclaude, F., Ricordeau, G., Martin, P., Remeuf, F., Vassal, L. and Bouillon, J. (1994) Du gène au fromage: le polymorphisme de la caséine α_{s1} caprine, ses effets, son évolution. *INRA Prod. Anim.*, **7**, 3–19.

Grosclaude, F. and Martin, P. (1997) Casein polymorphisms in the goat. In *Milk Protein Polymorphism, Special Issue*, **9702**, Brussels, Belgium, PP 241–53.

Guillou, H., Miranda, G. and Pélissier, J.P. (1987) Analyse quantitative des caséines dans le lait de vache par chromatographie liquide rapide d'échange d'ions (FPLC). *Lait*, **67**, 135–48.

Haenlein, G.F.W., Gonyon, D.S., Mather, R.E. and Hines, H.C. (1987) Associations of bovine blood and milk polymorphisms with lactation traits. Guernseys. *J. Dairy Sci.*, **70**, 2599–609.

Halliday, J.M., Bell, K., McKenzie, H.A. and Shaw, D.C. (1990) Feline whey proteins: identification, isolation and initial characterization of α-lactalbumin, β-lactogtobulin and lysozyme. *Comp. Biochem. Physiol.*, **95B**, 773–9.

Han, S.K., Chang, K.J. and Chung, E.Y. (1986) Relationship between milk protein polymorphisms and bacterial mastitis in Holstein-Friesian cattle. *Kor. J. Anim. Sci.*, **28**, 37–42.

Han, S.K., Chung, E.Y. and Lee, K.M. (1983) Studies on the genetic polymorphism of milk proteins in Korean cattle. *Proc. 5th World Conf. Anim. Prod.*, **2**, 51–2.

Han, S.K., Lee, K.M., Chung, E.Y. and Jang, K.J. (1984) Studies on the genetic polymorphism of milk proteins. II. Genetic variants of β-casein and β-lactoglobulin. *Korean J. Anim. Sci.*, **26**, 217–24.

Hargrove, G.L., Kiddy, C.A., Young, C.W., Hunter, A.G., Trimberger, G.W. and Mather, R.E. (1980) Genetic polymorphisms of blood and milk and reproduction in Holstein cattle. *J. Dairy Sci.*, **63,** 1154–66.

Hartung, H and Gernandt, E. (1993) Influence of protein genotypes of β- and κ-casein in milk from German Black Pied dairy cows on milk composition and coagulation properties. *VDLUFA Kongress, Hamburg. Qualität und Hygiene von Lebensmittel in Produktion und Verarbertung*, pp. 429–32.

Hill, J.P. (1993) The relationship between β-lactoglobulin phenotypes and milk composition in New Zealand dairy cattle *J. Dairy Sci.*, **76**, 281–6.

Hill, J.P. and Paterson, G.R. (1994) The variation in milk composition from individual beta-lactoglobulin AA and BB phenotype cows. *Proc. N.Z. Soc. Anim. Prod.*, **54**, 293–5.

Hill, J.P., Paterson, G.R. and MacGibbon, A.K.H. (1997a) Joint effect of β-lactoglobulin and κ-casein variants on the heat stability of milk. In *Milk Protein Polymorphism, Special Issue*, **9702**, International Dairy Federation, Brussels, Belgium, pp. 231–6.

Hill, J.P., Boland, M.J. and Smith, A.F. (1997b) The effect of β-lactoglobulin variants on milk powder manufacture and properties. In *Milk Protein Polymorphism, Special Issue*, **9702**, International Dairy Federation, Brussels, Belgium, pp. 372–94.

Hillier, R.M., Lyster, R.L.J. and Cheeseman, G.C. (1979) Thermal denaturation of alpha-lactalbumin and beta-lactoglobulin in cheese whey: effect of total solids concentration and pH. *J. Dairy Res.*, **46**, 103–11.

Hines, H.C., Haenlein, G.F.W., Zikakis, J.P. and Dickey, H.C. (1977) Blood antigen, serum protein and milk protein gene frequencies and genetic interrelationships in Holstein cattle. *J. Dairy Sci.*, **60**, 1143–51.

Hoagland, P.D., Thompson, M.P. and Kalan, E.B. (1971) Amino acid composition of α_{s3}- α_{s4}- and α_{s5}-caseins. *J. Dairy Sci.*, **54**, 1103–10.

Hollar, C.M., Law, A.J.R., Dalgleish, D.G., Medrano, J.F. and Brown, R.J. (1991) Separation of β-casein A^1, A^2, and B using cation-exchange fast protein liquid chromatography. *J. Dairy Sci.*, **74**, 3308–13.

Holt, C., Waninge, R., Sellers, P., Paulsson, M., Bauer, R., Ogendal, L., Roefs, S.P.F.M., Vanmill, P., de Kruif, G.G., Leonil, J., Fauquant, J. and Maubois, J.L. (1998) Comparison of the effect of heating on the thermal denaturation of nine different β-lactoglobulin preparations of genetic variants A, B or A/B, as measured by microcalorimetry. *Int. Dairy J.*, **8**, 99–104.

Hoogendorn, M.P., Moxley, J.E., Hawes, R.O. and MacRae, H.F. (1969) Separation and gene frequencies of blood serum transferrin, casein and beta-lactoglobulin loci of dairy cattle and their effects on certain production traits. *Can. J. Anim. Sci.*, **49**, 331–41.

Horne, D.S. and Muir, D.D. (1994) Genetic polymorphism of κ-casein and rennet coagulation time. Effects of serum phase components. *Milchwissenschaft*, **49**, 446–9.

Horvath, I. and Meszaros, I. (1972) Casein types in Hungarian spotted cattle and other breeds playing a role in its development. In *Proc. XIIIth Europ. Conf. Anim. Blood Grps. Biochem. Polym.*, Budapest, 1970, (W. Junk ed.) The Hague, pp. 217–23.

Huang, X.L., Catignani, G.L. and Swaisgood, H.E. (1994) Comparison of the size and rate of formation of peptides released by limited proteolysis of β-lactoglobulin A and B with immobilized trypsin. *J. Agr. Food Chem.*, **42**, 1281–4.

Humphrey, R.S. and Newsome, L.J. (1984) High performance ion-exchange chromatography of the major bovine milk proteins. *N.Z. J. Dairy Sci. Technol.*, **19**, 197–204.

Ikonen, T., Ruottinen, O., Erhardt, G. and Ojala, M. (1996) Allele frequencies of the major milk proteins in Finnish Ayrshire and detection of a new κ-casein variant. *Anim. Genet.*, **27**, 179–81.

Imafidon, G.I. (1990) *Genetic Polymorphism and Physico-Chemical Properties of Milk Proteins*, PhD Thesis, McGill University.

Imafidon, G.I. and Farkye, N.Y. (1994) Influence of pH on chymosin action in solutions of different κ-casein variants. *J. Agr. Food Chem.*, **42**, 1598–601.

International Dairy Federation (1995) Implication of genetic polymorphism of milk proteins on production and processing of milk. *Bulletin No* **304**, International Dairy Federation, Brussels, Belgium, pp. 2–25.

International Dairy Federation (1997) Milk protein polymorphism. Proceedings of the IDF seminar, Palmerston North New Zealand, February 1997. Special Issue **9702**, International Dairy Federation, Brussels, Belgium.

Jacob, E. and Puhan, Z. (1986) The significance of genetic variants of kappa-casein for rennetability of milk. *Deutche Molk. Zeit.*, **107**, 833–4.

Jairam, B.T. and Nair, P.G. (1983) Genetic polymorphisms of milk proteins and economic characters in dairy animals. *Ind. J. Anim. Sci.*, **53**, 1–8.

Jairam, B.T., Vijayalakshmi, S.J., Reddy, Y.K. and Rao, C.H. (1994) Impact of milk protein polymorphism on the incidence of mastitis in cattle. *Ind. J. Dairy Sci.*, **47**, 435–7.

Jakob, E. (1994a) Genetic polymorphism of milk proteins, *Bulletin*, **298**, International Dairy Federation, pp. 17–27.

Jakob, E. (1994b) Genetic polymorphism of milk proteins. *Mljekarstvo*, **44**, 197–217.

Jakob, E. and Puhan, Z. (1992) Technological properties of milk as influenced by genetic polymorphism of milk proteins a review. *Int. Dairy J.*, **2**, 157–78.

Janicki, C. (1978) Beta-lactoglobulin and some milk production characteristics of Holstein Friesian cows in Poland imported from USA and Canada. *Rocz. Akad. Roln. Poz.*, **25**, 169–75.

Jansà Pérez, M., Leroux, M., Sánchez-Bonastre, A. and Martin, P. (1994) Occurrence of a LINE sequence in the 3′UTR of the goat α_{s1}-casein E- encoding allele associated with a reduced protein synthesis level. *Gene*, **147**, 179–87.

Jansen, N.E., Klastrup, O., Madsen, P.S., Madsen, P. and Nielsen, S.M. (1981) The influence of milk protein types on mastitis. 32nd Ann. Meeting. EAAP III, P6.

Jordana, J., Amills, M., Diaz, E., Angulo, C., Serradilla, J.M. and Sànchez, A. (1996) Gene frequencies of caprine α_{s1}-casein polymorphism in Spanish goat breeds. *Small Ruminant Research*, **20**, 215–21.

Josephson, R.V. (1972) Isoelectric focusing of bovine milk caseins. *J. Dairy Sci.*, **55**, 1535–43.

Kalan, E.B., Draeling, R.R. and Gerrits, R.J. (1971) Isolation and partial characterization of a polymorphic swine whey protein. *Int. J. Biochem.*, **2**, 232–44.

Kaminski, S. (1996) Dde I RFLP at the 5′ region of bovine κ-casein gene. *J. Appl. Genet.*, **37**, 173–8.

Kaminski, S. and Zabolewicz, T. (1997) Genotyping of β-lactoglobulin by PCR-SSCP technique. *J. Appl. Genet.*, **38**, 471–6.

Kawamoto, Y., Namikawa, T., Adachi, A., Amano, T., Shotake, T., Nishida, T., Hayeshi, Y., Kattel, B. and Rajubhandary, H.B. (1992) A population genetic study on yaks, cattle and their hybrids in Nepal using milk protein variations. *Anim. Sci. Technol. (Jpn)*, **63**, 563–75.

Kemmer, B. (1969) β-Lactoglobulin Typen in der Sauenmilch. *Züchtungskunde*, **41**, 331–4.

Kiddy, C.A. (1975) Gel electrophoresis in vertical polyacrylamide beds. In *Methods of Gel Electrophoresis of Milk Proteins*, (H.E. Swaisgood, B.L. Larson, E.B. Kalan, J.R. Brunner, C.V. Morr and P.M.T. Hansen eds.) American Dairy Science Association, Champaign, Illnois, p. 14.

Kiddy, C.A., McCann, R.E. and Thatcher, W.W. (1968) Gene frequencies in milk protein polymorphisms in dairy cattle. *Immunogenet. Letts.*, **5**, 150–2.

Kim, S., Ng-Kwai-Hang, K.F. and Hayes, J.F. (1996) The relationship between milk protein phenotypes and lactation traits in Ayrshires and Jerseys. *Asian Austral. J. Anim. Sci.*, **9**, 685–93.

Kim, S., Ng-Kwai-Hang, K.F. and Hayes, J.F. (1998) The relationship between milk protein phenotypes and lactation traits in Brown Swiss and Canadienne. *Asian Austral. J. Anim. Sci.*, **11**, 311–17.

King, J.W.B. (1966) The caseins of sheep's milk. In *Polymorphismes Biochimiques des Animaux. Proc. Xth Eur. Conf. Anim. Blood Grps. Biochem. Polym.*, Paris, 427–31.

King, J.W.B. (1969) The distribution of sheep β-lactoglobulins. *Anim. Prod.*, **11**, 53–7.

King, J.W.B., Aschaffenburg, R., Kiddy, C.A. and Thompson, M.P. (1965) Non-independent occurrence of α_{s1}- and β-casein variants of cow's milk. *Nature*, **206**, 424.

Kingsbury, E.T. and Gaunt, S.N. (1977) Heterogeneity in whey proteins of mare's milk. *J. Dairy Sci.*, **60**, 274–7.

Kirchmeier, O., Mehana, A., Graml, R., Buchberger, J. and Pirchner, F. (1983) Studies on physico-chemical properties of casein: genetic and seasonal effects. *Milchwissenschaft*, **38**, 589–91.

Koczan, D., Hobom, G. and Seyfert, H.M. (1993) Characterization of the bovine α_{s1}-casein gene C-allele, based on Mae III polymorphism. *Anim. Genet.*, **24**, 74.

Kolde, H.J. and Braunitzer, G. (1983) The primary structure of ovine β-lactoglobulin 1. Isolation of the peptides and sequence. *Milchwissenchaft*, **38**, 18–20.

Komatsu, M. and Abe, T. (1980) Milk casein haplotypes in nine Japanese and European cattle breeds. *Jap. J. Zootech. Sci.*, **51**, 799–802.

Komatsu, M., Abe, T. and Oishi, T. (1977) Relationships between beta-lactoglobulin types and the concentrations of beta-lactoglobulin and alpha-lactalbumin in milk. *Jap. J. Zootech. Sci.*, **48**, 237–42.

Konevtsova, M.I. (1978) Milk protein polymorphism in Red steppe cows and its connection with milk production indice *Nav. Tru. Lenin. Selsko. Inst.*, **342**, 46–8.

Kraeling, R.R and Gerrits, R.J. (1969) Polymorphism of a protein of cow's whey. *J. Dairy Sci.*, **52**, 2036–8.

Krause, I., Buchberger, J., Weiss, G., Pflügler, M. and Klostermeyer, H. (1988) Isoelectric focusing in immobilized pH gradients with carrier ampholytes added for high-resolution phenotyping of bovine β-lactoglobulin: characterization of a new genetic variant. *Electrophoresis*, **9**, 609–13.

Kreuzer, M., Schulz, J.P., Fry, C. and Abel, H. (1996) Rennet coagulation properties of milk from cows at three stages of lactation supplied with graded levels of an antimicrobial feed supplement. *Milchwissenschaft*, **51**, 243–7.

Kriventsov, Y.M. (1978) Relationship between the preweaning growth of crossbred heifers and the beta-lactoglobulin type of the milk of nurse cows. *Selskok. Nauk. Lenina.*, **7**, 29–30.

Kroeker, E.M., Ng-Kwai-Hang, K.F., Hayes, J.F. and Moxley, J.E. (1985) Effects of environmental factors and milk protein polymorphism on composition of casein fraction in bovine milk. *J. Dairy Sci.*, **68**, 1752–7.

Larsen, B. (1971) Blood groups and polymorphic protein in cattle and swine. *Ann. Génét. Sél. Anim.*, **3**, 5–70.

Larsen, B. and Thymann, W. (1966a) Studies on milk protein polymorphism in Danish cattle and the interaction of the controlling genes. *Acta Vet. Scand.*, **7**, 189–205.

Larsen, B. and Thymann, M. (1966b) Studies on milk protein polymorphism and the interaction of the controlling genes. In *Polymorphismes Biochimiques des Animaux. Proc. Xth Eur. Conf. Anim. Blood Grps. Biochem. Polym.*, Paris, pp. 421–5.

Larsen, B. Gruchy, C.L. and Moustgaard, J. (1974) Studies on blood groups and polymorphic protein systems in Jersey cattle on the Isle of Jersey. *Acta Agric. Scand.*, **24**, 99–110.

Leonhard-Kluz, I. and Kaminska, B. (1981) Beta-lactoglobulin polymorphism and udder health in cows. *Dairy Sci. Abstr.*, **44**, 2356.

Leroux, C., Mazure, N. and Martin, P. (1992) Mutations away from splice site recognition sequences might *cis*-modulate alternative splicing of goal α_{s1}-casein transcripts. Structural organization of the relevant gene. *J. Biol. Chem.*, **267**, 6147–157.

Levéziel, H., Méténier, L., Mahé, M.F., Chopin, J., Furet, J.P., Paboeuf, G., Mercier, J.C. and Grosclaude, F. (1988) Identification of the two common alleles of the bovine κ-casein locus by RFLP technique using the enzyme Hind III. *Génét. Sél. Evol.*, **20**, 247–54.

Levéziel, H., Méténier, L., Guérin, G., Cullen, P., Provot, C., Bertaud, M. and Mercier, J.-C. (1991) Restriction fragment length polymorphism of ovine casein genes. Close linkage between the α_{s1}-, α_{s2}-, β- and κ-Cn loci. *Anim. Genet.*, **22**, 1–10.

Levéziel, H., Rodellar, C., Leroux, C., Pépin, L., Grohs, C., Vairman, D., Mahé, M.F., Martin, P. and Grosclaude, F. (1994) A microsatellite within the bovine κ-casein gene reveals a polymorphism correlating strongly with polymorphism previously described at the protein as well as the DNA level. *Anim. Genet.*, **25**, 223–8.

Li, F.H.F. and Gaunt, S.N. (1972) Study of genetic polymorphisms of milk β-lactoglobulin, α_{s1}-casein, β-casein and κ-casein in five dairy breeds. *Bioch. Genet.*, **6**, 9–20.

Lien, S., Alestrom, P., Klungland, H. and Rogne, S. (1992) Detection of multiple β-casein (CASB) alleles by amplification created restriction sites (ACRS). *Anim. Genet.*, **23**, 333–8.

Lien, S., Kaminski, S., Aleström, P. and Rogne, S. (1993) A simple and powerful method for linkage analysis by amplification of DNA from single sperm cells. *Genomics*, **16**, 41–4.

Lien, S., Gomez-Raya, L., Steine, 1., Fimland, T. and Rogne, S. (1995) Associations between casein haplotypes and milk yield traits. *J. Dairy Sci.*, **78**, 2047–56.

Lien, S., Kantanen, J., Olsaker, I., Holm, L.E., Eythorsdottir, E., Sandberg, K., Dalsgard, B. and Adalsteisson, S. (1999) Comparison of milk protein allele frequencies in Nordic cattle breeds. *Anim. Genet.*, **30**, 85–91.

Lin, C.Y. (1995) Influence of genetic polymorphisms of milk proteins on dairy production-related traits. *Bulletin 304,* International Dairy Federation, Brussels, Belgium, pp. 21–2.

Lin, C.Y., McAllister, A.J., Ng-Kwai-Hang, K.F. and Hayes. J.F. (1986) Effects of milk protein loci on first lactation production in dairy cattle. *J. Dairy Sci.*, **69**, 704–12.

Lin, C.Y., McAllister, A.J., Ng-Kwai-Hang, K.F., Hayes, J.F., Batra, T.R., Lee, A.J., Roy, G.L., Vesely, J.A., Wauthy, J.M. and Winter, K.A. (1987) Association of milk protein types with growth and reproductive performance of dairy heifers. *J. Dairy Sci.*, **70**, 29–39.

Lin, C.Y., Sabour, M.P. and Lee, A.J. (1992) Direct typing of milk proteins as an aid for genetic improvement of dairy bull and cows: a review. *Anim. Breed. Abstr.*, **60**, 1–10.

Lodes, A., Buchberger, J., Krause, I., Aumann, J. and Klostermeyer, H. (1996) The influence of genetic variants of milk proteins on the compositional and technological properties of milk. 2. Rennet coagulation time and firmness of the rennet curd. *Milchwissenschaft*, **51**, 543–8.

Lodes, A., Buchberger, J., Krause, I., Aumann, J. and Klostermeyer, H. (1997a) The influence of genetic variants of milk proteins on the compositional and technological properties of milk. 3. Content of protein, casein, whey protein, and casein number. *Milchwissenschaft*, **52**, 3–8.

Lodes, A., Buchberger, J., Krause, I. and Klostermeyer, H. (1997b) The influence of rare genetic variants of milk proteins on compositional properties of milk rennetability, casein micelle size and content of non-glycosylated κ-casein. In *Milk Protein Polymorphism, Special Issue*, **9702**, International Dairy Federation, Brussels, Belgium, pp. 158–61.

Loftus, R.T., MacHugh, D.E., Bradley, D.G., Sharp, P.M. and Cunningham, P. (1994) Evidence for two independent domestications of cattle. *Proc. Natl. Acad Sci. USA*, **91**, 2757–61.

Losi, G., Capella, P., Castagnetti, G.B., Grazia, L., Zambonelli, C., Mariani, P. and Russo, V. (1973) Influenza delle variants genetiche della caseine κ sulla formazione e sulle caracteristiche della cagliata. *Scien. Tecn. Alimenti.*, **3**, 373–6.

Losi, G., Castagnetti, G.B. and Morini, D. (1975) Importanza delle varianti genetiche delle proteine del latte per l'industria lattiero-casearia. *Mondo del Latte.*, **29**, 727–39.

Lowe, R. and Hill, J.P. (1997) The denaturation and aggregation of β-lactoglobulin A, B and C in whey protein concentrate permeate. In *Milk Protein Polymorphism, Special Issue*, **9702**, International Dairy Federation, Brussels, Belgium, pp. 224–30.

Ma, C.Y., Ng-Kwai-Hang, K.F., Harwalkar, V.R. and Imafidon, G.I. (1990) Effect of genetic variants of β-lactoglobulin and κ-casein on thermostability of β-lactoglobulin in model systems. *Brief Communications and Abstracts of Posters. 23rd International Dairy Congr.* Montreal 2: 359.

Mácha, I. and Novackova (1975) Polymorphisms bilkovin Mlece klisen. *Zivoc. Vyroba.*, **20**, 73–8.

Mahé, M.F. and Grosclaude, F. (1982) Polymorphisme de la caséine α_{s2} des bovines: caracterisation de variant C du yak (*Bos grunniens*). *Ann. Génét. Sél. Anim.*, **14**, 401–16.

Máhé, M.F. and Grosclaude, F. (1989) α_{s1}-Cn^{D}, another allele associated with a decreased synthesis rate at the caprine α_{s1}-casein locus. *Génét. Sél. Evol.*, **21**, 127–9.

Mahé, M.F. and Grosclaude, F. (1993) Polymorphism of β-casein in the Creole goat of Guadeloupe: evidence for a null allele. *Génét. Sél. Evol.* **25**, 403–8.

Mahé, M.F., Miranda, G., Queval, R., Bado, A., Souvenir Zafindrajaona, P. and Grosclaude, F. (1999) Genetic polymorphism of milk proteins in African *Bos taurus* and *Bos indicus* populations. Characterization of variants α_{s1}-Cn H and κ-Cn J. *Génét. Sél. Evol.*, **31**, 239–53.

Manderson, G.A., Creamer, L.K. and Hardman, M.J. (1997) Electrophoretic examination of the heat-induced changes in β-lactoglobulin A, B and C. In *Milk Protein Polymorphism, Special Issue*, **9702**, International Dairy Federation, Brussels, Belgium, pp. 217–23.

Manson, W., Annan, W.D. and Barnes, G.K. (1976) α_{s0}-Casein: its preparation and characterization. *J. Dairy Res.*, **43**, 133–6.

Manson, W., Carolan, T. and Annan, W.D. (1977) Bovine α_{s0}-casein: a phosphorylated homologue of α_{s1}-casein. *Eur. J. Biochem.*, **78**, 411–17.

Mao, F.C. (1994) A bovine α-lactalbumin gene Mn1 I restriction fragment length polymorphism. *J. Anim. Sci.*, **72**, 259.

Mariani, P. (1983) Sulla presenza di una terza κ-caseina nel latte di vacche di razza Bruna. *Scien. Tecn. Latt.-Casear.*, **34**, 174–81.

Mariani, P. (1985) Casein content in Friesian cow's milk. *Scien. Tecn. Latt.-Casear.*, **36**, 191–209.

Mariani, P. and Leoni, M. (1982) Relazione genetica tra i loci α-lattalbumina e β-lattoglobulina nei bovini. *Ann. Fac. Med. Vet. Parma.*, **2**, 209–16.

Mariani, P. and Leoni, M. (1985) II tempo di coagulazione del latte in rapporto alla varianti genetiche delle casein β et κ. *Ann. Fac. Med. Vet. Univ. Parma*, **5**, 185–95.

Mariani, P. and Russo, V. (1977) Polimorfismo genetico della α-lattalbumina nelle razze bovine. *Riv. Zootec. Vet.*, **5**, 603–13.

Mariani, P., Losi, G., Russo, V., Castagnetti, G.B., Grazia, L., Morini, D. and Fossa, E. (1976) Caseification tests made with milk characterised by Variants A and B of κ-casein in the production of Parmigiano-Reggiano cheese. *Scien. Tecn. Latt.-Casear.*, **27**, 208–27.

Mariani, P., Morini, D., Losi, G., Castagnetti, G.B., Fossa, E. and Russo, V. (1979a) Ripartizione delle frazioni azotate del latte in vache caratterizatte da genetipo diverse nel locus β-lactoglobulina. *Scien. Tecn. Latt.-Casear.*, **30**, 153–76.

Mariani, P., Losi, G., Morini, D. and Castagnetti, G.B. (1979b) Citric acid content in milk of cows with different kappa-casein genotypes. *Scien. Tecn. Latt.-Casear.*, **30**, 375–84.

Mariani, P., Resmini, P. and Losi, G. (1984) Sialic acid content of total casein in relation to cow genotype at the kappa-Cn locus. *Att. Soc-Itnl. Sci. Vet.*, **38**, 410–12.

Mariani, P., Meazza, M., Resmini, P., Pagani, M.A., Pecorari, M. and Fossa, E. (1986) Observations on beta-casein types and renneting properties of milk. *Ind. del Latte*, **22**, 35–58.

Mariani, P., Bonatti, P. and Pecorari, M. (1988) Rennet coagulation properties of cow milk in relation to alpha S_1-casein genotype. *Scien. Tecn. Latt.-Casear.*, **39**, 431–8.

Marie, C. and Delacroix-Buchet, A. (1994) Comparison of the A and C variants of β-casein in milk from Tarentaise cows used in Beaufort-type cheese production. II. Proteolysis and cheese quality. *Lait*, **74**, 443–59.

Martin, P. and Leroux, C. (1994) Characterization of a further α_{s1}-casein variant generated by exon skipping. Abstracts, *Proc. XXIV International Conference on Animal Genetics*, Prague, 23–29 July, p. 86.

Marziali, A.S. and Ng-Kwai-Hang, K.F. (1986a) Effects of milk composition and genetic polymorphism on coagulation properties of milk. *J. Dairy Sci.*, **69**, 1793–8.

Marziali, A.S. and Ng-Kwai-Hang, K.F. (1986b) Relationships between milk protein polymorphisms and cheese yielding capacity. *J. Dairy Sci.*, **69**, 1193–201.

Marziali, A.S. and Ng-Kwai-Hang, K.F. (1986c) Effects of milk composition and genetic polymorphism on cheese composition. *J. Dairy Sci.*, **69**, 2533–42.

Matyukov, V.S. (1983) The value of β-casein polymorphism for selection in cattle. *Selsk. Khoz. Biol.*, **18**, 73–8.

Mauriello, R., Addeo, F., Pieragostini, E. and Bufano, G. (1990) Polimorfismo delle caseine in pecore di razza Altamurana. *Sci. Tecn. Latt.-Casear.*, **41**, 357–64.

Mayer, H.K., Ortner, M., Tschager, E. and Ginzinger, W. (1997) Composite milk protein phenotypes in relation to composition and cheesemaking properties of milk. *Int. Dairy J.*, **7**, 305–10.

McGillivray, R.T.A., Brew, K. and Barnes, K. (1979) The amino-acid sequence of goat α-lactalbumin. *Arch. Biochem Biophys.*, **197**, 404–14.

McKenzie, H.A. (1971) β-Lactoglobulins. In *Milk Proteins: Chemistry and Molecular Biology*, Vol. II, (H.A. McKenzie ed.) Academic Press, New York, pp. 257–330.

McKenzie, H.A. and Larson, B.L. (1978) Purification and characterization of rat alpha lactalbumin: apparent genetic variants. *J. Dairy Sci.*, **61**, 714–22.

McLean, D.M., Graham, E.R.B., Ponzoni, R.W. and McKenzie, H.A. (1984) Effects of milk protein genetic variants on milk yield and composition. *J. Dairy Res.*, **51**, 531–46.

McLean, D.M., Graham, E.R.B., Ponzoni, R.W. and McKenzie, H.A. (1987) Effects of milk protein genetic variants and composition on heat stability of milk. *J. Dairy Res.*, **54**, 219–35.

Medrano, J.F. and Aguilar-Cordova, E. (1990) Genotyping of bovine κ-casein loci following DNA sequence amplification. *Biotechnology*, **8**, 144–6.

Mercier, J.C. (1981) Phosphorylation of caseins, present evidence for an amino acid triplet code posttranslationally recognized by specific kinases. *Biochimie*, **63**, 1–17.

Mercier, J.C., Grosclaude, F. and Ribadeau-Dumas, B. (1971) Structure primaire de la caséine α_{s1}-bovine. *Eur. J. Biochem.*, **23**, 41–51.

Mercier, J.C., Brignon, G. and Ribadeau-Dumas, B. (1973) Structure primaire de la caséine κ B bovine. Séquence complète. *Eur. J. Biochem.*, **35**, 222–35.

Mercier, J.C., Chobert, J.M. and Addeo, F. (1976) Comparative study of the amino acid sequences of the caseino-macropeptides from seven species. *FEBS Lett.*, **72**, 208–14.

Merlin, P. and Di Stasio, L. (1982) Study on milk proteins loci in some decreasing Italian cattle breeds. *Ann. Génét. Sél. Anim.*, **14**, 17–28.

Micari, P., Chiofalo, L. and Michailidis, J. (1986) Studio elettroforetico dei loci caseinici e delle siero-proteine nel latte della pecora di Chios (Grecia): raffronti con la pecora Barbaresca-Siciliana. *Atti. Soc. Ital. Vet.*, **40**, 578–81.

Michalak, W. (1967) Anomalous electrophoretic pattern of milk proteins. *J. Dairy Sci.*, 50, 1319–20.

Michalak, W. (1969) Hereditary polymorphism of milk proteins in some breeds of cattle raised in Poland. Part II: *Biul Zaki. Hodowli Dosw. Zwierzat PAN*, **15**, 89–111.

Miranda, G., Anglade, P., Mahé, M.F. and Erhardt, G. (1993) Biochemical characterization of the bovine genetic κ-casein C and E variants. *Anim. Genet.*, **24**, 27–31.

Mitra, A., Schlee, P., Krause, I., Blusch, J., Werner, T., Balakrishnan, C.R. and Pirchner, F. (1998) κ-Casein polymorphisms in Indian dairy cattle and buffalo: a new genetic variant in buffalo. *Anim. Biotech.*, **9**, 81–7.

Mityukov, A.S. (1982) The content of genetically determined polymorphous proteins in cow's milk. *Proc. XXI Intern. Dairy Congr* (*Moscow*), Vol. 1, Book 1, p. 204.

Moody, D.E., Pomp, D., Newman, S. and MacNeil, M.D. (1996) Characterization of DNA polymorphism in three populations of Hereford cattle and their associations with growth and maternal EPD in line 1 Herefords. *J. Anim. Sci.*, **74**, 1784–93.

Morini, D., Losi, G., Castagnetti, G.B., Beneveilli, M., Resmini, P. and Volonteno, G. (1975) The influence of genetic variants of kappa-casein on the size of casein micelles. *Scien. Tecn. Latt.-Casear.*, **26**, 437–44.

Morini, D., Losi, G., Castagnetti, G.B. and Mariani, P. (1979) Cheesemaking experiments with milk characterized by kappa-casein variants A and B: characteristics of the ripened cheese. *Scien. Tecn. Latt.-Casear.*, **30**, 243–62.

Mountfort, M. (1984) Patient project building herd of rare casein type-A cows. *Dairy Exporter*, December 1984, 29–31.

Moustgaard, J., Moller, I. and Sorensen, P.H. (1960) Polymorphism of bovine β-lactoglobulin. *Aarsberetn. Inst. Sterilitetsforskn. K. Vet. Landbohojsk. Kbh.*, 111–23.

Munro, G.L. (1978) Effect of genetic variants of milk proteins on yield and composition of milk. *Proc. XX Intern. Dairy Congr.*, Paris, p10.

Nebola, M., Dvorak, J. and Szulc, T. (1996) κ-Casein gene polymorphism in cattle breeds in the Czech Republic and Poland. *Zivocisna Vyroba.,* **41**, 429–31.

Neelin, J.M. (1964) Variants of κ-casein revealed by improved starch gel electrophoresis. *J. Dairy Sci.*, **47**, 506–9.

Ng-Kwai-Hang, K.F. (1998) Genetic polymorphism of milk proteins: relationships with production traits, milk composition and technological properties. *Can. J. Anim. Sci.*, **78** (suppl), 131–47.

Ng-Kwai-Hang, K.F. and Dong, C. (1994) Semipreparative isolation of bovine casein components by high performance liquid chromatography. *Int. Dairy J.*, **4**, 99–110.

Ng-Kwai-Hang, K.F. and Grosclaude, F. (1992) Genetic polymorphism of milk proteins. In *Advanced Dairy Chemistry*, *Vol. I*, *Proteins*, 2nd edn., (P.F. Fox ed.) Elsevier Applied Science, London, pp. 405–55.

Ng-Kwai-Hang, K.F. and Imafidon, G.I. (1990) Effect of genetic variants of κ-casein and β-casein on stability of calcium caseinate model systems. *Brief*

Communications and Abstracts of Posters. 23rd Int. Dairy Congr., Montreal, **2**, 364 (abstr).

Ng-Kwai-Hang, K.F. and Kim, S. (1996) Different amounts of β-lactoglobulin A and B in milk from heterozygous AB cows. *Int. Dairy J.*, **6**, 689–95.

Ng-Kwai-Hang, K.F. and Pélissier, J.P. (1989) Rapid separation of bovine caseins by mass ion exchange chromatography. *J. Dairy Res.*, **56**, 391–7.

Ng-Kwai-Hang, K.F., Hayes, J.F., Moxley, J.E. and Monardes, H.G. (1984) Association of genetic variants of casein and milk serum proteins with milk, fat, and protein production by dairy cattle. *J. Dairy Sci.*, **67**, 835–40.

Ng-Kwai-Hang, K.F., Hayes, J.F., Moxley, J.E. and Monardes, H.G. (1986) Relationships between milk protein polymorphisms and major milk constituents in Holstein-Friesian cows. *J. Dairy Sci.*, **69**, 22–6.

Ng-Kwai-Hang, K.F., Hayes, J.F., Moxley, J.E. and Monardes, H.G. (1987) Variation in milk protein concentration associated with genetic polymorphism and environmental factors. *J. Dairy Sci.*, **70**, 563–70.

Ng-Kwai-Hang, K.F., Monardes, H.G and Hayes, J.F. (1990a) Association between genetic polymorphism of milk proteins and production traits during three lactations. *J. Dairy Sci.*, **73**, 3414–20.

Ng-Kwai-Hang, K.F., Monardes, H.G. and Hayes, J.F. (1990b) Association between genetic polymorphism of milk proteins and parameters of reproduction in Holstein cows. Research Report, Department of Animal Science, Macdonald College, McGill University, pp. 6–7.

Nielsen, B.T., Singh, H. and Latham, J.M. (1996) Aggregation of bovine β-lactoglobulin A and B on heating at 75°C. *Int. Dairy J.*, **6**, 519–27.

Nuyts-Petit, V. (1991) *Influence des Variants Génétiques des Caséines Bovines sur l'Aptitude Fromagère du Lait de Vaches de Races Traditionnelles*, Thesis, Univ. Compiègne, France.

Ojala, M., Famula, T.R. and Medrano, J.F. (1997) Effects of milk protein genotypes on the variation for milk production traits of Holstein and Jersey cows in California. *J. Dairy Sci.*, **80**, 1776–85.

Orita, M., Iwahana, H., Kanazawa, H., Hayashi, K. and Sekiya, T. (1989) Detection of polymorphism of human DNA by gel electrophoresis as single-strand conformation polymorphisms. *Proc. Nat. Acad. Sci.*, **86**, 2766–70.

Ortner, M., Essl, A. and Solkner, J. (1995) On the importance of various milk protein genotypes in cattle breeding. *Zuchtungskunde*, **67**, 353–67.

Osta, R., Marcos, S. and Rodellar, C. (1995a) A Mnl I polymorphism at the bovine α_{s2}-casein gene. *Anim. Genet.*, **26**, 213.

Osta, R., Garcia-Muro, E. and Zarazaga, I. (1995b) A Msp I polymorphism at the α-lactalbumin gene. *Anim. Genet.*, **26**, 204–5.

Otte, J.A.H.J., Kristiansen, K., Zakora, M. and Qvist, K.B. (1994) Separation of individual whey proteins and measurement of α-lactalbumin and β-lactoglobulin by capillary zone electrophoresis. *Neth. Milk Dairy J.*, **48**, 81–97.

Pabst, K. (1998) The significance of different types of milk protein (particularly κ-casein) for cheesemaking. *Archiv für Tierzucht.*, **41**, 269–76.

Pagnacco, G. and Caroli, A. (1987) Effect of casein and β-lactoglobulin genotypes on renneting properties of milks. *J. Dairy Res.*, **54**, 479–85.

Panicke, L., Freyer, G. and Erhardt, G. (1998) Effects of milk protein variant genes on fertility traits in Black Pied cattle. *Archiv für Tierzucht.*, **41**, 447–54.

Papoff, C.M., Delacroix-Buchet, A., Le Bars, D., Campus, R.L. and Vodret, A. (1995) Hydrolysis of bovine β-casein C by plasmin. *Italian J. Food Sci.*, **7**, 157–68.

Paterson, G.R., Otter, D.E. and Hill, J.P. (1995) Application of capillary electrophoresis in the identification of phenotypes containing the β-lactoglobulin C variant. *J. Dairy Sci.*, **78**, 2637–44.

Pearce, R.J. (1983) Analysis of whey proteins by high performance liquid chromatography. *Aust. J. Dairy Technol.*, **38**, 114–17.

Pearce, R.J. and Zadow, J.G. (1978) Isoelectric focusing of milk proteins. *Proc. XX Intern. Dairy Congr.* (*Paris*), pp. 217–18.

Perez-Beato, O. and Ponce, P. (1985) Concentrations of lactose, sodium and potassium in the milk of Zebu cows in Cuba and their relationship with genotype at the alpha-lactalbumin locus. *Rev. Sal. Anim.*, **7**, 211–15.

Persuy, M.A., Printz, C., Medrano, J.F. and Mercier, J.C. (1996) One mutation might be responsible for the absence of β-casein in two breeds of goats. *Anim. Genet.*, **27** (Suppl. 2), 96.

Peterson, R.F. (1969) Electrofocusing of milk proteins. *J. Dairy Sci.*, **52**, 903 (abstr.).

Peterson, R.F. and Kopfler, F.C. (1966) Detection of new types of α-casein by polyacrylamide gel electrophoresis at acid pH: a proposed nomenclature. *Biochem. Biophys. Res. Commun.*, **22**, 388–92.

Piletz, J.E. and Gansehow, R.E. (1981) Genetic variation of milk proteins in mice. *Biochem. Genet.*, **19**, 1023–30.

Pirisi, A., Piredda, G., Fraghi, F., Papoff, C.M. and Chianese. (1997) Influences of sheep AA, CC and DD α_{s1}-casein variants on milk composition and cheese yield. In *Milk Protein Polymorphism, Special Issue*, **9702**, International Dairy Federation, Brussels, Belgium, pp. 254–8.

Pokalov, V.P. (1975) Relationship of blood serum and milk proteins with body weight of cows. *Anim. Breeding Abstr.*, **44**, 4147.

Ponzone, A. and Voglino, G.F. (1976) Milk casein polymorphism in man. *Helv. Paediat. Acta*, **31**, 47–52.

Ponzone, A., Voglino, G.F. and Tognolo, A. (1975) Milk casein polymorphism in the Kikuyu population. *Ann. Genet.*, **18**, 203–5.

Poplin, F. (1979) Origine du mouflon de Corse dans une nouvelle perspective paléontologique: par marronage. *Ann. Génét. Sél. Anim.*, **11**, 133–43.

Prinzenberg, E.M. and Erhardt, G. (1994) Characterization of bovine α_{s1}-casein A by PCR-RFLP. *Anim. Genet.*, **26**, suppl II, 30–1.

Prinzenberg, E.M. and Erhardt, G. (1997) SSCP identifies bovine α_{s1}-casein D. *Anim. Genet.*, **28**, 460.

Prinzenberg, E.M. and Erhardt, G. (1998) High resolution SSCP analysis reveals new alleles at the κ-casein (CSN3) *locus in Bos taurus* and *Bos indicus* cattle. *Proc. XXVI Int. Conf. Animal Genetics. Auckland*, p. 17.

Prinzenberg, E.M., Hiendleder, S., Ikonen, T. and Erhardt, G. (1996) Molecular genetic characterization of new bovine κ-casein alleles CSN3F and CSN3G and genotyping by PCR-RFLP. *Anim. Genet.*, **27**, 347–9.

Prinzenberg, E.M., Anglade, P., Ribadeau-Dumas, B. and Erhardt, G. (1998) Biochemical characterization of bovine α_{s1}-casein F and genotyping with sequence-specific primers. *J. Dairy Res.*, **65**, 223–31.

Prinzenberg, E.M., Krause, I. and Erhardt, G. (1999) SCCP analysis of the bovine CSN3 locus discriminates six alleles corresponding to known protein variants (A, B, C, E, F, G) and three new DNA polymorphism (H, I, A^1). *Anim. Biotech.*, **10**, 49–62.

Pujolle, J., Ribadeau-Dumas, B., Garnier, J. and Pion, R. (1966) A study of κ-casein components. *Biochem. Biophys. Res. Commun.*, **25**, 285–90.

Pupkova, G.V. (1980) Milk protein polymorphism and milk production of Estonian Black Pied cows. *Dairy Sci. Abstr.*, **45**, 6620.

Raimondi, R. (1982) Marcatori genetici in popolazioni animali scarsamente valorizzate o in via di estinzione. Progetto finalizzato Difese delle risorse genetiche delle popolazioni animali; obiettivi e risultati. Milano, pp. 120–38.

Rampilli, M., Caroli, A., Bolla, P. and Pirlo, G. (1988) Relationships of milk protein genotypes with casein composition and rennet ability of milk during lactation. *Scien. Tecn. Latt-Casear.*, **39**, 262–79.

Rampilli, M., Cecchi, F., Giuliotti, L. and Cattaneo, T.M.P. (1997) The influence of β-lactoglobulin genetic polymorphism on protein distribution and coagulation properties in milk of Massese breed ewes. In *Milk Protein Polymorphism, Special Issue*, **9702**, International Dairy Federation, Brussels, Belgium, pp. 311–5.

Ramunno, L., Rando, A., Gregorio, P., di Capogreco, B. and Masina, P. (1994) Indagine popolazione sui geni a effetto maggiore sul contenuto de caseina α_{s1} e β nel latte di capra. *Zootec. e Nutriz. animale*, **20**, 107–11.

Rando, A., Di Gregorio, P. and Masina, P. (1988) Identification of bovine κ-casein genotypes at the DNA level. *Anim. Genet.*, **19**, 51–4.

Rando, A., Di Gregorio, P., Davoli, R., Dall'olio, S. and Masina, P. (1990) Identification of the two common alleles of the bovine α_{s1}-casein locus by means of RFLPs. *Proc. 4th World Cong. Genet. Appl. Livest. Prod*, Edinburgh, Scotland, **XIV**, 241–4.

Rando, A., Ramunno, L., di Gregorio, P., Davoli, R. and Masina, P. (1992) A rare insertion in the bovine α_{s1}-casein gene. *Anim. Genet.*, **23** (Suppl. 1), 25 (abstr).

Rando, A., Di Gregorio, P., Ramunno, L., Mariani, P., Fiorella, A., Senese, S., Marletta, D. and Masina, P. (1998) Characterization of the $CSN1A^{G}$ allele of the bovine α_{s1}-casein locus by the insertion of a relict of a long interspersed element. *J. Dairy Sci.*, **81**, 1735–42.

Rausch, W.H., Hines, H.C., Treece, J.M. and Tyrrell, H.F. (1967) A preliminary report on polymorphism frequencies of 15 genetic systems in Guernseys. *Immunogenet. Letts.*, **5**, 87–92.

Recio, I., Amigo, L. and Lopezfandino R. (1997a) Assessment of the quality of dairy products by capillary electrophoresis of milk proteins. *J. Chromatogr. B.*, **697**, 231–42.

Recio, I., Perez-Rodriguez, M.L., Ramos, M. and Amigo, L. (1997b) Capillary electrophoretic analysis of genetic variants milk proteins from different species. *J. Chromatogr.*, **768**, 47–56.

Recio, I., Perez-Rodriguez, M.L., Amigo, L. and Ramos M. (1997c) Study of polymorphism of caprine milk caseins by capillary electrophoresis. *J. Dairy Res.*, **64**, 515–23.

Recio, I., Ramos, M. and Amigo, L. (1997d) Study of polymorphism of ovine α_{s1}- and α_{s2}-caseins by capillary electrophoresis. *J. Dairy Res.*, **64**, 525–34.

Reichardt, W., Gernand, E. and Schuler, D. (1995) Relationships between protein content and casein number of cow milk. 1. Review on factors which influence casein number. *Archiv für Tierzucht.*, **38**, 263–76.

Ribadeau-Dumas, B., Brignon, G., Grosclaude, F. and Mercier, J.C. (1972) Structure primaire de la caseine β bovine. Séquence complète. *Eur. J. Biochem.*, **25**, 505–14.

Richardson, B.C. and Creamer, L.K. (1975) Comparative micelle structure. IV. The similarity between caprine α_s-casein and bovine α_{s3}-casein. *Biochim. Biophys. Acta*, **393**, 37–47.

Ricordeau, G., Mahé, M.F., Persuy, M.A., Leroux, C., François, V. and Amigues, Y. (1999) Fréquences alléliques des caséines chez les chèvres des Pyrénées. Cas particuliers de la caséine β nulle. *INRA Prod. Anim.*, **12**, 29–38.

Rogne, S., Lien, S., Vegarud, G., Steine, T., Lansgrud, T. and Alestrom, P. (1989) A method for κ-casein genotyping of bulls. *Anim. Genet.*, **20**, 317–21.

Ronda, R., Perez-Beato, O. and Granado, A. (1984) Relationship between milk proteins and quantitative parameters in 3/4 Holstein-Friesian × 1/4 Zebu cows. *Rev. Sol. Anim.*, **6**, 593–603.

Russo, V. and Mariani, P. (1974) Polymorfismo delle proteine del latte nelle razze bovine da carne. III. Chianina. *Riv. Zootech. Vet.*, **2**, 139–43.

Russo, V. and Mariani, P. (1978) Polimorfismo delle proteine del latte e relazioni tra varianti genetiche e caratteristiche di interesse zootecnico, tecnologico e caseario. *Riv. Zootec. Vet.*, **6**, 365–79.

Russo, V., Chiofalo, L. and Micari, P. (1977) Esame elettroforetico delle proteine del latte in popolazioni caprine allevate Sicilia. *Zootech. Nutr. Anim.*, **3**, 247–53.

Russo, V., Davoli, R., Dall'olio, S. and Tedeschi, M. (1986) Ricerche sul polimorfismo del latte caprino. *Zoot. Nutr. Anim.*, **12**, 55–62.

Saama, P.M., Mao, I.L., Lie, S., Steine, T. and Solbu, H. (1992) Effects of polymorphic milk protein genes or weight gain in Norwegian bull. *J. Dairy Sci.*, **75**, Suppl 1, 285 (abstr).

Sadler, A.M., Kiddy, C.A., McCann, R.E. and Mattingly, W.A. (1968) Acid production and curd toughness in milks of different α_{s1}-casein types. *J. Dairy Sci.*, **51**, 28–30.

Samarieanu, M., Stamazescu, E., Granciu, I. and Sotu, E. (1983) Genetic variations of β-caseins in cattle. *Anim. Breed. Abstr.*, (1986) Abst. No.2767, **54**, 362.

Savva, D., Mazhar, K., Heriz, A., Tejedor, M.T., Monteagudo, L.V., Skidmore, C.V. and Arruga M. (1994) Confirmation of the assignation of the bovine β-lactoglobulin gene and analysis of polymorphism by the PCR method. *Itea Produccion Animal*, **90**, 155–65.

Schaar, J. (1984) Effects of κ-casein genetic variants and lactation number on the renneting properties of individual milks *J. Dairy Res.*, **51**, 397–406.

Schaar, J. (1986) *Variation in Milk Protein Composition. Studies on κ-Casein and β-Lactoglobulin Genetic Polymorphism and on Milk Plasmin*, Thesis, University of Agricultural Sciences, Uppsala, Sweden.

Schaar, J., Hansson, B. and Petterson, H.E. (1985) Effects of genetic variants of κ-casein and β-lactoglobulin on cheesemaking. *J. Dairy Res.*, **52**, 429–37.

Schild, T.A., Wagner, V. and Geldermann, H. (1994) Variants within the 5′-flanking regions of bovine milk protein genes. I. κ-Casein encoding gene. *Theoret. Appl. Genet.*, **89**, 116–20.

Schild, T.A. and Geldermann, H. (1996) Variants within the 5′-flanking regions of bovine milk protein encoding genes. III. Genes encoding the Ca-sensitive caseins α_{s1}, α_{s2} and β. *Theoret. Appl. Genet.*, **93**, 887–93.

Schlee, P. and Rottman, O. (1992) Genotyping of bovine β-casein, β-lactoglobulin and α-lactalbumin using the polymerase chain reaction. *J. Anim. Breed. Genet.*, **109**, 456–64.

Schlieben, S., Erhardt, G. and Senft, G. (1991) Genotyping of bovine κ-casein (κ-CN^A, κ-CN^B, κ-CN^C, κ-CN^E) following DNA sequence amplification and direct sequencing of κ-CN^E PCR product. *Anim. Genet.*, **22**, 333–42.

Schmidt, D.G. (1964) Starch gel electrophoresis of κ-casein. *Biochim. Biophys. Acta*, **90**, 411–14.

Schroeder, W.A., Shelton, J.B., Shelton, J.R., Powars, D., Friedman, S., Baker, J., Finklestein, J.Z., Miller, B., Johnson, C.S., Sharpsteen, J.R., Sieger, L. and Kawaoka, E. (1982) Identification of eleven human haemoglobin variants by high-performance liquid chromatography: additional data on functional properties and clinical expression. *Biochem. Genet.*, **20**, 133–52.

Seibert, B., Erhardt, G. and Senft, B. (1985) Procedure for simultaneous phenotyping of genetic variants in cow's milk by isolectric focusing. *Anim. Blood Grps. Biochem. Genet.*, **16**, 183–91.

Seibert, B., Erhardt, G. and Senft, B. (1987) Detection of a new κ-casein variant in cow's milk. *Anim. Genet.*, **18**, 269–72.

Sen, A. and Chaudhuri, S. (1962) Non-casein proteins of goat's milk. *Nature*, **195**, 286–7.

Sherbon, J.W., Ledford, R.A. and Regenstein, J. (1967) Variants of milk proteins and their possible relation to milk properties. *J. Dairy Sci.*, **50**, 951.

Singh, H., Bhat, P.N. and Singh, R. (1981) Association of protein polymorphic genotypes with certain performance traits in crossbred cattle. *Ind. J. Anim. Sci.*, **51**, 5–10.

Smith, M.H., Hill, J.P., Creamer, L.K. and Plowman, J.E. (1997) Towards understanding the variant effect on the rate of cleavage by chymosin on κ-casein A and C. In *Milk Protein Polymorphism, Special Issue*, **9702**, International Dairy Federation, Brussels, Belgium, pp. 185–8.

Stambekov, S. Zh., Shapiro, Y.U.O., Mandrusova, E.E. and Romanyuk, N.A. (1977a) Polymorphism of milk proteins in Latvian Darkheaded ewes. *Dairy Sci. Abstr.*, **39** Abstract No. 1000.

Stambekov, S. Zh., Shapiro, Y.U.O. and Mandrusova, E.E. (1977b) Milk protein polymorphism in Soviet Merino and Latvian Darkheaded ewes and its connection with economically valuable characteristics. *Dairy Sci. Abstr.*, **39** Abstract No 6873.

Stur, I., Neumeister, F., Mayr, B., Glawisehnig, F. and Schleger, W. (1976) Relationships between mastitis resistance, degree of heterozygosis and milk proteins in Austrian Brown cattle. *Wien. Tier. Monat.*, **63**, 352–6.

Sulimova, G.E., Badagueva, Y.N. and Udina, I.G. (1996) Polymorphism of the κ-casein gene in subfamilies of the Bovidae. *Genetika* (*Moskva*), **32**, 1576–82.

Szmelik, L., Zagulski, T. and Michalak, W. (1971) Investigations on genetic polymorphism of β-casein in cattle bred in Poland. *Biul. Inst. Genet, Hodowli Zwierzat Pol. Akad. Nauk.*, **23**, 51–65.

Tanev, G. (1967) Thin layer starch gel electrophoretic separation of the principal proteins of ewe's milk. *Zhivotnovdi Nauki.*, **41**, 23–9.

Tervala, H.L., Antila, V., Syvaejaervi, J. and Lindstroem, U.B. (1983) Variations in the renneting properties of milk. *Meij. Aika.*, **41**, 24–33.

Tervala, H.L., Antila, V. and Syvaejaevi, J. (1985) Factors affecting the renneting properties of milk, *Meij. Aika.*, **43**, 16–25.

Thompson, M.P. (1966) DEAE-cellulose-urea chromatography of casein in the presence of 2-mercaptoethanol. *J. Dairy Sci.*, **49**, 792–5.

Thompson, M.P. (1970) Phenotyping milk proteins: A review. *J. Dairy Sci.*, **53**, 1341–8.

Thompson, M.P. and Pepper, L. (1964) Genetic polymorphism in caseins of cow's milk. IV. Isolation and properties of β-casein A, B, and C. *J. Dairy Sci.*, **47**, 633–6.

Thompson, M.P., Kiddy, C.A., Pepper, L. and Zittle, C.A. (1962) Variations in the α-casein fraction of individual cow's milk. *Nature*, **195**, 1001–2.

Thompson, M.P., Kiddy, C.A., Johnston, J.O. and Weinberg, R.M. (1964) Genetic polymorphism in caseins of cow's milk. II. Confirmation of the genetic control of β-casein variation. *J. Dairy Sci.*, **47**, 378–81.

Thompson, M.P., Gordon, W.G., Boswell, R.T. and Farrell, H.M. (1969a) Solubility, solvation, and stabilization of α_{s1}- and β-caseins. *J. Dairy Sci.*, **52**, 1166–73.

Thompson, M.P., Gordon, W.G., Pepper, L. and Greenberg, R. (1969b) Amino acid composition of β-caseins from the milks of *Bos indicus* and *Bos taurus* cows: a comparative study. *Comp. Biochem. Physiol.*, **30**, 91–8.

Threadgill, D.W. and Womack, J.E. (1990) Genomic analysis of the major bovine milk protein genes. *Nucleic Acids Res.*, **18**, 6935–42.

Thresher, W.C. (1997) Interactions of β-lactoglobulin variants with κ-casein and DNTB: formation of a covalent complex. In *Milk Protein Polymorphism, Special Issue*, **9702**, International Dairy Federation, Brussels, Belgium, pp. 138–45.

Tong, P.S., Vink, S., Farkye, N.Y. and Medrano, J.F. (1994) Effect of genetic variants of milk proteins on the yield of Cheddar cheese. In *Cheese Yield and Factors Affecting its Control, Special Issue*, **9402**, International Dairy Federation, Brussels, Belgium, pp. 179–87.

van den Berg, G. (1994) Genetic polymorphism of κ-casein and β-lactoglobulin in relation to milk composition and cheesemaking properties. In *Cheese Yield and Factors Affecting its Control, Special Issue*, **9402**, International Dairy Federation, Brussels, Belgium, pp. 123–33.

van den Berg, G., Escher, J.T.M., de Koning, P.J. and Bovenhuis H. (1992) Genetic polymorphism of κ-casein and β-lactoglobulin in relation to milk composition and processing properties. *Neth. Milk Dairy J.*, **46**, 145–68.

Van Eenennaam, A.L. and Medrano, J.F. (1991) Differences in allelic protein expression in the milk of heterozygous κ-casein cows. *J. Dairy Sci.*, **74**, 1491–6.

Van Riel, J. and Olieman, C. (1995) Determination of caseinomacropeptide with capillary zone electrophoresis and its application to the detection and estimation of rennet whey solids in milk and buttermilk powder. *Electrophoresis*, **16**, 529–33.

Velmala, R., Grohs, C., Mahé, M.F., Lien, S. and Levéziel, H. (1994) Genotyping of the bovine β-casein C allele by amplification created restriction site. *Anim. Genet.*, **25**, 429.

Velmala, R., Vilkki, J., Elo, K. and Maki-Tanila, A. (1995) Casein haplotypes and their association with milk production traits in Finnish Ayrshire cattle. *Anim. Genet.*, **26**, 419–25.

Visser, S., Jenness, R. and Mulin, R.J. (1982) Isolation and characterization of β- and γ-caseins from horse milk. *Biochem. J. Mol. Asp.*, **203**, 131–9.

Visser, S., Slangen, K.J. and Rollema, H.S. (1986) High performance liquid chromatography of bovine caseins with the application of various stationary phases. *Milchwissenschaft*, **41**, 559–62.

Visser, S., Slangen, C.J. and Rollema, H.S. (1991) Phenotyping of bovine milk proteins by reversed-phase high-performance liquid chromatography. *J. Chromatogr.*, **548**, 361–70.

Visser, S., Slangen, C.J., Lagerwerf, F.M., van Dongen, W.D. and Haverkamp, J. (1995) Identification of a new genetic variant of bovine β-casein using reversed-phase high-performance liquid chromatography and mass spectrometric analysis *J. Chromatogr.* A., **711**, 141–50.

Voelker, G.R., Bleck, G.T. and Wheeler, M.B. (1997) Single-base polymorphism within the 5′-flanking region of the bovine α-lactalbumin gene. *J. Dairy Sci.*, **80**, 194–7.

Voglino, G.F. (1972) A new β-casein variant in Piedmont cattle. *Anim. Blood Grps. Biochem. Genet.*, **3**, 61–2.

Voglino, G.F. and Ponzone, A. (1972) Polymorphism in human casein. *Nature New Biol.*, **238**, 149–50.

Voglino, G.F., Caone, M. and Ponzone, A. (1975) A new beta-casein variant in human milk. *Biomedicine*, **32**, 342–5.

Wagner, V.A., Schild, T.A. and Geldermann, H. (1994) DNA variants within the 5′-flanking region of milk protein encoding genes. II. The β-lactoglobulin encoding gene. *Theoret. Appl. Genet.*, **89**, 121–6.

Wake, R.G. and Baldwin, R.L. (1961) Analysis of casein fractions by zone electrophoresis in concentrated urea. *Biochim. Biophys. Acta*, **47**, 225–39.

Walsh, C.D., Guinee, T., Harrington, D., Mehra, R., Murphy, J., Connoly, J.F. and Fitzgerald, R.J. (1995) Cheddar cheesemaking and rennet coagulation characteristics of bovine milks containing κ-casein AA or BB genetic variants. *Milchwissenschaft*, **50**, 492–6.

Watterson, G.A. and Guess, H.A. (1977) Is the most frequent allele the oldest? *Theor. Popul. Biol.*, **11**, 141–60.

Wilkins, R.J. and Xie T. (1997a) Two distinct gene mutation-one milk protein polymorphism: the example of the α_{s1}-casein A variant. In *Milk Protein Polymorphism, Special Issue*, **9702**, International Dairy Federation, Brussels, Belgium, pp. 330–3.

Wilkins, R.J. and Xie, T. (1997b) Designing DNA tests for the rarer milk protein polymorphism. in *Milk Protein Polymorphism, Special Issue*, **9702**, International Dairy Federation, Brussels, Belgium, pp. 345–9.

Willige, R.W.G. and Fitzgerald, R.J. (1995) Tryptic and chymotryptic hydrolysis of β-lactoglobulin A, B, and AB at ambient and high pressure. *Milchwissenschaft*, **50**, 183–6.

Winkelman, A.M. and Wickham, B.W. (1996) Associations between milk protein genetic variants and production traits in New Zealand dairy cattle. *Proc. N.Z. Soc. Anim. Prod. Hamilton, New Zealand*, **56**, 24–7.

Woychik, J.H. (1964) Polymorphism in κ-casein of cow's milk. *Biochem. Biophys. Res. Commun.*, **16**, 267–71.

Yaguchi, M. and Rose, D. (1971) Chromatographic separation of milk proteins: A review. *J. Dairy Sci.*, **54**, 1725–43.

Yebrovski, L.S. and Komissarenko, A. (1982) Effect of genotype on intra-breed variation in the amino acid content of milk proteins. *Proc. XXI Intern. Dairy Congr.* (*Moscow*). Vol. 1, Book 1, 46 (abstr).

Zadworny, D., Kuhnlein, U. and Ng-Kwai-Hang, K.F. (1990) Determination of kappa-casein alleles in Holstein dairy cows and bulls using the polymerase chain reaction. *Proc. 4th World Congr. Genet. Appl Livestock Prod.* **XIV**: 251–4.

Zebrovskii, L.S., Babukov, A.V. and Mityutko, V.E. (1977) The relationship of polymorphic proteins with productivity in Black Pied cattle. *Zhivotnovodstvo*, **7**, 25–7.

Zikakis, J.P., Haenlein, G.F.W., Hines, H.C., Mather, R.E. and Tung, S. (1974) Gene frequencies of electrophoretically determined polymorphisms in Guernsey blood and milk. *J. Dairy Sci.*, **57**, 405–10.

17

GENETIC ENGINEERING OF MILK PROTEINS

J. LEAVER AND A.J.R. LAW

17.1 INTRODUCTION

Changes in the productivity, composition or appearance of plants and animals have, traditionally, been achieved by means of selective breeding, drawing upon naturally occurring variations within the gene pool of a species. Many commercial plant species such as cereals and members of the brassica family bear very little resemblance to their ancestral forms and their productivity is very substantially higher. Changes achieved with domesticated animals are less dramatic but, in conjunction with improved animal husbandry, selective breeding has been so successful in increasing the milk yield of dairy cattle that within the European Union, quotas have had to be imposed in order to limit surpluses. By the use of artificial insemination and embryo transplants, an individual bull with the desired characteristics can produce thousands of offspring.

However, in contrast to the dramatic increases which have been achieved in yield, changes in the actual composition of the milk, such as the ratios of fat to protein and whey protein to casein and in the overall solids content, have been much more limited. In addition, within a species, the amino acid sequences of the milk proteins, which largely dictate their functionality as components in processed foods, are generally well conserved. In cattle, for example, with the exception of α_{s1}-casein (α_{s1}-CN) where there is a deletion of a sequence of 13 amino acids in the A variant, variation is restricted to single amino residues, severely limiting the scope for changing the functionality of individual proteins.

These restrictions can, potentially, be overcome by the application of a relatively new area of biotechnology, transgenesis. This technique utilises recent advances in molecular biology to provide an alternative method for

Advanced Dairy Chemistry Volume 1: Proteins, 3rd edn.
Edited by P.F. Fox and P.L.H. McSweeney, Kluwer Academic/Plenum Publishers, 2003.

producing much more dramatic, but equally much more specific, genomic changes. Genes from one species can be introduced into the genome of organisms of the same or different species and subsequently transmitted to their progeny in a stable fashion. Transgenic techniques were used originally to modify microorganisms, such as yeasts and bacteria, in order to produce large quantities of valuable, pharmacologically-active proteins and peptides usually present in only minute amounts in higher organisms. Palmiter *et al.* (1982) extended this technique to mammals and reported the successful

Table 17.1
Expression of some pharmaceutical proteins in transgenic milks

Protein	*Regulatory sequence*	*Species*	*Maximum level* ml^{-1}	*Reference*
Human α_1-antitrypsin	ovine βLG	sheep mouse	33 mg 21 mg	Carver *et al.* (1993)
Human factor IX	ovine βLG	sheep	25 ng	Clark *et al.* (1989)
Human tissue plasminogen activator	bovine αs1-cn	mouse, rabbit	0.05 mg	Riego *et al.* (1993)
Human gamma interferon	ovine βLG	mouse	1800 I.U.	Dobrovolsky *et al.* (1993)
Cystic fibrosis transmembrane conductance regulator	goat β-cn	mouse	1 mg	Ditullio *et al.* (1992)
Human superoxide dismutase	mouse WAP ovine βLG	mouse		Hansson *et al.* (1994)
Human protein C	mouse WAP	mouse, pig	0.7 mg 1 mg	Drohan *et al.* (1994) Velander *et al.* (1992)
FSH	rat β-cn	mouse	60 I.U. (15ng)	Greenberg *et al.* (1991)
Human CD6 antibody	rabbit WAP	mouse		Limonta *et al.* (1995)
Human glycosyltransferase	mouse WAP	mouse		Prieto *et al.* (1995)
Human parathyroid hormone	mouse WAP	mouse		Rokkones *et al.* (1995)
Human insulin-like growth hormone	bov α_{s1}-cn	rabbit	1 mg	Brem *et al.* (1994)
Human growth hormone	rabbit WAP rabbit WAP bov α_{s1}-cn	mouse mouse mouse	10 mg 22 mg 11 μg	Stinnakre *et al.* (1993) Devinoy *et al.* (1994) Ninomiya *et al.* (1994)
Bovine growth hormone	rabbit WAP	mouse	16 mg	Thepot *et al.* (1995)
Ovine trophoplastin	bovine αLA	mouse	1 μg	Stinnakre *et al.* (1991)

expression of rat growth hormone in mice. This was achieved by the direct microinjection of rat DNA, fused to the regulatory sequence of the metallothionein gene, into fertilized mouse eggs. The growth hormone was produced in the liver, and plasma levels up to 100-fold higher than normal were detected. However, the foreign protein interfered with the normal development of the host and transgenic mice were significantly larger than normal. It was quickly realised that a better option would be to target expression specifically to the mammary gland.

Expressing therapeutic proteins and peptides in milk is an attractive proposition. The mammary gland itself is capable of very high rates of protein expression and secretion and, since the stored milk is relatively isolated from the general circulation of the animal, transgenic proteins are much less likely to influence the physiology of the host. Milk is a well defined substance, relatively free from proteolytic enzymes and infective organisms such as viruses and prions and can be harvested in a non-invasive fashion using normal milking procedures and subsequently processed on a very large scale, if necessary.

In this article, after briefly considering changes achieved by site-directed mutagenesis in the structure and functionality of milk proteins expressed in microorganisms, we will concentrate mainly on the use of transgenic techniques to modify the composition and properties of milk both as a food and also as a means of producing valuable pharmaceutical proteins.

17.2 Expression of milk proteins in non-mammalian systems

17.2.1 Design of expression systems

The prokaryotic bacterium, *Escherichia coli*, and the eukaryotic yeasts, *Saccharomyces cerevisiae*, *Kluyveromyces lactis* and *Pichia pastoris*, are the principal microorganisms in which native and mutated milk proteins have been expressed. Typically, the gene coding for the milk protein is fused to an inducible promoter sequence (although constitutive expression is becoming more common) and inserted into a suitable vector such as a plasmid or phage. The vector is then used to transform the host and after screening, transformed cells are cultured. After the cell mass reaches a sufficiently high level, the appropriate inducer is added and expression of the milk protein gene begins. By incorporating a secretion signal peptide, it is possible to direct export of the gene product into the medium.

17.2.2 Whey proteins

Bovine β-lactoglobulin (β-lg) has been expressed in both prokaryotic and eukaryotic organisms. In *E. coli*, levels of 8 to 10 mg L^{-1} of culture were reported (Batt *et al.*, 1990). However, the protein was deposited within the cells in the form of inclusion bodies and required solubilisation and refolding. The same protein was expressed using the glyceraldehyde-3-phosphate promoter in the eukaryotic yeast, *S. cerevisiae*, as a soluble, secreted product at a concentration of about 1.1 mg L^{-1} (Totsuka *et al.*, 1990). Ovine β-lg has also been reported to be expressed in *S. cerevisiae* at 3 to 4 mg L^{-1} but was secreted in cultures of another yeast, *K. lactis*, using the phosphoglycerate kinase promoter, at a concentration of 40 to 50 mg L^{-1} (Rocha *et al.*, 1996). Introducing an additional disulphide bridge into bovine β-lg by site-directed mutagenesis in *E. coli*, increased the thermal stability of the protein by 8 to 10°C (Cho *et al.*, 1994). In contrast, introducing two additional thiol groups into the protein decreased its thermostability by more than 15°C, increased gel strength and decreased syneresis in yogurts made from milk fortified with this novel β-lg (Lee *et al.*, 1994).

Human lactoferrin has been expressed in *S. cerevisiae* and in the Baculovirus expression system. In *S. cerevisiae*, using the yeast chalatin promoter, the recombinant protein was secreted at levels up to 2 mg L^{-1} in a glycosylated form which bound iron and copper (Liang *et al.*, 1993). The protein was purified from the Baculovirus expression system at a concentration of 10 to 15 mg L^{-1} of culture supernatant (Salmon *et al.*, 1997). The iron-binding properties of this transgenic protein were identical to those of the native protein, but the two differed in the extent of glycosylation. Recombinant truncated variants of bovine lactoferrin have been expressed in *E. coli* at a level of 10 mg L^{-1} (Sitaram *et al.*, 1998). By determining the ability of these truncated proteins to bind to the Ca^{2+}-dependent receptors on hepatocytes, it was established that lactoferrin does so *via* its C-lobe.

Recombinant α-lactalbumin (α-la) expressed in *E. coli* was shown to have an additional methionine at the N-terminus. The effects of this modification on the physical properties of recombinant caprine (Chaudhuri *et al.*, 1999) and human (Ishikawa *et al.*, 1998) α-la were very pronounced, with the Met-α-la form of the protein being considerably more heat labile than the native form. Comparison of the X-ray structures of the caprine Met-α-la with that of the native protein showed the structural differences were confined to the N-terminal region.

17.2.3 Caseins

Human β-CN expressed in *E. coli* at a level of 300–500 mg L^{-1}, using an inducible T7-based expression system (Hansson *et al.*, 1993), was shown to be intracellular and to co-migrate with authentic full-length, unphosphorylated human β-CN. Co-expression of human casein kinase II in the same plasmid resulted in *in vivo* phosphorylation of the transgenic protein (Thurmond *et al.*, 1997).

Bovine β-CN expressed in *S. cerevisiae* under the control of the hexokinase P1 promoter at levels as high as 10 mg L^{-1} was believed to be located either in the periplasmic space or in the cytoplasm (Jiminez-Flores *et al.*, 1990). Detailed analysis of the protein showed that, in addition to being multiply phosphorylated, some had also undergone *O*-glycosylation. Bovine β-CN has also been expressed in *E. coli*, again using the T7 expression system (Simons *et al.*, 1993). The level of expression was estimated to be 20% of the total soluble proteins, with most of it being located in the periplasmic fraction. Amino acid sequence analysis showed that the protein had an additional methionine at the N-terminal, and the glutamine residue at position 175 had been mutated to an arginine residue. The β-CN cDNA was modified in order to remove the principal chymosin cleavage site at position 192–193, which gives rise to bitter peptides during cheese maturation. This was achieved by changing the Leu-Tyr sequence at this point to a chymosin-insensitive Pro-Pro bond, which was not cleaved, or to Leu-stop which produced a β-CN molecule truncated at position 192.

κ-CN has been expressed in *E. coli* at 2 mg mL^{-1} using the secretion vector pIN-III-ompA and mutated to increase both its nutritional quality (Oh and Richardson, 1991a) and rate of proteolysis by chymosin (Oh and Richardson, 1991b). In the first instance, three methionine residues were inserted between Ala_{167} and Thr_{168} in order to increase the content of sulphur-containing amino acids. In the latter instance, the Phe_{105}–Met_{106} bond was changed to Phe–Phe and the rate of chymosin-catalysed hydrolysis of the novel protein was shown to be 80% higher than that of the wild-type.

17.3 Genetic engineering of milk proteins in animals

17.3.1 Production of transgenic animals

Currently, the method most commonly used to incorporate foreign genes into host animals is pro-nuclear injection. DNA is microinjected into the pro-nucleus of one-cell embryos obtained from superovulated females and the eggs are then surgically implanted into hormonally-primed foster mothers. This technique, originally developed for the generation of transgenic mice, has been extended, with minor adaptations, to larger animals, including ruminants (see review by Clark, 1998). Generally, the success rate of this technique, as measured by the number of live births of transgenic animals, is about 1 to 2%.

An alternative method for generating transgenic animals is by modifying the genome of embryonic stem (ES) cells. These are obtained early in the development of the embryo and can be cultured in a suitable medium whilst retaining their totipotency. Following the introduction of foreign DNA, usually including a marker gene in order to enable positive cells to be identified, these cells can be injected into a host blastocyst. Here, the cells become integrated into the tissues of the developing embryo, including the germline, producing transgenic animals capable of transferring the transgene to their progeny. While this technique has been used widely to produce transgenic mice, this is the only species from which embryonic stem cells have been obtained.

Nuclear transfer has caused enormous excitement in the popular press in recent years. A review of this technique and an assessment of its potential for modifying livestock has been published recently (Woolliams and Wilmut, 1999). It has great potential to produce cloned copies of mammals, including dairy animals, whose milk composition has been modified in specific ways. Basically, this technique involves fusing cultured cells obtained either from embryonic, fetal or adult tissues with enucleated, unfertilized egg cells, using electrical pulses. The reconstructed embryos are cultured further and those that develop to morula or blastocyst are implanted into recipient females and allowed to develop to full-term. An obvious extension to this technique is to use cloned transgenic cell lines as donors in order to produce cloned transgenic offspring. Transgenic sheep, produced by nuclear transfer, expressing human factor IX in their milk were reported by Schnieke *et al.* (1997). The authors estimated that the efficiency of production of these animals was more than twice as high as could be attained by pro-nuclear microinjection.

Table 17.2
Expression of milk proteins in transgenic milks

Protein	*Host species*	*Max. Level ($mg\ ml^{-1}$)*	*Reference*
Ovine βLG	mouse	23	Simons *et al.* (1987)
Ovine βLG (phosphorylated)	mouse	c. 30	unpublished
Caprine κ-casein	mouse	3	Persuy *et al.* (1995)
Bovine αLA	mouse	1.5	Bleck & Bremel (1994)
Bovine β-casein	mouse	10	Bleck *et al.* (1995)
Bovine β-casein	mouse	5	Hitchin *et al.* (1996)
Caprine β-casein	mouse	40	Persuy *et al.* (1992)
Human serum albumin	mouse	10	Hurwitz *et al.* (1994)
Human lactoferrin	mouse	0.04	Platenburg *et al.* (1994)
Human α-lactalbumin	mouse	1.4	Stacey *et al.* (1995)
Mouse WAP	pig	1.5	Shamay *et al.* (1991)
	sheep	0.5	
			Wall *et al.* (1996)

17.3.2 Targeting expression to the mammary gland

To direct the expression of transgenic proteins specifically to the mammary gland, a gene construct is formed by coupling the gene coding for the foreign protein to a milk protein regulatory sequence. Promoter elements which have been used for this purpose include those from various caseins, whey acidic protein (WAP) from both mouse and rabbit, ovine β-lg and α-la genes. Both the efficiency of expression and tissue specificity are affected by the nature of the promoter. WAP expressed in transgenic sheep under the control of the mouse WAP promoter was detected in the liver, lung, heart, kidney and bone marrow as well as in the milk, but not in skeletal muscle or intestine (Wall *et al.*, 1996). Ectopic expression of transgene products from constructs containing the ovine β-lg 5′-regulatory sequences has also been detected but usually in the salivary gland (Carver *et al.*, 1992) and at very low levels (0.1% of that in milk). The caprine β-CN transgene with 3 kb of 5′ and 6 kb of 3′ flanking region has been expressed in mice (Persuy *et al.*, 1992). High levels of caprine β-CN were detected in 5 lines but analysis of the mRNA from a variety of tissues by Northern blotting showed that caprine β-CN mRNA was present only in mammary gland. The exception to this was the detection of some caprine β-CN mRNA in the skin of a few animals but this may have been due to contamination with mammary tissue.

17.3.3 Post-translational processing of transgenic proteins

One of the major benefits of expressing transgenic proteins in mammalian cell cultures and milk is the ability of these systems to perform correctly a variety of post-translational modifications such as glycosylation, phosphorylation, γ-carboxylation, proteolytic processing and assembly of multicomponent complexes.

(a) Glycosylation
Human extracellular superoxide dismutase, a glycosylated, tetrameric metalloprotein, has been expressed in mouse milk under the control of the mouse WAP promoter, at a maximum concentration of 3 mg mL^{-1} (Stromqvist *et al.*, 1997). Analysis of the purified protein showed that it had the correct polypeptide backbone and glycosylation pattern and was tetrameric. In contrast, human bile salt-stimulated lipase, also expressed in mouse milk under the control of murine WAP regulatory elements at a concentration up to 1 mg mL^{-1}, was found to have a significantly lower degree of glycosylation than either the native protein or recombinant protein isolated from mouse cell cultures (Stromqvist *et al.*, 1996). However, the lipolytic activity and the physical properties of the recombinant protein isolated from milk were found to be very similar to those of the native protein. Human erythropoietin has been expressed in transgenic mouse and

rabbit milks as a bovine β-lg fusion protein (Korhonen *et al.*, 1997). After specific proteolytic cleavage of the purified fusion protein, the erythroprotein portion of the molecule was found to have a lower molecular mass than the native protein. The difference in molecular mass was shown to be due to different types of glycosylation. Human granulocyte-macrophage colony-stimulating factor expressed in mouse milk was found to have the same glycosylation-derived size heterogeneity as the native protein. Expressing a human glycosyltransferase enzyme in mouse milk resulted in the production of not only the active enzyme but also of a series of novel glycoconjugates and milk oligosaccharides resulting from its *in vivo* activity (Prieto *et al.*, 1995). By varying the specificity of the glycosyltransferase it may be possible to use the mammary gland to synthesise different series of glycoconjugates and oligosaccharides.

(b) Phosphorylation

Bovine β-CN expressed in mouse milk was shown by tryptic digestion and mass spectrometry to have the correct polypeptide backbone and to be fully phosphorylated (Hitchin *et al.*, 1996). Phosphatase digestion was also used to demonstrate that transgenic bovine β-CN had the correct degree of phosphorylation (Jeng *et al.*, 1997). Bovine α_{s1}-CN expressed in mouse milk was indistinguishable from the native protein on the basis of specific immunoblotting and radioimmunoassay (Rijnkels *et al.*, 1997). Caprine κ-CN expressed in mouse milk was found to have four additional amino acid residues at its N-terminus, presumably as a result of incorrect cleavage of the signal peptide (Persuy *et al.*, 1995). A modified form of ovine β-lg in which the native gene had been altered to encode a casein kinase recognition

TABLE 17.3

Peptides detected by MALDI-TOF mass spectrometry in tryptic digests of transgenic bovine β-casein and authentic bovine β-casein, phenotypes A[1] and B

Measured peptide mass (Da)	*Identity of peptide (calculated mass: Da)*	*Transgenic β-casein*	*β-casein A[1]*	*β-casein B*
1138.3	114–122 (1137.3)	–	–	+
1382.7	191–202 (1382.7)*	+	+	+
1768.1	53–68 (1769.0)*	+	–	–
2107.7	191–209 (2107.6)*	+	+	+
2184.5	184–202 (2185.5)	+	+	+
2910.3	184–209 (2909.5)	+	+	+
3123.1	1–25 (3123.3)§	+	+	+
5306.7	123–169 (5312.2)	–	–	+
5359.3	49–97 (5359.3)	+	+	+
6361.7	114–169 (6361.7)	+	+	–
7396.3	33–97 (7402.2)§	+	+	+

+ = detected; * = non-specific clevages; § = phosphopeptides.

site on an exposed loop of the molecule, was expressed in mouse milk and shown to be partially phosphorylated *in vivo* (McClenaghan *et al.*, 1999).

(c) γ-Carboxylation

In order to bind calcium, which is an essential cofactor, some plasma proteins possess clusters of carboxylated glutamic acid residues. The ability of the mammary gland to perform γ-carboxylations of these proteins is therefore crucial if they are to be expressed in milk in an active form, particularly since none of the endogenous milk proteins are modified in this fashion. Human protein C (hPC) has been expressed in transgenic mice but only trace amounts were γ-carboxylated and therefore active (Drohan *et al.*, 1994). In contrast, Yull *et al.* (1995) reported that most of the human factor IX recovered from transgenic mouse milk was active and therefore the γ-carboxylation process was relatively efficient.

(d) Proteolytic processing

Maturation of hPC precursor requires endoproteolytic cleavage at two sites in the molecule to remove the propeptide from the N-terminal and to convert the single chain to a heterodimer by excising a dipeptide. Most of the recombinant hPC expressed in the milk of transgenic swine was shown to be inefficiently processed by the endoproteases of the mammary epithelial cells (Lee *et al.*, 1995).

(e) Assembly of complexes

Assembly of transgenic protein complexes has also been achieved in milk. Fibrinogen is a heterodimeric glycoprotein composed of two subunits of three different polypeptide chains. Three expression cassettes, each encoding one of the polypeptide chains under the control of ovine β-lg promoter sequences, were co-injected into fertile mouse eggs (Prunkard *et al.*, 1996). The fibrinogen subunits were expressed in the milk at levels up to 2 mg mL^{-1} and in some animals up to 100% of the individual subunits were incorporated into active, fully-assembled hexamers.

17.4 Modification of the nutritional properties of milk

In industrialised countries, milk proteins are increasingly being consumed not in liquid milk but as components in processed foods ranging from cheese and yoghurt to low-fat spreads and sauces. Since the composition of bovine milk has evolved to enable it to be consumed directly, from the mammary gland by neonatal calves, it is not surprising that food technologists can suggest ways in which its processing behaviour could be improved (see Table 17.1 and reviews by Karatzas and Turner, 1997; Batt, 1997; Wall *et al.,* 1997; Murray, 1999).

Despite the technology being available to produce ruminants with modified milks, very few actual modifications have been made. This is largely due to the expense of producing large transgenic animals given the low success rate of the current transgenic techniques, their long gestation period and the time required for them to reach sexual maturity. Much of the reported research has therefore been performed on small animals such as the mouse, which produce milk very different from that of dairy animals. The amino acid sequences of the caseins of mouse milk, for example, are different from those of the corresponding proteins in ruminants and the protein content of mouse milk is two to three times higher than that of bovine milk. In addition, mouse milk contains no β-lg and very little α-la. The major whey protein in mouse milk, WAP, is not present in ruminant milks. While over-extrapolation of the effects of changes in mouse milk to the effect of similar changes in the milk of dairy animals is not wise, some of the changes which have been achieved have had interesting effects on the properties of the milk (Table 17.2).

17.4.1 Altering protein content and casein/whey protein ratio

It may be possible to increase the protein content of milk by inserting multiple copies of milk protein genes. Obviously, this depends on the full protein synthetic capacity of the mammary gland not already being utilised. Expression of high levels of ovine β-lg in mouse milk was achieved at the expense of endogenous protein production with the production of WAP being more suppressed than that of the caseins. No significant increase in the total protein content of the transgenic milk was detected (McClenaghan *et al.*, 1995). The suppression of endogenous protein production was matched by a decrease in the corresponding steady-state mRNA levels. In contrast, expression of human α-antitrypsin (hAAT) in ovine milk at a level up to 50 mg mL^{-1} was not achieved at the expense of endogenous milk proteins (Colman, 1996).

17.4.2 Changing lactose content

It has been estimated that approximately 70% of the adult human population has a deficiency of the intestinal enzyme, lactase, and therefore cannot digest lactose. This leads to a variety of intestinal problems arising from the malabsorption of lactose and an understandable tendency to avoid consuming milk. Reducing the lactose content of milk should therefore enable its consumption by some people who suffer from lactose intolerance. Since lactose also plays a major role in regulating the osmolarity of milk, reducing its concentration should also reduce the water content of the modified milk, increasing the solids contents and reducing transport and processing costs.

Two different genetic engineering approaches have been used in an attempt to reduce the lactose content of milk. Since α-la is involved in the synthesis of lactose in milk, the most obvious method for reducing lactose content is by disrupting the α-la gene, thus preventing the expression of this protein. This has been achieved by homologous recombination in mouse embryonic stem cells (Stinnakre *et al.*, 1994). Homozygous mutant mice produced very viscous milk, rich in proteins and fat. Unfortunately, due to its viscosity, pups were unable to remove this milk from the mammary gland. Heterozygous mutant mice showed a 40% decrease in the level of α-la but only a 10–20% decrease in the lactose content of the milk. Stacey *et al.* (1995) observed the same effect when they deleted the murine α-la gene. However, when the endogenous murine α-la gene was replaced by the human α-la gene, the level of α-la detected in homozygous females, which is very low in the native strains, was found to be 14 times greater than in the native mice and these animals lactated normally.

An alternative method for obtaining low-lactose milk is to produce transgenic mice expressing rat intestinal lactase in both the apical side of mammary alveolar cells and the outer membrane of fat globules in the milk (Jost *et al.*, 1999). This enzyme, which hydrolyses lactose to glucose and galactose, reduced the lactose content of the milk by up to 85% with no effect on fat or protein content or on pup development. The authors proposed that if the same technique was applied to dairy cattle, the low-lactose milk would satisfy the needs of lactose maldigesters since a 50 to 70% decrease in lactose content is sufficient to prevent the intestinal disorders associated with this condition.

Bovine α-la has been expressed in the milk of transgenic pigs in order to determine the effect of over-production of α-la on milk production and piglet growth (Bleck *et al.*, 1998). Levels of bovine α-la as high as 0.9 g L^{-1} were attained, resulting in a 50% increase in the total level of α-la produced throughout a lactation. The concentration of bovine α-la varied, being highest in the early stages of lactation; the ratio of the bovine and porcine α-la also changed throughout the lactation; indicating that the transgene and the endogenous gene were under different control elements. Although there was a 3.8% increase in the lactose content of the transgenic milks at day 0 of lactation (which may indicate an increase in milk output), by days 5 and 10, the difference in lactose content between transgenic and control milks was no longer significant. The authors speculated that if the increase in lactose concentration at the start of lactation was accompanied by an increase in milk production, this might lead to more efficient growth of offspring.

17.4.3 Changes in micellar properties

Milk is not simply a solution of proteins and salts. Under physiological conditions, the caseins and much of the calcium and phosphate are organised into colloidal particles, the casein micelles. The micelles in milk are stabilised by a layer of κ-CN and there is a negative correlation between κ-CN content and micellar size. It has been proposed that if the relative concentration of κ-CN could be increased by inserting multiple copies of this gene, the effect would be to reduce the mean diameter of the micelles. This should increase the heat stability of the milk during processing and produce cheese with a firmer texture. Bovine κ-CN has been expressed in mouse milk with the mean concentration in individual lines being as high as 3.8 mg mL^{-1} (Gutierrez-Adan *et al.*, 1996). This resulted in a decrease in the range of mean micelle diameters from 250–274 nm in the controls to 141–211 nm in homozygous lines. There was no effect on rennet coagulation time, but the curd formed from the transgenic milks was significantly stronger than that produced from control milk. The total protein content of the milk was not affected, indicating that the synthesis of the bovine κ-CN was at the expense of other endogenous proteins. A relative increase in the κ-CN content of mouse milk was achieved by disrupting the β-CN gene by gene targeting in embryonic stem cells (Kumar *et al.*, 1994). This resulted in a decrease in both the total casein and total protein content of the β-CN-deficient milks. However, since β-CN was absent from the milk of homozygous animals, the proportion of κ-CN increased and the mean micelle diameter decreased by 10%. The antimicrobial enzyme, lysozyme, is present in human milk at a concentration of 400 μg mL^{-1} compared to 0.13 μg mL^{-1} in cows' milk. Expressing human lysozyme in mouse milk at a concentration ranging from 0.25 to 0.71 μg mL^{-1} caused a decrease in mean micelle size from 172 to 157 nm, a 35% decrease in rennet clotting tune and a 2.5 to 5-fold increase in gel strength (Maga *et al.*, 1995). The authors speculated that these changes were due to the association of the positively charged lysozyme with the negatively charged micelles which would reduce electrostatic repulsion between individual micelles.

Bovine β-CN has been expressed in mouse milk under the control of both ovine β-lg and bovine α-la 5'-flanking regions. In the former case, the protein was shown to be incorporated into the micelles and the appearance of the milk and the development of the pups were normal (Hitchin *et al.*, 1996). In the latter case, the transgenic milk was very viscous and lactation stopped at between 1 and 18 days post-partum, depending on the concentration of the bovine protein (Bleck *et al.*, 1995). This may have been due to some, or all, of the transgenic protein being non-micellar, but this was not investigated.

17.4.4 Improving infant formulae

The composition of bovine milk has evolved to act as the optimum food for the development of calves, rather than for human neonates. A great deal of effort has therefore been applied to adapting the composition of infant formulae based on bovine milk, in order to promote growth and development equivalent to that in infants fed on human milk. A variety of ways for improving the composition of infant formula have been suggested, particularly for the nutrition of premature infants and those with special dietary requirements as a result of allergic, inflammatory and metabolic disorders (see review by Lo and Kleinman, 1996). These include incorporation of antibodies and cytokines, increasing the level of lactoferrin, lysozyme and metallothionein and the removal of allergenic epitopes from proteins. Some of these changes may be achievable through the use of genetic engineering.

In an attempt to provide passive immunisation against enteric infections, transgenic mice secreting a monoclonal antibody (Mab) against transmissible gastroenteritis coronavirus (TGEV) were generated (Catilla *et al.*, 1998). The Mab was a chimera consisting of the constant modules of a human IgG and the variable modules of the murine TGEV-specific Mab 6A.C3 expressed under the control of WAP regulatory sequences. Antibody expression titers up to 10^6 that reduced TGEV infectivity by 10^6-fold were measured in some transgenic milks. Antibody levels were independent of transgene copy number but were related to the site of integration.

Human lysozyme has been expressed in mouse milk at a concentration up to 0.7 $\mu g\ mL^{-1}$ (Maga *et al.*, 1995). The transgenic protein was shown to have antimicrobial activity. Human lactoferrin has been successfully expressed in mouse milk under the control of regulatory elements of the bovine α_{s1}-CN gene (Nuijens *et al.*, 1997). The transgenic protein was 90% saturated with iron whereas natural lactoferrin is only 3% saturated. The transgenic protein was immunologically identical to the natural protein, as were other properties such as iron release and ligand binding.

The hypothesis that milk-borne growth factors might stimulate gastrointestinal growth and development was tested by determining the growth of pups suckling from mice expressing des(l–3) Human Insulin-Like Growth Factor-l (hIGF-l) in their milk (Burrin *et al.*, 1999). Although the concentration of des(l–3) hIGF-1 was 40- to 200-fold higher than mouse IGF, there was little evidence of stimulation of growth or of the transgenic protein being adsorbed into the circulation after ingestion of the milk.

17.5 The mammary gland as a factory for the manufacture of therapeutic proteins

A pharmacologically active protein, human tissue plasminogen activator, has been produced in mouse milk (Gordon *et al.*, 1987). Since then, a variety of other proteins with potential pharmaceutical interest have been expressed in the milk of different species and a new industry has been created based on this technology (see reviews by Houdebine, 1994 and Clark, 1998).

17.5.1 Choice of species

Whereas food proteins are high volume, low value products, therapeutic proteins are low volume and high value. Two factors need to be considered when deciding which species of animal to use for the expression of therapeutic proteins. These are the demand for the protein and the cost of producing the transgenic animal. The mouse, the species on which most research in this area has been performed, has the advantages of small size (reducing feeding and housing costs), short generation time and large numbers of pups in each litter. The major disadvantages are that a typical female produces less than 1 mL of milk per milking and repeated milking is very difficult. It has been estimated that the production of 100 kg of a transgenic protein which is expressed in milk at a concentration of 1 mg mL^{-1} would require 6,000,000 mice, 2000 rabbits, 200 sheep or 10 cows (Lee and de Boer, 1994). However, the cost of producing a transgenic cow has been estimated to be about $300,000, with the delay time from microinjection to first milking being 2 years compared to 3 months for the mouse. The world requirement for some therapeutic proteins is very low and large-scale production is unnecessary.

17.5.2 Expression of therapeutic proteins

Pharmaceutical proteins which have been expressed in milk are listed in Table 17.3, although this is not exhaustive. As expected, the most popular host is, again, the mouse, although the rabbit, sheep and pig have been used also. Most of the transgenic animals have been produced by pro-nuclear injection although antithrombin III and human factor IX have been produced in goats (Baguisi *et al.*, 1999) and sheep (Schnieke *et al.*, 1997), respectively, using the nuclear transfer cloning technique.

Expression levels of therapeutic proteins are very variable, ranging from approximately 3 μg to 33 mg mL^{-1} and higher in the case of hAAT expressed in sheep (Carver *et al.*, 1993). Generally, transgenic proteins were found to be active although some problems have been reported with respect to the authenticity of their post-translational modifications (see Section 17.3.3). Some proteins, such as erythropoietin, have been expressed as fusion proteins in order to reduce their activity and prevent side effects caused by ectopic expression or leakage from the mammary gland (Korhonen *et al.*, 1997). These proteins require specific proteolytic cleavage after collecting the milk to produce an active protein. As mentioned previously, hPC, a vitamin K-dependent anti-coagulant plasma protein, has been expressed in the milk of transgenic mice and pigs (Drohan *et al.*, 1994). The precursor protein undergoes a variety of post-translational modifications during maturation, including glycosylation, β-hydroxylation, γ-carboxylation and proteolytic processing prior to forming the active heterodimer. Since the mammary epithelial cells perform the proteolytic processing inefficiently (Lee *et al.*, 1995), the hPC precursor was co-expressed with human furin in mouse milk (Paleyanda *et al.*, 1997). Furin is an enzyme which cleaves paired basic amino acids from proteins and is involved in proteolytic processing of proteins in the secretory pathway. Human furin, both as the expected intracellular form and also a smaller secreted form, was detected at levels more than 100 times greater than the corresponding endogenous mouse enzyme. Whereas most of the hPC in the original transgenic mice consisted of the unproteolysed precursor, the hPC in the mice expressing human furin was processed to the mature, active protein.

17.6 Future developments

While the dream of herds of transgenic cows producing modified milks for specific manufactured food products is still some way from reality, it is potentially achievable. The success rate of producing transgenic animals is increasing and costs are decreasing. However, the political problems associated with attaining this goal may well outweigh the scientific ones. Will the idea of genetically modified milks produced by cloned animals prove acceptable to the consumer? At the time of writing, genetically modified crops such as maize and soya bean are a topic of great debate within the UK and elsewhere. Some pressure groups are calling for their cultivation, even on well-defined tests sites, to be banned and foods containing material derived from such crops to be clearly labelled. An increasing number of manufacturers are seeking, under consumer pressure, guarantees from their suppliers that raw materials are not derived from these sources. Whether milk, a product that has long benefited from its natural, wholesome image, would continue to do so if it underwent any form of genetic modification of the types suggested is debatable.

The production of therapeutic proteins in milk is politically much less contentious. Producing these proteins in transgenic milk rather than by fermentation or cell culture is considerably less demanding, both financially and environmentally when feeding, waste disposal and product isolation costs are considered. A number of these proteins were at various stages of clinical evaluation in early 2000. Human lactoferrin, for heparin neutralisation, was expected to start Phase II clinical trials. hAAT, produced in kilogram-sized batches from milk obtained from flocks of sheep, had almost completed Phase II trials for the treatment of cystic fibrosis and congenital emphysema which affects one child in every 2000 in the Western Hemisphere. Human alpha-glucosidase was undergoing Phase II trials for the treatment of Pompe's disease, a lethal lysosomal storage disorder. Recombinant human antithrombin III, which helps regulate blood clotting, was in Phase III trials. If these products are successful, more will undoubtedly follow. It will be interesting to see the status of both these areas by the time the next edition of this book is published.

ACKNOWLEDGEMENTS

Core funding for the Hannah Research Institute is provided by The Scottish Executive Rural Affairs Department.

REFERENCES

Baguisi, A., Behboodi, E., Melican, D.T., Pollock, J.S., Destrempes, M.M., Cammuso, C., Williams, J.L., Nims, S.D., Porter, C.A., Midura, P., Palacios, M.J., Ayres, S.L., Denniston, R.S., Hayes, M.L., Ziomek, C.A., Meade, H.M., Godke, R.A., Gavin, W.G., Overstrom, E.W. and Echelard, Y. (1999) Production of goats by somatic cell nuclear transfer. *Nat. Biotechnol.*, **17**, 456–61.

Batt, C.A. (1997) Genetic engineering of food proteins. In *Food Proteins and Their Applications*, (S. Damodoran and A. Paraf ed.) Marcel Dekker Inc, New York, pp. 425–41.

Batt, C.A., Rabson, L.D., Wong, D.W.S. and Kinsella, J.E. (1990) Expression of recombinant bovine β-lactoglobulin in *Escherichia coli*. *Agric. Biol. Chem.*, **54**, 949–55.

Bleck, G.T. and Bremel, R.D. (1994) Variation in the expression of a bovine α-lactalbumin transgene in milk of transgenic mice. *J. Dairy Sci.*, **77**, 1897–904.

Bleck, G.T, Jiminez-Flores, R. and Bremel, R.D. (1995) Abnormal properties of milk from transgenic mice expressing bovine β-casein under the control of the bovine α-lactalbumin 5'-flanking region. *Int. Dairy J.*, **5**, 619–32.

Bleck, G.T., White, B.R., Miller, D.J. and Wheeler, M.B. (1998) Production of bovine α-lactalbumin in the milk of transgenic pigs. *J. Anim. Sci.*, **76**, 3072–8.

Brem, G., Hartl, P., Besenfelder, U., Wolf, E., Zinovieva, N. and Pfaller, R. (1994) Expression of synthetic cDNA sequences encoding human-insulin-like growth-factor-I (IGF-1) in the mammary-gland of transgenic rabbits. *Gene*, **149**, 351–5.

Burrin, D.G., Fiorotto, M.L. and Hadsell, D.L. (1999) Transgenic hypersecretion of des(l–3) human insulin-like growth factor I in mouse milk has limited effects on the gastrointestinal tract in suckling pups. *J. Nutr.*, **129**, 51–6.

Carver, A.S., Dalrymple, M.A., Wright, G., Cottom, D.S, Reeves, D.B., Gibson, Y.H., Keenan, J.L., Barrass, J.D., Scott, A.R., Colman, A. and Garner, I. (1993) Transgenic livestock as bioreactors: Stable expression of human alpha-1-antitrypsin by a flock of sheep. *Bio/Technology*, **11**, 1263–70.

Castilla, J., Pintado, B., Sola, I., Sanchez-Morgado, J.M. and Enjuanes, L. (1998) Engineering passive immunity in transgenic mice secreting virus-neutralizing antibodies in milk. *Nat. Biotechnol.*, **16**, 349–54.

Cho, Y., Gu, W., Watkins, S., Lee, S-P., Kim, T-R., Brady, J.W. and Batt, C.A. (1994) Thermostable variants of bovine β-lactoglobulin. *Protein Eng.*, **7**, 263–70.

Chaudhuri, T.K., Horii, K., Yoda, T., Arai, M., Nagata, S., Terada, T.P., Uchiyama, H., Ikura, T., Tsumoto, K., Kataoka, H., Matsushima, M., Kuwajima, K. and Kumagai, I. (1999) Effect of extra-N-terminal methionine residue on the stability and folding of recombinant α-lactalbumin expressed in *Escherichia coli*. *J. Mol. Biol.*, **285**, 1179–94.

Clark, A.J. (1998) The mammary gland as a bioreactor: Expression, processing, and production of recombinant proteins. *J. Mammary Gland Biol.*, **3**, 337–50.

Colman, A. (1996) Production of proteins in the milk of transgenic livestock: problems, solutions and successes. *Am. J. Clin. Nutr.*, **63**, 639S.

Devinoy, E., Thepot, D., Stinnakre, M.G., Fontaine, M.L., Grabowski, H., Puissant C., Pavirani, A. and Houdebine, L.M. (1994) High-level production of human growth hormone in the milk of transgenic mice – The upstream region of the rabbit whey acidic protein (WAP) gene targets transgene expression to the mammary gland. *Transgenic Res.*, **3**, 79–89.

DiTullio, P., Cheng, S.H., Marshall, J., Gregory, R.J., Ebert, K.M., Meade, H.M. and Smith, A.E. (1992) Production of cystic-fibrosis transmembrane regulator in the milk of transgenic mice. *Bio/Technology*, **10**, 74–7.

Drobovolsky, V.N., Lagutin, O.V., Vinagradova, T.V., Frolova, I.S., Kuznetsov, V.P. and Larionov, O.A. (1993) Human gamma-interferon expression in the mammary-gland of transgenic mice. *FEBS Lett.*, **319**, 181–4.

Drohan, W.N., Zhang, D.W., Paleyanda, R.K., Chang, R.L., Wroble, M., Velander, W. and Lubon, H. (1994) Inefficient processing of human protein-C in the mouse mammary-gland. *Transgenic Res.*, **3**, 355–65.

Greenberg, N.M., Anderson, J.W., Hsueh, A.J.W., Nishimori, K., Reeves, J.J., Deavila, D.M., Ward, D.N. and Rosen, J.M. (1991) Expression of biologically-active heterodimeric bovine follicle-stimulating-hormone in milk of transgenic mice. *Proc. Natl. Acad. Sci. USA*, **88**, 8327–31.

Guitierrez-Adan, A., Maga, E.A., Meade, H., Shoemaker, C.F., Medrano, J.F., Anderson, G.B. and Murray, J.D. (1996) Alterations in the physical characteristics milk from transgenic mice producing bovine κ-casein. *J. Dairy Sci.*, **79**, 791–9.

Hansson, L., Bergstrom, S., Heraell, O., Lonnerdal, B., Nilson, A.K. and Stromqvist, M. (1993) Expression of human milk β-casein in *Escherichia coli*-comparison of recombinant protein with native isoforms. *Protein Expres. Purif.*, **4**, 373–81.

Hansson, L., Edlund, M., Edlund, A., Johansson, T., Marklund, S.L., Fromm, S., Stromqvist, M. and Tornell, J. (1994) Expression and characterization of biologically-active human extracellular-superoxide dismutase in milk of transgenic mice. *J. Biol. Chem.*, **269**, 5358–63.

Hitchin, E., Stevenson, E.M., Clark, A.J., McClenaghan, M. and Leaver, J. (1996) Bovine β-casein expressed in transgenic mouse milk is phosphorylated and incorporated into micelles. *Protein Expres. Purif.*, **7**, 247–52.

Houdebine, L-M. (1994) Production of pharmaceutical proteins from transgenic animals. *J. Biotechnol.*, **34**, 269–87.

Hurwitz, D.R., Nathan, M., Barash, I., Ilan, H. and Shani, M. (1994) Specific combinations of human serum-albumin introns direct high-level expression of albumin in transfected cos cells and in the milk of transgenic mice. *Transgenic Res.*, **3**, 365–75.

Ishikawa, N., Chiba, T., Len, L.T., Shimizu, A., Ikeguchi, M. and Sugai, S. (1998) Remarkable destabilization of recombinant α-lactalbumin by an extraneous N-terminal methionyl residue. *Protein Eng.*, **11**, 333–5.

Jeng, S-Y., Bleck, G.T., Wheeler, M.B. and Jiminez-Flores, R. (1997) Characterization and partial purification of bovine α-lactalbumin and β-casein produced in milk of transgenic mice. *J. Dairy Sci.*, **80**, 3167–75.

Jost, B., Vilotte, J.-L., Duluc, I., Rodeau, J.-L. and Freund, J.-N. (1999) Production of low-lactose milk by ectotopic expression of intestinal lactase in the mouse mammary gland. *Nat. Biotechnol.*, **17**, 160–4.

Jiminez-Flores, R., Richardson, T. and Bisson, L.F. (1990) Expression of bovine β-casein in *Saccharomyces cerevisiae* and characterization of the protein produced *in vivo*. *J. Agric. Food Chem.*, **38**, 1134–41.

Karatzas, C.N. and Turner, J.D. (1997) Toward altering milk composition by genetic manipulation: Current status and challenges. *J. Dairy Sci.*, **80**, 2225–32.

Korhonen, V.-P., Tolvanen, M., Hyttinen, J.-M., Uusi-Oukari, M., Sinervirta, R., Alhonen, L., Jauhiainen, M., Janne, O.A. and Janne, J. (1997) Expression of bovine β-lactoglobulin/human erythropoietin fusion protein in the milk of transgenic mice and rabbits. *Eur. J. Biochem.*, **245**, 482–9.

Kumar, S., Clarke, A.R., Hooper, M.L., Horne, D.S., Law, A.J.R., Leaver, J., Springbett, A., Stevenson, E. and Simons, J.P. (1994) Milk-composition and lactation of β-casein-deficient mice. *Proc. Natl. Acad. Sci. USA*, **91**, 6138–42.

Lo, C.W. and Kleinman, R.E. (1996) Infant formula, past and future: opportunities for improvement. *Am. J. Clin. Nutr.*, **63**, 646S–50S.

Lee, S.H. and de Boer, H.A. (1994) Production of biomedical proteins in the milk of transgenic dairy cows: the state of the art. *J. Control. Release*, **29**, 213–21.

Lee, S-P., Kim, D-S., Watkins, S. and Batt, C.A. (1994) Reducing whey syneresis in yoghurt by the addition of a thermolabile variant of β-lactoglobulin. *Biosci. Biotech. Biochem.*, **58**, 309–13.

Lee, T.K., Drohan, W.N. and Lubon, H. (1995) Proteolytic processing of human Protein C in swine mammary gland. *J. Biochem.*, **118**, 81–7.

Liang, Q.W. and Richardson, T. (1993) Expression and characterization of human lactoferrin in yeast *Saccharomyces cerevisiae*. *J. Agric. Food Chem.*, **41**, 1800–7.

Limonta, J., Pedraza, A., Rodriguez, A., Freyre, F.M., Barral, A.M., Castro, F.O., Lleonart, R., Gracia, C.A., Gavilondo, J.V. and Delafuente, J. (1995) Production of active anti-CD6 mouse-human chimeric antibodies in the milk of transgenic mice. *Immunotechnol.*, **1**, 107–13.

Limonta, J.M., Castro, F.O., Martinez, R., Puentes, P., Ramos, B., Aguilar, A., Lleonart, R.L. and Delafuente, J. (1995) Transgenic rabbits as bioreactors for the production of human growth-hormone. *J. Biotechnol.*, **40**, 49–58.

McClenaghan, M., Springbett, A., Wallace, R.M., Wilde, C.J. and Clark, A.J. (1995) Secretory proteins compete for production in the mammary-gland of transgenic mice. *Biochem. J.*, **310**, 637–41.

McClenaghan, M., Hitchin, E., Stevenson, E.M., Clark, A.J., Holt, C. and Leaver, J. (1999) Insertion of a casein kinase recognition sequence induces phosphorylation of ovine β-lactoglobulin in transgenic mice. *Protein Eng.*, **12**, 259–64.

Murray, J.D. (1999) Genetic modification of animals in the next century. *Theriogenol.*, **51**, 149–59.

Ninomiya, T., Hirabayashi, M., Sagara, J. and Yuki, A. (1994) Functions of milk protein gene 5′-flanking regions on human growth-hormone gene. *Mol. Reprod. Dev.*, **37**, 276–83.

Nuijens, J.H., van Berkel, H.C., Geerts, M.E.J., Hartevelt, P.P., de Boer, H.A., van Veen, H.A. and Pieper, F.R. (1997) Characterization of recombinant human lactoferrin secreted in milk of transgenic mice. *J. Biol. Chem.*, **272**, 8802–7.

Oh, S. and Richardson, T. (1991a) Genetic-engineering of bovine κ-casein to improve its nutritional quality. *J. Agric. Food Chem.*, **39**, 422–7.

Oh, S. and Richardson, T. (1991b) Genetic-engineering of bovine κ-casein to enhance proteolysis by chymosin. *ACS Symposium Series*, **454**, 195–211.

Paleyanda, R.K., Velander, W.H., Lee, T.K., Scandella, D.H., Gwazdauskas, F.C., Knight, J.W., Hoyer, L.W., Drohan , W.N. and Lubon, H. (1997) Transgenic pigs produce functional human factor VIII in milk. *Nat. Biotechnol.*, **15**, 971–5.

Paleyanda, R.K., Drews, R., Lee, T.K. and Lubon, H. (1997) Secretion of human furin into mouse milk. *J. Biol. Chem.*, **272**, 15270–4.

Persuy, M.A., Stinnakre, M.G., Printz, C., Mahe, M.F. and Mercier J.C. (1992) High expression of the caprine β-casein gene in transgenic mice. *Eur. J. Biochem.*, **205**, 887–93.

Persuy, M.A., Legrain, S., Printz, C., Stinnakre, M.G., Lepourry, L., Brignon, G. and Mercier, J.C. (1995) High-stage and mammary-tissue-specific expression of a caprine κ-casein-encoding minigene driven by a β-casein promoter in transgenic mice. *Gene*, **165**, 291–6.

Platenburg, G.J., Koowijk, E.P.A., Kooiman, P.M., Woloshuk, S.L., Nuijens, J.H., Krimpenfort, P.J.A., Pieper, F.R., de Boer, H.A. and Strijker, R. (1994) Expression of human lactoferrin in milk of transgenic mice. *Transgenic Res.*, **3**, 99–108.

Prieto, P.A., Mukerji, P., Kelder, B., Erney, R., Gozalez, D., Yun, J.S., Smith, D.F., Moremen, K.W., Mardelli, C., Pierce, M., Li, Y.S., Chen, X., Wagner, T.E., Cummings R.D. and Kopchick, J.J. (1995) Remodelling of mouse milk glycoconjugates by transgenic expression of a human glycosyltransferase. *J. Biol. Chem.*, **270**, 29515–9.

Prunkard, D., Cottingham, I., Garner, I., Bruce, S., Dalrymple, M., Lasser, G., Bishop, P. and Foster, D. (1996) High-level expression of recombinant human fibrinogen in the milk of transgenic mice. *Nat. Biotechnol.*, **A4**, 867–71.

Riego, E., Limonta, J., Aguilar, A., Perez, A., Dearmas, R., Solano, R., Ramos, B., Castro, F.O. and Delafuente, J. (1993) Production of transgenic mice and rabbits that carry and express the human tissue plasminogen-activator cDNA under the control of a bovine α_{s1}-casein promoter. *Theriogenol.*, **39**, 1173–85.

Rijnkels, M., Kooiman, P.M., Platenburg, G.J., van Dixhoorn, M., Nuijens, J.H., de Boer, H.A. and Pieper, F.R. (1997) High-level expression of bovine α_{s1}-casein in milk of transgenic mice. *Transgenic Res.*, **7**, 5–14.

Rocha, T.L., Paterson, G., Crimmins, K., Boyd, A., Sawyer, L. and Fothergill-Gilmore, L.A. (1996) Expression and secretion of recombinant ovine β-lactoglobulin in *Saccharomyces cerevisiae* and *Kluyveromyces lactis*. *Biochem. J.*, **313**, 927–32.

Rokkones, E., Fromm, S.H., Kareem, B.M., Klungland, H., Olstad, O.K., Hogset, A., Iversen, J., Bjoro, K. and Gautvik, K.M. (1995) Human parathyroid hormone as a secretory peptide in milk of transgenic mice. *J. Cell. Biochem.*, **59**, 168–76.

Salmon, V., Legrand, D., Georges, B., Slomianny, M.-C., Coddeville, B. and Spik, G. (1997) Characterization of human lactoferrin produced in the baculovirus expression system. *Protein Expres. Purif.*, **9**, 203–10.

Schnieke, A.E., Kind, A.J., Ritchie, W.A., Mycock, K., Scott, A.R., Ritchie, M., Wilmut, I., Colman, A. and Campbell, K.H.S. (1997) Human factor IX transgenic sheep produced by transfer of nuclei from transfected fetal fibroblasts. *Science*, **278**, 2130–3.

Shamay, A., Solinas, S., Pursel, V.G., McKnight, R.A., Alexander, L., Beattie, C., Hennighausen, L. and Wall, R.J. (1991) Production of the mouse whey acidic protein in transgenic pigs. *J. Anim. Sci.*, **69**, 4552–62.

Shani, M., Barash, I., Nathan, M., Ricca, G., Searfoss, G.H., Dekel, I., Faerman, A., Givol, D. and Hurwitz, D.R. (1992) Expression of human serum albumin in the milk of transgenic mice. *Transgenic Res.*, **1**, 195–208.

Simons, J.P., McClenaghan, M. and Clark, A.J. (1987) Alteration of the quality of milk by expression of sheep β-lactoglobulin gene in transgenic mice. *Nature*, **328**, 530–2.

Simons, G., van den Heuvel, W., Reynen, T., Frijters, A., Rutten, G., Slangen, C.J., Groenen, M., de Vos, W.M. and Siezen, R.J. (1993) Overproduction of bovine β-casein in *Escherichia coli* and engineering of its main chymosin cleavage site. *Protein Eng.*, **6**, 763–70.

Sitaram, M.P., Moloney, B. and McAbee, D.D. (1998) Prokaryotic expression of bovine lactoferrin deletion mutants that bind to the Ca^{2+}-dependent lactoferrin receptor on isolated rat hepatocytes. *Protein Expres. Purif.*, **14**, 229–36.

Stacey, A., Schnieke, A., Kerr, M., Scott, A., McKee, C., Cottingham, I., Binas, B., Wilde, C. and Colman A. (1995) Lactation is disrupted by α-lactalbumin deficiency and can be restored by human α-lactalbumin gene replacement in mice. *Proc. Natl. Acad. Sci. USA*, **92**, 2835–9.

Stinnakre, M.G., Vilotte, J.L., Soulier, S., Lharidon, R., Charlier, M., Gaye, P. and Mercier, J.C. (1991) The bovine α-lactalbumin promoter directs expression of ovine trophoblast interferon in the mammary-gland of transgenic mice. *FEBS Lett.*, **284**, 19–22.

Stinnakre, M.G., Devinoy, E., Thepot, D., Chene, N., Bayatsamardi, M., Grabowski, H. and Houdebine, L.M. (1993) The synthesis of a recombinant protein in the milk of transgenic mice. *Livest. Prod. Sci.*, **36**, 35–8.

Stinnakre, M.G., Vilotte, J.L., Soulier, S. and Mercier, J.C. (1994) Creation and phenotypic analysis of α-lactalbumin-deficient mice. *Proc. Natl. Acad. Sci. USA*, **91**, 6544–8.

Stromqvist, M., Tornell, J., Edlund, M., Edlund, A., Johansson, T., Lindgren, K., Lundberg, L. and Hansson, L. (1996) Recombinant human bile salt-stimulated lipase: an example of defective *O*-glycosylation of a protein produced in milk of transgenic mice. *Transgenic Res.*, **5**, 475–85.

Stromqvist, M., Houdebine, L.-M., Andersson, J.-O., Edlund, A., Johansson, T., Viglietta, C., Puissant, C. and Hansson, L. (1997) Recombinant human extracellular superoxide dismutase produced in milk of transgenic rabbits. *Transgenic Res.*, **6**, 271–8.

Thepot, D., Devinoy, E., Fontaine, M.L., Stinnakre, M.G., Massoud, M., Kann, G. and Houdebine, L.M. (1995) Rabbit whey acidic protein gene upstream region controls high-level expression of bovine growth hormone in the mammary-gland of transgenic mice. *Mol. Reprod. Dev.*, **42**, 261–7.

Thurmond, J.M., Hards, R.G., Seipelt, C.T., Leonard, A.E., Hansson, L., Stromqvist, M., Bystrom, M., Enquist, K., Xu, B.X.C., Kopchick, J.J. and Mukerji, P. (1997) Expression and characterization of phosphorylated recombinant human β-casein in *Escherichia coli*. *Protein Expres. Purif.*, **10**, 202–8.

Totsuka, M., Katakura, Y., Shimizu, M., Kumagai, I., Miura, K. and Kaminogawa, S. (1990) Expression and secretion of bovine β-lactoglobulin in *Saccharomyces cerevisiae*. *Agric. Biol. Chem.*, **54**, 3111–6.

Velander, W.H., Johnson, J.L., Page, R.L., Russell, C.G., Subramanian, A., Wilkins, T.D., Gwasdaukas, F.C., Pittius, C. and Drohan, W.N. (1992) High-level expression of a heterologous protein in the milk of transgenic swine using the cDNA-encoding human protein-C. *Proc. Natl. Acad. Sci. USA*, **89**, 12003–7.

Wall, R.J., Rexroad, C.E., Powell, A., Shamay, A., McKnight, R. and Hennighausen, L. (1996) Synthesis and secretion of the mouse whey acidic protein in transgenic sheep. *Transgenic Res.*, **5**, 67–72.

Wall, R.J., Kerr, D.E. and Bondioli, K.R. (1997) Transgenic dairy cattle: Genetic engineering on a large scale. *J. Dairy Sci.*, **80**, 2213–24.

Woolliams, J.A. and Wilmut, I. (1999) New advances in cloning and their potential impact on genetic variation in livestock. *Anim. Sci.*, **68**, 245–56.

Yull, F., Harold, G., Cowper, A., Percy, J., Cottingham, I. and Clark, A.J. (1995) Fixing human factor IX (fIX): Correction of a cryptic RNA splice enables the production of biologically active human factor IX in the mammary gland. *Proc. Natl. Acad. Sci. USA*, **92**, 10899–903.

18

ENZYMATIC COAGULATION OF MILK

D.B. Hyslop

18.1 Introduction

The first step in the cheesemaking process involves the coagulation of milk. For the principal family of cheeses (≈ 75% of total cheese), this is achieved by adding a small amount of rennet or some other milk clotting enzyme or combination of enzymes to the milk; the alternative mechanism of coagulation is *via* acid (isoelectric) precipitation (see Chapter 22). Rennet is an extract from calfs' stomach and contains as the clotting factor, calf chymosin. The term "rennet" is often used loosely to designate any particular milk clotting enzyme preparation.

When rennet is added to milk, nothing seems to happen initially. However, after a lag period, the milk rather abruptly begins to coagulate to form a coagulum, that at some point becomes a three-dimensional gel. Actually, a process is set in motion that continues through the aging of the cheese product.

Numerous changes occur during this process and it becomes a somewhat arbitrary matter to decide how many steps are involved. However, it is convenient to divide the process into three steps. First, there is the enzymatic step, or primary phase, which leads to activation of the milk clotting particles, the casein micelles. This is followed by a secondary phase during which the enzyme-modified casein micelles begin to aggregate with each other. It is also possible to consider a third phase that involves changes in the properties and structure of the coagulum once it has formed. This chapter will largely neglect the third phase and focus on the first two. Strictly speaking, the primary and secondary phases are sequential in nature, i.e., casein micelles cannot coagulate until they are activated by rennet. One of the more difficult features of the clotting reaction to deal with when trying to study and understand it, is that there is overlap between the

Advanced Dairy Chemistry Volume 1: Proteins, 3rd edn.
Edited by P.F. Fox and P.L.H. McSweeney, Kluwer Academic/Plenum Publishers, 2003.

primary and secondary phases, i.e., the secondary phase begins before the primary phase is complete.

The casein micelle itself has been the subject of an enormous amount of study over the years and its structure is still a matter of debate (see Chapters 1 and 5). It is known, however, that casein micelles are relatively large, roughly spherical aggregates of individual casein molecules. These aggregates have colloidal dimensions and form stable suspensions in aqueous environments, until destabilized. Any one micelle may have literally thousands of individual casein molecules, predominantly α_{s1}-, α_{s2}-, β- and κ-caseins. In addition, the micelle has a significant non-protein content. Most important in this regard is colloidal calcium phosphate. The micelle seems to be held together largely by hydrophobic and ionic forces. It has long been established that κ-casein is the critical casein fraction when considering micelle stability. The κ-casein molecule can be divided into two parts. The critical activity of chymosin, or other proteolytic enzyme, during the primary phase of the clotting reaction, is the hydrolysis of the Phe_{105}–Met_{106} peptide bond. This divides κ-casein into *para*-κ-casein (residues 1–105), which is quite hydrophobic, and casein-macropeptide (CMP) (residues 106–169), which is very hydrophilic. Some macropeptides have bound sialic acid; glycosylated macropeptides are often referred to as glycomacropeptides (GMP).

It is now generally agreed that κ-casein is situated at the micelle surface and that the macropeptide portion exists as segments, or "hairs," which protrude from the surface and provide the micelle with a steric layer, that stabilizes it. This is the basis of the "hairy micelle" depiction of casein micelles (see Chapter 5). The role of rennet is to cut away or shave off the hairy layer so that the subsequently exposed micelle cores, which are hydrophobic in nature, begin to aggregate. This aggregation constitutes the secondary phase of the clotting reaction. The focus of this chapter is largely on the nature of the primary and secondary phases of the clotting reaction.

In terms of form and content, this chapter relies extensively on the previous review in this series (Dalgleish, 1993). However, an attempt has been made both to review more recent research and to discuss in more detail some areas of research, both old and new. In particular, a whole section (Section 18.10) on the measurement of aggregation is included.

Holt and Horne (1996) recently reviewed developments pertaining to the stability and structure of casein micelles. Included is a discussion of functionality theory as applied to casein micelle coagulation. Use of functionality theory is further elaborated upon here.

Walstra (1990) reviewed the factors that affect the stability of casein micelles. The present author has found many of the interpretations of Walstra (1990) valuable. Probably, the most extensive review concerned with modelling of the total clotting reaction is still that of Payens (1989), and it has been read carefully by the present author. Payens (1979) and McMahon and Brown (1984) have also been valuable.

18.2 Measurement of enzyme activity

Historically, the enzymatic clotting activity on a milk substrate has been defined in terms of how long it takes the clotting enzyme to clot a standard milk sample. In the past, this activity was often measured in terms of Soxhlet Units (SU) (Soxhlet, 1877a,b); today, it is more commonly measured in terms of Rennet Units (RU) (Berridge, 1945). Although there is an approximately linear relationship between clotting time and the reciprocal of enzyme concentration (see Section 18.11), so that enzyme concentration can be determined fairly easily from clotting time measurements, clotting time itself cannot be defined unambiguously, as it contains contributions from both proteolysis and aggregation.

In order to remove the above ambiguity, it is necessary to focus on the proteolytic nature of the primary phase of the clotting reaction. The proteolytic reaction can be depicted as follows:

$$\kappa\text{-casein} + \text{rennet} \rightarrow para\text{-}\kappa\text{-casein} + \text{CMP}$$

This mechanism suggests that the best way to measure rennet activity is to measure the time-dependent disappearance of intact κ-casein, or the production of either *para*-κ-casein or CMP.

There are a number of problems associated with quantifying whole κ-casein. Firstly, there are three genetic variants (A, B and C) of κ-casein (Jakob and Puhan, 1992; Chapter 16). This problem, of course, can be overcome by using milk from individual cows. κ-Casein also has a strong tendency to self-associate. The protein has two SH groups, at positions Cys_{11} and Cys_{88}, and each is capable of disulfide bond formation (Rasmussen *et al.*, 1992; de Kruif and May, 1991; Slattery and Evard, 1973; Thurn *et al.*, 1987; Vreeman *et al.*, 1981, 1986; Farrell *et al.*, 1991; Dalgleish, 1998). However, disulfide bonds can be reduced and thiol groups can be blocked by using thiol blocking agents like mercaptoethanol or dithiothreitol.

Some macropeptide residues are glycosylated but to different extents (Pujolle *et al.*, 1966; Schmidt *et al.*, 1966; Armstrong *et al.*, 1967). This causes an additional, possibly more serious, problem when trying to quantify the extent of proteolysis. Pujolle *et al.* (1966) found 10 fractions of both κ-casein A and κ-casein B on DEAE-cellulose columns and starch electrophoretograms in the presence of mercaptoethanol. More recently, Humphrey and Newsome (1984) obtained multiple peaks from κ-casein A when using high performance liquid chromatography (HPLC). Using high performance cation-exchange chromatography, Dalgleish (1986) obtained a single, skewed peak for κ-casein in the presence of mercaptoethanol. However, the technique was not useful for studying rennet-induced proteolysis because there was little difference in retention time between whole κ-casein and *para*-κ-casein. On an anion exchange column, whole κ-casein was distributed amongst several peaks. Despite this, the rate of disappearance of κ-casein was quantified successfully.

Assaying *para*-κ-casein has some advantages compared to whole κ-casein. Although *para*-κ-casein is insoluble in most buffers, it is soluble in urea and its concentration can be quantified on polyacrylamide gels (Chaplin and Green, 1980) or cellulose acetate strips (Dalgleish, 1979). *Para*-κ-casein is positively charged at pH 6.6 and migrates in the opposite direction from the other caseins, which are negatively charged at this pH. In addition, glycosylation is not a problem because all the carbohydrates in κ-casein are bound to the macropeptide segment. Although both thiol groups are on the *para* portion of κ-casein (residues 1–105), they can be blocked with thiol blocking agents.

The most common way to measure the extent of proteolysis has been to isolate and quantify CMP. In the presence of 2% trichloroacetic acid (TCA), virtually all of the macropeptides are soluble (Dalgleish, 1993). With the exception of ~75% of β-lactoglobulin, which is soluble in 2% TCA, all other milk proteins are insoluble (Fox *et al.*, 1967). As the concentration of TCA is increased, both less β-lactoglobulin and less CMP remain soluble. At 12% TCA, only the carbohydrate-rich GMP fraction of CMP is soluble (Armstrong *et al.*, 1967). Peptides soluble in 2 and 12% TCA have been used to distinguish between glycosylated and non-glycosylated fractions of CMP (Hindle and Wheelock, 1970). About 8% TCA is required to precipitate all whey proteins (van Hooydonk and Olieman, 1982).

The macropeptide has been quantified by a number of methods. Since it has no aromatic residues, near UV spectroscopy is useless. Far UV spectroscopy, however, has been used successfully (Castle and Wheelock, 1972; van Hooydonk and Olieman, 1982). Beeby (1980) used a fluorescent technique (Udenfriend *et al.*, 1972) to quantify CMP. Van Hooydonk and Olieman (1982) used HPLC to measure the extent of proteoysis. CMP soluble in 8% TCA eluted as a single peak from the column. In recent years, immunochemical techniques have been developed to quantify CMP; both enzyme-linked-immuno-sorbant assay (ELISA) (Bitri *et al.*, 1993; Picard *et al.*, 1994) and nephelometric (Prin *et al.*, 1996) methods have been used.

The methods discussed above are based on taking aliquots of sample at different times during proteolysis and subsequently analyzing them. An ideal method would be one in which it is possible to monitor continuously the extent of reaction in the sample itself during real time (i.e., while the reaction is taking place). An attempt to develop such a technique was made by Saputra *et al.*, (1994), who monitored diffuse reflectance of near IR radiation during the proteolysis of κ-casein. They were able to generate data which resembled peptide release data, but only if the concentrations of κ-casein and chymosin were known beforehand. Perhaps more importantly, it seems likely to the present author that changes in diffuse reflectance are related only indirectly to the hydrolysis of the Phe_{105}–Met_{106} bond.

18.3 MICHAELIS-MENTEN KINETICS

It was demonstrated many years ago (Ernstrom, 1974) that the primary action of chymosin, the proteolytic enzyme in calf rennet, is to cleave the Phe_{105}–Met_{106} bond of κ-casein. Previously, Waugh (1958) had shown that the rate for this hydrolysis was 1000 times faster than the action on any other peptide bond in the casein complex. Based on these observations, it was suggested that the Michaelis-Menten theory would be appropriate in order to model the primary phase of the clotting reaction. For one substrate we have:

$$E + S \underset{k_2}{\overset{k_1}{\rightleftharpoons}} ES \xrightarrow{k_{cat}} P + M + E$$

where
E = enzyme (rennet),
S = substrate (κ-casein),
ES = enzyme-substrate complex,
P = *para*-κ-casein,
M = CMP,
k_1, k_2 = pre-steady state rate constants
k_{cat} = catalytic rate constant.

Michaelis-Menten velocity (v) is given by

$$v = \mathrm{d}[P]/\mathrm{d}t = -\mathrm{d}[S]/\mathrm{d}t = V_{max}[S]/(K_M + [S]) \tag{1}$$

where
$V_{max} = k_{cat}\, E_0$,
E_0 = total enzyme concentration
$K_M = (k_2 + k_{cat})/k_1$ = Michaelis-Menten constant.

Michaelis-Menten theory is based on the assumption of steady-state kinetics. A ramification of this is that $\mathrm{d}[P]/\mathrm{d}t = -\mathrm{d}[S]/\mathrm{d}t$. The usual way to study Michaelis-Menten kinetics is to measure the initial rate of substrate depletion or product formation at different substrate concentrations, and then plot data according to one of several formats in order to calculate K_M and V_{max}. Since steady-state is reached very quickly ($\ll$1 s), this approach seems quite reasonable. K_M has units of substrate concentration and specifically refers to the concentration of substrate when the Michaelis-Menten velocity is half its maximum value (i.e., $0.5V_{max}$). For the reaction between κ-casein and chymosin, a wide range of K_M values (*ca.* 1×10^{-6} to 5×10^{-4} M) have been calculated.

K_M is a measure of the affinity of a substrate for an enzyme. In principle, its value is independent of enzyme and substrate concentrations. Actually, it is a composite parameter quantifying the ability of ES to dissociate back to its reactants (k_2/k_1), and the ability of the same complex to turn over to P (k_{cat}/k_1). Lower values of K_M are usually associated with more active enzyme systems.

V_{max}, unlike K_M, is very dependent on enzyme concentration. However, it is independent of substrate concentration. As shown in Equation 1, $V_{max} \propto k_{cat}$. The range of values for k_{cat} in the renneting reaction is from *ca.* 2 to 100 s^{-1}. Larger values indicate more active systems. From Equation 1 it is evident that if E_0 is known, k_{cat} can be calculated from V_{max}.

The ratio, k_{cat}/K_M, is often used as a measure of catalytic activity. Large values for this ratio suggest greater enzyme activity. Consideration of this ratio may shed light on the controversy concerning whether or not glycosylation affects the kinetics of the primary phase of renneting (Gibbons and Cheeseman, 1962; Chaplin and Green, 1980; van Hooydonk *et al.*, 1984; Dalgleish, 1986). Vreeman *et al.* (1986) made a careful study of the subject and found that K_M decreased with increased glycosylation, suggesting an increased affinity between κ-casein and chymosin, but k_{cat} also decreased, suggesting a slower turnover of ES to P. These effects tend to offset each other, suggesting that glycosylation has little overall effect on proteolysis.

A critical issue when studying the renneting reaction concerns the nature of κ-casein. In order for information to be relevant to cheesemaking, it is desirable to use conditions as similar as possible to those found in the cheese vat. If primary phase kinetics are studied by way of initial velocity measurements, substrate concentration must be altered. A question may be posed concerning what effect dilution might have on micelle structure and consequently on rennet activity. In short, it may be advantageous to study the renneting reaction at a single substrate concentration. This can be done mathematically by integrating the Michaelis-Menten equation:

$$([S_0] - [S])/t = V_{max} - K_M[\ln([S_0]/[S])]/t \qquad (2)$$

where $[S_0]$ = initial substrate concentration.

This expression is linear and can be used to calculate K_M and V_{max} at a single S_0 value. A problem in using this approach is that the Michaelis-Menten theory is based on the assumption of a steady state, an assumption which slowly breaks down during the course of the enzymatic reaction. This can lead to calculated K_M values being larger than those obtained from initial velocity studies. Dalgleish (1979) and Chaplin and Green (1980) calculated K_M values of 5×10^{-4} and 2.3×10^{-4} M, respectively, using the integration method. These are probably the largest K_M values reported to date.

A number of years ago, van Hooydonk *et al.* (1984) reported the results of a study which has lead to a re-evaluation of the nature of the proteolytic reaction. They reported that the primary phase of the reaction, at least when κ-casein was intact in casein micelles, followed first order kinetics. This does not necessarily mean that primary phase kinetics fail to follow Michaelis-Menten theory, which predicts zero or first order, depending on the relative values of $[S]$ and K_M. When $[S] \gg K_M$, Equation 1 reduces to:

$$-d[S]/dt = k_{cat}E_0 = V_{max} \tag{3}$$

so that the rate of transformation of substrate is zero order with respect to [S]. On the other hand, when $[S] \ll K_M$, Equation 1 reduces to:

$$-d[S]/dt = k_1[S] \tag{4}$$

$$k_1 = h_{cat}E_0/K_M$$

These are classic first-order kinetics. In all cases, $[S]$ becomes $< K_M$ late in the reaction, so Michaelis-Menten kinetics always approach first order kinetics as the reaction proceeds. However, if $S_0 \ll K_M$, the reaction is first order from beginning to end. The concentration of κ-casein in milk is *ca.* 1–1.5 × 10^{-4} M. Since reported K_M values range from 1×10^{-6} to 5×10^{-4} M, kinetics could be closer to zero order (if K_M is at the lower end of the range), or closer to first order (if K_M is at the higher end). In this regards, it is significant that lower K_M values have been obtained for isolated κ-casein (K_M range from 3.3×10^{-5} to 6.6×10^{-5} M) (Garnier, 1963; Azuma *et al.*, 1984; Vreeman *et al.*, 1986), and higher K_M values have been obtained for whole casein micelles (K_M range from 1×10^{-4} to 5×10^{-4} M) (Beeby 1979; Dalgleish, 1979; Chaplin and Green, 1980). It would seem that chymosin has a greater affinity for isolated κ-casein than for κ-casein in intact casein micelles, presumably because of the inaccessibility of the latter.

For isolated κ-casein, the magnitude of k_{cat} is in the range *ca.* 25 to 100 s^{-1} (Garnier, 1963; Garnier *et al.*, 1968; Vreeman *et al.*, 1986). For whole micelles, reported values for k_{cat} are < 25 s^{-1} (Castle and Wheelock, 1972), between 25 and 100 s^{-1} (Beeby, 1979) or > 100 s^{-1} (calculated by the present author from data of van Hooydonk *et al.*, 1984, assuming that $K_M = 5 \times 10^{-4}$ M). More research would be useful here.

It would seem that it can be concluded that the kinetics of the primary phase are different for micellar and isolated κ-casein, and it would appear that the hydrolysis of micellar κ-casein, at least, follows first order kinetics.

18.4 Surface enzyme kinetics

Since intact κ-casein seems to be located at the surface of casein micelles, it is possible that chymosin must first bind to this surface before it can cause proteolysis. At the very least, it would seem that the enzyme would have to penetrate the hairy layer before it could react with κ-casein. With isolated κ-casein, such binding or penetration would seem less necessary. These considerations may explain some of the differences in the kinetic parameters when the hydrolysis of micellar κ-casein is compared to that of isolated κ-casein.

Verger *et al.* (1973) proposed a model for the action of soluble enzymes at interfaces. The model was designed to explain the behaviour of lipolytic enzymes at oil-water interfaces, but, in principle, there is no reason why it cannot be applied to the casein micelle surface. The model consists of two successive reversible steps: the enzyme must first penetrate to the surface and then form an *ES* complex.

The general steady-state equation for enzyme activity following this model can be expressed as:

$$v = V_{max} S^* \sigma / (\sigma K_M^* (1 + K_i/\sigma) + \sigma S^*) \tag{5}$$

In this equation, v has the same meaning and units it does in conventional Michaelis-Menten theory, as does V_{max}. However, the parameters S^* and K_M^* are surface concentration and surface Michaelis-Menten constant, respectively, and have units mol per unit surface area. The parameter K_i is the ratio of a desorption rate constant to a penetration rate constant, so that a large value of K_i indicates that the enzyme has difficulty penetrating the interface. The units of K_i are surface area per unit volume. The parameter σ is the ratio of total surface area to total volume and as such has the same units as K_i. The product σS^* is equal to normal substrate concentration (mol per unit volume). By the same token, $\sigma K_M^* = K_M$, the usual Michaelis-Menten constant, with units mol per unit volume.

Given all this information, Equation 5 may be expressed as:

$$v = V_{max}[S]/(K'_M + [S]) \tag{6}$$

$$K'_M = K_M(1 + K_i/\sigma)$$

The parameter, K'_M, is an apparent K_M, possibly that when κ-casein is incorporated in casein micelles, and K_M itself possibly that of free κ-casein. In any event, a relatively large value for K_i would indicate that the ability of the enzymes to penetrate the hairy layer is inhibited. If K_i is large enough, Equation 6 reduces to a first order expression. As we have already seen, renneting of whole micelles follows first order kinetics (van Hooydonk *et al.*, 1984).

In Equation 6, the parameter K'_M is predicted to increase in magnitude as the micelles increase in size. This is because the surface area to volume ratio decreases as the micelle size increases. Vreeman *et al.* (1986) found that when the size of κ-casein aggregates increased (sedimentation coefficient increased from 12 to 22 S), the Michaelis-Menten constant increased *ca.* two-fold. On the other hand, Dalgleish (1986) found little difference in the rate of peptide release from micelles of different size.

There has long been evidence that chymosin binds to casein, and that this binding becomes greater at lower pH values. Holmes *et al.* (1977) reported that 31% of the rennet added to milk remained in the curd after renneting at pH 6.6; this increased to 86% when the pH was lowered to 5.2. Recent work has confirmed these results (de Roos *et al.*, 1995). Moreover, there is now evidence that chymosin specifically binds to *para*-κ-casein when the latter is adsorbed on emulsion droplets (de Roos *et al.*, 1995; de Roos, 1999), is free in solution (Dunnewind *et al.*, 1996), and to a lesser extent is associated with α_s- and β-caseins in micelles (de Roos, 1999). In the latter case, it is argued that α_s- and β-caseins compete with chymosin for κ-casein (de Roos, 1999). Adsorption is inhibited at high ionic strength and this may indicate that binding is electrostatic in nature. This can be explained as an ionic shielding effect (Daniels and Alberty, 1966). Reactions between ions of the same charge sign are facilitated by increasing ionic strength, while reactions between ions of opposite charge are inhibited. The critical idea behind this phenomenon is the fact that a charged entity in solution will attract a counter-ion atmosphere, which becomes more dense at higher ionic strength.

A critical matter in regard to the use of the interface enzyme kinetics model is that binding discussed above occurs subsequent to the splitting of the CMP from κ-casein (de Roos, 1999). The interface model is based on the assumption that binding precedes the formation of the *ES* complex. Enzyme penetration of the hairy layer certainly precedes proteolysis and it is arguable that this penetration could take the place of explicit binding. If penetration is slow enough, the parameter K_i in Equation 5 becomes large, as does the apparent K'_M in Equation 6. This is consistent with the fact that the K_M for whole micelles is larger than that for isolated κ-casein.

18.5 Diffusion kinetics

The casein micelle is such that the substrate for the renneting reaction is on the surface of a much larger particle. We have seen that the interface enzyme kinetics model (i.e., the model of Verger *et al.*, 1973) predicts first order kinetics if the size of the particle having the interface is large enough. However, it has been argued that enzyme kinetics of a heterogeneous system, consisting of soluble enzyme and insoluble substrate, cannot normally be described by the classical Michaelis-Menten theory (Gatt and Bartfai, 1977). As it turns out, we can also arrive at first order kinetics from diffusion considerations. Since the diffusion coefficient is inversely related to molecular mass, the casein micelle ($M_w \approx 10^8$ to 10^9 Da) has a much lower diffusion coefficient than chymosin ($M_w \approx 35{,}000$ Da). Relatively speaking, the micelle is practically standing still compared to the enzyme, as the latter moves about. So the enzyme must come to the substrate, rather than the other way around. For this reason, *inter alia*, it is not difficult to imagine why there has been little success in using immobilized rennet in order to coagulate casein. In such systems, the micelle must go to the enzyme. Even if the micelle can do this, there is still the question of whether the enzyme can penetrate the hairy layer; it seems that it can not (Beeby, 1979).

Although there have been reports of the successful use of immobilized rennets to coagulate milk, it is likely that few, if any, of these studies actually achieved proteolysis by way of bound enzyme, because of the possibility of having some free enzyme present. Few reports of success have in fact tested for the presence of free enzyme (Carlson *et al.*, 1986). At present, most authors probably agree that immobilizing rennet for milk coagulation is not feasible.

For situations involving free enzyme, i.e., where a fast-moving enzyme diffuses to a slow-moving substrate, van Hooydonk *et al.* (1984) proposed a diffusion model to describe primary phase kinetics:

$$-\mathrm{d}[S]/\mathrm{d}t = cpA\exp(-E_a/RT) \tag{7}$$

where

A = frequency factor
p = steric factor
c = constant
E_a = activation energy
R = gas constant
T = absolute temperature.

The steric factor (p) accounts for the fraction of collisions that do not lead to proteolysis. This should be proportional to unhydrolyzed κ-casein, so $p = a[S]$, where a is a constant. Insertion of this relationship into Equation 7 yields first order kinetics. Saputra *et al.* (1994) have further argued that the frequency factor (A) should be dependent on collisions between enzyme and micelle and, as a consequence, should be directly proportional to total enzyme concentration (E_0) and inversely proportional to micelle concentration (c_0). An inverse relationship is argued because at a constant enzyme concentration, the number of enzyme molecules per micelle decreases as micelle concentration increases. This leads to:

$$-\mathrm{d}[S]/\mathrm{d}t = k_1[S] \tag{8}$$

$$k_1 = b(E_0/c_0)\exp(-E_a/RT)$$

$$b = \text{constant}$$

This suggests that k_1 should increase with E_0 and decrease with c_0, and this is, indeed, the case (van Hooydonk *et al.*, 1984; Hyldig, 1993; Lomholt and Qvist, 1997; Lomholt *et al.*, 1998). It should be pointed out that Michaelis-Menten theory, with (Equation 6) or without (Equation 1) surface binding, also predicts these results.

If the reaction is diffusion-controlled, its rate should be controlled by collision frequency. Such processes yield relatively low values of E_a because the temperature coefficient for collision frequency is generally less than those for processes where chemical bonds are formed or broken. One of the earliest studies of the kinetics of proteolysis of κ-casein was that of Nitschmann and Bohren (see Waugh, 1971), who obtained a value for E_a of *ca.* 10 kcal/mol (41.9 J/mol) when Na caseinate was used as substrate. Later, Garnier (1963) reported a value of 6.5 kcal/mol (27.2 J/mol) for k_{cat} when isolated κ-casein was substrate. van Hooydonk *et al.* (1984) obtained a value of 6.2 kcal/mol (26.0 J/mol) for k_1 in skim milk. All these values indicate a $Q_{10} < 2$; the latter two indicate a value of *ca.* 1.4. van Hooydonk and Walstra (1987) also obtained a $Q_{10} \approx 1.4$ for the effect of temperature on the reciprocal of milk viscosity during renneting. Diffusion (D) and viscosity (η) in solutions are inversely related by the Stokes-Einstein relationship:

$$D = kT/6\pi\eta R_h \tag{9}$$

where

k = Boltzmann constant

η = solvent viscosity

R_h = radius of diffusing particle (hydrodynamic radius)

These results support the contention that proteolysis of κ-casein during renneting is diffusion-controlled.

18.6 The enzyme-substrate complex

Chymosin is an aspartyl proteinase, i.e., a proteolytic enzyme characterized by two aspartic acid residues (Asp_{34} and Asp_{215}) at the active site (Foltmann, 1993). The molecule has an irregular kidney shape, consisting of two lobes and an extensive cleft with well-defined hydrophobic pockets along its walls (Foltmann, 1993). The active site is inside the cleft.

Studies on the action of chymosin on synthetic peptides based on the amino acid sequence of κ-casein surrounding its cleavage site (Phe_{105}–Met_{106}) have expanded our knowledge of the primary phase reaction. Probably, most observers now agree that the sequence His_{98} to Lys_{112} is most critical for proteolytic activity:

His_{98}–Pro_{99}–His_{100}–Pro_{101}–His_{102}–Leu_{103}–Ser_{104}–Phe_{105}–Met_{106}–Ala_{107}–Ile_{108}–Pro_{109}–Pro_{110}–Lys_{111}–Lys_{112}

At least 5 amino acids, including the sequence Ser_{104}–Phe_{105}–Met_{106}–Ala_{107}, are needed in order for chymosin to cleave the Phe_{105}–Met_{106} bond (Visser *et al.*, 1976). Visser (1981) summarized many of the important developments that led to a better understanding of the action of chymosin on κ-casein. Perhaps the most important observation is that a peptide representing the sequence His_{98} to Lys_{112} of κ-casein gives kinetic parameters nearly the same as those for whole κ-casein.

Visser *et al.* (1987) proposed that the sequence Leu_{103} to Ile_{108} is accomodated within the active site cleft of chymosin and is brought into position by electrostatic binding near the cleft entrance. This would be mediated by His_{98}, His_{100} and His_{102} in the sequence His_{98}–His_{102}, and possibly by Lys_{111}. These interactions are strongly supported by Pro residues at positions 99, 101, 109 and 110 of κ-casein. The positive charges on His and Lys would be attracted by the negative charges on the enzyme.

Since the active site of chymosin is within a cleft, it would seem necessary that the region of κ-casein containing the susceptible bond (Phe_{105}–Met_{106}) would be able to fit into this cleft. It appears that chymosin can accommodate the κ-casein sequence, Leu_{103} to Ile_{108}, in its cleft (Gillialand *et al.*, 1991). More specifically, the residues Leu_{103} to Ile_{108} are hydrophobic and it seems they can fit into the hydrophobic pockets in the chymosin cleft (Plowman and Creamer, 1995). In addition, negatively charged residues in chymosin may pair up with positively charged residues in κ-casein, in the sequences His_{98} to His_{102} and Ile_{108} to Lys_{112} (Plowman and Creamer, 1995). Aspartate and glutamate residues of chymosin may be involved here.

These latter results support the proposal by Visser (1987, 1980) that electrostatic attractions play an important role in the formation of the *ES* complex. The above studies were based largely on the behaviour of synthetic peptides simulating the critical sequence, His_{98} to Lys_{112}, of κ-casein. The results of other studies have also suggested an electrostatic contribution to the formation of the *ES* complex. The effect of ionic strength on the clotting time of casein micelles is to yield a minimum when the total ionic strength is *ca.* 0.125 M (Hamdy and Edelstein, 1970; Alais and Lagrange, 1972). Likewise, Payens and Both (1980) found a minimum for the clotting time of κ-casein, but at a lower ionic strength (*ca.* 0.05 M). Possibly, this result may be explained in terms of two opposing ionic shielding effects (Payens and Both, 1980; Visser, 1981). Both chymosin and κ-casein have a net negative charge, so their approach would be facilitated by increasing ionic strength, which would make the rate of complex formation more rapid. Assuming that the positively charged groups near the active site of κ-casein must complex with their negatively charged counterparts near the active site of chymosin, complex formation would be inhibited by increasing ionic strength. When *E* and *S* are broadly separated, the first effect probably dominates, but when closely spaced, the second effect probably dominates. Both effects pertain to the primary phase of the clotting reaction, so if the above ideas are correct, there must be evidence of a maximum in the rate of proteolysis as a function of ionic strength.

Bringe and Kinsella (1986) found a maximum rate of proteolysis after addition of *ca.* 0.008 M $CaCl_2$ (ionic strength, $\mu = 0.024$ M). Green and Marshall (1977) reported that the rate of proteolysis increased on addition of $CaCl_2$, but they increased the concentration to only 0.005 M. On the other hand, van Hooydonk *et al.* (1986c) found that Ca^{2+} had no effect on proteolysis. Dalgleish (1993) discussed reasons for the possible discrepancy. van Hooydonk *et al.* (1986c) also reported that NaCl inhibited proteolysis at concentrations >0.01 M ($\mu > 0.01$ M). This latter result is supported by J. Johnson (personal communication), although she found a maximum value for k_1 at *ca.* 0.01 M NaCl, so that at lower concentrations of NaCl, proteolysis was slower. The same researcher found that Mg^{2+} and Ca^{2+} also showed a maximum value for k_1 at $\mu \approx 0.01$ M. It seems likely from the above results that ionic strength plays an important role in defining chymosin activity.

18.7 Rennet substitutes

By about 1960, it became evident that the demand for calf rennet was beginning to exceed its supply (de Koning, 1978), and a search for substitutes was begun. It has long been felt that only acid proteinases, of which chymosin is an example, are suitable to properly coagulate milk. Until a few years ago, only two types of substitute, the so-called traditional substitutes, were used. First were the pepsins, from the stomachs of adult ruminants and other animals, and the second were acid proteinases from bacterial and fungal sources.

However, advances in recombinant DNA technology have made it possible to clone chymosin in microorganisms. This cloned chymosin apparently is identical to that found in the calf's stomach. Considerable development was made in this area during the 1980s (Foltmann, 1993). Early studies indicated that cloned chymosin could be used to make cheese essentially identical to that made using calf chymosin (Hicks *et al.*, 1988; Bines *et al.*, 1989). In the United States and elsewhere, cloned chymosins are used widely today.

However, the more traditional substitutes are still used, so it is worth discussing these animal and microbial rennets. In addition, calf rennet is often used in combination with other acid proteinases, probably most commonly with bovine pepsin. The commonly used traditional substitutes are the pepsins from cow, sheep, goat, pig and chicken, and acid proteinases from *Rhizomucor miehei*, *R. pusillus* and *Cryphonectria parasitica*.

Although all these rennet substitutes, with the exception of *C. parasitica* proteinase (which hydrolyzes Ser_{104}–Phe_{105}), hydrolyze the critical Phe_{105}–Met_{106} bond, their overall activities are not identical to that of chymosin. The general pH optimum for fungal proteinases seems to be similar to that of chymosin (*ca.* pH 4) (Hamme, 1972; Martin *et al.*, 1980; Foltmann, 1993); however, that for pepsin is lower (*ca.* pH 2–3) (Bohak, 1969; Foltmann, 1993). Since renneted milk is coagulated at *ca.* pH 6.6 in cheesemaking, it is significant that calf chymosin (Dalgleish, 1993) and *R. miehei* proteinase (Alais and Lagrange, 1972) are stable at pH values below 6.7, while porcine pepsin is unstable above pH 6.0 (Fox, 1969), and is extensively, possibly completely, inactivated during cheesemaking (Green and Foster, 1974).

Chymosin exhibits maximum proteolytic activity at *ca.* 40°C, whereas the proteinases from *R. pusillus* and *R. miehei* are maximally active at 55°C and >63°C, respectively (Martin *et al.*, 1980). Also, fungal rennets are more resistant to heat inactivation than chymosin or pepsin and, with the exception of *C. parasitica* proteinase (Thunell *et al.*, 1979), become more resistant when the pH is lowered (Thunell *et al.*, 1979; Hyslop *et al.*, 1979). With the possible exception of *R. miehei* proteinase, all fungal rennets can be inactivated by HTST pasteurization, provided the pH is high enough (Thunell *et al.*, 1979; Hyslop *et al.*, 1979). *R. miehei* produces more than one proteinase (Lagrange *et al.*, 1980), some of which are heat sensitive (Dalgleish, 1993).

In order for a rennet substitute to be used successfully to make cheese, it is critical that it have weak general proteolytic activity and narrow specificity. For example, trypsin is inappropriate because of its excessive proteolytic activity. Indeed, one of the most significant problems in the use of rennet substitutes is the occurrence of bitter flavor defects in the final cheese caused by excessive proteolysis.

One way to assess the usefulness of a rennet substitute is to measure the clotting-to-proteolytic activity ratio (CPR). Clotting activity is generally expressed in terms of RU (de Koning, 1978). Proteolytic activity is measured on synthetic peptides which mimic the sequence of κ-casein around the Phe_{105}–Met_{106} bond. Studying chymosin, bovine pepsin and proteinases from *R. miehei* and *R. pusillus*, Martin *et al.* (1980) found that CPR is significantly higher for chymosin than for the fungal proteinases; however, bovine pepsin had the lowest ratio. On the other hand, the values of k_{cat}/K_M for fungal proteinases on synthetic peptides were considerably higher than those for chymosin or pepsin.

On individual caseins, the rennet substitutes exhibit less specific activity than chymosin (Tam and Wittaker, 1972; Shammet *et al.*, 1972; Paquet and Alais, 1978).

18.8 PRETREATMENT OF MILK

When milk is acidified, the rate of the primary phase of coagulation increases (van Hooydonk *et al.*, 1986b). In addition, the degree of proteolysis at gelation increases (van Hooydonk *et al.*, 1986b), indicating that pH affects the rate of the two phases differently. This is not surprising, given the complexity of the changes that occur in the casein micelle on acidification.

A decrease in pH leads to dissolution of calcium phosphate. Micellar phosphate seems to exist primarily as seryl phosphate, although some phosphate seems to be esterified to threonine. Seryl phosphate residues tend to have a pK_a value in the pH range *ca.* 5.65 to 6.40. It would be expected that as the pH is lowered below the pK_a value, calcium would tend to dissociate so that free [Ca^{2+}] would increase and probably remove colloidal calcium phosphate that is cross-linked by calcium to seryl phosphate. It also might be expected that dissolution of colloidal calcium phosphate might be fairly abrupt with respect to pH but it tends to be continuous down to *ca.* pH 5.3, at which virtually all inorganic phosphorus is removed from the micelle (van Hooydonk *et al.*, 1986b). Some calcium persists in the micelle $<$pH 5.3, and this might be bound to carboxyl groups (van Hooydonk *et al.*, 1986a). In any event, a decrease in pH does result in increased [Ca^{2+}].

When the temperature is increased, the pH of milk decreases. However, [Ca^{2+}] does not increase; rather, calcium phosphate begins to precipitate (Fox, 1981). The favoured form of phosphoric acid in milk is $H_2PO_4^-$ as long as pH is between *ca.* 2 and 7. Since monoprotonated and unprotonated forms of phosphate seem to be required for precipitation, precipitation is accompanied by the liberation of protons. pH is also lowered by breakdown of lactose (Fox, 1981; Dalgleish, 1989). On cooling, some calcium phosphate dissolves, causing an increase in [Ca^{2+}] and pH. However, the process is not completely reversible (Kannan and Jenness, 1961; Dalgleish, 1989).

If heating is severe enough to denature β-lactoglobulin and other whey proteins (*ca.* 75°C for a few minutes or 90°C for 1 min), another process occurs that significantly affects the casein micelles. Denaturation of β-lactoglobulin leads to its aggregation in which disulfide interchange has been implicated (Sawyer, 1969). These aggregates then react irreversibly with κ-casein; again disulfide interchange has been implicated (Smits and van Brouwershaven, 1980; Pearse *et al.*, 1985). Assuming that the linkage between β-lactoglobulin and κ-casein is entirely *via* disulfide bonds (which may not be so), the linkage must be with the *para* portion of κ-casein, because it contains the thiol groups of the protein.

Binding of whey protein aggregates to κ-casein suggests the possibility of inhibition of the enzyme-induced clotting of milk. Early studies (Zittle *et al.*, 1962; Morrisey, 1969) indicated that inhibition does occur. More recently, van Hooydonk *et al.* (1987) found that heating skim milk sufficiently to denature all β-lactoglobulin caused a *ca.* 20% reduction in the rate of proteolysis compared to unheated milk. When *ca.* 25% of whey protein was denatured (15 s at 90°C), the rate of proteolysis was inhibited only slightly, but the time required for viscosity to increase was increased significantly. These results suggest that the secondary phase of coagulation is affected more than the primary phase.

Since denaturation of whey proteins leads to binding with κ-casein, it is worthwhile questioning whether all κ-casein molecules are available for proteolysis subsequent to heating; it appears that significant amounts of κ-casein cannot be hydrolyzed (Wilson and Wheelock, 1972; Wheelock and Kirk, 1974; Damicz and Dziuba, 1975; Reddy and Kinsella, 1990).

Although chymosin hydrolyzes micelles at refrigeration temperatures, these rennet-altered micelles fail to coagulate, at least within a reasonable time. As already indicated, lowering the temperature increases the pH and solubilizes colloidal calcium phosphate; both these changes adversely affect rennet coagulation. Probably more importantly, however, is the fact that individual caseins, especially β-casein, partially dissociate from the micelles on cooling (Roefs *et al.*, 1985). Although all the caseins have a considerable hydrophobic character, β-casein is the most hydrophobic, at least in terms of the Bigelow Index (Hill and Wake, 1969). Since hydrophobic molecules become more soluble at low temperature, it is generally believed that hydrophobic considerations account for a significant amount of the dissociation. It is likely that the hairy layer on the micelle is "thickened" by the partial dissociation of β-casein (Walstra, 1990). The idea is that, at low temperature, the hairy layer includes both the C-terminal region of κ-casein and a portion of β-casein. The β-casein may extend further away from the micelle surface than the C-terminal region of κ-casein. Only the κ-casein portion of the hairy layer would be expected to be susceptible to hydrolysis. It seems likely that steric stabilization prevents coagulation subsequent to proteolysis. When milk is reheated, the micelle reassembles, and in fact coagulates.

More frequently, milk for cheese is concentrated by ultrafiltration (UF). Although clotting time increases when milk is diluted (Fox and Morrissey, 1972; Dalgleish, 1980a), it is not so clear whether it increases (Culioli and Sherman, 1978; Garnot *et al.*, 1982) or decreases (Garnot and Corre, 1980; Darling and van Hooydonk, 1981; Reuter *et al.*, 1981; Mehai and Cheryan, 1983b) when milk is concentrated. In either case, the effect is not very great (Dalgleish, 1980a, 1981). Payens (1984) argued that clotting time may increase, decrease or remain unchanged when milk is concentrated by UF, depending on whether the primary phase obeys zero order, first order, or Michaelis-Menten kinetics. The argument, however, is based, at least in part, on the assumption that the rate constant for the secondary phase of coagulation is independent of both substrate concentration and time after addition of rennet. It is questionable whether either is true.

Lomholt *et al.* (1998) used an energy barrier model (see Section 18.12) to study the effects of concentration on the secondary phase of the clotting reaction. Assuming that the energy barrier decreases in proportion to the extent of proteolysis, a two-fold increase in casein concentration caused the initial energy barrier (V) to decrease from 26.5 kT to 15.5 kT. This decrease may be due in part to changes in [Ca^{2+}] caused by ultrafiltration (UF) (Mehai and Cheryan, 1983a,b). In any event, it seems clear that the secondary phase of coagulation is facilitated, possibly significantly, by UF. The only slight change in clotting time may be due to the fact that the enhancement of the secondary phase is at least partially offset by inhibition of the primary phase discussed above (see Section 18.5).

The fat phase of milk is an oil-in-water emulsion in which the oil phase is made up of milk fat globules (MFG), surrounded by a natural surfactant system, referred to as the milk fat globule membrane (MFGM). Homogenization causes an approximate 10-fold increase in interfacial area (Mulder and Walstra, 1974). There is not enough MFGM material available to cover this increased area and additional surfactant material is supplied by serum proteins, but more importantly, by caseins. Casein micelles and micelle subunits have been found on the surface of homogenized MFGs (Oortuwijn *et al.*, 1977) but there is also evidence that micelles undergo disruption and disintegration at the surface of MFGs (Oortuwijn and Walstra, 1979).

Although homogenized milk coagulates faster than un-homogenized milk (Zollikofer, 1949; Humbert *et al.*, 1980), Robson and Dalgleish (1984) found the rate of proteolysis of κ-casein in homogenized milk is not greater than in un-homogenized milk. In addition, the Smoluchowski rate constant for homogenized milk was *ca.* 100 times less than that for un-homogenized milk. Although these results seemed inconsistent with the shorter clotting times for homogenized milk, the authors observed that the onset of aggregation began at a lower level of proteolysis when milk was homogenized. κ-Casein should be spread out over a larger surface area in homogenized milk and because of this, there should be a lower critical extent of proteolysis for the formation of active (i.e., coagulable) "particles" of casein. The idea of the critical extent of proteolysis and active patches will be discussed in Sections 18.11 and 18.12.

18.9 Casein stability

Although it is unlikely that anyone now seriously believes that the stability of casein micelles can be explained completely by the Theory of the Stability of Lyophobic Colloids (DLVO theory), it is likely that it plays some part. It also seems a good starting place when trying to understand the forces that render casein micelles stable or unstable. According to DLVO theory (Hiemenz, 1986), coagulation of colloidal particles occurs if the free energy of attraction is greater than that of repulsion. Repulsion arises from surface charge and an accompanying counter-ion atmosphere, the so-called electrical double layer. Because of surface charge, there is a surface potential and because of the counter-ion atmosphere, the potential decays as a function of distance from the surface. The so-called double layer thickness or shielding length is the distance over which the electrical potential decreases to 1/e, or *ca.* 37% of the surface potential, and can be calculated if the ionic strength is known. Milk has an ionic strength of *ca.* 0.075 M (van Kreveld and van Minnen, 1955), and this yields a shielding thickness of *ca.* 1 nm. If the surface potential is also known, the repulsive free energy can be calculated as a function of inter-particle distance. Surface potential is usually estimated from the zeta (ζ)-potential, which is actually less than surface potential, because it is the potential at the "slip plane" created by the colloidal particle when it moves in an electric field and is some distance away from the particle. Probably, the main reason why attempts have been made to apply DLVO theory to the stability of casein micelles is because the ζ-potential is reduced to *ca.* 50% of its initial value (from *ca.* −20 to −10 mV) by renneting (Green and Crutchfield, 1971; Pearse, 1976; Darling and Dickson, 1979; Dalgleish, 1984).

Attraction arises from internal dipoles (the so-called London-van der Waals' forces), and is governed by the Hamaker constant, which is dependent on dipole density. The voluminosity (volume of solute/mass of solute) of casein micelles is *ca.* 10 times greater than that of typical globular proteins (Bloomfield and Morr, 1973; Payens, 1979). This means that within a micelle there are 3–4g H_2O/g protein. As a result, the Hamaker constant for casein micelles is low compared to inorganic colloidal suspensions.

An energy barrier against coagulation is obtained by taking the difference between repulsion and attraction as a function of inter-particle distance. The classic way to achieve coagulation is to increase the ionic strength, which, if large enough, eliminates the barrier. However, this does not work for casein micelles. Increasing the ionic strength, at least with NaCl, inhibits, rather than promotes, coagulation (Payens, 1979; Dalgleish, 1983). Another problem in applying DLVO theory to casein micelles is that the energy barrier develops at very short inter-particle distances (< 1 nm) where surface roughness and short-range interactions, both of which are not accounted for in the DLVO theory, are likely to be important. This is well inside the hairy layer, which may extend as far as 12 nm from the micelle surface (Holt and Dalgleish, 1986). Hydrolysis of the hairy layer leads to a decrease of *ca.* 5 nm in its thickness (Walstra *et al.*, 1981).

The presence of this very hydrophilic hairy layer and the shortcomings of the DLVO theory suggest the importance of steric stabilization (Napper, 1983; Tadros, 1984). It is likely that most observers today feel that steric stabilization is more relevant than the DLVO theory in explaining the stability of casein micelles. Steric stabilization arises when a core particle is surrounded by a segmented protective coat or layer. With casein micelles, the CMPs provide the segmental steric layer, or as it is usually referred to, the hairy layer. This hairy layer has a significant voluminosity (Walstra,

1979; van Hooydonk and Walstra, 1987). In addition, it is glycosylated and has sites for phosphorylation. If the CMP was a random coil, the hairy layer would be *ca.* 4.5 nm thick (van Hooydonk and Walstra, 1987), i.e., smaller than the thickness measured experimentally (*ca.* 6 to 12 nm) (van Hooydonk *et al.*, 1986c; Holt and Dalgleish, 1986). The macropeptide appears to have both α-helix and β-sheet configurations, as well as β-turns (Raap *et al.*, 1983; Kumosinski *et al.*, 1991, 1993; Farrell *et al.*, 1993). Nevertheless, CMP may be fairly flexible.

As with the DLVO theory, attraction arises from London-van der Waals' forces when considering steric stabilization. Since double layer forces are not involved, there is no repulsion until the steric layers touch and interpenetrate. Steric "mixing" is involved until interpenetration is far enough so that the segment of one particle touches the core of another. Further penetration means that the segments must be compressed. This leads to "volume restriction," which is always repulsive in nature. de Kruif and May (1991) concluded that κ-casein layers interpenetrate < 1 nm, suggesting that volume restriction effects, as well as DLVO barrier effects, are not important. Late during proteolysis, however, volume restriction effects may be important (van Hooydonk and Walstra, 1987).

The mixing effect is often referred to as the "osmotic" effect, because it is caused by generation of an osmotic pressure in the overlap region of interpenetrating steric layers. The mixing effect is then tied to the second virial coefficient (B_2), which is an interaction parameter. This parameter can be negative (poor solvent conditions), positive (good solvent conditions), or zero (θ-conditions). Then, steric mixing can be repulsive ($B_2 > 0$), attractive ($B_2 < 0$) or ideal ($B_2 = 0$). Horne (1984) found that the hairy layer shrank by *ca.* 10 nm in the presence of 20% ethanol. The collapse of the hairy layer was gradual as ethanol concentration increased, and as a result, Horne (1984) cautioned against the notion that there is a well-defined θ-point governing the stability of casein micelles. de Kruif *et al.* (1992) have shown that proteolysis leads to a decrease in the second virial coefficient from a positive to a negative value.

Conventional activation energy studies have been used to gain insights into the nature of the forces involved when casein micelles coagulate. Dalgleish (1983) subjected micelles to cold hydrolysis and then measured the early rate of coagulation in the temperature range 15 to 60°C. An Arrhenius plot showed that the Smoluchowski rate constant for coagulation increased with temperature in a non-linear manner: activation energy decreased continuously with temperature until at a high temperature (*ca.* 60°C) it became nearly independent of temperature. These results suggest that more than one process is observed. At low temperature, some β-casein dissociates from casein micelles (Roefs *et al.*, 1985); since this dissociation seems to be as a consequence of the weakening of hydrophobic interactions, and since renneted micelles do not coagulate at a low temperature (Berridge, 1942), it would seem that as the temperature is increased, the change in activation energy might be due to at least a couple of actions: reincorporation of β-casein into the micelle must occur and hydrophobic interactions between micelles may occur. Other interactions might be important also. Once gels are formed, cooling increases the magnitude of the shear modulus (Zoon *et al.*, 1988); if coagulation involved only hydrophobic interactions, the gel would soften at low temperature. Dalgleish (1983) found that Na^+ inhibited, whereas Ca^{2+} accelerated, coagulation. This suggests that the effect of Na^+ is an ionic strength effect, as discussed above (Payens, 1979), while that of Ca^{2+} is due to ion pairing. In short, Ca^{2+} may be involved in cross-linking micelles during coagulation. Part of the effect of Ca^{2+} could be to help neutralize micelles which are negatively charged. Charge neutralization would seem to explain the influence of polycations which facilitate coagulation (Baki and Wolfe, 1971; Pearse, 1976; Green and Marshall, 1977). Neutralization of micelles would reduce charge repulsion and favour hydrophobic interactions.

18.10 Measurement of aggregation

A critical problem when trying to quantify aggregation is to match a measurement technique (usually a physical one) with a mathematical modelling equation. With this in mind, a convenient place to begin is with the equation of Drake (1972):

$$\frac{\mathrm{d}\mu_n}{\mathrm{d}t} = \frac{1}{2}\sum_{i=1}^{\infty}\sum_{j=1}^{\infty}[(i+j)^n - i^n - j^n]k_{ij}c_ic_j \tag{10}$$

where

$\mu = \sum i^n c_i = n$th moment of particle distribution,
i,j = number of subunits in aggregates,
k_{ij} = aggregation rate constant
c_i, c_j = molar concentration of particles of i and j subunits.

From the standpoint of aggregation, the most important statistical moments are the zeroth, first and second. The zeroth moment is the total number of particles per unit volume, which for casein micelles, at least, can be counted from electron micrographs. It also may be noted that if Equation 10 is solved for the zeroth moment, the Smoluchowski equation (Overbeek, 1952) is produced. Proper interpretation of light scattering data will yield weight-average molecular mass (M_w). The ratio μ_2/μ_1 gives the ratio M_w/M_0, where M_0 is the initial molecular mass. This ratio is usually referred to as weight-average degree of polymerization ($\bar{x}_w$):

$$\bar{x}_w = M_w/M_0 \tag{11}$$

What are called z-average parameters come from the ratio μ_3/μ_2.

Using an electron microscope (EM), it is, in principle at least, possible to detect particles that have dimensions of *ca.* 1 nm. Since the average diameter of casein micelles is *ca.* 120 nm, not only can micelles be detected, but so can much of their microstructure. There have been numerous studies on the casein micelle using EM (Holt, 1985), mainly in order to study its structure and size distribution. The number of individual micelles in aggregates can, literally, be counted if the aggregates are not too large, so that all statistical moments can be calculated. In short, both μ_0 and $\bar{x}_w$ are accessible from EM data. Probably, the most significant problem in using EM data pertains to the effects of "fixing" samples during the preparation of micrographs, as this can change the nature of the micelles. Instrumentation for EM has been improved significantly during the last several years (McMahon and McManus, 1998).

At least one EM study has been performed during the renneting reaction (Green *et al.*, 1978). It was found that during coagulation, M_w increased *ca.* 10 to 15 times, and then decreased slowly. A maximum value for $\bar{x}_w$ followed by a decrease, has also been observed with light scattering. This decrease may be due to sedimentation (Horne, 1987; Bauer *et al.*, 1995). Electron micrographs show that initially, linear strands of micelles are formed; later, branching is observed (Green *et al.*, 1978). These results appear to be consistent with aggregation mediated by random collisions of spherical particles (Sutherland, 1970).

Dark field microscopy (DFM) is capable of giving a resolution of *ca.* 5 nm under ideal conditions. This technique has a definite advantage over EM in that it is non-invasive. In both EM and DFM, particles are seen and can be counted, and do not need the numerous assumptions made in light scattering experiments. Ruettimann and Ladish (1991) monitored enzyme-induced coagulation of casein micelles using DFM. However, these authors did not count particles, presumably because of insufficient resolution.

A very sophisticated, but controversial, way for following aggregation is to use light scattering. Traditional scattering is often referred to as "static" scattering. Zimm plots (van Holde, 1971; Hiemenz, 1986) make it possible to evaluate M_w in addition to B_2 and the radius of gyration (R_g). A couple of restrictions associated with scattering are that solutions must not be too concentrated and scattering particles must not be too large. Zimm plots make use of the Debye scattering factor, which, as a rule of thumb, seems fairly valid up to particle sizes approaching the wavelength of a given light source, and this is normally in the visual region of the electromagnetic spectrum. Visual light can be used to study casein micelles since they have diameters in the range *ca.* 20 to 600 nm (Schmidt, 1982). The problem is that during coagulation, particle size increases. In early studies (Dalgleish *et al.*, 1981; Payens and Brinkhuis, 1986), micelles were diluted 500 to 1000 times, and $\bar{x}_w$ was measured up to a value of 3 or 4. Resolution in these early studies was probably *ca.* 500 nm (Bauer *et al.*, 1995). Bauer *et al.* (1995) also diluted micelles 500 to 1000 times, but used an indirect Fourier Transform method to analyze data. This method may give reasonable results for particles up to *ca.* 1500 nm (Glatter *et al.*, 1985; Hofer, 1991). Bauer *et al.* (1995) obtained data for values of $\bar{x}_w$ up to *ca.* 100, and argued (based on the slope of double log plots of $\bar{x}_w$ *vs* R_g) that early aggregation involved the formation of "extended chains", and that later there was a structural change in aggregates which involved "internal mass filling". This sequence of events seems consistent with EM data (Green *et al.*, 1978).

Zimm plots are based on measuring the angular dependence of scattered light. It is also possible to determine M_w and B_2 from the wavelength dependence of transmitted light (Doty and Steiner, 1950). This method has been used to study the aggregation of casein micelles (Horne, 1987; Worning *et al.*, 1998; Lomholt *et al.*, 1998) and in at least two studies (Lomholt *et al.*, 1998; Worning *et al.*, 1998) micelles were investigated at normal concentration, two-fold dilution and two-fold concentration. The concentrated solutions were studied by reducing the light path in the cuvette to 0.05–0.25 mm.

The effectiveness of dynamic scattering has improved markedly recently. Rather than measure the average of scattered light by a given sample, as in static scattering, fluctuations in scattered light are incorporated into an autocorrelation function. Although this technique is normally applied to dilute solutions, it can, in the back-scattering mode, be applied to concentrated, in fact opaque, solutions. A *z*-average diffusion coefficient (D_z) is calculated from dynamic scattering data. This can be used to calculate the hydrodynamic radius (R_h) using equation 9.

Walstra *et al.* (1981) demonstrated that the R_h of casein micelles decreased by *ca.* 5 nm during proteolysis when milk was diluted 100 times. More recently, Horne and Davidson (1993) have demonstrated a similar decrease in UF retentate by use of back-scattering.

Static and dynamic scattering can be combined to calculate the ratio R_g/R_h, referred to as ρ in the literature. The hydrodynamic radius of a hard sphere is the actual radius (R), so $\rho = 0.78$, since $R_g = 0.78R$ for a hard sphere. For a flexible polymer, $\rho = 1.73$ (Schmidt *et al.*, 1978). More interesting, from the standpoint of aggregation of casein micelles, is the fact that polydispersity causes ρ to increase and branching causes it to decrease (Burchard, 1978; Burchard *et al.*, 1980). Bauer *et al.* (1995) calculated ρ during the course of enzyme-induced coagulation of casein micelles. Initially, $\rho = 1$, indicating that intact micelles are not hard spheres; ρ increased during the early phases of aggregation and then decreased. Clearly, the early increase was due to polydispersity created by early aggregation. The later decrease was likely caused by branching which was a consequence of the combining of smaller aggregates.

One of the easier ways to monitor continuously enzyme-induced coagulation of milk is to measure viscosity. Although it is possible to combine viscosity measurements with those from dynamic light scattering to calculate the M_w of intact casein micelles (Dewan and Bloomfield, 1972), it is less certain whether this could be done successfully during coagulation owing to changes in shape and hydration of particles during the reaction.

Many years ago, Scott Blair and Oosthuizen (1961) demonstrated that the relative viscosity (η_r) of milk decreased by *ca.* 5% during early proteolysis. Viscosity then increased, leaving a minimum in the η_r *vs.* time curve. Although part of the decrease may be due to the decrease in molecular mass caused by removal of CMP from casein micelles, most observers believe that changes in hydration are more important: the maximum decrease in molecular mass would be *ca.* 8%, while the actual decrease in voluminosity is *ca.* 35% (Walstra, 1978). In addition, the comments of de Kruif (1998) imply an increase in viscosity, as well as in scattered light, depend not on aggregation, but on "increased osmotic compressibility of the colloidal system." Be that as it may, experimental evidence (Green *et al.*, 1978) seems to indicate that a 2-fold increase in η_r corresponds to a 5 to 10 fold increase in molecular mass. It has been suggested also that viscosity increases before aggregation (de Kruif *et al.*, 1992). This argument is based on Coulter counter data (Guthy and Novak, 1977). The Coulter counter in question detected only particles larger than 2 to 3 μm in diameter. In short, aggregation had to proceed far enough so that the size of the aggregates increases to >20-fold the size of the casein micelle in order to be detected. Rheological and EM data (Green *et al.*, 1978) suggest that the increase in viscosity and aggregation begin at about the same time, i.e., when proteolysis is *ca.* 60 to 80% complete. However, there are scattering data (Dalgleish, 1980b) which suggest that aggregation begins when proteolysis is closer to 90% complete.

Before aggregation begins, it is likely the Adhesive Hard Sphere (AHS) model of de Kruif *et al.* (1992) applies to viscosity changes:

$$\eta_r = 1 + 2.5\phi + k_h\phi^2 \tag{12}$$

where

ϕ = volume fraction of casein micelles
k_h = Huggins' coefficient.

The second term on the right (2.5ϕ) is the Einstein term for dilute spheres. The third term ($k_h\phi^2$) accounts for particle interactions and as such is a function of the second virial coefficient (B_2). Since both k_h, and ϕ are set as functions of the extent of proteolysis (α) (de Kruif *et al.*, 1992), the model predicts that α is completely dependent on η_r, which has been shown experimentally to be untrue (Lomholt and Qvist, 1997). It would seem that the model should break down when aggregation begins. Thus, the model may be used to advantage to pin-point the initiation of aggregation. In the next section, models will be presented which include explicit aggregation terms.

18.11 Modelling the Total Clotting Reaction

Probably the first mathematical model proposed for the enzyme-induced coagulation of milk was that of Segelke and Storch (see Foltmann, 1971), who suggested that the product of clotting time (t_c) and enzyme concentration (E_0) is a constant. This expression implies that a plot of t_c *vs.* E_0^{-1} is linear with a slope equal to the constant and a y-intercept equal to zero. Actual y-intercepts fall in the range of *ca.* 30–300 s (Qvist, 1979a,b). This discrepancy is due to the fact the model fails to account for coagulation. Holter (1932) modified the relationship by making a distinction between the duration of the primary phase (t_1) and that of secondary phase (t_2). If the Segelke-Storch relationship is modified to read $t_1\ E_0 = C$, where C is constant, and if $t_c = t_1 + t_2$, then the following results:

$$t_c = t_2 + C/E_0 \tag{13}$$

which is the Holter equation. Although this expression describes data fairly well if the range of E_0 is not too large, it has limitations. Firstly, it is assumed that there is no overlap between the primary and secondary phases of the clotting reaction. Experimental evidence indicates that the secondary phase begins when the primary phase is *ca.* 60 to 80% complete (Green *et al.*, 1978), at least under normal conditions. Perhaps more important is the fact (implicit in the first deficiency) that the Holter equation is based on the assumption that the extent of proteolysis (α) is always the same at t_c. Otherwise, C in equation 13 would not be constant. Experimental evidence (Lomholt and Qvist, 1997) indicates that α at t_c increases as E_0 increases. This at least partly explains the non-linearity of Holter plots if the range of E_0 is too large. The secondary phase of coagulation must be dealt with in a more realistic way to overcome this problem. Nevertheless, Holter plots are used widely, and Hyslop and Qvist (1996) argued that for a clotting model to be viable, it should predict a correct t_2 value (*ca.* 30 to 300 s) from a Holter plot.

Because of the non-linearity of Holter plots, Payens *et al.* (1977) argued for the use of a power law (PL) relationship in order to model t_c:

$$t_c E_0^{\gamma} = \text{constant} \tag{14}$$

where γ is the slope of a double log plot of t_c *vs.* E_0^{-1}. It is possible to derive an explicit relationship between the Holter expression and the PL. Let ln t_c $= y$ and ln $E_0^{-1} = x$. Substitute into Equation 13 and differentiate dy/dx, which is equal to γ. The final result is $\gamma = (t_c - t_2)/t_c$. Thus, γ can be estimated from Holter law data. For example, in the cheese vat, $t_c \approx 1200$ s. A typical t_2 value is *ca.* 100 s. The magnitude of γ then should be *ca.* 1100/1200 $\approx$ 0.92. Actual values fall in the range *ca.* 0.75 to 1 (Payens *et al.*, 1977). In this chapter, it is assumed that a viable model must predict γ to be in the range of *ca.* 0.75 to 1.

Clotting time measurements give no direct information concerning the time course of aggregation. In fact, there is a slight decrease in $\bar{x}_w$ before coagulation proceeds rapidly. The best way to model the time course of coagulation is to use the Drake equation, but also to include a term for the velocity (v) of the primary phase:

$$\frac{d\mu_n}{dt} = v + \frac{1}{2}\sum_{i=1}^{\infty}\sum_{j=1}^{\infty}([i+j]^n - i^n - j^n)k_{ij}c_ic_j \tag{15}$$

Payens (1989), in his last review, made a distinction between "source" and "gradual destabilization" models. The former relies on Equation 15 and the latter on Equation 10. If the primary phase is complete before aggregation begins, there is no need for v in Equation 15. This is the case for salt-induced coagulation of a classic DLVO colloidal suspension in which activation of clotting particles is essentially instantaneous.

Payens (1989) used source models when k_{ij} was taken as time independent, and gradual destabilization models when k_{ij} was taken as time dependent. Dalgleish (1980b, 1988) has done likewise. Hyslop (1989, 1993) preferred to use source models regardless of the time dependence of k_{ij}.

Payens and co-workers (Payens 1976, 1977; Payens *et al.*, 1977) made a significant step forward in modelling by introducing Smoluchowski kinetics to the clotting reaction. For the first time, the secondary phase of coagulation was put in the main stream of modern physical chemistry. From Smoluchowski kinetics (Overbeek, 1952) comes the notion of a diffusion-controlled coagulation reaction; if every collision between particles leads to adhesion, the rate constant (k_{ij}) is given by:

$$k_{ij} = k_s = 8kTN_A/3\eta \tag{16}$$

where N_A = Avagadro's number and kT is the energy of Brownian movement.

In aqueous solution at room temperature, $k_s \approx 6.6 \times 10^{12}$ ml/mol s.

In his first model, Payens (1976, 1977) assumed that both v and k_{ij} were constant. The former was set equal to V_{max} in Michaelis-Menten theory (see Equation 1), and the latter to a second order rate constant (k_2). Making use of Equation 15 to solve for μ_1 and μ_2, followed by substitution of these latter parameters into Equation 11, yields:

$$\bar{x}_w = 1 + (k_2V_{max}/3)t^2 \tag{17}$$

Letting t_c be t when $\bar{x}_w$ is some constant value, the following accrues:

$$t_c(k_2V_{max})^{1/2} = \text{constant} \tag{18}$$

This expression indicates that t_c is equally dependent on the primary and secondary phases of the clotting reaction, as would be expected. Since $V_{max} \propto E_0$ (see Equation 1), Equation 18 also predicts that γ in Equation 14 should be equal to 0.5. Since γ falls in the range *ca.* 0.75 to 1, the above model is inconsistent with experimental data.

One problem with Payens' model is that it fails to account for the effects of proteolysis on the mass of intact micelles. In order to deal with this, Payens (1977) developed Equation 19:

$$\bar{x}_w = 1 - (1 - \delta^2)\mu_1/c_0 + (1 - \delta)^2\mu_2/c_0 \tag{19}$$

where

δ = mM_1/M_0 = weight fraction of CMP in intact casein micelles
m = number of κ-casein residues in a micelle
M_0 = molecular mass of intact micelles
M_1 = molecular mass of CMP
c_0 = molar concentration of casein micelles.

This expression is the counterpart of Equation 11. Assuming $v = V_{max}$ and $k_{ij} = k_2$, combination of the Drake equation with Equation 19 yields:

$$\bar{x}_w = 1 - 2\delta(1 - \delta)V_{max}/c_0 + (1 - \delta)^2(k_2V_{max}^2/3c_0)t^3 \tag{20}$$

It can be shown that Equation 20 predicts $\gamma \approx 0.67$. Equation 19, therefore, makes modelling more realistic. When k_2 is calculated from experimental data, a value of *ca.* 5×10^5 ml/mol s is obtained (Payens, 1977). This value is much less than the diffusion-controlled value (*ca.* 6.6×10^{12} ml/mol s).

The above models of Payens are based on the assumption that v and k_{ij} are true constants. Certainly, v is not as its magnitude decreases during proteolysis. Can things be improved upon by using a more realistic expression for v? The present author has attempted to answer this question by assuming that proteolysis follows first order kinetics (Hyslop, 1993; Hyslop and Qvist, 1996). An analytical solution for $\bar{x}_w$ is obtained from the combination of equations 15 and 19:

$$\bar{x}_w = 1 - 2\delta(1 - \delta)(1 - \exp(-k_1t)) + (1 - \delta)^2k_2c_0\{t - (2/k_1) \times (1 - \exp(-k_1t)) + (1/2k_1)(1 - \exp(-2k_1t))\} \tag{21}$$

This expression predicts that $\gamma \approx 3/2\ k_1\ t_c$. In the cheese vat, $k_1 \approx 0.0025\ \mathrm{s}^{-1}$ and $t_c \approx 1200$ s. Using these numbers, $\gamma = 0.5$. In short, using a realistic expression for v fails to improve upon the model of Payens. Payens (1989) referred to Equations 17, 18 and 20 as "catch-and-razor" models, the idea being "that unstable micelles are created only after being caught and completely razored by the enzyme."

In any event, Dalgleish (1979, 1980b) made a significant contribution to our understanding of the clotting reaction when he accounted for, in a sophisticated way, the time dependence of proteolysis by applying a binomial distribution to the process. The idea is that the clotting enzyme would randomly cleave peptides from the micelle surface. The binomial distribution was incorporated into the Michaelis-Menten theory, and the latter into a source model for coagulation with a constant k_{ij} value. Although an analytical solution for $\bar{x}_w$ was not obtained, realistic values for γ were. This so-called "random attack" model permitted Dalgleish (1979, 1980b) to calculate a k_2 value *ca.* 100 times greater than Payens calculated with his "catch-and-razor" models. Significantly, Dalgleish (1979, 1980b) found virtually no aggregation until $\alpha \approx 0.88$. In addition, the binomial expression indicated that ≈97% of the κ-casein on an average micelle had to be cleaved before it was able to aggregate. This led Dalgleish to propose a "step-function" approach to the clotting reaction (Dalgleish, 1980a). The idea is that there is a critical extent of proteolysis (α_c) before aggregation is possible.

If there is any drawback to the model of Dalgleish (1980b), it is in its complexity: no analytical solution is attained. In unpublished calculations, the present author has derived a relatively simple step-function model which yields an analytical solution. Proteolysis is assumed to follow first order kinetics, and aggregation is not permitted until α_c is reached. After submitting the Drake equation to a La Place Transform technique, the following can be obtained from Equation 19:

$$\begin{aligned}\bar{x}_w &= 1 - 2\delta(1-\delta)(1-\exp(-k_1 t)) + (1-\delta)^2(k_2 c_0/k_1)\\ &\quad \times u_c\{k_1 t - \ln(1-\alpha_c)^{-1} - 2(1-(\exp(-k_1 t))/\\ &\quad \times (1-\alpha) + (1/2)(1-(\exp(-2k_1 t))/(1-\alpha_c)^2)\}\end{aligned} \tag{22}$$

$$u_c = \begin{cases} 0 & \alpha < \alpha_c \\ 1 & \alpha \geq \alpha_c \end{cases}$$

This equation reduces to Equation 21 when $\alpha_c = 0$. It seems to predict $\alpha_c \approx$ 0.8–0.9, as does the "random proteolysis" model of Dalgleish (1979, 1980b), and like the latter model, it gives a viable value for γ. Unlike the Dalgleish model, it has no explicit assumption concerning the randomness of proteolysis.

The step-function model is based on the premise that there is no aggregation until α_c. If α_c is *ca.* 0.8–0.9, then it is likely that viscosity increases before aggregation, as maintained by de Kruif *et al.* (1992).

From Equation 22, an expression may be found for t_c:

$$\begin{aligned}t_c &= (\bar{x}_{w,t_c} - 1 + 2\delta(1-\delta))/(1-\delta)^2 k_2 c_0 - 2\delta\exp(-k_1 t_c)/(1-\delta)k_2 c_0\\ &\quad + \{3/2 + \ln(1-\alpha_c)^{-1} - 2\exp(-k_1 t_c)/(1-\alpha_c)\\ &\quad + \exp(-2k_1 t_c)/2(1-\alpha_c)^2\}1/k_1\end{aligned} \tag{23}$$

where

$$\bar{x}_{w,t_c} = \bar{x}_w \text{ at } t_c$$

This will not predict a linear relationship between t_c and E_0^{-1} ($k_1 \propto E_0$; see Equation 8), but neither do experimental data if the E_0 range is large enough. The value of δ is ≈ 0.04. If this is considered small enough and, in addition if $\exp(-k_1 t_c) \approx 0$, which is true only at a relatively high E_0, and finally if Equation 8 is used for k_1, Equation 23 can be expressed as

$$t_c = (\bar{x}_{w,t_c} - 1)/k_2 c_0 + \{3/2 + \ln(1 - \alpha_c)^{-1}\} c_0 \exp(E_a/RT)/bE_0 \quad (24)$$

This expression is the result of an analytical solution from the total integration of a step-function source model. Besides being consistent with the Holter rule, it appears to be capable of predicting the correct dependence of t_c on micelle concentration (see Section 18.8).

18.12 Time dependent rate constants

The step-function describes data well, at least up to t_c. However, it does not really explain why a critical amount of proteolysis is required before coagulation can occur. In order to deal with this issue, probably most observers now agree that it is best to consider k_{ij} as a time-dependent rate constant. The question then arises as to how to best define the time dependency. It seems that the first idea published in this regards was that of Darling and van Hooydonk (1981). This is the idea, now well tested (van Hooydonk and Walstra, 1987; Dalgleish, 1988; Hyslop, 1989; Payens, 1989; Hyslop and Qvist, 1996), of an energy barrier which the enzyme removes before micelles can coagulate. The rate parameter, k_{ij}, can be expressed as:

$$k_{ij} = k_s \exp(-V/kT) \quad (25)$$

where V is the energy barrier and k_s, is given by Equation 16. The ratio k_{ij}/k_s gives the fraction of effective collisions and is sometimes referred to as the "stability ratio." Most observers have assumed that $V \propto 1-\alpha$, but there may be no good reason for this. Indeed, the results of van Hooydonk and Walstra (1987) suggest that the barrier cannot change linearly with $1-\alpha$ throughout the course of the total reaction. In addition, the present writer, at least, feels that a more sound theoretical function would be $V \propto \ln(1-\alpha)$, as has been used by de Kruif *et al.* (1992). It is thus surprising that Lomholt *et al.* (1998) found that the use of the idea $V \propto 1-\alpha$ gave a better fit to light scattering data than other models tested.

Although the present writer (Hyslop, 1989, 1993; Hyslop and Qvist, 1996) has applied a source modelling mechanism (Equation 15) when using Equation 25, most others have applied a gradual destabilization modelling mechanism (Equation 10). Using the latter, Dalgleish (1988) obtained reasonable γ values if the initial barrier was *ca.* 15 to 20 kT. On the other hand, using the former, Hyslop and Qvist (1996) found that reasonable γ values required a higher initial barrier (≥ 30 kT). In both studies, it was assumed that $V \propto 1-\alpha$. Using the model of Hyslop and Qvist (1996), Lomholt *et al.* (1998) obtained from light scattering data an initial barrier of *ca.* 25 kT when micelle concentration was that in normal milk. At the end of proteolysis, the magnitude of the barrier was *ca.* 10 kT. From rheological data, van Hooydonk and Wastra (1987) predicted that the barrier for fully renneted micelles was closer to 7 or 8 kT. From light scattering data, Dalgleish (1983) and Dalgleish and Holt (1988) found that fully renneted micelles had an even lower barrier of 6 or 7 kT at ambient temperature. This yields a value for k_{ij} of *ca.* 1/500 of k_S; at 60°C, k_{ij} was *ca.* 0.5 k_s (Dalgleish, 1983). These values for k_{ij} seem more realistic than those obtained using the earlier time-independent rate constants (Payens, 1977; Dalgleish, 1980b).

Energy barrier models imply uniform micelle surfaces. From the standpoint of the clottability of micelles, their surfaces seem to be more discrete (Dalgleish, 1998). It is questionable whether the best way to define a flocculation rate parameter is with an exponential energy barrier. Therefore, Dalgleish and Holt (1988) developed a "geometric model" based on steric stabilization, such that micelle cores must be able to touch each other in order to coagulate. In this model, $k_{ij} \propto \alpha^N$. The parameter N refers to the number of CMP "hairs" which must be hydrolyzed before micelle cores can touch. Strictly speaking, the model deals only with dimerization, and calculations indicate that *ca.* 20 peptides must be cut in order to generate a sufficiently large "patch" on the surface of micelle cores. The need for these patches explains the necessity of a critical extent of proteolysis which must be achieved before aggregation can begin. Using a gradual destabilization model approach, Dalgleish (1988) found that the geometric model yielded realistic γ values.

The geometric model suggests the importance of functionality theory. This has led the present writer (Hyslop, 1993; Hyslop and Qvist, 1996) to investigate Flory-Stockmayer theory, and more specifically, the use of the Stockmayer (1943) rate parameter:

$$k_{ij} = K(4 + 2(f - 2)(i + j) + (f - 2)^2 ij) \tag{26}$$

where

K = proportionality factor
f = number of functional sites (functionality)
$i\,j$ = number of particle subunits.

If f = 1, only dimers are possible, and k_{ij} reduces to K. If f = 2, linear polymers are predicted, and k_{ij} reduces to $4K$. This latter case could be construed as a Smoluchowski case: k_{ij} would be a flocculation rate constant that is time independent. In this case, three-dimensional gels are impossible: if k_{ij} = constant, $\bar{x}_W$ becomes infinite only in infinite time; for gelation kinetics, $\bar{x}_W$ must become infinite in finite time. Experimental evidence suggests that early in the micelle clotting reaction, linear strands are formed

(Green *et al.*, 1978; Bauer *et al.*, 1995), and $f \approx 2$ (Green and Morant, 1981; Bauer *et al.*, 1995; Lomholt *et al.*, 1998).

Gelation kinetics are not possible unless $f > 2$. If $f \gg 2$, then Equation 26 becomes $K(f-2)^2ij$. Payens and Brinkhuis (1986) proposed a model based on the idea that $k_{ij} \propto ij$, but such a model fails to yield viable γ values (Hyslop and Qvist, 1996).

A realistic functionality theory model must account for the fact that at $t = 0$, $f = 0$, but then increases during the course of the reaction. Experimental evidence indicates that when $\bar{x}_w$ approaches 10, $f \approx 2$ (Bauer *et al.*, 1995; Lomholt *et al.*, 1998). At larger $\bar{x}_w$ values, aggregation slows down. Although this would seem to contradict any increase in f, which should speed up aggregation, there is also evidence of a "filling in" process, or, to say it the other way around, aggregates begin to branch out (Bauer *et al.*, 1995). Branching suggests a larger value for f. The slowing of aggregation could be explained by a decrease in mobility caused by the bulkiness of aggregates and increased viscosity. This would be reflected in a smaller value of K in Equation 26. Such interpretations, however, should be made with caution, owing to the questionable validity of data when aggregates become large and of uncertain shape.

Another issue pertains to the nature of the time dependent change in f. The general assumption has been that $f \propto \alpha$ (Hyslop, 1993; Hyslop and Qvist, 1996; Lomholt *et al.*, 1998), but this implies that f develops sequentially. In light of the geometric model, it seems more likely that functional sites, which would actually be patches of micelle surface, would be developed simultaneously. In such a situation, it is arguable (Hyslop, unpublished calculations) that

$$f/f_\infty = 1 - \{1 + k_f \ln(1-\alpha)^{-1}\}^{-1/(f_\infty - 1)} \tag{27}$$

where

$f_\infty = f$ when $\alpha = 1$
k_f = constant.

It remains to be seen whether this expression can describe actual clotting data.

Use of functionality theory can be criticized. Although the Stockmayer expression (Equation 26) includes no explicit geometric assumptions, it is usually assumed that the aggregating entities are large extended polymers in which many functional sites are possible, and in which steric problems are unlikely. With spherical particles, like casein micelles, it is likely that steric problems could easily manifest themselves once two such spheres react. It would then be expected that K, or f, or both in Equation 26 would be low. But the Stockmayer expression would still be valid.

18.13 Conclusion

Considerable attention has been paid here to modelling the rennet-induced coagulation of milk. Despite the fact that the primary phase can be dealt with more directly than the secondary phase, both from a theoretical and experimental stand point, we have found that, depending on circumstances and assumptions, at least three different kinetic models (Michaelis-Menten, Section 18.3; surface Michaelis-Menten, Section 18.4 or diffusion, Section 18.5) appear to provide a reasonable description of the proteolysis of κ-casein. Our present understanding of this reaction could probably be improved if there was a more direct and continuous way to collect and monitor proteolysis data (Section 18.2).

Although the two phases of the coagulation of milk are theoretically separable, it is debatable whether they are separable experimentally. They are not separable during normal clotting in the cheese vat. This leads to serious modelling problems. The most sophisticated models have been based on solving differential equations which simultaneously include information on both phases of the clotting reaction. Depending on circumstances and assumptions, all 3 models, i.e., step-function (Section 18.11), energy barrier (Section 18.12) and functionality theory (Section 18.12) may be used successfully to describe the aggregation phase of the clotting reaction. A serious problem here is how to relate experimental data to the actual aggregation of casein micelles. Because of advances in experimental techniques (eg., EM and LS), it is arguable that recent data (Section 18.10) are of higher quality than earlier data.

It is arguable that further information concerning the mechanism of coagulation is unnecessary in order to make fine cheese, which has been made for centuries. However, history, especially the history of science, surely teaches us that the pursuit of fundamental problems can lead to good results, sometimes in unexpected ways. The enzymatic coagulation of milk follows a mechanism which involves numerous fundamental problems from the stand point of physics and chemistry. We should continue to try to solve these problems.

References

Alais, C. and Lagrange, A. (1972) Etude biochemique d'une protease coagulante produit par *Mucor miehi*. I. Active coagulante et activite proteolytique. *Lait*, **52**, 407–27.

Armstrong, C.E., MacKinlay, A.G., Hill, R.J. and Wake, R.G. (1984) The action of rennin on κ-casein: the heterogenity and origin of the soluble product. *Biochim. Biophys. Acta*, **140**, 123–31.

Azuma, N., Kaminnnogawa, S. and Yamauchi, K. (1984) Properties of glyco-macropeptide and para-κ-casein derived from human κ-casein and comparison of human and bovine κ-casein as to susceptibility to chymosin and pepsin. *Agr. Biol. Chem.*, **48**, 2025–31.

Baki, M. and Wolfe, F.H. (1971) Destabilization of casein micelles by lysozyme. *Can. J. Biochem.*, **49**, 882–4.

Bauer, R., Hansen, M., Hansen, S., Ogendal, L., Lomholt, S.B., Qvist, K.B. and Horne, D.S. (1995) The structure of casein aggregates during renneting studied by indirect Fourier transformation and inverse Laplace transformation of static and dynamic light scattering data, respectively. *J. Chem. Phys.*, **103**, 2725–37.

Beeby, R. (1979) The proteolysis of casein by immobilized preparations of α-chymotrypsin, chymosin and a fungal protease. *N.Z. J. Dairy Sci. Technol.*, **14**, 1–11.

Beeby, R. (1980) The use of fluorescamine at pH 6.0 to follow the action of chymosin on κ-casein and to estimate this protein in milk. *N.Z. J. Dairy Sci. Technol.*, **15**, 99–108.

Berridge, N.J. (1942) The second phase of rennet coagulation. *Nature*, **149**, 194–5.

Berridge, N.J. (1945) The purification and crystallization of rennin. *Biochem. J.*, **39**, 179–86.

Bines, V., Young, P. and Law, J.A. (1989) Comparison of Cheddar cheese made with a recombinant calf chymosin with standard calf rennet. *J. Dairy Res.*, **56**, 657–64.

Bitri, L., Rolland, M.P. and Besancon, P. (1993) Immunological detection of bovine caseinomacropeptide in ovine and caprine dairy products. *Milchwissenschaft*, **48**, 367–70.

Bloomfield, V.A. and Morr, C.V. (1973) Structure of casein micelles: physical methods. *Neth. Milk Dairy J.*, **27**, 103–20.

Bohak, Z. (1969) Purification and characterization of chicken pepsinogen and chicken pepsin. *J. Biol. Chem.*, **244**, 4638–48.

Bringe, N.A. and Kinsella, J.E. (1986) Influence of calcium chloride on the chymosin-initiated coagulation of casein micelles. *J. Dairy Res.*, **53**, 371–9.

Burchard, W. (1978) Application of cascade theory to calculation of quasielastic scattering functions. I. Polydisperse, ideal, linear chains in dilute solution. *Macromolecules*, **11**, 455–9.

Burchard, W., Schmidt, M. and Stockmayer, W.H. (1980) Information on polydispersity and branching from combined quasielastic and integrated scattering. *Macromolecules*, **13**, 1265–72.

Carlson, A., Hill, G.C. and Olson, N.F. (1986) The coagulation of milk with immobilized enzymes: a critical review. *Enz. Microb. Technol.*, **8**, 642–50.

Castle, A.V. and Wheelock, J. (1972) Effect of varying enzyme concentration on the action of rennin on whole milk. *J. Dairy Res.*, **39**, 15–22.

Chaplin, B. and Green, M.L. (1980) Determination of the proportion of κ-casein hydrolyzed by rennet on coagulation of skim milk. *J. Dairy Res.*, **47**, 351–8.

Culioli, J. and Sherman, P. (1978) Rheological aspects of the renneting of milk concentrated by ultrafiltration. *J. Text. Studies*, **9**, 257–81.

Dalgleish, D.G. (1979) Proteolysis and aggregation of casein micelles treated with immobilized or soluble chymosin. *J. Dairy Res.*, **46**, 653–61.

Dalgleish, D.G. (1980a) Effect of milk concentration on the rennet coagulation time. *J. Dairy Res.*, **47**, 231–5.

Dalgleish, D.G. (1980b) A mechanism for the chymosin-induced flocculation of casein micelles. *Biophys. Chem.*, **11**, 147–55.

Dalgleish, D.G. (1981) Effect of milk concentration on the nature of curd formed during renneting – a theoretical discussion. *J. Dairy Res.*, **48**, 65–9.

Dalgleish, D.G., Brinkhuis, J. and Payens, T.A.J. (1981) The coagulation of differently sized casein micelles by rennet. *Eur. J. Biochem.*, **119**, 257–61.

Dalgleish, D.G. (1983) Coagulation of renneted casein micelles: dependence on temperature, calcium ion concentration and ionic strength. *J. Dairy Res.*, **50**, 331–40.

Dalgleish, D.G. (1984) Measurement of electrophoretic mobilities and zeta-potentials of particles from milk using laser doppler electrophoresis. *J. Dairy Res.*, **51**, 425–38.

Dalgleish, D.G. (1986) Analysis of fast protein liquid chromatography of variants of κ-casein and their relevance to micellar structure and renneting. *J. Dairy Res.*, **53**, 43–51.

Dalgleish, D.G. (1988) A new calculation of the kinetics of the renneting reaction. *J. Dairy Res.*, **55**, 221–8.

Dalgleish, D.G. (1989) The behavior of minerals in heated milks, *Bulletin 238*, International Dairy Federation, Brussels, pp. 31–4.

Dalgleish, D.G. (1993) The enzymatic coagulation of milk, in *Cheese: Chemistry, Physics and Microbiology*, Vol. 2, (P.F. Fox ed.) Chapman and Hall, London, pp. 69–100.

Dalgleish, D.G. (1998) Casein micelles as colloids: surface structures and stabilities. *J. Dairy Sci.*, **81**, 3013–8.

Dalgleish, D.G. and Holt, C. (1988) A geometric model to describe the initial aggregation of partly renneted casein micelles. *J. Colloid Interf. Sci.*, **123**, 80–4.

Damicz, W. and Dziuba, J. (1975) Studies on casein proteolysis. I. Enzymatic phase of casein coagulation as influenced by heat treatment of milk proteins. *Milchwissenschaft*, **30**, 399–405.

Daniels, F. and Alberty, R.A. (1966) *Physical Chemistry*, 3rd edn, John Wiley, New York, pp. 360–2.

Darling, D.F. and Dickson, J. (1979) The determination of the zeta potential of casein micelles. *J. Dairy Res.*, **46**, 329–32.

Darling, D.F. and van Hooydonk, A.C.M. (1981) Derivation of a mathematical model for the mechanism of casein micelle coagulation by rennet. *J. Dairy Res.*, **48**, 189–200.

de Koning, P.J. (1978) Coagulating enzymes in cheese making. *Dairy Ind. Int.*, **43** (7), 7–13.

de Kruif, C.G. and May, R.P. (1991) Structure, interaction and gelling studied by small-angle neutron scattering. *J. Biochem.*, **200**, 431–6.

de Kruif, C.G. Jeurinik, T.J.M. and Zoon, P. (1992) The viscosity of milk during the initial stages of renneting. *Neth. Milk Dairy J.*, **46**, 123–37.

de Kruif, C.G. (1998) Supra-aggregates of casein micelles as a prelude to coagulation. *J. Dairy Sci.*, **81**, 3019–28.

de Roos, A.L. (1999) *The Adsorption of Lysozyme and Chymosin onto Emulsion Droplets and their Association with Caseins*, Thesis, Agricultural University of Wageningen, Netherlands.

de Roos, A.L., Walstra, P. and Guerts, T.J. (1995) Association of chymosin with adsorbed caseins, in *Food Macromolecules and Colloids*, (E. Dickinson and D. Lorient eds.) The Royal Society of Chemistry, Cambridge, pp. 50–7.

Dewan, R.A.J. and Bloomfield, V.A. (1972) Molecular weight of bovine milk casein micelles from diffusion and viscosity measurements. *J. Dairy Sci.*, **56**, 66–8.

Doty, P. and Steiner, R.F. (1950) Light scattering and spectrophotemetry of colloidal solutions. *J. Chem. Phys.*, **18**, 1211–20.

Drake, R.L. (1972) The scalar transport equation of coalescence theory: moments and kernals. *J. Atm. Sci.*, **29**, 537–47.

Dunnewind, B., de Roos, A.L. and Guerts, T.J. (1996) Association of chymosin with caseins in solution. *Neth. Milk Dairy J.*, **50**, 121–33.

Ernstrom, C.A. (1974) Milk clotting enzymes and cheese chemistry, in *Fundamentals of Dairy Chemistry*, 2nd edn, (B.H. Webb, A.H. Johnson and J.A. Alford eds.) AVI Publishing Co, Inc., Westport, CT, pp. 694–700.

Farrell, H.M., Jr., Brown, E.M. and Kumosinski, T.F. (1993) Three-dimensional molecular modeling of bovine caseins. *Food Structure*, **12**, 235–50.

Farrell, H.M., Wickham, E.D. and Groves, M.L. (1998) Environmental influences on purified κ-casein: disulfide interactions. *J. Dairy Sci.*, **81**, 2974–84.

Foltmann, B. (1971) The biochemistry of prorennin (prochymosin) and rennin (chymosin), in *Milk Proteins; Chemistry and Molecular Biology*, Vol. 2, (H.A. McKenzie ed.) Academic Press, New York, pp. 218–21.

Foltmann, B. (1993) General and molecular aspects of rennets, in *Cheese: Chemistry, Physics and Microbiology*, Vol. 2, 2nd edn. (P.F. Fox ed.) Chapman and Hall, London, pp. 37–68.

Fox, K.K., Holsinger, V.H., Polsati, L.P. and Pallansch, M.J. (1967) Separation of β-lactoglobulin from other milk serum protein by trichloroacetic acid. *J. Dairy Sci.*, **50**, 1363–7.

Fox, P.F. (1969) Milk – clotting and proteolytic activities of rennet, and of bovine pepsin and porcine pepsin. *J. Dairy Res.*, **36**, 427–33.

Fox, P.F. (1981) Heat-induced changes in milk preceding coagulation. *J. Dairy Sci.*, **6**, 2127–37.

Fox, P.F. and Morrissey, P.A. (1972) Casein micelle structure: location of κ-casein. *J. Dairy Res.*, **39**, 387–94.

Garnier, J. (1963) Etude cinetique d'une protolyse limitee: action de la pressure sur la casein κ. *Biochim. Biophys. Acta*, **66**, 366–77.

Garnier, J., Mocquot, G., Ribadeau-Dumas, B. and Maubois, J.-L. (1968) Coagulation du lait par la pressure: aspects scientifiques et technologiques. *Ann. Nutr. Alim.*, **22**, B495–552.

Garnot, P. and Corre, C. (1980) Influence of milk protein concentration on the gelling activity of chymosin and bovine pepsin. *J. Dairy Res.*, **47**, 103–11.

Garnot, P., Rank, T.C. and Olson, N.F. (1982) Influence of protein and fat contents on ultrafiltrated milk and rheological properties of gels formed by chymosin. *J. Dairy Sci.*, **65**, 2267–73.

Gatt, S. and Bartfai, T. (1977) Rate equations and simulation curves for enzymatic reactions which utilize lipids as substrates, II. Effect of adsorption of the substrate or enzyme on the steady-state kinetics. *Biochim. Biophys. Acta*, **488**, 13–24.

Gibbons, R.A. and Cheeseman, G.C. (1962) Action of rennin on casein: The function of neuraminic acid residues. *Biochim. Biophys. Acta*, **56**, 354–56.

Gillialand, G.L., Oliva, M.T. and Dill, J. (1991) Functional implications of the three dimension structure of bovine chymosin, in *Structure and Function of Aspartic Proteinases. Genetics, Structures and Mechanisms*, (B.M. Dunn ed.) Plenum Press, New York, pp. 23–37.

Glatter, D., Hofer, M., Jorde, C. and Eigner, W.D. (1985) Interpretation of elastic light-scattering data in real space. *J. Colloid Interf. Sci.*, **105**, 577–86.

Green, M.L. and Crutchfield, G. (1971) Density gradient electrophoresis of nature and rennet-treated casein micelles. *J. Dairy Res.*, **38**, 151–64.

Green, M.L. and Foster, P.M.D. (1974) Comparison of the rates of proteolysis during ripening of Cheddar cheeses made with calf and swine pepsin as coagulants. *J. Dairy Res.*, **41**, 269–82.

Green, M.L. and Marshall, R.F. (1977) The acceleration by cationic materials of the coagulation of casein micelles by rennet. *J. Dairy Res.*, **44**, 521–31.

Green, M.L., Hobbs, D.G., Morant, S.V. and Hill, V.A. (1978) Intermicellar relationships in rennet treated separated milk. II. Process of gel assembly. *J. Dairy Res.*, **45**, 413–22.

Green, M.L. and Morant, S.V. (1981) Mechanism of aggregation of casein micelles in rennet-treated milk. *J. Dairy Res.*, **48**, 57–63.

Guthy, K. and Novak, G. (1977) Observations on the primary phase of milk coagulation by rennet under standardized conditions. *J. Dairy Res.*, **44**, 363–6.

Hamdy, A. and Edelstein, D. (1970) Some factors affecting the coagulation strength of three different microbial rennets. *Milchwissenschaft*, **25**, 450–3.

Hicks, C.L., O'Leary, J. and Bucy, J. (1988) Use of recombinant chymosin in the manufacture of Cheddar and Colby cheese. *J. Dairy Sci.*, **71**, 1127–31.

Hiemenz, P.C. (1986) *Principles of Colloid and Surface Chemistry*, 2nd edn, Marcel Dekker, New York, pp. 223–86.

Hill, R.J. and Wake, R.G. (1969) Amphiphile nature of κ-casein as the basis for its micelle stabilizing property. *Nature*, **221**, 635–9.

Hindle, E.J. and Wheelock, J.V. (1970) The primary phase of rennin action in heat stabilized milk. *J. Dairy Res.*, **37**, 389–96.

Holmes, D.G., Duersch, J.W. and Ernstrom, G.A. (1977) Distribution of milk – clotting enzymes between curd and whey and their survival during Cheddar cheesemaking. *J. Dairy Sci.*, **60**, 862–9.

Holt, C. (1985) The size distribution of bovine casein micelles: a review. *Food Microstructure*, **4**, 1–10.

Holt, C. and Dalgleish, D.G. (1986) Electrophoretic and hydrodynamic properties of bovine casein micelle interpreted in terms of particles with an outer hairy layer. *J. Colloid Interf. Sci.*, **114**, 513–24.

Holt, C. and Horne, D.S. (1996) The hairy casein micelles: evolution of the concept and its implication for dairy technology. *Neth. Milk Dairy J.*, **50**, 85–111.

Holter, H. (1932) Uber die labwirkung. *Biochem. Z.*, **255**, 160–88.

Horne, D.S. (1984) Steric effects in the coagulation of casein micelles by ethanol. *Biopolymers*, **23**, 989–93.

Horne, D.S. (1987) Determination of the fractal dimension using turbidimetric techniques. *Faraday Disc. Chem. Soc.*, **83**, 259–70.

Horne, D.S. and Davidson, C.M. (1993) Direct observation of decrease in size of casein micelles during the initial stages of renneting of skim milk. *Int. Dairy J.*, **3**, 61–71.

Humbert, G., Driou, A., Guerin, J. and Alais, C. (1980) Effect of high pressure homogenzation on the properties of milk and its enzyme clotting characteristics. *Lait*, **60**, 574–94.

Humme, H.E. (1972) The optimum pH for the limited specific proteolysis of κ-casein by rennin (primary phase of milk clotting). *Neth. Milk Dairy J.*, **26**, 180–5.

Humphrey, R.S. and Newsome, L.J. (1984) High performance ion-exchange chromatography of the major bovine milk proteins. *N.Z. J. Dairy Sci. Technol.*, **19**, 197–204.

Hyldig, G. (1993) *Effect of Technological Factors on the Enzymatic Reaction and Gel Formation in Milk and Ultrafiltration Concentrates*, Thesis, The Royal Veterinary and Agricultural University, Copenhagen, Denmark.

Hylsop, D.B. (1989) Enzymatically initiated coagulation of casein micelles: a kinetic model. *Neth. Milk Dairy J.*, **43**, 163–70.

Hyslop, D.B. (1993) Enzyme-induced coagulation of casein micelles: a number of different kinetic models. *J. Dairy Res.*, **60**, 517–33.

Hyslop, D.B. and Qvist, K.B. (1996) Application of numerical analysis to a number of models for chymosin-induced coagulation of casein micelles. *J. Dairy Res.*, **63**, 223–32.

Hyslop, D.B., Swanson, A.M. and Lund, D.B. (1979) Heat inactivation of milkclotting enzymes at different pH. *J. Dairy Sci.*, **62**, 1227–32.

Jakob, E. and Puhan, Z. (1992) Technological properties of milk as influenced by genetic polymorphism of milk proteins-a review. *Int. Dairy J.*, **2**, 157–78.

Kannan, A. and Jenness, R. (1961) Relation of milk serum proteins and milk salts to the effects of heat treatment clotting. *J. Dairy Sci.*, **44**, 808–22.

Kumosinski, T.F. Brown, E.M. and Farrell, H.M., Jr. (1991) Three-dimensional molecular modelling of bovine caseins: κ-casein. *J. Dairy Sci.*, **74**, 2879–87.

Kumosinski, T.F., Brown, E.M. and Farrell, H.M., Jr. (1993) Three-dimensional molecular modeling of bovine caseins: a refined, energy minimized κ-casein structure. *J. Dairy Sci.*, **76**, 2507–20.

Lagrange, A., Paquit, P.D. and Alais, C. (1980) Comparative study of two *Mucor miehei* acid proteinases. Purification and some molecular properties. *Int. J. Biochem.*, **11**, 347–52.

Lomholt, S.B. and Qvist, K.B. (1997) Relationship between rheological properties and degree of κ-casein proteolysis during renneting of milk. *J. Dairy Res.*, **64**, 541–9.

Lomholt, S.B., Worning, P., Ogendal, L., Qvist, K.B., Hyslop, D.B. and Bauer, R. (1998) Kinetics of the renneting reaction followed by measurement of turbidity as a function of wavelength. *J. Dairy Res.*, **65**, 545–54.

Martin, P., Raymond, M.N., Bricas, E. and Ribadeau Dumas, B. (1980) Kinetic studies on the action of *Mucor pusillus*, *Mucor miehei* acid proteinases and chymosins A and B on synthetic chromophoric hexapeptide. *Biochim. Biophys. Acta*, **612**, 410–20.

McMahon, D.J. and Brown, R.J. (1984) Enzymatic coagulation of casein micelles: a review. *J. Dairy Sci.*, **67**, 919–29.

McMahon, D.J. and McManus, W.R. (1998) Rethinking casein micelle structure using electron microscopy. *J. Dairy Sci.*, **81**, 2985–93.

Mehai, M.A. and Cheryan, M. (1983a) The secondary phase of milk coagulation. Effect of calcium, pH and temperature on clotting activity. *Milchwissenschft*, **38**, 137–40.

Mehai, M.A. and Cheryan, M. (1983b) Coagulation studies of ultrafiltraion – concentrated skin milk. *Milchwissenschaft*, **38**, 708–10.

Morrissey, P.A. (1969) The rennet hysteresis of heated milk. *J. Dairy Res.*, **36**, 333–41.

Mulder, H. and Walstra, P. (1974) *The Milk Fat Globule. Emulsion Science as Applied to Milk Products and Comparable Foods*, Commonwealth Agricultural Bureau, Farnham Royal, UK.

Napper, D.H. (1983) *Polymeric Stabilization of Colloidal Dispersions*. Academic Press, London, pp. 197–215.

Oortwijn, H., Walstra, P. and Malder, H. (1977) The membranes of recombined fat globules. I. Electron microscopy. *Neth. Milk Dairy J.*, **31**, 134–47.

Oortwijn, H. and Walstra, P. (1979) The membranes of recombined fat globules. II. Composition. *Neth. Milk Dairy J.*, **33**, 134–54.

Overbeek, J.T.G. (1952) Kinetics of flocculation, in *Colloid Science*, Vol. 1, (H.R. Kruyt ed.) Elsevier, Amsterdam, pp. 278–323.

Paquet, P.D. and Alais, C. (1978) Action de proteases foniqiques sur la casein bovine et ses constituants. *Milchwissenschqft*, **33**, 87–90.

Payens, T.A.J. (1976) On the enzyme – triggered clotting of casein; a preliminary account. *Neth. Milk Dairy J.*, **30**, 55–9.

Payens, T.A.J., Wiersma, A.K. and Brinkhuis, J. (1977) On enzymatic clotting processes. I. Kinetics of enzyme triggered coagulation reactions. *Biophys. Chem.*, **6**, 253–61.

Payens, T.A.J. (1977) On enzymatic clotting processes, II. The colloidal instability of chymosin-treated casein micelles. *Biophys. Chem.*, **6**, 263–70.

Payens, T.A.J. (1979) Casein micelles: The colloidal chemical approach. *J. Dairy Res.*, **46**, 291–306.

Payens, T.A.J. and Both, P. (1980) On enzymatic coagulation processes. IV. The chymosin-triggered clotting of para-κ-casein, in *Bioelectrochemistry: Ions, Surfaces, Membranes*. (M. Blank ed.) *Adv. Chem. Series No 188* American Chemical Society, Washington, DC, pp. 129–41.

Payens, T.A.J. and Visser, S. (1981) What determines the specificity of chymosin towards κ-casein? *Neth. Milk Dairy J.*, **35**, 387–9.

Payens, T.A.J. (1984) The relationship between milk concentration and rennet coagulation time. *J. Appl. Biochem.*, **6**, 22–239.

Payens, T.A.J. and Brinkhuis, J. (1986) Mean field kinetics of the enzyme – triggered gelation of casein micelles. *Colloid. Surf.*, **20**, 37–50.

Payens, T.A.J. (1989) The enzyme – triggered coagulation of casein micelles. *Adv. Colloid Interf. Sci.*, **30**, 31–69.

Pearse, K.N. (1976) Moving boundry electrophoresis of native and rennet-treated casein micelles. *J. Dairy Res.*, **43**, 27–36.

Pearse, M.J., Linklater, P.M., Holl, R.J. and MacKinlay, A.G. (1985) Effect of heat induced interaction between β-lactoglobulin and κ-casein on syneresis. *J. Dairy Res.*, **52**, 159–65.

Picard, C., Plard, I., Rondaux - Gaida, D. and Collin, J.-C. (1994) Detection of proteolysis in raw milk stored at low temperature by inhibition ELISA. *J. Dairy Res.*, **61**, 395–404.

Plowman, J.E. and Creamer, L.K. (1995) Restrained molecular dynamics study of the interaction between bovine κ-casein peptide 98–111 and bovine chymosin and porcine pepsin. *J. Dairy Res.*, **62**, 451–67.

Prin, C., El Bari, N., Montagne, P., Cuilliere, M-C,B., Faure, G., Humbert, G. and Linden, G. (1996) Microparticle - enhanced nephelometric immunoassay for caseinomacropeptide in milk. *J. Dairy Res.*, **63**, 73–81.

Pujolle, J., Ribadeau Dumas, B., Garnier, J. and Pion, R. (1966) A study of κ-casein components. I. Preparation. Evidence of a common C-terminal sequence. *Biochem. Biophys. Res. Comm.*, **25**, 285–90.

Qvist, K.B. (1979a) Reestablishment of the original rennetability of milk after cooling. I. The effect of cooling and LTST pasteurization of milk and renneting. *Milchwissenschaft*, **34**, 467–9.

Qvist, K.B. (1979b) Reestablishment of the original rennetability of milk after cooling, II. The effect of some additives. *Milchwissenschaft*, **34**, 600–3.

Raap, J., Kerling, H.J., Vreeman, H.J. and Visser, S. (1983) Peptide substrates for chymosin (rennin): conformational studies of κ-casein and κ-casein-related oligopeptides by circular dichroism and secondary structure prediction. *Arch. Biochem. Biophys.*, **221**, 117–24.

Rasmussen, L.K., Hojrup, P. and Petersen, T.E. (1992) The multimeric structure and disulfide – boning pattern of bovine κ-casein. *Eur. J. Biochem.*, **207**, 215–22.

Reddy, I.M. and Kinsella, J.E. (1990) Interaction of β-lactoglobulin with κ-casein in micelles as assessed by chymosin hydrolysis: effect of temperature, heating time, β-lactoglobulin concentration and pH. *J. Agr. Food Chem.*, **38**, 50–8.

Reuter, H., Hisserich, D. and Prokopek, D. (1981) Beitragzus formalkinetic der labqerinnung von durch ultrafiltration konzentrierter milch. *Milchwissenschaft*, **36**, 13–8.

Robson, E.W. and Dalgleish, D.G. (1984) Coagulation of homogenized milk particles by rennet. *J. Dairy Res.*, **51**, 417–24.

Roefs, S.P.F.M., Walstra, P., Dalgleish, D.G. and Horne, D.S. (1985) Preliminary note on the change in casein micelles caused by acidification. *Neth. Milk Dairy J.*, **39**, 119–22.

Ruettimann, K.W. and Ladish, M.R. (1991) *In situ* observation of casein micelle coagulation. *J. Colloid Interf. Sci.*, **146**, 276–87.
Saputra, D., Payne, F.A. and Hicks, G.L. (1994) Analysis of enzymatic hydrolysis of κ-casein in milk using diffuse reflectance near-infrared radiation. *Trans. ASAE*, **37**, 1947–55.
Sawyer, W.H. (1969) Complex between β-lactoglobulin and κ-casein. A review. *J. Dairy Sci.*, **52**, 1347–55.
Schmidt, D.G. (1982) Association of caseins and casein micelle structure, in *Developments in Dairy Chemistry*. Vol. 1, (P.F. Fox ed.) Applied Science, London, pp. 61–86.
Schmidt, D.G., Both, P. and de Koning, P.J. (1966) Fractionation of some properties of kappa-casein variants. *J. Dairy Sci.*, **49**, 776–82.
Schmidt, M., Dittman, N. and Burchard, W. (1978) Quasi-elastic light scattering from branched polymers: I. Polyvinyl acetate and polyvinyl acetate-microgels prepared by emulsion polymerization. *Polymer*, **20**, 582–8.
Scott-Blair, G.W. and Oosthuizen, J.C., (1961) A viscometric study of the breakdown of casein in milk by rennin and rennet. *J. Dairy Res.*, **28**, 165–73.
Shammet, K.M., Brown, R.J. and McMahon, D.J., (1992) Proteolytic activity of some milk-clotting enzymes on κ-casein. *J. Dairy Sci.*, **75**, 1373–9.
Slattery, C.W. and Evard, R., (1973) A model for the formation and structure of casein micelles from subunits of variable composition. *Biochim. Biophys. Acta*, **317**, 529–38.
Smits, P. and van Brouwershaven, J.H. (1980) Heat-induced association of β-lactoglobulin and casein micelles. *J. Dairy Res.*, **47**, 313–25.
Soxhlet, F. (1877a) Die darftelung halt baren labtliiffigteiten. *Milch. Z.*, **37**, 495–501.
Soxhlet, F. (1877b) Die darftellung halt baren labtliiffigteiten. *Milch. Z.*, **38**, 513–4.
Stockmayer, W.H. (1943) Theory of molecular size distribution and gel formation in branched chain polymers. *J. Chem. Phys.*, **11**, 45–55.
Sutherland, D.N. (1970) Chain formation of fine particle aggregates. *Nature*, **226**, 1241–2.
Tadros, T.F. (1984) Suspensions, in *Surfactants*, (T.F. Tadros ed.) Academic Press, London, pp. 197–220.
Tam, J.J. and Whitaker, J.R. (1972) Rates and extents of hydrolysis of several caseins by pepsin, rennin, *Endothia parasitica* proteinase and *Mucor pusillus* proteinase. *J. Dairy Sci.*, **55**, 1523–31.
Thunell, R.K., Duersch, J.W. and Ernstrom, C.A. (1979) Thermal inactivation of residual milk-clotting enzymes in whey. *J. Dairy Sci.*, **62**, 373–7.
Thurn, A., Blanchard, W. and Niki, R. (1987) Structure of casein micelles. I. Small angle neutron scattering from β- and κ-casein. *Colloid Polym. Sci.*, **265**, 653–66.
Undenfriend, S., Stein, S., Bohlen, P., Dairman, W., Leimgruber, W. and Weigele, M. (1972) Fluorescamine: a reagent for assay of amino acids, peptides, proteins and primary amines in the picomole range. *Science*, **178**, 871–2.
van Holde, K.E. (1971) *Physical Biochemistry*. Prentice Hall, London, pp. 180–201.
van Hooydonk, A.C.M. and Olieman, C. (1982) A rapid and sensitive high-performance liquid chromatography method following the action of chymosin on milk. *Neth. Milk Dairy J.*, **36**, 153–8.
van Hooydonk, A.C.M., Olieman, C. and Hagedoorn, H.G. (1984) Kinetics of the chymosin-catalyzed proteolysis of κ-casein in milk. *Neth. Milk Dairy J.*, **38**, 207–22.
van Hooydonk, A.C.M., Hagedoorn, H.G. and Boerringter, I.J. (1986a) pH-induced physico-chemical changes of casein micelles in milk and their effect on

renneting. I. Effect of acidification on physico-chemical properties. *Neth. Milk Dairy J.*, **40**, 281–96.

van Hooydonk, A.C.M., Hagedoorn, H.G. and Boerringter, I.J. (1986b) pH-induced physico-chemical changes of casein micelles in milk and their effect on renneting. II. Effect of pH on renneting of milk. *Neth. Milk Dairy J.*, **40**, 297–313.

van Hooydonk, A.C.M., Hagedoorn, H.G. and Boerringter, I.J. (1986c) The effect of various cations on the renneting of milk. *Neth. Milk Dairy J.*, **40**, 369–90.

van Hooydonk, A.C.M. and Walstra, P. (1987) Interpretation of the kinetics of the renneting reaction in milk. *Neth. Milk Dairy J.*, **41**, 19–47.

van Hooydonk, A.C.M., de Koster, P.G. and Boerringter, I.J. (1987) The renneting properties of heated milk. *Neth. Milk Dairy J.*, **41**, 3–18.

van Kreveld, A. and van Minnen, G. (1995) Calcium and magnesium ion activity in raw milk and processed milk. *Neth. Milk Dairy J.*, **9**, 1–29.

Verger, R., Mieras, M.C.E. and de Hass, H.G., (1973) Action of phospholipase A at interfaces. *J. Biol. Chem.*, **248**, 4024–34.

Visser, S., Van Rooijen, P.J., Schattenlzerk, C. and Kerling, K.E.T. (1976) Peptide substrates for chymosin (rennin). Kinetic studies with peptides of different chain length including parts of the sequence 101–112 of bovine κ-casein. *Biochim. Biophys. Acta*, **438**, 265–72.

Visser, S., van Rooijen, P.J. and Slangen, C.J. (1980) Peptide substrates for chymosin (rennin). Isolation and substrate behavior of two tryptic fragments of bovine κ-casein. *Eur. J. Biochem.*, **108**, 415–421.

Visser, S. (1981) Proteolytic enzymes and their action on milk proteins. A review. *Neth. Milk Dairy J.*, **35**, 65–88.

Visser, S., Slangen, C.J. and van Rooijen, P.J. (1987) Interaction sites in κ-casein-related sequence located outside the (103–108)-hexapeptide region that fits into the enzyme's active-site cleft. *Biochem. J.*, **244**, 553–8.

Vreeman, H.J., Brinkhuis, J.A. and van der Spek, C.A. (1981) Some association properties of bovine SH-κ-casein. *Biophys. Chem.*, **14**, 185–93.

Vreeman, H.J., Visser, S., Slanger, C.J. and van Riel, A. (1986) Characterization of bovine κ-casein fractions and kinetics of chymosin-induced macropeptide release from carbohydrate-free and carbohydrate-containing fractions determined by high-performance gel-permeation chromatography. *Biochem. J.*, **240**, 87–97.

Walstra, P. (1979) The voluminosity of bovine casein micelles and some of its implications. *J. Dairy Res.*, **46**, 317–23.

Walstra, P., Bloomfield, V.A., Wei, G.J. and Jenness, R. (1981) Effect of chymosin on the hydrodynamic diameter of casein micelles. *Biochim. Biophys. Acta*, **669**, 258–9.

Walstra, P. (1990) On the stability of casein micelles. *J. Dairy Sci.*, **73**, 1965–79.

Waugh, D.F. (1958) The interactions of α_s-, β- and κ-caseins in micelle formation, *Disc. Faraday Soc.*, **25**, 186–96.

Waugh, D.F. (1971) Formation and structure of casein micelles, in *Milk Proteins. Chemistry and Molecular Biology*, Vol. 2, (H.A. McKenzie ed.) Academic Press, New York, pp. 3–85.

Wheelock, J.V. and Kirk, A. (1974) The role of β-lactoglobulin in the primary phase of rennin action on heated casein micelles and heated milk. *J. Dairy Res.*, **41**, 367–72.

Wilson, G.A. and Wheelock, J.V. (1972) Factors affecting the action of rennin in heated milk. *J. Dairy Res.*, **39**, 413–9.

Worning, P., Bauer, R., Ogendal, L. and Lomholt, S.B. (1998) A novel approach to turbidity of dense systems: an investigation of the enzymatic gelation of casein micelles. *J. Colloid Interf. Sci.*, **203**, 265–77.

Zittle, C.A. Thompson, M.P., Custer, J.H. and Cerbulis, J. (1962) κ-Casein-β-lactoglobulin interaction in solution when heated. *J. Dairy Sci.*, **45**, 807–10.

Zollikofer, E. (1949) Influence of homogenization on the rennet coagulation time of milk. *Proc. XII Int. Dairy Congr.* (Stockholm) Vol. 3, pp. 129–31.

Zoon, P.T., van Vliet, T. and Walstra, P. (1988) Rheological properties of rennet-induced skim milk gels. II. The effect of temperature. *Neth. Milk Dairy J.*, **42**, 271–94.

19

HEAT-INDUCED COAGULATION OF MILK

J.E. O'Connell and P.F. Fox

19.1 Abstract

In this chapter, methods for assessing the heat stability of unconcentrated and concentrated milk are discussed. The effects of compositional factors, processing conditions and additives on the heat stability of milk are reviewed and the changes that occur on heating are considered in relation to the heat stability of milk. Three possible explanations for the pH-dependence and mechanism of thermal coagulation of milk, which incorporate the research on the heat stability of milk over the last century, are presented.

19.2 Heat stability of milk

19.2.1 Introduction

The utilization of milk by humans as a readily digestible source of proteins, lipids and carbohydrate, dates back to antiquity. Historically, milk was preserved by fermentation in the form of fermented milk or cheese or the lipid phase was conserved as butter or ghee. However, with developments in thermal processing, milk can now be stored in liquid form or as powders. The first heat treatment of milk with a specific objective has been attributed to Louis Pasteur, *ca.* 1860. Today, the majority of milk, regardless of its final use, is subjected to at least one heat treatment. Common heating regimes and their specific objectives are listed in Table 19.1.

The effect of heat-treating milk on its nutritional and sensory quality, inactivation of enzymes, the technology of thermal processing and methods for assessing the intensity of heat treatments have been reviewed comprehensively (Burton, 1984; Calvo and La Hoz, 1992; Andersson and

Advanced Dairy Chemistry Volume 1: Proteins, 3rd edn.
Edited by P.F. Fox and P.L.H. McSweeney, Kluwer Academic/Plenum Publishers, 2003.

TABLE 19.1
Common heat treatments applied in the dairy industry, their parameters and specific objectives

Heating regime	*Conditions*	*Objective*
Thermization	65°C × 15 min	Killing of spoilage microbes
Pasteurization		
LTLT[1]	63°C × 30 min	Killing of pathogenic microbes
HTST[2]	72°C × 15 sec	
Forewarming	90°C × 2–10 min	Preparatory step for sterilizaton
	120°C × 20 sec	
Sterilization		
UHT[3]	130–140°C × 3–5 sec	Sterilization
In-container	110–115°C × 10–20 min	
Production of specific products	85–90°C × 5–15 min	Yoghurts and protein coprecipitates

[1]Low temperature long time.
[2]High temperature short time.
[3]Ultra-high temperature.

Öste, 1995a,b; Hinrichs and Kessler, 1995; Farkye and Imafidon, 1995; Nursten, 1995; Pellegrino *et al.*, 1995; Stepaniak and Sørhaug, 1995; Recio *et al.*, 1997).

Relative to other food systems, milk is extremely heat stable and can tolerate the conditions applied in most thermal processes. Indeed, if the pH of milk is readjusted periodically to its original value, it may be heated at 140°C for at least 3 h without coagulating (Fox, 1981a). The very high thermal stability of milk is due to the loose, ill-defined three-dimensional structure of its principal proteins, the caseins. Conversely, the whey proteins, which have compact globular structures with unique native conformations, are quite heat labile (see Singh, 1995). For a detailed overview of the colloidal stability of milk considered in relation to the tertiary and quaternary structures of milk proteins, the reader is referred to reviews by Holt (1992) and Wong *et al.* (1996).

However, under certain conditions, milk and milk products may be unstable to the processing conditions imposed. Coagulation during sterilization and gelation or sedimentation of sterilized milk products during storage are encountered occasionally, e.g., sterilization of reconstituted/recombined concentrated milk often results in partial or complete coagulation (see Singh and Creamer, 1992). Although, the heat coagulation of milk proteins is generally a deleterious effect, the production of a special product in South America, *dulce de leche*, entails the controlled thermal coagulation of concentrated milk containing a high level of sucrose (Pauletti *et al.*, 1996).

The mechanism and pH-dependence of the heat-induced coagulation of milk is one of the classical physico–chemical phenomena that has intrigued dairy chemists for about a century and consequently has been the subject of a considerable amount of research. A very extensive literature on the heat stability of milk has accumulated which has been reviewed periodically (Rose, 1963; Fox and Morrissey, 1977; Fox, 1982; Walstra and Jenness, 1984; Muir, 1985; Singh, 1988; van Boekel *et al.*, 1989a,b; Singh and Creamer, 1992; McCrae and Muir, 1995; Singh, 1995). It is apparent from these reviews that steady progress has been made on explaining the variability and mechanism of heat-induced coagulation of milk since the first paper on the subject was published by Sommer and Hart (1919). However, the phenomenon has not been completely explained at the molecular level and there is still significant academic interest in the subject. Problems related to the heat stability of milk are still encountered by the dairy industry, especially with concentrated milk products.

This review will concentrate on recent studies on the heat stability of milk but a substantial part of the older literature is included to explain the basis for the more recent work.

19.2.2 Assessment of the heat stability of milk

The phenomenon of heat-induced coagulation of milk, first reported by Hammersten (1874), is apparent as flocculation, gelation or changes in protein sedimentability (Fox, 1982). In the past, an array of indicators of heat stability were used, such as checking for coagulation in milk that had been autoclaved in test tubes at 136°C for 20 min, or heated in sealed ampoules in a xylene vapour bath (see Fox and Morrissey, 1977). Other early, usually indirect, methods included the ethanol stability assay (Sommer and Binney, 1923), the whitening test (Burton, 1956), the phosphate test (Ramsdell *et al.*, 1931) and viscosity tests (Ball, 1955).

A 'subjective heat stability assay' was developed by Miller and Sommer (1940) and refined by Davies and White (1966). Stability is expressed as the **time** (heat coagulation time, HCT) which elapses between placing a sample of milk, contained in a small glass tube, in an oil bath at a definite temperature, usually 140°C for unconcentrated milk or 120°C for concentrated milk, and the onset of coagulation. The reaction has a Q_{10} value of approximately 3, an activation energy of $\sim$140 kJ mol^{-1} and shows a slight curvilinear relationship between log HCT and assay temperature (Parker and Dalgleish, 1977; Hyslop and Fox, 1981, O'Connell and Fox, 1999c).

Less frequently, heat stability is expressed as the temperature at which coagulation occurs within a short time (usually 2 min). The heat coagulation **temperature** (HCTemp) is determined by intrapolating the HCTime at a range of temperatures to give the value at 2 min. The HCTemp indicates the intrinsic heat stability of milk because it is effectively a measure of instantaneous coagulation and is unaffected by changes that occur on prolonged heating, i.e., dephosphorylation, covalent polymerization of proteins and acid production due to the thermal oxidation of lactose (Miller and Sommer, 1940). Since it is time-consuming to obtain the necessary data, HCTemp is rarely measured and the effects of various compositional and processing parameters thereon have not been studied. The correlation between HCT measured by the subjective assay and the stability of milk towards commercial sterilization is often poor. It appears that pilot plant or laboratory methods which simulate commercial sterilization conditions are better predictors of the stability of milk to commercial sterilization (see Singh and Creamer, 1992).

An objective heat stability assay was developed by Davies and White (1966), in which the percentage of total nitrogen/protein sedimentable by low gravitational forces (~400 g) after various heating intervals is measured (Figure 19.1). The resulting **nitrogen/protein depletion curve** shows a sudden break at the onset of coagulation. Coagulation times as determined by the 'objective assay' correlate well with the subjective heat stability assay (Davies and White, 1966).

An objective method for determining the heat stability of reconstituted milk powders, based on viscosity measurements, was developed by Kieseker and Aitken (1988). Automated methods for assessing heat stability based on a falling ball viscometer (de Wit *et al.*, 1986) or an electromagnetic device (Foissy and Kneifel, 1984) have been developed and found to correlate well with the classical subjective heat stability assay. The viscosity of sterilized milk products is an important quality attribute and hence analytical methods which provide information on the viscosity of the product are especially useful for the quality of such products.

19.3 Compositional factors that affect the heat stability of milk

19.3.1 Introduction

The heat stability of milk and concentrated milk is very variable, indeed the stability of milk from different quarters of the same udder can differ considerably (Benton, 1929). Stability is affected by a number of compositional factors and the significance of any individual factor is difficult to assess as it can not be considered in isolation because variations in one constituent are likely to influence other compositional factors.

19.3.2 pH

(a) Bovine milk

Early studies on the effect of pH on the heat stability of milk were inconclusive. Some researchers reported that stability decreased on slight

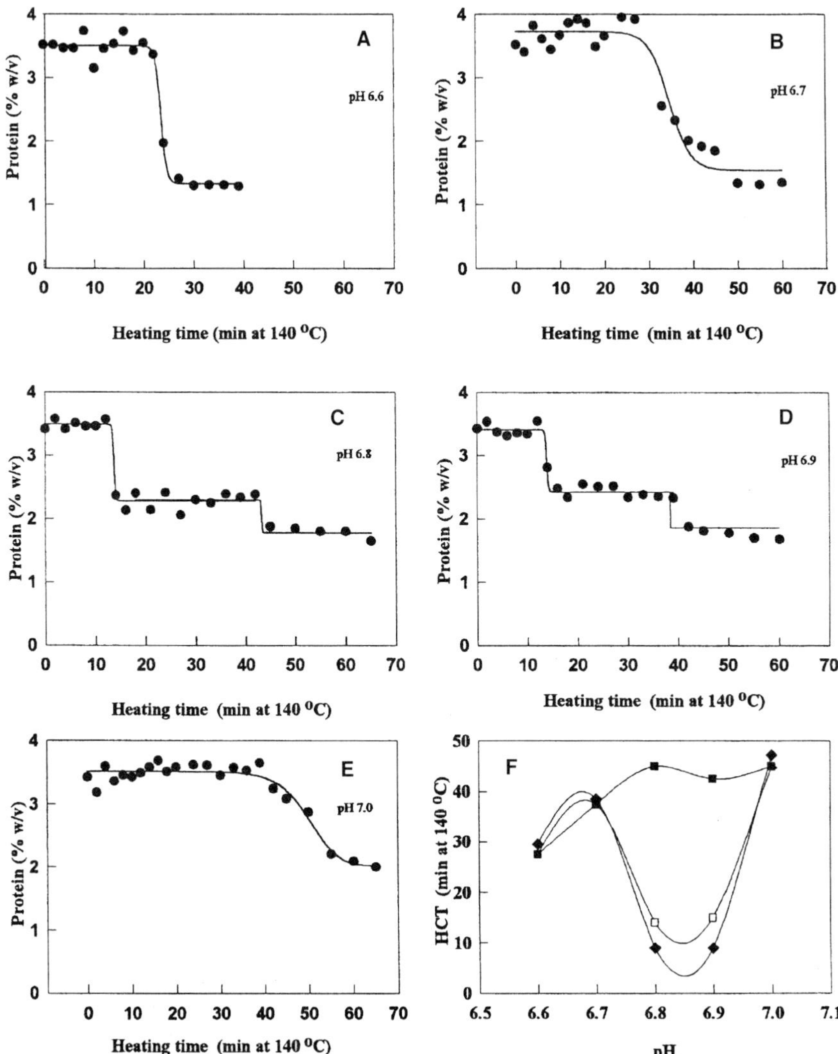

Figure 19.1 (A to E) Protein depletion curves for skimmed milk in the pH range 6.6 to 7.0 on heating at 140°C and (F) the heat coagulation time-pH profile of skimmed milk as determined by the subjective heat stability assay (◆), the objective heat stability assay taking the first inflexion points within the minimum (□), the objective heat stability assay taking the second inflexion points within the minimum (■).

acidification while others reported the opposite effect (Sommer and Hart, 1922; Frazier, 1925; Webb, 1928). The true significance of pH on the heat stability of milk was elucidated by Rose (1961a,b) and thereafter, stability has been considered in relation to pH. Two principal HCT-pH profiles, A and B, are observed. Type A milks exhibit a marked maximum at pH ~ 6.7 and a minimum at pH ~ 6.9 (Figure 19.2); for most type A milks, the natural pH coincides with or is close to the pH of maximum stability during most of the year (Holt *et al.*, 1978a). Milks that show a type A HCT-pH profile also exhibit a maximum and minimum in the HCTemp-pH profile (Figure 19.3).

Although the minimum in the HCTemp-pH profile is less marked than that in the HCTime-pH profile, its occurrence indicates that the minimum is either an intrinsic instability feature of the protein system or is due to a very rapid heat-induced change rather than to a slow heat-induced change, e.g., acid production or dephosphorylation (see Sections 19.5.6 "Dephosphorylation and proteolysis" and 19.5.9 "Heat-induced acidification"). The HCT-pH profile of type A milk as determined by the objective heat stability assay also has a maximum and minimum (Figure 19.1F). Outwith the region of the minimum, coagulation proceeds by a single-stage process (Figure 19.1A, B and E) while within the region of the minimum, coagulation occurs in two stages (Figure 19.1C and 19.1D), the first of which coincides with the HCT observed in the subjective heat stability assay while the second occurs at a time that would be expected if the minimum did not exist (Figure 19.1F; Sweetsur and White, 1974; O'Connell and Fox, 2000a).

The stability of type B milk increases as a function of pH (Figure 19.2; Rose, 1961a,b). It would be expected that the heat stability of all milks should exhibit type B characteristics as the stability of proteins should increase as the pH moves away from the isoelectric point and there is a

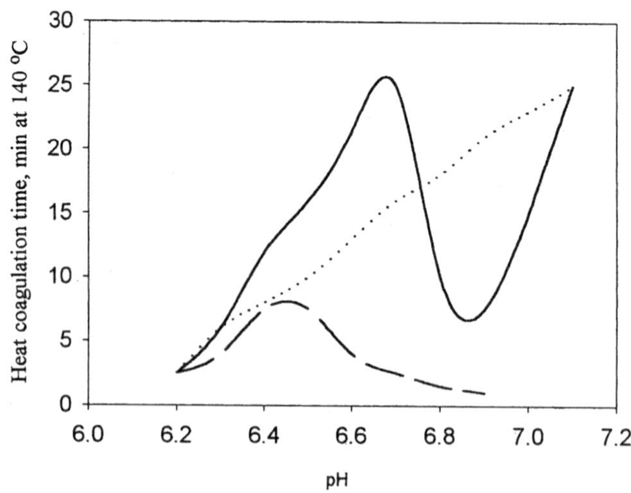

Figure 19.2 Heat coagulation time-pH profile of unconcentrated type A milk, —, unconcentrated type B milk ··· and 3-fold concentrated type A milk - - -.

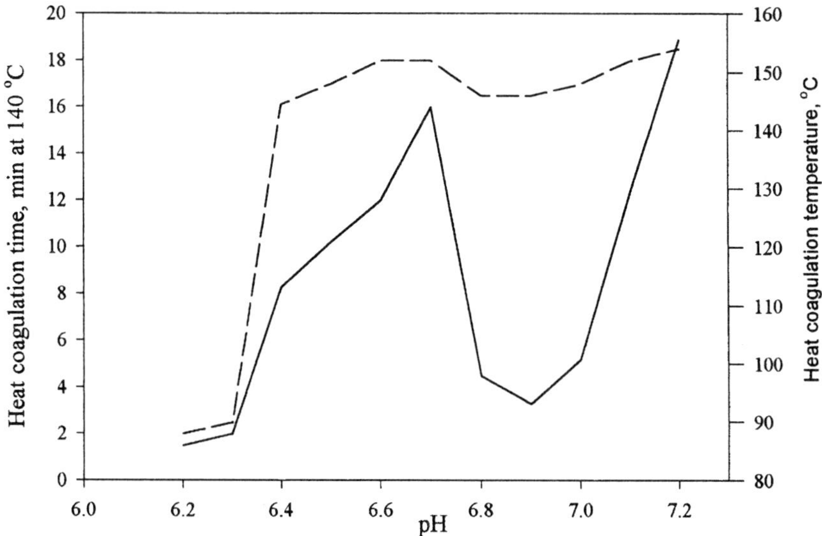

Figure 19.3 Effect of pH on the heat coagulation time —— and heat coagulation temperature - - - of skimmed (type A) milk (O'Connell and Fox, unpublished).

concomitant increase in protein net charge, hydration and voluminosity. Furthermore, the calcium ion activity, which is a key determinant of milk protein stability, decreases with increasing pH (Geerts *et al.*, 1983). Type B milks appear to coagulate by a single-stage process throughout the pH range (Sweetsur and White, 1974).

Type A milks predominate in most countries (with the exception of Japan), accounting for 70% of Scottish, 80% of Canadian and ~100% of Irish and Australian individual-cow milks, while all bulk milks exhibit type A behaviour (Tessier and Rose, 1964; Feagan *et al.*, 1972; Fox, 1982; Muir, 1985). The reasons for the regional differences in the pH-dependence of heat stability are unknown although dietary, environmental and genetic factors are likely to be significant (see Fox, 1982). Sweetsur and White (1974) reported that the type of HCT-pH profile is dependent on assay temperature, with type A milks assuming type B characteristics as the assay temperature is reduced; however, Fox and Hearn (1978b) and Hyslop and Fox (1981) could not confirm this. Methods for converting a type A milk to a type B milk and *vice versa* are listed in Table 19.2.

Concentrated milk is less stable than unconcentrated milk throughout the pH range, it has maximum stability at pH 6.4–6.6 and stability does not recover at higher pH values, i.e., on the alkaline side of the maximum (Figure 19.2; Singh and Creamer, 1992). The coagulation of concentrated milk at the pH of maximum stability is a two-stage process (Muir and Sweetsur, 1978). Bovine Na caseinate dissolved in milk diffusate is very

TABLE 19.2
Methods for inter-converting type A and B HCT-pH profiles

Conversion of type A → B	*Reference*
Reducing assay temperature (150 to 120°C)	Sweetsur and White (1974)
Addition of κ-casein	Tessier and Rose (1964)
Reducing the concentration of inorganic phosphate	Morrissey (1969)
Dialysis of type A milk against type B milk	Morrissey (1969) Fox and Morrissey (1977)
Interchange of ultracentrifugal supernatants of type A and B milks	Rose (1962)
Conversion of type B → A	
Increasing assay temperature (120 to 150°C)	Sweetsur and White (1974)
Pre-heating milk at 80°C for 30 min	Sweetsur and White (1974)
Addition of lactose	Van Boekel *et al.* (1989b)
Addition of β-lactoglobulin	Tessier and Rose (1964)
Addition of divalent cations	Morrissey (1969) Fox and Morrissey (1977)
Dialysis against excess type A milk	Morrissey (1969)
Interchange of ultracentrifugal supernatants of type A and B milks	Rose (1962)

stable (HCT at pH 6.7 and 140°C is ~45 min) and stability increases as a function of pH (Fox and Hearn, 1978b).

Buttermilk has a type B HCT-pH profile owing to its low concentration of calcium and β-lactoglobulin (β-Lg; O'Connell and Fox, 2000c). The addition of buttermilk to milk prior to evaporation and drying increases the stability of reconstituted concentrated milk (Singh and Tokley, 1990). Colostrum has very low heat stability (see Section 19.3.4 "Lactation, season and health").

(b) Interspecies differences

Large interspecies differences in the HCT and shape of the HCT-pH profile have been reported. Human milk is quite heat stable and shows type B behaviour (McLean and McKenzie, 1978). Buffalo milk shows type A heat stability characteristics but is less stable than bovine milk, perhaps because of its high concentrations of casein and calcium (Ganguli, 1979; Ghatak *et al.*, 1980). The heat stability of porcine milk is relatively low and increases as a function of pH (Hoynes and Fox, 1975; Gallagher *et al.*, 1996). Ovine and caprine milks show a maximum in the HCT-pH profile at pH 6.8–7.0 (HCT at 140°C ~10 min) but are very unstable at more alkaline pH values (Fox and Hoynes, 1976; Muir *et al.*, 1993; Tziboula, 1997; Anema and Stanley, 1998). Equine milk is very unstable and coagulates rapidly (<10 min) at temperatures as low as 100°C (Fox and Hoynes, 1976). Farah and

Atkins (1992) reported that camel milk is very unstable to heat, with a HCT <1 min at 140°C at pH values 6.5 to 7.2; when the assay temperature is reduced to 100°C, camel milk has a type B HCT-pH profile.

19.3.3 Milk salts

For many years, it was considered that differences in the heat stability of milk were due to variations in the composition of milk salts and, consequently, the area was researched intensely. Sommer and Hart (1919) proposed the 'salt balance theory' which embraced the concept that there is an optimum value for the ratio of total calcium plus magnesium to total phosphate plus citrate and that any deviation from this value reduces heat stability. However, this theory was based on data from experiments involving the deliberate addition of salts to milk. Many subsequent investigators have endeavoured to correlate heat stability with natural variations in the composition of milk salts. Holt *et al.* (1978a) failed to establish any statistically significant correlation between heat stability and any mineral in milk, while Morrissey *et al.* (1981) found only weak correlations between heat stability and milk salt parameters, alone or in various combinations.

More recently, Ghatak *et al.* (1989) reported that the heat stability of both bovine and buffalo milk is significantly correlated with soluble calcium content while Pouliot and Boulet (1995) correlated the heat stability of concentrated milk with the colloidal Ca to colloidal phosphate ratio.

It must be concluded that although artificial alterations in the concentrations of various salts affect the heat stability of milk, natural variations in milk salts are likely to be too small to be significant, especially in bulk milk.

Nevertheless, changes in the heat stability of milk resulting from removal or addition of either colloidal or soluble salt components are important. An equilibrium exists in the distribution of calcium and phosphate between the colloidal and soluble phases. Colloidal calcium phosphate (CCP) plays an integral role in maintaining the integrity of the casein micelles and consequently removal of CCP markedly affects heat stability. Fox and Hoynes (1975) reported that removal of 40% of the CCP increased heat stability throughout the pH range 6.4 to 7.4, while removal of 60 to 100% of CCP increased stability in the pH range 6.4 to 7.0 but had a destabilizing effect at pH values >7.1 (Figure 19.4A). Doubling of the CCP content had a slight destabilizing effect throughout the pH range 6.4 to 7.4. The reason for the effect of CCP on the pH-dependence of heat stability is unclear, Fox and Hoynes (1975) proposed that the removal of CCP increased heat stability at pH values <6.7 by increasing protein charge but caused destabilization on the alkaline side of the minimum due to a reduction in micellar integrity.

Soluble salts, particularly calcium and phosphate, also play an important role in the heat stability of milk. If the total concentration of Ca^{2+} plus Mg^{2+} is reduced by 2 mM, from 13 to 11 mM, a type A HCT-pH profile is converted to a type B profile, i.e., stability increases as a function of pH (Figure 19.4B; Morrissey, 1969). Reducing the soluble phosphate content shifts the HCT-pH profile to more alkaline values (Figure 19.4C; Morrissey, 1969). Conversely, addition of phosphate increases heat stability due to

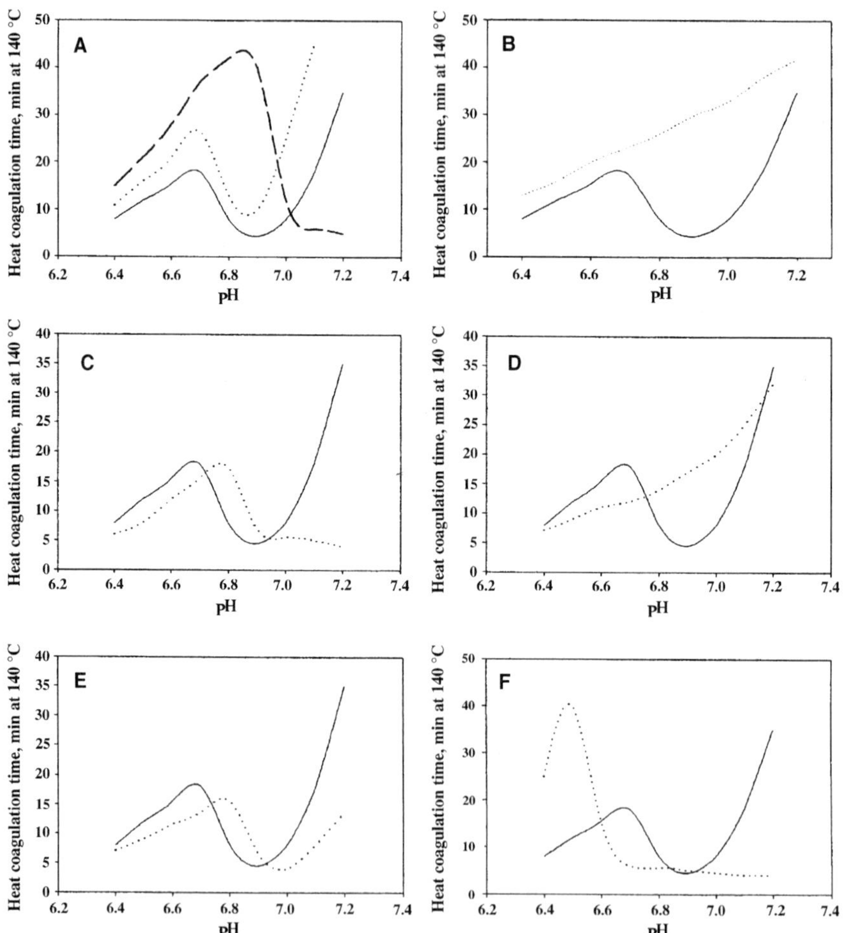

Figure 19.4 Effect of altering the milk salt system on the heat stability of milk; — skimmed milk and (A) removal of 40% ··· or >60% - - of CCP, (B) with 2 mM reduction in calcium content···, (C) reduced phosphate content ···, (D) demineralized skimmed milk ···, (E) with 20 mM added NaCl ···, (F) with 20 mM added Na citrate ···.

chelation of calcium and perhaps to a lesser extent to its buffering capacity (Sweetsur and White, 1974; see Fox and Morrissey, 1977).

Demineralization of type A milk by dialysis against H_2O for a short period (~3 h) converts it to a type B milk (Figure 19.4D; Fox and Hearn, 1978a) while demineralization of a type B milk by electrodialysis reduces stability at pH values >6.6 (Okonogi and Tomita, 1977). Conversely, dialysis of milk against concentrated synthetic milk ultrafiltrate (SMUF; concentrated 2 fold) reduced stability in the region of the minimum, but increased stability at the pH of the maximum stability (Sweetsur and White, 1975).

The effect of other serum salts is not as pronounced as that of calcium or phosphate. Addition of KCl to milk shifts the HCT-pH profile to more alkaline values and markedly reduces stability at pH values >7.0 (Morrissey, 1969). NaCl has a similar effect but higher concentrations are required (Figure 19.4E; Fox and Nash, 1979; Aoki *et al.*, 1999). The addition of 10 mmol L^{-1} Na citrate shifts the HCT maximum to more acid pH values and stability does not recover on the alkaline side of the minimum (Figure 19.4F; Mohammad and Fox, 1983). The effect of Na citrate is presumably due to chelation of both soluble and colloidal calcium; at pH values <6.6, solubilization of CCP increases stability, while at pH values >6.6, stability is reduced as chelation of Ca^{2+} enhances dissociation of κ-casein (CN) (see Section 19.5.4 "Dissociation").

The heat stability of concentrated milk is strongly affected by salt composition (Newstead *et al.*, 1977; Hardy *et al.*, 1984). Concentrates with a lower soluble salt content (i.e., concentrates prepared by ultrafiltration or evaporation of milk which had been dialysed prior to concentration by evaporation) were markedly more stable than those prepared by standard evaporation (Newstead, 1977; Muir and Sweetsur, 1978; Sweetsur and Muir, 1980b).

19.3.4 Lactation, season and health

The HCT-pH profile of milk is relatively constant throughout the lactation period, i.e., HCT_{max} and HCT_{min} do not differ greatly, with the exception of very late lactation, when stability is very low (Rose, 1961a). However, the heat stability at the natural pH of milk does vary throughout the lactation, with variations being more pronounced in countries where milk production is seasonal (Sweetsur and Muir, 1982; Kelly *et al.*, 1982). The HCT of bulk milk at its natural pH and at the pH of maximum stability is lowest from December to May and highest from July to November (de Koning *et al.*, 1974; Holt *et al.*, 1978a).

Also, recombined evaporated milk prepared from powder produced from late lactation milk has poor heat stability properties (Newstead *et al.*, 1976);

as a result, the production of powders in New Zealand is often limited to the mid-lactation period, October to March (Singh *et al.*, 1992). Powders produced from late-lactation milk have low heat stability due to compositional factors and a high somatic cell count (Auldist *et al.*, 1996).

Colostrum is very unstable to heat (Kruk, 1979b) and its addition to milk results in a commensurate decrease in the heat stability of the mixture (Ibrahim *et al.*, 1990). The low heat stability of colostrum may be related to its high protein content; initially, colostrum has a protein content of ~15% but within 15 days *post partum*, its protein content decreases to that of normal milk (Ibrahim *et al.*, 1990).

The stability of concentrated milk also varies with season (Muir *et al.*, 1978). Concentrates prepared during late autumn and early winter from Scottish milk were unstable; high stability was directly related to protein content and inversely to the concentration of diffusible calcium (Sweetsur and Muir, 1982; Muir and Sweetsur, 1992a). Changes in the nitrogen content of milk in September and April have been attributed to changes in feeding patterns (Sweetsur and Muir, 1982).

Although it is difficult to distinguish between lactational and dietary effects, it appears that the type of feed affects heat stability (Feagan *et al.*, 1972) and it is likely that the high stability of milk from cows fed on pasture is associated with an increased concentration of urea (see Fox, 1982).

The health status of dairy cattle also affects the heat stability of milk; mastitic infection reduces heat stability (Feagan *et al.*, 1966; Kruk, 1979b; Heeschen, 1996).

19.3.5 Proteins

(a) Whey proteins

Whey proteins affect the heat stability of milk throughout the pH range 6.4 to 7.4 (Rose, 1961b; Tessier and Rose, 1964; Morrissey, 1969; Fox and Hoynes, 1975; Dziuba and Bochenek, 1989). The heat stability of serum protein-free casein micelle (SPFCM) dispersions in milk diffusate increases as a function of pH (Figure 19.5). The addition of β-Lg to a SPFCM dispersion results in the development of type A HCT-pH characteristics, with stability increasing and decreasing at the pH of maximum and minimum stability of a type A milk, respectively (Tessier and Rose, 1964). Addition of β-Lg to a type B milk converts its heat stability characteristics to type A (Tessier and Rose, 1964).

The addition of porcine β-Lg to a bovine SPFCM dispersion does not introduce a maximum and minimum (Gallagher and Mulvihill, 1997), perhaps because, unlike bovine β-Lg, porcine β-Lg does not have a sulphydryl group and can not interact with κ-CN *via* a disulphide bond; the significance of this interaction is discussed in Section 19.5.10 "Heat-

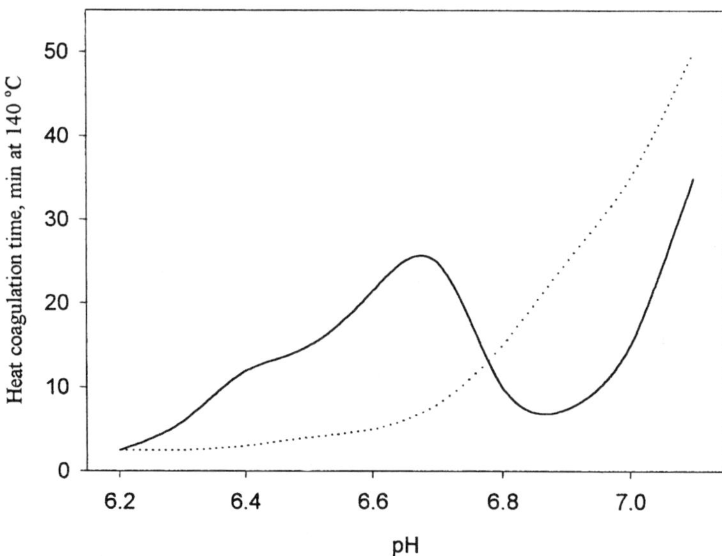

Figure 19.5 Heat coagulation time-pH profile of a serum protein-free casein micelle (SPFCM) dispersion in milk ultrafiltrate with, —, or without, ···, the addition of β-lactoglobulin to 0.4%, w/v.

induced whey protein–casein interactions". Fox and Hearn (1978b) reported that α-lactalbumin (α-La) also introduces a maximum and minimum in the HCT-pH profile of a SPFCM dispersion in milk diffusate while Schmidt and Poll (1986) reported that α-La has a destabilizing effect throughout the pH range 6.4 to 7.1. It appears that the purity and isolation method employed determines the effect of α-La on the stability of a SPFCM dispersion (O'Connell and Fox, unpublished data).

The other whey proteins also affect heat stability. Bovine serum albumin (BSA) destabilizes milk and SPFCM dispersions throughout the pH range, as does lysozyme (Fox and Hearn, 1978a). Variations in the concentration of BSA in normal bovine milk are probably too small to have a significant effect on the heat stability of milk. Human milk contains a very high level of lysozyme (Hennart *et al.*, 1991) which may effect the heat stability but there are no data on this. Human milk has a low concentration of Ca^{2+} and no CCP which may offset the destabilizing effect of a high level of lysozyme. The effect of lysozyme on the heat stability of milk may play a role in the production of an ancient curdled milk product prepared in the Netherlands by boiling milk containing eggs which have a high lysozyme content (van Winter, 1994).

In contrast to its effect on unconcentrated milk, β-Lg (at values of 10 to 150% of the original level) destabilizes concentrated milk throughout the pH range (Newstead *et al.*, 1977). The addition of β-Lg to milk prior to concentration also reduces the heat stability the concentrate; this effect may be reduced, but not eliminated, by pre-heating the milk at 85°C for 30 min prior to adding β-Lg (Newstead *et al.*, 1977; Pearce, 1979). The heat stability of concentrated milk is inversely correlated to its concentration of β-Lg (Newstead *et al.*, 1977; Muir and Sweetsur, 1978).

(b) Caseins

Type A milk may be converted to a type B milk by adding κ-CN (Tessier and Rose, 1964) [note: the effect of κ-CN and β-Lg on heat stability are not opposite, as the latter affects both maximum and minimum stability while the former affects only minimum stability]. Tessier and Rose (1964) proposed that the amount of κ-CN on the surface of the micelles and the κ-CN:β-Lg ratio (rather than the concentration of κ-CN *per se*) determines the shape of the HCT-pH profile and that type B milks lack a minimum because they contain an excess of κ-CN. Stability on the alkaline side of the minimum is markedly affected by κ-CN, as shown by the fact that limited proteolysis of κ-CN by chymosin markedly lowers HCT at pH values > 7.0 (Fox and Hearn, 1978c). According to Robitaille and Ayers (1995), natural variations in the level of glycosylation or enzymatic desialyation of κ-CN have little effect on the heat stability of milk, but it has also been reported that the heat stability of skim milk (Dzuiba *et al.*, 1991) or a SPFCM dispersion (Minkiewicz *et al.*, 1993) is reduced by enzymatic desialyation of κ-CN.

The other casein components also affect the heat stability of milk, albeit to a lesser extent than κ-CN. Schmidt and Koops (1977) showed that the ratio of α:β:κ-CN had a marked effect of the stability of artificial casein micelles. Addition of α_{s2}-CN to milk reduces stability in the region of the minimum and extends the pH range over which the minimum exists (Snoeren *et al.*, 1978); the effect occurs in a SPFCM dispersion whether β-Lg is present or not (Snoeren and van Riel, 1979) and may be related to the high sensitivity of α_{s2}-CN to calcium (Kudo, 1980b).

The heat stability of milk is inversely related to casein micelle size (Zbikowska *et al.*, 1989, 1992; Dziuba *et al.*, 1991; O'Connell and Fox, 2000a) which is presumably related to the protein profile of micelles of varying size (the proportion of κ-CN increases as micelle size decreases; McGann *et al.*, 1980; Zbikowska *et al.*, 1989, 1992; O'Connell and Fox, 2000a). The fact that smaller micelles exhibit weaker van der Waal's attractive forces is also likely to be significant (see van Boekel *et al.*, 1989a). It appears that the two-stage coagulation of type A milk within the region of the minimum is due to the premature coagulation of the larger casein micelles (O'Connell and Fox, 2000a).

As discussed above, the heat stability of milk is affected not only by β-Lg or κ-CN or the ratio of these two proteins but also by the other caseins and whey proteins (Fox, 1982). However, as with milk salts, natural variations in protein concentration or profile appear to be too small to be responsible for the natural variations in the heat stability of milk (de Koning *et al.*, 1974; Holt *et al.*, 1978a; Kelly *et al.*, 1982).

Standardization of the protein content of milk by addition of the ultrafiltrate of skim milk or sweet whey leads to a progressive increase in heat stability, particularly in the region of the minimum (Rattray and Jelen, 1996). Conversely, standardization through the addition of the ultrafiltrate of acid whey reduces stability, presumably due to an increase in soluble calcium content (Peter *et al.*, 1996; Rattray and Jelen, 1996).

19.3.6 Genetic polymorphism

Early attempts to correlate genetic polymorphism of the milk proteins and heat stability yielded rather inconclusive results, with only an erratic tendency for stability at pH 6.7 to increase if κ-CN AA and β-Lg BB were present (Feagen *et al.*, 1972; McLean *et al.*, 1987). Robitaille (1995) proposed that genetic polymorphism of κ-CN and β-Lg affects heat stability; at the pH of maximum stability, milk containing κ-CN AB was more stable than milk with κ-CN AA, with the genotype of β-Lg becoming significant only when κ-CN AA was present (milk with β-Lg AA was more stable than milk with the BB variant). This may be related to the fact that β-Lg AA is more heat stable than the BB variant (Klostermeyer and Reimerdes, 1977; Allmere *et al.*, 1998). The minimum stability of type A milk containing κ-CN AB and β-Lg BB was greater than that of milk containing any other combination of genetic polymorphs examined (Robitaille, 1995). A considerable proportion of milks containing κ-CN BB have a type B HCT-pH profile (Robitaille, 1995). Paterson *et al.* (1999) reported results that were similar in many respects to those of Robitaille (1995): milk with the highest maximum stability had a β-Lg/κ-CN combination of AA/BB, followed by AA/AA, BB/BB and BB/AA, while minimum stability was lowest in milks with a β-Lg/κ-CN combination of AA/AA, followed by AA/BB, BB/AA and BB/BB. The effect of genetic variants of κ-CN on the heat stability of milk may be related to the fact that the degree of gylcosylation of κ-CN and casein micelle size appear to be related to the genetic variant of κ-CN present (Lodes *et al.*, 1996).

van den Berg *et al.* (1992), who studied the effect of genetic polymorphism on the heat stability of concentrated milk, proposed that the genetic variant of β-Lg is of most significance (with milk containing BB being the least stable) and that genetic polymorphism of κ-CN is less important.

The effect of genetic polymorphism *per se* on the heat stability of milk is complicated by the fact that genetic variants are correlated with other properties, e.g., κ-CN BB is correlated to higher total protein, κ-CN and ionic calcium content while β-Lg variant AA has been correlated with a high concentration of β-Lg in milk (Banks *et al.*, 1984; van den Berg *et al.*, 1992).

19.3.7 Urea

The ability of urea to enhance the heat stability of milk at its natural pH is well established and the degradation of indigenous urea through the action of urease reduces the heat stability of milk (Pyne, 1958; Robertson and Dixon, 1969; Muir and Sweetsur, 1976). The addition of urea at 1 to 2% to the basal diet of cattle enhances the heat stability of milk, among other changes, e.g., a reduced sensitivity to Ca^{2+} and rennet and an increase in milk yield and protein content (Mousa and Ibrahim, 1991) and manipulation of cattle feed to produce milk with a high urea content increases heat stability (Banks *et al.*, 1984). Natural variations in urea content, 190 to 450 mg L^{-1} (Muir and Sweetsur, 1976), are highly correlated ($p < 0.001$) with heat stability (Holt *et al.*, 1978a; Kelly *et al.*, 1982) and it is estimated that natural variations in the concentration of urea are responsible for 72 to 90% of the natural variability in the heat stability of bulk milk in SW Scotland during the period June to May (Holt *et al.*, 1978a).

Type A milks are stabilized by the addition of urea (~6 mM) in the region of the maximum and on the alkaline side of the minimum, while type B milks are stabilized throughout the pH range 6.4 to7.4 (Muir and Sweetsur, 1977). Urea does not affect the heat stability of concentrated milk at all or of unconcentrated milk in the region of the HCT minimum, unless added at a very high level, ~1000 mg L^{-1}, or if *N*-ethyl malemide (NEM) is added also (Muir and Sweetsur, 1977, 1978). However, urea does act synergistically with carbonyls to stabilize concentrated milk (Muir *et al.*, 1979).

The apparent activation energy for the heat-induced coagulation of milk is not affected by urea, which implies that it has no effect on the mechanism of heat-induced coagulation (Muir and Sweetsur, 1977). The alcohol stability and rennetability of milk are not affected by urea, suggesting that some heat-induced reaction is a requisite for its stabilizing effect (Muir and Sweetsur, 1977). The thermal degradation of urea has been reported to be a first (Metwalli *et al.*, 1996) or a pseudo-first (Saidi and Warthesen, 1993) order reaction with an activation energy of ~80 kJ mol^{-1} and a Q_{10} of 2 to 2.2 (Fox *et al.*, 1980; Saidi and Warthesen, 1993). Fox *et al.* (1980) reported that other amides and related compounds such as biuret, triuret, methyl urea and ethyl urea, as well as urea, increase heat stability and attributed the effect of urea (and related compounds) to the thermal degradation with the production of ammonia (Figure 19.6, reaction 1) which then enhances stability due to its ability to reduce the decrease in pH of milk caused by heating (i.e., buffering capacity, see Section 19.5.9). The ammonia content in milk increases from ~15 to 100 μg mL^{-1} after 18 min at 140°C (O'Connell and Fox, unpublished data) of which ~70 and 30% is a consequence of heat-induced degradation of urea and deamidation (see Section 19.5.6), respectively (Metwalli and van Boekel, 1998; Metwalli *et al.*, 1998; O'Connell and Fox, unpublished data). This view was supported by Rajput *et al.* (1984a) who showed that the buffering ability of urea increased linearly on heating at 140°C and appeared to be greater in milk than in a protein-free milk salt solution. The fact that the pH of milk at the point of coagulation is roughly the same, regardless of whether urea is added or not, even though the HCT may differ markedly, suggests that the buffering capacity of urea is critically important in its stabilizing action (Metwalli and van Boekel, 1996).

Considering that ammonium carbonate has little effect on the heat stability of milk, Sweetsur and Muir (1981) suggested that the buffering effect is not exclusively responsible for the stabilizing action of urea. O'Reilly and Kelly (1982), who used radiolabelled urea to study its fate in heated milk, showed that a significant proportion of urea interacted with proteins on heating (120°C for 60 min). Preheat treatment of milk with NEM or formaldehyde inhibited the interaction of urea with proteins which suggests that urea interacts specifically with nucleophilic amino acid residues, i.e., lysine and cysteine (O'Reilly and Kelly, 1982). Sweetsur and Muir (1981) proposed that on heating, urea is degraded to cyanate which may then interact with milk proteins through cyanate-lysine interactions resulting in the production of homocitrulline (Figure 19.6, reaction 2). However, Manson *et al.* (1985) and Metwalli *et al.* (1998) suggested that carbamylation of lysine residues is limited and the former authors proposed that carbamylation of the cysteine residues (Figure 19.6, reaction 3), especially of β-Lg, is more likely. Metwalli and van Boekel (1996) suggested that urea may interact with lactose to form lactosyl-urea and that cyanate, which is formed on the thermal degradation of urea, competes with lactose for the interaction with lysine residues, to form homocitrulline. It is postulated that, in combination, these two reactions reduce the rate of the Mailllard reaction in heated milk and consequently inhibit covalent polymerization of proteins (Metwalli and van Boekel, 1996).

$$\underset{\text{(urea)}}{H_2N-CO-NH_2} \rightarrow NH_4CNO \rightarrow NH_3 + \underset{\text{(cyanate)}}{HNCO} \qquad (1)$$

$$\underset{\text{(cyanate)}}{HCNO} + \text{lysine} \rightarrow \underset{\text{(homocitrulline)}}{R-(CH_2)_4-NH-CO-NH_2} \qquad (2)$$

$$\underset{\text{(cyanate)}}{HCNO} + H_2O + \text{cysteine} \rightarrow \underset{\text{(}S\text{-carbamyl cysteine)}}{R-CH_2-S-CO-NH_2} + OH^- \qquad (3)$$

Figure 19.6 Thermal decompositon of urea and subsequent interaction of cyanate with lysine and cysteine residues in proteins.

However, urea enhances the extent of the Maillard reaction in milk on heating (Fox *et al.*, 1980; O'Connell and Fox, 2000b). Tan-Kintia (1996) proposed that urea stabilizes milk through its thermal decomposition to ammonia and CO_2, both of which have a buffering effect; ammonia directly buffers against heat-induced acidification while CO_2 inhibits the thermal oxidation of reductones to organic acids and consequently reduces the decrease in pH on heating (see Section 19.5.9; Tan-Kintia, 1996). Secondly, cyanate may interact with lysine residues to form homocitrulline which is then degraded (or perhaps regenerated to lysine), resulting in the release of an additional mole of ammonia. Tan-Kintia (1996) suggested that these two reactions explain why ammonium carbonate has little effect on heat stability (it does not decompose to CO_2 on heating) and the observation that urea has a greater buffering capacity in milk than in a synthetic milk salts solution (Rajput *et al.*, 1984a). The greater buffering capacity of urea in milk than in synthetic milk salts solution may also be due simply to the inherent buffering capacity of proteins. It is noteworthy that the presence of lactose or another reducing sugar, such as glucose, but not sucrose (which is a non-reducing sugar), is necessary for the stabilizing effect of urea (Kudo, 1980a; Shalabi and Fox, 1982a).

Urea also causes swelling of the casein micelles (Dalgleish *et al.*, 1987b) which may inhibit their aggregation by increasing steric hinderance (Metwalli *et al.*, 1996). This may be related to the fact that urea appears to reduce the extent of covalent bond formation on heating (Dalgleish *et al.*, 1987b).

19.3.8 Lactose

Lactose (4-*O*-β-D-galactopyranosyl-D-glucopyranose), a reducing disaccharide composed of glucose and galactose, is present in bulk bovine milk at ~4.8%, w/v. Attempts to correlate indigenous levels of lactose with variations in heat stability have been unsuccessful (Holt *et al.*, 1978a). However, altering the concentration of lactose in milk has a considerable effect on stability, e.g., increasing lactose to 150% of its normal level destabilized type A milk throughout the pH range 6.4 to 7.4 and shifted the minimum to more alkaline pH values (Sweetsur and White, 1975). These authors reported that increasing the lactose content of a type B milk to 150% of its original value reduced stability throughout the pH range, but stability still increased as a function of pH, while van Boekel *et al.* (1989b) showed that doubling the concentration of lactose in a type B milk introduced a minimum at pH 6.9, but maximum stability was unaffected. The addition of lactose to lactose-free milk progressively increases maximum stability up to approximately 1%, w/v, but higher concentrations reduce stability (Shalabi and Fox, 1982a).

Enzymatic hydrolysis of lactose prior to heating enhances the heat stability of fresh milk (particularly in the region of the minimum) and also of concentrated milk prepared from low- or medium-heat skim milk powder. However, pre-heating at 90°C for 10 min eliminates the ability of lactose hydrolysis to increase heat stability, which may be due to whey protein denaturation and a shift in the salts equilibria (Tan-Kintia and Fox, 1996).

Replacement of air in the headspace by O_2 reduces the heat stability of milk throughout the pH range 6.4–7.0, but has little effect on lactose-free milk (Sweetsur and White, 1975). Conversely, replacement of air in the headspace by N_2 increases stability throughout and decreases the pH range over which the minimum exists (Sweetsur and White, 1975). The effect of headspace composition on the heat stability of milk is presumably related to the thermal oxidation of lactose (see Section 19.5.7 "Thermal Degradation of Lactose").

The significance of lactose in relation to covalent polymerization and heat-induced acidification is discussed in Sections 19.5.8 "Heat-induced crosslinking" and 19.5.9 "Heat-induced acidification"; in addition, lactose increases the calcium ion activity in milk which would be expected to be significant also (see van Boekel *et al.*, 1989b).

19.4 Processing conditions and additives that affect stability

19.4.1 Pre-heating

Early work showed that pre-heating (forewarming) milk at 80 to 90°C reduces the heat stability of unconcentrated milk (Leighton and Deysher, 1923; Davies, 1959). Pre-heating a type A milk shifts the HCT-pH profile (at 140°C) to more acid values without markedly affecting maximum stability (Rose, 1962). When the pH of unconcentrated milk is not adjusted to its initial pH after heat treatment, the effect of pre-heating is minimal (Darling, 1980). Tan-Kintia and Fox (1999) supported this view but also showed that, when assayed at a lower temperature (120°C), stability was reduced markedly by preheat treatment (90°C × 10 min). Preheat treatment at UHT temperatures [120°C × 20 min (Fox, 1981a), 150°C × 1.5 min (Sweetsur and Muir, 1981) or 140°C × 6–12 min (Tan-Kintia and Fox, 1999) enhances stability throughout the pH range. The heat stability of type B milk is unaffected by preheat treatment (Morrissey and O'Mahony, 1976), while pre-heating (90°C for 10 min) a SPFCM dispersion in SMUF increases heat stability. The heat stability of a SPFCM dispersion to which β-Lg was added prior to pre-heating was lower than when β-Lg was added after pre-heating (Tan-Kintia and Fox, 1999).

The pH of milk prior to pre-heating appears to be significant: pre-heating milk increases its heat stability when the pH of the milk is on the acid side of the maximum, but has a destabilizing effect if the pH of the milk is on the alkaline side of the maximum (Griffin *et al.*, 1976; Newstead, 1977).

The stability of concentrated milk is markedly enhanced by pre-heating prior to concentration. Webb and Holm (1932) proposed that a solids non-fat content of 13% is critical in determining whether pre-heating has a stabilizing (>13% SNF) or a destabilizing (<13% SNF) effect. Preheat treatment is thought to stabilize concentrated milk by causing whey protein denaturation and precipitation of calcium phosphate under relatively mild conditions (see Fox, 1982).

The heat stability of recombined milk is also improved if the milk is preheated at 120°C for 2 min prior to concentration (Webb and Bell, 1942; Newstead *et al.*, 1975; Newstead and Baucke, 1983; Sweetsur and Muir, 1980a; van Mil and de Koning, 1992). The effect of pre-heating milk on the stability of concentrates prepared therefrom is least in early and late lactation (van Mil and de Koning, 1992).

The possibility that pre-heating enhances stability by glycosylating proteins has not been examined; it has been shown that extensive glycosylation can occur at pre-heating temperatures (Leonil *et al.*, 1997) and that glycosylation enhances the heat stability of β-Lg (Kitabatake *et al.*, 1985). Conventionally, glycosylation is achieved by incubating proteins with sugars (or sugar acids) in the presence of a catalyst, such as dimethylamine borane or cyanoborohydride. This method is complicated by the fact that most common catalysts also activate a proportion of carboxylic acid residues which then interact with amine residues, causing protein crosslinking. With the advent of electro-assisted glycosylation (Tainturier *et al.*, 1992), this problem should be eliminated and thus the significance of glycosylation (and also of the Maillard reaction) on the heat stability of milk could be elucidated.

19.4.2 Lipids and homogenization

Lipids *per se* have little effect on heat stability as illustrated by the fact that the stability of unhomogenized whole milk and its corresponding skimmed milk are similar (Sweetsur and Muir, 1983a; McCrae, 1999).

Passage of skimmed milk through a homogenizer at up to 31.0 MPa has little effect on its heat stability, while homogenization of whole milk reduces stability (Sweetsur and Muir, 1983a). The extent of destabilization is related to the temperature (Rose, 1963) and pressure of homogenization (Sweetsur and Muir, 1983b) and the season, winter milk being most affected (Sweetsur and Muir, 1983a).

Homogenization entails the disruption of fat globules into smaller ones, resulting in an increase in the area of the lipid/water interface and a concomitant increase in the amount of protein, including micellar casein, adsorbed at the interface (the proportion of proteins on the milk fat globule membrane after homogenization at 30 MPa is approximately 13, 68, 7 and 5% native membrane proteins, caseins, β-Lg and α-La, respectively; Cano-Ruiz and Richter, 1997). van Boekel *et al.* (1989b) proposed that some micelles disintegrate into submicelles on homogenization and that milk fat globules covered with submicelles with a low κ-CN content behave in a manner analogous to κ-CN-depleted casein micelles and are, consequently, inherently less stable. This view is supported by Sharma and Singh (1999), who showed that stability is affected markedly by the concentration of surface protein in homogenised milk.

Sulphydryl blocking agents, oxidizing agents or alterations in serum protein concentration have similar effects on homogenized whole milk and skimmed milk (Sweetsur and Muir, 1983b), which suggests that the coagulation mechanism of homogenized milk is similar to that for skimmed milk. McCrae *et al.* (1994) showed that the heat stability of SPFCM dispersions is not affected by homogenization but addition of serum proteins to an homogenized SPFCM dispersion reduced stability relative to the corresponding unhomogenized SPFCM dispersion containing serum proteins and consequently proposed that the decrease in stability on homogenization is not due to adsorption of whey proteins.

Destabilization due to homogenization poses major problems in dairy processing, e.g., it is a limiting factor in the development of sterile full-fat evaporated milk products (see Singh and Creamer, 1992). The deleterious effect of homogenization may be reduced by using two-stage homogenization, i.e., 20.7 followed by 3.5 MPa, multiple-pass homogenization, preheating at 90°C for 10 min or at a higher temperature (120°C for 120 s, 145°C for 5 s) or the addition of phosphates, Na-caseinate or lecithin (Newstead *et al.*, 1979; McCrae and Muir, 1992; Muir and Sweetsur, 1992b; McCrae and Muir, 1995, Whiteley and Muir, 1996; Sharma and Singh, 1999).

19.4.3 Ultrafiltration and microfiltration

Sweetsur and Muir (1980b) and Muir and Sweetsur (1984) reported an extensive study on the heat stability of concentrates prepared by ultrafiltration under various conditions; concentrates prepared by ultrafiltration are markedly more stable than concentrates produced by conventional evaporation, which has been attributed to a reduction in the concentration of lactose and soluble calcium (Sweetsur and Muir, 1980b).

Muir and Sweetsur (1984) attempted to reduce the colloidal calcium phosphate content of milk by ultrafiltration at pH 5.5. However, ultrafiltration at this pH was unsuccessful because extensive membrane fouling occurred and the concentrates produced were very viscous on readjustment of the pH to 6.6 and had poor heat stability unless reneutralized.

Manipulation of the protein profile by ultrafiltration using a diaporous membrane (nominal pore size, 0.45 μm) rather than conventional UF membranes produced a concentrate with a lower β-Lg content and higher stability (Muir and Sweetsur, 1984). Dialysis of milk against H_2O or dilution with H_2O prior to ultrafiltration enhanced the stability of the concentrate; diafiltration had a similar effect (Muir and Sweetsur, 1984).

Permeates produced by filtration through 1.4 or 0.8 μm membranes were more stable than the corresponding retentates, presumably due to their lower total solids content. Adjustment of the total solids content of the permeate to the level of the retentate reduced the stability of the former (Tziboula *et al.*, 1998). After standardization (of total solids), the retentates were less stable than permeates on the acid side of the minimum but more stable in the region of the minimum. Changes in stability due to microfiltration were attributed to fractionation of large micelles (which have low κ-CN content) and small micelles (which have high κ-CN content) into the retentate and permeate, respectively (Tziboula *et al.*, 1998).

19.4.4 Storage

Davies and White (1966) suggested that milk can be stored at 4°C for up to one week without any significant change in heat stability, while Ghatak *et al.* (1990) reported that the heat stability of milk increased slightly on storage at 5°C for 48 h. The heat stability of pre-heated milk has been reported to decrease on storage (Vujicic and Vulic, 1991). However, stability was reduced when milk containing an antimicrobial agent was stored at ambient temperature (Feagan *et al.*, 1966). Late lactation milk, which contains high levels γ-CNs, products of β-CN hydrolysis by the indigenous alkaline proteinase, plasmin, has low heat stability (Rose, 1961a) and milk to which proteose peptone fractions were added were less stable than unmodified milk, although the high concentration of Ca^{2+} and protein of late lactation milk are also likely to contribute to the reduced stability (Kelly, 1982; Ram and Jushi, 1990).

Heat stability is unaffected by pyschrotrophic bacteria until the population reaches $\sim 10^9$ cfu mL^{-1} (see Fox, 1982), while a population in excess of 10^7 cfu mL^{-1} causes an increase in fouling of heat exchangers (Jeurnink, 1991). The destabilizing effect of pyschrotrophic bacteria is due to the production of proteinases, rather than to the possibility that the bacterial cells act as nuclei for the coalescence of proteins (Jeurnink, 1991).

19.4.5 Sugars and aldehydes

In addition to lactose, which, as stated in Section 19.3.8, has a large effect on the heat stability of milk, other sugars also affect heat stability. Reducing sugars such as glucose, galactose, lactulose, maltose, mannose and fructose at 100 mM stabilize lactose-free milk; the minimum is eliminated and stability increases as a function of pH (Holt *et al.*, 1978b; Tan-Kintia, 1996). 2-Deoxyribose, which markedly enhances stability at 5 mM, is more effective than hexoses because the hemi-acetal structure of the latter makes them less reactive (Holt *et al.*, 1978b; Tan-Kintia, 1996). Non-reducing sugars, e.g., sucrose, trehalose, and the sugar alcohol, lactitol, have little effect on the heat stability of milk (Tan-Kintia, 1996).

Formaldehyde, at 250 mg L^{-1}, increases stability throughout the pH range 6.4 to 7.4, particularly in the region of the minimum, which is eliminated (Nelson, 1954; Holt *et al.*, 1978b; Singh and Fox, 1985a). It has been suggested that formaldehyde stabilizes milk by interacting with lysine residues to form methylol groups (Figure 19.7, reaction 1) which then interact with the amide group of asparagine or glutamine (Figure 19.7, reaction 2) or with the guanidyl group of arginine residues (Figure 19.7, reaction 3; Metwalli and van Boekel, 1995). The formation of such crosslinks inhibits the heat-induced dissociation of κ-CN from the casein micelle (Holt *et al.*, 1978b; Singh and Fox, 1985a; Metwalli and van Boekel, 1995). Aldehydes and urea act synergistically in stabilizing milk (Shalabi and Fox, 1982a) and concentrated milk (Muir *et al.*, 1979) against heat-induced coagulation. Aoki and Kako (1984) showed that formaladehyde, at 600 mg L^{-1}, stabilizes concentrated milk and concentrated SPFCM dispersion, crosslinks protein and inhibits the heat-induced dissociation of protein from the casein micelle. Holt *et al.* (1978b) and Metwalli and van Boekel (1995) proposed that formaldehyde stabilizes milk proteins by interacting with lysine residues which results in an increase in the net negative charge on the protein and a concomitant increase in stability to calcium; Horne (1983) and O'Connell and Fox (1999e) presented evidence which supports this theory. Surprisingly, the possibility that formaldehyde increases the heat stability of milk through its interaction with sulphydryl groups has not been investigated.

$$\text{Lysine} + \text{HCHO} \rightarrow \underset{\text{(methylol)}}{\text{R–NH–CH}_2\text{OH}} \qquad (1)$$

$$\text{R–NH–CH}_2\text{OH} + \text{R–CO–NH}_2 \rightarrow \text{R–NH–CH}_2\text{–NH–CO–R} + \text{H}_2\text{O} \qquad (2)$$

$$\text{R–NH–CH}_2\text{OH} + \text{R–NH–CNH–NH}_2 \rightarrow \text{R–NH–CH}_2\text{–NH–CNH–NH–R} + \text{H}_2\text{O} \qquad (3)$$

Figure 19.7 Interaction of formaldehyde with lysine residues (1) and subsequent interaction of the methylol group with glutamine or asparagine (2) or arginine (3) groups.

19.4.6 Other additives

Several other additives enhance the heat stability of milk. H_2O_2 increases stability, probably by oxidising sulphydryl groups (Koops and Westerbeck, 1970). κ-Carragennan, at 0.05%, w/v, eliminated the HCT minimum without significantly altering maximum stability but at higher levels it markedly destabilized milk; κ-carrageenan also enhances the stability of concentrated bovine (Shalabi and Fox, 1982b) and buffalo (Prasad and Balachandran, 1987) milk. Starch destabilizes both regular and concentrated milk; the destabilizing effect is reduced, but not eliminated, by acid hydrolysis and crosslinking of starch prior to addition to milk (Tziboula and Muir, 1993).

Sodium dodecyl sulphate at 0.2%, w/v, increases the maximum heat stability and shifts it to more acid values and when added to a SPFCM dispersion has an effect analogous to β-Lg, creating a maximum at pH ~6.7 and minimum at pH ~6.9 (Shalabi and Fox, 1982c).

Oxidising and reducing agents affect the heat stability of milk. β-Mercaptoethanol, a reducing agent, at 5 mM markedly destabilizes milk while oxidising agents such as $KBrO_4$ and iodobenzoate at 5 mM convert a type A milk to a type B milk, i.e., increased stability in the region of the minimum but reduce stability in the region of the maximum in the HCT-pH profile (Singh and Fox, 1987a). Another oxidising agent, potassium iodate, at 5 mM, markedly enhances stability throughout the pH range 6.3 to 7.2 (Singh and Fox, 1987a).

Several amino acids affect the heat stability of milk. Cystine and cysteine at 2 or 4 mM reduced the maximum stability without affecting stability in the region of the minimum or at pH values on the alkaline side of the minimum but at 8 mM, cysteine destabilized milk throughout the pH range 6.2 to 7.2 (Rajput *et al.*, 1984b). The basic amino acids, lysine, arginine and histidine, when added to milk at 2.5 mmol L^{-1} destabilized milk at pH values on the alkaline side of the minimum without affecting maximum or minimum stability. Proline and hydroxyproline, at 4 mM, had little effect on the heat stability of milk (Rajput *et al.*, 1984b).

Diketones, e.g., diacetyl and 1,2-cyclohexanedione, at 5 mM, stabilize milk in the region of the maximum without affecting minimum stability, while dialdehydes such as glyoxal have a strong stabilizing effect throughout the pH range (Shalabi and Fox, 1982d,e); these authors attributed the effect of diketones to their interaction with arginine residues which they proposed play an important role in the heat stability of milk.

Dziuba (1979) and O'Connell and Fox (2000b) have also reported that modification of arginine residues markedly enhances the heat stability of milk protein systems. Modification of carboxyl and tryptophan residues also increases the heat stability of milk protein systems (Dziuba, 1979; O'Connell and Fox, 2000b), while nitration of tyrosine residues reduced stability (Dziuba, 1979).

Ethanol, at 3 to 10%, v/v, markedly destabilizes milk. The destabilizing effect of alcohols increases with increasing hydrocarbon chain length, i.e., the destabilizing effect is in the order of butanol > propanol > ethanol > methanol (Mohammad and Fox, 1986) and is presumably due to a reduction in the dielectric constant of milk which causes the hairy polypeptide chains on the casein micelles (glycomacropeptide segment of κ-CN) to collapse, making the casein micelles susceptible to calcium-induced precipitation (Horne, 1992). The effect of alcohols on the heat stability of milk is partially offset by the addition of EDTA (Mohammad and Fox, 1986). The effect of alcohols on the heat stability of milk may have been used unwittingly in the production of 'poshotte', a milk curd prepared by thermally curdling milk containing ale (van Winter, 1994).

Polyphenol-rich extracts of tea, coffee, cocoa, wine, oak and pine bark and oak leaves and several pure phenolic compounds increase the heat stability of unconcentrated milk (particularly in the region of the minimum) and concentrated milk (O'Connell *et al.*, 1998; O'Connell and Fox, 1999a,b); caffeic acid is the most effective of the polyphenols examined (O'Connell and Fox, 1999b). Interaction of phenolic compounds and thermally-generated quinones with nucleophilic amino residues has been implicated in their stabilizing effect (O'Connell and Fox, 1999c).

Addition of the Na or K salts of orthophosphoric, citric or carbonic acids is permitted in many countries and increases the stability of concentrated milk by shifting the HCT-pH profile to more acid values (Sweetsur and Muir, 1980a; Augustin and Clarke, 1990; van Mil and de Koning, 1992). The addition of commercial phosphates (mixture of NaH_2PO_4, Na_2HPO_4 and Na_3PO_4 at 10 g L^{-1}) prevents the coagulation of goats' milk on UHT processing (Montilla and Calvo, 1997).

Lecithin at 0.2%, w/w, enhances the heat stability and subsequent shelf-life of concentrated full-cream milk (Hardy *et al.*, 1985, 1986). The effect appears to be more pronounced if lecithin is added to milk before drying (Singh *et al.*, 1992). The stabilizing effect of lecithin, which has not been explained fully, appears to be more complicated than the displacement of protein from the lipid-serum interface (Singh *et al.*, 1992).

19.5 Heat-induced changes

19.5.1 Introduction

Heat-induced changes in milk occur as a function of temperature, varying with respect to rate but not nature. Up to ~90°C, reactions occur at a relatively slow rate and are largely reversible (with the exception of whey protein denaturation), while at higher temperatures, reactions occur rapidly and most are irreversible. Concerning heat-induced changes, it remains unclear which particular changes are directly responsible for coagulation and which are only predisposing, incidental factors or, in fact, are caused by the coagulation process.

19.5.2 Heat-induced reactions which may increase the heat stability of milk

Although the vast majority of research has been concerned with changes that may cause or contribute to coagulation, there is also a body of evidence which suggests that certain heat-induced reactions occur which may promote or enhance the heat stability of milk. This is supported by the fact that the pH of milk decreases to 5.5–5.0 on heating, yet if a milk sample is preadjusted to pH 5.5 prior to heating it coagulates instantaneously at a temperature as low as 66°C (Fox, 198la). Possible antagonistic reactions may include reduced sensitivity of casein to Ca^{2+} (Pyne, 1958; Mohammad, 1985; O'Connell and Fox, 2000b) or the ability of heating to reduce the Ca^{2+} binding capacity as a consequence of thermal dephosphorylation or alternatively by other non-specific interactions (Mohammad, 1985; Pappas, 1992). Also, heating causes a significant decrease in the Ca^{2+} activity, e.g., Ca^{2+} was reduced from 0.82 to 0.40 mM after heating at 115°C for 10 min (Geerts *et al.*, 1983). Thermal degradation of lactose may also indirectly stabilize milk, as breakdown products of $carbon_5/carbon_6$ compounds have been reported to enhance stability (Tan-Kintia, 1996).

19.5.3 Aggregation of casein micelles

Much work has been done on heat-induced aggregation of model casein systems. Kresheck *et al.* (1964), using light scattering techniques, showed that there was little change in the average molecular weight or gyration constant of whole Na caseinate when heated up to 90°C; when studied individually, only κ-CN showed any change in molecular weight and polydispersity when heated to 90°C, suggesting that complex formation between α_s- and κ-CN stabilizes κ-CN against aggregation.

Turbidity measurements indicate that aggregation of Na caseinate at higher temperatures (120 to 150°C) for 60 min increased with temperature up to 140°C and decreased thereafter, possibly due to thermal degradation of casein (Guo *et al.*, 1989). κ- and α_{s2}-CN are very prone to heat-induced aggregation, presumably due to the presence of two cysteine residues in each protein, as modification of κ-CN by the addition of β-mercaptoethanol or other reducing agents inhibit aggregation (Fox, 1981b). Aggregation of κ-CN is affected by the presence of salts; an aqueous solution of the protein is unaffected by heating at 100°C for 5 min, but when heated under the same conditions in 0.05 M NaCl it becomes heat labile, as indicated by its inability to protect α_s-CN against calcium (Zittle, 1969a,b).

Heating at up to 90°C has little effect on casein micelle size, but when milk is subjected to higher temperatures there is an increase in casein micelle size and a decrease in the range of micelle size (Carroll *et al.*, 1971; Mohammad and Fox, 1987a). Fox (1981b) and Mohammad and Fox (1987b) showed, using ultracentrifugation, viscosity measurements and gel permeation chromatography, that on heating, casein micelles in milk aggregate initially but then dissociate before very extensive aggregation occurs at the onset of coagulation. This was partly confirmed by Dalgleish *et al.* (1987b) who, using photon correlation spectroscopy, followed aggregation at 130°C and showed that the average diameter of casein micelles increased slowly initially and that the micelles underwent rapid aggregation just before coagulation.

Harwalkar *et al.* (1989), who heated milk at 200°C for 3 min and fixed the casein micelles at the assay temperature for examination by electron microscopy, showed that the micelles did not disintegrate on heating but did become enlarged and appeared clustered and distorted, van Boekel *et al.* (1989a) also showed that casein micelles do not disintegrate at elevated temperatures. Aggregation appears to progress simultaneously with an increase in the number of small particles which are possibly formed on disaggregation of micelles (Morr, 1973; Creamer *et al.*, 1978; Creamer and Matheson, 1980; Mohammad and Fox, 1987a).

Aggregation of casein micelles, which appears to be caused by the exposure of hydrophobic binding sites (Ono *et al.*, 1999), is influenced by a number of factors, including pH and Ca^{2+}. On heating at 130°C at pH values outside the minimum in the HCT-pH profile (i.e., pH 6.7 and 7.2), the size of casein micelle increases initially and then remains constant until just before the onset of coagulation, at which point micelle size increases drastically (Dalgleish *et al.*, 1987b). At the pH of minimum stability (pH 6.9), the increase in micelle size on heating at 130°C is more gradual (Dalgleish *et al.*, 1987b). A decrease in [Ca^{2+}] reduces the initial rate of aggregation, possibly by affecting the formation or stability of κ-CN-β-Lg complexes (Jeurnink, 1992). Surprisingly, initial aggregation does not appear to be affected by whey proteins, i.e., the rate of aggregation was the same for milk and a SPFCM dispersion (Mohammad and Fox, 1987b), although the presence of whey proteins increases the extent of micelle growth on heating (Ono *et al.*, 1999; Beaulieu *et al.*, 1999).

Aggregation also occurs in concentrated milk. However, as in the case of unconcentrated milk on heating, it is not possible to relate changes in casein micelle size to other colloidal properties, i.e., micellar integrity and zeta potential (Aoki *et al.*, 1977; Singh and Creamer, 1991c).

19.5.4 Dissociation

The non-sedimentable casein increases on heating milk (Fox *et al.*, 1967) or SPFCM dispersions (Aoki *et al.*, 1974, 1975) at temperatures >110°C.

Approximately 40% of total dissociated protein is κ-CN (Aoki *et al.*, 1974, 1975) and the Q_{10} for the heat-induced dissociation of κ-CN in skim milk at its natural pH is 1.3 from 90 to 120°C (see van Boekel *et al.*, 1989a). Heat-induced dissociation of κ-CN-rich protein from the casein micelles in a SPFCM dispersion was pH-dependent, increasing as a function of pH from pH 6.4 to 7.3 (Kudo, 1980c). Dissociated κ-CN formed on heating (130–140°C for 15 sec) a concentrated whey protein-free milk had a higher sialic acid and hexosamine content and had a higher capacity to stabilize α_{s1}-CN to calcium than non-dissociated κ-CN (Aoki and Kako, 1985). Heat-induced dissociation of κ-CN-rich protein also occurs in caprine milk (Anema and Stanley, 1998). Kudo (1980c) proposed that the pH-dependent dissociation of κ-CN is responsible for the shape of the HCT-pH profile and this work was extended and supported by Singh and Fox (1987b) who showed that the addition of β-Lg to a SPFCM dispersion has a marked effect on dissociation, reducing dissociation at pH values <6.7 but having the opposite effect at higher pH values. Singh *et al.* (1996) showed that the pH-dependence of the distribution of β-Lg in heated milk was similar to that of κ-CN, i.e., after heating at 130°C for 4 min, ~15 and 25% of β-Lg and κ-CN, respectively, were in the serum phase at pH 6.3 and 90 and 70%, respectively, at pH 7.1. Singh and Fox (1985b, 1986) proposed that dissociated κ-CN may reassociate with the micelles on cooling while van Boekel *et al.* (1989a) proposed that a net association of dissociated protein occurs after approximately 2 min of heating at 140°C and pH 7.05, i.e., dissociated κ-CN reassociates with the micelles. However, Anema and Klostermeyer (1997a) could not confirm this in either unconcentrated or concentrated milk heated at a temperature in the range 80 to 120°C and pH 7.10 for up to 20 min.

The protein profile of dissociated protein also varies as a function of pH. At pH 6.7, dissociated protein is comprised of peptides and non-aggregated monomeric caseins, while at pH 7.3, the dissociated protein is comprised of small whey protein-κ-CN complexes and monomeric caseins (Singh and Lantham, 1993).

Fox *et al.* (1967) proposed that on heating milk, calcium, which had been previously associated with citrate, precipitates as calcium phosphate, rendering the citrate free. The citrate then solubilizes indigenous colloidal calcium phosphate, resulting in micellar integrity being compromised and dissociation of micellar κ-CN. This view was supported by the fact that intermolecular protein interactions and solubilization of micellar calcium affect dissociation (Aoki *et al.*, 1975). Altering the CCP content to 40 or 120% of its original value has little effect of the dissociation of κ-CN at pH 7.3, while increasing the level of soluble Ca^{2+} reduced dissociation, possibly due to reduced electrostatic repulsion (Singh and Fox, 1987c; Jeurnink and de Kruif, 1995). Conversely, the addition of phosphate or citrate increases dissociation (Fox *et al.*, 1967). Electrostatic repulsion may be responsible for dissociation (Singh and Fox, 1987b,c); on heating, the net negative charge on the casein micelles increases to such an extent that hydrophobic bonds (which are severely weakened at higher temperatures) are no longer able to maintain micellar integrity. This proposal is supported by the effect of ionic detergents and protein modification on heat-induced dissociation; addition of anionic detergents, e.g., SDS, or succinylation of basic residues enhanced dissociation, while cationic detergents, e.g., CTAB and amidation of proteins had the opposite effect (Singh and Fox, 1986, 1987c).

The fact that dissociated proteins can be resolved by SDS-PAGE and gel filtration chromatography but not by urea-PAGE suggests that proteins undergo modification of charged groups and/or structural changes (Singh and Creamer, 199la; Singh and Lantham, 1993). Heat-induced dissociation of α_{s1}-CN, α_{s2}-CN and β-CN is likely to result from a weakening of micellar protein-protein interactions, possibly as a result of dephosphorylation (Dalgleish *et al.*, 1987a).

The dissociation of κ-CN at temperatures <100°C is also pH dependent (Anema and Klostermeyer, 1997a); little dissociation occurs up to pH 6.7 while up to 57% of total κ-CN dissociates when milk is heated at 90°C for 15 min at pH 7.1. Using isoelectric focusing, it was shown that heating at temperatures <100°C caused little change in the charge on dissociated κ-CN which led the authors to suggest that dissociation at temperatures <100°C involves weakening of CCP links in the micelle so that at higher pH values, CCP links are not strong enough to prevent dissociation of κ-CN due to electrostatic repulsion (Anema and Klostermeyer, 1997b). This hypothesis appears to be consistent with the results of Aoki *et al.* (1990) who reported that the linkage of CCP to protein is weakened on heating.

Heat-induced dissociation of κ-CN increases directly with the degree of glycosylation of the protein (Aoki and Kato, 1984) and inversely with casein micelle size (Aoki and Kato, 1984) but the significance of these factors has not been established.

Dissociation of κ-CN in concentrated milk is also pH-dependent; increasing concentration (10–25%, SNF) leads to a progressive decrease in the pH at which κ-CN dissociates from the casein micelles and to an increase in the amount of κ-CN that dissociates; up to 65% of total κ-CN dissociates when milk at pH 6.85 is heated at 120°C for 8 min (Singh and Creamer, 1991a,b,c,d; Anema, 1998). Differences in protein charge, as determined by resolution in urea-PAGE, between the native and dissociated κ-CN suggest that electrostatic repulsion causes dissociation in concentrated milk also (Singh and Creamer, 1991b). Singh and Creamer (1991a,b,c) suggested that heat-induced interaction of ε-amino groups and lactose, in addition to dephosphorylation of the caseins and hydrolysis of κ-CN, may contribute to charge modification.

19.5.5 Zeta potential and hydration

Unfortunately, it is not possible to measure the zeta potential of casein micelles at the high assay temperature and the effect of heating on zeta potential and related properties has been measured after cooling. Darling and Dickson (1979) showed that heating (135°C for 50 min) caused little change in zeta potential. Heating reconstituted milk (at 140°C) at pH 6.5 caused an initial increase in the zeta potential which then decreased on further heating but heating at pH 7.1 markedly reduced zeta potential initially which remained relatively constant thereafter (Anema and Klostermeyer, 1996, 1997b). The zeta potential of casein micelles in a SPFCM dispersion in SMUF at pH 6.7 is largely unaffected by heat treatment (Schmidt and Poll, 1986). However, the addition of whey proteins to a SPFCM dispersion before heating led to an increase in zeta potential when heated at pH 6.6 to 6.9 and a reduction in the zeta potential when heated at pH 7.0 to 7.1 (Figure 19.8). Anema and Klostermeyer (1997b) proposed that the initial changes in zeta potential were due to association of whey proteins with casein micelles and precipitation of calcium phosphate onto the micelles while subsequent changes in zeta potential are caused by dissociation of κ-CN and dephosphorylation of caseins.

The effect of heat treatments on the zeta potential of casein micelles is intriguing because of the similarities between the effect of serum proteins on

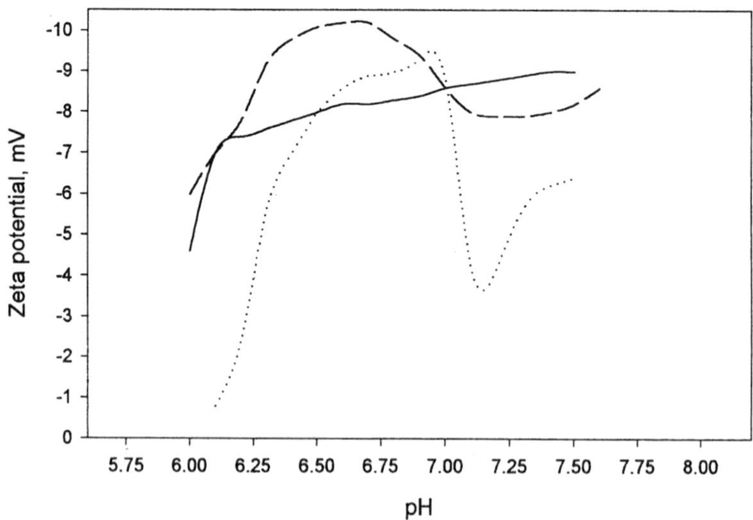

Figure 19.8 Zeta potential-pH profile of a serum protein-free casein micelle dispersion in SMUF, containing 0,—, 0.3 - - - or 0.8· · ·, %, w/v, β-lactoglobulin and heated at 120°C for 10 min (Schmidt and Poll, 1986).

the HCT-pH profile and zeta potential-pH profile of a SPFCM dispersion (Rose, 1961b; Schmidt and Poll, 1986).

Thompson *et al.* (1969) and Kruk (1979) reported that variations in the heat stability of bulk and individual-cow milk is correlated with casein hydration, but this correlation could not be confirmed by Morrissey *et al.* (1981). There also appears to be a correlation between the heat stability of milk as a function of pH and casein hydration (Kruk, 1979a,b); the concentrations of ionic calcium and κ-CN are considered to be key determinants of the latter.

The hydration and equilibrium water content of casein micelles as determined by ultracentrifugation (Fox, 1981b) or water-sorption isotherms (Rüegg *et al.*, 1979) is considerably reduced by heating. However, Creamer and Matheson (1980) reported that heat treatment does not greatly affect casein micelle hydration if the pH is readjusted to the original value after heat treatment. Considering the above, it would appear that heat-induced changes in the hydration of milk proteins are reversible and a consequence of the decrease in pH on heating (see "Heat-induced acidification") rather than changes the physico–chemical properties on the proteins. However, a reduction in protein hydration (regardless of the cause) is likely to be significant in affecting the colloidal stability of the milk protein system.

The addition of β-Lg to a SPFCM dispersion at the pH of maximum stability leads to an increase in hydration on heating, while the opposite effect is evident on heating at pH 7.3, i.e., the addition of β-Lg reduces hydration on heating (Singh and Fox, 1987b).

Heat-induced coagulation is unlikely to emanate directly from changes in either zeta potential or hydration but decreases in these physico-chemical properties are important secondary reactions which may increase the susceptibility of proteins to coagulation.

19.5.6 Dephosphorylation and proteolysis

Howat and Wright (1934) estimated that Na caseinate is completely dephosphorylated after 5 h at 120°C, with 50% dephosphorylation occurring within 60 min; dephosphorylation proceeds at a slower rate in Ca caseinate (80% dephosphorylation within 5 h at 120°C; see Fox, 1982) and micellar casein (Table 19.3; Belec and Jenness, 1962a,b; Davies and White; 1959; Walstra and Jenness, 1984). Whole casein is dephosphorylated at approximately the same rate as α_s-CN which is more susceptible than β-CN (Belec and Jenness, 1962a).

Dephosphorylation in Na caseinate solutions and milk follows first order kinetics in the temperature range 110 to 140°C, with an activation energy of ~105 to 120 kJ mol^{-1} and a Q_{10} of 2.6 to 2.8 at 100°C; the reaction is independent of pH from 6.0 to 7.0 (Belec and Jenness, 1962a,b; Walstra and

TABLE 19.3
Effect of heat treatment on the thermal dephosphorylation of micellar casein and Na caseinate

Heating conditions	*% dephosphorylation*		*Reference*
	Micellar casein	*Na caseinate*	
120°C for 20 min	10	15	Belec and Jenness (1962a,b)
135°C for 1 h	80	90 to 95	Walstra and Jennes (1984)

Jenness, 1984; Mohammad, 1985). Concentration of milk increases the rate of dephosphorylation which is reduced by pre-heat treatment prior to concentration, although no explanation has been provided (Belec and Jenness, 1962a,b). Polyvalent cations, e.g., Ca^{2+} and La^{3+}, are reported to reduce the rate of dephosphorylation of Na caseinate, presumably due to ion exchange of Ca^{2+} or La^{3+} for Na^{+} (as stated above, Ca caseinate is dephosphorylated at a slower rate than Na caseinate); however, the reason why the valency of the caseinate counter-ion might affect the rate of dephosphorylation is unclear (see Belec and Jenness, 1962a).

Howat and Wright (1934) suggested that dephosphorylation of casein progresses in a similar manner to alkali hydrolysis, i.e., by splitting the phosphate esters at the —C—O— bond, resulting in the formation of serine (Figure 19.9 reaction 1; Anderson and Kelley, 1959). Belec and Jenness (1962a) proposed that dephosphorylation involves fission of —O—P— bonds by β-elimination, a view supported by the fact that the activation energy for hydrolysis of casein phosphate is similar to that for glycerol-1-phosphate and serine-*O*-phosphate (Figure 19.9, reaction 2; Belec and Jenness, 1962a).

Recently, van Boekel (1999) presented evidence to suggest that dephosphorylation may proceed by both β-elimination and hydrolysis, but that the latter is principally responsible.

The true significance of thermal dephosphorylation has yet to be elucidated. The ability of κ-CN to protect α_s-CN (α_{s1}- + α_{s2}-CN) against Ca^{2+} is markedly reduced if the former is enzymatically dephosphorylated prior to interaction (Pepper and Thomson, 1963). Dephosphorylation would be expected to reduce heat stability by reducing the net negative micellar charge and disrupting the native micellar structure which is dependant

$$\underset{\text{(phosphoserine)}}{R-CH_2-O-PO_3^{2-}} + H_2O \rightarrow \underset{\text{(serine)}}{R-CH_2-OH} + HPO_4^{2-} \quad (1)$$

$$\underset{\text{(phosphoserine)}}{R-CH_2-O-PO_3^{2-}} \rightarrow \underset{\text{(dehydroalanine)}}{R=CH_2} + HPO_4^{2-} \quad (2)$$

Figure 19.9 Possible mechanisms for the thermal dephosphorylation of casein.

on CCP links (Fox, 1982; Dalgleish *et al.*, 1987a). Also, dephosphorylation of phosphoserine residues may generate the reactive intermediate, dehydroalanine, which may then react with nucleophilic amino acids (Manson and Carotan, 1980; see Singh and Creamer, 1995), the significance of which is discussed in Section 19.5.8 "Heat-induced crosslinking".

At high temperatures, the peptide bonds of proteins undergo partial isomerization from the *trans* to the *cis* configuration; the latter configuration is inherently unstable and more susceptible to heat-induced hydrolysis (Walstra and Jenness, 1984). It is likely that the mechanism of thermal proteolysis is similar to alkaline hydrolysis, which as outlined by Whitaker (1980), involves the peptide bond being attacked by hydroxyl ions, resulting in the formation of an anionic tetrahedral intermediate which is followed by the expulsion of the R—NH_2 moiety (Whitaker, 1980; Figure 19.10).

Early studies showed that proteolysis of caseins in milk on heating [which was determined by quantifying the formation of non-protein nitrogen (NPN)] is linear with time; NPN accounts for 10 to 20% of total nitrogen after 5 h at 120°C (Howat and Wright, 1934) or 1 h at 135°C (White and Davies, 1966), while heating for 40 min at 140°C or under sterilization conditions (120°C for 30 min) increases the NPN content from 5.5 to 7.5% of total nitrogen (Walstra and Jenness, 1984; Saidi and Warthesen, 1993; the NPN content of unheated milk is ~5% of total nitrogen and is composed of urea, orotic acid, creatinin, creatin, free amino acids, ammonia, etc). Proteolysis has an activation energy of ~70 kJ mol^{-1}, a Q_{10} of 2.3, follows zero order kinetics (see Saidi and Warthesen, 1993) and is independent of pH, from pH 6.5 to 7.2 (Guo *et al.*, 1989), whey proteins and lactose (Gaucheron *et al.*, 1999).

Hustinx *et al.* (1997) showed that the heat-induced formation of pH 4.6- and 2% TCA-soluble peptides from Na-caseinate was linear with time at pH 7.0 for the duration of heating at 140°C (100 min). Morales and Jiménéz-Pérez

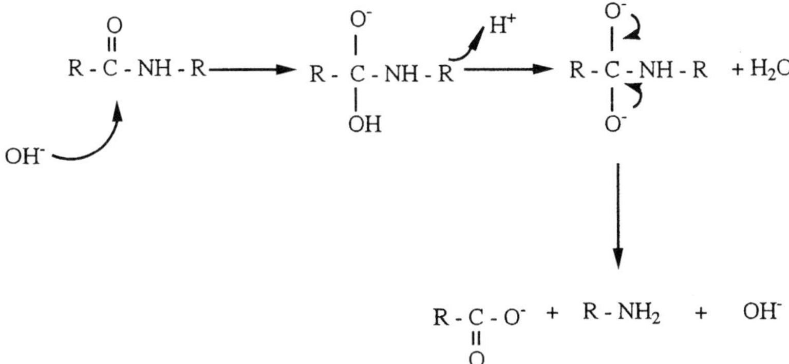

Figure 19.10 Proposed mechanism for the heat-induced hydrolysis of peptide bonds (Whitaker, 1980).

(1998), who monitored the formation of two 12% TCA-soluble peptides, showed that proteolysis followed zero-order kinetics with the formation of the two peptides having activation energies of ~55 and 90 kJ mol^{-1}.

The increase in NPN on heating milk proteins may be due to hydrolysis of peptide bonds or to the release of ammonia by deamidation of amide groups (asparagine and glutamine). Metwalli and van Boekel (1998), who distinguished between proteolysis and deamidation, showed that in caseinate systems, the increase in NPN was a first and second order reaction with respect to concentration and time, respectively, with an activation energy of 107 kJ mol^{-1}. In caseinate systems, deamidation is a first order reaction with respect to concentration but tends towards a second order reaction with respect to time and has an activation enthalpy of 92 kJ mol^{-1} (Metwalli and van Boekel, 1998). Deamidation proceeds by a β-aspartyl/glutamyl shift mechanism through a succinimide/glutarimide intermediate, to form aspartic/glutamic acid (van Boekel, 1999). It appears that deamidation of asparagine occurs more readily than glutamine, possibly because the amide group of the former is closer to adjacent amino acid residues which may catalyse deamidation (van Boekel, 1999). It is estimated that heating for 90 min at 140°C results in a 20% increase in NPN and that ammonia, the product of deamidation, accounts for 10 to 15% of total NPN (Metwailli and van Boekel, 1998).

The nature of proteolysis in heated milk remains unclear. α_s-CN (particularly α_{s2}-CN) and κ-CN are more susceptible to thermal hydrolysis than β-CN and peptide bonds containing aspartic acid residues are hydrolysed especially rapidly (Guo *et al.*, 1989; Hustinx *et al.*, 1997; Gaucheron *et al.*, 1999). Some of the peptide bonds cleaved on heating are shown in Table 19.4. Peptides similar to those produced by the action of chymosin on κ-CN have been isolated from heated milk (Alais *et al.*, 1967), although it appears that the bond(s) which is cleaved is located further towards the N-terminal than the peptide bond cleaved specifically by chymosin (Phe_{105}-Met_{106}; Kindle and Wheelock, 1970). The formation of 12% TCA-soluble sialic acid (a component of the glycomacropeptide) is linear with time up to ~ 90% hydrolysis and has a Q_{10} value of ~3 from 110 to 150°C; approximately, 20 to 30% of κ-CN is degraded at the point of coagulation, regardless of assay temperature (Fox and Hearn, 1978c), suggesting that there may be a critical level of κ-CN hydrolysis before coagulation occurs. The significance of κ-CN hydrolysis is supported by the finding that hydrolysis of ~20% of total κ-CN by chymosin causes instantaneous coagulation on heating at 140°C on the alkaline side of the minimum (Fox and Hearn, 1978c). Proteolysis of κ-CN would be expected to affect the zeta potential significantly and hydration of casein micelles. Deamidation of micellar casein may affect heat stability by altering the net protein charge and the release of ammonia may promote the Maillard reactions (Metwalli and van Boekel, 1998).

TABLE 19.4
Peptide bonds hydrolysed on heating

Casein	*Bond hydrolysed*
α_{s1}-casein	Asn_{74}–$SerP_{75}$(a)
	Asp_{175}–Ala_{176}(a,b)
	Asp_{181}–Ile_{182}(a,b)
	Asn_{184}–Pro_{185}(a,b)
	Pro_2–Lys_3(a)
	Pro_{177}–Ser_{178}(a,b)
	Phe_{23}–Phe_{24}(a)
α_{s2}-casein	Gln_{94}–Tyr_{95}(a)
β-casein	Glu_{14}–$SerP_{15}$(b)
	Leu_{16}–$SerP_{17}$(b)
κ-casein	Leu_{50}–Ile_{51}(a)
	Glu_{154}–Ser_{155}(a)
	Arg_{10}–Cys_{11}(b)
	Asn_{160}–Thr_{161}(b)
	Pro_{150}–Glu_{151}(b)

[a]Hustinx *et al.* (1997); pH 4.6-soluble peptides from Na caseinate in H_2O, heated at 140°C for 120 min, pH 7.0.
[b]Gaucheron *et al.* (1999); 12% TCA-soluble peptides from SPFCM dispersion in milk salts solution, heated at 120°C for 30 min, pH 6.7.

Heat-induced formation of amino acids does not appear to occur extensively in milk on heating at 110 to 125°C for up 40 s (Meisel and Schlimme, 1995), although a more thorough study at higher temperatures and longer holding times is warranted before the significance of free amino acids formation can be elucidated.

19.5.7 Thermal degradation of lactose

The thermal degradation of lactose has been reviewed comprehensively by O'Brien (1995); only a brief overview is provided here. When heated at an elevated temperature, lactose may be transformed and degraded *via* two pathways, i.e., isomerization/degradation reactions which result in the formation of organic acids (Figure 19.11, pathway A) or *via* the Maillard reaction (Figure 19.11, pathway B).

It is estimated that after heating at 140°C for 20 min, approximately 30% of total lactose has been degraded, of which ~80% is isomerized to lactulose through the Lobry de Bruyn-Alberda van Ekenstein reaction, which is catalyzed by the milk salts system (Andrews and Prasad, 1987) and subsequently degraded to galactose, formic acid and C_5/C_6 compounds such as deoxyribose, 3-deoxypentulose and furfurol (Berg and van Boekel, 1994;

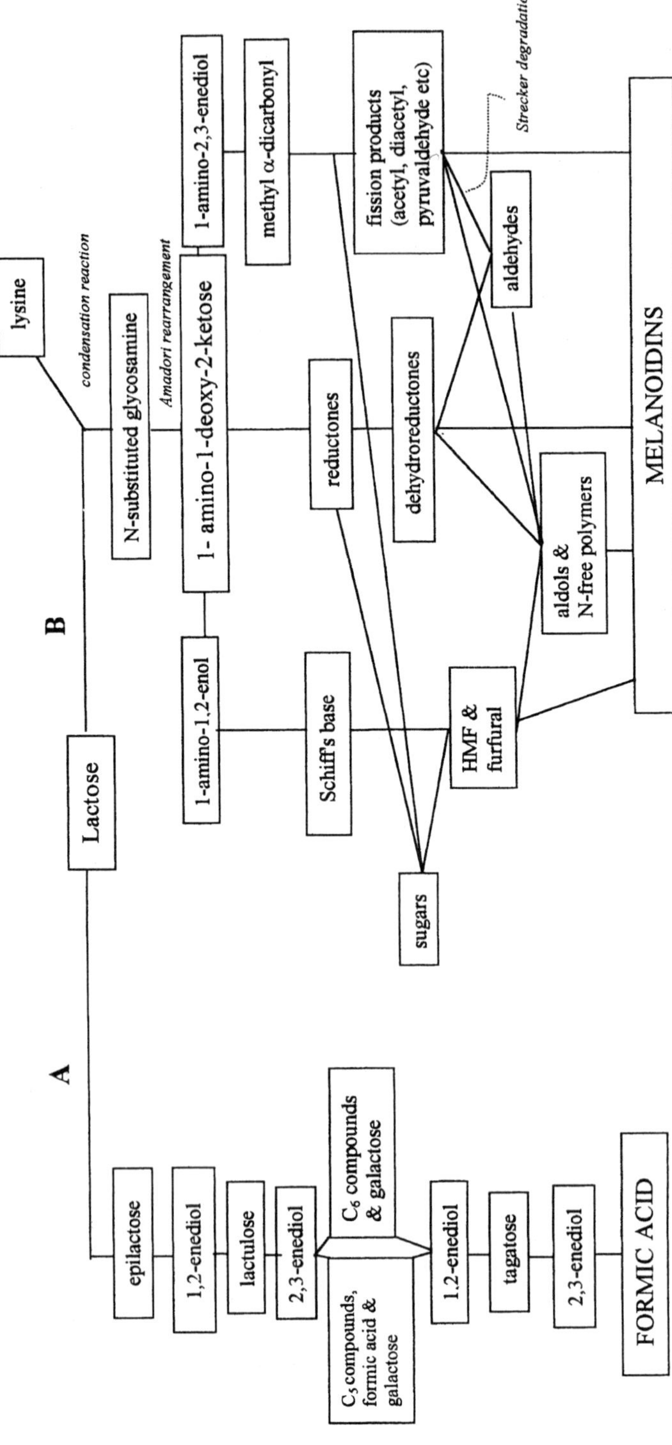

Figure 19.11 Pathways for the degradation of lactose in heated milk: A, isomerization and formation of organic acid and B, the Maillard reaction (see chemical structures below).

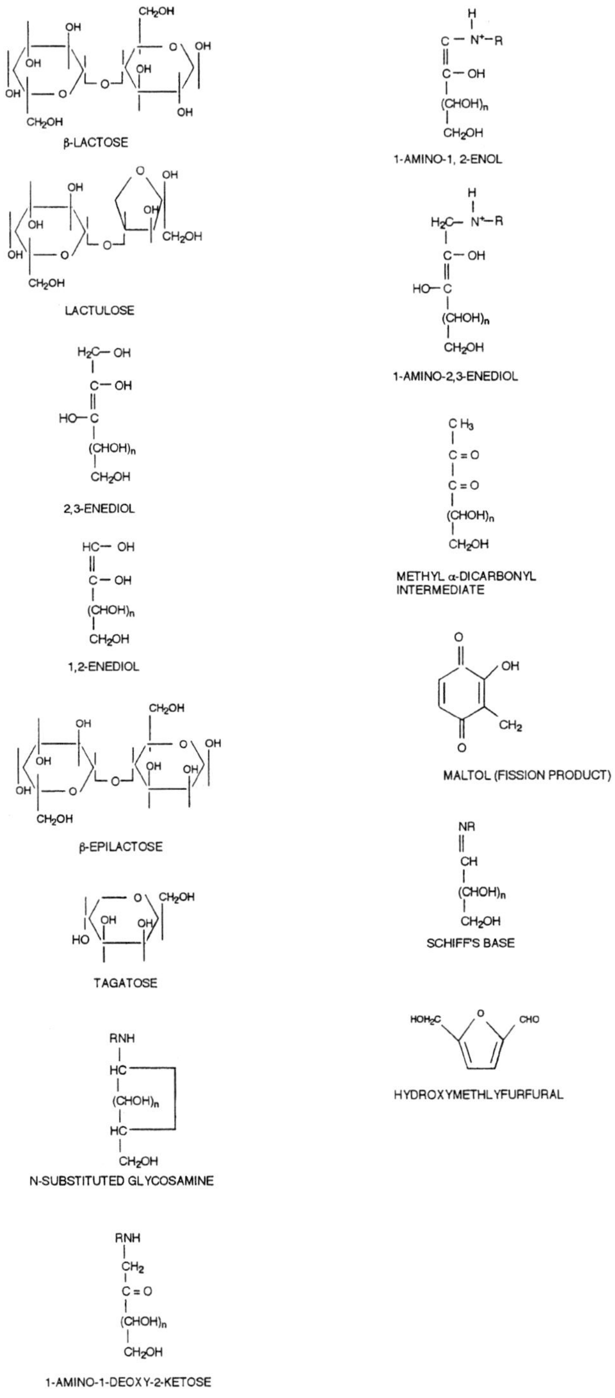

β-LACTOSE

LACTULOSE

2,3-ENEDIOL

1,2-ENEDIOL

β-EPILACTOSE

TAGATOSE

N-SUBSTITUTED GLYCOSAMINE

1-AMINO-1-DEOXY-2-KETOSE

1-AMINO-1, 2-ENOL

1-AMINO-2,3-ENEDIOL

METHYL α-DICARBONYL INTERMEDIATE

MALTOL (FISSION PRODUCT)

SCHIFF'S BASE

HYDROXYMETHLYFURFURAL

Figure 19.11 (Continued)

Figure 19.11, pathway A). The activation energy for the isomerization of lactose to lactulose has been reported to be 102 (de Rafael *et al.*, 1997) or 125 (Olano and Calvo, 1989) kJ mol^{-1} and to follow zero (de Rafael *et al.*, 1997) or first (Olano and Calvo, 1989) order reaction kinetics. The concentration of lactulose in milk heated at 120°C for 50 s and in direct UHT-treated milk is approximately 100 and 500 mg L^{-1}, respectively (Fink and Kessler, 1988; de Rafael *et al.*, 1997).

On heating at 110 to 150°C for 20 min, approximately 20% of the lactose, which has been thermally degraded, reacts with proteins, principally with the casein micelles, through the Maillard reaction (Figure 19.11, pathway B; Turner *et al.*, 1978; Berg and van Boekel, 1994). The Maillard reaction involves a condensation reaction between proteins (predominantly the ε-amino group of lysine but also to a lesser extent the indolyl group of tryptophan, the imidazole group of histidine, the guanidino group of arginine and N-terminal α-amino group of proteins) and sugars (in an open chain configuration) to form a glycosylamine. This is followed by an Amadori rearrangement, resulting in the formation of a N-substituted-1-amino-1-deoxy-2-ketose. The N-substituted-1-amino-1-deoxy-2-ketose may then enolize to either 1-amino-1,2-enol and 1-amino-2,3-enediol or dehydrate to form reductones. 1-Amino-1,2-enol is dehydrated to hydroxymethylfurfural and furfural, while 1-amino-2,3-enediol is dealdolized to an α-dicarbonyl intermediate. The α-dicarbonyl intermediate may then be broken down to an array of fission compounds, such as 2-oxo-propanal (also referred to as methylglyoxal or pyruvaldehyde), acetol, diacetyl, hydroxydiacetyl, maltol and C-methyl reductones. Fission products polymerize to form melanoidins or alternatively react, *via* the Strecker degradation pathway, with amine groups to form aldehydes, CO_2 and pyrazines.

Compounds produced through the dehydration, dealdolization or Strecker degradation pathways then interact and polymerize to form melanoidins through an ill-defined mechanism and the products formed have not been isolated or identified (Nursten, 1981; Walstra and Jenness, 1984).

The Maillard reaction in milk follows zero order kinetics, as determined by furosine formation and fluorescence (associated with products of the Maillard reaction); these two reactions have an activation energy of 104 and 84 kJ mol^{-1}, respectively (Morales *et al.*, 1996; de Rafael *et al.*, 1997). The production of hydroxymethyl furfural (HMF) and the loss of reactive lysine residues are also used to quantify the extent of the Maillard reaction in heated milk; Morales *et al.* (1995) reported that the activation energy and Q_{10} for these reactions are 93.04–118.5 and 66.67 kJ mol^{-1} and 2.55 and 2.0, respectively, in the temperature range 35 to 160°C. Kessler and Fink (1986) reported that the formation of HMF and the loss of lysine have an activation energy of 109 and 139 kJ mol^{-1} and follow zero and first order reaction kinetics, respectively, from 20 to 160°C.

The significance of the heat-induced degradation of lactose to the heat stability of milk is obscured by the complexity of its breakdown pathways. The possible significance of organic acid formation (pathway A) and the Maillard reaction (pathway B) in terms of heat stability are discussed in Sections 19.5.9 and 19.5.8, respectively. In addition to changes in acidity and the Maillard reaction, lactose, when heated with casein, reduces the calcium binding capacity of caseins, which would be expected to affect stability (Pappas and Rothwell, 1991; Pappas, 1992). Tan-Kintia (1996) proposed that reductones, which are products of the Maillard reaction, promote heat stability. It is suggested that urea acts synergistically with reducing sugars to stabilize milk proteins against coagulation; on heating, urea is broken down to ammonia and CO_2. The formation of the CO_2 inhibits the degradation of reductones to organic acids, thereby enhancing stability by inhibiting heat-induced acidification and also by the fact that the degradation of reductones, which are known to enhance the heat stability, is inhibited (Tan-Kintia, 1996).

19.5.8 Heat-induced crosslinking

When heated at a high temperature, nucleophilic amino acid residues of milk proteins (e.g., lysine and cysteine) become very reactive and, in the presence of lactose (which is reactive due to its carbonyl group), can form inter- and intra-molecular bonds. The nature, factors affecting and significance of these covalent bonds have yet to be elucidated.

Heat-induced polymerization has been reported in SPFCM dispersions heated at temperatures as low as 80°C (Zin El-Din and Aoki, 1993), UHT-treated milk (Andrews 1975; Zin El-Din and Aoki, 1993) and Na-caseinate solutions heated at 140°C (Lorient, 1979; Fox, 1981b; Guo *et al.*, 1989).

Polymerization may result from many types of crosslinks, including disulphide bonds (Section 19.5.10), crosslinks arising from the Maillard reactions (e.g., pentosidine and lysylpyraline, Pellegrino *et al.*, 1999; Section 19.5.7), dehydroalanine-derived crosslinks and the formation of isopeptide bonds (see van Boekel *et al.*, 1989a) or through the formation of thermally-generated CCP links (Singh, 1994). Dehydroalanine, which is formed on the β-elimination of cysteine, phosphoproteins or glycoproteins, is extremely reactive and interacts readily with nucleophilic amino acid residues such as lysine, cysteine, arginine, ornthinine and histidine to crosslink proteins (Figure 19.12, Manson and Carolan, 1980; Henje *et al.*, 1993; see Creamer and Singh, 1995). Isopeptide bonds are formed by amidation, i.e., the reaction of the ε-amino group of lysine with the carboxyl group of either glutamic acid or aspartic acid (Figure 19.13, reaction 1) or alternatively by transamidation; interaction of lysine with the amide bond of asparagine or glutamine (Figure 19.13, reaction 2) and have been reported in heated milk protein systems (Otterburn *et al.*, 1977).

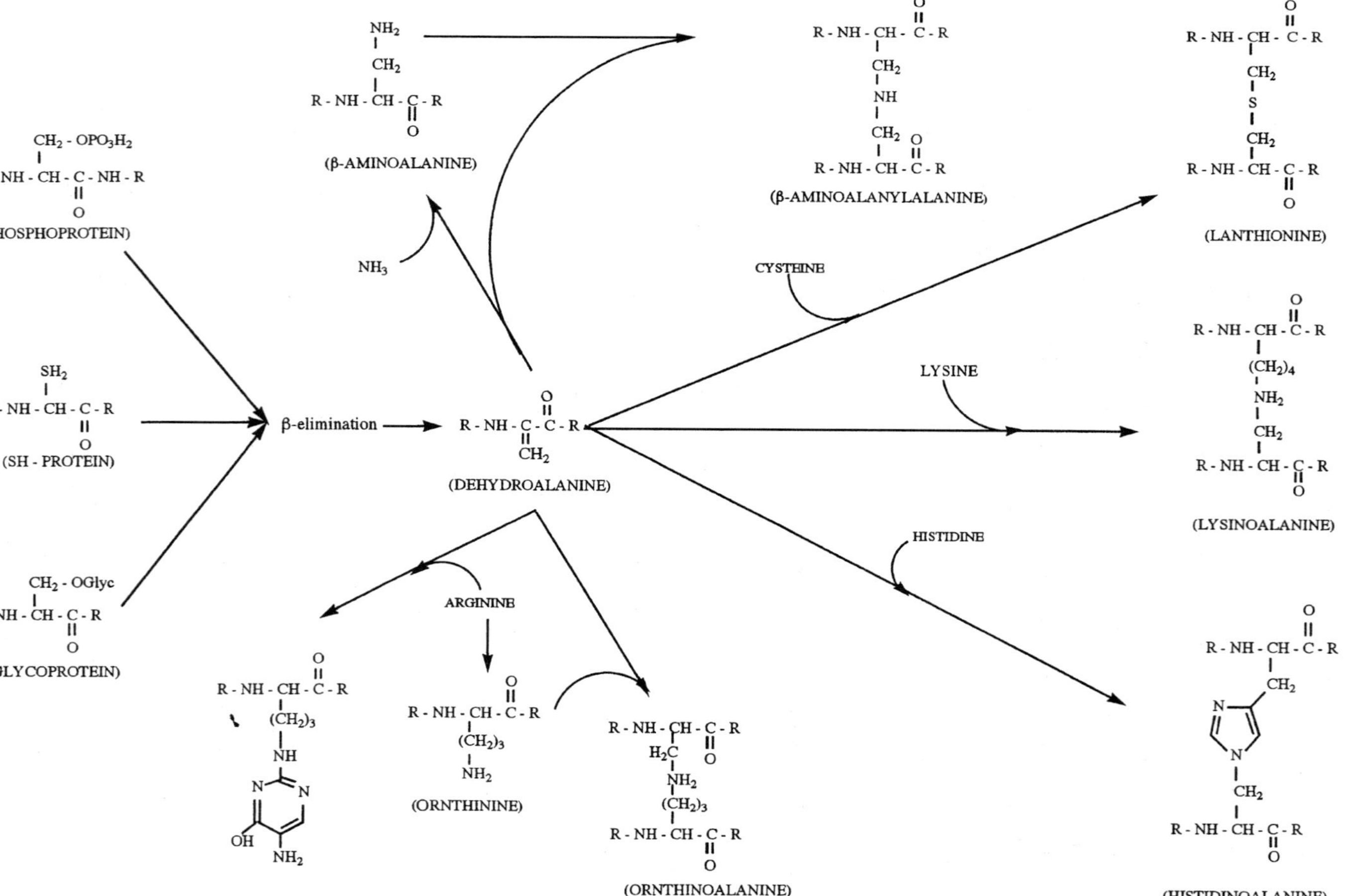

Figure 19.12 Heat-induced formation of dehydroalanine and its subsequent interaction with nucleophilic amino acids and ammonia.

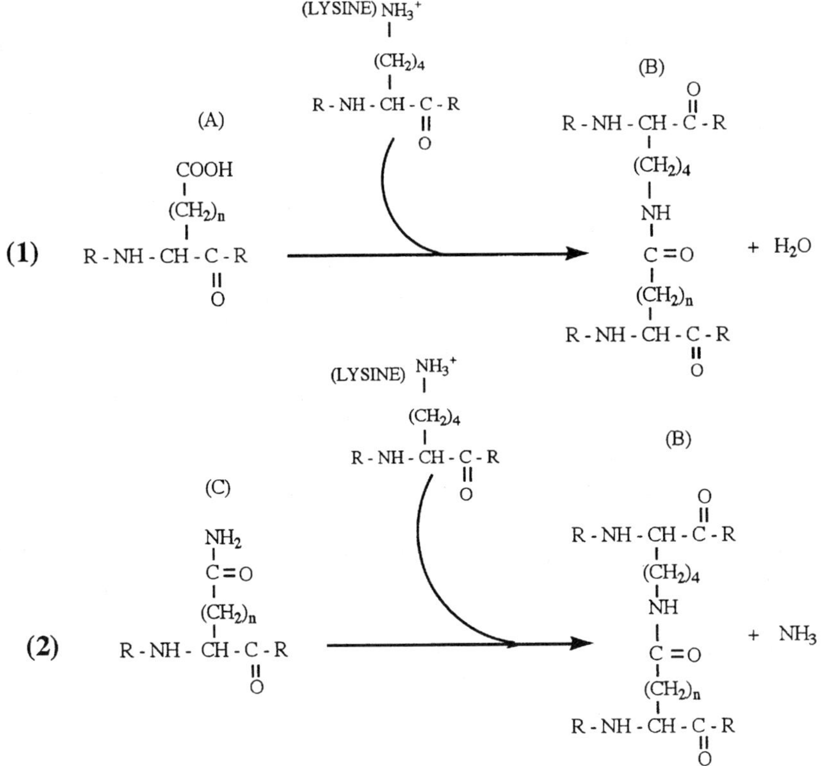

Figure 19.13 Possible reaction pathways for the formation of isopeptide bonds A: n = 2, aspartic acid; n = 3, glutamic acid, B: n = 2, ε-N-(β-aspartyl)-lysine; n = 3, ε-N-(γ-glutamyl)-lysine,C: n = 2, asparagine; n = 3, glutamine.

It remains unclear whether these crosslinks are formed intra- or intermolecularly and if the latter, whether the crosslinks are inter- or intramicellar. van Boekel *et al.* (1989a) suggested that polymerization is more likely to occur within micelles, owing to the fact that the protein molecules are closer together. Supporting this conclusion, Zin El-Din and Aoki (1993) showed that polymerization was reduced if the micelles were disaggregated into submicelles prior to heating.

Polymerization is affected by a number of factors, temperature being the most important. The percentage of total casein polymerized in whey protein-free milk after 75 s at 80 or 140°C was approximately 2.5 and 24%, respectively (Zin El-Din and Aoki, 1993). van Boekel *et al.* (1989a) reported that heat-induced polymerization shows an initial lag period, but Zin El-Din and Aoki (1993) proposed that polymerization progresses linearly with time. Covalent crosslinking of proteins appears to occur most rapidly at lower pH (pH 6.7 to 7.2; Dalgleish *et al.*, 1987b), possibly due to a reduction in electrostatic repulsion and a concomitant increase in protein-protein interactions (van Boekel *et al.*, 1989a). Other factors known to affect covalent polymerization include lactose (Zin El-Din and Aoki, 1993) and urea (Dalgleish *et al.*, 1987b).

Covalent polymerization has been implicated in the loss of quality of rennet casein (Ennis *et al.*, 2000) and in the age gelation of UHT-treated milk (Andrews, 1975). Although its significance in heat-induced coagulation has yet to be determined, it has been postulated that polymerization is ultimately responsible for coagulation outwith the region of the minimum in the HCT-pH profile (Darling, 1980; Walstra and Jenness, 1984; van Boekel *et al.*, 1989a); this proposal was based on the fact that coagula formed outwith the region of the minimum are not dispersible in solutions containing urea and/or oxalate/EDTA (Dalgleish *et al.*, 1987b; van Boekel *et al.*, 1989a). However, it is quite possible that polymerization is merely a concurrent reaction to the reaction(s) responsible for coagulation.

The fact that the size of casein particles does not increase after initial heating up to the point of coagulation, while turbidity increases linearly with time (Dalgleish *et al.*, 1987b), suggests that heat-induced polymerization is predominantly intramolecular or intramicellar while intermicellar polymerization at the onset of coagulation may be a consequence of, rather than a cause of, coagulation.

19.5.9 Heat-induced acidification

Heat-induced acidification plays a key role in creating an environment that favours the coagulation of milk proteins; this is illustrated by the fact that coagulation at 140°C may be prolonged for at least 3 h if the pH is readjusted occasionally to its original value (Pyne, 1958; Fox, 1981a). Pyne (1958) showed that the pH of milk decreased from about pH 6.7 to 5.5–6.0 on heating at 140°C to the point of coagulation. Fox (1981a) estimated, from pH reversal curves, that the pH of milk at the assay temperature (140°C) at the point of coagulation is ~4.9. Dalgleish *et al.* (1987b) reported that the pH of milk at the point of coagulation was the same regardless of the initial pH, but van Boekel *et al.* (1989a) suggested that the pH at coagulation depends on the initial pH.

It has been claimed (Darling, 1980) that the rate of acidification may be pH dependent, decreasing as a function of the initial pH up to the pH of maximum stability and within the pH range of the minimum there is a sharp increase in the rate of acidification. However, van Boekel *et al.* (1989a), who did not support this view, maintained that differences in the decrease in pH as a function of coagulation time were due to the rate of acidification being non-linear, with a rapid decrease in pH during the first 2 min, after which the rate of acidification is reduced; the rate of hydrogen ion formation on heating at 140°C and pH 6.3 was ~5.0 mmol H^+ min^{-1} initially and decreased to ~1.75 and ~1.0 mmol H^+ min^{-1} after heating for 5 and 10 min, respectively (van Boekel *et al.*, 1989a).

The decrease in pH on heating is caused by a number of reactions. The thermal oxidation of lactose to organic acids is thought to account for up to 50% of heat-induced acidification. Formic acid, which accounts for up to 80% of total acids formed, is formed at a rate of 0.4 mol m^{-3} min^{-1} at 140°C in skim milk, resulting in a decrease of 0.03 pH units min^{-1}; its formation has a Q_{10} of 1.8 (van Boekel *et al.*, 1989a). Other organic acids formed on heating include acetic, lactic, pyruvic and propionic acids (Morr *et al.*, 1957). The second reaction responsible for acidification is the precipitation of primary and secondary calcium phosphate as tertiary phosphate, with the concomitant release of H^+. This reaction is considered to contribute ~30% of total acidity on heating and occurs within the first 5 min of heating at 140°C (Rose and Tessier, 1959; Nieuwenhuijse *et al.*, 1988; Tan-Kintia, 1996). Dephosphorylation of casein and the subsequent precipitation of phosphate as tertiary calcium phosphate is also considered to contribute to acidification though its relative importance remains unclear. However, it has been established that increasing the protein concentration results in a concomitant increase in acid production on heating (Sweetsur and White, 1975). If proteins were not a source of heat-induced acidity (through dephosphorylation), then it would be expected that the heat-induced decrease in pH would be reduced with increasing protein concentration due to the inherent buffering capacity of proteins. Other reactions that may contribute to acidification include the Maillard reaction (Sweetsur and White, 1975) and proton dissociation from a buried carboxyl group in β-Lg on the N → R conformational change (de Wit, 1981). van Boekel *et al.* (1989a) concluded that denaturation of whey proteins does not contribute to heat-induced acidification as there was no difference in the rate of acid production between skim milk and whey protein-free milk.

Although heat-induced acidification plays a critical role in coagulation, the true extent of its contribution has not been elucidated. The pH of milk at the point of coagulation (pH ~4.9) approaches that at which acid coagulation occurs (pH ~4.6) but heat-induced coagulation is not due merely to an indirect acid coagulation mechanism. This is illustrated by the fact that the Q_{10} for thermal acidification and coagulation differ, being ~ 2 and 3, respectively; also, the pH at coagulation is not constant and the coagula formed on heating are not redispersible if the pH is increased (van Boekel *et al.*, 1989a). Heat-induced acidification, by reducing the pH, decreases electrostatic replusion which is likely to promote coagulation.

19.5.10 Heat-induced whey protein–casein interactions

β-Lg is denatured on heating at temperatures > 70°C. Thermal denaturation of β-Lg has been reported to follow pseudo-first order kinetics (Dalgleish, 1990) and is affected by Ca^{2+}, lactose, casein and whey protein concentration, and pH (de Wit, 1981; Relkin, 1996; Law and Leaver, 1999).

The ability of thermally denatured β-Lg to interact with casein on heating has been demonstrated (Tobias *et al.*, 1952; Slatter and van Winkle, 1952); the rate at which it occurs depends on a number of factors such as temperature (Smits and Brouwerschaven, 1980; Noh and Richardson, 1989), β-Lg to casein ratio, genetic polymorphism (Allmere *et al.*, 1998), κ-CN glycosylation (Dziuba, 1979; Doi *et al.*, 1983) and pH. Interaction at 70°C is slow but increases rapidly at higher temperatures; ~75 and ~95% of total β-Lg complexes with casein micelles when heated at 90 or 120°C for 20 min, respectively (Smits and van Brouwershaven, 1980). Interaction decreases as the pH is increased from 6.8 to 7.3 (Smits and Brouwershaven, 1980). Ionic strength also affects the interaction: when an aqueous solution of β-Lg and κ-CN was heated at 100°C for 30 min, only 14% of the β-Lg and 17% of the κ-CN complexed but when heated in synthetic milk ultrafiltrate under the same conditions, 77% of β-Lg and 83% of κ-CN complexed (El Negoumy, 1974).

The molecular size and stiochiometry of β-Lg complexes have not been determined in milk, although Long *et al.* (1963) reported that, in a model system, the extent and nature of complex formation is dependent on the β-Lg:κ-CN ratio. Haque *et al.* (1987) reported when present at equimolar concentrations, 3 moles of β-Lg interact with 1 mole of κ-CN when heated at 70°C [both κ-CN and β-Lg are present at 180 mol m^{-3} in milk; Walstra and Jenness, 1984].

Other milk proteins, apart from κ-CN and β-Lg, are also capable of complex formation on heating. Although it appears that α-La does not interact directly with κ-CN (Smits and Brouwershaven, 1980; Noh *et al.*, 1989 a,b), when in the presence of β-Lg, complexes containing all 3 proteins are formed on heating (Table 19.5; Baer *et al.*, 1976; Snoeren and van der Speek, 1977; Creamer, 1984; Noh and Richardson, 1989; Singh and Creamer, 1991b; Singh and Lantham, 1993; Corredig and Dalgleish, 1999). Corredig and Dalgleish (1999) suggested that β-Lg and α-La interact in the serum phase and then complex with κ-CN while Fairise *et al.* (1999) proposed that α-La interacts with β-Lg-κ-CN complexes at the micelle surface. Of the other caseins, α_{s2}-CN has been shown to form complexes with κ-CN (Snoeren and van der Speek, 1977).

At low temperatures, ~70°C, hydrophobic bonding is principally responsible for complex formation between κ-CN and β-Lg (Haque and Kinsella, 1988; Jang and Swaisgood, 1990), but at higher temperatures, it appears that sulphydryl-disulphide interchange reactions are involved

TABLE 19.5
Proportions of β-lactoglobulin (β-Lg) and α-lactalbumin (α-La) complexed with casein micelles after various heat treatment

Heating conditions	*% β-Lg*	*% α-La*
70°C × 45 min	2	0.3
95°C × 0.5 min	58	8
95°C × 20 min	85	55
140°C × 2 sec	43	9
140°C × 4 sec	54	12

Measurements were made on skimmed milk.

(Trautman and Swanson, 1958; Sawyer *et al.*, 1963; Jang and Swaisgood, 1990). This is supported by the fact that copper and NEM prevent complexation and that β-mercaptoethanol disrupts complexes (Sawyer *et al.*, 1963; Jang and Swaisgood, 1990). However, Sawyer (1969), Dziuba (1979) and Wong *et al.* (1996) expressed reservations concerning complex formation *via* sulphydryl-disulphide interchange reactions and suggested that non-specific association between aggregates of β-Lg and κ-CN may be responsible for complex formation; this view supports the results of Hartman (1967) who showed that NEM failed to prevent complex formation between denatured β-Lg and κ-CN and of Long *et al.* (1963) who showed that denatured β-Lg was capable of complex formation with κ-CN at 20°C. It has been proposed that on heating at 70°C, β-Lg forms a trimer *via* hydrophobic bonding which then interacts with κ-CN, initially through hydrophobic bonding but eventually through sulphydryl-disulphide interchange reaction (Haque *et al.*, 1987).

Once the β-Lg-κ-CN complex has formed, its distribution between the serum and colloidal phases appears to be pH-dependent. Using electron microscopy, Creamer *et al.* (1978), Mohammad and Fox (1987b) and Singh *et al.* (1996) showed that small thread-like appendages are formed on heating milk; these appendages remain on the surface of the micelles when heated at pH 6.7, but when heated at pH 7.3 they are in the serum phase and the micelle surface appears eroded. Creamer *et al.* (1978) isolated protein complexes from heated milk by gel filtration chromatography. These complexes had a similar appearance to the appendages located on the micelle surface when milk is heated at pH 6.7 when examined by electron microscopy and had an amino acid composition intermediate between κ-CN and the whey proteins; electrophoretic analysis showed the complexes were comprised of β-Lg and κ-CN. This is consistent with the effect of whey proteins on the dissociation of protein from the casein micelle on heating at ~pH 7.3 (Section 19.5.4).

19.6 pH DEPENDENCE AND MECHANISM OF HEAT-INDUCED COAGULATION

19.6.1 Introduction

The pH-dependence of and mechanism for the heat-induced coagulation of milk is intriguing and has been intensely researched for the last 40 years. An integrated theory for heat-induced coagulation over the pH range 6.3 to 7.4 has been developed and largely accepted (van Boekel *et al.*, 1989a,b; Singh and Creamer, 1992).

19.6.2 Current hypothesis on the heat stability of milk

The theory commonly accepted for the pH-dependence and coagulation mechanism, throughout the pH range 6.3 to 7.4, splits the HCT-pH profile into 4 discrete regions, as shown in Figure 19.14. In region 1, milk, presumably due to the low pH and high calcium activity (the calcium activity of skimmed milk at pH 6.25 and 7.25 is 1.386 and 0.511 mM, respectively; Geerts *et al.*, 1983), is inherently unstable and coagulates rapidly on heating at an elevated temperature. The inherent instability of milk within this region is highlighted by a HCTemp-pH profile which shows that there is a sigmoidal relationship between HCTemp and pH from pH 5.7

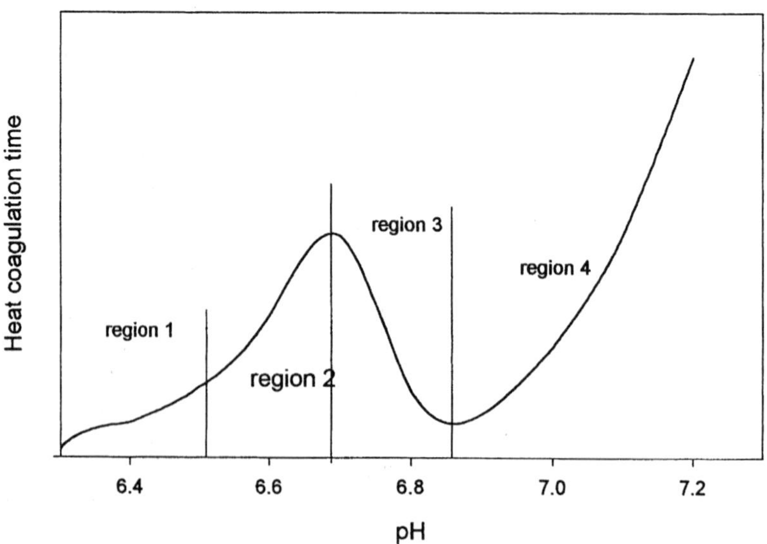

Figure 19.14 Heat coagulation time-pH profile of unconcentrated type A milk.

to 6.7 (see Figure 19.3). Coagulation within this region may be due to calcium-induced destabilization (see Horne and Muir, 1990).

In region 2, stability increases as the pH is increased. As described earlier in "Whey proteins", addition of β-Lg to a SPFCM dispersion introduces a maximum and minimum in the HCT-pH profile. It is postulated that β-Lg (and possibly other whey proteins also) complex with κ-CN *via* sulphydryl-disulphide interchange reactions in region 2 (Section 19.5.10) and increases stability by inhibiting heat-induced dissociation of micellar κ-CN and by increasing zeta potential, hydration and steric hindrance (Section 19.5.5) (Singh and Fox, 1987b). It should be noted that the precipitation of β-Lg and its non-specific interaction with casein has also been suggested (de Wit, 1981; van Boekel *et al.*, 1989b; Wong *et al.*, 1996). In region 2, milk is very stable and can withstand heating at 140°C for up to 25 min; the coagulation mechanism within this region has been suggested to involve covalent polymerization of proteins (van Boekel *et al.*, 1989a).

In region 3, which contains the minimum in the HCT-pH profile, milk proteins exhibit relatively low heat stability. In this region, β-Lg, through the formation of complexes with κ-CN [Section 19.5.10], promotes dissociation of κ-CN from the casein micelles (Section 19.5.4), making them inherently unstable. Singh and Fox (1986) isolated casein micelle heated within the region of the minimum and showed that they were extremely unstable to calcium, ethanol and enzymatic coagulation. It is postulated that the coagulation mechanism within the region of the minimum is due to calcium-induced precipitation of κ-CN-depleted casein micelles (Singh and Fox, 1986, 1987b). The decrease of zeta potential and hydration (see "Zeta Potential and Hydration") on heating would also be expected to predispose micelles to premature coagulation. van Boekel *et al.* (1989a) suggested that within the region of the minimum, β-Lg does not interact with micellar κ-CN, but remains in the serum phase and interacts with the dissociated κ-CN.

In region 4, i.e., the pH range on the alkaline side of the minimum, stability increases as a function of pH. Increased stability in this region may be due to a decrease in calcium ion activity and increases in hydration and zeta potential at higher pH values (Singh, 1995). van Boekel *et al.* (1989a) proposed that increased stability on the alkaline side of the minimum is due partly to reassociation of κ-CN with the micelles. Net reassociation, i.e., when reassociation occurs to a greater extent than dissociation, of protein onto the casein micelles was suggested by Kudo (1980c) and Dalgleish *et al.* (1987a) to occur after 10 min at 140°C or 20 min at 130°C, while van Boekel *et al.* (1989a) proposed that dissociation climaxes after ~2 min and thereafter, reassociation occurs, particularly after the pH has decreased below pH 6.6. The reversible nature of the dissociation reaction appears to be supported by the fact that milk pre-heated at the pH of minimum stability for up to 80% of its coagulation time exhibited a normal type A HCT-pH profile following cooling and readjustment of pH (see Fox, 1982).

The coagulation mechanism in region 4 may involve covalent polymerization (van Boekel *et al.*, 1989a).

The reason why the κ-CN-β-Lg complex remains at the micelle surface at pH values <pH 6.7 but dissociates at higher pH values remains unclear. de Wit (1981) showed that when heated at pH values <6.8, β-Lg precipitates but remains in solution when heated at a higher pH value. This is possibly due to N → R conformational changes in β-Lg at pH 6.8. Conformational changes at ~pH 6.8 may also affect the reactivity of the thiol group of β-Lg and consequently the extent of β-Lg-κ-CN complex formation, through sulphydryl-disulphide interchange reactions.

Problems encountered when trying to apply the above theory to the heat-induced coagulation include the fact that if the dissociation of κ-CN is so critically important in explaining the heat stability of milk and pH-dependence of the HCT-pH profile, why does urea, which promotes dissociation of κ-CN from the micelle (Dalgleish *et al.*, 1987b; Metwalli *et al.*, 1996) stabilize milk? The fact that extensive dissociation κ-CN occurs at low temperatures (~80°C) (Anema and Klostermeyer, 1997a) while milk can withstand heating at this temperature for up to 15 h without coagulating (Tan-Kintia, 1996) would also appear to mitigate against the current hypothesis. This hypothesis is also complicated by the fact the maximum stability of milk occurs at pH 6.7 while dissociation of κ-CN becomes significant only at pH values >6.8.

An alternative explanation for the pH dependence of the heat coagulation of unconcentrated milk has been presented by Horne and Muir (1990), who proposed that urea and β-Lg compete for interaction with the -S-S- groups of κ-CN. Outwith the pH of the minimum, urea-κ-CN interactions are favoured, while within the pH of the minimum, β-Lg is more reactive, possibly due to exposure of a buried thiol group at pH values >6.8 on heating (de Wit, 1981; Kella and Kinsella, 1988) and consequently the formation of the β-Lg-κ-CN complex predominates. It is postulated that the formation of β-Lg-κ-CN complexes on the surface of the micelles destabilizes milk by crosslinking proteins and making the protruding κ-CN hairs on the surface of the micelle more rigid, as a result of which the casein micelles are more susceptible to calcium-induced precipitation (Horne and Muir, 1990). This hypothesis would explain why low levels of urea do not affect minimum stability whereas high levels do, since at the higher levels, urea may out-compete β-Lg for interaction with κ-CN within the region of the minimum (Horne and Muir, 1990).

However, this hypothesis would appear unable to explain the pH-dependence of the heat stability of milk as it fails to explain a number of phenomena:

- whey proteins stabilize SPFCM dispersions at pH 6.7 (Rose, 1961 b,c)
- urea stabilizes SPFCM dispersions (Metwalli and van Boekel, 1996)

- degradation of indigenous urea in a type A milk does not alter the pH-dependence of heat stability (Muir and Sweetsur, 1977)
- cyanate (a thermal breakdown product of urea) interacts only with a fraction of the –SH groups in milk (Manson *et al.*, 1985)
- inhibition of sulphydryl interactions does not eliminate the stabilizing effect of urea (Muir and Sweetsur, 1977).

19.6.3 New hypothesis for the mechanism of heat-induced coagulation of milk

A third hypothesis on the pH-dependence and coagulation mechanism of milk is introduced here. It is proposed that the pH dependence of the thermal denaturation of β-Lg plays a critical role. From pH ~6.3 to the pH of maximum stability, β-Lg, as a dimer with the thiol group inaccessible and unreactive to serum phase constituents, denatures through a calcium-mediated mechanism and enhances the thermal stability of the casein micelles by chelating calcium (Zitue *et al.*, 1957). At pH values >6.9, the disulphide bonds of β-Lg and κ-CN are reduced on heating, which facilitates complex formation. Concurrently, the hydrophobic β-barrel of β-Lg is exposed which results in an increase in the surface hydrophobicity of the β-Lg-casein micelle complexes, thereby sensitizing the casein micelles to the destabilizing effect of heat-induced calcium phosphate precipitation. At pH values on the alkaline side of the minimum, stability increases as a function of pH, possibly because of an increase in protein stability with increasing pH (micellar zeta potential and hydration increase as a function of pH) and a decrease in calcium activity with pH. It is also proposed that the extent of heat-induced precipitation of calcium phosphate, which is affected by the initial concentration of calcium phosphate in the serum, and the rate of precipitation is markedly pH dependent. Extensive precipitation of calcium phosphate occurs at pH values <6.5, due to the high level of serum-phase calcium phosphate, and at the pH of minimum stability, due to the relatively high serum-phase calcium phosphate content and rate of precipitation. However, at the pH of maximum stability and at pH values on the alkaline side of the minimum, the extent of calcium phosphate precipitation is low, due to a low serum-phase calcium phosphate content and a low rate of precipitation, respectively (Figure 19.15).

19.6.4 Heat coagulation of concentrated milk

Concentrated milk is inherently less stable than unconcentrated milk and will coagulate within 25 min at 120°C (unconcentrated milk can tolerate up to 70 min at 120°C without coagulating). Because of the lower assay temperature

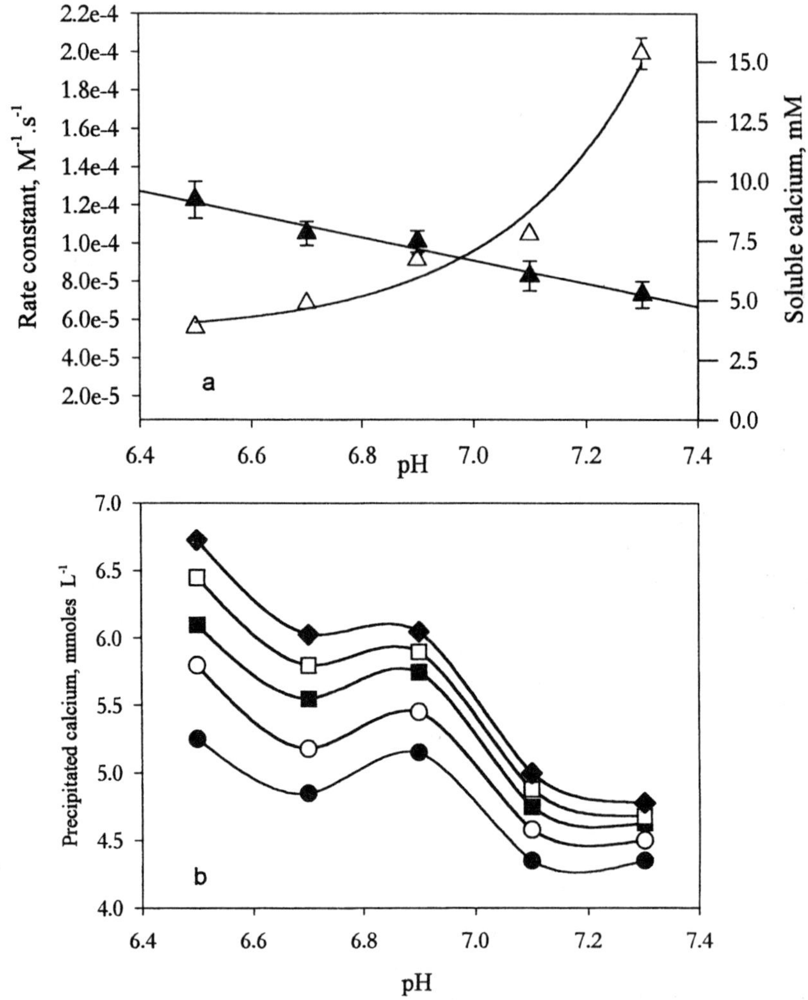

Figure 19.15 Effect of pH on (A) the serum phase calcium content of milk (▲) and the rate constant for heat-induced precipitation of calcium (as calcium phosphate) in milk ultrafiltrate at 140°C (△); (B) amount of calcium precipitated after heating at 140°C for 4 (●), 5 (○), 6 (■), 7 (□), or 8 (◆) min, as calculated from the rate equation and the data in (A).

and shorter coagulation times, many of the changes (e.g., covalent polymerization, dephosphorylation, proteolysis), that occur in unconcentrated milk either occur at reduced rate or do not occur at all in concentrated milk.

It has been suggested that the coagulation mechanism of concentrated milk is in many respects similar to the coagulation mechanism of unconcentrated milk in the region of the minimum (Table 19.6). Considering these

Table 19.6
Apparent similarities between coagulation mechanisms for concentrated milk and unconcentrated milk at pH 6.7 and 6.9

Conditions	*Unconcentrated milk at pH 6.7*	*Unconcentrated milk at pH 6.9*	*Concentrated*	*Reference*
Effect of β-Lg additon to SPFCM dispersion or concentrated milk	Stabilizes	Destabilizes	Destabilizes	Rose (1961b,c), Newstead *et al.* (1977)
Effect of urea addition to ∼6 mM	Stabilizes	No effect	No effect	Muir and Sweetsur (1977)
Dispersibility of coagulation 6 M urea buffer	No	Yes	Yes	Delgleish *et al.* (1987b), van Boekel *et al.* (1989a)
One stage or two stage coagulation process	One	Two	Two	Sweetsur and White (1974) Muir and Sweetsur (1978)
κ-CN dissociation	Limited	Extensive	Extensive	Kudo (1980c), Singh and Creamer (1991a,b)

similarities, an integrated model for the coagulation process of concentrated milk throughout the pH range 6.2 to 7.1 has been proposed (Singh and Creamer, 1992; Singh *et al.*, 1995). At pH values <6.4, heating causes denaturation and attachment of whey proteins to the casein micelles, resulting in coagulation through the formation of a gel network. Ionic interactions are likely to play a key role in gelation/coagulation within this pH range. The coagulation of concentrated milk from pH 6.4 to 6.7 occurs in a manner analogous to coagulation of unconcentrated milk within the region of the minimum. Heat-induced dissociation of κ-CN from the casein micelles occurs and leads to calcium-induced precipitation of the κ-CN-depleted micelles. At pH values greater than pH 6.7, heating causes extensive dissociation of casein micelles and coagulation proceeds rapidly by a 'gelation-like' mechanism (Nieuwenhuijse *et al.*, 1991; Singh *et al.*, 1995).

19.7 Suggestions for further work

Albert Einstein is attributed to have said 'We live in a world of perfect means and confused objectives'. It is suggested that the converse is true for the dairy scientist attempting to elucidate the pH-dependence and mechanism of coagulation of milk on heating, as many of the factors which affect the stability of milk proteins, e.g., pH, zeta potential, hydration, redox-potential and colloidal characteristics, cannot be measured at assay temperature. Three possible models for the coagulation mechanism and pH dependence of milk on heating have been presented here, although it must be appreciated that, at best, any model is a simplification of what actually occurs. In order to improve our understanding of the heat-induced changes in milk, an in-depth and thoroughly integrated kinetic study is needed. The nature and significance of covalent bond formation also warrants further research, as does the pH-dependence and denaturation mechanism of β-Lg in milk. The application of nuclear magnetic resonance (to study changes in the secondary structure of caseins), electro-acoustics (to study the colloidal properties of micelles) and confocal scanning microscopy (to study the surface hydrophobicity of micelles as affected by κ-CN dissociation) may also yield valuable information.

References

Alais, C., Kiger, N. and Jollès, P. (1967) Action of heat on cow κ-casein. Heat caseino-glycopeptide. *J. Dairy Sci.*, **50**, 1738–43.

Allmere, T., Andrén, A., Lundén, A. and Björck, L. (1998) Interaction in heated skim milk between genetic variants of β-lactoglobulin and κ-casein. *J. Agric. Food Chem.*, **46**, 3004–8.

Anderson, L. and Kelley, J.J. (1959) The dephosphorylation of casein by alkalies. *J. Am. Chem. Soc.*, **81**, 2275–6.

Andersson, I. and Öste, R. (1995a) Sensory quality of UHT milk, in *Heat-induced Changes in Milk*, (P.F. Fox ed.) Special Issue 9501, International Dairy Federation, Brussels, pp. 279–301.

Andersson, I. and Öste, R. (1995b) Nutritional quality of heat processed liquid milk, in *Heat-induced Changes in Milk*, (P.F. Fox ed.) Special Issue 9501, International Dairy Federation, Brussels, pp. 318–28.

Andrews, A.T. (1975) Properties of aseptically packed ultra-high-temperature milk. III. Formation of polymerized protein during storage at various temperatures. *J. Dairy Res.*, **42**, 89–99.

Andrews, G.R. and Prasad, S.K. (1987) Effect of protein, citrate and phosphate content of milk on formation of lactulose during heat treatment. *J. Dairy Res.*, **54**, 207–18.

Anema, S.G. (1998) Effect of milk concentration on heat-induced, pH-dependent dissociation of casein from casein micelles in reconstituted skim milk at temperatures between 20 and 120°C. *J. Agric. Food Chem.*, **46**, 2299–305.

Anema, S.G. and Klostermeyer, H. (1996) ζ-Potential of casein micelles from reconstituted skim milk heated at 120°C. *Int. Dairy J.*, **6**, 673–87.

Anema, S.G. and Klostermeyer, H. (1997a) Heat-induced, pH-dependent dissociation of casein micelles on heating reconstituted skim milk at temperatures below 100°C. *J. Agric. Food Chem.*, **45**, 1108–15.

Anema, S.G. and Klosyermeyer, H. (1997b) The effect of pH and heat treatment on the κ-casein content and ζ potential of particles in reconstituted skim milk. *Milchwissenschaft*, **52**, 217–23.

Anema, S.G. and Stanley, D.J. (1998) Heat-induced, pH dependent behaviour of protein in caprine milk. *Int. Dairy J.*, **8**, 917–23.

Aoki, T. and Kako, Y. (1984) Effect of formaldehyde on the heat stability of concentrated milk and the formation of soluble casein. *Agric. Biol. Chem.*, **48**, 1017–21.

Aoki, T. and Kako, Y. (1985) The soluble and micellar κ-casein in heated concentrated whey protein-free milk. *Jap. J. Zootech. Sci.*, **56**, 347–52.

Aoki, T., Hatanaka, C. and Imamura, T. (1977) Large casein micelles and soluble casein in heated concentrated whey protein-free milk. *Agric. Biol. Chem.*, **41**, 2349–55.

Aoki, T., Suzuki, H. and Imamura, T. (1974) Formation of soluble casein in whey protein-free milk heated at high temperature. *Milchwissenchaft*, **29**, 589–94.

Aoki, T., Suzuki, H. and Imamura, T. (1975) Some properties of soluble casein in heated concentrated whey protein-free milk. *Milchwissenchaft*, **30**, 30–5.

Aoki, T., Umeda, T. and Kako, Y. (1990) Cleavage of the linkage between colloidal calcium phosphate and casein on heating milk at high temperature. *J. Dairy Res.*, **57**, 349–54.

Aoki, T., Umeda, T. and Nakano, T. (1999) Effect of sodium chloride on the properties of casein micelles. *Milchwissenschaft*, **54**, 91–3.

Augustin, M.A. and Clarice, P.T. (1990) Effects of added salts on the heat stability of recombined concentrated milk. *J. Dairy Res.*, **57**, 213–26.

Auldist, M.J., Coats, S.T., Sutherland, B.J., Clarke, P.T., McDowell, G.H. and Rogers, G.L. (1996) Effect of somatic cell count and state of lactation on the quality of full cream milk powder. *Aust. J. Dairy Technol.*, **51**, 94–8.

Baer, A., Oroz, M. and Blanc, B. (1976) Serological studies on heat-induced interaction of α-lactalbumin and milk proteins. *J. Dairy Res.*, **43**, 419–32.

Ball, C.O. (1955) Studies on high temperature sterilization of evaporated milk with emphasis on stability aspects. *Food Res.*, **20**, 351–70.

Banks, W., Clapperton, L.J., Muir, D.D., Powell, A.K. and Sweetsur, A.W.M. (1984) The effect of dietary-induced changes in milk urea levels on the heat stability of milk. *J. Sci. Food Agric.*, **35**, 165–72.

Beaulieu, M., Pouliot, Y. and Pouliot, M. (1999) Composition and microstructure of casein: whey protein aggregates formed by heating model solutions at 95°C. *Int. Dairy J.*, **9**, 393–4.

Belec, J. and Jenness, R. (1962a) Dephosphorization of casein by heat treatment. I. In caseinate solutions. *J. Dairy Sci.*, **45**, 12–9.

Belec, J. and Jenness, R. (1962b) Dephosphorization of casein by heat treatment. II. In skim milks. *J. Dairy Sci.*, **45**, 20–6.

Benten, A.G. (1929) Variations in the milk from different quarters of the same udder: Their significance in studies of coagulability. *J. Dairy. Sci.*, **12**, 481–3.

Berg, H.E. and van Boekel, M.A.J.S. (1994) Degradation of lactose during heating of milk. I. Reaction pathways. *Neth. Milk Dairy J.*, **48**, 157–75.

Burton, H. (1956) A new approach to the problem of heat stability in evaporated milk. *Proc. 14th Int. Dairy Congr.* (*Rome*), **1**, 342—9.

Burton, H. (1984) Reviews in Progress of Dairy Science: The bacteriological, chemical, biochemical and physical changes that occur in milk at temperatures of 100–150°C. *J. Dairy Res.*, **51**, 341–63.

Calvo, M.M. and de la Hoz, L. (1992) Flavour of heated milks: a review. *Int. Dairy J.*, **2**, 69–81.

Cano-Ruiz, M.E. and Richter, R.L. (1997) Effect of homogenization pressure on the milk fat globule membrane. *J. Dairy Sci.*, **80**, 2732–9.

Carroll, R.J., Thompson, M.P. and Melnychyn, P. (1971) Gelation of concentrated skim milk: Electron microscopic study. *J. Dairy Sci.*, **54**, 1245–52.

Corredig, M. and Dalgleish, D.G. (1999) The mechanisms of heat-induced interaction of whey proteins with casein micelles in milk. *Int. Dairy J.*, **9**, 233–6.

Creamer, L.K. (1984) Fractionation of casein micelles and whey protein aggregates on Sephacryl S-1000. *J. Chromat.*, **291**, 460–3.

Creamer, L.K. and Matheson, A.R. (1980) Effect of heat treatment on the proteins of pasteurized skim milk. *N.Z. J. Dairy Sci. Technol.*, **15**, 37–49.

Creamer, L.K., Berry, G.P. and Matheson, A.R. (1978) The effect of pH on protein aggregation in heated skim milk. *N.Z. J. Dairy Sci. Technol.*, **13**, 9–15.

Dalgleish. D.G. (1990) Denaturation and aggregation of serum proteins and caseins in heated milk. *J. Agric. Food Chem.*, **38**, 1995–9.

Dalgleish, D.G., Pouliot, Y. and Paquin, P. (1987a) Studies on the heat stability of milk. I. Behaviour of divalent cations and phosphate in milk heated in a stainless steel system. *J. Dairy Res.*, **54**, 29–37.

Dalgleish, D.G., Pouliot, Y. and Paquin, P. (1987b) Studies on the heat stability of milk. II. Association and dissociation of particles and the effect of added urea. *J. Dairy Res.*, **54**, 39–49.

Darling, D.F. (1980) Heat stability of milk. *J. Dairy Res.*, **47**, 199–210.

Darling, D.F. and Dickson, J. (1979) The determination of the zeta potential of casein micelles. *J. Dairy Res.*, **46**, 329–32.

Davies, J.M. (1959) Decrease in the heat stability of milk on forewarming. *Proc. 15th Int. Dairy Congr.* (*London*), **3**, 1859–64.

Davies, D.T. and White, J.C.D. (1959) Determination of heat-induced changes in the protein stability and chemical composition of milk. *Proc. 15th Int. Dairy Congr.* (*London*), **3**, 1677–85.

Davies, D.T. and White, J.C.D. (1966) The stability of milk protein to heat I. Subjective measurement of heat stability of milk. *J. Dairy Res.*, **33**, 67–81.

de Koning, P.J., Koops, J. and van Rooijen, P.J. (1974) Some features of the heat stability of concentrated milk. III. Seasonal effects on the amounts of casein, individual whey proteins and NPN and their relation to variations in heat stability. *Neth. Milk Dairy J.*, **28**, 186–202.

de Rafael, D., Villumiel, M. and Olano, A. (1997) Formation of lactulose and furosine during heat treatment of milk at temperatures 100–120°C. *Milchwissenschaft*, **52**, 76–8.

de Wit, J.N. (1981) Structure and functional behaviour of whey proteins. *Neth. Milk Dairy J.*, **35**, 47–64.

de Wit, J.N., Klarenbeek, G. and de Graaf, C. (1986) Klaro-graph: a new approach for the measurement of the viscosity, density and heat stability in milk and milk concentrates at temperatures up to 140°C. *Voedingsmiddelentechnologie*, **6**, 25–7.

Doi, H., Ideno, S., Ibuki, F. and Kanamori, M. (1983) Participation of the hydrophobic bond in complex formation between κ-casein and β-lactoglobulin. *Agric. Biol. Chem.*, **47**, 407–9.

Zin El-Din, M. and Aoki, T. (1993) Polymerization of casein on heating milk. *Int. Dairy J.*, **3**, 581–8.

El-Negoumy, A.M. (1974) β-Lactoglobulin-κ-casein interaction in various salt solutions. *J. Dairy Sci.*, **57**, 1302–8.

Ennis, M.P., Mahon, P. and Mulvihill, D.M. (2000) Heat-induced modification of rennet casein hydration behaviour. *Irish. J. Agric. Food Res.*, **39**, 133 (abstr).

Fairise, J.-F., Cayot, P. and Lorient, D. (1999) Characterisation of the protein composition of casein micelles after heating. *Int. Dairy J.*, **9**, 249–54.

Farah, Z. and Atkins, D. (1992) Heat coagulation of camel milk *J. Dairy Res.*, **59**, 229–31.

Farkye, N.Y. and Imafidon, G.I. (1995) Thermal denaturation of milk enzymes, in *Heat-induced Changes in Milk*, (P.F. Fox ed.) Special Issue 9501, International Dairy Federation, Brussels, pp. 331–45.

Feagan, J.T., Griffin, A.T. and Lloyd, G.T. (1966) Effects of subclinical mastitis on the heat stability of fluid milk. *J. Dairy Sci.*, **49**, 933–9.

Feagan, J.T., Bailey, L.F., Hehir, A.F., McLean, D.M. and Ellis, N.J.S. (1972) Coagulation of milk proteins. 1. Effect of genetic variants of milk proteins on rennet coagulation and heat stability of normal milk. *Aust. J. Dairy Technol.*, **27**, 129–34.

Fink, R. and Kessler, H.G. (1988) Comparison of methods for distinguishing UHT-treatment and sterilization of milk. *Milchwissenschaft*, **43**, 275–80.

Foissy, H. and Kneifel, W. (1984) An automatic method for measuring the heat coagulation time of milk powder solutions. *J. Dairy Res.*, **51**, 325–9.

Fox, K.K., Harper, M.K., Holsinger, V.H. and Pallansch, M.J. (1967) Effect of high-heat treatment on stability of calcium caseinate aggregates in milk. *J. Dairy Sci.*, **50**, 443–50.

Fox, P.F. (1981a) Heat stability of milk: significance of heat-induced acid formation in coagulation. *Irish J. Food Sci. Technol.*, **5**, 1–11.

Fox, P.F. (1981b) Heat-induced changes in milk preceding coagulation. *J. Dairy Sci.*, **64**, 2127–37.

Fox, P.F. (1982) Heat-induced coagulation of milk in *Developments in Dairy Chemistry-I. Proteins.* (P.F. Fox ed.) Applied Science Publishers, London, pp. 189–228.

Fox, P.F. and Hearn, C.M. (1978a) Heat stability of milk: influence of dilution and dialysis against water. *J. Dairy Res.*, **45**, 149–57.

Fox, P.F. and Hearn, C.M. (1978b) Heat stability of milk: influence of denaturable proteins and detergents on pH sensitivity. *J. Dairy Res.*, **45**, 159–72.

Fox, P.F. and Hearn, C.M. (1978c) Heat stability of milk: influence of κ-casein hydrolysis. *J. Dairy Res.*, **45**, 173–81.

Fox, P.F. and Hoynes, M.C.T. (1975) Heat stability of milk: influence of colloidal calcium phosphate and β-lactoglobulin. *J. Dairy Res.*, **42**, 427–35.

Fox, P.F. and Hoynes, M.C.T. (1976) Heat stability characteristics of ovine, caprine and equine milk. *J. Dairy Res.*, **43**, 433–42.

Fox, P.F. and Morrissey, P.A. (1977) Reviews of the Progress of Dairy Science: The heat stability of milk. *J. Dairy Res.*, **44**, 627–46.

Fox, P.F. and Nash, B.M. (1979) Physico–chemical characteristics of casein micelles in dilute aqueous media *J. Dairy Res.*, **46**, 357–63.

Fox, P.F., Nash, B.M., Horan, T.J., O'Brien, J. and Morrissey, P.A. (1980) Effect of selected amides on heat-induced changes in milk. *J. Dairy Res.*, **47**, 211–9.

Frazier, W.C. (1925) The influence of some bacterial enzymes on the heat coagulation of milk. *J. Dairy Sci.*, **8**, 370–89.

Gallagher, D.P. and Mulvihill, D.M. (1997) Heat stability and renneting characteristics of milk systems containing bovine casein micelles and porcine or bovine β-lactoglobulin. *Int. Dairy J.*, **7**, 221–8.

Gallagher, D.P., Lucey, J.A. and Mulvihill, D.M. (1996) Heat stability characteristics of porcine milk and mixed porcine-bovine milk systems. *Int. Dairy J.*, **6**, 597–611.

Ganguli, N.C. (1979) Stability of buffalo casein micelles. *J. Dairy Res.*, **46**, 401–6.

Gaucheron, F., Mollé, D., Briard, V. and Léonil, J. (1999) Identification of low molecular weight peptides released during sterilization of milk. *Int. Dairy J.*, **9**, 515–21.

Geerts, J.P., Bekhof, J.J. and Scherjon, J.W. (1983) Determination of calcium ion activities in milk with an ion-selective electrode. A linear relationship between the logarithm of time and the recovery of the calcium ion activity after heat treatment. *Neth. Milk Dairy J.*, **37**, 197–211.

Ghatak, P.K., Bandyopadhyay, A.K. and Gupta, M.P. (1990) Influence of chilling treatment on heat stability and proteolysis of cow milk. *Asian J. Dairy Res.*, **9**, 231–4.

Ghatak, P.K., Bandyopadhyay, A.K. and Gupta, M.P. (1989) The relation between the heat stability of milk and its chemical composition. *Asian J. Dairy Res.*, **8**, 165–8.

Ghatak, P.K., Singh, A., Bhavadasan, M.K. and Ganguli, N.C. (1980) The effect of the addition of aldehydes or sugars on the heat stability of buffalo milk. *N.Z. J. Dairy Sci. Technol.*, **15**, 159–65.

Guo, M., Fox, P.F., Flynn, A. and Mohammad, K.S. (1989) Heat-induced changes in sodium caseinate. *J. Dairy Res.*, **56**, 503–12.

Griffin, A.T., Hickey, M.W., Bailey, L.F. and Feagan, J.T. (1976) The significance of preheat and pH adjustment in the manufacture of recombined evaporated milk. *Aust. J. Dairy Technol.*, **31**, 134–7.

Hammersten, O. (1874) Uber den Chemischen Verlauf bein der Gerinnung des Caseins mit Lab. *Jahres-Bericht uber die Fortschnitte Der Thier-Chemie*, **4**, 135–54. [cited from McCrae and Muir, 1995].

Haque, Z. and Kinsella, J.E. (1988) Interaction between heated κ-casein and β-lactoglobulin: predominance of hydrophobic interactions in the initial stages of complex formation. *J. Dairy Res.*, **55**, 67–80.

Haque, Z., Kristjansson, M.M. and Kinsella, J.E. (1987) Interaction between κ-casein and β-lactoglobulin: possible mechanism. *J. Agric. Food Chem.*, **35**, 644–9.

Hardy, E.E., Muir, D.D., Sweetsur, A.W.M. and West, I.G. (1984) Changes of calcium phosphate partition and heat stability during manufacture of sterilized concentrated milk. *J. Dairy Sci.*, **67**, 1666–73.

Hardy, E.E., Sweetsur, A.W.M., West, I.G. and Muir, D.D. (1985) Heat stability of concentrated milk: enhancement of initial heat stability by incorporation of food grade lecithin. *J. Food Technol.*, **20**, 97–105.

Hardy-Lloyd, E.E., Sweetsur, A.W.M., West, I.G. and Muir, D.D. (1986) Preparation and properties of sterilized concentrated milk incorporating lecithin. *Milchwissenschaft*, **41**, 470–3.

Hartman, G.H. (1967) Heat treatment of mixtures of β-lactoglobulin and κ-casein. *Diss. Abstr.*, **28**, 474.

Harwalkar, V.R., Allan-Wojtas, P. and Kaláb, M. (1989) Effect of heating to 200°C on casein micelles in milk: a metal shadowing and negative staining electron microscope study. *Food Microstructure*, **8**, 217–24.

Heeschen, W. (1996) Influence of mastitis on quality and hygienic characteristics of milk. *Praktische-Tierarzt*, **77**, 223–6. (cited from *CAB Abstr*. 96040996).

Henle, T., Walter, A.W. and Klostermeyer, H. (1993) Detection and identification of the crosslinking amino acids N^{τ} and N^{π}-(2′-amino-2′-carboxy-ethyl)-L-histidine ("Histidinoalanine", HAL) in heated milk products. *Z. Lebensm. Unters. Forsch.*, **197**, 114–7.

Hennart, P.F., Brasseur, D.J., Delogne-Desnoeck, J.B., Dramaix, M.M. and Robyn, C.E. (1991) Lysozyme, lactoferrin and secretory immunoglobulin A content in breast milk: Influence of duration of lactation, nutrition status, prolactin status and parity of mother. *Amer. J. Clin. Nut.*, **53**, 32–39.

Hindle, E J. and Wheelock, J.V. (1970) The release of peptides and glycopeptides by the action of heat on cow's milk. *J. Dairy Res.*, **37**, 397–405.

Hinrichs, J. and Kessler, H.G. (1995) Thermal processing of milk-processes and equipment, in *Heat-induced Changes in Milk*, (P.P. Fox ed.) Special Issue 9501, International Dairy Federation, Brussels, pp. 8–21.

Holt, C. (1992) Structure and stability of bovine casein micelles. *Adv. Protein Chem.*, **43**, 63–151.

Holt, C., Muir, D.D. and Sweetsur, A.W.M. (1978a) Seasonal changes in the heat stability of milk from creamery silos in South-West Scotland. *J. Dairy Res.*, **45**, 183–90.

Holt, C., Muir, D.D. and Sweetsur, A.W.M. (1978b) The heat stability of milk and concentrated milk containing added aldehydes and sugars. *J. Dairy Res.*, **45**, 47–52.

Horne, D.S. (1983) Calcium-induced precipitation of α_{s1}-casein: Effect of modification of lysine residues. *Int. J. Biol. Macromol.*, **5**, 296–300.

Horne, D.S. (1992) Ethanol stability, in *Advanced Dairy Chemistry-1*: *Proteins*, (P.F. Fox ed.) Elsevier Applied Science, London, pp. 657–90.

Horne, D.S. and Muir, D.D. (1990) Alcohol and heat stability of milk protein. *J. Dairy Sci.*, **73**, 3613–26.

Howat, G.R. and Wright, N.C. (1934) The heat coagulation of caseinogen. I. The role of phosphorus cleavage. *Biochem. J.*, **28**, 1336–45.

Hoynes, M.C.T. and Fox, P.F. (1975) Some physico–chemical properties of porcine milk. *J. Dairy Res.*, **42**, 43–56.

Hustinx, J.C.A., Singh, T.K. and Fox, P.F. (1997) Heat-induced hydrolysis of sodium caseinate. *Int. Dairy J.*, **7**, 207–12.

Hyslop, D.B. and Fox, P.F. (1981) Heat stability of milk: interrelationship between assay temperature, pH and agitation. *J. Dairy Res.*, **48**, 123–9.

Ibrahim, E.M., Mohan, M.A. and Hanafy, N.E. (1990) Physico–chemical characteristics of colostrum and the influence of its addition to some technological properties of normal milk. *Assiut J. Agric. Sci.*, **21**, 221–39.

Jang, H.D. and Swaisgood, H.E. (1990) Disulphide bond formation between thermally denatured β-lactoglobulin and κ-casein in casein micelles. *J. Dairy Sci.*, **73**, 900–4.

Jeurnink, T.J.M. (1991) Effect of proteolysis in milk on fouling in heat exchangers. *Neth. Milk Dairy J.*, **45**, 23–32.

Jeurnink, T.J.M. (1992) Changes in milk on mild heating: turbidity measurements. *Neth. Milk Dairy J.*, **46**, 183–96.

Jeurnink, T.J.M. and de Kruif, K.G. (1995) Calcium concentration in milk in relation to heat stability and fouling. *Neth. Milk Dairy J.*, **49**, 151–65.

Kella, N.K.D. and Kinsella, J.E. (1988) Structural stability of β-lactoglobulin in the presence of kosmotropic salts. *Int. J. Peptide Protein Res.*, **32**, 396–405.

Kelly, P.M. (1982) The effect of preheat temperature and urea addition on the seasonal variation in the heat stability of skim milk powder. *J. Dairy Res.*, **49**, 187–96.

Kelly, P.M., O'Keeffe, A.M., Keogh, M.K. and Phelan, J.A. (1982) Studies of milk composition and its relationship to some processing criteria. III: Seasonal variation in heat stability of milk. *Irish J. Food Sci. Technol.*, **6**, 29–38.

Kessler, H.G. and Fink, R. (1986) Changes in heated and stored milk with an interpretation by reaction kinetics. *J. Food Sci.*, **51**, 1105–11.

Kieseker, F.G. and Aitken, B. (1988) An objective method for determination of heat stability of milk powders. *Aust. J. Dairy Technol.*, **43**, 26–31.

Kitabatake, N., Cuq, J.L. and Cheftel, J.C. (1985) Covalent binding of glycosyl residues to β-lactoglobulin: effects on solubility and heat stability. *J. Agric. Food Chem.*, **33**, 125–30.

Klostermeyer, H. and Reimerdes, E.H. (1977) Heat-induced crosslinks in milk proteins and consequences for the milk system, in *Advances in Experimental Medicine and Biology, Vol 86 B: Protein Crosslinking,* (M. Friedman ed.) Plenum Press, New York, pp. 263–76.

Koops, J. and Westerbeek, D. (1970) Some features of the heat stability of concentrated milk. II. Effect of hydrogen peroxide. *Neth. Milk Dairy J.*, **24**, 52–60.

Kresheck, G.C., van Winkle, Q. and Gould, L.A. (1964) Physical changes in milk proteins at elevated temperatures as determined by light scattering. I. Casein fractions. *J. Dairy Sci.*, **47**, 117–25.

Kudo, S. (1980a) Influence of lactose and urea on the heat stability of artificial milk systems. *N.Z. J. Dairy Sci. Technol.*, **15**, 197–200.

Kudo, S. (1980b) The influence of α_s-casein on the heat stability of artificial milks. *N.Z. J. Dairy Sci. Technol.*, **15**, 245–54.

Kudo, S. (1980c) The heat stability of milk: Formation of soluble proteins and protein-depleted micelles at elevated temperatures. *N.Z. J. Dairy Sci. Technol.*, **15**, 255–63.

Law, A.J.R. and Leaver, J. (1999) Factors affecting the heat denaturation of whey proteins in cow's milk. *Int. Dairy J.*, **9**, 407–8.

Leighton, A. and Deysher, E.R. (1923) Factors influencing the heat coagulation of milk and the thickening of condensed milk. *Proc. World's Dairy Congr.*, **2**, 1276–84. [cited from Singh and Creamer, 1992].

Leonil, J., Molle, D., Fauquant, J., Maubois, J.L., Pearse, R.J. and Bouhallab, S. (1997) Characterisation by ionisation mass spectroscopy of lactosyl β-lactoglobulin conjugates formed during heat treatment of milk and whey and identification of one lactose-binding site. *J. Dairy Sci.*, **80**, 2270–81.

Lodes, A., Krause, I., Buchberger, J., Aumann, J. and Klostermeyer, H. (1996) The influence of genetic variants of milk proteins on the compositional and technological properties of milk. 1. Casein micelle size and content of non-glycosylated κ-casein. *Milchwissenchaft*, **51**, 368–73.

Long, J.E., van Winkle, Q. and Gould, I.A. (1963) Heat-induced interaction between crude κ-casein and β-lactoglobulin. *J. Dairy Sci.*, **46**, 1329–34.

Lorient, D. (1979) Covalent bonds formed in proteins during milk sterilization: studies on caseins and casein peptides. *J. Dairy Res.*, **46**, 393–6.

Manson, W. and Carolan, T. (1980) Formation of lysinoalanine from individual bovine caseins. *J. Dairy Res.*, **47**, 193–8.

Manson, W., Carolan, J. and Annan, W.D. (1985) Products of heat-promoted reaction between urea and the protein fraction of bovine milk. *J. Dairy Res.*, **52**, 401–7.

McCrae, C.H. (1999) Heat stability of milk emulsions: phospholipid-protein interactions. *Int. Dairy J.*, **9**, 227–31.

McCrae, C.H. and Muir, D.D. (1992) The influence of phospholipid classes of crude lecithin on the heat stability of recombined milk. *Milchwissenschaft*, **47**, 755–9.

McCrae, C.H. and Muir, D.D. (1995) Heat stability of milk, in *Heat-induced Changes in Milk*, (P.F. Fox ed.) Special Issue 9501, International Dairy Federation, Brussels, pp. 206–30.

McCrae, C.H., Hirst, D., Law, A.J.R. and Muir, D.D. (1994) The heat stability of homogenised milk: role of interfacial protein. *J. Dairy Res.*, **61**, 507–16.

McGann, T.C.A., Donnelly, W.J., Kearney, R.D. and Buchheim, W. (1980) Composition and size distribution of bovine casein micelles. *Biochim. Biophys. Acta*, **630**, 261–70.

McLean, D.M. and McKenzie, H.A. (1978) Heat stability of milk from several species. *Proc. 20th Int. Dairy Congr.* (Paris), p. 247 (*abstr*).

McLean, D.M., Graham, E.R.B., Ponzoni, R.W. and McKenzie, H.A. (1987) Effects of milk protein genetic variants and composition on heat stability of milk. *J. Dairy Res.*, **54**, 219–35.

Meisel, H. and Schlimme, E. (1995) Casein-bound phosphorus and contents of free amino acids in milk subjected to different heat treatments. *Kieler Milchwirtschaftliche Forschungsberichte*, **47**, 289–95.

Metwalli, A.A.M. and van Boekel, M.A.J.S. (1995) Effect of formaldehyde on heat stability of milk. *Neth. Milk Dairy J.*, **49**, 177–89.

Metwalli, A.A.M. and van Boekel, M.A.J.S. (1996) Effect of urea on the heat coagulation of milk. *Neth. Milk Dairy J.*, **50**, 459–76.

Metwalli, A.A.M. and van Boekel, M.A.J.S. (1998) On the kinetics of heat-induced deamidation and breakdown of caseinate. *Food Chem.*, **61**, 53–61.

Metwalli, A.A.M., Lammers, W.L. and van Boekel, M.A.J.S. (1998) Formation of homocitrulline during heating of milk. *J. Dairy Res.*, **65**, 579–89.

Metwalli, A.A.M., Metwalli, N.H. and van Boekel, M.A.J.S. (1996) Effect of urea on heat-induced changes in milk. *Neth. Milk Dairy J.*, **50**, 427–57.

Miller, P.G. and Sommer, H.H. (1940) The coagulation temperature of milk as affected by pH, salts, evaporation and previous heat treatment *J. Dairy Sci.*, **23**, 405–21.

Minkiewicz, P., Dziuba, J. and Muzínska, B. (1993) The contribution of N-acetlyneuraminic acid in the stabilization of micellar casein. *Polish J. Food Nut. Sci.*, **2**, 39–48.

Mohammad, K.S. (1985) *Heat-induced Changes in Micellar and Soluble Casein*, PhD thesis, National University of Ireland, Cork.

Mohammad, K.S. and Fox, P.F. (1983) Influence of some polyvalent organic acids and salts on the colloidal stability of milk. *J. Soc. Dairy Technol.*, **36**, 112–7.

Mohammad, K.S. and Fox, P.F. (1986) Heat and alcohol-induced coagulation of casein micelles. *Irish J. Food Sci. Technol.*, **10**, 47–55.

Mohammad, K.S. and Fox, P.F. (1987a) Heat-induced microstructural changes in casein micelles before and after heat coagulation. *N.Z. J. Dairy Sci. Technol.*, **22**, 191–203.

Mohammad, K.S. and Fox, P.F. (1987b) Heat-induced association-dissociation of casein micelles preceding coagulation. *J. Dairy Res.*, **54**, 377–87.
Montilla, A. and Calvo, M.M. (1997) Goat's milk stability during heat treatment: Effect of pH and phosphates. *J. Agric. Food Chem.*, **45**, 931–4.
Morales, F.J. and Jiménez-Pérez, S. (1998) Monitoring of heat-induced proteolysis in milk and milk resembling systems. *J. Agric. Food Chem.*, **46**, 4391–7.
Morales, F.J., Romero, C. and Jiménez-Pérez, S. (1995) New methodologies for the kinetic study of 5-(hydroxymethyl) furfural formation and reactive lysine blockage in heat-treated milk and model systems. *J. Food Prot.*, **58**, 310–5.
Morales, F.J., Romero, C. and Jiménez-Pérez, S. (1996) Fluoresence associated with Maillard reaction in milk and milk resembling systems. *Food Chem.*, **57**, 423–8.
Morr, C.V. (1973) Milk ultracentrifugal opalescent layer 2: Physico-chemical properties. *J. Dairy Sci.*, **56**, 1258–66.
Morr, C.V., Harper, N.J. and Gould, I.A. (1957) Some organic acids in raw and heated milk. *J. Dairy Sci.*, **40**, 964–72.
Morrissey, P.A. (1969) The heat stability of milk as affected by variations in pH and milk salts. *J. Dairy Res.*, **36**, 343–51.
Morrissey, P.A. and O'Mahony, F. (1976) Heat stability of forewarmed milks: Influence of κ-casein, serum proteins and divalent cations. *J. Dairy Res.*, **43**, 267–74.
Morrissey, P.A., Murphy, M.F., Hearn, C.M. and Fox, P.F. (1981) Composition and stability of mid-lactation milks. *Irish J. Food Sci. Technol.*, **5**, 117–27.
Mousa, S.M. and Ibrahim, E.M. (1991) Effect of urea addition to dairy cattle rations with low protein contents on milk yield, composition and some technological properties. *Assiut J. Agric. Sci.*, **23**, 179–96.
Muir, D.D. (1985) Heat stability of milk and concentrated milk. *Int. J. Biochem.*, **17**, 291–9.
Muir, D.D. and Sweetsur, A.W.M. (1976) The influence of naturally occurring levels of urea on the heat stability of bulk milk *J. Dairy Res.*, **43**, 495–9.
Muir, D.D. and Sweetsur, A.W.M. (1977) Effect of urea on the heat coagulation of the caseinate complex in skim milk. *J. Dairy Res.*, **44**, 249–57.
Muir, D.D. and Sweetsur, A.W.M. (1978) The effect of concentration on the heat stability of skim milk. *J. Dairy Res.*, **45**, 37–45.
Muir, D.D. and Sweetsur, A.W.M. (1984) Optimization of the heat stability of protein-rich concentrates prepared by ultrafiltration of skim-milk. *J. Food Technol.*, **19**, 263–71.
Muir, D.D. and Sweetsur, A.W.M. (1992a) Production and properties of in-can sterilised concentrated milk with 39% solids: seasonal effects. *Milchwissenschaft*, **47**, 8–11.
Muir, D.D. and Sweetsur, A.W.M. (1992b) Production and properties of in-can sterilised concentrated milk with 39% solids: process optimisation. *Milchwissenschaft*, **47**, 80–3.
Muir, D.D., Abbot, J. and Sweetsur, A.W.M. (1978) Changes in the heat stability of milk protein during the manufacture of dried skim-milk. *J. Food Technol.*, **13**, 45–53.
Muir, D.D., Sweetsur, A.W.M. and Holt, C. (1979) The synergic effect of urea and aldehydes on the heat stability of concentrated skim-milk. *J. Dairy Res.*, **46**, 381–4.
Muir, D.D., Horne, D.S., Law, A.J.R. and Sweetsur, A.W.M. (1993) Ovine milk. 2. Seasonal changes in indices of stability. *Milchwissenschaft*, **48**, 442–5.
Nelson, V. (1954) The effects of formaldehyde and copper salts on the heat stability of evaporated milk. *J. Dairy Sci.*, **37**, 825–9.
Newstead, D.F. (1977) Effect of protein and salt concentration on the heat stability of evaporated milk. *N.Z. J. Dairy Sci. Technol.*, **12**, 171–5.

Newstead, D.F. and Baucke, A.G. (1983) Heat stability of recombined evaporated milk and reconstituted concentrated skim milk. Effects of temperature and time of preheating. *N.Z. J. Dairy Sci. Technol.*, **18**, 1–11.

Newstead, D.F., Conaghan, E.F. and Sanderson, W.B. (1976) Effects of whey protein concentration on the heat stability of evaporated milk. *N.Z. J. Dairy Sci. Technol.*, **11**, 223–30.

Newstead, D.F., Conaghan, E.F. and Baldwin, A.J. (1979) Studies on the induction of heat stability in evaporated milk by preheating: Effects of milk concentration, homogenization and whey proteins. *J. Dairy Res.*, **46**, 387–91.

Newstead, D.F., Sanderson, W.R. and Baucke, A.G. (1975) The effect of heat treatment and pH on the heat stability of recombined evaporated milk. *N.Z. J. Dairy Sci. Technol.*, **10**, 113–8.

Newstead, D.F., Sanderson, W.B. and Conaghan, E.F. (1977) Effects of whey protein concentrations and heat treatment on the heat stability of concentrated and unconcentrated milk. *N.Z. J. Dairy Sci. Technol.*, **12**, 29–36.

Nieuwenhuijse, J.A., Timmermans, W. and Walstra, P. (1988) Calcium and phosphate partitions during the manufacture of sterilized concentrated milk and their relations to the heat stability. *Neth. Milk Dairy J.*, **42**, 387–421.

Nieuwenhuijse, J.A., Sjollema, A., van Boekel, M.A.J.S., van Vliet, T. and Walstra, P. (1991) The heat stability of concentrated milk. *Neth. Milk Dairy J.*, **45**, 193–224.

Noh, B. and Richardson, T. (1989) Incorporation of radiolabelled whey proteins into the casein micelles by heat processing. *J. Dairy Sci.*, **72**, 1724–31.

Noh, B., Creamer, L.K. and Richardson, T. (1989a) Thermally induced complex formation in an artificial milk system *J. Agric. Food Chem.*, **37**, 1395–400.

Noh, B., Richardson T. and Creamer, L.K. (1989b) Radiolabelling study of the heat-induced interactions between α-lactalbumin, β-lactoglobulin and κ-casein in milk and in buffer solutions. *J. Food Sci.*, **54**, 889–93.

Nursten, H.E. (1981) Recent developments in studies of the Maillard reaction. *Food Chem.*, **6**, 263–77.

Nursten, H.E. (1995) Heat-induced changes in the flavour of milk, in *Heat-induced Changes in Milk*, (P.F. Fox ed.) Special Issue 9501, International Dairy Federation, Brussels, pp. 308–15.

O'Brien, J. (1995) Heat-induced changes in lactose: isomerization, degradation, Maillard reaction, in *Heat-induced Changes in Milk*, (P.F. Fox ed.) Special Issue 9501, International Dairy Federation, Brussels, pp. 134–62.

O'Connell, J.E. and Fox, P.F. (1999a) Effect of extracts of oak (*Quercus petraea*) bark, oak leaves, aloe vera (*Curacao aloe*), coconut shell and wine on the colloidal stability of milk and concentrated milk. *Food Chem.*, **66**, 93–6.

O'Connell, J.E. and Fox, P.F. (1999b) Effect of phenolic compounds on the heat stability of milk and concentrated milk. *J. Dairy Res.*, **66**, 399–407.

O'Connell, J.E. and Fox, P.F. (1999c) Proposed mechanism for the effect of polyphenols on the heat stability of milk. *Int. Dairy J.*, **9**, 523–30.

O'Connell, J.E. and Fox, P.F. (2000a) The two-stage coagulation of milk proteins in the minimum of the heat coagulation time-pH profile of milk: effect of casein micelle size. J. *Dairy Sci.*, **83**, 378–86.

O'Connell, J.E. and Fox, P.F. (2000b) Heat-induced changes in the calcium sensitivity of caseins. *Int. Dairy J.*, **9**, 839–47.

O'Connell, J.E. and Fox, P.F. (2000c) The heat stability of buttermilk. *J. Dairy Sci.*, **83**, 1728–32.

O'Connell, J.E., Fox, P.D., Tan-Kintia, R. and Fox, P.F. (1998) Effect of extracts of tea, coffee and cocoa on the colloidal stability of milk *Int. Dairy J.*, **8**, 689–93.

Okonogi, S. and Tomita, M. (1977) Effect of pH on the heat stability of skim-milk subjected to demineralization by means of electrodialysis with ion permselective membranes. *Jap. J. Zootech. Sci.*, **48**, 437–8.

Olano, A. and Calvo, M.M. (1989) Kinetics of lactulose, galactose and epilactose formation during heat treatment of milk. *Food Chem.*, **34**, 239–48.

Ono, T., Yoshida, M., Tanaami, H. and Ohkosi, H. (1999) Changes in the casein micelle size induced by heating. *Int. Dairy J.*, **9**, 405–6.

O'Reilly, M. and Kelly, P.M. (1982) Heat induced urea interaction and their heat stabilizing effects in milk. *Irish J. Food Sci. Technol.*, **6**, 202 (*abstr*).

Otterburn, M.S., Healy, M. and Sinclair, W. (1977) The formation, isolation and importance of isopeptides in heated proteins, in *Advances in Experimental Medicine and Biology*, *Vol 86 B*: *Protein Crosslinking*, (M. Friedman ed.) Plenum Press, New York, pp. 239–62.

Pappas, C.P. (1992) Interactions between milk proteins: Influence of heat, calcium and lactose. 1. Interactions between (I) whole casein and β-lactoglobulin and (II) κ-casein and β-lactoglobulin. *Lebensm. Wiss. u. Technol.*, **25**, 102–12.

Pappas, C.P. and Rothwell, J. (1991) The effect of heating, alone or in the presence of calcium or lactose, on the calcium binding to milk proteins. *Food Chem.*, **42**, 183–201.

Parker, T.G. and Dalgleish, D.G. (1977) The theory of branching process applied to milk proteins. *J. Dairy Res.*, **44**, 85–92.

Paterson, G.R., MacGribbon, A.K.H. and Hill, J.P. (1999) Influence of κ-casein and β-lactoglobulin phenotype on the heat stability of milk. *Int. Dairy J.*, **9**, 375–6.

Pauletti, M.S., Castelao, E.L. and Sequro, E. (1996) Kinetics of heat coagulation of concentrated milk proteins containing high sucrose contents. *J. Food Sci.*, **61**, 1207–10.

Pearce, R.J. (1979) Heat stability in concentrated and non-concentrated milks: the effect of urea and β-lactoglobulin levels and the influence of preheating. *J. Dairy Res.*, **46**, 385–6.

Pellegrino, L., Resmini, P. and Luf, W. (1995) Assessment (indices) of heat treatment of milk, in *Heat-induced Changes in Milk*, (P.F. Fox ed.) Special Issue 9501, International Dairy Federation, Brussels, pp. 419–53.

Pellegrino, L., van Boekel, M.A.J.S., Gruppen, H., Resmini, P. and Pagani, M.A. (1999) Heat-induced aggregation and covalent linkages in β-casein model systems. *Int. Dairy J.*, **9**, 255–60.

Pepper, L. and Thompson, M.P. (1963) Dephosphorylation of α_s- and κ-caseins and its effect on micelle stability in the κ-α_s-casein system. *J. Dairy Sci.*, **46**, 764–7.

Peter, S., Rattray, W. and Jelen, P. (1996) Heat stability and sensory quality of protein standardized 2% fat milk. *Milchwissenschaft*, **51**, 611–6.

Pouliot, Y. and Boulet, M. (1995) Observations on the seasonal variations in the salt balance of concentrated milk. *Int. Dairy J.*, **5**, 75–85.

Prasad, C. and Balachasandran, R. (1987) Effect of chemical additives on the pH and heat stability of buffalo milk at different concentrations. *N.Z. J. Dairy Sci. Technol.*, **22**, 123–30.

Pyne, G.T. (1958) The heat coagulation of milk. II. Variations in sensitivity of casein to calcium ions. *J. Dairy Res.*, **25**, 467–74.

Rajput, Y.S., Bhavadasan, M.K. and Ganguli, N.C. (1984a) Effect of urea on heat-induced acidity and milk coagulation. *N.Z. J. Dairy Sci. Technol.*, **19**, 49–54.

Rajput, Y.S., Bhavadasan, M.K. and Ganguli, N.C. (1984b) Influence of amino acids on heat stability of milk. *Milchwissenschaft*, **39**, 598–600.

Ram, M. and Joshi, V.K. (1990) Heat stability of milk as affected by proteose peptone components during cold storage. *Indian J. Dairy Sci.*, **43**, 217–9.

Ramsdell, G.A., Johnson, W.T. and Evans, F.R. (1931) A test for the detection of a milk unstable to heat. *J. Dairy Sci.*, **14**, 93–106.

Rattray, W. and Jelen, P. (1996) Thermal stability of skim milk with protein content standardized by the addition of ultrafiltration permeate. *Int. Dairy J.*, **6**, 157–70.

Recio, I., Amigo, L. and López-Fandiño, R. (1997) Assessment of the quality of dairy products by capillary electrophoresis of milk proteins. *J. Chrom. B*, **697**, 231–42.

Relkin, P. (1996) Thermal unfolding of β-lactoglobulin, α-lactalbumin and bovine serum albumin. A thermodynamic approach. *CRC Crit. Rev. Food Sci. Nut.*, **36**, 565–601.

Robertson, N.H. and Dixon, A. (1969) The nitrogen fractions and heat stability of bovine milk. *Agroanimnalia*, **1**, 141–4.

Robitaille, G. (1995) Influence of κ-casein and β-lactoglobulin genetic variants on the heat stability of milk. *J. Dairy Res.*, **62**, 593–600.

Robitaille, G. and Ayers, C. (1995) Effects of κ-casein glycosylation on the heat stability of milk. *Food Res. Int.*, **28**, 17–21.

Rose, D. (1961a) Variations in the heat stability and composition of milk from individual cows during lactation. *J. Dairy Sci.*, **44**, 430–41.

Rose, D. (1961b) Factors affecting the pH-sensitivity of the heat stability of milk from individual cows. *J. Dairy Sci.*, **44**, 1405–13.

Rose, D. (1961c) β-Lactoglobulin and pH-sensitivity of the heat stability of evaporated milk. *J. Dairy Sci.*, **44**, 1763.

Rose, D. (1962) Factors affecting the heat stability of milk. *J. Dairy Sci.*, **45**, 1305–11.

Rose, D. (1963) Heat stability of bovine milk: a review. *Dairy Sci. Abstr.*, **25**, 45–52.

Rose, D. and Tessier, H. (1959) Composition of ultrafiltrates from milk heated at 80 to 230°F in relation to heat stability. *J. Dairy Sci.*, **42**, 969–80.

Rüegg, M., Blanc, B. and Lüscher, M. (1979) Hydration of casein micelles: kinetics and isotherms of water sorption of micellar casein isolated from fresh and heat-treated milk. *J. Dairy Res.*, **46**, 325–8.

Saidi, B. and Warthesen, J.J. (1993) Heat and fermentation effects on total nonprotein nitrogen and urea in milk *J. Food Sci.*, **58**, 548–51.

Sawyer, W.H. (1969) Complex between β-lactoglobulin and κ-casein. A review. *J. Dairy Sci.*, **52**, 323–9.

Sawyer, W.H., Coulter, S.T. and Jenness, R. (1963) Role of sulfydryl groups in the interaction of κ-casein and β-lactoglobulin. *J. Dairy Sci.*, **46**, 564–5.

Schmidt, D.G. and Koops, J. (1977) Properties of artificial casein micelles. 2. Stability towards ethanol, dialysis, pressure and heat in relation to casein composition. *Neth. Milk Dairy J.*, **31**, 342–57.

Schmidt, D.G. and Poll, J.K. (1986) Electrokinetic measurements on unheated and heated casein micelle systems. *Neth. Milk Dairy J.*, **40**, 269–80.

Shalabi, S.I. and Fox, P.F. (1982a) Heat stability of milk: synergic action of urea and carbonyl compounds. *J. Dairy Res.*, **49**, 197–207.

Shalabi, S.I. and Fox, P.F. (1982b) Effect of κ-carrageenan on the heat stability and rennet coagulation of milk. *Irish J. Food Sci. Technol.*, **6**, 183–8.

Shalabi, S.I. and Fox, P.F. (1982c) Heat stability of milk: Influence of cationic detergents on the pH sensitivity. *J. Dairy Res.*, **49**, 597–605.

Shalabi, S.I. and Fox, P.F. (1982d) Heat stability of milk: influence of modification of lysine and arginine on the heat stability-pH profile. *J. Dairy Res.*, **49**, 607–17.

Shalabi, S.I. and Fox, P.F. (1982e) Effect of diacetyl on the heat stability of concentrated milks. *J. Food Technol.*, **17**, 753–60.

Sharma, R. and Singh, H. (1999) Heat stability of recombined milk systems as influenced by the composition of fat globule surface layers. *Milchwissenschaft*, **54**, 193–6.

Singh, H. (1988) Effects of high temperatures on casein micelles. *N.Z. J. Dairy Sci. Technol.*, **23**, 257–73.

Singh, H. (1994) Crosslinking of milk proteins on heating concentrated milk at 120°C. *Int. Dairy J.*, **4**, 477–89.

Singh, H. (1995) Heat-induced changes in caseins including interactions with whey proteins, in *Heat-induced Changes in Milk*, (P.F. Fox ed.) Special Issue 9501, International Dairy Federation, Brussels. pp. 86–99.

Singh, H. and Creamer, L.K. (1991a) Influence of concentration of milk solids on the dissociation of micellar κ-casein on heating reconstituted milk at 120°C. *J. Dairy Res.*, **58**, 99–105.

Singh, H. and Creamer, L.K. (1991b) Aggregation and dissociation of milk protein complexes in heated reconstituted skim milk, *J. Food Sci.*, **56**, 238–46.

Singh, H. and Creamer, L.K. (1991c) Changes in size and composition of protein aggregates on heating reconstituted concentrated skim milk at 120°C. *J Food Sci.*, **56**, 671–77.

Singh, H. and Creamer, L.K. (1991d) Denaturation, aggregation and heat stability of milk protein during the manufacture of skim milk powder. *J. Dairy Res.*, **58**, 269–83.

Singh, H. and Creamer, L.K. (1992) Heat stability of milk, in *Advanced Dairy Chemistry-I Proteins*, (P.F. Fox ed.) Elsevier Applied Science Publishers, London. pp. 621–56.

Singh, H. and Fox, P.F. (1985a) Heat stability of milk: the mechanism of stabilization by formaldehyde. *J. Dairy Res.*, **52**, 65–76.

Singh, H. and Fox, P.F. (1985b) Heat stability of milk: pH-dependent dissociation of micellar κ-casein on heating milk at ultra high temperatures. *J. Dairy Res.*, **52**, 529–38.

Singh, H. and Fox, P.F. (1986) Heat stability of milk: further studies on the pH-dependent dissociation of micellar κ-casein. *J. Dairy Res.*, **53**, 237–48.

Singh, H. and Fox, P.F. (1987a) Heat stability of milk: Influence of modifying sulphydryl-disulphide interactions on the heat coagulation time-pH profile. *J. Dairy Res.*, **54**, 347–59.

Singh, H. and Fox, P.F. (1987b) Heat stability of milk: role of β-lactoglobulin in the pH-dependent dissociation of micellar κ-casein. *J. Dairy Res.*, **54**, 509–21.

Singh, H. and Fox, P.F. (1987c) Heat stability of milk: influence of colloidal and soluble salts and protein modification on the pH-dependent dissociation of micellar κ-casein. *J. Dairy Res.*, **54**, 523–34.

Singh, H. and Lantham, J.M. (1993) Heat stability of milk: aggregation and dissociation of protein at ultra-high temperatures. *Int. Dairy J.*, **3**, 225–37.

Singh, H. and Tokley, R.P. (1990) Effects of preheat treatments and buttermilk addition on the seasonal variations in the heat stability of recombined evaporated milk and reconstituted concentrated milk. *Aust. J. Dairy Technol.*, **45**, 10–6.

Singh, H., Creamer, L.K. and Newstead, D.F. (1995) Heat stability of concentrated milk, in *Heat-induced Changes in Milk*, (P.F. Fox ed.) Special Issue 9501, International Dairy Federation, Brussels, pp. 256–74.

Singh, H., Sharma, R. and Tokley, R.P. (1992) Influence of incorporation of soya lecithin into skim milk powder on the heat stability of recombined evaporated milk. *Aust. J. Dairy Technol.*, **47**, 33–7.

Singh, H., Sharma, R., Taylor, M.W. and Creamer L.K. (1996) Heat-induced aggregation and dissociation of protein and fat particles in recombined milk. *Neth. Milk Dairy J.*, **50**, 149–66.

Slatter, W.L. and van Winkle, Q. (1952) An eletrophoretic study of the proteins in skim milk. *J. Dairy Sci.*, **35**, 1083–93.

Smits, P. and van Brouwershaven, J.H. (1980) Heat-induced association of β-lactoglobulin and casein micelles. *J. Dairy Res.*, **47**, 313–25.

Snoeren, T.H.M. and van der Spek, C.A. (1977) The isolation of a heat-induced complex from UHTST milk. *Neth. Milk Dairy J.*, **31**, 352–5.

Snoeren, T.H.M. and van Riel, J.A.M. (1979) The properties of α_{s1}-group casein. *Zuivelziht*, **71**, 766–8.

Snoeren, T.H.M., Koops, J. and Westerbeek, D. (1978) Some features of the heat stability of (concentrated) milk. 4. The effect of the α_{S2}-caseins on the heat stability. *Neth. Milk Dairy J.*, **32**, 255–7.

Sommer, H.H. and Binney, T.H. (1923) A study of the factors that influence the coagulation of milk in the alcohol test. *J. Dairy Sci.*, **6**, 176–97.

Sommer, H.H. and Hart, E.B. (1919) The heat coagulation of milk. *J. Biol. Chem.*, **40**, 137–51.

Sommer, H.H. and Hart, E.B. (1922) The heat coagulation of milk. *J. Dairy Sci.*, **5**, 525–43.

Stepaniak, L. and Sørhaug, T. (1995) Thermal denaturation of bacterial enzymes in milk, in *Heat-induced Changes in Milk*, (P.F. Fox ed.) Special Issue 9501, International Dairy Federation, Brussels. pp. 349–59.

Sweetsur, A.W.M. and Muir, D.D. (1980a) The use of permitted additives and heat-treatment to optimize the heat-stability of skim milk and concentrated skim milk. *J. Soc. Dairy Technol.*, **33**, 101–5.

Sweetsur, A.W.M. and Muir, D.D. (1980b) Effect of concentration by ultrafiltration on the heat stability of skim-milk. *J. Dairy Res.*, **47**, 327–35.

Sweetsur, A.W.M. and Muir, D.D. (1981) Role of cyanate ions in the urea-induced stabilization of the caseinate complex in skim-milk. *J. Dairy Res.*, **48**, 163–6.

Sweetsur, A.W.M. and Muir, D.D. (1982) Natural variation in heat stability of concentrated milk before and after homogenization. *J. Soc. Dairy Technol.*, **35**, 120–6.

Sweetsur, A.W.M. and Muir, D.D. (1983a) Effect of homogenization on the heat stability of milk. *J. Dairy Res.*, **50**, 291–300.

Sweetsur, A.W.M. and Muir, D.D. (1983b) Influence of sulphydryl group interactions on the heat stability of homogenized concentrated milk. *J. Dairy Res.*, **50**, 301–8.

Sweetsur, A.W.M. and White, J.C.D. (1974) Studies on the heat stability of milk protein. I. Interconversion of type A and type B milk heat stability curves. *J. Dairy Res.*, **41**, 349–58.

Sweetsur, A.W.M. and White, J.C.D. (1975) Studies on the heat stability of milk protein. III. Effect of heat-induced acidity in milk. *J. Dairy Res.*, **42**, 73–88.

Tainturier, G., Roullier, L., Martenot, J.P. and Lorient, D. (1992) Electroassisted glycosylation of bovine casein: An alternative to the use of reducing chemicals in N-alkylation of proteins. *J. Agric. Food Chem.*, **40**, 760–3.

Tan-Kintia, R.H. (1996) *Heat-induced Changes and the Heat Stability of Milk*, Ph.D thesis, National University of Ireland, Cork.

Tan-Kintia, R.H. and Fox, P.F. (1996) Effect of enzymic hydrolysis of lactose on the heat stability of milk or reconstituted milk. *Neth. Milk Dairy J.*, **50**, 267–77.

Tan-Kintia, R. and Fox, P.F. (1999) Effect of various preheat treatments on the heat stability of unconcentrated milk. *Int. Dairy J.*, **9**, 219–25.

Tessier, H. and Rose, D. (1964) Influence of κ-casein and β-lactoglobulin on the heat stability of skim milk. *J. Dairy Sci.*, **47**, 1047–51.

Thompson, M.P., Boswell, R.T., Martin, V., Jenness, R. and Kiddy, C.A. (1969) Casein-pellet-solvation and heat stability of individual cow's milk. *J. Dairy Sci.*, **52**, 796–8.

Tobias, J., McWhitney, R.M. and Tracy, P.H. (1952) Electrophoretic properties of milk proteins. II. Effect of heating at 300°F by means of the Mallory small tube heat exchanger. *J. Dairy Sci.*, **35**, 1036–45.

Trautman, J.C. and Swanson, A.M. (1958) Additional evidence of a stable complex between β-lactoglobulin and α-casein. *J. Dairy Sci.*, **41**, 715.

Turner, L.G., Swaisgood, H.E. and Hansen, A.P. (1978) Interaction of lactose and proteins of skim milk during ultrahigh-temperature processing. *J. Dairy Sci.*, **61**, 384–92.

Tziboula, A. (1997) Casein diversity in caprine milk and its relation to technological properties: heat stability. *Int. J. Dairy Technol.*, **4**, 134–7.

Tziboula, A. and Muir, D.D. (1993) Effect of starches on the heat stability of milk *Int. J. Food Sci. Technol.*, **28**, 13–24.

Tziboula, A., Steele, W., West, I. and Muir, D.D. (1998) Microfiltration of milk with ceramic membranes: influence of casein composition and heat stability. *Milchwissenschaft*, **53**, 8–11.

van Boekel, M.A.J.S. (1999) Heat-induced deamidation, dephosphorylation and breakdown of caseinate. *Int. Dairy J.*, **9**, 237–41.

van Boekel, M.A.J.S., Nieuwenhuijse, J.A. and Walstra, P. (1989a) The heat coagulation of milk: Mechanisms. *Neth. Milk Dairy J.*, **43**, 97–127.

van Boekel, M.A.J.S., Nieuwenhuijse, J.A. and Walstra, P. (1989b) The heat coagulation of milk. 3. Comparison of theory and experiment. *Neth. Milk Dairy J.*, **43**, 147–62.

van den Berg, G., Esher, J.T.M., de Koning, P.J. and Bovenhuis, H. (1992) Genetic polymorphism of κ-casein and β-lactoglobulin in relation to milk composition and processing properties. *Neth. Milk Dairy J.*, **46**, 145–68.

van Mil, P.J.J.M. and de Koning, J. (1992) Effect of heat treatment, stabilizing salts and seasonal variation on heat stability of reconstituted concentrated skim milk. *Neth. Milk Dairy J.*, **46**, 169–82.

van Winter, J.M. (1994) The consumption of dairy products in the Netherlands in the fifteenth and sixteenth century, in *Milk and Milk Products from Medieval to Modern Times*, (P. Lysght ed.) Canongate Academic, Edinburgh, pp. 3–13.

Vujicic, I.F. and Vulicic, M. (1991) The effects of forewarning on the heat stability of milk. *Mljekarstvo*, **41**, 59–64. [cited form *CAB Abstr.* 930462460].

Walstra. P. and Jenness, R. (1984) *Dairy Chemistry and Physics*, Wiley and Sons, New York.

Webb, B.H. (1928) Heat coagulation of evaporated milk as affected by mixing different grades of raw milk. *J. Dairy Sci.*, **11**, 471–8.

Webb, B.H. and Bell, R.W. (1942) The effect of high-temperature short-time forewarming of milk upon the heat stability of its evaporated product. *J. Dairy Sci.*, **25**, 301–11.

Webb, B.H. and Holm, G.E. (1932) The heat coagulation of milk. II. The influence of various added salts upon the heat stability of milks of different concentrations. *J. Dairy Sci.*, **15**, 345–66.

Whitaker, J.R. (1980) Changes occurring in proteins in alkaline solution, in *Chemical Deterioration of Proteins*. (J.R. Whitaker and M. Fujimaki eds.) American Chemical Society, Washington, DC, pp. 145–63.

White, J.C.D. and Davies, D.T. (1966) The stability of milk protein to heat. III. Objective measurement of heat stability of milk. *J. Dairy Res.*, **36**, 93–102.

Whiteley, A.J. and Muir, D.D. (1996) Heat stability of homogenised concentrated milk. 2. Synergic effect of addition of sodium caseinate and urea. *Milchwissenchaft*, **51**, 385–90.

Wong, D.W.S., Camirand, W.M. and Pavlath, A.E. (1996) Structure and functionalities of milk proteins. *CRC Crit. Rev. Food Sci. Nut.*, **36**, 807–44.

Zbikowska, A., Dziuba, J., Jaworska, H. and Zaborniaic, A. (1992) The influence of casein micelle size on selected functional properties of bulk milk proteins. *Polish J. Food Nut. Sci.*, **42**, 23–32.

Zbikowska, A., Dziuba, J., Kostyra, H. and Smoczynska, K. (1989) A study of the relationship between the molecular state and the physico-chemical properties of milk proteins. *Milchwissenschaft*, **44**, 25–8.

Zittle, C.A. (1969a) Influence of heat on κ-casein. *J. Dairy Sci.*, **52**, 12–16.

Zittle, C.A. (1969b) Influence of heat on κ-casein: effect of α_s-casein and concentration of calcium chloride and sodium chloride. *J. Dairy Sci.*, **52**, 1356–8.

Zittle, C.A., Dellamonicia, E.S., Rudd, R.K. and Custer, J.H. (1957) The binding of calcium ions by β-lactoglobulin both before and after aggregation by heating in the presence of calcium ions. *J. Am. Chem. Soc.* **79**, 4661–46.

20

PROTEIN STABILITY IN STERILISED MILK AND MILK PRODUCTS

JOHANNES A. NIEUWENHUIJSE AND MARTINUS A.J.S. VAN BOEKEL

20.1 INTRODUCTION

The principal aim of heat-treating foods is to produce microbiologically safe products and to extend shelf-life. Shelf-life may be defined as the period over which a product retains acceptable bacteriological, physical and (bio) chemical characteristics, which mostly comes down to more or less retaining the properties it had shortly after manufacture. Extension of shelf-life may be by some days only, even if the product is refrigerated, as for pasteurised milk, or by several years at ambient temperature, as for retort sterilised concentrated milk or sweetened condensed milk. Product composition, heat-treatment, storage conditions and, in addition, consumer attitude, all play a part in the attainable storage period.

During the storage of sterilised milk products, proteins change due to chemical or biochemical reactions and, in addition, the salt equilibria between serum and proteinaceous particles may change. These changes are treated in Section 20.3. Sometimes, changes in proteins result in age-gelation of the product, which gives the name to many studies and reviews on storage stability, but perhaps even more frequently they result in thinning rather than gelation, or in no physical change at all but in a flavour defect. These aspects are treated in Section 20.4. The chapter starts with a short overview of processing methods (Section 20.2).

We will focus on mechanisms rather than on covering all aspects of processing and composition; an excellent coverage of this is already included in the previous edition of this book (Harwalkar, 1992).

Advanced Dairy Chemistry Volume 1: Proteins, 3rd edn.
Edited by P.F. Fox and P.L.H. McSweeney, Kluwer Academic/Plenum Publishers, 2003.

20.2 Methods of preservation

Producing a liquid milk product with a shelf-life of more than a few weeks in the refrigerator requires the killing of all vegetative microorganisms and inactivation of spores of microorganisms that are capable of growing in the product at the relevant storage temperature. In addition, enzymes must be inactivated. Heating is by far the most widely applied method to achieve this. Two basic methods can be distinguished. One is continuous-flow heating, characterised by heating and cooling times of less than about 2 min, and holding at the top temperature of about 140°C mostly for a few seconds. The most common processes are direct and indirect ultra-high temperature (UHT)-heating. The former involves instantaneous heating from about 80°C to the sterilisation temperature accompanied by 10–15% dilution of the milk with water in the sterilisation holder, while the latter involves more gradual heating and no dilution. Ultrapasteurisation and ESL (Extended Shelf Life) heating are essentially variations on UHT. The other basic method involves batch heating, characterised by heat-treatment after packaging, using heating, holding and cooling times of about 10 min each. The most common containers are cans and glass or plastic bottles, and the holding temperature for sterilisation is about 120°C.

An essential additional processing step in the manufacture of concentrated products is preheating, i.e., heating of the milk before evaporation or reverse osmosis. The standard preheating for the production of evaporated milk is 0.5–3 min at 110–130°C (Walstra *et al.*, 1999).

Alternative methods to heating, like high-pressure, pulsed electric fields and microfiltration, have been studied for decades, but so far only microfiltration has found a limited application for the production of ESL-milk. The storage stability of products treated by these methods may be totally different from that of a conventionally heated product having similar microbiological characteristics, because chemical and physical changes caused by these processes may be very different from those caused by heating. We will not discuss products treated by these methods any further.

20.3 Physical and chemical changes caused by heating and storage

20.3.1 Thermodynamic and kinetic aspects

Physical and chemical reactions occur during heating and storage of milk. These reactions are driven by thermodynamics and are (mostly) kinetically controlled; it thus seems useful to discuss briefly the thermodynamic and kinetic aspects of such changes. [see van Boekel and Walstra (1995) for an extensive discussion of thermodynamics and kinetics, with examples from milk products.]

Thermodynamics is important because it tells us whether or not a reaction is possible at all, and in what direction it will go. Thermodynamics is concerned with the equilibrium position and the effect of temperature and pressure on it. It is also important in understanding the concept of activity, because ultimately it is the activity, not concentration, that determines the reactivity or effective concentration of a component. A useful formula in this respect is:

$$a_i = \gamma_i c_i$$

in which a_i is the activity of a component i, γ_i is the activity coefficient (related to molar scale) and c_i is the molar concentration. The activity coefficient takes non-ideal behaviour into account. Unfortunately, concentrations are much easier to determine than activities, with the notable exception of ion-selective electrodes and pH electrodes which measure activities directly. The activity coefficient in milk for mono-, di- and trivalent ions is estimated to be 0.8, 0.4 and 0.13, respectively (Walstra *et al.*, 1999); thus, there can be an enormous difference between concentration and activity. In milk and milk products, activities are important with respect to solubility (salts, lactose), crystallisation (fat, lactose, salts), partition equilibrium between fat and plasma (free fatty acids) and adsorption phenomena.

The driving force for a reaction is the difference in free energy between reactants and products. If the free energy decreases (either by enthalpy or entropy changes), the reaction will proceed spontaneously. The change in free energy is determined by the chemical potential of reactants and products, which is in turn governed by the activity of the reactants and products. If the composition of a reaction mixture is governed by the equilibrium position, the reaction is said to be thermodynamically controlled. Such situations can occur with salt equilibria in milk.

However, if it follows from thermodynamic considerations that a reaction can occur, this does not mean that it will take place. This brings us into the realm of kinetics: it is the driving force of a reaction divided by a possible resistance which determines what happens. In other words, it may take a long time before equilibrium is reached. If the composition of a reaction mixture is not determined by its equilibrium position, the reaction is said to be kinetically controlled. This is the case for most reactions in foods. In fact, it is a most important task for the food technologist to prevent reactions from reaching equilibrium because that would mean total spoilage in many cases.

This chapter on storage stability is thus mostly on the kinetics of changes that occur. A quantitative description of this would be ideal, using the appropriate kinetic equations. A very important effect in relation to storage is that of temperature. The effect of temperature on rate constants is usually described *via* the Arrhenius equation, or its more theoretically based counterpart, the Eyring equation (van Boekel and Walstra, 1995). These equations are certainly useful in describing changes during storage. In a way, storage can be considered as a heat-treatment at rather low temperature but with a very long holding time, so that changes do become noticeable. However, the changes that are described in this chapter are not yet known in such a way, one exception being the Maillard reaction as studied by the formation of hydroxymethylfurfural (Kessler and Fink, 1988). We will therefore not go into a quantitative treatment of kinetics, although we would like to make a plea for future research in this direction with respect to the storage stability of milk products.

20.3.2 Salts

Milk is supersaturated with respect to calcium, magnesium, phosphate and citrate, but their salts are kept in dispersion because a large part of the calcium and phosphate are incorporated into the casein micelles. A useful model with respect to storage stability proposes the presence of a calcium-phosphate-casein phase ('Micellar Calcium Phosphate'; MCP), which mostly behaves as if in thermodynamic equilibrium with the serum phase (Holt, 1995). Thus, heat-induced changes in salts are largely reversible on cooling, as observed by, e.g., Rose and Tessier (1959) and Geerts *et al.* (1981). Concentrating milk results in the association of additional calcium and phosphate with the MCP, apparently without greatly changing the composition of the MCP (Nieuwenhuijse *et al.*, 1988). This study indicated that sterilisation, especially of concentrated milk with added phosphate, may cause some irreversible changes.

However, at least one change during heating and storage may induce irreversible changes in salts: acids (mainly formic) are formed by degradation of lactose (Berg and van Boekel, 1994). For milk, this results in a slight decrease in pH during UHT-sterilisation and a decrease by 0.2–0.3 units during batch sterilisation. During storage at 5–10°C, the pH of UHT-sterilised milk remains constant, but it decreases at a higher storage temperature, e.g., from 6.75 to 6.5 in 6 months at 30°C (Kocak and Zadow, 1985a; Kondal Reddy *et al.*, 1991). For higher storage temperature and longer times, much lower values were found, e.g., pH 5.6 for milk containing a gelled sediment after 34 mo at 37°C (Andrews *et al.*, 1977). Evaporated milk has a lower pH than milk, due mainly to the liberation of H^+ by association of calcium phosphate, and the pH mostly decreases further during sterilisation (Nieuwenhuijse *et al.*, 1988). For UHT-treated evaporated milk, the pH remains more or less constant at 30°C (Aoki and Imamura, 1974a; Harwalkar and Vreeman, 1978a), whereas that of retort-sterilised evaporated milk decreases to a similar extent as that of milk, pH becoming as low as 5.5 after 1 year at 40°C without gelation or coagulation of the sample (Nieuwenhuijse, unpublished results).

A second change is the dephosphorylation of casein, but it can be estimated from the data of Belec and Jenness (1962) that the extent during UHT- or retort-sterilisation of milk is still very limited. This is in accordance with the results by Aoki *et al.* (1990) for milk and Nieuwenhuijse *et al.* (1988) for evaporated milk. Dephosphorylation was not observed during the storage of UHT-sterilised evaporated milk (Harwalkar and Vreeman, 1978a).

The effect of storage of heated milk on the partition of salts appears to have been studied only by Aoki and Imamura (1974b). These authors investigated changes in the ultracentrifugal supernatant (110,000 × g at 4°C) of milks stored at 30°C. UHT-heated samples had an initial pH of 6.6, while that of autoclaved samples was 6.3. Ca in the supernatant of the UHT-milk decreased by about 1.2 mM (or 10%) during the initial 2 months of storage, and by an additional 0.7 mM during the subsequent 13 months. Inorganic P (which is presumably in the form of inorganic PO_4) in the supernatant decreased by about 0.8 mM during the total storage period, while Mg remained virtually constant. In the autoclaved milk, Ca in the ultracentrifugal supernatant increased from about 10.5 to about 11.5 mM during the first 3 months of storage and remained constant thereafter. Inorganic P increased by 0.4 mM, while Mg remained constant. Summarising, changes in these salts during storage were small, indicating a high stability of the MCP in milk during long storage. It is remarkable, however, that the decrease in pH during storage does not induce solubilisation of Ca and PO_4 – but the same is true for heating.

In UHT-heated evaporated milk, the concentration of Ca in the supernatant decreased from 21 mM immediately after sterilisation to 16 mM after 60 days, inorganic P decreased from 35 to 25 mM, while Mg increased from 7.3 to 9.6 mM (Aoki and Imamura, 1974a). During the remainder of the storage period, concentrations remained more or less constant. Deysher and Webb (1952) reported the formation of large (several mm) calcium citrate crystals in retort-sterilised evaporated milk; if storage temperature was 30 or 40°C, crystals were formed within a few months. Addition of orthophosphate was found to prevent their formation, at least for storage up to 1 year. Fox *et al.* (1967) reported that commercial samples, presumably with added orthophosphate, contained granules (size 1–8 mm) containing mainly $CaHPO_4$ after storage at 4°C for at least 3 years. Samples stored at room temperature contained calcium citrate crystals after about 10 months. These observations indicate that in evaporated milk, MCP may change into some other thermodynamically more stable form.

Changes in salts are relevant for the storage stability of the proteins in milk products, because salts affect colloidal interactions and because changes in the MCP may result in a less effective association of casein with MCP (van Dijk, 1992). Aoki *et al.* (1990) showed that cross-linking of casein by MCP is reduced after UHT heating of 'whey protein-free' milk. A longer heating time gave less MCP cross-linked casein, and evaporated milk had less cross-linked casein after the same heat-treatment than milk. Such data are not yet available for any effect of storage, but the dissociation of casein molecules from the micelles during storage (see Section 20.3.3 "Protein Aggregation and Dissociation") point to such a change.

20.3.3 Protein aggregation and dissociation

The bovine casein micelle is a very stable, and dynamic, colloidal particle. After heating above about 75°C, these particles are altered due to association of denatured serum proteins with the casein micelles, and to the dissociation of casein molecules (or small aggregates of molecules) from the micelles. Very briefly, at least 50% of the serum proteins are denatured in direct UHT-treated milk, and virtually all in retort-sterilised products (Burton, 1988). In these products, denatured serum proteins are associated among themselves, in small particles with dissociated (mainly) κ-casein, and with the casein micelles (e.g., Smiths and van Brouwershaven, 1980). Different heating processes will cause differently aggregated proteins, as can be concluded for results of, e.g., Metwalli *et al.* (1996) – a longer heating time or a higher heating temperature generally yield less protein in the casein 'micelles', with κ-casein dissociating fastest and most extensively. At the natural pH and after the same heating, evaporated milk contains a lower proportion of protein in the casein 'micelles' than milk (Fox *et al.*, 1966; Nieuwenhuijse *et al.*, 1991).

As observed in electron micrographs, the volume-average size of the casein micelles in direct UHT milk is about the same as in raw milk (Rollema *et al.*, 1987). However, micrographs of UHT milk showed a large number of small particles, presumably consisting of aggregated serum proteins and/or dissociated caseins. Particles in UHT-sterilised evaporated skim milk were considerably larger (roughly 250 nm) than those in raw milk (Schmidt, 1968; Harwalkar and Vreeman, 1978b) These 'micelles' were almost spherical, had a spiky surface, and few small particles were present. The size and appearance of the particles in UHT-sterilised, UF-concentrated milks (Venkatachalam *et al.*, 1993; McMahon, 1996) were very similar to those in UHT-sterilised evaporated milk. Apparently, the heat-induced coagulation of casein micelles in UF-concentrated milk is similar to that in evaporated milk rather than to that in milk, as can also be concluded from the shape of the HCT-pH plot for UF-concentrated milk, which resembles that of evaporated milk (Muir and Sweetsur, 1984). In retort-sterilised evaporated milk, 'micelles' were also roughly 250 nm, but were aggregated into small clusters and were less spiky, and many small particles were present (Schmidt, 1968).

Summarising, proteinaceous particles in sterilised milk products are, from the beginning of storage, different from the casein micelles in raw milk.

Association and dissociation of proteins continue during storage. Studies on this appear to have been pioneered in the United States: Melnychyn (unpublished results mentioned in Carroll *et al.*, 1971) found that 35% of the protein was not ultracentrifugable in evaporated milk immediately after sterilisation and 65% after prolonged storage. Similarly, Aoki and Imamura (1974a,b) reported that for UHT-treated evaporated milk, total N minus non-casein N (NCN) in the supernatant increased from l to 3.5%, i.e., about 50% of the total protein, in 60 days at 30°C. Since the content of organic phosphorus in the supernatant did not change, while calcium and inorganic phosphate decreased, it can be tentatively concluded that mainly κ-casein and whey proteins dissociated, similar to the observations by McMahon (1996, see below) or UHT-sterilised UF-concentrated milk.

For UHT unconcentrated milk, Aoki and Imamura (1974a,b) found no change for total N minus NCN, i.e., dissociation did not occur. In autoclaved milk, 'not-sedimentable' casein increased from about 15 to 25% in 1 year at 30°C. Thus, dissociation in milk appears to be much less than in evaporated milk or UF-concentrated milk.

Electron micrographs by Schmidt (1968), Harwalkar and Vreeman (1978b) and McMahon (1996) also show changes that strongly indicate dissociation of proteinaceous material in UHT-sterilised evaporated or UF-concentrated milk: immediately after sterilisation, the 'micelles' were spiky and the serum was almost devoid of particles, while the micelles were smooth and the serum contained many small particles after 6-10 weeks of storage at 28°C, or 4–8 mo at 22°C for UF-concentrate. The results of Schmidt (1968) indicate that dissociation is less in retort-sterilised evaporated milk, but more small particles were already present in the serum shortly after sterilisation (see above). This author also showed that the dissociated protein contains a high proportion of κ-casein. From the work by McMahon (1996), using immunogold labelling and electron microscopy on UF-concentrated UHT-milk, β-lactoglobulin and κ-casein appear to be the main proteins that dissociate, whereas α_s- and β-caseins dissociate only after 10 or more months of storage.

The above results agree reasonably well with some observations on milk and evaporated milk immediately after sterilisation, which indicate that in heat-treated milk κ-casein is much more susceptible to dissociation than in unheated milk. For example, van Hooydonk *et al.* (1987) found that 40% of the κ-casein had dissociated in milk heated for 5 min at 120°C and stored at 4°C for 1 day, whereas Nieuwenhuijse *et al.* (1991) found that only 8% of the κ-casein had dissociated in milk heated for 3 min at 120°C and stored overnight at 20°C. Thus, presuming that 5 min heating gives no great increase in dissociated casein compared to 3 min heating, the difference between 40% and 8% is caused largely by storing the milk in the cold.

Similarly, Nieuwenhuijse *et al.* (1991) reported 14% and 21% of dissociated κ-casein in (not heated) evaporated milk made from preheated milk at pH 6.5 and 7.0, respectively, whereas evaporated milk made from pasteurised milk contained only 9% of dissociated κ-casein at all pH values. Thus, increasing the pH of preheated evaporated milk at room temperature resulted in more dissociated κ-casein.

20.3.4 Covalent cross-linking of proteins

Heating and storage of milk products cause covalent cross-linking of the proteins. Obviously, S—S cross-linked aggregates of κ-casein and β-lactoglobulin (and presumably α-lactalbumin also) make up part of the covalently cross-linked proteins in heated milks (Singh, 1995). By gel permeation chromatography in dissociating solvents, Andrews (1975) estimated that about 25% of the proteins in milk immediately after indirect UHT-sterilisation (1.5 s at 139°C) were present in aggregates cross-linked by S—S bridges, which is less than the sum of cysteine-containing proteins. The fraction of protein in aggregates that dissociated in the presence of 2-mercaptoethanol (i.e., in aggregates that were linked by S—S bridges only) decreased during storage of UHT-milk. Presumably, this indicates that aggregates that are linked by S-S bridges after heating become cross-linked by other bonds as well, the S-S bridges remaining intact, since Wilson *et al.* (1963) found that titratable SH groups also decreased during storage of evaporated milk.

Andrews and Cheeseman (1972) reported that 5% of the proteins in UHT milk obtained from a retail outlet were covalently cross-linked by non-S—S bridges, and 11% in retort-sterilised milk. Similar values were found by Zin El-Din *et al.* (1991) using gel permeation HPLC in the presence of 2-mercaptoethanol: 7.5% of the proteins in milk after indirect UHT-sterilisation for 2 s at 138°C were covalently cross-linked by non S—S bridges, compared to 2.5% of the proteins in raw milk. During storage of UHT milk, more protein molecules become cross-linked by covalent bonds other than S—S bridges. Andrews (1975) found for UHT milk that per month of storage at 4°C, about 0.7% of the total protein became cross-linked in large aggregates, and at 30°C about 5.8%. Zin El-Din *et al.* (1991) found the increase to be the same at 4°C, but to be 3.7% per month at 30°C. Analytical differences may have caused part of this difference. More important perhaps was the extensive proteolysis which occurred during storage of the milk at 30°C in the latter investigation, since polymerisation with concomitant proteolysis is likely to result in smaller aggregates than polymerisation only.

Since casein heated in the absence of lactose contained hardly any polymerised protein, Andrews and Cheeseman (1972) presumed that the

Maillard reaction was responsible for the non S—S type of cross-linking during heat treatment. This is in agreement with the results of Pellegrino *et al.* (1999) which showed for lactose-free model systems that covalent aggregation of β-casein did occur in the presence of glucose (i.e., a reducing sugar) but not in its absence. Covalent cross-linking during storage was also attributed to the Maillard reaction by Andrews and Cheeseman (1972). Möller *et al.* (1977) and Jimenez-Perez *et al.* (1992) demonstrated the formation of Maillard reaction products (lactulosylysine and fructosylysine, and 5-hydroxymethyl-furfural, respectively) in milk during storage at 30°C and higher. On the other hand, Venkatachalam *et al.* (1993) concluded that the concentration of lactose in UF concentrated milk (<0.05, 3 and 6%) and the formation of high molecular weight aggregates as detected by sodium dodecylsulphate polyacrylamide gel electrophoresis (SDS-PAGE) in the presence of β-mercaptoethanol were not related. However, according to the present authors, the gel for the sample with <0.05% lactose seems to contain less high molecular weight material than the others, and SDS-PAGE may be less sensitive in detecting aggregated proteins than gel permeation methods. Still, other covalent cross-linking reactions appear to be at least as important as the Maillard reaction, but nothing is known about the nature of these.

As occurs during heat treatment, covalent cross-links may be formed during storage between proteins in the same casein micelle, or between molecules in different particles. Intra-micellar bonds seem to predominate during storage, since flocculation of the micelles is usually not observed until shortly before gelation (if this occurs), even at high temperatures, where the extent of cross-linking is highest.

20.3.5 Proteolysis

Milk contains many enzymes and, in addition, enzymes produced by contaminating microorganisms may be present. Some of the enzymes of either origin are not completely inactivated by sterilisation. Indigenous enzymes are discussed in Chapter 11; in particular, plasmin is sufficiently heat-stable to play a part in the 'age-gelation' of UHT milk.

Proteinases produced by some psychrotrophic bacteria have a much higher heat stability than plasmin. For instance, Driessen (1983) found D values at 130°C of 8 and 11 min for the extracellular proteinase of *Achromobacter* spp. 1–10 and *P. fluorescens* 22F, respectively. Such D values result in residual activity even after retort sterilisation. Consequently, the production of the enzymes before heat-treatment rather than the intensity of heat treatment will determine residual activity. A special heat treatment, the so-called low temperature inactivation (LTI) treatment, i.e., heating for, e.g., 30 min at about 50°C, is worthwhile mentioning here. The LTI treatment results in inactivation of the enzyme by autoproteolysis; however, in the presence of other proteins this occurs to only a limited extent (Schokker and van Boekel, 1998a,b). It should also be noted that such a heat treatment does not result in an appreciable inactivation of plasmin (Metwalli *et al.*, 1998). Production of proteinases by psychrotrophic bacteria occurs mainly at the end of the exponential growth phase (e.g., Law *et al.*, 1977). Therefore, enzymes should not be present in products produced from good quality milk, but several studies indicated the presence of bacterial proteinases in commercial UHT milk (López-Fandiño *et al.*, 1993; García-Risco *et al.*, 1999).

Residual proteolytic enzymes will result in proteolysis during storage, if conditions allow activity. Two approaches for investigating proteolysis during storage may be distinguished. One is to investigate changes in enzyme activity (the 'amount' of enzyme present) during storage using a (usually chromogenic) substrate under conditions of optimum activity. By using a method specific for plasmin/plasminogen, de Koning *et al.* (1985) and Manji *et al.* (1986) found that plasmin activity was low immediately after sterilisation and that the plasminogen present after sterilisation was converted to plasmin during the first few months of storage. Some loss of plasmin + plasminogen occurred during these months. Plasmin activity did not decrease upon longer storage, in agreement with Kohlman *et al.* (1991) who found no change in activity if plasmin was added aseptically after sterilisation. Activation of plasminogen was slower at 4 than at 24 or 37°C and full conversion appeared not to occur at 4°C (Manji *et al.*, 1986). Results by Kelly and Foley (1997) indicate that the rate and extent of conversion of plasminogen to plasmin is higher in milk having a high somatic cell count, these cells being a source of, apparently heat stable, plasminogen activators. The most important effect of this gradual conversion of plasminogen to plasmin during the storage of UHT milk is that the maximum rate of proteolysis by plasmin will occur only after some time of storage, as was confirmed by Kelly and Foley (1997). Specific measurement of the activity of bacterial proteinases by the Azocoll method was possible immediately after UHT-heating, but not after some months of storage (Rollema *et al.*, 1990).

A second approach is to determine the hydrolysis of milk proteins in the product during storage. Fractionation of the proteins into total N, NCN, and non-protein N (NPN), determination of trichloroacetic acid (TCA)-soluble NH_2 groups, electrophoresis and HPLC methods have been used most often. Fractionation of proteins is simple and gives, for low degrees of conversion, some information about the type of enzyme involved: plasmin gives an increase in NCN but hardly any in NPN, whereas both NCN and NPN increase following proteolysis by (at least some frequently occurring) bacterial proteinases (Driessen, 1983). The method is not very sensitive, however, and does not show which types of protein are broken down.

Measurement of proteolysis as the increase in TCA-soluble free amino groups shows overall breakdown of protein, and proteolysis will become apparent only when (small) TCA-soluble free peptides have been produced. Consequently, a lag phase is to be expected for analytical reasons only. At a higher storage temperature, the extent of proteolysis is likely to be underestimated by this method, since hydrolysis of covalently cross-linked proteins and covalent cross-linking of peptides will yield less TCA-soluble material, which explains results obtained by Kocak and Zadow (1985a). In addition, free amino groups will participate in the Maillard reaction. Electrophoresis shows the decrease of (almost) unchanged proteins, as well as the formation of the large fragments, like γ-casein and *para*-κ-casein; small peptides are not detected. Quantification of the extent of proteolysis is possible, at least for a low extent of conversion (Driessen, 1983). At a high storage temperature, and for intensely heated products, quantification is less accurate or even impossible due to loss of sharpness of the bands and the appearance of streaky patterns, which are caused by chemical reactions (e.g., Venkatachalam *et al.*, 1993). The pros and cons of HPLC methods are similar to those for electrophoresis; the main advantage of HPLC is that the detection of small fragments is possible. Both electrophoretic and (especially reversed-phase-) HPLC methods provide some insight into whether plasmin or bacterial proteinases are involved in proteolysis (Driessen, 1983; López-Fandiñio *et al.*, 1993).

Evidence for proteolysis during the storage of UHT-milk is abundant. Only for investigations in which aseptically drawn milk was used, as by Snoeren *et al.* (1979), Driessen (1983) and de Koning *et al.* (1985), is it certain that proteolysis by plasmin only was involved. For direct UHT-sterilised (2 s at 142°C) milk, Snoeren *et al.* (1979) reported that complete breakdown of β- and α_{s2}-caseins occurred in about 60 days at 28°C, and of α_{s1}-casein in about 100 days; about 50% of the κ-casein was, however, still intact after 100 days. Similar rates of proteolysis, except for κ-casein that was hardly hydrolysed, were reported by de Koning *et al.* (1985). Proteolysis in indirect UHT milk (2 s at 143°C) could not be detected by electrophoresis and NPN in this milk increased only very slightly during storage (Snoeren and Both, 1982). In UHT sterilised evaporated milk, made from milk preheated for 3 min at 120°C, proteolysis was absent (de Koning *et al.*, 1985). All of this is in agreement with the inactivation kinetics of plasmin during heat treatment.

For some other investigations, in which fresh milk of good bacteriological quality was sterilised shortly after reception, one may safely assume that proteolysis by bacterial proteinases must have been virtually absent, considering that the aforementioned enzymes are produced at the end of the exponential growth phase. Examples are the studies by Corradini (1975), Guthy *et al.* (1983) and Manji *et al.* (1986). The last named authors found that the difference in the extent of proteolysis between direct and indirect

UHT milk correlated with the difference in plasmin activity. Activity was measured by using a chromogenic substrate and proteolysis was determined by quantifying 6% TCA-soluble NH_2 groups. Even for the study by Kocak and Zadow (1985a), in which the raw milk was used to study the temperature dependency of various changes during storage had a (psychrotrophic) plate count of about 1×10^5 cfu ml^{-1} after 116 h storage at 2°C, one may assume that proteolysis was due to plasmin only. The optimum temperature for proteolysis was 30–40°C (Kocak and Zadow, 1985a), but appreciable proteolysis (as measured by the increase of NCN) occurred at 20°C (Guthy *et al.*, 1983).

For studies on proteolysis by bacterial proteinases in which indirect UHT sterilised milk was used, it is fairly certain that bacterial proteinases only were involved, and not proteolysis by plasmin as well, because this is inactivated during sterilisation. Examples are Richardson and Newstead (1979) and Mitchell and Ewings (1985), in which indirect UHT milk was inoculated with isolated bacterial proteinases, Law *et al.* (1977), in which milk was inoculated with psychrotrophic bacteria, and Snoeren and Both (1981), in which raw milk was aged until the bacterial count was higher than 1×10^7 cfu ml^{-1} before heating. Rapid conversion of κ-casein to *para*-κ-casein-like material was found in all studies. Breakdown of the other caseins was not observed by, e.g., Snoeren and Both (1981), but other authors (e.g., Law *et al.*, 1977; Mitchell and Ewings, 1985) did find breakdown of β-casein and some α_s-casein as well. These observations suggest that all of the highly heat stable bacterial proteinases attack κ-casein, but only some hydrolyse β-casein.

From the results of several other investigations, it is not clear whether proteolysis was due to plasmin, bacterial proteinases, or both. López-Fandiño *et al.* (1993), who investigated proteolysis in milk produced on commercial direct or indirect UHT plants, found up to 58% degradation of κ-casein and up to 13% of β-casein in direct UHT milk; degradation was about half as much in indirect UHT milk, indicating proteolysis by predominantly bacterial proteinases. Judging from the denaturation of serum proteins in the milk, residual plasmin should be present in the direct UHT milk but hardly in the indirectly heated milk. Similar results were reported by García-Risco *et al.* (1999). Kocak and Zadow (1985a) found a different rate of proteolysis (measured as TCA-soluble amino acids) for 21 batches of aged raw milk that was subsequently direct UHT sterilised. Plate counts ranged between 9.1×10^4 and 7.2×10^7 cfu ml^{-1}. Clearly, proteolysis by plasmin must have occurred in all batches, whereas proteolysis by bacterial proteinases is likely to have occurred only in batches with a plate count higher than about 1×10^7 cfu ml^{-1}.

Proteolysis may result in deterioration of the flavour of heat-treated milk products and in physical instability; the latter will be discussed in Section 20.4 "Physical Instability". In cases where plasmin was the only proteinase involved, bitterness was reported to develop only after extensive hydrolysis of the caseins (Driessen, 1983). A clearly perceptible off-flavour usually occurred somewhat earlier than physical instability. For proteolysis by bacterial proteinases, bitterness frequently developed at a much lower level of protein breakdown (Mitchell and Ewings, 1985); the level of protein breakdown at which bitterness became perceivable differed considerably for the various proteinases investigated. For most enzymes, bitterness was perceptible before physical instability.

20.4 Physical instability

20.4.1 General aspects

Sterilised milk products are usually considered stable during storage for as long as they remain more or less homogeneous. However, even if they are homogenised and contain casein micelles of the same size as in raw milk, they are not stable over a period of many months: fat globules rise to the top of the container, and (especially the large) casein micelles sink to the bottom. This is not considered unacceptable, generally. Storage stability becomes unacceptable if some type of aggregation of protein, or of protein-covered fat globules, occurs, leading to accelerated creaming/sedimentation or to a gel.

Below, we will discuss physical instability in two sections: one on 'gelation', in which physical instability (i.e., gelation, sedimentation, flocculation, etc.) due to physico-chemical changes in the product during storage will be discussed, and one on sedimentation and/or creaming, in which long-term effects of aggregation during the sterilisation process will be treated briefly.

20.4.2 Age-'gelation'

If particles change during storage of the product (e.g., due to proteolysis and/or aggregation), this is usually presumed to result in a gel. Harwalkar (1992) defined age gelation as being 'characterised by loss of fluidity of the product as a result of changes during storage'. However, many authors used 'coagulation' or 'destabilisation' to describe the instability, and for UHT-sterilised whole milk, Visser (1981) reported that, 'the presence of bacterial proteinase in whole milk leads to the formation of a sediment, while the natural proteinase causes creaming with a more or less clear serum layer at the bottom'.

Viscometry, using Brookfield, torsion or falling ball viscometers, was used in a number of investigations to measure the progress of the aggregation reaction leading to gelation. Graphs showed an initial decrease in viscosity, followed by a period in which it hardly changed and then a sharp increase shortly before gelation (e.g., Tarassuk and Tamsma, 1956; de Koning *et al.*, 1985; Kocak and Zadow, 1985a). Such figures are, however, an over-simplification, since various types of gel were reported to form, ranging from weak and transparent (de Koning *et al.*, 1985) to rennet-like (Snoeren *et al.*, 1979). For some (evaporated) milks, it was even observed that age thickening was followed by thinning (Kocak and Zadow, 1985a; de Koning *et al.*, 1992). These observations clearly show the need for more sophisticated rheological studies on age gelation.

Below, an attempt is made to arrange the various types of age-'gelation' by cause.

(a) Gelation induced by proteolysis of κ-casein

Several investigations showed that UHT-milk gelled during storage if it was produced from milk having a high count of psychrotrophic bacteria before sterilisation. Most of the κ-casein was broken down at the time of gelation. One of the earliest studies which clearly showed this was that by Law *et al.* (1977), who found that gelation of UHT-sterilised milk occurred after 12 days or 9 weeks at 20°C if (inoculated) *Ps. fluorescens* had grown in the milk before sterilisation to 5×10^7 or 8×10^6 cfu ml^{-1}, respectively, but not if the colony count was 8×10^5 ml^{-1} or lower. Breakdown of κ-casein to *para*-κ-casein occurred only in the milks having the higher bacterial counts. Similarly, Snoeren *et al.* (1979) found that direct UHT-sterilised milk produced from milk to which a culture of psychrotrophic bacteria had been added gelled after 3 weeks of storage at 28°C, while a similarly-treated indirect UHT-sterilised milk gelled after 12 weeks (Snoeren and Both, 1981). The gel was 'reminiscent of a rennet curd', i.e., likely to be a particulate gel. Driessen (1983) reported that milk in which psychrotrophic bacteria had been allowed to produce proteolytic enzymes before direct UHT-sterilisation, 'destabilised' after 6 weeks at 20°C. Neither milk sterilised fresh nor milk sterilised before bacteria had produced a measurable enzyme activity 'destabilised' upon storage. Also, UHT-milk to which cell-free supernatants of various *Pseudomonas* or *Serratia* cultures had been added before (Richardson and Newstead, 1979) or after (Mitchell and Ewings, 1985) sterilisation, invariably gelled upon storage, κ-casein being extensively degraded at the point of gelation.

Thus, there is ample evidence that age-gelation of UHT-sterilised milk occurs as a result of hydrolysis of κ-casein similar to that by rennet. The subsequent aggregation of the denuded particles is, however, different from that of renneted casein micelles. A major difference is that it is heated milk which gels during storage, whereas the aggregation of *para*-κ-casein 'micelles' in heated milk is known to proceed very slowly (van Hooydonk *et al.*, 1987) – but on the other hand, age-gelation is fast if it happens in weeks, and renneting slow if it takes hours. Also, the percentage of κ-casein converted at the point of age-gelation varies considerably, and may be as low as 50%, although it is usually higher (e.g., Mitchell and Ewings, 1985). The effect of added hexametaphosphate (HMP) is also different: 0.5 g HMP/1 (Snoeren *et al.*, 1979; Snoeren and Both, 1981) delayed age-gelation only slightly, whereas 0.3 g HMP/l prolonged renneting time tenfold (Odagiri and Nickerson, 1964). Since data on particle aggregation during age-gelation, e.g., by light scattering techniques or rheological studies, are lacking, one can only speculate on possible causes for these differences. Perhaps, the very slow rate of hydrolysis of κ-casein during age-gelation, compared to that during renneting, plays a part, allowing slow aggregation of partly-denuded casein particles, not only by colloidal interactions, but also by bond formation between reactive groups on the remaining κ-casein hairs in a depleted hairy layer, as suggested by Walstra (1990).

This type of gelation does not occur very often in practice, even though cold-storage of milk is common practice, because enzyme production is virtually absent during the exponential growth phase of psychrotrophic bacteria (Law *et al.*, 1977; Driessen, 1983). As already discussed, this explains the absence of a relation between psychrotrophic count, proteolysis and gelation time in, e.g., the study by Kocak and Zadow (1985a). Finally, it should be noted that even retort-sterilised products may be destabilised by this mechanism, because of the high heat stability of some bacterial proteinases.

(b) Destabilisation due to proteolysis by plasmin

A clear example of this type of age-'gelation' is in a paper by de Koning *et al.* (1985). These authors observed gelation of aseptically drawn, direct (2 s at 142°C) UHT-sterilised, skim milk after 10 weeks at 20°C. The gel was weak, transparent and easily redispersible. In electron micrographs, it appeared to be built of threads and irregular knots of protein rather than of casein micelles. After 3 weeks, all α_{s2}- and β-caseins were degraded, and at the time of gelation (10 weeks) most of the α_{s1}-casein as well; κ-casein was still intact, however. Thus, polypeptides must be the main building blocks of the gel. The same milk containing plasmin inhibitor did not gel, proteolysis of casein did not occur, the integrity of the casein micelles was maintained and a small amount of sediment was observed after 9 months of storage. Many other papers (e.g., Snoeren *et al.*, 1979; Guthy *et al.*, 1983; Manji *et al.*, 1986) report similar observations for (direct-) UHT milk produced from good quality milk. Again, both the rate of proteolysis and the rate of aggregation have to be taken into account when discussing this type of storage instability. Several aspects can be distinguished:

1) Proteolysis is fast at a high residual or added enzyme activity or at a temperature close to the optimum. However, this appears to result in a solution of (poly)peptides and small particles, while gelation occurs at lower rates of proteolysis. Clear examples of the effect of the amount of plasmin are the studies by Kohlmann *et al.* (1988, 1991). UHT-sterilised

milk to which 0.3 or 1.5 mg plasmin/l had been added aseptically after sterilisation became yellow after a few weeks at 25°C and clots (presumably consisting mainly of fat globules) formed on the surface of the milk. However, a gel was formed (initially only at the bottom of the container, and later in a larger part) if 0.15 mg plasmin/l were added, or after addition of 0.3 mg plasminogen/l. A similar relation between plasmin/plasminogen activity and gelation/destabilisation was observed by Kelly and Foley (1997). The effect of temperature is best observed in the various studies using direct UHT-sterilised milk, which still contains an appreciable level of plasmin. Gelation was usually found to be most rapid at 20–25°C; at a lower temperature, gelation was slower, while it did not, or only after very long storage, occur above 30°C (Zadow and Chituta, 1975; Guthy *et al.*, 1983; Kocak and Zadow, 1985a; Manji *et al.*, 1986). Presumably, at a high storage temperature, gelation does not occur in direct UHT milk because proteolysis by plasmin is too fast. Still, it appears that fast proteolysis at these higher temperatures can result in aggregates since Guthy *et al.* (1983) observed that the increase in the volume of aggregates (as measured by the Coulter counter) was slightly more rapid at a storage temperature of 35°C than at 20°C. At 20°C, gelation occurred a few weeks after the increase in particle volume, while at 35°C a voluminous sediment formed after about 9 months.

2) For a 'fine-stranded'gel to be formed, polypeptides must diffuse into the serum, or at least protrude from the remaining particles. From the observation by Aoki *et al.* (1974b) that the NCN, but not the casein N, content of ultracentrifugal supernatant increased during storage of UHT milk (see Section 20.3.3), it can be concluded that pH 4.6-soluble polypeptides diffuse into the serum, but not the pH 4.6-insoluble material (which is intact casein and presumably also breakdown products).
3) The type of bonds involved and the rate of bond formation must be important, but very little is known about either. From the study by Manji and Kakuda (1988), it can be concluded that the aggregation of the polypeptides (and residual protein particles) is faster at a higher temperature. These authors applied a high level of induced proteolysis, which was stopped before the start of the storage temperature, and observed faster gelation at a higher temperature. Thus, the rate of bond formation is higher at a higher temperature, as is expected for chemical cross-linking, formation of hydrophobic bonds, or formation of salt bridges. Added salts may change gelation time *via* their effect on the rate of bond formation and on the physico-chemical properties of the casein micelles, but not by an effect on proteolysis (Kocak and Zadow, 1985b); for an overview, we refer to the previous edition of this book (Harwalkar, 1992). Typically, addition of orthophosphate (e.g., Snoeren *et al.*, 1979) and of citrate (Kocak and Zadow, 1985b) slightly reduces the gelation time of milk, but the latter authors also observed that although 0.1%

citrate reduced gelation time, 0.3% did not. From these observations, it can tentatively be concluded that Ca-bridges are not important for the aggregation of the polypeptides and remaining particles. Addition of HMP to the milk strongly delayed gelation of portions in which extensive proteolysis by plasmin (Snoeren *et al.*, 1979) or by plasmin and perhaps also bacterial proteinases (Kocak and Zadow, 1985b, see Section 20.3.5) occurred. HMP was observed to associate with the casein micelles (Ogadiri and Nickerson, 1965), thereby causing considerable dissociation of protein molecules from the micelles, without dissolution of MCP (Morr, 1967). Presumably, HMP associates with the peptides derived from casein as well, thus giving additional negative charge to the peptides, which greatly reduces encounter frequency and thus rate of aggregation.

Proteolysis by plasmin is not involved in the gelation of evaporated milk, if only because the preheating applied in almost all investigations was sufficient to completely inactivate this enzyme and its zymogen. For instance, Leviton and Pallansch (1962), who preheated at 100°C for 17 min, found that the evaporated milk gelled in as short as 2–3 weeks at 30°C after UHT-sterilisation (5 s at 137°C). Similarly, de Koning *et al.* (1985), who applied a preheating of 3 min at 120°C, could detect no plasmin/plasminogen activity nor breakdown of protein, while gelation occurred in about 7 weeks at 30°C. Moreover, de Koning and Kaper (1985) observed gelation after 5 weeks at 30°C irrespective of the presence of proteinase inhibitor, 20% of α_s- and β-caseins being degraded in the control portion at the time of gelation. Only if autoclaved evaporated milk was subjected to proteolysis by plasmin before it was stored did it gel more rapidly than a control portion (Manji and Kakuda, 1988). Casein micelles must have been largely degraded in the treated portion, however, and this study merely indicates that (poly)peptides in a batch-sterilised evaporated milk system gel more rapidly during storage than the mostly very stable casein particles in such a product.

(c) 'Gelation' due to physico-chemical changes of the casein micelles

If proteolysis is (almost) absent, aggregation of the casein micelles in unconcentrated milk is slow. Still, after storage for at least 5 months but generally one year or longer, milk may destabilise or gel (e.g., Samel *et al.*, 1971; Andrews *et al.*, 1977; Auldist *et al.*, 1996). Mostly, a voluminous gelled sediment has been reported – it should be noted that the gelled cream layer, about 5 mm thick, reported for some samples by Auldist *et al.* (1996) may be due to partial coalescence of the fat rather than to gelation of the protein. In their extensive studies on the storage stability of UHT-milk, Andrews *et al.* (1977) observed that a gelled sediment was present after 3 years, but not yet after 14 months. At 4°C, the gel was voluminous, comprising the bottom 80% of the carton, but the higher the storage temperature, the more compact it was, the gel at 37°C occupying only the bottom 10% of the carton. The gel was a particle-gel at all temperatures, at 4°C being built up of particles of roughly 0.1–0.2 μm connected by large (up to several hundred nm) 'tendrils', and of large (about 0.5 μm) roughly spherical particles at 37°C. After storage at low temperature, casein particles were relatively 'loose', but after storage at high temperature they were relatively 'compact'.

As mentioned in the previous section, gelation of UHT-sterilised evaporated milk occurs within a few months, proteolysis not being involved. It is thus the casein 'micelles' that aggregate during the storage of evaporated milk, which usually causes the product to become more viscous or to gel. Only Aoki and Immamura (1974a) reported the 'formation of a visible sediment and whey-off' during the storage of evaporated skimmed milk. These authors found a rapid increase of protein sedimentable at 500 × g between 20 and 60 days of storage at 30°C, until after 60 days all N was either sedimentable at 500 × g, or dissolved, i.e., not sedimentable at 110,000 × g. All other investigators found that gelation occurred. As measured by viscometry and judged from electron micrographs, aggregation of particles in UHT-sterilised evaporated milk occurs only shortly before gelation (Schmidt, 1968; Carrol *et al.*, 1971; Harwalkar and Vreeman, 1978b; de Koning *et al.*, 1985). Aggregation in retort-sterilised evaporated milk appears to proceed more gradually, if it occurs at all (Harwalkar *et al.*, 1983; de Koning *et al.*, 1992). In samples observed around the time of gelation, aggregation *via* 'strands of protein' appears to predominate, but flocculation and fusion of the particles also occurred (Harwalkar and Vreeman, 1978b, de Koning *et al.*, 1992).

Venkatachalam *et al.* (1993) reported that UHT-sterilised UF-concentrated milk gelled after 5 months at 20 or 4°C, but not during the storage at 35°C. Again, the gel consisted of particles connected by strands of protein. Somewhat surprisingly, the viscosity of retort-sterilised UF-concentrated milk increased considerably during 7 mo of storage at 15°C, while after 7 mo at 37°C a precipitate had formed (Muir *et al.*, 1984), whereas retort-sterilised milk and evaporated milk are stable for years (Samel *et al.*, 1971; de Koning *et al.*, 1992).

Thus, a kind of 'bridging flocculation' leads to the gelation of UHT milk if stored for many months at room temperature or below, and of UHT-sterilised evaporated milk after storage for a few months, the higher the storage temperature, the faster the gelation process. Such a 'bridging flocculation' model was first proposed by Wilson *et al.* (1963) for evaporated milk, and extensively studied by McMahon (1996, and the references therein) for UHT-sterilised UF-concentrated milk. For (unconcentrated) UHT milk stored at about 30°C or higher and for retort-sterilised milk and evaporated milk, sedimentation of proteinaceous particles predominates, which after long storage times may result in a gelled sediment due to flocculation and fusion of the close-packed sedimented particles. Aspects of this simplified model for the 'gelation' due to physico–chemical changes in the casein micelles are:

1) From the work of McMahon (1996) it can be concluded that in UHT-sterilised, UF-concentrated milk stored at 4 or 20°C, the 'bridges' are made of dissociated proteins (mostly β-lactoglobulin and κ-casein) which reaggregate upon long storage. By using immunogold labelling and transmission electron microscopy, these authors showed that β-lactoglobulin and κ-casein gradually dissociated from the surface of the casein micelles and appeared as linear patterns between the micelles after some 8–12 months of storage. As discussed in Section 20.3.3 "Protein Aggregation and Dissociation", dissociation in UHT-sterilised products is faster in evaporated milk, and apparently also in UF-concentrated milk, than in milk. This is a principal factor in the much faster gelation of evaporated and UF-concentrated milk as compared to that of milk. Similarly, the effect of temperature on the gelation of unconcentrated milk, in which gelation was not slower at 4°C than at 20 or even 30°C (Samel *et al.*, 1971), may be tentatively explained by the more extensive dissociation of proteins, especially κ-casein, in heat-treated milk stored in the cold, which compensates for the presumably lower rate of bond formation between the dissociated proteins at low temperature. During storage at 30–40°C, hydrophobic interactions hold the proteins in the 'micelles' and, in addition, intra-micellar chemical cross-links are formed during storage. The result is compact spherical particles that sediment slowly, cross-linking of particles in the sediment finally giving a compact gelled sediment.
2) The gelation time is shorter if the concentration factor of the milk is higher (Leviton and Pallansch, 1962; Ellerston and Pearce, 1964). This is likely to be due largely to the effect of concentration on dissociation, as discussed in (1). In addition, the concentration factor of protein and (homogenised) fat plays a part, because gelation occurs sooner at a higher volume fraction of particles, if the rate of aggregation and shape of the aggregates are the same (Bremer, 1992).
3) Added citrate or phosphate give faster gelation of UHT-sterilised evaporated milk, and added HMP much slower gelation (Leviton and Pallansch, 1962; Harwalkar and Vreeman, 1978a). The effect of phosphate and citrate may well be due to a more extensive dissociation of protein in heat-treated milk containing these additives (Fox *et al.*, 1967; Morr, 1967; Nieuwenhuijse *et al.*, 1988). Correspondingly, electron micrographs of samples of sterilised coffee cream containing phosphate or citrate stabilisers showed more small particles than those of samples sterilised without additives and, moreover, the fraction of small particles increased during storage (Buchheim *et al.*, 1986). The gel that finally

formed in the coffee cream was built up of strands of protein and small protein aggregates.

From the electron micrographs of Harwalkar and Vreeman (1978b), it appears that added HMP has no effect on the dissociation of proteins. Apparently, 'micelles' and dissociated proteins in evaporated milk with added HMP aggregate much more slowly, presumably due to the additional negative charge brought about by the adsorbed HMP, as discussed in Section 20.4.2(*b*) "Destabilisation due to Proteolysis by Plasmin".

4) A higher heating intensity appears to slow this type of gelation. For unconcentrated milk, Samel *et al.* (1971) observed that UHT milk which was subsequently retort-sterilised remained stable for more than 2 years, while the UHT milk samples had gelled after 13 mo at 4, 20 or 30°C, but not at 37°C. Retort-sterilised evaporated milk does not gel (although sedimentation and creaming may be considerable), unless the evaporated milk is cold stored for several days before sterilisation, as was shown by Harwalkar *et al.* (1983) and de Koning *et al.* (1992). Also, additional, or more intense, heat treatments in the concentrated state, e.g., a pre- or post-sterilisation heat treatment at, e.g., 80–90°C, slows down gelation of UHT-sterilised evaporated milk (Ellerston and Pearce, 1964). A longer ('UHT') sterilisation, or a higher sterilisation temperature also slow gelation, but a more intense preheating of the unconcentrated milk does not. We tentatively conclude that dissociation of protein is less in more intensely heated products (although the fraction of dissociated proteins is higher immediately after heating), presumably because during storage the proteins are held within the 'micelles' by intra-micellar covalent cross-links formed during heating. We suggest that the faster gelation of retort-sterilised evaporated milk which is cold stored before sterilisation may be due to dissociation of mainly κ-casein and β-lactoglobulin during the cold storage and that cross-linking of these proteins in the 'micelles' during the subsequent sterilisation occurs to a lesser extent than for concentrate that is sterilised shortly after evaporation.
5) Again, hardly anything is known about the type of bonds involved. Wilson *et al.* (1963) observed that the 'short chains involved in gelation' of UHT-sterilised evaporated milk dissociated in reducing agents like Na_2SO_3 or ascorbic acid. This indicates that S—S cross-links are involved in these 'bridges'. However, H_2O_2 was found to reduce the gelation time of UHT-sterilised evaporated milk (Harwalkar and Vreeman, 1978b) and retort-sterilised evaporated milk cold stored before sterilisation (Harwalkar *et al.*, 1983) and H_2O_2 oxidises SH-groups, thereby preventing the formation of S—S cross-links. Oxidation of SH-groups may also increase the dissociation of κ-casein and β-lactoglobulin, thereby enhancing gelation if these proteins subsequently aggregate by other bonds. Andrews *et al.* (1977) observed that particles in un-gelled milk after

1 year of storage could be dissolved in EDTA, but the gel formed after 3 years of storage could not. Once the gel is formed, salt bridges are apparently not required for it to remain intact. Summarising, much remains to be elucidated about the type of bonding involved in aggregation, especially of the 'bridges'; it is conceivable that more than one reaction is involved.

20.4.3 Creaming and sedimentation

The rate of creaming and sedimentation in long-life milk products depends strongly on the size of the particles and on the difference in mass density between particles and (concentrated) milk serum. Theoretical considerations of creaming of fat globules and sedimentation of casein micelles have been discussed by Walstra and Oortwijn (1975) and Dalgleish (1992b), respectively.

Creaming and sedimentation in heat-treated homogenised milk products often results in a cream layer that contains, besides fat, a relatively high concentration of protein, and in a sediment layer that contains, besides protein, a relatively high concentration of fat. Extreme examples of this were reported by, e.g., Schmidt *et al.* (1971): evaporated milk after 6 months of storage had a higher fat and a higher protein content in both cream and sediment than in the middle layer. This can be explained by fat globules being largely covered by (partly spread) casein micelles during homogenisation, the smallest fat globules having the highest protein load (Walstra and Oortwijn, 1982). Thus, the density of large protein-covered fat globules is lower than that of serum, but that of most small protein-covered fat globules may well be higher. If during heating (and storage) only dissociation of caseins occurs but no aggregation, as during UHT heating of milk, the cream layer will contain relatively little protein, and sedimentation will be slight. If, on the other hand, during heating aggregation of protein occurs, as in evaporated milk, the density of protein-covered fat globules becomes even higher, and sedimentation will be faster. However, if aggregation of fat and protein is extensive, this may retard sedimentation and creaming: Maxy and Sommer (1954) found that sedimentation and creaming in retort-sterilised evaporated milk were slower in product having a high rather than a low viscosity after sterilisation.

For some products, processing conditions can control the fast creaming/sedimentation due to heat-induced aggregation of particles. One example is the formation of a compact cream layer in coffee cream containing large clusters of fat globules, which can be prevented by proper homogenisation (Buchheim *et al.*, 1986). Another is the sedimentation and creaming of aggregates of protein and fat (which also cause the 'chalkiness' defect) which occurs in direct UHT-sterilised milk that is homogenised before sterilisation

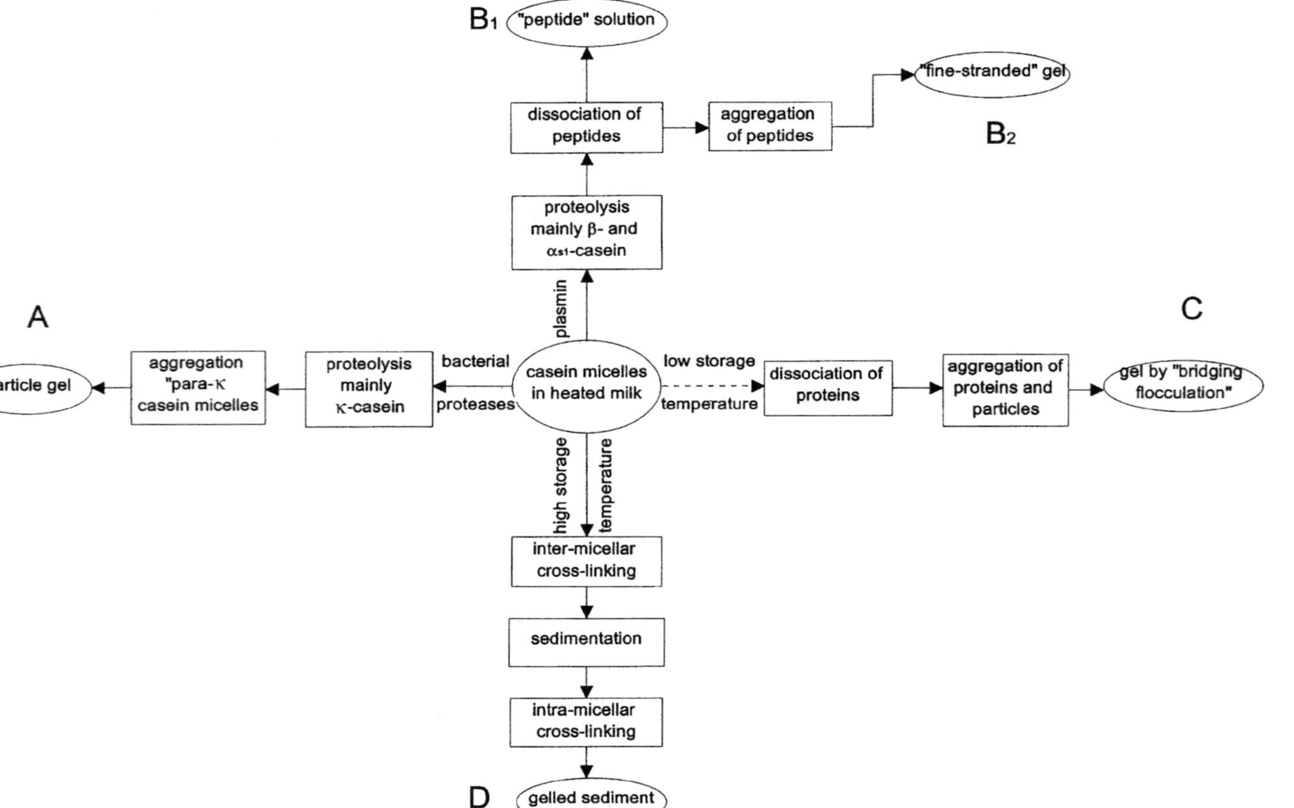

Figure 20.1 Simplified scheme of the various pathways for destabilisation of the protein in milk and milk products. 'Gelation' time; A: fast for milk, evaporated milk, B_1: fast for milk, if plasmin activity is 'high', not for evaporated milk, B_2: fast for milk, if plasmin activity is 'moderate', not for evaporated milk, C: very slow for UHT sterilised milk, fast for UHT sterilised evaporated milk, D: slow for all products, unless sterilisation produced large proteinaeous particles.

and which can be prevented by homogenisation after sterilisation (Hostettler, 1972). If, however, part of the aggregates that are formed during sterilisation are sufficiently strong to withstand the subsequent homogenisation, as in UHT-sterilised evaporated milk, this will only retard, but not prevent, this defect.

From the above, it is clear that general rules for the control of sedimentation and creaming by control of heat stability cannot be given.

20.5 Concluding remarks

The proteinaceous particles in heat-treated milk products are remarkably stable during storage, as long as the system resembles the casein micelles in the product immediately after heating. Further changes, like proteolysis or dissociation of part of the protein, decrease stability and may eventually result in gelation or destabilisation. A simplified scheme is shown in Figure 20.1.

Proteolysis has been investigated amply; physico-chemical changes of the casein micelles during storage have been studied very little, and neither has the aggregation of particles and peptides during storage. The latter two need to be investigated for understanding of storage stability to advance significantly.

References

Andrews, A.T. (1975) Properties of aseptically packed ultra-high-temperature milk. III. Formation of polymerised protein during storage at various temperatures. *J. Dairy Res.*, **42**, 89–99.

Andrews, A.T. and Cheeseman, G.C. (1972) Properties of aseptically packed ultra-high-temperature milk. II Molecular weight changes of the casein components during storage. *J. Dairy Res.*, **39**, 395–408.

Andrews, A.T., Brooker, B.E. and Hobbs, D.G. (1977) Properties of aseptically packed ultra-high-temperature milk. IV. Electron microscopic examination of changes occurring during storage. *J. Dairy Res.*, **44**, 283–92.

Aoki, T. and Imamura, T. (1974a) Changes of the casein complex during storage of sterilised concentrated skim milk. *Agr. Biol. Chem.*, **38**, 309–14.

Aoki, T. and Imamura, T. (1974b) Changes of the casein complex during storage of sterilised skim milk. *Agr. Biol. Chem.*, **38**, 1929–34.

Aoki, T., Umeda, T. and Kako, Y. (1990) Cleavage of the linkage between colloidal calcium phosphate and casein on heating milk at higher temperature. *J. Dairy Res.*, **57**, 349–54.

Auldist, M.J., Coats, S.J., Sutherland, B.J., Hardham, J.F., McDowell, G.H. and Rogers, G.L. (1996) Effect of somatic cell count and stage of lactation on the quality and storage life of ultra high temperature milk. *J. Dairy Res.*, **63**, 377–86.

Belec, J. and Jenness, R. (1962) Dephosphorylation of casein by heat treatment, II. In skimmilks. *J. Dairy Sci.*, **45**, 20–6.

Berg, H.E. and van Boekel, M.A.J.S. (1994) Degradation of lactose during heating of milk. 1. Reaction pathways. *Neth. Milk Dairy J.*, **48**, 157–75.

Bremer, L.G.B. (1992) *Fractal Aggregation in Relation to Formation and Properties of Particle Gels*, Doctoral Thesis, Wageningen University.

Buchheim, W., Falk, G. and Hinz, A. (1986) Ultrastructural aspects and physico-chemical properties of ultra-high-temperature (UHT)-treated coffee cream. *Food Microstructure*, **5**, 181–92.

Burton, H. (1988) *Ultra-High Temperature Processing of Milk and Milk Products*. Elsevier Applied Science Publishers, London.

Carrol, R.J., Thompson, M.P. and Melnychyn, P. (1971) Gelation of concentrated skimmilk: electron microscopic study. *J. Dairy Sci.*, **54**, 1245–52.

Corradini, C. (1975) Gel formation behaviour in UHT-sterilised milk. *Milchwissenschaft*, **30**, 413–6.

Dalgleish, D.G. (1992b) Sedimentation of casein micelles during the storage of ultra-high temperature milk products–a calculation. *J. Dairy Sci.*, **75**, 371–9.

Dalgleish, D.G., Pouliot, Y. and Paquin, P. (1987) Studies on the heat stability of milk. I. Behaviour of divalent cations and phosphate in milks heated in stainless steel system. *J. Dairy Res.*, **54**, 29–37.

de Koning, P.J. and Kaper, J. (1985) Effects of some proteinase inhibitors and the Maillard reaction on the process of age-thinning and gelation of UHTST-sterilised concentrated casein micelle dispersions. *Neth. Milk Dairy J.*, **39**, 37–47.

de Koning, P.J., Kaper, J., Rollema, H.S. and Driessen, F.M. (1985) Age-thinning and gelation in concentrated and unconcentrated UHT-sterilised skim milk. *Neth. Milk Dairy J.*, **39**, 71–87.

de Koning, P.J., de Wit, J.N. and Driessen, F.M. (1992) Process conditions affecting age-thickening and gelation of sterilised canned evaporated milk. *Neth. Milk Dairy J.*, **46**, 3–18.

Deysher, E.F. and Webb, B.H. (1952) Factors that affect the formation of a crystalline deposit in evaporated milk. *J. Dairy Sci.*, **35**, 106–15.

Driessen, F.M. (1983) *Lipases and Proteinases in Milk. Occurrence, Heat Inactivation and their Importance for the Keeping Quality of Milk Products*, Doctoral thesis, Wageningen University.

Ellerston, M.E. and Pearce, S.J. (1964) Some observations on the physical-chemical stability of sterile concentrated milks. *J. Dairy Sci.*, **47**, 564–72.

Fox, K.K., Harper, M.K., Holsinger, V.H. and Pallansch, M.J. (1966) Effects of high-heat treatment on stability of calcium caseinate aggregates in milk. *J. Dairy Sci.*, **50**, 443–50.

Fox, K.K., Harper, M.K., Holsinger, V.H. and Pallansch, M.J. (1967) Composition of granules in evaporated milk stored at low temperatures. *J. Dairy Sci.*, **50**, 1032–7.

García-Risco, M.R., Ramos, M. and López-Fandiño R. (1999) Proteolysis, protein distribution and stability of UHT milk during storage at room temperature. *J. Sci. Food. Agr.*, **79**, 1171–8.

Geerts, J.P., Bekhof, J.J and Scherjon, J.W. (1983) Determination of calcium ion activities in milk with an ion-selective electrode. A linear relationship between the logarithm of time and the recovery of the calcium ion activity after heating. *Neth. Milk Dairy J.*, **37**, 197–211.

Guthy, K., Hong, Y.H. and Klostermeyer H. (1983) Gelation of UHT milk during storage–a comparison of some chemical and physical methods of analysis. *Milchwissenschaft*, **38**, 654–7.

Harwalkar, V.R. (1992) Age gelation of sterilised milks, in *Advanced Dairy Chemistry-1: Proteins*, 2nd edn, (P.F. Fox ed.) Elsevier Applied Science Publishers, London, pp. 691–734.

Harwalkar, V.R. and Vreeman, H.J. (1978a) Effect of added phosphates and storage on changes in ultra-high temperature short-term sterilised concentrated skim-milk. 1. Viscosity, gelation, alcohol stability, chemical and electrophoretic analysis of proteins. *Neth. Milk Dairy J.*, **32**, 94–111.

Harwalkar, V.R. and Vreeman, H.J. (1978b) Effect of added phosphates and storage on changes in ultra-high temperature short-term sterilised concentrated skim-milk. 2. Micelle structure. *Neth. Milk Dairy J.*, **32**, 204–16.

Harwalkar, V.R., Beckett, D.C., McKellar, R.C., Emmons, D.B. and Doyle, G.E. (1983) Age-thickening and gelation of sterilised evaporated milk. *J. Dairy Sci.*, **66**, 735–42.

Holt, C. (1995) Effect of heating and cooling on the milk salts and their interaction with casein, in *Heat Induced Changes in Milk*, 2nd edn, (P.F. Fox ed.) Special Issue, 9501, International Dairy Federation, Brussels, pp. 105–33.

Hostettler, H. (1972) Appearance, flavour and texture aspects. History of the development of UHT processes, in *Monograph on UHT Milk*, Annual Bulletin, Part V, International Dairy Federation, Brussels, pp. 169–74.

Jimenez-Perez, S., Corzo, N., Morales, F.J., Delgado, T. and Olano, A. (1992) Effect of storage temperature on lactulose and 5-hydroxymethyl-furfural formation in UHT milk. *J. Food Protect*, **55**, 304–6.

Kelly, A.L. and Foley, J. (1997) Proteolysis and storage stability of UHT milk as influenced by milk plasmin activity, plasmin/β-lactoglobulin complexation, plasminogen activation and somatic cell count. *Int. Dairy J.*, **7**, 411–20.

Kessler, A.G. and Fink, R. (1986) Changes in heated and stored milk with an interpretation by reaction kinetics. *J. Food Sci.*, **51**, 1105–11.

Kocak, H.R. and Zadow, J.G. (1985a) Age gelation of UHT whole milk as influenced by storage temperature. *Aust. J. Dairy Technol.*, **40**, 14–21.

Kocak, H.R. and Zadow, J.G. (1985b) Controlling age gelation of UHT milk with additives. *Aust. J. Dairy Technol.*, **40**, 58–64.

Kohlmann, K.L., Nielsen, S.S. and Ladisch, M.R. (1991) Effects of a low concentration of added plasmin on ultra-high temperature processed milk. *J. Dairy Sci.*, **74**, 1151–6.

Kohlmann, K.L., Nielsen, S.S. and Ladisch, M.R. (1988) Effects of serine proteolytic enzymes (trypsin and plasmin), trypsin inhibitor, and plasminogen activator addition to ultra-high temperature processed milk. *J. Dairy Sci.*, **71**, 1728–39.

Kondal Reddy, K., Nguyen, M.H., Kailasapathy, K. and Zadow, J.G. (1991) The effects of some treatments and storage temperatures on UHT whole milk. *Aust. J. Dairy Technol.*, **46**, 57–63.

Law, B.A., Andrews, A.T. and Sharpe, M.E. (1977) Gelation of ultra-high temperature-sterilised milk by proteinases from a strain of *Pseudomonas fluorescens* isolated from raw milk. *J. Dairy Res.*, **44**, 145–8.

Leviton, A. and Pallansch, M.J. (1962) High temperature short-time sterilised evaporated milk. IV. The retardation of gelation with condensed phosphates, manganous ions, polyhydric compounds and phosphatides. *J. Dairy Sci.*, **45**, 1045–56.

López Fandiño, R., Olano, A., San José, C. and Ramos, M. (1993) Application of reversed phase HPLC to the study of proteolysis in UHT milk. *J. Dairy Res.*, **60**, 111–6.

Manji, B. and Kakuda, Y. (1988) The role of protein denaturation, extent of proteolysis and storage temperature of the mechanism of age gelation in a model system. *J. Dairy Sci.*, **71**, 1455–63.

Manji, B., Kakuda, Y. and Arnott, D.R. (1986) Effect of storage temperature on age gelation of ultra-high temperature milk processed by direct and indirect heating systems. *J. Dairy Sci.*, **69**, 2944–3001.

Maxy, R.B. and Sommer, H.H. (1954) Fat separation in evaporated milk. III. Gravity separation and heat stability. *J. Dairy Sci.*, **37**, 1061–70.

McMahon D.J. (1996) Age-gelation of UHT milk: changes that occur during storage, their effect on shelf life and the mechanisms by which age-gelation occurs, in *Heat Treatments and Alternative Methods*, Special Issue, 9602, International Dairy Federation, Brussels, pp. 315–26.

Metwalli, A.A.M., de Jongh, H.H.J. and van Boekel, M.A.J.S. (1998) Heat inactivation of bovine plasmin. *Int. Dairy J.*, **8**, 47–56.

Metwalli, A.A.M., Metwalli, N.H. and van Boekel, M.A.J.S. (1996) Effect of urea on heat-induced changes in milk. *Neth. Milk Dairy J.*, **50**, 427–57.

Mitchell, G.E. and Ewings, K.N. (1985) Quantification of bacterial proteolysis causing gelation in UHT-treated milk. *N.Z. J. Dairy Sci. Technol.*, **20**, 65–76.

Möller A.B., Andrews, A.T. and Cheeseman, G.C. (1977) Chemical changes in ultra-heat-treated milk during storage, II. Lactuloselysine and fructoselysine formation by the Maillard reaction. *J. Dairy Res.*, **44**, 267–75.

Morr, C.V. (1967) Some effects of pyrophosphate and citrate ions upon the colloidal caseinate-phosphate micelles and ultrafiltrate of raw and heated skimmilk. *J. Dairy Sci.*, **50**, 1038–44.

Muir, D.D., Banks, W., Golightly, S.J and Sweetsur, A.W.M. (1984) Preparation and properties of in-can sterilised skim-milk based concentrates. *J. Food Technol.*, **19**, 369–76.

Muir, D.D. and Sweetsur, A.W.M. (1984) Optimization of the heat stability of protein-rich concentrates prepared by ultrafiltration of skim-milk. *J. Food Technol.*, **19**, 263–71.

Nakai, S., Wilson, H.K and Herreid, E.O. (1965) Effect of changes in sulphur compounds on stability and gelation of caseins and of sterilised concentrated milk. *J. Dairy Sci.*, **48**, 431–7.

Nieuwenhuijse, J.A., Timmermans, W. and Walstra, P. (1988) Calcium and phosphate partitions during the manufacture of concentrated milk and their relations to heat stability. *Neth. Milk Dairy J.*, **42**, 387–421.

Nieuwenhuijse, J.A., van Boekel, M.A.S.J. and Walstra P. (1991) Association and dissociation of proteins in concentrated skim milk. *Neth. Milk Dairy J.*, **45**, 3–22.

Odagiri, S. and Nickerson, T.A. (1964) Complexing of calcium by hexametaphosphate, oxalate, citrate, and EDTA in milk. I. Effects of complexing agents on turbidity and rennet coagulation. *J. Dairy Sci.*, **47**, 1306–9.

Odagiri, S. and Nickerson, T.A. (1965) Complexing of calcium by hexametaphosphate, oxalate, citrate, and EDTA in milk. II. Dialysis of milk containing complexing agents. *J. Dairy Sci.*, **48**, 19–22.

Pellegrino, L., van Boekel, M.A.J.S., Gruppen, H., Resmini, P. and Pagani, M.A. (1999) Heat-induced aggregation and covalent linkages in β-casein model systems. *Int. Dairy J.*, **9**, 255–61.

Richardson, B.C. and Newstead, D.F. (1979) Effect of heat stable proteinases on the storage life of UHT milk. *N.Z. J. Dairy Sci. Technol.*, **14**, 273–9.

Rollema, H.S., Poll, J.K. and Brinkhuis, J.A. (1990) Improvement of the keeping of UHT-milk products. *NIZO Report NOV 1386*.

Rollema, H.S., Vreeman, H.J., Schmidt, D.G., Siezen, R.J., Both, P., Brinkhuis, J.A. and van Marwijk, B.W. (1987) Improvement of the structure and properties of dairy products: studies of the microstructure of casein micelles. *NIZO Report NOV 1265*.

Rose, D. and Tessier, H. (1959) Composition of ultrafiltrates from milk heated at 80 to 230°F. in relation to heat stability. *J. Dairy Sci.*, **42**, 969–80.

Samel, R., Weaver, R.W.V. and Gammack, D.B. (1971) Changes on storage in milk processed by ultra-high temperature sterilisation. *J. Dairy Res.*, **38**, 323–32.

Schmidt, D.G. (1968) Electron-microscopic studies on gelation of UHTST sterilised concentrated skim milk. *Neth. Milk Dairy J.*, **22**, 40–9.

Schmidt, D.G., Buchheim, W. and Koops, J. (1971) An electron-microscopial study of the fat-protein complexes in evaporated milk, using the freeze-etching technique. *Neth. Milk Dairy J.*, **25**, 200–16.

Schokker, E.P. and van Boekel, M.A.J.S. (1998a) Mechanism and kinetics of inactivation at 40–70°C of the extracellular proteinase from *Pseudomonas fluorescens* 22F. *J. Dairy Res.*, **65**, 261–72.

Schokker, E.P. and van Boekel, M.A.J.S. (1998b) Effect of protein content on low temperature inactivation of the extracellular proteinase from *Pseudomonas fluorescens* 22F. *J. Dairy Res.*, **65**, 347–52.

Singh, H. (1995) Heat-induced changes in casein, including interactions with whey proteins, in *Heat-Induced Changes in Milk*, (P.F. Fox ed.) Special Issue 9501, International Dairy Federation, Brussels, pp. 86–104.

Smits, P. and van Brouwershaven, J.H. (1980) Heat-induced association of β-lactoglobulin and casein micelles. *J. Dairy Res.*, **47**, 313–25.

Snoeren, T.H.M. and Both, P. (1981) Proteolysis during the storage of UHT-sterilised whole milk. 2. Experiments with milk heated by the indirect system for 4 seconds at 142°C. *Neth. Milk Dairy J.*, **35**, 113–9.

Snoeren, T.H.M., van Riel, J.A.M. and Both, P. (1979) Proteolysis during the storage of UHT-sterilised whole milk. 1. Experiments with milk heated by the direct system for 4 seconds at 142°C. *Neth. Milk Dairy J.*, **33**, 31–9.

Tarassuk, N.P. and Tamsma, A.F. (1956) Milk sterilisation. Control of gelation in evaporated milk. *J. Agr. Food Chem.*, **4**, 1033–5.

van Boekel, M.A.J.S. and Walstra, P. (1995) Use of kinetics in studying heat-induced changes in foods, in *Heat-Induced Changes in Milk*, (P.F. Fox ed.) Special Issue 9501, International Dairy Federation, Brussels, pp. 22–50.

van Dijk, H.J.M. (1990) The properties of casein micelles. 3. Changes in the state of the micellar calcium phosphate and their effect on other changes in the casein micelles. *Neth. Milk Dairy J.*, **44**, 125–41.

van Hooydonk, A.C.M., de Koster, P.G. and Boerrigter, I.J. (1987) The renneting properties of heated milk. *Neth. Milk Dairy J.*, **41**, 3–18.

Venkatachalam, N., McMahon, D.J. and Savello, P.A. (1993) Role of protein and lactose in the age gelation of ultra-high temperature processed concentrated skim milk. *J. Dairy Sci.*, **76**, 1882–94.

Visser S. (1981) Proteolytic enzymes and their action on milk proteins. A review. *Neth. Milk Dairy J.*, **35**, 65–88.

Walstra, P. (1990) On the stability of casein micelles. *J. Dairy Sci.*, **73**, 1965–79.

Walstra, P., Geurts, T.J., Noomen, A., Jellema, A. and van Boekel, M.A.J.S. (1999) *Dairy Technol*, Marcel Dekker, Inc., New York.

Walstra, P. and Oortwijn, H. (1975) Effect of globule size and concentration on creaming in pasteurised milk. *Neth. Milk Dairy J.*, **36**, 103–13.

Walstra, P. and Oortwijn, H. (1982) The membranes of recombined milk fat globules. 3. Mode of formation. *Neth. Milk Dairy J.*, **36**, 103–13.

Wilson, H.K., Vetter, J.L., Sasago, K. and Herreid, E.O. (1963) Effects of phosphates added to concentrated milk before sterilisation at ultra-high temperatures. *J. Dairy Sci.*, **46**, 1038–43.

Zadow, J.G. and Chituta, F. (1975) Age gelation of ultra-high-temperature milk. *Aust. J. Dairy Technol.*, **30**, 104–6.
Zadow, J.G. and Hardman, J.F. (1981) Studies on the stability of concentrated and reconstituted concentrated skim milks towards UHT processing. *Aust. J. Dairy Technol.*, **36**, 30–3.
Zin El-Din, M., Aoki, T. and Kako, Y. (1991) Polymerisation and degradation of casein in UHT milk during storage. *Milchwissenschaft*, **46**, 284–7.

21

ETHANOL STABILITY

D.S. Horne

21.1 Introduction

In the previous edition of Advanced Dairy Chemistry, Horne (1992) reviewed what was mainly his work and that of colleagues at the Hannah Research Institute on the effect of ethanol addition on the stability of the casein micelle system. These studies were part of the Institute's wider programme on micellar stability, particularly on the effects of renneting or heating milk. In that context experiments were also carried out at sub-critical ethanol levels and contributed to the effective exploitation of the theories of steric stabilization as applied to the casein micelle, the so-called 'hairy micelle' model of Holt (1975) and Walstra (1979). Holt and Horne (1996) recently reviewed such work in this area. Models of internal micelle structure have also to be consistent with the observations of micellar behaviour in the presence of ethanol and here we ask how well the latest suggestions stand up to this test.

In this revised chapter, emphasis is placed on the role of milk composition and processing factors in controlling the alcohol stability of milk. The single-point alcohol stability test was used, and is still being used, as a test of the suitability of milk for processing. It is cheap and easy to apply, but it is also apparent that the possibilities of other causes of failure are either not understood or are being overlooked in the blanket application of this test, indicating the necessity for a further, alternative, confirmatory test before final condemnation of the milk.

Advanced Dairy Chemistry Volume 1: Proteins, 3rd edn.
Edited by P.F. Fox and P.L.H. McSweeney, Kluwer Academic/Plenum Publishers, 2003.

21.2 Alcohol stability test

21.2.1 Historical introduction

The earliest accounts of the alcohol stability test for milk appeared in the scientific literature almost a century ago. Particularly in central Europe, the test was often enshrined in city statutes or public health regulations and provided an objective assessment of the keeping quality of milk for retail sale. A review of this early work and the value of the test in this context is contained in a 1915 Bulletin of the United States Department of Agriculture (Ayers and Johnson, 1915).

Most people employed the simple unsophisticated single alcohol test. If a precipitate formed when an equal volume of ethanol solution, usually 70%, v/v, was added to a milk sample, the batch of milk was rejected. Dahlberg and Garner (1921) concluded that "the alcohol test shows good possibilities as a practical and reliable test for determining the quality of milk for quality of milk for condenseries making evaporated milk". Its use for this purpose continued until the 1930s, with such alcohol tests being accepted as indicators of whether the milk was going sour, or was mixed with colostrum or was contaminated with milk from a cow suffering from mastitis (Padmos, 1930). Thereafter, following a report of Ramsdell *et al.* (1931), who stated that no available test, including the alcohol test, was sufficiently "definite for advantageous use in grading milk to sterilization purpose", the alcohol test fell into disfavour and its usage declined in the USA and Europe. However, recent evidence suggests that the single-point alcohol test is still widely applied in Central and South America and in the Far East as a means of rapidly defining the acceptability of milk on arrival at the dairy plant (Guo *et al.*, 1998; M.A.V.P. Brito, Personal Communication; P. Tangkawattana, Personal Communication).

21.2.2 Ethanol stability/pH profile

When the alcohol test first came into use, it was regarded largely as a measure of acidity, particularly acidity produced by bacterial fermentation. However, it was soon observed that fresh milk from individual cows frequently coagulated and that the test was dependent on factors other than acidity. In view of the fact that rejection of a farmer's milk output at the dairy can lead to a considerable drop in income, it is crucial that the potential causes of failure, other than bacterial contamination, are fully understood by the practitioners and advocates of the test. This information is available in the literature from the results of early studies and from the more recent developments.

The early major studies relating to the alcohol test concentrated on the role of the inorganic components of the milk serum. Sommer and Binney

(1923) concluded that coagulation by alcohol occurred as a result of an excess of calcium and magnesium over citrate and phosphate in the serum, in keeping with the salt-balance theory for heat stability proposed by Sommer and Hart (1919). Mitamura (1937) confirmed the importance of bivalent cations. Davies and White (1958) took this a stage further when they observed that the strength of the ethanol solution required to induce coagulation in an equal volume of milk was inversely related to the concentration of ionic calcium.

Horne and Parker (1980) demonstrated that artificial adjustment of the pH of a milk provided an ethanol stability/pH profile characteristic of that milk. This plot is analogous, but in no way identical, to the coagulation time/pH profiles first observed by Rose (1961a,b) in heat stability studies.

These ethanol stability/pH profiles (Figure 21.1) were sigmoidal in shape, with a minimum stability at low pH values and a maximum stability at high pH values. Because increasing pH increased the stability of milk to such an extent that pure ethanol often failed to induce precipitation in the single alcohol test with one-to-one volume addition, Horne and Parker (1980) modified the test, adding two volumes of aqueous ethanol solution to one volume of milk. The ethanol stability of the milk at a specified pH was then reported as the concentration of ethanol solution which induced instantaneous precipitation of the milk protein.

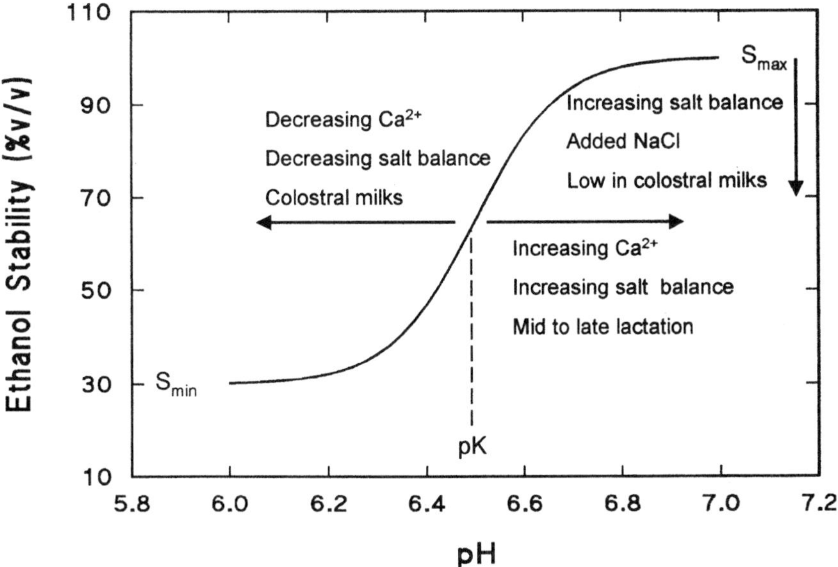

Figure 21.1 The ethanol stability phase diagram based on the ethanol stability/pH profile. Parameters defining the profile shape and position are indicated, together with listings of identified factors influencing their values.

Unlike the heat stability coagulation time/pH profile, the ethanol stability /pH profile can be assigned a functional form, the parameters of which can be related to the composition of milk or the processing conditions to which the milk has been subjected. The equation relating ethanol stability (S) to pH takes the form

$$\frac{S - S_{min}}{S_{max} - S} = 10^{(pH - pK)/b} \tag{1}$$

which may be more usefully transformed into the linear equation:

$$pH = pK + b \log_{10}[(S - S_{min})/(S_{max} - S)] \tag{2}$$

Note that Equation (1) was incorrectly transcribed in Horne (1992).

Industry favours a single-point alcohol stability test in which the milk is mixed with an equal volume of alcohol solution of a pre-selected ethanol concentration, thus fixing the value of S in Equation 2. Using this value and Equation (2), we can calculate an equivalent pH at which the milk would be unstable at this ethanol concentration, S. As long as the true pH of the milk is greater than this equivalent pH, the milk will be stable and pass the test. Conversely, if milk pH is less than the equivalent pH, the milk will fail the test. Four parameters are involved in defining the summation making up the right-hand side of Equation (2). These serve to characterize fully the ethanol stability/pH profile. They are the asymptotic values, S_{min} and S_{max}, of the two arms of the sigmoid at low and high pH, respectively, an inflexion point or effective pK value denoting the position of the profile along the pH-axis and b, the gradient of the linear transform (Equation 2). Fitting is usually carried out by varying S_{min} and S_{max} until a maximum correlation coefficient is achieved for the linear equation form, pK being the intercept of this straight line fit on the pH-axis and b the gradient of this straight tine.

21.2.3 Factors that affect the values of profile parameters

Unlike heat-induced coagulation, where two forms of the coagulation time/ pH curve are encountered (Rose, 1961a,b), the sigmoidal shape of the ethanol stability/pH profile is invariant. Though the majority of determinations have been made on bovine milk from various breeds (generally unknown) and locations, the profiles obtained for goat (Horne and Parker, 1982; Guo *et al.*, 1998) and buffalo (Unnikrishnan *et al.*, 1988) milks show the same characteristic shape and are located at similar positions along the pH axis.

The ethanol stability/pH profile defines a phase diagram linking ethanol stability and pH (Figure 21.1). For all combinations of ethanol and pH lying to the left and above the sigmoidal curve, the milk is unstable and precipitation occurs on mixing. Stability is assured for all combinations lying to the right of the curve. In bovine milk, at least, the location of this curve along the pH-axis, defined mainly by the pK parameter, is not fixed but varies from milk to milk from individual cows. The profile pK value also shifts systematically during lactation (Horne *et al.*, 1986).

The pK of the profile can be shifted to more alkaline values, thus destabilizing the milk, by increasing the concentration of ionic calcium, and conversely to more acidic values (increasing stability) by addition of sequestrants such as phosphate, citrate or EDTA (Horne and Parker, 1981a). A survey of the ethanol stability of Irish creamery milks (Donnelly and Horne, 1986) also noted a strong positive correlation between profile pK values and the soluble salt balance ($r = 0.82$), whether soluble salt balance was expressed as the ratio of (Ca + Mg) to (P_i + Citrate), or as the net difference between the summed concentrations of these cations and anions. Since it can be readily demonstrated algebraically that the concentration of soluble free calcium is governed by the soluble salt balance ratio (Donnelly and Horne, 1986), these relationships provide confirmation, from natural compositional variations, of the previous observations on milks dosed with calcium salts or sequestrants. Milks with a high concentration of free ionic calcium have higher pK values, and are consequently less stable, than those with low Ca^{2+} activity.

Horne *et al.* (1986) monitored lactational changes in the ethanol stability of the milks of 8 individual cows. The pK value of the profile decreased in early lactation but increased thereafter throughout lactation for all 8 cows monitored. This alkaline shift was the most obvious effect observed in direct measurement and translated into a decline in ethanol stability through lactation from an early lactation maximum which was characteristic of each animal. This change in the pK value of the profile mirrored the known lactational behaviour of the soluble salt balance ratio (recalculated from the original data of White and Davies, 1958), and again emphasizes the previously quoted findings on the dominant role of free ionic calcium.

As is apparent in Figure 21.1, low values of S_{min} and S_{max} are indicative of less stable milk. No changes in S_{min} and S_{max} were observed in the cross-dialysis experiments of Horne and Parker (1981a), suggesting that diffusible components were not important in fixing their values. S_{max}, the asymptotic value at high pH, was, however, observed to correlate negatively ($r \sim -0.72$) with the soluble salt balance ratio in the Irish creamery survey (Donnelly and Horne, 1986). Low values of S_{max} were also encountered in colostral milks from Friesian cows in the Hannah Research Institute herd (Horne *et al.*, 1986) but these increased rapidly in the first week of lactation. Thereafter, in this group of cows, no further lactational trends in the value of S_{max} could be discerned (Horne *et al.*, 1986). The value of S_{min} showed no obvious lactational trend in the milk samples studied by Horne *et al.* (1986), although a stepped decrease (and therefore instability) was observed when the autumn and winter-calving cows went over to summer grazing.

The slope parameter, b, controls the rotation of the profile about the inflexion point at the pK value. Little information is available on milk compositional factors that affect its value, although a gradual slow increase throughout lactation was noted for all 8 cows monitored by Horne *et al.* (1986).

Lactational effects on alcohol stability at the natural pH of milk had been noted by Mitamura (1937) and Davies and White (1958). Mitamura (1937) found that milk was very unstable to alcohol in early lactation, but thereafter increased in stability to a level which remained fairly constant for most of the lactation and which was specific for each cow. He found that towards the end of lactation, the milk from some cows became more stable but the stability of most samples decreased. Davies and White (1958) confirmed the observation that early lactation milks were very unstable to ethanol. They also found that mid-lactation milks were much more stable than those in early lactation, but late lactation milks showed a wide range of stability. Such variations in behaviour during lactation were confirmed by Horne *et al.* (1986) based on complete ethanol stability/pH profiles. In early lactation, the most important factor appears to be a low natural pH of milk, so that the natural stability is low on the acid arm of the profile. In late lactation, the most important contributory factor is a high soluble salt balance ratio. A shift in the profile to more alkaline pH values is associated with this and, if the natural milk pH does not undergo a similar shift, a decrease in alcohol stability, as measured by the single point test, must ensue. If, however, the natural pH of the milk increases sufficiently to over-compensate for the shift in the profile, then an increase in alcohol stability at the natural pH will be noted. Hence, the reason for the inconsistent behaviour of the alcohol stability of late lactation milk seen by Mitamura (1937) and Davies and White (1958). On the other hand, if the pH of the milk is adjusted to a chosen, fixed value, decreased stability of the late lactation milk, as a result of the increasing soluble salt balance ratio, should be clearly and consistently observed. Thus, Horne *et al.* (1986) found that the ethanol stability calculated at a fixed pH of 6.6 from measured profile parameters passed in early lactation through a maximum value, characteristic for each cow, but declined steadily thereafter. When the ethanol stability for the milk of each cow was normalized to this maximum value, a normalized decay rate of 1×10^{-3} per day, constant to within 10%, fitted the data from all eight cows in the study (Horne *et al.*, 1986). This implies that from a few measurements during the first few weeks of lactation, it should be possible to predict the ethanol stability of the milk from any individual cow or synchronously calving herd to within a few percent ethanol for the remainder of lactation.

21.2.4 Effects of processing milk on its alcohol stability

Increasingly, all milk reaching the consumer is processed in some manner. Some operations, such as pasteurization and cooling, are relatively simple;

others, such as concentration and conversion to milk powder, are relatively complex in their effects on the properties of the finished product. The effects of two such processes, forewarming and concentration, on the ethanol stability behaviour of milk have been studied in depth (Horne and Parker, 1981d, 1983).

(a) Effects of forewarming milk

Forewarming is a technique frequently used by milk processors to improve the heat stability of milk to be concentrated subsequently. It is also used extensively in yoghurt manufacture to improve texture and reduce syneresis and, incidentally, to kill indigenous bacteria prior to inoculation with the fermentative organisms. In this operation, the milk is heated to 90 or 95°C and held for 10 to 30 min. Horne and Parker (1981d) studied the effects of forewarming at 90°C for 30 min on the ethanol stability of unconcentrated skim milk. They found that:

- the sigmoidal shape of the ethanol stability/pH profile was retained,
- the asymptotic maximum and minimum stabilities were largely unchanged from the values of the original milk,
- the entire profile was shifted laterally to more acid pH values by approximately 0.15 pH units, giving an apparently increased ethanol stability at the natural pH of milk.

This alteration in the position of the pH profile was found to be insensitive to the presence of sulphydryl-blocking agents, thus apparently ruling out the involvement of denatured serum protein in the effect (Horne and Parker, 1981d). Among other reactions that occur in heated milk is the precipitation of calcium phosphate (Hilgemann and Jenness, 1951). That such reactions occurred under the conditions of forewarming used by Horne and Parker (1981d) was confirmed by direct measurement of the concentration of soluble calcium as a function of pH. The observed effects on the ethanol stability of the forewarmed milk were fully consistent with such a reduction in the concentration of Ca^{2+} by precipitation, as viewed from the model suggested by Horne and Parker (1981b).

Buffalo milk showed a similar response to forewarming but also demonstrated an increase in the value of S_{max} (Unnikrishnan *et al.*, 1988). On storage, however, these effects on profile behaviour were partially reversed, particularly the behaviour of the alkaline arm of the profile. Moreover, the extent of the improvement in stability and of the rate of reversion varied widely between milk samples. Such behaviour was considered to be consistent with slow resolubilization of the calcium phosphate precipitated on heating, as reported by Jenness and Patton (1959).

(b) Effects of concentrating milk

Concentration also produced marked changes in the ethanol stability/pH profile of milk. The main effect was seen in the stability at high pH values, which decreased progressively as the milk was concentrated by evaporation (Horne and Parker, 1983). At low pH values, on the acid side of the natural pH, ethanol stability was largely unaffected by concentration. At pH values of 7.0 or above, Horne and Parker (1983) found that the ethanol stability decreased linearly with the chloride ion content of the concentrate. By adding NaCl to the original milk, they reproduced this linear decay behaviour. In concentrated samples, the chloride ion content of which had been reduced subsequently by dialysis, ethanol stability values were found to be as predicted by the remaining chloride level. By adding other monovalent salts, the effect was shown to be non-specific (Horne and Parker, unpublished). Empirically, therefore, the concentration-dependent behaviour of the alkaline arm of the ethanol stability/pH profile was found to be governed largely by the ionic strength of the system. Because of the multiplicity of ionic strength-dependent reactions possible in milk, it is difficult to reconcile these observations within a simple theoretical framework. Nevertheless, they do appear to be useful from two contrasting viewpoints: first, predictive, since by adding NaCl to milk, the ethanol stability of its evaporative concentrate can be predicted; secondly, conservative, since if that ethanol stability is found to be unacceptably low, it can be increased by incorporating a short dialysis stage to reduce the chloride concentration, prior to concentration.

21.2.5 Ethanol stability and heat stability

A detailed comparison of the ethanol and heat stability tests formed the basis of a review by Horne and Muir (1990). The main thrust of their argument may be summarized as follows. The ethanol stability test requires the addition of ethanol solutions of increasing concentration to milk samples until instantaneous precipitation is induced. In the conventional heat stability test, the milk sample is heated at a pre-selected temperature until coagulation occurs, the actual duration of heating, or clotting time, being defined as the measure of heat stability. This extended heating period allows a multiplicity of concurrent and sequential reactions to occur, as summarized by Fox (198l) and in Chapter 19. Some of these reactions are deleterious while some are advantageous from the heat stability viewpoint, but each has its own concentration- and temperature-dependent characteristics. It is not surprising that no correlation has been observed between the alcohol stability and the heat coagulation time.

However, when the basis of the heat stability test is altered to one in which the temperature required to cause instantaneous coagulation is measured, a different picture emerges. Miller and Sommer (1940) measured the heat stability of milk as a function of the degree of acidification using just such a procedure. The coagulation temperature as a function of pH showed sigmoidal behaviour, remarkably similar to the ethanol stability/pH profile. Indeed, Horne and Muir (1990) were able to demonstrate that an equation of the form used to fit ethanol stability profiles could also be applied to the heat stability data of Miller and Sommer (1940). Moreover, sufficient results were contained in the paper of Miller and Sommer (1940) to demonstrate that these heat stability profiles responded to the addition of calcium and phosphate in a manner similar to ethanol stability/pH profiles, i.e., an addition of calcium shifted the profile to more alkaline pH values whereas addition of a calcium sequestrant, phosphate or citrate, shifted the profile to more acidic values. The lack of correlation between ethanol and heat stability seems to be due to test protocols rather than in a fundamental difference in the destabilization reactions. The exact nature of those reactions is now reasonably certain in the case of ethanol stability but is still speculative for heat stability and must await further research, although the mechanism proposed by Horne and Muir (1990) is not unreasonable.

21.3 Ethanol stability and casein micelle structure

21.3.1 Casein micelle stability

The technological properties of milk are strongly influenced by the stability of the casein micelle system or rather by the fact that this stability can be lost as a result of any of several treatments, including acidification, heating, addition of ethanol, limited proteolysis and addition of calcium. Each of these processes modifies the structure of the micelle in such a fashion as to render it unstable. Understanding how this is brought about provides a valuable contribution to our fundamental knowledge of casein micelle structure and stability which ultimately must reap dividends in optimizing process control and manufacturing efficiency. Although controversy still rages, the application of new concepts, notably the adhesive sphere approach of de Kruif and co-workers (see de Kruif, 1998 or de Kruif and Holt, Chapter 5, for a review) has assisted in rationalizing much of the data.

(a) Micellar interactions

All current discussions of casein micelle stability begin with the tacit assumption that the micelles can be treated as hard-sphere, colloidal particles. Their treatment as colloids is adequately based on their particle size. Their diameter ranges from 20 to 600 nm and consistently averages around 200 nm (McGann *et al.*, 1980; Horne and Dalgleish, 1985; Holt, 1992). De Kruif (1998) has reviewed the evidence justifying their treatment as hard spheres, citing evidence of size and polydispersity from neutron scattering (Hansen *et al.*, 1996), supported by results from sedimentation field-flow fractionation (de Kruif, 1998). This picture also ties in with the voluminosity (de Kruif; 1998), viscosity (Griffin *et al.*, 1989) and diffusivity (de Kruif; 1992) behaviour of casein micelles.

It is now almost universally accepted that casein micelles are sterically stabilized by a diffuse layer of flexible hydrophilic polypeptide chains, mainly the C-terminal portion of κ-casein (Holt, 1975, 1992; Walstra, 1979, 1990; Holt and Horne, 1996). This so-called 'hairy layer' presents a virtually impregnable barrier against aggregation unless the hairs are removed enzymatically (as in the renneting of milk) or if the solvent quality is reduced (as in the presence of added ethanol in the ethanol stability test) or on acidification. De Kruif (1998, 1999) has interpreted such removal as imparting a stickiness to the micelle surface, the so-called adhesive sphere theory, which predicts phase separation when a critical level of adhesiveness is exceeded. Although such theories provide generic mechanisms for micellar destabilization, being thermodynamic in nature, they can be criticized for conveying little knowledge of the kinetics and dynamics of these processes. Although important in rennet-induced aggregation, the latter issues of kinetics seldom are important in discussions of ethanol stability where catastrophic and rapid stability failure ensues at the critical alcohol concentration and where rapid equilibration apparently occurs at sub-critical ethanol levels (Horne, 1984, 1986; Horne and Davidson, 1986).

Convincing verification of the 'hairy-layer' model was provided by Walstra *et al.* (1981), who directly measured changes in the hydrodynamic size of casein micelles as a result of chymosin treatment. A reduction in diameter of about 10 nm was observed, consistent with the removal of a macropeptide hairy layer. At longer times, they observed an increase in apparent size as coagulation ensued. Independent confirmation of the depilatory action of chymosin has since been provided by several laboratories (Horne, 1984; Holt and Dalgleish, 1986; Griffin, 1987) and has included direct observations on undiluted milk using a fibre optic dynamic light scattering technique (Horne and Davidson, 1993). Other evidence in support of this steric stabilization model comes from proton NMR studies which reveal considerable flexibility and high levels of mobility in the C-terminal of the κ-casein polypeptide chain on the casein micelle (Griffin and Roberts, 1985; Rollema *et al.*, 1988) and in the N-terminal of adsorbed β-casein (Ter Beek *et al.*, 1996). Yet more evidence in favour of this model is presented in Section 21.3.3 when the response of casein micelles to sub-critical levels of ethanol is considered.

21.3.2 Ethanol addition and steric stabilization of casein micelles

It is well known that, in an unfavourable solvent environment, proteins and other polymers undergo a conformational transition to a collapsed form, accompanied by a dramatic reduction in the effective volume of the molecule. Solvent conditions that lead to collapse are often similar to those that lead to aggregation (Post and Zimm, 1982). It was therefore postulated that the action of ethanol in micellar aggregation might be to collapse the 'hairs' and remove the steric stabilizing component from the system. To test this hypothesis, Horne (1984) used photon correlation spectroscopy to measure the hydrodynamic radius of micelles in the presence of ethanol at concentrations up to those that induce coagulation. At these sub-critical ethanol levels, the radius of micelles, diluted from a micellar fraction of narrow size distribution, exhibited a maximum decrease of approximately 10 nm from the value measured in the absence of ethanol. The hydrodynamic radius passed through this minimum and increased dramatically as the concentration of ethanol was increased through the critical level, the time-dependent increase in radius indicating micellar aggregation.

Horne (1984) recognized that the decrease in radius could be the result either of a uniform global shrinkage of the micelle or of a conformational change in the 'hairy' outer layer. Support for the occurrence of a purely surface phenomenon came from an experiment similar to that of Walstra *et al.* (1981), in which aggregation was induced by chymosin addition. In the absence of ethanol, a decrease in radius of approximately 5 nm was observed, confirming the observations of Walstra *et al.* (1981). In the presence of 16% (v/v) ethanol, with no rennet added, the micelle radius was reduced by approximately 8 nm. On adding rennet to this solution, no further change in radius was recorded before aggregation induced by enzyme action ensued (Horne, 1984). The reaction still exhibited a delay period, or clotting time, dependent on enzyme concentration, indicating that although the 'hairs' had been collapsed by the ethanol and the steric component thus removed from the interaction equation, reduction of some other repulsive component (presumably electrostatic) through enzyme action was still required to allow aggregation to proceed.

Protein desorption or dissociation from the micelle would be an equally applicable explanation for the measured decrease in hydrodynamic size. However, no increases were observed in serum protein levels, even at micellar concentrations 20 times those employed in the dynamic light scattering experiments (Horne, 1986). It was recognized, however, that this was not totally convincing evidence of the absence of desorption, since only a few hairs on the micellar surface would be sufficient to produce a significant increase in hydrodynamic radius whilst their putative loss in ethanol solution could easily lie within detection limits of serum protein levels.

Further arguments in favour of hair collapse rather than desorption were advanced as follows. The observed change in micelle radius occurred very rapidly and showed no time dependence. The value of the radius, measured immediately after mixing, remained constant at that value over a substantial period of time. Indeed, rennet action was required to induce further change in micellar radius (Horne, 1984). The reversibilty of the decrease in radius was equally rapid and could be achieved simply by reducing the ethanol concentration (Horne, 1996). No hysteresis was detected and the radius measured at any given sub-critical ethanol level was the same, no matter which path was followed to that concentration Equally important, the value of the radius measured at any particular ethanol level remained constant with time. The immediacy of response and its lack of time dependence, and the absence of hysteresis on cycling the ethanol levels are evidence against desorption, since the rate of re-adsorption ought to be demonstrably a function of the micellar concentration and hence the extent of dilution employed.

The constancy of the micelle radius with time applied only to measurements in what were termed sub-critical levels of ethanol. Beyond a critical ethanol concentration, which was dependent on buffer pH and Ca^{2+} concentration, a time-dependent increase in radius was observed. Since this increase in radius was, at least in its initial stages, accompanied by an increase in light scattering intensity, and since the rate of increase was also dependent on both the micellar and ethanol concentrations (the latter being above the critical level), this behavior could only be the result of an aggregation process. The change in micelle size on exposure to ethanol was thus due to a collapse of the protein hairs onto the micelle surface, subsequent aggregation beyond the critical ethanol concentration clearly demonstrating the role of the steric stabilizing component in micellar interactions and stability.

21.3.3 Micellar core radius and its consequences

By decreasing the concentration of casein micelles and reducing the rate of aggregation to a negligible level, Horne and Davidson (1986) found that increasing ethanol concentrations did not cause a continued decrease in micelle size. Instead, it produced a range of ethanol concentrations over which the collapsed micelles were stable. This suggests that with total collapse of the 'hairs', the micelle radius can decrease only to a certain minimum value. This was defined as a 'core' radius to distinguish it from a simple minimum, and its measurement allowed an estimate of the hydrodynamic thickness of the barrier layer to be deduced.

The casein micelle system is very polydisperse, with particle sizes ranging from 20 to 600 nm (McGann *et al.*, 1980). Since only an average hydrodynamic radius can be measured using photon correlation spectroscopy, in order to eliminate effects due to possible changes in the micelle size distribution, the majority of barrier thickness measurements were carried out on 'monodisperse' micellar fractions, prepared by resuspension of pellets produced in a series of successive centrifugation steps, always using the supernatant of the previous stage (Horne and Dalgleish, 1985). In this manner, a series of 'monodisperse' slices of the micellar size distribution can be obtained. The barrier layer thickness measured for such fractions increased monotonically with micelle size, i.e., small micelles had thin barriers whereas larger micelles had thicker barriers. Such an observation does not at first sight seem consistent with the simple picture of an outer layer of κ-casein providing a stabilizing sheath. In that model, the same barrier thickness might be anticipated, no matter the size of the particle. Faced with similar behaviour from the polymer polyvinyl alcohol, when adsorbed onto polystyrene latex particles, Garvey *et al.* (1976) developed a model which took the curvature of the particle surface into account. When more extensive studies were made on the response of barrier thickness to change in buffer conditions using the same micellar fraction, it was soon recognized that such a model was not applicable to the casein micelle situation (Horne, 1986).

All theories of steric stabilization highlight the essential role of the stabilizing polymer and the thickness of the stabilizing layer. Solvent quality, which defines that conformation and thickness, is a function also of the ionic constituents of the buffer system. This is of particular importance where the polymeric stabilizer is a polyelectrolyte species such as a protein. Horne and Davidson (1986) found that the correlation between layer thickness and the concentration of ethanol required to achieve total collapse to the core, previously observed for varying micelle size (Horne, 1986), was maintained when changes in barrier thickness were induced by varying the pH, ionic strength or calcium ion concentration of the ethanolic buffer system into which the casein micelles were diluted.

Qualitatively, the observed changes in critical ethanol concentration could be explained on the basis of a mechanism involving electrostatic repulsion between charged polyelectrolyte hairs (Horne and Davidson, 1986). This is a major increase in complexity from the simple model for micellar interaction where the various contributions are treated as additive. Rather, now we envisage the introduction of cross-terms into the interaction equation. Charge and its magnitude influence the strength of the steric stabilization contribution although present theories are as yet incapable of quantitatively estimating such effects. The observed variations with calcium concentration, pH and ionic strength are, however, consistent with the qualitative predictions of a charged hair model. Thus, the effect of calcium ion concentration, where increasing the calcium level decreases the critical ethanol concentration for total collapse, could be envisaged as resulting from specific calcium ion binding reducing the net negative charge on the protein, decreasing inter-hair repulsion and making them more susceptible to collapse (Horne and Davidson, 1986), similar effects to those observed previously in calcium-induced aggregation of individual caseins (Horne and Dalgleish, 1980; Dalgleish *et al.*, 1981). Similarly, the increase in barrier strength with pH, its increasing resistance to collapse, was viewed as being consistent with the increase in charge brought about by greater ionization of the casein phosphate groups which have a pK value of around 6.5 (Horne and Davidson, 1986). The increase in stability recorded with increasing ionic strength required a more convoluted argument to fit this model, since it was to be expected that increased ionic strength would reduce the range of the putative inter-hair repulsion, thus weakening the steric stabilization barrier—the exact opposite of the behaviour observed. Recourse was made to the observations of Dalgleish and Parker (1980) that calcium binding by individual caseins decreases as ionic strength increases. Decreases in calcium binding would mean an increase in protein charge and result in an increase in inter-hair repulsion and in barrier strength. Domination of this binding effect over the contraction of the Debye-Huckel range of electrostatic repulsion between the hairs would therefore lead to an increase in the strength of the barrier with ionic strength. The reversal in the ethanol stability behaviour at the highest ionic strength employed could then be interpreted as the contraction process eventually becoming the dominant component in the balance (Horne and Davidson, 1986). Later studies of the hydrodynamic thickness of β-casein layers adsorbed onto polystyrene latex spheres showed a decrease in thickness on addition of ionic calcium (as calcium chloride) or sodium chloride (Brooksbank *et al.*, 1993), consistent with these observations on the effect of ethanol on micellar size in these environments.

Whilst such a 'charged hair' model accommodates the behaviour observed when ionic species in the buffer solution are varied, it does not address the changes in barrier thickness that also result from these variations. The mechanism envisaged has nothing to contribute on the subject of barrier thickness and would be equally valid in a situation where barrier thickness remained constant. Barrier thickness did not remain constant, however, and consequently a more refined model was required. Moreover, when detailed plots of core radius were included as a function of the ionic variable, it became apparent that the decrease in barrier thickness arose not only from the decrease in hydrodynamic radius from the value observed in alcohol-free solutions, but also from an increase in the actual core radius (Horne and Davidson, 1986). Indeed, increases in core radius were the main contributor to decreases in barrier thickness when the concentration of calcium was increased (Horne and Davidson, 1986). In such situations, the micelle was apparently being hardened and its steric stabilizing coat was being thinned from within.

21.3.4 Implications for structural models of the casein micelle

It is of interest to consider the various models of the casein micelle in the light of the steric stabilization concept, and in particular of the behaviour of the micelle in the presence of ethanol. The principal shortcomings of the sub-unit model were highlighted by Horne (1992) and will not be repeated here. Much of the early support for the sub-unit model came from electron microscopy images (Shimmin and Hill, 1964; Schmidt and Buchheim, 1970; Buchheim and Welsch, 1973; Schmidt *et al.*, 1973; Kalab *et al.*, 1982; Carroll *et al.*, 1985), where the micelles showed a corpuscular, raspberry-like appearance. In effect, these electron microscopy studies were only indicative of heterogeneity in the internal structure of the casein micelle. The presence of such heterogeneity is also the conclusion from a number of neutron (Stothart and Cebula, 1982; Hansen *et al.*, 1996) and X-ray (Pessen *et al.*, 1989) scattering studies. Complementary to these observations are NMR relaxation measurements on casein micelles, which suggest that the water within them is not in a homogeneous continuum which would permit rapid molecular motion but is trapped in discrete volumes of diameter approximately 7 nm, a value close to, but somewhat smaller than, other estimates of subunit size (Kumosinski *et al.*, 1987; Farrell *et al.*, 1989). Against this evidence of heterogeneity are the latest electron microscopy images of McMahon and McManus (1998) which show no electron dense areas greater than 2–3 nm. Discounting this latter study, what all of these techniques are unable to demonstrate is that these submicelles are capable of a discrete, separate existence or that their formation is a necessary step in casein micelle production in the mammary gland.

In a review questioning the existence of sub-micelles, Walstra (1999) pointed out several experimental observations which were difficult to reconcile within the sub-micelle model. These included the presence of small calcium phosphate particles (now referred to as nanoclusters) seen in electron micrographs (Knoop *et al.*, 1973) or in proteolytic digests (Holt *et al.*, 1986) and the longevity of calcium exchange between micelles and solution. Such observations had led Holt (1992) to propose what was effectively an internal structure model of the casein micelle. In this model, the micelle is envisaged as a mineralized cross-linked protein gel with the colloidal calcium phosphate nanoclusters as the agents responsible for cross-linking the proteins and holding the network together. However, Walstra (1999) drew attention to a major deficiency in the Holt model in that it does not explain in any simple way how the casein micelles acquire an outer layer of κ-casein. It provides no mechanism for limiting the growth of casein micelles to the size range observed. Horne (1998) raised further objections and dealt with them in a new dual-binding model for the casein micelle, described briefly below.

Based on ideas encapsulated in a computer simulation of micelle assembly from individual casein molecules (Horne *et al.*, 1989), assembly and growth in this dual-binding model takes place by a polymerization process, as previously, but now involving and identifying, as the name suggests, two distinct and strictly separate forms of bonding. These are cross-linking through hydrophobic regions of the caseins and neutralization and bridging of negatively charged regions, mainly phosphoserine clusters (phosphate centres), by mineral calcium phosphate nanoclusters (Figure 21.2). Central to the model is the concept that micelle integrity, and hence stability, is maintained by a local excess of non-covalent attraction over electrostatic repulsion. In the main, it is anticipated that the majority of this non-covalent attraction will be provided by hydrophobic interactions between regions of the casein molecules. The self-association tendencies of the caseins are well documented (Schmidt, 1982). Thus, α_{s1}-casein, as a BAB block copolymer, forms a worm-like chain polymer (Figure 21.2), the degree of polymerization of which is limited and controlled by solution pH or ionic strength (Payens and Schmidt, 1966; Schmidt, 1970a,b). In the dual-binding model, it has been suggested that the protein cross-linking occurs through interaction between the hydrophobic regions, B, at each end of the molecule (Horne, 1998). Effectively, further growth is limited by the negative charge carried by the central hydrophilic region, denoted A. In the casein micelle, this charge can also be neutralized by the positively charged calcium phosphate nanocluster, which can then also cross-link protein chains, providing the second form of bonding, as shown schematically in Figure 21.2.

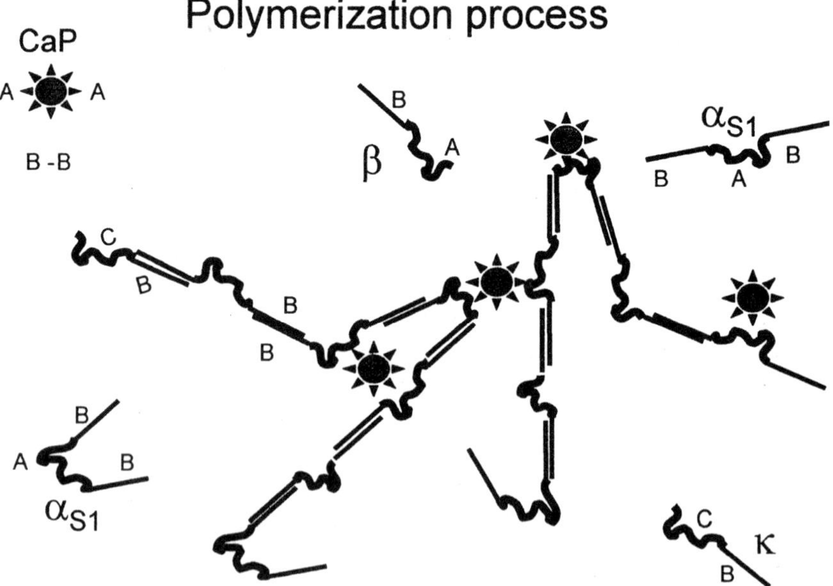

Figure 21.2 Illustration showing polymerization processes which may occur during micellar assembly. Paths within the network can be traced along the protein chains through hydrophobic interaction links or through CaP nanocluster cross-links. Growth along these paths is infinite until terminated by a κ-casein molecule hydrophobically bonding to the (n−1)th casein molecules.

The second essential concept in this dual-binding model is the idea that the individual caseins possess differing functionalities, different numbers of hydrophobic regions and phosphate clusters capable of entering bonding links. A minimum of two functional groups of any combination of types is necessary to sustain polymer growth and three per molecule allows branching and network formation. As mentioned above, chains of α_{s1}-casein can be viewed as cross-linked through nanoclusters as branch points. Viewing the clusters as multi-functional, this same cluster could be linked to the phosphoserine residue cluster of a β-casein molecule through which the polymerization pathway could be sustained by linking of its hydrophobic region to that of another β-, α_s- or κ-casein molecule. κ-Casein, however, is seen as mono-functional only. Its N-terminal half is hydrophobic but the molecule has no cluster of phosphoserine residues to cross-link to nanoclusters. The particular polymerization pathway is terminated by a κ-casein molecule. The ending of growth in this fashion leaves κ-casein molecules on the surface of the micellar aggregates, the location found in nature and a consequence of the unique properties of this particular casein.

In the original computer simulation of Horne *et al.* (1989), the growing cluster was permitted to circumvent the blocked or terminating growth site, allowing some significant fraction of κ-casein to be buried in the micellar interior. In this new model, chain branching is also permitted by allowing more than two hydrophobic regions to come together in a stack of three or more. This is how the micellization of β-casein is considered to occur. Such variable functionality is likely to, indeed should, depend on the level of charge carried by the hydrophilic regions in the immediate neighbourhood. Possibilities of reducing this are low for κ-casein, reducing the hydrophobic functionality and in turn the likely opportunities for easy circumvention of the blocked growth site.

A major failing of earlier micelle models was their lack of a plausible mechanism for assembly, growth and more importantly termination of growth. All such elements are in place in this dual-binding model. κ-Casein naturally finds itself on the micelle surface as extended hairs to sterically stabilize the micelle. The potential for removal of these hairs by chymosin action or collapse in the presence of ethanol is immediately evident. Furthermore, the concept of a localized excess of non-covalent attraction over electrostatic repulsion, as well as allowing the growth of micelles to be modeled, successfully accommodates the response of the casein micelle to changes in pH, temperature, the presence of denaturants such as urea, and the removal of calcium phosphate by acidification or sequestrants (Horne, 1998).

In the present context of ethanol stability, this model of the casein micelle is particularly attractive because it retains the main ideas of the microgel structure postulated by Horne and Davidson (1986) to explain their observations on the influence of solution environment on the response of the casein micelle to sub-critical concentrations of ethanol. The varying reactivity of the monomers not only gives rise to internal heterogeneity necessary to explain many of the physical observations, but also creates a more physically meaningful picture of steric stabilization in which the hairs are linked, not to a solid core surface, but also to each other. The effects of calcium addition in those experiments, which could be viewed as simply reducing inter-hair repulsion, can now be seen as promoting increased cross-linking of the surface layers into the more rigid structure that is the micelle core. In other words, this model allows for increases in the radius of the core through the formation of further bonds in the region of the outer layer, bonds which are unlikely to differ in chemical or physical nature from those already within the micelle interior. Similarly, the effects of increased charge, with pH for example, would be explicable as the promotion of greater repulsion leading to a reduction in cross-linking, the protein hairs now being free to extend and give rise to the increased hydrodynamic radii observed in alcohol-free buffers (Horne and Davidson, 1986).

The microgel model, and now its equivalent the dual-binding model, also predicts that in situations where high levels of cross-linking are operative, the stabilizing sheath should be less responsive to the action of ethanol and *vice versa*. When the stability layer is thin, and the system is consequently less stable, the presence of the cross-links renders the hairs less readily collapsible and less susceptible to the effects of added ethanol, compared to the converse situation where the system is sterically stable but has few inter-hair cross-links so that paradoxically the hairs collapse more readily in the presence of the ethanol. Such behaviour was in fact observed (Horne and Davidson, 1986), assuming that the initial gradient of the plot of hydrodynamic radius against ethanol concentration indicated the susceptibility of the steric-stabilizing sheath to alcohol-induced collapse. The magnitude of this gradient increased linearly with buffer pH. As qualitatively predicted by the microgel model, the steric stabilizing barrier became softer due to the decreasing number of cross-links weakening the gel sheath, as pH was increased. The increased softness was not accompanied by reduced stability, however, because as the barrier became softer it also became thicker and total thickness was the controlling factor which regulated the overall stability of the micelles. In the final analysis, higher concentrations of ethanol were required to totally collapse that thicker barrier and allow aggregation to proceed. This microgel model was also used to explain the observations noted when the concentration of ionic calcium was varied. The ethanol response gradient or collapse susceptibility decreased as the calcium concentration was increased (Horne and Davidson, 1986). Therefore, the gel sheath became more rigid as the increased calcium level neutralized greater amounts of charge, reduced localized inter-hair repulsion and promoted cross-links between hairs. The barrier became thinner as the inner layers became more rigid and were incorporated into the micelle core. Although it became progressively more difficult to induce conformational collapse in the thinning gel sheath as the concentration of calcium was increased, the overall stability of the micelle system decreased, because again the overriding parameter controlling that stability was the barrier thickness alone.

21.4 Mechanism for alcohol destabilization of milk protein

The experiments described in the last section confirm the contribution of steric stabilizaiton to micelle stability but relate to the behaviour of the casein micelle in an artificial environment, one devoid of inorganic phosphate. Attention was drawn previously (Horne, 1992) to the differences between the stability of micelles in this environment and the alcohol stability of milk, where the presence of phosphate proved to be the cause of the sigmoidal ethanol stability/pH profile (Horne, 1987). This mechanism for alcohol destabilization of milk proposed by Horne (1987, 1992) has not been challenged and fits well with the proposed dual-binding model of the casein micelle outlined above.

The mechanism suggests two effects of ethanol on the milk system: destabilization of micelles through loss of the 'hairy layer' and precipitation of calcium phosphate, as noted by Pierre (1985). This later precipitation would reduce the concentration of free calcium, drawing away caseinate-bound calcium, increasing the thickness of the steric layer and micellar charge, all of which increase micelle stability. The higher the alcohol level, the faster the precipitation of calcium phosphate and the greater its extent. The ensuing adjustments in protein charge and conformation, although relatively rapid, as observed in 'hairy layer' studies (Horne, 1986), still require a finite time to respond. Countering them, however, are the destabilizing effects on the micelles themselves, promoting cross-linking and causing the collapse of the steric stabilizing layers. This removal of a major repulsive contribution to the energy barrier might speed up the rate of aggregation of micelles by orders of magnitude. When the coagulation reaction occurs faster than the response time for the adjustment of charge and conformation resulting from calcium phosphate precipitation, the aggregation reaction dominates and precipitation of micelles follows.

According to this mechanism, the origin of the sigmoidal pH profile can also explained through the effect of pH on calcium phosphate solubility (Horne, 1987). The argument may be stated as follows. It is known that on acidification of milk, micellar calcium phosphate is solubilized. It is also known that this colloidal calcium phosphate is composed mainly of small crystallites or nanoclusters (Lyster *et al.*, 1984). The large exposed surface area of these small crystals would account for the ease of dissolution and, perhaps of greater importance here, of the re-precipitation of calcium phosphate on exposure to ethanol. Increasing pH promotes the calcium phosphate precipitation reaction, enhancing its stabilizing effect by requiring more ethanol to overcome the increasing energy barrier being erected. Similarly, the observed decrease in the ethanol stability of milk as ionic strength is increased can be accounted for by the inhibition of calcium phosphate precipitation at higher ionic strengths (Horne, 1987), leading to a reduced demand for calcium from caseinate and a consequent decrease in the level of ethanol required to destabilize the system. These proposals are based qualitatively on the effects of ionic strength on the solubility of minerals and the pK values of phosphate in the absence of ethanol. As was pointed out by Pierre (1985), the solubility of salts is also dependent on the dielectric constant of the mixture. The domination of mineral precipitation behaviour would then satisfactorily explain the relation observed by Horne and Parker (1981c) between ethanol stability and the dielectric constant of the precipitation mixture.

The proposed role of calcium phosphate precipitation as a stabilizing influence would provide a route by which the composition of the serum phase could exert control over micelle stability, consistent with the relationships found by Donnelly and Horne (1986) and earlier by Davies and White (1958). Similarly, when milk is forewarmed, the resulting precipitation of calcium phosphate would enhance the effectiveness of ethanol-induced mineral precipitation by shifting the stability profile in the direction of reduced availability of calcium, thereby producing the observed increases in micellar stability (Horne and Parker, 1981d).

Finally, what is effectively a competition between mineral precipitation and micellar aggregation reactions for calcium in the system could also explain the anomalous destabilization behaviour of trifluoroethanol (Horne and Davidson, 1987). With this alcohol, it was found that after passing through a critical range of trifluoroethanol concentration which caused protein coagulation, higher levels gave rise to micellar dissociation. In the mechanism proposed, this would be the result of the mineral precipitation behaviour proceeding so fast under these high alcohol concentrations that the micelle disintegrated before the aggregation reaction could occur, so much calcium was being drawn off the casein.

21.5 Practical applications

Whilst the main reason for pursuing research on the ethanol stability of milk was to gain greater insight into the inherent stability of the casein micelle, there have been some eminently practical benefits from these studies.

21.5.1 Cream liqueurs

The main product to benefit from the increased knowledge of ethanol stability was cream liqueur, sales of which continue to be a major source of income for many dairy and liquor companies, aside from the excise revenue generated for the exchequer. The base cream liqueur is an oil-in-water emulsion of fat droplets stabilized by sodium caseinate in an aqueous dispersion containing 10–20% ethanol, usually with up to 20% added sucrose. The alcohol source and other added flavours and colourings produce the distinguishing features of the various brands, although some variants may emphasize the source of the milk fat, for example.

In the early varieties, two major technical problems contributed to a limited shelf-life. In the first, a fat plug formed in the neck of bottles of cream liqueur during storage. Muir and Banks (1986) recognized that this was caused by poor homogenization of the emulsion which could be solved by greater control of conditions and an increase in the severity of homogenization. Subsequently, Dickinson *et al.* (1989) suggested that other factors, notably low pH, high calcium content and a low concentration of caseinate, all of which contribute to low emulsion stability, are likely to enhance and aggravate the problems of fat plug formation. The second major defect manifested itself as complete destabilization of the emulsion and was attributed to residual amounts of calcium in the cream source. This was overcome by incorporating a calcium sequestrant, such as sodium citrate, into the formulation (Banks *et al.*, 1981), or by using washed cream or anhydrous milk fat (Banks and Muir, 1988).

21.5.2 Single point stability tests and milk quality

The single point alcohol stability test is still used widely in South and Central America (M.A.P.V. Brito, Personal Communication) and in parts of Asia (P. Tangkawattana, Personal Communication). Failure in this stability test indicates that a sample of milk is of inferior quality and unsuitable for further processing, and represents a substantial loss of income to the presenting farmer. The ethanol stability/pH profile demonstrates that reducing the pH of milk from its natural value reduces the alcohol stability of the milk. Presumably, such milks are being rejected because the acidification is attributed to bacterial contamination. There are, however, a number of reasons which cause the ethanol stability profile to shift along the pH axis to a higher pK value, producing the same reduction in ethanol stability as acidification. These have been summarized in Section 21.2.3. In recognition of these other possible causes of failure, it would be more fair to subject doubtful milks which fail the alcohol-screening test to further testing before final rejection.

References

Ayers, S.H. and Johnson Jr., W.T. (1915) The alcohol test in relation to milk. *Bulletin No. 202, US Dept. Agric.*, Government Printing Office, Washington.

Banks, W., Muir, D.D. and Wilson, A.G. (1981) Extension of the shelf-life of cream-based liqueurs at high ambient temperatures. *J. Food Technol.*, **16**, 587–95.

Banks, W. and Muir, D.D. (1988) Stability of alcohol-containing emulsions, in *Advances in Food Emulsions and foams*, (E. Dickinson and G. Stainsby eds.) Elsevier Applied Science, London, pp. 257–83.

Brooksbank, D.V., Davidson, C.M., Horne, D.S. and Leaver, J. (1993) Influence of electrostatic interactions on β-casein layers adsorbed on polystyrene latices. *J. Chem Soc. Faraday Trans.*, **89**, 3419–25.

Buchheim, W. and Welsch, U. (1973) Evidence for the submicellar composition of casein micelles on the basis of electron microscopical studies. *Neth. Milk Dairy J.*, **27**, 163–80.

Carroll, R.J., Basch, J.J., Phillips, J.G. and Farrell, H.M., Jr. (1985) Ultrastructural and biochemical investigations of mature human milk. *Food Microstruct.*, **4**, 323–31.

Dahlberg, A.O. and Garner, H.S. (1921) The alcohol test as a means of determining quality of milk for condenseries. *Bulletin No. 944*, *US Dept. Agric.*, Government Printing Office, Washington.

Dalgleish, D.G. and Parker, T.G. (1980) Binding of calcium ions to bovine α_{s1}-casein and precipitability of the protein-calcium ion complex. *J. Dairy Res.*, **47**, 113–22.

Dalgleish, D.G., Paterson, E. and Horne, D.S. (1981) Kinetics of aggregation of α_{s1}-casein/Ca^{2+} mixtures: charge and temperature effects. *Biophys. Chem.*, **13**, 307–14.

Davies, D.T. and White, J.C.D. (1958) The relation between the chemical composition of milk and the stability of the casein complex II. Coagulation by ethanol. *J. Dairy Res.*, **25**, 256–66.

De Kruif, C.G. (1992) Casein micelles: diffusivity as a function of renneting time. *Langmuir*, **8**, 2932–7.

De Kruif, C.G. (1998) Supra-aggregates of casein micelles as a prelude to coagulation. *J. Dairy Sci.*, **81**, 3019–28.

De Kruif, C.G. (1999) Casein micelle interactions. *Int. Dairy J.*, **9**, 183–8.

Dickinson, E., Narham, S.K. and Stainsby, G. (1989) Factors affecting the properties of cohesive creams formed from cream liqueurs. *J. Sci. Food Agric.*, **48**, 225–34.

Donnelly, W.J. and Horne, D.S. (1986) Relationship between ethanol stability of bovine milk and natural variations in milk composition. *J. Dairy Res.*, **53**, 23–33.

Farrell, H.M., Jr., Pessen, H. and Kumosinski, T.F. (1989) Water interactions with bovine caseins by hydrogen-2. Nuclear magnetic resonance relaxation studies: structural implications. *J. Dairy Sci.*, **72**, 562–74.

Fox, P.F. (1981) Heat-induced changes in milk preceding coagulation. *J. Dairy Sci.*, **64**, 2127–37.

Garvey, M.J., Tadros, Th.F. and Vincent, B. (1976) A comparison of the adsorbed layer thickness obtained by several techniques of various molecular weight fraction of poly (vinyl alcohol) on aqueous polystyrene latex particles. *J. Colloid Interf. Sci.*, **55**, 440–53.

Griffin, M.C.A. (1987) Structural studies of casein micelles using photon correlation spectroscopy. *J. Colloid Interf. Sci.*, **115**, 499–506.

Griffin, M.C.A. and Roberts, G.C.K. (1985) A ^{1}H-NMR study of casein micelles. *Biochem. J.*, **228**, 273–6.

Griffin, M.C.A., Price, J.C. and Griffin, W.G. (1989) Variation of the viscosity of a concentrated, sterically stabilized colloid: effect of ethanol on casein micelles of bovine milk. *J. Colloid Interf. Sci.*, **128**, 223–9.

Guo, M.R., Wang, S., Li, Z., Qu, J., Jin, L. and Kindstedt, P.S. (1998) Ethanol stability of goat's milk. *Int. Dairy J.*, **8**, 57–60.

Hansen, S., Bauer, R., Lomholt, S.B., Qvist, K.B., Pedersen, J.S. and Mortensen, K. (1996) Structure of casein micelles studied by small angle neutron scattering. *Eur. Biophys. J.*, **24**, 143–7.

Hilgemann, M. and Jenness, R. (1951) Observations on the effect of heat treatment upon the dissolved calcium and phosphorus in skim milk. *J. Dairy Sci.*, **34**, 483–4.

Holt, C. (1975) The stability of casein micelles, in *Proceedings of an International Conference on Colloid and Surface Science, Budapest*, Vol. 1, (E. Wolfram ed.) Akademia Kiado, Budapest, pp. 641–4.

Holt, C. (1992) Structure and stability of the bovine casein micelle, in *Advances in Protein Chemistry*, Vol. 43, (C.B. Anfinsen, J.D. Edsall, F.K. Richards and D.S. Eisenberg eds.) Academic Press, New York, pp. 63–151.

Holt, C. and Dalgleish, D.G. (1986) Electrophoresis and hydrodynamic properties of bovine casein micelles interpreted terms of particles with an outer hairy layer. *J. Colloid Interf. Sci.*, **114**, 513–24.

Holt, C. and Horne, D.S. (1996) The hairy casein micelle: evolution of the concept and its implications for dairy technology. *Neth. Milk Dairy J.*, **50**, 85–111.

Holt, C., Davies, D.T. and Law, A.J.R. (1986) The effects of colloidal calcium phosphate content and milk serum free calcium ion concentration on the dissociation of bovine casein micelles. *J. Dairy Res.*, **53**, 557–72.

Horne, D.S. (1984) Steric effects in the coagulation of casein micelles by ethanol. *Biopolymers*, **23**, 989–93.

Horne, D.S. (1986) Steric stabiliztion and casein micelle stability. *J. Colloid Interf. Sci.*, **111**, 250–60.

Horne, D.S. (1987) Ethanol stability of casein micelles-a hypothesis concerning the role of calcium phosphate. *J. Dairy Res.*, **54**, 389–95.

Horne, D.S. (1992) Ethanol stability, in *Advanced Dairy Chemistry 1: Proteins*, 2nd edn., (P.F. Fox ed.) Elsevier Applied Science, London, pp. 657–81.

Horne, D.S. (1998) Casein interactions: casting light on the black boxes, the structure of dairy products. *Int. Dairy J.*, **8**, 171–7.

Horne, D.S. and Dalgleish, D.G. (1980) Electrostatic interaction and the kinetics of protein aggregation: α_{s1}-casein. *Int. J. Biol. Macromol.*, **2**, 154–60.

Horne, D.S. and Dalgleish, D.G. (1985) A photon correlation spectroscopy study of size distributions of casein micelle suspensions. *Eur. Biophys. J.*, **11**, 249–58.

Horne, D.S. and Davidson, C.M. (1986) The effect of environmental conditions on the steric stabilization of casein micelles. *Colloid Polym. Sci.*, **264**, 727–34.

Horne, D.S. and Davidson, C.M. (1987) Alcohol stability of bovine skim milk. Anomalous effects with trifluoroethanol. *Milchwissenschaft*, **42**, 509–12.

Horne, D.S. and Davidson, C.M. (1993) Direct observation of decrease in size of casein micelles during initial stages of renneting of skim milk. *Int. Dairy J.*, **3**, 61–71.

Horne, D.S. and Muir, D.D. (1990) Alcohol and heat stability of milk protein. *J. Dairy Sci.*, **73**, 3613–26.

Horne, D.S. and Parker, T.G. (1980) The pH sensitivity of the ethanol stability of individual cow milks. *Neth. Milk Dairy J.*, **34**, 126–30.

Horne, D.S. and Parker, T.G. (1981a) Factors affecting the ethanol stability of bovine milk. I. Effect of serum phase components *J. Dairy Res.*, **48**, 273–84.

Horne, D.S. and Parker, T.H. (1981b) Factors affecting the ethanol stability of bovine milk. II. The origin of the pH transition. *J. Dairy Res.*, **48**, 285–91.

Horne, D.S. and Parker, T.G. (1981c) Factors affecting the ethanol stability of bovine casein micelles: 3. Substitution of ethanol by other organic solvents. *Int. J. Biol. Macromol.*, **3**, 399–402.

Horne, D.S. and Parker, T.G. (1981d) Factors affecting the ethanol stability of bovine milk. IV. Effect of forewarming. *J. Dairy Res.*, **48**, 405–15.

Horne, D.S. and Parker, T.G. (1982) Some aspects of the ethanol stability of caprine milk. *J. Dairy Res.*, **49**, 459–68.

Horne, D.S. and Parker, T.G. (1983) Factors affecting the ethanol stability of bovine milk. VI. Effect of concentration. *J. Dairy Res.*, **50**, 425–32.

Horne, D.S., Parker, T.G. and Dalgleish, D.G. (1989) Casein micelles, polycondensation and fractals, in *Food Colloids*, (R.D. Bee, P. Richmond and J. Mingins eds.) Special publication No. 75, Royal Society of Chemistry, London, pp. 400–5.

Horne, D.S., Parker, T.G., Donnelly, W.J. and Davies, D.T. (1986) Factors affecting the ethanol stability of bovine skim milk. VII., Lactational and compositional effects. *J. Dairy Res.*, **53**, 407–17.

Jenness, R. and Patton, S. (1959) *Principles of Dairy Chemistry*, Wiley, New York.

Kalab, M., Phibbs-Todd, B.E. and Allan-Wojtas, P. (1982) Milk gel structure XIII. Rotary shadowing of casein micelles for electron microscopy. *Milchwissenschaft*, **37**, 513–8.

Knoop, A.M., Knoop, E. and Weichen, A. (1973) Electron microscopical investigations on the structure of casein micelles. *Neth. Milk Dairy J.*, **27**, 121–7.

Kumosinski, T.F., Pessen, H., Prestrelski, J.T. and Farrell, H.M., Jr. (1987) Water interactions with varying molecular states of bovine casein: ^{2}H-NMR relaxation studies. *Arch. Biochem. Biophys.*, **257**, 259–68.

Lyster, R.L.J., Mann, S., Parker, S.B. and Williams, R.J.P. (1984) Nature of micellar calcium phosphate in cows' milk as studied by high-resolution electron microscopy. *Biochim. Biophys. Acta*, **801**, 315–7.

McGann, T.C.A., Donnelly, W.J., Kearney, R.D. and Buchheim, W. (1980) Composition and size distribution of bovine casein micelles. *Biochim. Biophys. Acta*, **630**, 261–70.

McMahon, D.J. and McManus, W.R. (1998) Rethinking casein micelle structure using electron microscopy. *J. Dairy Sci.*, **81**, 2985–93.

Miller, P.G. and Sommer, H.H (1940) The coagulation temperature of milk as affected by pH, salts, evaporation and previous heat treatment. *J. Dairy Sci.*, **23**, 405–21.

Mitamura, K. (1937) Studies on the alcohol coagulation of fresh cow milk. *J. Fac. Agric. Hokkaido Univ.*, **41**, 97–362.

Muir, D.D. and Banks, W. (1986) Multiple homogenization of cream liqueurs. *J. Food Technol.*, **21**, 229–32.

Padmos, W.H. (1930) On the value of the alcohol test for the appraisal of milk. *J. Dairy Res.*, **1**, 201 (abst.).

Payens, T.A.J. and Schmidt, D.G. (1966) Boundary spreading of rapidly polymerizing α_{s1}-casein B and C during sedimentation. Numerical solutions of the Lamm-Gilbert-Fujita equation. *Arch. Biochem. Biophys.*, **115**, 136–45.

Pessen, H., Kumosinski, T.F. and Farrell, H.M., Jr. (1989) Small-angle X-ray scattering investigation of the micellar and submicellar forms of bovine casein. *J. Dairy Res.*, **56**, 443–51.

Pierre, A. (1985) Milk coagulation by alcohol. Studies on the solubility of the milk calcium and phosphate in alcoholic solutions. *Lait*, **65**, 201–12.

Post, C.B. and Zimm, B.H. (1982) Theory of DNA condensation: Collapse versus aggregation. *Biopolymers*, **21**, 2123–37.

Ramsdell, G.A., Johnson, W.T., Jr. and Evans, F.R. (1931) A test for the detection of milk unstable of heat. *J. Dairy Sci.*, **14**, 93–106.

Rollema, H.S., Brinkhuis, J.A. and Vreeman, H.J. (1988) ^{1}H-NMR studies of bovine κ-casein and casein micelles. *Neth. Milk Dairy J.*, **42**, 233–48.

Rose, D. (1961a) Variations in the heat stability and composition of milk from individual cows during lactation. *J. Dairy Sci.*, **44**, 430–41.

Rose, D. (1961b) Factors affecting the pH-sensitivity of the heat stability of milk from individual cows. *J. Dairy Sci.*, **44**, 1405–13.

Shimmin, P.D. and Hill, R.D. (1964) An electron microscope study of the internal structure of casein micelles. *J. Dairy Res.*, **31**, 121–3.

Schmidt, D.G. (1970a) The association of α_{s1}-casein at pH 6.6. *Biochim. Biophys. Acta*, 207, 130–8.

Schmidt, D.G. (1970b) Differences between the association of the genetic variants B,C and D of α_{s1}-casein. *Biochim. Biophys. Acta*, **221**, 140–2.

Schmidt, D.G. (1982) Association of caseins and casein micelle structure, in *Developments in Dairy Chemistry 1: Proteins*, (P.F. Fox ed.) Elsevier Applied Science, London, pp. 61–86.

Schmidt, D.G. and Buchheim, W. (1970) An electron-microscopical investigation of the sub-structure of the casein micelles in cows' milk. *Milchwissenschaft*, **25**, 596–600.

Schmidt, D.G., Walstra, P. and Buchheim, W. (1973) The size distribution of casein micelles in cow's milk. *Neth. Milk dairy J.*, **27**, 128–42.

Sommer, H.H. and Binney, T.H. (1923). A study of the factors that influence the coagulation of milk in the alcohol test. *J. Dairy Sci.*, **6**, 176–97.

Sommer, H.H. and Hart, E.B. (1919) The heat coagulation of milk. *J. Biol. Chem.*, **40**, 137–51.

Stothart, P.R. and Cebula, D.J. (1982) Small-angle neutron scattering study of bovine casein micelles and submicelles. *J. Mol. Biol.*, **160**, 391–5.

Ter Beek, L.C., Ketelaars, M., McCain, D., Smulders, P.E.A., Walstra, P. and Hemminga, M.A. (1996) Nuclear magnetic resonance study of the conformation and dynamics of β-casein at the oil/water interface in emulsions. *Biophys. J.*, **70**, 2396–402.

Unnikrishnan, V., Bhavadasan, M.K. and Rama Murthy, M.K. (1988) Alcohol stability of buffalo milk. *Ind. J. Dairy Sci.*, **41**, 421–6.

Walstra, P. (1979) The voluminosity of bovine casein micelles and some of its implications. *J. Dairy Res.*, **49**, 317–23.

Walstra, P. (1990) On the stability of casein micelles. *J. Dairy Sci.*, **73**, 1965–79.

Walstra, P. (1999) Casein sub-micelles: Do they exist? *Int. Dairy J.*, **9**, 189–92.

Walstra, P., Bloomfield, V.A., Wei, G.J. and Jenness, R. (1981) Effects of chymosin action on the hydrodynamic diameter of casein micelles. *Biochim. Biophys. Acta*, **669**, 258–9.

White, J.C.D. and Davies, D.T. (1958) The relation between the chemical composition of milk and the stability of the caseinate complex. I. General introduction, description of samples, methods and chemical composition of samples. *J. Dairy Res.*, **25**, 236–55.

22

ACID COAGULATION OF MILK

J.A. Lucey and H. Singh

22.1 Introduction

Acidified milk products are one of the oldest and most popular foodstuffs and are produced throughout the world. Worldwide production of fermented milk products (such as yoghurt) probably exceeds 20 million tonnes (IDF, 1995). The popularity of fermented milks and yoghurts is due, at least in part, to various health claims and therapeutic benefits that have been associated with some of these products. In spite of the commercial importance of acidified milk products, much less is known about the formation, structures and physico-chemical properties of acid-coagulated products than is known about cheese. The formation and physical properties of acidified milk gels have been reviewed recently (Lucey and Singh, 1997; Horne, 1999).

A wide variety of acidified milk products are produced; some of the main types are described briefly. Yoghurt is formed by the slow fermentation of lactose to lactic acid by the thermophilic starter bacteria, *Streptococcus thermophilus* and *Lactobacillus delbrueckii* subsp. *bulgaricus*. In set-style yoghurt, gels are formed (undisturbed) in the retail pot. Stirred-type yoghurt is made by breaking a set gel before mixing with fruit and filling into retail containers. The microbiology of the starter culture used, as well as the technologies involved in yoghurt and fermented milks, have been reviewed extensively (e.g., Robinson and Tamime, 1993; Tamime and Marshall, 1997; Tamime and Robinson, 1999). The effects of compositional and processing parameters on the textural properties of acid milk gels were reviewed by Lucey and Singh (1997). The manufacture and technologies involved in the production of fresh acid cheeses have been reviewed also (Guinee *et al.*, 1993; Puhan *et al.*, 1994; Kosikowski and Mistry, 1997). In fresh acid cheeses, the coagulation of milk, cream or whey is achieved *via* a combination of acid and

Advanced Dairy Chemistry Volume 1: Proteins, 3rd edn.
Edited by P.F. Fox and P.L.H. McSweeney, Kluwer Academic/Plenum Publishers, 2003.

heat. The production of fresh acid cheeses generally involves pre-treatments (which may include heat treatment and/or homogenisation of the milk), slow acidification and gelation, whey separation and perhaps curd treatment (Guinee *et al.*, 1993). A small amount of rennet may be used in the production of Quarg, Cottage cheese and *Fromage frais* (Guinee *et al.*, 1993). Processing parameters, such as fat content and heat treatment of the milk, method of whey separation, heat treatment of the curd and addition of hydrocolloids to the curd are varied to produce different types of fresh cheese. The use of rennet in acid milk gels has a dramatic influence on the rheological properties of these "combined" gels (Lucey *et al.*, 2000, 2001). Acid casein is produced by the acidification of milk by starter (lactic casein) or acids (e.g., HCl), followed by heating, whey separation, multiple-stage washing with water, mechanical dewatering and thermal drying of the resultant precipitate (see Mulvihill, 1992 and Chapter 26). A common factor in all of these different acidified milk products is that the initial step involves the formation of an acid-induced gel, which is then further processed. This chapter focuses primarily on the formation of acid milk gels and their physical, rheological and microstructural properties.

22.2 Acidification of milk

22.2.1 Method of acidification of milk

Milk can be acidified by bacterial cultures, which ferment lactose to lactic acid, the direct addition of acids, such as HCl, or by the use of glucono-δ-lactone (GDL) which hydrolyses in solution to gluconic acid. Most studies on the formation of acid milk gels have involved the use of GDL (Lucey and Singh, 1997). An extensive study on the formation and rheological properties of gels formed by cold acidification of milk with HCl and subsequent heating to the gelation temperature has been reported (Roefs, 1986; Roefs *et al.*, 1990; Roefs and van Vliet, 1990). The rate of acidification is different between milk acidified with GDL or bacterial cultures; GDL is hydrolysed rapidly to gluconic acid (especially at high temperatures) whereas after the addition of starter bacteria, the pH changes little initially, but then decreases gradually with time (Figure 22.1). The final pH attained in GDL-induced gels is a function of the amount of GDL added to the milk whereas starter bacteria can continue to produce acid until a very low pH (e.g., <4.1) is attained when the bacteria are inhibited by the low pH; in practice, bacterially-acidified gels are cooled when sufficient acidity has been attained (Tamime and Robinson, 1999). In the manufacture of casein, skim milk is mixed with mineral acid at room temperature and the decrease in pH is very rapid. The rate of pH change during fermentation or addition of acid is controlled by the acid-base buffering properties of milk (Lucey *et al.*, 1993).

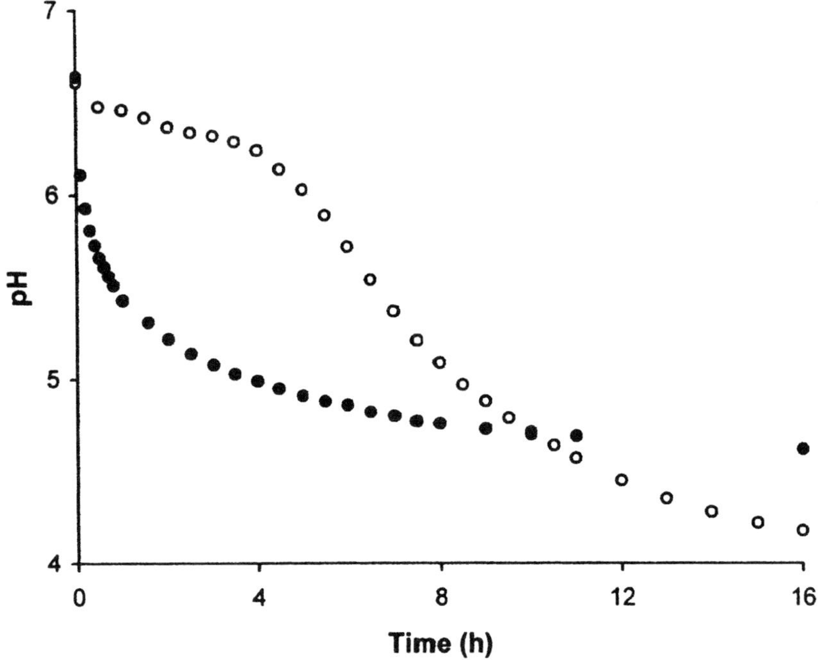

Figure 22.1 Changes in pH during the acidification of milk at 30°C with 1.3% glucono-δ-lactone (GDL) (●) or 2% (w/w) starter culture (○). Milk was heated at 85°C for 30 min. Reprinted from *Food Research International 31*, Lucey *et al.*, A comparison of the formation, rheological properties and microstructure of acid skim milk gels made with a bacterial culture or glucono-δ-lactone, pp. 147–155, 1998, with permission from Elsevier Science.

The rheological and physical properties of gels made with GDL differ from those of fermentation-produced gels (Lucey *et al.*, 1998d), probably due to the different rates of acidification during the critical stage of aggregation of the casein particles and the concomitant physico-chemical changes in casein micelles.

22.2.2 Effect of acidification on the properties of casein micelles

Caseins constitute approximately 80% of the protein in bovine milk, with four main types (α_{s1}-, α_{s2}-, β- and κ-caseins) in combination with appreciable quantities of micellar or colloidal calcium phosphate (CCP) in the form of aggregates called casein micelles (see Chapter 5). At least two models for the structure of casein micelles have been proposed; one model proposes that the micelle core is divided into discrete subunits (submicelles) with distinctly different properties from an outside hairy layer (Schmidt, 1982; Walstra,

1990); the other model suggests that the internal substructure resembles a mineralized, entangled or cross-linked web of chains of casein molecules (Holt, 1992) or is a result of polymerization of casein molecules (Horne, 1998, see Chapter 21). Originally, it was assumed that submicelles were held together ("cemented") in the micelle by bridges of CCP (Schmidt, 1982) but it is now clear that a number of factors are responsible for maintaining the integrity of the micelle. The formation of CCP is an effective means of "burying" a considerable amount of Ca and phosphate within casein micelles (which is important for milk to achieve its primary role of nutrition of the young calf) but CCP can be removed without completely disrupting the micelle if milk is acidified at a temperature >20°C (Dalgleish and Law, 1988; Lucey *et al.*, 1997a). Hydrophobic and hydrogen bonding are important for micelle integrity since the addition of urea disrupts the micelle structure (McGann and Fox, 1974); Ca^{2+} also play a role in the integrity of the micelle (Holt *et al.*, 1986).

During the acidification of milk, many of the physico-chemical properties of the casein micelles undergo considerable change, especially in the pH range 5.5 to 5.0, including voluminosity, solvation and dissociation of the caseins (Roefs *et al.*, 1985; Walstra, 1990). As the pH of unheated milk is reduced, CCP is dissolved (Pyne and McGann, 1960) and the caseins are liberated into the serum phase (Roefs *et al.*, 1985; Dalgleish and Law, 1988). The extent of liberation of caseins depends on the temperature at acidification; at 30°C, a decrease in pH causes virtually no liberation of protein but at 4°C, about 40% of the caseins are liberated into the serum at pH ~ 5.5 (Dalgleish and Law, 1988). The temperature at acidification has no effect on the solubilization of CCP. Apparently, little change in the average hydrodynamic diameter of casein micelles occurs during acidification of (unheated) milk to pH ~ 5.0 (Roefs *et al.*, 1985). Aggregation of casein occurs as the isoelectric point (pH 4.6) is approached.

Milk used for the manufacture of yoghurt is subjected to a severe heat treatment. Heat treatment of milk at a temperature above 70°C causes denaturation of whey proteins, some of which associate with casein micelles, involving κ-casein, *via* hydrophobic interactions and intermolecular disulphide bonds (Haque and Kinsella, 1988; Singh, 1995). Moderate heating does not appear to affect the size of casein micelles although these treatments cause the whey proteins to denature and bind to micellar κ-casein; more extensive heat treatment causes some degree of micellar aggregation and an increase in particle size (Dalgleish *et al.*, 1987). Heat treatment of milk markedly affects the formation and properties of acid milk gels, as described in detail in Section 22.3.1b "Possible physico-chemical mechanisms involved in the formation of gels from unheated and heated milk."

The effect of heat treatment on the solubilization of CCP and the release of caseins during the acidification of milk has been studied recently (Law, 1996; Singh *et al.*, 1996); heat treatment (in the range 70 to 90°C) prior to acidification had little effect on the extent of solubilization of calcium and P_i from the micelles.

From this discussion, it appears that during the acidification of unheated or heated milk to pH $\sim$ 5.1 at a temperature above 22°C, most of the CCP in the micelles has been solubilized, the charge on individual caseins has been altered and the ionic strength of the solution increased. As a result, the forces responsible for the integrity of these "micelle-like" CCP-depleted casein particles are considerably different from those in native micelles even if their average hydrodynamic diameter appears largely unchanged. The balance between the intermolecular attractive and repulsive forces, which is important for gelation properties, is also modified (Horne, 1999). Lucey *et al.* (1997a) showed that CCP-depleted casein particles, formed at pH $\sim$ 5.2, disintegrate if the pH is restored to normal values, demonstrating the important role of electrostatic forces in maintaining the integrity of casein particles.

22.3 Formation and properties of acid-induced milk gels

22.3.1 Mechanisms involved in the formation of acid-induced gels

(a) Theoretical models

Acid milk gels like yoghurt are examples of particle gels (Horne, 1999). At least three theoretical models, namely fractal, adhesive hard spheres and percolation models, have been used to describe the formation of acidified milk gels.

Fractal aggregation theory has been applied to the formation of various casein gels (Bremer *et al.*, 1989, 1990, 1993) and is described in this section. In particle gels, a fractal scaling regime may occur only over small length scales, which are of the order of the aggregating clusters. At longer length scales, the microstructure appears homogeneous. Fractal behaviour is not expected in gels made from high volume fraction systems (Dickinson, 1997a; Horne, 1999). Fractal aggregation assumes that (hard) spherical particles of radius, a, can move by Brownian motion and that they can aggregate when they encounter each other. The aggregates formed then also aggregate with each other. If no further changes occur among the particles in an aggregate, once they are incorporated, this cluster-cluster aggregation process leads to aggregates which obey the scaling relation:

$$\frac{N_p}{N_0} = \left(\frac{R}{a_{eff}}\right)^{D-3} \tag{1}$$

where N_p is the number of particles in an aggregate of radius R, N_0 is the total number of primary particles that could form such a floc, D is a constant called the fractal dimensionality ($D < 3$) and a_{eff} is the radius of the effective building blocks forming the fractal cluster. The volume fraction of particles in an aggregate, ϕ_A, decreases as aggregation proceeds and R increases. At a certain radius, R, the average ϕ_A will equal the volume fraction of particles in the system, ϕ, and all aggregates will touch, forming a continuous gel network:

$$\phi_A = \left(\frac{R_g}{a_{eff}}\right)^{D-3} \tag{2}$$

Rearranging, gives the radius of the aggregates at the point of gel formation, R_g:

$$R_g = a_{eff}\phi^{1/(D-3)} \tag{3}$$

R_g is a measure of the upper cut-off length, i.e., the largest length scale at which the fractal regime exists (van Vliet *et al.*, 1997). The structure of a gel can be characterized by the parameters a, ϕ, R_g and D. For casein gels (both rennet- and acid-induced), $D \approx 2.3$ has been observed generally (Bremer *et al.*, 1989, 1993). However, in gels with a similar D, both a and ϕ can vary and, as a result, the properties of the gel can differ substantially (Lucey *et al.*, 1991d). van Vliet and Keetels (1995) reported that D is similar in acid milk gels made from low or high heat-treated milk although the rheological properties of these gels differ markedly.

This simple fractal approach, although it has described successfully semi-quantitative features in irreversibly aggregating systems, appears to have some deficiencies, including the lack of any allowance for aggregate rearrangement (before, during and after gelation) and interpenetration, and the assumption that all aggregates have the same size at the gel point (Dickinson, 1997a). Dickinson (1997a) suggested that in reality, in a fairly concentrated dispersion, when the biggest clusters are just beginning to join together to form a system-spanning network, a significant proportion of the particles could still exist as very small clusters and that these clusters may be incorporated into the developing network at times well beyond the gel point. It is possible that the slow increase in the storage modulus (G') of acid milk gels after gelation (Lucey and Singh, 1997) reflects the incorporation of additional protein clusters into the gel network as well as the continued fusion and rearrangement of bonds and strands in the network. Horne (1999) also questioned the fractal definition of the gelation point and asked what happens to D if there are rearrangements of particles and the re-making of bonds in the network. After some rearrangements, D probably increases but it is likely that after severe rearrangements a fractal description of the clusters will no longer hold (van Vliet, 1999). Horne (1999) concluded that until these questions are answered, the fractal description of particle gels may be more applicable to fully-formed mature gels and may not be as useful for providing insights into the dynamics of gel development.

The aggregation of casein particles during the acidification of milk has also been modelled using the adhesive hard sphere theory (de Kruif *et al.*,

1995; de Kruif and Roefs, 1996; de Kruif, 1997, 1999). In this model, it is proposed that the glycomacropeptide (GMP) part of κ-casein sterically stabilizes casein micelles and the GMP is considered as a polyelectrolyte brush which collapses on the surface of the micelle as the pH of the system approaches the pK_a of the charged groups on the brush (reduced charge density). Horne (1999) pointed out that the adhesive hard sphere model could be applied only to situations where there is weak attraction between particles and it has no inherent time scale. This model provides no information on the dynamics of gel formation or on the mechanical properties of the gel (Horne, 1999).

Dickinson (1997b) has attempted to incorporate the concept of fractal aggregate formation into the sticky hard sphere model, thereby including the simultaneous effects of strong bonding and weak reversible interactions into the same description of particle gels.

Horne (1999) reviewed the suitability of percolation models for acid milk gels. In this theory, it is assumed that percolation clusters form bonds between close neighbours in a lattice. Bond formation is assumed to be random and as the number of bonds between neighbours increases, the clusters increase in size. In computer simulations, these functions can be calculated (de Kruif *et al.*, 1995). Above a certain threshold, a large cluster occurs which extends throughout the lattice. Analogies between percolation and gelation can be drawn with particles establishing an increasing number of links as they aggregate until, at a certain threshold, a cluster is created which spans the container. At this stage, only a fraction of bonds have been joined and even more importantly not all individual fractions have been incorporated into the system-spanning cluster and this mechanism would explain the continued increase in G' after gelation (Horne, 1999). However, Horne (1999) suggested that the percolation model may be applicable only at the gel point and suggested that it is difficult to use this theory to model the mechanical properties of acid gels. Recently, a new dual-binding model has been proposed for the formation of casein micelles (Horne, 1998) and this model appears to explain some of the properties of acid milk gels (Lucey, 2002).

(b) Possible physico-chemical mechanisms involved in the formation of gels from unheated and heated milk

Native casein micelles (in milk of normal pH) are stabilized by a negative charge and steric repulsion (Walstra, 1990; Mulvihill and Grufferty, 1995). On acidification, casein particles aggregate as a result of (mainly) charge neutralization, leading to the formation of chains and clusters that are linked together to form a three-dimensional network (Mulvihill and Grufferty, 1995). Acid casein gels can be formed from sodium caseinate and these gels can have generally similar properties to acid gels made from milk (Lucey *et al.*, 1997c,d), which suggests that CCP is not essential for acid gelation. Hydrophobic interactions are unlikely to play a direct role in the strength of acid milk gels as the G' of these gels increases with decreasing assay temperature (Roefs and van Vliet, 1990; Lucey *et al.*, 1997b,c). Cooling gels to a low temperature results in an increase in G', probably due to swelling of casein particles and an increase in the contact area between particles (Lucey *et al.*, 1997c). Roefs and van Vliet (1990) reported that increasing the concentration of NaCl added to cold-acidified skim milk samples resulted in a decrease in the dynamic moduli of gels formed on subsequent heating, confirming that electrostatic interactions are important for particle interactions. With increasing ionic strength, charged groups on casein would be screened, thereby weakening interactions between particles (Roefs and van Vliet, 1990; Lucey *et al.*, 1997c).

Lucey *et al.* (1997b) suggested that gels made from unheated milks undergo extensive particle rearrangements during the gel formation, resulting in the formation of dense clusters of aggregated casein particles, which aggregate to form a gel. From these dense clusters, it would be expected that many particles would hardly contribute to cross linking of the network; thus, unheated milk gels would probably have a low G' value, which has indeed been reported (Lucey *et al.*, 1997b).

Acid-induced gels made from heated milk have a higher pH at gelation (Horne and Davidson, 1993; Heertje *et al.*, 1995) and are considerably more firm than those from unheated milk (Lucey *et al.*, 1997b, 1998c). High heat treatment of milk causes denaturation of whey proteins and subsequently, a proportion of denatured whey proteins associates with the casein micelles, involving κ-casein (see review by Singh, 1995). These whey proteins appear as appendages or filaments on the casein micelle surface in electron micrographs (Davies *et al.*, 1978; Kalab *et al.*, 1983; Mottar *et al.*, 1989). When heated milk is acidified, the denatured whey proteins associated with the casein micelles become susceptible to aggregation, as the repulsive charge on the proteins is reduced. The isoelectric pH of the major whey protein, β-lactoglobulin, is ~5.3, which is higher than that of the caseins (Kinsella and Whitehead, 1989). This would explain the high pH of gelation of heated milk. Denatured whey proteins associated with casein micelles act as bridging material by interacting with other denatured whey proteins associated with casein micelles (Lucey *et al.*, 1998c). This increases the number and strength of bonds between protein particles.

The different physico-chemical mechanism involved in the formation of gels from acidified heated milk would probably result in the formation of gels with a different microstructure from those formed from unheated milk. Indeed, a very different microstructure has been reported; there appears to be more branching or interconnectivity in the gel network in heated milk gels than in unheated milk gels (Lucey *et al.*, 1998e). The presence of denatured whey proteins on the surface of casein particles may hinder the close approach of other casein particles and reduce the likelihood that dense clusters of casein particles could be formed. Cross-linking of aggregating particles may occur *via* the denatured whey proteins associated with the casein micelle surface (Lucey *et al.*, 1998c) instead of between charged residues or hydrophobic groups on casein molecules. van Vliet and Keetels (1995) suggested that in acid gels made from heated milk, aggregates were linked by "straight" strands compared to unheated milk, which may have tortuous or "bent" strands. The altered orientation/arrangement of aggregating particles (due to the denatured whey protein associated with casein micelles) is responsible mainly for the different mechanical properties of acid-induced gels, produced from unheated or heated milk (Lucey *et al.*, 1998c).

Based on the results of an electron microscopy study of acidified heated milk, Heertje *et al.* (1985) proposed that at pH ~ 5.5 there is preferential dissociation of β-casein and that at pH ~ 5.2 it reassociates with the micelles, which coincides with a "stage of contraction and rearrangement". However, recent studies (Law, 1996; Singh *et al.*, 1996) have shown that at the temperature (e.g., 30°C) used for the formation of acid gels from heated milk, no preferential dissociation of β-casein from the micelles occurs during the acidification of milk.

In this section, a number of theories which have been used to describe the formation and properties of acidified milk gels were described briefly. Each has some attractions but also limitations. Any useful theory should not only adequately describe the gelation kinetics but should also provide useful information on the mechanical properties and should be consistent with the very different mechanism which occurs in heated milk gels. Theories that incorporate rearrangements at both the particle and cluster level could be important if we are to better understand the formation of acid milk gels. Computer simulation of aggregating particles is a powerful tool in probing the effects of various types and strengths of interactions and how they influence microstructure and small and large deformation properties of simulated particle gels.

22.3.2 Physical properties of acid-induced milk gels

(a) Rheological properties of acid milk gels

In most studies, the textural properties of set-style acid milk gels have been measured empirically as firmness or viscosity (e.g., Parnell-Clunies *et al.*, 1986; Dannenberg and Kessler, 1988b). The relevance of viscosity measurements in a set gel is unclear, although it is important in stirred-style yoghurts. Fundamental large deformation studies of acid milk gels are, in our opinion, a much more powerful and useful tool for understanding the effects of processing procedures such as stirring.

Most rheological parameters characterizing casein gels depend on the number and strength of bonds between the casein particles, on the structure of the latter and the spatial distribution of the strands making up these particles (Roefs *et al.*, 1990; Lucey and Singh, 1997).

Dynamic testing, which involves an oscillatory applied strain or stress, can provide very useful information on the gel formation process (Lopes da Silva and Rao, 1999). Some of the main parameters that are usually determined from these tests include the G', which is a measure of the energy stored per oscillation cycle, the viscous or loss modulus (G''), which is a measure of the energy dissipated as heat per cycle, and the loss tangent (tan δ), which is the ratio of the viscous to elastic properties (Lopes da Silva and Rao, 1999). These parameters are defined as follows:

$$G' = \left(\frac{\tau_0}{\gamma_0}\right) \cos \delta \tag{4}$$

$$G'' = \left(\frac{\tau_0}{\gamma_0}\right) \sin \delta \tag{5}$$

$$\tan \delta = \frac{G''}{G'} \tag{6}$$

where τ_0 is the amplitude of the shear stress, γ_0 is the amplitude of the strain and δ is the phase angle.

The rheological properties of acid gels made from unheated milk at 30°C with GDL are shown in Figure 22.2. Unheated milk forms a weak gel and

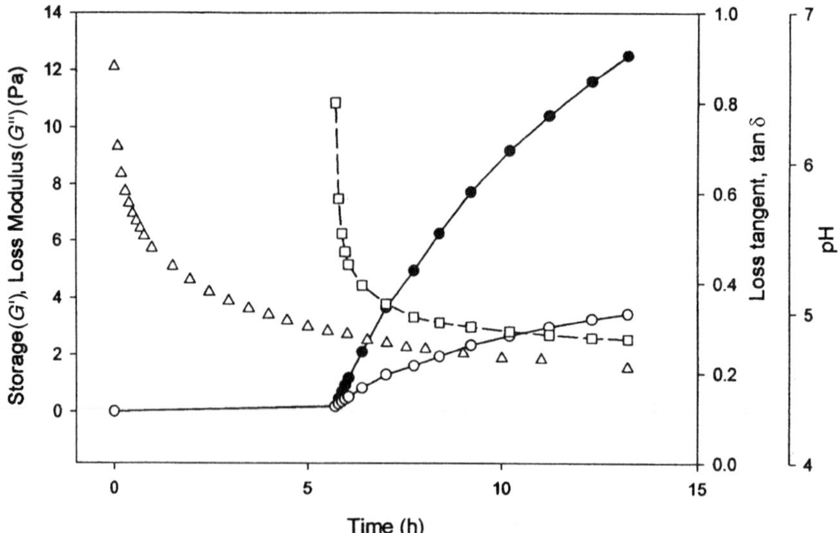

Figure 22.2 Rheological properties of an acid milk gel made by acidification with 1.3% glucono-δ-lactone (GDL) at 30°C from unheated milk. Storage modulus (G') (●), loss modulus (G″) (○), loss tangent (tan δ) (□) and pH (△).

the pH at gelation is generally ~ 4.8 (Lucey *et al.*, 1997b). After gelation, G' increases rapidly but starts to plateau during ageing of the gel; tan δ decreases to <0.4 soon after gelation and to ~ 0.25 during ageing of the gel. Roefs (1986) demonstrated that for acid milk gels made by cold acidification and quiescent heating, G' continues to increase for a period of up to several days, due presumably to slow ongoing fusion of casein particles. In contrast to these typical results, Gastaldi *et al.* (1996) reported that unheated milk has a "stiffness" or "modulus" of ~ 1000 Pa and that after acid-induced gelation, this "stiffness" increases to ~ 5000 Pa. They also suggested that the loss tangent of milk is ~ 8 and that after gelation it increased to ~16. In liquids, such as milk, the loss tangent is >1 and after gelation it decreases to values such as 0.2 to 0.3 for milk gels (see Figure 22.3). The rheological properties of acid milk gels reported by Gastaldi *et al.* (1996) appear to be unrealistic.

There have been a number of reports on the effects of heat treatment on the rheological properties of acid milk gels determined by dynamic low amplitude (strain) oscillation (van Vliet and Keetels, 1995; Lucey *et al.*, 1997b, 1998c). van Vliet and Keetels (1995) reported that acid-induced gels made from reconstituted low-heat skim milk powder (SMP) had much lower

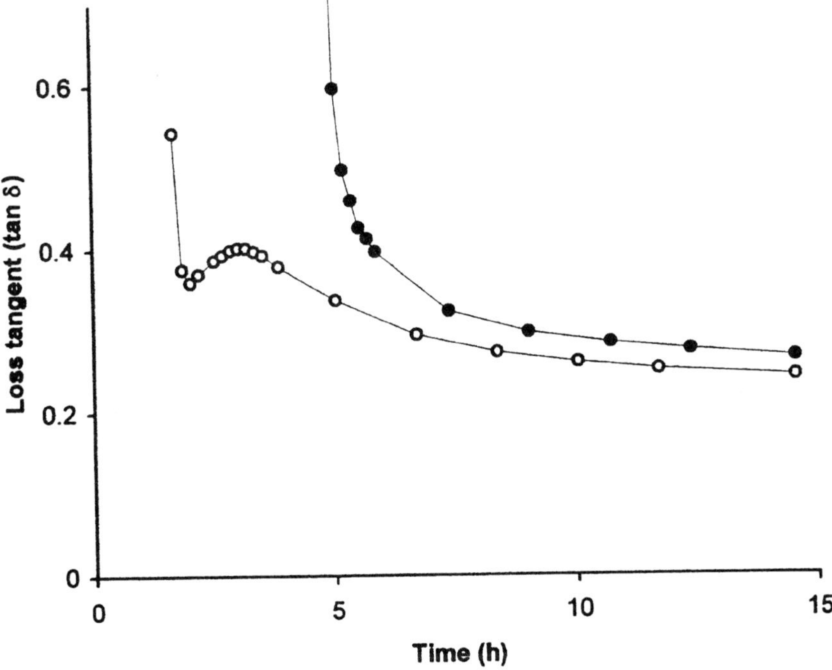

Figure 22.3 Loss tangent (tan δ) as a function of time for acid milk gels made at 30°C by acidification of unheated milk (●), or heated (85°C for 30 min) milk (○) with 1.3% glucono-δ-lactone (GDL).

dynamic moduli than gels made from high-heat SMP. Lucey *et al.* (1997b) reported that heating milk at a temperature ⩾78°C greatly (350 to 450 Pa) increased the G' of resulting acid-induced gels compared to those from unheated milk (~15 Pa). Increased cross-linking or bridging, by denatured whey proteins, within gels made from heated milk may be responsible for the increased rigidity and G' of the network (Lucey *et al.*, 1997b, 1998c).

An unusual rheological phenomenon is observed soon after the formation of an acid gel from heated milk: tan δ decreases initially but then increases to a maximum value before decreasing again (Figure 22.3) (Biliaderis *et al.*, 1992; Rönnegård and Dejmek, 1993; van Marle and Zoon, 1995; Lucey *et al.*, 1998c). A high tan δ indicates an increased susceptibility of bonds and strands in the gel to break or relax, thus facilitating further rearrangements of the gel (van Vliet *et al.*, 1991). The maximum in tan δ may be a consequence of a partial loosening of the weak initial gel network due to the solubilization of CCP, while at lower pH values there would be increased protein-protein attractions between casein particles as the net charge decreases with the approach of the isoelectric point (Lucey *et al.*, 1998c). The maximum in tan δ does not occur in acid-induced gels made from unheated milk or from milk heated in the presence of a sulphydryl blocking agent because gelation in these systems occurs at low pH values (<5.2), which is below the pH range where most of the dramatic changes in the physico-chemical properties of casein micelles (e.g., solubilization of CCP) occur (Lucey *et al.*, 1998c).

Horne (1999) reported that the rheological properties of acidified milk gels exhibit a form of scaling behaviour. In this procedure, the complex shear modulus (G^*) for individual milk samples are replotted as a function of reduced time, defined as the reaction time, t, divided by the gelation time for that profile. Then, each of these individual reduced time plots is normalized against the value of its own shear modulus at three times the gelation time. For acid gels made from unheated or heated (90°C for 10 min) milk, there are two distinct "master curves", (Figure 22.4). Horne (1999) suggested that since gels produced from heated or unheated milk had different "master curves", this implies that there are fundamental differences in the kinetics and dynamics of the gel formation process in these two types of gel. It would appear that the rate at which bonds form between protein particles and the mechanism by which clusters grow into a network are altered by heat treatment. This observation provides further support for the suggestion by Lucey *et al.* (1998c) that an altered physico-chemical mechanism is involved in the formation of acid gels from heated milk.

The effects of fat on the fundamental rheological properties of acid milk gels have been reported (van Vliet and Dentener-Kikkert, 1982; van Vliet, 1988; Lucey *et al.*, 1998b; Cho *et al.*, 1999). The nature of the fat globule membrane determines the types of interaction that can occur between fat globules and the protein matrix. Fat globules act as an inert filler if the

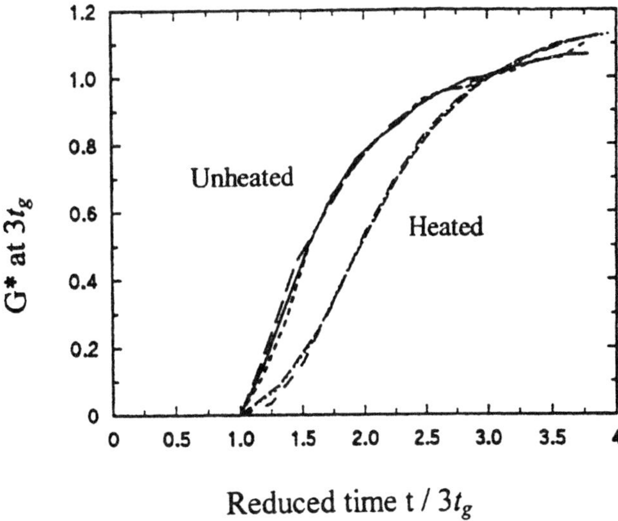

Figure 22.4 Complex shear modulus (G^*) of unheated and heated (90°C for 10 min) milk scaled by the procedure of Horne (1999). The time scale for various gel profiles were reduced by the gelation time (t_g) for that profile and then the profiles were normalized against its shear modulus value at three times the t_g (G^* at $3t_g$). Reprinted from *International Dairy Journal 9*, Horne, Formation and structure of acidified milk gels, pp. 261–268, 1999 with permission from Elsevier Science.

native fat globule membrane is intact since this membrane does not interact with casein particles (van Vliet and Dentener-Kikkert, 1982; van Vliet, 1988). The G' value of acid milk gels decreases with an increasing volume fraction of fat, if the fat globules have an intact native membrane (van Vliet and Dentener-Kikkert, 1982; van Vliet, 1988). In homogenized or recombined milk, the native membrane is replaced largely by casein and some whey protein so that the surface of fat particles can interact with the protein matrix (largely casein but some denatured whey proteins when the gel is made from heated milks) of acid milk gels (van Vliet and Dentener-Kikkert, 1982; van Vliet, 1988). In acid gels made from recombined milk, G' increases with an increasing volume fraction of fat (van Vliet and Dentener-Kikkert, 1982; van Vliet, 1988; Lucey *et al.*, 1998b). Cho *et al.* (1999) showed that the G' of acid gels made from either heated or unheated milk is influenced by the nature of fat globule membrane; gels containing fat globules stabilized by sodium caseinate or denatured whey proteins have very high G' values compared with those stabilised with skim milk powder or native whey proteins.

In experiments where the time-scale of the applied deformation was varied (frequency sweeps), log G' versus log angular frequency gave linear curves with a slope of ~ 0.15 for various types of acid casein gels (Roefs and van

Vliet, 1990; Lucey *et al.*, 1997b). This suggests that similar structural components are present in all types of acid casein gels.

Large deformation studies provide information on properties that may be related to the consistency of the gel during consumption or shearing (which is used in the production of stirred-style yoghurt). There is little information on the large scale deformation properties of yoghurt (Rönnegård and Dejmek, 1993). Mixing and stirring of set gels prior to rheological testing means that many reported (e.g., Dannenberg and Kessler, 1988b) yield properties are not those of the original 'set' gel. Fundamental large deformation rheological properties of acid casein gels have been reported (Bremer *et al.*, 1990; van Vliet *et al.*, 1991; van Vliet and Keetels, 1995; Lucey *et al.*, 1997b,c). Gross fracture of acid casein gels made with GDL was observed at a strain of 0.5 to 0.6 (Roefs, 1986; van Vliet *et al.*, 1991; van Marle and Zoon, 1995). Lucey *et al.* (1997c), using a low constant shear rate technique for acid casein gels made *in situ*, found that the apparent shear stress at fracture increased with decreasing gelation temperature. The apparent shear stress at fracture of acid casein gels increased with ageing while the strain at fracture decreased (Lucey *et al.*, 1997c). Lucey *et al.* (1997b) reported that heat treatment of milk prior to acidification resulted in a large reduction in the strain at fracture, from ~1.5 for gels made from unheated milk to 0.5 to 0.8 for gels made from milks heated at a temperature ⩾80°C (Figure 22.5). The influence of different shearing rate on the fracture behaviour of acid milk gels using the low constant shear rate technique proposed by Lucey *et al.* (1997b,c) has not been reported and it would be interesting to compare a range of shearing rates.

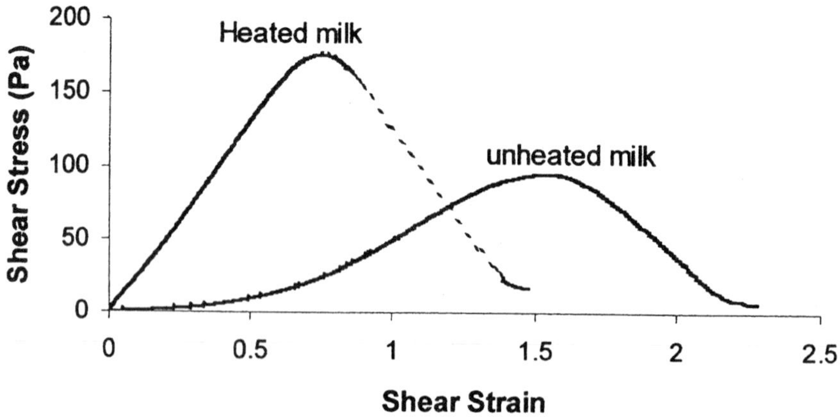

Figure 22.5 Shear stress as a function of applied deformation in a constant shear rate (~0.002 s^{-1}) experiment on acid milk gels made at 30°C by acidification with 1.3% glucono-δ-lactone (GDL). Milk was unheated (—) or heated at 80°C for 30 min (····). Redrawn from the results of Lucey *et al.* (1997b) with kind permission by *Journal of Dairy Research.*

(b) Microstructure of acid milk gels

Electron microscopy (EM) studies on acid milk gels, such as yoghurt, have shown that these gels consist of a coarse particulate network of casein particles linked together in clusters, chains and strands (Kalab *et al.*, 1983). The network has pores or void spaces in which the aqueous phase is confined. The diameter of these pores varies considerably, with larger pores in gels made at a high gelation temperature or from milk with a low protein content. There have been several EM studies on the microstructure of gels formed by acidification of heated milk (Davies *et al.*, 1978; Parnell-Clunies *et al.*, 1987; Mottar *et al.*, 1989). Harwalkar and Kalab (1980) proposed, based on the examination of electron micrographs, that yoghurt gels made from unheated milk had larger protein clusters than gels made from heated milk, which they described as highly branched.

Many of the preparation steps required in EM, including dehydration, fixation, embedding, sectioning and staining, can disrupt the native structure of milk products and lead to the creation of artefacts, unless great care is taken. Confocal laser scanning microscopy (CSLM) is a relatively new technique which enables samples to be observed with minimal preparation procedures due to its unique optical sectioning capabilities and high spatial resolution (Brooker, 1995) and is very suitable for observing the overall microstructure of milk gels (Hassan *et al.*, 1995; Lucey *et al.*, 1997d, 1998b, e). A confocal micrograph of an acid gel made from unheated milk is shown in Figure 22.6. The effects of heat treatment on the microstructure of GDL-induced milk gels have been reported (Lucey *et al.*, (1998e). Confocal images are very amenable to image analysis since the images are already in a digital form and they can be used to calculate the fractal dimensionality (D) of acid casein gels (Bremer *et al.* 1993).

Many of the preparation steps used in EM of whole milk yoghurt can result in partial extraction of fat globules (Allan-Wojtas and Kalab, 1984). Barrantes *et al.* (1996) reported that in yoghurt made from recombined milk, the fat globules were not noticeable using scanning EM but could be observed using transmission EM. The microstructure of acid gels of different fat content was investigated using CSLM by Lucey *et al.* (1998b). In acid-induced gels made from recombined milk, fat globules appeared to be dispersed throughout the gel and the microstructure was quite different from that of skim milk gels (Lucey *et al.*, 1998b). No large pores were visible in acid-induced gels made from recombined milk; probably, fat globules obscured the finer details of pores and strands.

(c) Permeability of acid milk gels

Permeability measurements give information about inhomogenities at the level of the gel network (van Dijk and Walstra, 1986; Roefs *et al.*, 1990; Lucey *et al.*, 1998e). The permeability coefficient of acid milk gels can be calculated as follows:

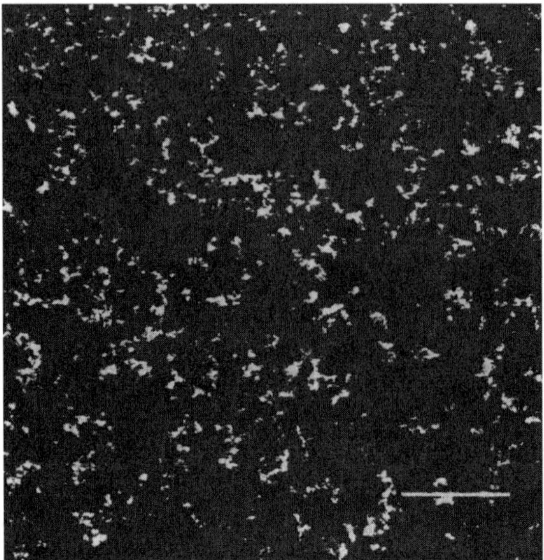

Figure 22.6 Confocal scanning laser micrograph of acid milk gels made at 30°C by acidification with 1.3% glucono-δ-lactone (GDL) from unheated milk. Scale bar = 20 μm. The protein matrix appears white. Reprinted from *Food Hydrocolloids 12*, Lucey *et al.*, Microstructure, permeability and appearance of acid gels made from heated skim milk, pp. 159–165, 1998, with permission from Elsevier Science.

$$B = -\left[\ln \frac{(h_\infty - h_{t_2})}{(h_\infty - h_{t_1})}\right] \eta H / [\rho g(t_2 - t_1)] \tag{7}$$

where B is the permeability coefficient (m^2), h_∞ is the height of the whey in the reference tube (m), h_{t_1} is the height of the whey in the gel tube at t_1 (m), h_{t_2} is the height of the whey in the gel tube at t_2, η is the viscosity of the whey, H is the height of the gel (m), ρ is the density of the whey and g is acceleration due to gravity. In acid milk gels made at 30°C with GDL, the value of B is usually in the range 1 to 2×10^{-13} m^2 (Roefs *et al.*, 1990; Lucey *et al.*, 1998e).

In rennet-induced milk gels, B increases with time, which has been taken as evidence of "microsyneresis" or breakage of strands in the network resulting in the formation of larger pores (Walstra, 1993). Studies on the permeability of acid-induced gels have shown that the value of B does not change with time (Roefs *et al.*, 1990; Lucey *et al.*, 1997d); however, in these studies, B was determined in aged gels. It is possible that the B of acid milk gels could change with time at least for a short period after gel formation.

(d) Appearance

Set-style yoghurt gels should have a smooth, semi-solid consistency with no surface whey (Lucey and Singh, 1997). The appearance of a set gel should be smooth with no cracks or 'blemishes'. Gels made from severely heated milks with GDL had a "rough" surface with visible cracks and some whey separation (Lucey *et al.*, 1998a,e). It was suggested that rearrangement of the network just after gel formation might be responsible for these defects. Gels made from severely heated milk had a low strain at fracture compared to gels made from unheated milk (Figure 22.5) and this may make heated gels more susceptible to localized fracturing of strands in the network (Lucey *et al.*, 1997b). The "transition" in the rheological properties (as indicated by the maximum in loss tangent) may increase the susceptibility of protein-protein bonds to relax and if these bonds have a relatively short lifetime, this may lead to yielding or breaking of strands (van Vliet *et al.*, 1991).

(e) Whey separation

Whey separation refers to the appearance of liquid (whey) on the surface of a milk gel and is a common defect in fermented milk products. Syneresis is defined as shrinkage of a gel and this occurs concomitantly with expulsion of whey. It is useful to define spontaneous syneresis as the contraction of a gel without the application of an external force (e.g., centrifugation) and this is related to instability of the gel network (i.e., due to large-scale rearrangements) (Walstra, 1993). In practice, yoghurt manufacturers try to prevent whey separation by adding stabilizers (e.g., pectin or gelatin) or whey protein concentrate (WPC).

A simple method for quantifying spontaneous whey separation in acid milk gels has been proposed (Lucey *et al.*, 1998a). Previous studies had determined the quantity of whey expelled from yoghurt as result of high speed centrifugation or drainage through a screen (Harwalkar and Kalab, 1983, 1986; Dannenberg and Kessler, 1988a). The drainage of whey from a broken gel distributed over a screen measures whey separation when a very large surface area is available and is more relevant to products such as Cottage cheese or casein than to set gels, like yoghurt (Lucey *et al.*, 1998a). High-speed centrifugation measures the water-holding capacity of the gels under relatively high forces. Therefore, both of these methods for measuring whey expulsion are not relevant to the spontaneous separation of whey from set-style gels. The approach of Lucey *et al.* (1998a) was to make gels in containers and determine the amount of surface whey that was expelled during gelation. Whey separation (in volumetric flasks) was increased significantly by heat treatment and gelation temperature (Lucey *et al.*, 1998a); coincidentally, a high heat treatment and high gelation temperature are used commonly in the manufacture of yoghurt.

Rapid acidification of milk and a high incubation temperature may be two of the main causes of whey separation in acid gels, like yoghurt (Lucey and Singh, 1997). Acid-induced milk gels formed by slow acidification of milk at a low temperature and quiescent heating exhibit little wheying-off or spontaneous syneresis (Roefs, 1986).

It has been shown (van Dijk and Walstra, 1986) that the one dimensional syneresis of milk gels is related to the flow of liquid (whey) through the network and is governed by the equation of Darcy:

$$v = \frac{B}{\eta}\frac{p}{x} \tag{8}$$

where v is the superficial flow velocity of the syneresing liquid, B is the permeability coefficient, η is the viscosity of the liquid, p is the pressure acting on the liquid and x the distance over which the liquid must flow. In milk gels, an endogenous syneresis pressure can occur if there is a tendency of the casein network to rearrange after its formation (Lucey *et al.*, 1997d). It has been shown that endogenous syneresis pressure is generally small in acid sodium caseinate gels and this results in a smaller tendency for shrinkage of these gels compared to rennet-induced gels (Lucey *et al.*, 1997d). One-dimensional syneresis of acid milk gels made with GDL increased at high gelation temperature and high pH values (van Vliet *et al.*, 1997).

Differences in the ease of water loss from different casein gels have been related to the susceptibility of the network to rearrangements just after gel formation (van Vliet and Walstra, 1994). Parameters that affect the time-scale for rearrangements of bonds in a gel include the dynamic moduli, which indicates the strength and number of bonds in the network, the yield stress and shear deformation at yielding, which determine the susceptibility of the strands to breakage, and tan δ, with higher values favouring the relaxation of bonds (van Vliet *et al.*, 1991; Lucey *et al.*, 1997b,d). In freshly made gels, the number of bonds between each junction is not yet very high, as indicated by the low dynamic moduli and tan δ is higher than in aged gels; these factors might explain why wheying-off occurs sometimes in young but not in aged gels. The "maximum in tan δ" which has been observed in acid gels made from heated milk (Figure 22.3) would indicate an even greater likelihood of relaxation of bonds during the initial period after gel formation. The relationship between the mechanical properties of acid milk gels and the spontaneous whey separation defect were discussed recently (Lucey, 2001).

Acid-induced milk gels that were first cooled to a low temperature (e.g., 5°C) before wetting their surface actually increased slightly in height, possibly due to the absorption of water by casein particles, which swell at low temperatures (Lucey *et al.*, 1997d). Surface whey that was expelled during gelation is sometimes reabsorbed by the gel on cooling and storage at low temperature (Lucey and Singh, 1997).

(f) Textural defects

A wide range of cultured milk products is now on the market and each has a very different texture or consistency. The textural properties of acid milk gels can be assessed by a range of fundamental and empirical (instrumental) methods such as dynamic low amplitude oscillation, large amplitude oscillatory shear, penetration, rotational viscometry and flow through an orifice such as a Posthumus funnel. An excessively firm texture can be caused by factors such as a very high total solids content of the mix or an excessive amount of added stabilizers (this is exploited in the manufacture of "custard-style" yoghurt). A weak body can be caused by factors such as a low solids (fat) content of the mix, insufficient heat treatment of the milk, low acidity (high pH) and a high gelation temperature (>40°C). Textural defects described as 'lumpiness' or 'granular' are objectionable as consumers usually expect a smooth, fine-bodied product (Bodyfelt *et al.*, 1988; Tamime and Robinson, 1999). Lumpiness usually refers to the presence of large protein aggregates in yoghurt that can often range in size from 1 to 5 mm (Lucey and Singh, 1997). Excessive production of acid at a high incubation temperature, the use of rennet and the use of an excessive amount of starter have been associated with these defects (Humphreys and Plunkett, 1969; Lucey and Singh, 1997). Some of these defects may be caused by conditions which favour the formation of dense protein clusters during gelation. Excessive heat treatment of milk and the addition of a high level of WPC have also been associated with other textural defects. In yoghurt samples in which ⩾20% of milk solids-non-fat (SNF) were replaced by WPC, a 'grainy' texture was observed (Greig and Van Kan, 1984). Substituting WPC for skim milk powder (SMP) to elevate the total solids content of yoghurt mixes increased the 'lumpy' or 'granular' defect (Guirguis *et al.*, 1988), while replacement of casein by WPC resulted in a yoghurt with a 'less smooth and clumpy' appearance (Jelen *et al.*, 1987). Addition of WPC to milk, followed by heat treatment, resulted in acid gels becoming more brittle (Lucey *et al.*, 1999). In stirred-type yoghurt, stabilizers are added to control textural defects and prevent whey separation but stabilizers are not always added to plain, set-style yoghurt (Lucey and Singh, 1997). Stabilizers, such as pectin, gelatin, starch and various gums, have been used to increase the viscosity of stirred-type yoghurt products. Generally, the level of stabilizers used in stirred yoghurt is ⩽0.7%. Parameters such as milk heat treatment, total solids level, addition of WPC, incubation temperature and amount of starter added are all process variables which manufacturers can exploit to improve/modify the textural properties of acidified milk products.

The sensory or flavour attributes of yoghurts and other cultured milk products are very important (Bodyfelt *et al.*, 1988). In many markets, such as the USA, fruits and sweeteners are added, which to a large extent, determine the sensory properties of cultured products, especially if low acid- and flavour-producing cultures are used for fermentation.

22.4 Concluding remarks

Although acidified milk gels have been made for thousands of years, only the technological and microbiological aspects of these products were studied in any detail until relatively recently. Considerable progress has been made during the past 15 years on the formation and rheological properties of acidified milk gels. Recent developments were reviewed, including the use of techniques, such as dynamic low amplitude oscillatory rheology, to monitor the gel formation process; confocal scanning laser microscopy to examine the gel microstructure non-destructively; and various models including fractal aggregation were discussed in terms of possible mechanism involved in the formation of acid-induced milk gels. We believe further work is needed in the following areas:

- Comparisons of acid milk gels made by different acidification techniques, e.g., GDL or bacterial fermentation, as they can have different rheological properties. This is one of the major difficulties in translating the progress made on model systems (GDL-induced gels) into "real" yoghurt manufacturing circumstances.
- In a study (Lucey *et al.*, 1998d) using GDL-induced gels, surface whey separation was found to increase with high heat treatment of milk and high gelation temperature in contrast to the generally accepted belief that these two variables help reduce wheying-off in yoghurts. Further studies are needed in this area using bacterially-acidified gels and more "realistic" methods for assessing surface whey separation.
- Acid milk gels to which rennet is also added have unusual rheological properties (Lucey *et al.*, 2000, 2001); little is known about these types of "combined" gels although they are related to fresh cheese products such as Cottage or Quarg. More work is needed in this area in order to understand the physico-chemical changes which occur in casein particles and how these influence the formation of milk gels.
- There is a growing demand by the consumer for more "natural" products in which less or no additives/stabilizers are used. Fermented milk products have a positive health image and with increasing emphasis on probiotic products there is a growing need to be able to produce acid milk gels that do not whey-off during storage without added stabilizers.
- Studies that probe possible linkages between fundamental rheological properties and empirical tests and their correlation with sensory perception of texture and taste.
- Developing or modifying existing theoretical models for the formation of acid milk gels; a successful theory should be able to describe the kinetics of gelation, the dynamics of the rheological properties, help explain the different gelation behaviour of unheated and heated milk, be consistent with rearrangements of the aggregating particles before during and after gelation and explain the continued increase in G' after visual gelation. Computer (Brownian dynamic) simulation of particle aggregation (Bijsterbosch *et al.*, 1995; Mellema *et al.*, 1999; Wijmans *et al.*, 1999) could assist in achieving this goal.

References

Allan-Wojtas, P. and Kalab, M. (1984) Milk gel structure. XIV. Fixation of fat globules in whole milk yoghurt for electron microscopy. *Milchwissenschaft*, **39**, 323–7.

Barrantes, E., Tamime, A.Y., Sword, A.M., Muir, D.D. and Kalab, M. (1996) The manufacture of set-style natural yoghurt containing different oils—2: Rheological properties and microstructure. *Int. Dairy J.*, **6**, 827–37.

Bijsterbosch, B.H., Bos, M.T.A., Dickinson, E., van Opheusden, J.H.J. and Walstra, P. (1995) Brownian dynamic simulation of particle gel formation: from Argon to yoghurt. *Faraday Discuss.*, **101**, 51–64.

Biliaderis, C.G., Khan, M.M. and Blank, G. (1992) Rheological and sensory properties of yogurt from skim milk and ultrafiltered retentates. *Int. Dairy J.*, **2**, 311–23.

Bodyfelt, F.W., Tobias, J. and Trout, G.M. (1988) *The Sensory Evaluation of Dairy Products*, Van Nostrand Reinhold, New York.

Bremer, L.G.B., van Vliet, T. and Walstra, P. (1989) Theoretical and experimental study of the fractal nature of the structure of casein gels. *J. Chem. Soc. Faraday Trans.*, 1, **85**, 3359–72.

Bremer, L.G.B., Bijsterbosch, B.H., Walstra, P. and van Vliet, T. (1993) Formation, properties and fractal structure of particle gels. *Adv. Colloid Interface*, **46**, 117–28.

Bremer, L.G.B., Bijsterbosch, B.H., Schrijvers, R., van Vliet, T. and Walstra, P. (1990) On the fractal nature of the structure of acid casein gels. *Colloid and Surface*, **51**, 159–70.

Brooker, B.E. (1995) Imaging food systems by Confocal Laser Scanning Microscopy, in *New Physico-Chemical Techniques for the Characterization of Complex Food Systems*, (E. Dickinson ed.) Blackie Academic & Professional, Glasgow, pp. 53–68.

Cho, Y.H., Lucey, J.A. and Singh, H. (1999) Rheological properties of acid milk gels as affected by the nature of the fat globule surface material and heat treatment of milk. *Int. Dairy J.*, **9**, 537–45.

Dalgleish, D.G. and Law, A.J.R. (1988) pH-induced dissociation of casein micelles. 1. Analysis of liberated caseins. *J. Dairy Res.*, **55**, 529–38.

Dalgleish, D.G., Pouliot, Y. and Paquin, P.A. (1987) Studies on the heat stability of milk. II. Association and dissociation of particles and the effects of added urea. *J. Dairy Res.*, **54**, 39–49.

Dannenberg, F. and Kessler, H.-G. (1988a) Effect of denaturation of β-lactoglobulin on texture properties of set-style nonfat yoghurt. 1. Syneresis. *Milchwissenschaft*, **43**, 632–35.

Dannenberg, F. and Kessler, H.-G. (1988b) Effect of denaturation of β-lactoglobulin on texture properties of set-style nonfat yoghurt. 2. Firmness and flow properties. *Milchwissenschaft*, **43**, 700–4.

Davies, F.L., Shankar, P.A., Brooker, B.E. and Hobbs, D.G. (1978) A heat-induced change in the ultrastructure of milk and its effect on gel formation in yoghurt. *J. Dairy Res.*, **45**, 53–8.

de Kruif, C.G. (1997) Skim milk acidification. *J. Colloid Interf. Sci.*, **185**, 19–27.

de Kruif, C.G. (1999) Casein micelle interactions. *Int. Dairy J.*, **9**, 183–8.

de Kruif, C.G. and Roefs, S.P.F.M. (1996) Skim milk acidification at low temperatures: a model for the stability of casein micelles. *Neth. Milk Dairy J.*, **50**, 113–20.

de Kruif, G.G., Hoffman, M.A.M., van Marle, M.E., van Mil, P.J.J.M., Roefs, S.P.F.M., Verheul, M. and Zoon, N. (1995) Gelation of proteins from milk. *Faraday Discuss.*, **101**, 185–200.

Dickinson, E. (1997a) Aggregation processes, particle interactions, and colloidal structure, in *Food Colloids; Proteins, Lipids and Polysaccharides*, (E. Dickinson and B. Bergenståhl eds.) Royal Society of Chemistry, Cambridge, pp. 107–26.

Dickinson, E. (1997b). On gelation kinetics in a system of particles with weak and strong interactions. *J. Chem. Soc. Faraday Trans.*, **93**, 111–4.

Gastaldi, E., Lagaude, A. and Tarado de La Fuente, B. (1996) Micellar transition state in casein between pH 5.5 and 5.0. *J. Food Sci.*, **61**, 59–64, 68.

Greig, R.I.W. and Van Kan, J. (1984) Effect of whey protein concentrate on fermentation of yogurt. *Dairy Ind. Int*, **49** (10), 28–9.

Guinee, T.P., Pudja, P.D. and Farkye, N.Y. (1993) Fresh acid-curd cheese varieties, in, *Cheese: Chemistry, Physics and Microbiology - Volume 1, General Aspects*, 2nd edn, (P.F. Fox ed.) Chapman & Hall, London, pp. 363–419.

Guirguis, N., Hickey, M.W. and Freeman, R. (1988) Some factors affecting nodulation in yoghurt. *Aust. J. Dairy Technol.*, **43**, 45–7.

Haque, Z. and Kinsella, J.E. (1988) Interaction between κ-casein and β-lactoglobulin: predominance of hydrophobic interactions in the initial stages of complex formation. *J. Dairy Res.*, **55**, 67–80.

Harwalkar, V.R. and Kalab, M. (1980) Milk gel structure. XI. Electron microscopy of glucono-δ-lactone-induced skim milk gels. *J. Texture Stud.*, **11**, 35–49.

Harwalkar, V.R. and Kalab, M. (1983) Susceptibility of yoghurt to syneresis. Comparison of centrifugation and drainage methods. *Milchwissenschaft*, **38**, 517–22.

Harwalkar, V.R. and Kalab, M. (1986) Relationship between microstructure and susceptibility to syneresis in yoghurt made from reconstituted nonfat dry milk. *Food Microstruct.*, **5**, 287–94.

Hassan, A.N., Frank. J.F., Farmer, M.A., Schmidt, K.A. and Shalabi, S.I. (1995) Formation of yogurt microstructure and three-dimensional visualization as determined by confocal scanning laser microscopy. *J. Dairy Sci.*, **78**, 2629–36.

Heertje, I., Visser, J. and Smits, P. (1985) Structure formation in acid milk gels. *Food Microstruct.*, **4**, 267–77.

Holt, C. (1992) Structure and stability of bovine casein micelles. *Adv. Protein Chem.*, **43**, 63–151.

Holt, C., Davies, D.T. and Law, A.J.R. (1986) Effects of colloidal calcium phosphate content and free calcium ion concentration in milk serum on the dissociation of bovine casein micelles. *J. Dairy Res.*, **53**, 557–72.

Horne, D.S. (1998) Casein interactions: casting light on the black boxes, the structure in dairy products. *Int. Dairy J.*, **8**, 171–7.

Horne, D.S. (1999) Formation and structure of acidified milk gels. *Int. Dairy J.*, **9**, 261–8.

Horne, D.S. and Davidson, C.M. (1993) Influence of heat treatment on gel formation in acidified milks, Special Issue, 9303, International Dairy Federation, Brussels, pp. 267–76.

Humphreys, C.L. and Plunkett, M. (1969) Yoghurt: a review of its manufacture. *Dairy Science Abstracts*, **31**, 607–22.

IDF (1995) Consumption statistics for milk and milk products (1993), Bulletin 301, International Dairy Federation, Brussels.

Jelen, P., Buchheim, W. and Peters, K.-H. (1987) Heat stability and use of milk with modified casein: whey protein content in yogurt and cultured milk products. *Milchwissenschaft*, **42**, 418–21.

Kalab, M., Allan-Wojtas, P. and Phipps-Todd, B.E. (1983) Development of microstructure in set-style nonfat yoghurt. *Food Microstruct.*, **2**, 51–66.

Kinsella, J.E. and Whitehead, D.M. (1989) Proteins in whey: chemical, physical, and functional properties. *Adv. Food and Nutrition Res.*, **33**, 343–438.

Kosikowski, F.V. and Mistry V.V. (1997) *Cheese and Fermented Milk Foods*, 3rd edn, F.V. Kosikowski LLC, Westport, CT.

Law, A.J.R. (1996) Effects of heat treatment and acidification on the dissociation of bovine casein micelles. *J. Dairy Res.*, **63**, 35–48.

Lopes da Silva, J. and Rao, M.A. (1999) Rheological behavior of food gel systems, in *Rheology of Fluid and Semisolid Foods*, (M.A. Rao ed.) Aspen Publishers, Inc., Gaithersburg, MD, pp. 319–68.

Lucey, J.A. (2001) The relationship between rheological parameters and whey separation in acid milk gels. *Food Hydrocolloid*, **15**, 603–8.

Lucey, J.A. (2002) Formation and physical properties of milk protein gels. *J. Dairy Sci.*, **85**, 281–94.

Lucey, J.A., Hauth, B., Gorry, C. and Fox, P.F. (1993) Acid-base buffering of milk. *Milchwissenschaft*, **48**, 268–72.

Lucey, J.A. and Singh, H. (1997) Formation and physical properties of acid milk gels: a review. *Food Res. Int.*, **30**, 529–42.

Lucey, J.A., Dick, C., Singh, H. and Munro, P.A. (1997a) Dissociation of colloidal calcium-phosphate depleted casein particles as influenced by pH and concentration of calcium and phosphate. *Milchwissenschaft*, **52**, 603–6.

Lucey, J.A., Teo, C.T., Munro, P.A. and Singh, H. (1997b) Rheological properties at small (dynamic) and large (yield) deformations of acid gels made from heated milk. *J. Dairy Res.*, **64**, 591–600.

Lucey, J.A., van Vliet, T., Grolle, K., Geurts, T. and Walstra, P. (1997c) Properties of acid casein gels made by acidification with glucono-δ-lactone. 1. Rheological properties. *Int. Dairy J.*, **7**, 381–8.

Lucey, J.A., van Vliet, T., Grolle, K., Geurts, T. and Walstra, P. (1997d) Properties of acid casein gels made by acidification with glucono-δ-lactone. 2. Syneresis, permeability and microstructural properties. *Int. Dairy J.*, **7**, 389–97.

Lucey, J.A., Munro, P.A. and Singh, H. (1998a) Whey separation in acid skim milk gels made with glucono-δ-lactone: effects of heat treatment and gelation temperature. *J. Texture Stud.*, **29**, 413–26.

Lucey, J.A., Munro, P.A. and Singh, H. (1998b) Rheological properties and microstructure of acid milk gels as affected by fat content and heat treatment. *J. Food Sci.*, **63**, 660–4.

Lucey, J.A., Tamehana, M., Singh, H. and Munro, P.A. (1998c) Effect of interactions between denatured whey proteins and casein micelles on the formation and rheological properties of acid skim milk gels. *J. Dairy Res.*, **65**, 555–67.

Lucey, J.A., Tamehana, M., Singh, H. and Munro, P.A. (1998d) A comparison of the formation, rheological properties and microstructure of acid skim milk gels made with a bacterial culture or glucono-δ-lactone. *Food Res. Int.*, **31**, 147–55.

Lucey, J.A., Teo, C.T., Munro, P.A. and Singh, H. (1998e) Microstructure, permeability and appearance of acid gels made from heated skim milk. *Food Hydrocolloid*, **12**, 159–65.

Lucey, J.A., Munro, P.A. and Singh, H. (1999) Effects of heat treatment and whey protein addition on the rheological properties and structure of acid skim milk gels. *Int. Dairy J.*, **9**, 275–9.

Lucey, J.A., Tamehana, M., Singh, H. and Munro, P.A. (2000) Rheological properties of acid milk gels made by a combination of rennet action and acidification. *J. Dairy Res.*, **67**, 415–27.

Lucey, J.A., Tamehana, M., Singh, H. and Munro, P.A. (2001) Effect of heat treatment on the physical properties of milk gels made with both rennet and acid. *Int. Dairy J.*, **11**, 559–65.

Mellema, M., van Opheusden, J.H.J. and van Vliet, T. (1999) Brownian dynamic simulation of colloidal aggregation and gelation, in *Food Emulsions and Foams; Interfaces, Interactions and Stability*, (E. Dickinson and J.M. Rodriguez Patino eds.) Royal Society of Chemistry, Cambridge, pp. 177–91.

McGann, T.C.A. and Fox, P.F. (1974) Physico-chemical properties of casein micelles reformed from urea-treated milk. *J. Dairy Res.*, **41**, 45–53.

Mottar, J., Bassier, A., Joniau, M. and Baert, J. (1989) Effect of heat-induced association of whey proteins and casein micelles on yogurt texture. *J. Dairy Sci.*, **72**, 2247–56.

Mulvihill, D.M. (1992) Production, functional properties and utilisation of milk protein products, in *Advanced Dairy Chemistry-1-Proteins*, (P.F. Fox ed.) Elsevier Applied Science, London, pp. 369–404.

Mulvihill, D.M. and Grufferty, M.B. (1995) Effect of thermal processing on the coagulability of milk by acid, in *Heat-induced Changes in Milk*, 2nd edn, (P.F. Fox ed.) Special Issue, 9501, International Dairy Federation, Brussels, pp. 188–205.

Parnell-Clunies, E.M., Kakuda, Y. and de Man, J.M. (1986) Influence of heat-treatment of milk on the flow properties of yoghurt. *J. Food Sci.*, **51**, 1459–62.

Parnell-Clunies, E., Kakuda, Y. and Smith, A.K. (1987) Microstructure of yogurt as affected by heat treatment of milk. *Milchwissenschaft*, **42**, 413–17.

Puhan, Z., Dreissen, F.M., Jelen, P. and Tamime, A.Y. (1994) Fresh products—yoghurt, fermented milks, Quarg and fresh cheese. *Mljekarstvo*, **44(4)**, 285–98.

Pyne, G.T. and McGann, T.C.A. (1960) The colloidal calcium phosphate of milk. II. Influence of citrate. *J. Dairy Res.*, **27**, 9–17.

Robinson, R.K. and Tamime, A.Y. (1993) Manufacture of yoghurt and other fermented milks, in *Modern Dairy Technology, Volume 2, Advances in Milk Products*, 2nd edn, (R.K. Robinson ed.) Elsevier Applied Science, London, pp. 1–48.

Roefs, S.P.F.M. (1986) *Structure of Acid Casein gels. A Study of Gels Formed After Acidification in the Cold*, PhD thesis, Wageningen Agricultural University, Netherlands.

Roefs, S.P.F.M., de Groot-Mostert, A.E.A. and van Vliet, T. (1990) Structure of acid casein gels. 1. Formation and model of gel network. *Colloid Surface*, **50**, 141–59.

Roefs, S.P.F.M. and van Vliet, T. (1990) Structure of acid casein gels. 2. Dynamic measurements and type of interaction forces. *Colloid Surface*, **50**, 161–75.

Roefs, S.P.F.M., Walstra, P., Dalgleish, D.G. and Horne, D.S. (1985) Preliminary note on the change in casein micelles caused by acidification. *Neth. Milk Dairy J.*, **39**, 119–22.

Rönnegård, E. and Dejmek, P. (1993) Development and breakdown of structure in yoghurt studied by oscillatory rheological measurements. *Lait*, **73**, 371–9.

Schmidt, D.G. (1982) Association of casein and casein micelle structure, in *Developments in Dairy Chemistry - 1 Proteins*, (P.F. Fox ed.) Applied Science Publishers, London, pp. 61–86.

Singh, H. (1995) Heat-induced changes in casein, including interactions with whey proteins, in *Heat-induced Changes in Milk*, 2nd edn, (P.F. Fox ed.) International Dairy Federation, Brussels, pp. 86–104.

Singh, H., Roberts, M.S., Munro, P.A. and Teo, C.T. (1996) Acid-induced dissociation of casein micelles in milk: effects of heat treatment. *J. Dairy Sci.*, **79**, 1340–6.

Tamime, A.Y. and Marshall, V.M.E. (1997) Microbiology and technology of fermented milks, in *Microbiology and Biochemistry of Cheese and Fermented Milk*, 2nd edn, (B.A. Law ed.) Blackie Academic & Professional, London, pp. 57–152.

Tamime, A.Y. and Robinson, R.K. (1999) *Yoghurt*: *Science and Technology*, 2nd edn, CRC Press, Boca Raton, FL.

van Dijk, H.J.M. and Walstra, P. (1986) Syneresis of curd. 2. One-dimensional syneresis of rennet curd in constant conditions. *Neth. Milk Dairy J.*, **40**, 3–30.

van Marle, M.E. and Zoon, P. (1995) Permeability and rheological properties of microbially and chemically acidified skim-milk gels. *Neth. Milk Dairy J.*, **49**, 47–65.

van Vliet, T. (1988) Rheological properties of filled gels. Influence of filler matrix interaction. *Colloid Polym. Sci.*, **266**, 518–24.

van Vliet, T. (1999) Factors determining small-deformation behaviour of gels, in *Food Emulsions and Foams*; *Interfaces*, *Interactions and Stability*, (E. Dickinson and J.M. Rodriguez Patino eds.) Royal Society of Chemistry, Cambridge, pp. 307–17.

van Vliet, T. and Dentener-Kikkert, A. (1982) Influence of the composition of the milk fat globule membrane on the rheological properties of acid milk gels. *Neth. Milk Dairy J.*, **36**, 261–5.

van Vliet, T. and Keetels, C.J.A.M. (1995) Effect of preheating of milk on the structure of acidified milk gels. *Neth. Milk Dairy J.*, **49**, 27–35.

van Vliet, T., Lucey, J.A., Grolle, K. and Walstra, P. (1997) Rearrangements in acid-induced casein gels during and after gel formation, in *Food Colloids*; *Proteins*, *Lipids and Polysaccharides*, (E. Dickinson and B. Bergenståhl eds.) Royal Society of Chemistry, Cambridge, pp. 335–45.

van Vliet, T., van Dijk, H.J.M., Zoon, P. and Walstra, P. (1991) Relation between syneresis and rheological properties of particle gels. *Colloid Polym. Sci.*, **269**, 620–7.

van Vliet, T. and Walstra, P. (1994) Water in casein gels; how to get it out or keep it in. *J. Food Eng.*, **22**, 75–88.

Walstra, P. (1990) On the stability of casein micelles. *J. Dairy Sci.*, **73**, 1965–79.

Walstra, P. (1993) The syneresis of curd, in *Cheese*: *Chemistry*, *Physics and Microbiology - Volume 1*, *General Aspects*, 2nd edn, (P.F. Fox ed.) Chapman & Hall, London, pp. 141–91

Wijmans, C.M., Whittle, M. and Dickinson, E. (1999) Structure and rheology of simulated particle gel systems, in *Food Emulsions and Foams*; *Interfaces*, *Interactions and Stability*, (E. Dickinson and J.M. Rodriguez Patino eds.) Royal Society of Chemistry, Cambridge, pp. 342–55.

23

MANUFACTURE AND PROPERTIES OF MILK POWDERS

A.L. Kelly, J.E. O'Connell and P.F. Fox

23.1 Introduction

The first recoded reference to the manufacture of milk powder as a method for preserving milk was by Marco Polo, who observed the use of milk powder by Mongol soldiers in the 13th Century (Hall and Hedrick, 1975). The earliest modern commercial concentrated dairy products were air-dried concentrated milk tablets, developed in 1809, and vacuum-concentrated sweetened and unsweetened condensed milks, produced by Gail Borden in 1856. The second half of the 19th Century saw the production of solidified high-total solids milk products (Caríc and Kaláb, 1987). Roller drying was introduced around 1902 and rapidly became the predominant method for producing dried dairy products, such as infant formulae.

Spray drying, or atomisation of a feed into a hot air stream, was developed in 1872 for use in the chemical industry. Shortly thereafter, it was applied to dairy products and gradually replaced roller drying as the method of choice for the production of high-quality milk powders. Spray-dryers have evolved in structure and application over the past 50 years, and now a suite of dryer types is available, suitable for the production of a wide variety of dairy products. One of the key driving forces in dryer development has been the need of consumers for convenient, easily reconstituted ('instant') powders, produced from whole or skimmed milk, whey, buttermilk and a range of other dairy products, such as cream or cheese. Increased understanding of the chemical and physical changes which occur in milk during drying have paralleled developments in drying technology and enable control of the properties, and hence potential applications, of milk powders.

Advanced Dairy Chemistry Volume 1: Proteins, 3rd edn.
Edited by P.F. Fox and P.L.H. McSweeney, Kluwer Academic/Plenum Publishers, 2003.

The modern processes used for the production of dairy powders, the effects of these processes on the proteins and other constituents of milk, and the properties of milk powders will be considered in this review.

23.2 Technology of milk powder manufacture

Flow diagrams for the production of a number of common milk powders are summarised in Figure 23.1.

23.2.1 Milk pre-treatment

Milk of very high microbiological quality is required; bactofugation or microfiltration may be used to remove both bacterial cells and spores from milk, and thus ensure a milk powder of very high microbiological quality. Whole milk is generally standardised, usually to a fat to solids non-fat ratio of 1:2.76, to control the fat content of the final powder.

23.2.2 Preheating and concentration

Preheating of milk just before evaporation ensures the microbiological quality of the concentrate and of the final powder, but it is also a critical step in the control of the functional properties of the powder. Preheating is usually the highest temperature step applied during manufacture and is therefore the step at which most whey protein denaturation occurs (Singh and Creamer, 1991). Preheating may be performed using any of a range of heat exchangers, including plate heat exchangers, spiral heat exchangers wrapped around the tubes in the evaporator itself or very short-time steam injection heating systems. Direct heat exchangers are preferred over indirect systems, as biofilms of thermophilic bacteria may develop within indirect heat exchangers (Early, 1998).

As will be discussed in Section 23.3, skim milk powder (SMP) is often classified according to the heat treatment applied during preheating. There are 3 principal heat categories: low heat (typically heated at 75°C for 15 sec); medium heat (typically heated at 75°C for 1–3 min) and high heat (heated at 80°C for 30 min or 120°C for 1 min). Whole milk powder (WMP) is generally not heat classified, but is heated at 85–95°C for several minutes to ensure inactivation of indigenous lipase and to expose antioxidant sulphydryl groups (Hols and van Mil, 1991).

After pre-heating, the milk is concentrated to 45–50% or 42–48% total solids (TS) for whole or skim milk, respectively (Kyle, 1993). Typically, evaporation is performed in a multiple effect (stage) falling-film evaporator, where thermal efficiency is maximised by maintaining each subsequent effect

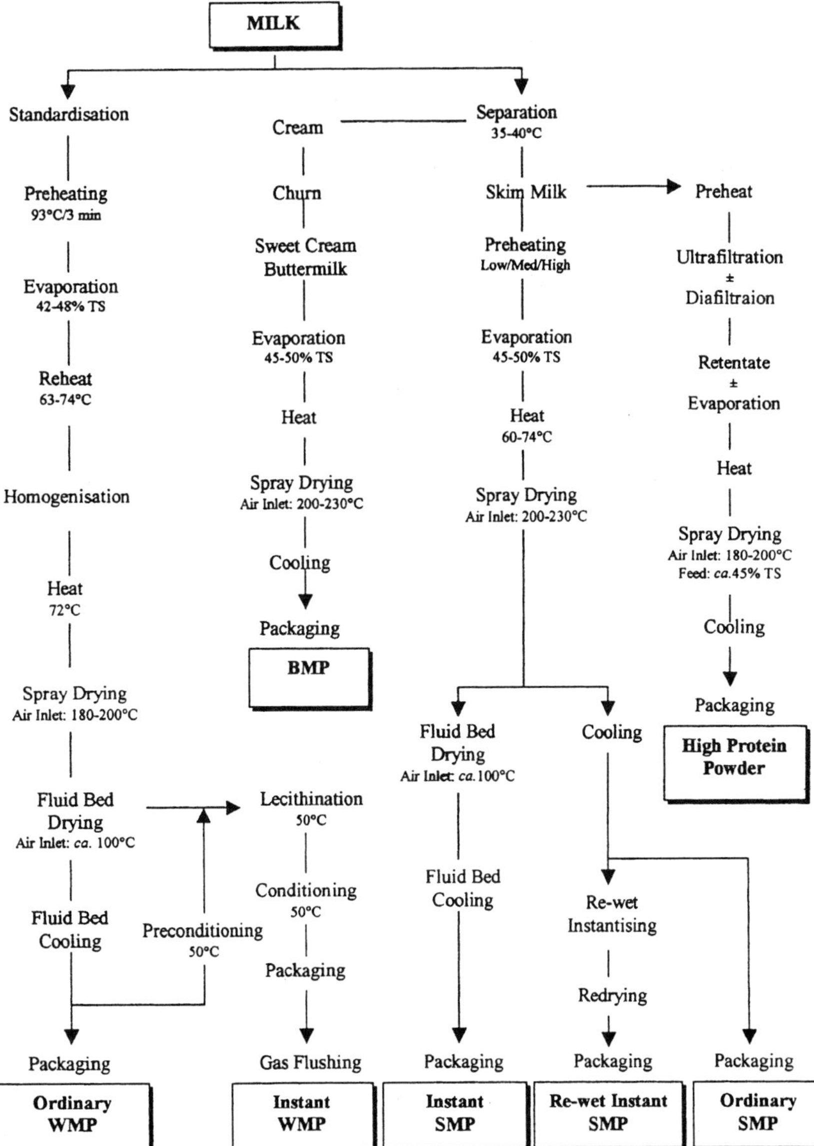

Figure 23.1 Flowchart of the technology for production of a range of milk powders (modified from Kyle, 1993).

at an increasingly low pressure, and thus boiling points, allowing the vapour from each effect to be used as the heating medium for the next effect (Figure 23.2). Steam consumption is reduced further, and the economy of

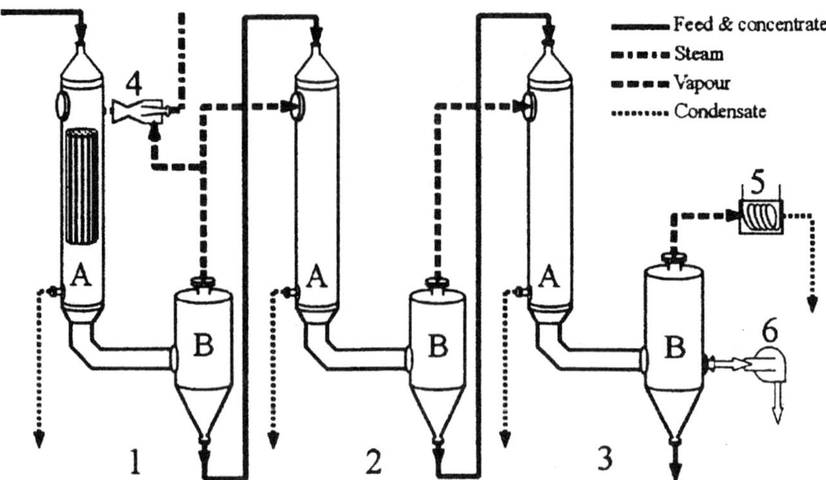

Figure 23.2 Schematic diagram of a triple effect evaporator showing (1) first effect, comprising calandria (A) and vapour-liquid separating cyclone (B), (2) second effect, (3) third effect, (4) thermocompressor, (5) vapour condenser and (6) vacuum pump. The function of the thermocompressor is to mix vapour from the first effect with steam and compress the mixture, thereby increasing its temperature and allowing its use as heating medium for the first effect and reducing the overall consumption of steam.

operation increased, by use of thermal or mechanical vapour recompression to increase the temperature of some or all of the vapour and allow it to be used as an additional heating medium. Conventionally, concentration is monitored using an on-line refractometer, although the use of an on-line viscometer has been recommended (O'Donnell *et al.*, 1996).

As an alternative to evaporation, membrane processing may be used for specific applications. While processes such as ultrafiltration (UF) fractionate milk constituents, reverse osmosis (RO) or hyperfiltration or nanofiltration (NF) remove essentially only water, and thus can serve as a pre-concentration step (Harding, 1985). RO is limited to a low achievable solids level (<20% TS) and a medium flow rate, but has lower operation costs and is far less thermally intensive than evaporation (Písecký, 1997). The concentration of milk by UF membranes, which are far less susceptible to fouling than RO membranes (El-Gazzar and Marth, 1991), has been suggested also. Concentration by UF presents advantages in that heat treatment during concentration is avoided and the levels of protein and lactose in the powder may be standardised and controlled (Sweetsur and Muir, 1980; Muir and Sweetsur, 1984; Mistry and Pulgar, 1996; Horton, 1997). The significance of these alterations in powder composition is discussed in "Section 23.5".

Crystallisation of lactose prior to drying is desirable for many products, particularly high-lactose products such as whey powders and, consequently, the concentrate may be held before drying under conditions which promote crystallisation and/or may be seeded with finally-ground lactose crystals, which promote crystallisation by acting as nuclei (Sanderson, 1978).

After concentration, milk may be homogenised. Conventionally, two-stage homogenisation is used at a temperature in the range of 60–70°C; typical homogenisation pressures are 15 MPa, followed by 5 MPa.

23.2.3 Drying

The first commercial milk powder plants used roller dryers, consisting typically of two co- or counter-rotating steam-heated drums, with or without a preliminary concentration step. Roller drying entails the direct transfer of heat from the drum to a thin film of milk (Figure 23.3). The water is evaporated off and the dried solids are removed and pulverised, which produces very irregularly-shaped powder particles, which are more wettable than finely divided spray-dried powder. This method of drying results in severe thermal damage to milk constituents, with adverse effects on flavour and quality. Modifications to the basic roller dryer to reduce damage to the product include drying in a vacuum chamber at a lower drum temperature. Generally, roller drying has been replaced by spray-drying, although it is still used for specific applications, such as the production of WMP with a high free fat content, which is desirable for chocolate manufacture.

Today, the majority of milk powder is produced by spray-drying. The following section will briefly summarise the processing steps involved in spray-drying of milk powders. For more detail on technological and engineering aspects of this process, comprehensive descriptions have been given by Hansen (1985), Masters (1991), Westergaard (1994) and Písecký (1998).

In spray-dryers, concentrate is taken from the evaporator, using a positive displacement pump, to an atomiser at the top of a spray drying chamber, which produces a spray of droplets which contact hot air and are dried to individual powder particles. The concentrate is generally heated to ~72°C before atomisation, to reduce viscosity and obtain optimal atomisation. The objective of atomisation is to convert the liquid feed to a spray of droplets 10–400 μm in diameter. Particle size and, in particular, the distribution of particle sizes is very important in determining the properties of milk powder, as will be discussed in Section 23.5. Atomisation may be achieved using nozzles under pressure or by centrifugal force using a rotating disk or wheel. Pressure nozzle atomisers are either high pressure/low capacity or low capacity/high pressure, and either have grooved core inserts or swirl chambers to impart rotary motion to the feed (Masters, 1991). The selection

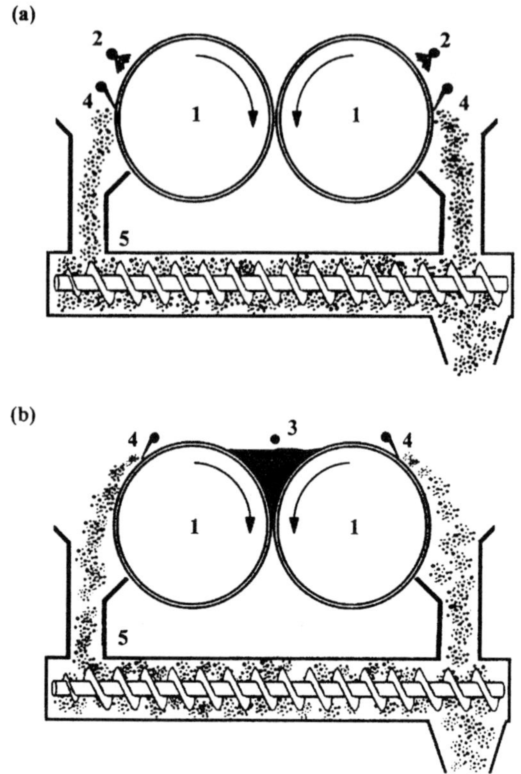

Figure 23.3 Principle of operation of roller dryers fed either by (a) spray applicators or (b) feed sump, showing (1) steam-heated drums (may be co- or counter-rotating); (2) spray applicator; (3) sump; (4) knife for removing dry product after completion of one (partial) rotation; (5) conveyor for grinding and transporting dry product.

of atomiser type is critical in determining the properties of the powder produced. For maximum flexibility, different types of atomiser can be used interchangeably in many spray-dryers. The influence of pressure nozzle and centrifugal atomisers on the properties of milk powder are compared in Table 23.1.

There are three principal classes of spray-dryer, based on the number of separate drying stages (one, two or three) used to achieve the final moisture content of the powder (typically 3 to 5%; Woodhams and Murray, 1978). These spray-dryer types are illustrated in Figure 23.4.

In single-stage dryers, the whole drying process takes place in the main cylindro-conical chamber of the spray dryer itself. Critical parameters which determine both the quality of the final powder and the efficiency of the drying process are the temperature of the drying air at the points of entry to

TABLE 23.1
Characteristics of different types of atomisation

Pressure nozzle atomisation	*Centrifugal atomisation*
Advantages	Advantages
Low occluded air in powder	Good flexibility
High powder bulk density	Handles high feed rates
Good powder flowability	Handles viscous concentrates
Simple and low cost	Wheel speed controls droplet size
Low energy consumption	Can steam sweep to control bulk density
Good control over spray flow	Insensitive to concentration variation
Low deposit levels in chamber	Low risk of blocking during run
Can agglomerate with angled nozzles	Can handle crystalline feeds
	Can co-dry separate feeds
Disadvantages	Disadvantages
Susceptible to blocking	High energy consumption
	High air incorporation
	High capital costs
	May give deposits in drying chamber

and exit from the main chamber [T_{inlet} (typically 160–220°C) and T_{outlet} (typically 70–90°C)]. During drying, milk droplets are cooled continuously by loss of the latent heat of evaporation, and generally never reach a temperature greater than 70°C during drying. Initially, droplets lose moisture at a constant rate, while saturation conditions exist at the surface, but eventually saturation can no longer be maintained and the rate of drying decreases as a hard dry shell forms at the droplet surface (falling rate period). If this shell becomes too solid or thick, case-hardening occurs, preventing further drying. If exposed to a high air temperature at the end of drying (T_{outlet}), steam and air within powder particles may expand, forming large vacuoles and, potentially, fracturing the powder particles, resulting in an increased level of small, light powder fragments (fines). The rate of drying should be controlled so that the end of the falling rate drying period coincides with the end of heating and drying.

Air exiting the chamber carries a significant amount of light powder particles and fines, which are removed from the air by passage through cyclone separators or, occasionally, by use of bag filters or wet scrubbing systems. Recovered powder is added to the main bulk of powder, which is removed from the base of the drying chamber, either by a pneumatic conveying system or a fluidised bed using cold air to cool and transport the powder.

Completion of drying in a single stage necessitates the use of a high air temperature, which results in poor-quality non-instant, dusty powder. In two- and three-stage drying the powder exits the main drying chamber at a higher moisture content (~10–15%), and drying is completed in additional drying

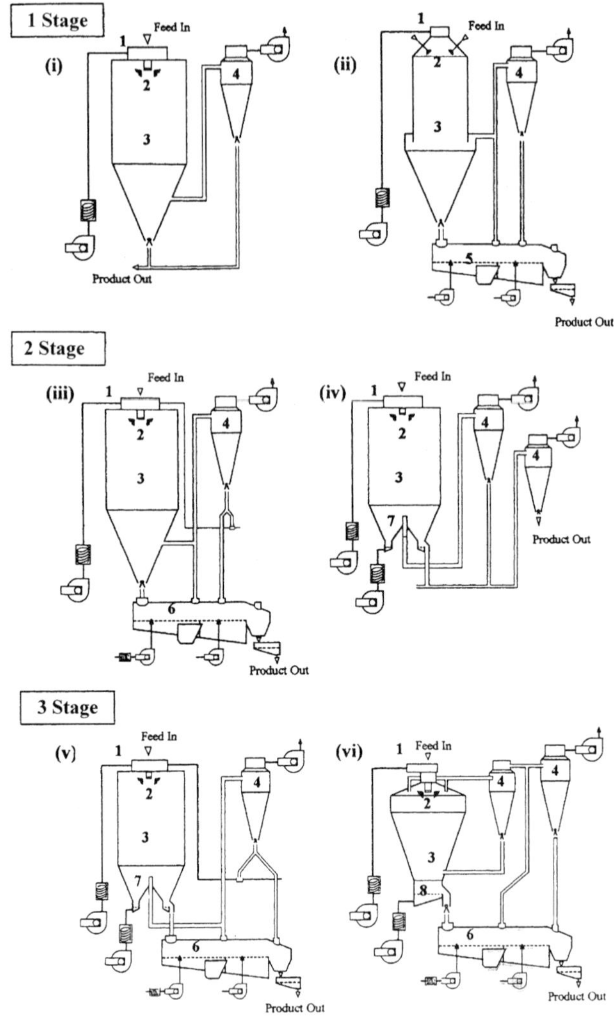

Figure 23.4 Common spray-dryer configurations used for the manufacture of milk powder: (i) single-stage dryer with pneumatic powder conveying; (ii) single stage dryer with fluidised bed cooling and conveying; (iii) two-stage dryer with external fluidised bed dryer; (iv) two-stage dryer with integrated fluidised bed dryer; (v) compact spray dryer (CSD, three stage) with integrated annular and external fluidised beds; (vi) multi-stage spray dryer (MSD, three stage) with integrated circular and external fluidised beds. Dryer components indicated include (1) hot air inlet, (2) atomiser (may be nozzle or rotary), (3) main drying chamber, (4) cyclone, (5) external fluidised bed cooler, (6) external fluidised bed dryer, (7) integrated annular bed dryer and (8) integrated circular bed dryer.

stages. Thus, in two-stage drying, the main chamber is followed by a fluidised bed dryer while in three-stage spray dryers, the spray-dryer chamber is followed sequentially by an integrated fluidised bed and an external plug-flow fluidised bed (Boersen, 1990). The two main categories of three-stage spray dryers used for milk powders are those with an internal annular fluidised bed surrounding the air outlet (compact spray dryers, CSD) or dryers with an internal circular fluidised bed (multi-stage dryers, MSD).

Separation of drying into two or three stages allows for improved control of powder properties, greater efficiency of the drying process and, because the rate of heat introduction is adjusted to the rate of evaporation, the process is milder than single-stage processes (Písecký, 1978). Another advantage of two- or three-stage spray drying is that such processes permit the drying of high TS concentrates without deterioration of product solubility. Two- and three-stage drying systems also have reduced stack losses (Písecký, 1978).

Movement of powder within the drying plant may be achieved either using a fluidised bed or pneumatic system, although the latter method will break up agglomerates, and is suitable only for powders of high bulk density. Milk powder may be sifted for size classification, and then either packaged directly, filled into intermediate transportation containers, or transferred to silos for storage before packaging. Milk powder may be packaged into wholesale, catering or retail packs (Warren, 1980). Generally, milk powder is packaged in multi-layer paper bags with a polyethylene inner lining. WMP may be packaged in an inert atmosphere, where air is displaced by N_2 or a mixture of N_2 and CO_2, to improve oxidative stability during storage. Occasionally, WMP may be packed in tins or plastic containers, or in bags consisting of multi-layer laminates incorporating metal foil and high density plastic films (Varnam and Sutherland, 1994).

23.2.4 Instantization

Single-stage drying produces dusty non-agglomerated non-instant powders of high bulk density. Production of milk powders that reconstitute well when dispersed in cold water (instant powders) has necessitated development of modified spray-drying processes.

For SMP, instantization is achieved by agglomeration, i.e., the production of porous clusters of particles, 250 to 750 μm in diameter, with a high level of entrapped air (Neff and Morris, 1967; Sanderson, 1978; Písecký, 1997). Instanisation reduces the bulk density of milk powder (produces lighter particles) and enhances its wettability, sinkability and solubility (see Section 23.5.2).

Agglomeration of SMP may be achieved by returning fines to the atomisation zone, where they act as nuclei for the growth of agglomerates (Refstrup, 1995), by removal of the powder from the spray-drying chamber at 8 to 15% moisture (and hence use of two- or three-stage drying), which favours agglomerate formation during fluidised bed drying, or by rewetting processes applied at some point after drying (Neff and Norris, 1967). Other methods of agglomeration include the use of multiple atomiser systems where nozzles are arranged so that sprays cross each other (Boersen, 1990), or the use of curved-vane rotary atomisers or gases such as CO_2 or N_2 which are introduced into the feed and expand towards the end of drying, increasing vacuole size. In order to produce agglomerates with optimum reconstitution properties, it has been proposed that low- or medium-heat milk be used and that the initial powder particles should have a high particle density and a diameter in the range 25 to 50 μm (Sanderson, 1978).

Rewet agglomeration systems involve wetting non-instant powder using either a gaseous phase (humidified air or steam) or a liquid (milk or water) wetting agent, holding for a certain period under conditions which favour the interaction of moist particles to form agglomerates, drying to remove the added moisture, and using size classification to screen agglomerates of the required size. Rewet processes produce better agglomeration and instant properties than straight-through systems, where agglomeration is achieved during the powder manufacturing process itself, but the rewet process is more costly.

To overcome the hydrophobic nature of milk fat, instantization of WMP requires a combination of agglomeration and lecithinization. Agglomeration of WMP may be achieved by the same procedures as for SMP. Lecithin, which aids instantization by virtue of its surfactant properties, is usually added to WMP at a level of ~0.2% between the spray-drying stage and the fluidised bed drying or, alternatively, in a rewet process (Jensen, 1975; Sanderson, 1978). When lecithin is added to WMP, the mixture should be held at ~50°C for 5 min to ensure complete coating of the particles with lecithin (Jensen, 1975; Sanderson, 1978). Alternative techniques for producing instant WMP, including co-drying of whole milk and crystalline lactose solutions and co-drying of cream and skim milk concentrate, were reviewed by Sanderson (1978).

23.3 Types of milk powders

A wide range of dry dairy products is produced worldwide, a selection of which are listed in Table 23.2. The two principal commercial milk powders are SMP and WMP, which are generally classified as either regular (non-instant) or instant. Food applications of SMP and WMP are summarised in Table 23.3, and global production trends are shown in Figure 23.5. Commercial SMP is also routinely classified according to the pre-heat treatment applied (heat classification) as low-, medium- or high-heat powders. Fat-filled, or filled, milk powders are products with a fat content close to that of WMP, produced by drying a blend of non-milk fat and skim milk. Encapsulated milk fat powders (40–60% fat) may be made using a

TABLE 23.2
Range of dried dairy products

Skim milk powder	Butter powder
Instant	Cheese powders
Regular	Buttermilk powder
Low-, medium-, high-heat	Whey powders
	Normal
Whole milk powder	Demineralised
Instant	Delectosed
Regular	Caseinates (sodium, potassium, calcium)
High free fat	Rennet casein
	Acid casein
Filled milk powder	Total milk proteinates
Infant formulae	Casein coprecipitates

blend of emulsifying salts, SMP and flour, starch or sucrose (Holsinger *et al.*, 2000). Such powders, which have good flow properties and are resistant to oxidation, compared to other high-fat powders, may be used as substitutes for vegetable shortenings in a range of food products.

High-fat powders, which may be defined as powders with a fat content in the range 42–65% (Early, 1990; Munns, 1991), present certain problems in drying due to the crystallisation of fat during handling and storage. Pneumatic transport from dryers may not cool the powder sufficiently to crystallise all the fat therein and some fat may crystallise subsequently on cooling in the package, releasing the latent heat of crystallisation and thus

TABLE 23.3
Principal food applications of skim and whole milk powders

Skim milk powder
Reconstitution
Cheesemaking (low heat SMP)
Confectionery products
Ice cream and other desserts
Hot and cold beverages
Recombined sweetened condensed milk
Bakery products (high heat SMP)
Calf milk replacers
Recombined milk production
Chocolate manufacture
Meat products
Whole milk powder
Reconstitution
Convenience soups and sauces
Milk chocolate (high free fat WMP)

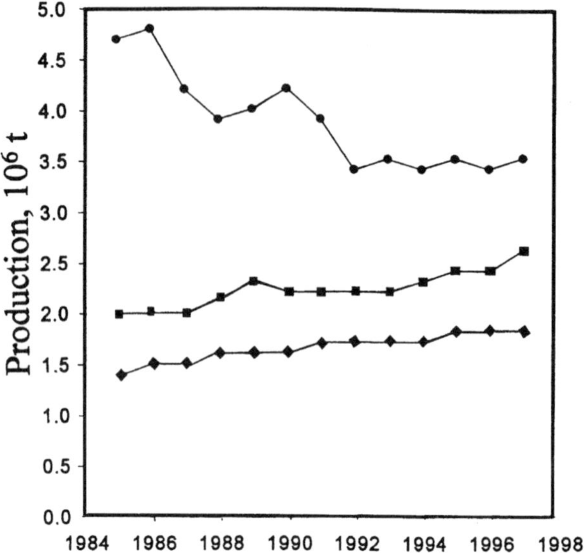

Figure 23.5 Worldwide production of skim milk powder (●), whole milk powder (■) and whey powder (◆).

causes an increase in the temperature causing fat to melt and caking, i.e., the formation of a solid mass, of the powder. This may be avoided by ensuring complete fat crystallisation, using a fluidised bed cooler, before packaging the powder (Early, 1998).

Infant formulae are a diverse group of products manufactured to high microbiological and hygienic standards which must reconstitute in warm water to give a homogeneous liquid free of undissolved lumps (Písecký, 1997). The composition of infant formulae may be modified to achieve a profile consistent with the nutritional needs of new-born babies. Such adjustment may be made during manufacture, by processes such as electrodialysis and ultrafiltration, which reduce the ionic content of milk (demineralisation). Full replacement of milk fat by vegetable fat, alteration of the casein:whey protein ratio or addition of lactose and vitamins and minerals are also common. As well as milk constituents, specific sources of protein (e.g., demineralised whey powders), fat (e.g., vegetable oils), carbohydrates, minerals and vitamins may be added during manufacture (Guo *et al.*, 1996). Infant formulae may be classified on the basis of the proportions of casein and whey proteins, and such products are frequently fortified with whey products (Mettler, 1980). Fortification with whey proteins, particularly with α-lactalbumin-rich preparations, is particularly important to adjust both the casein:whey protein ratio and the protein density (g/100 kcal) of cows' milk to resemble human milk more closely (Jost *et al.*, 1999).

In recent years, a number of fractionated or semi-refined milk protein powders have been developed for specific functional applications, with many products arising from selective use of membrane technologies, such as microfiltration, ultrafiltration and diafiltration (Mistry and Hassan, 1991; Kelly *et al.*, 2000). Kelly *et al.* (2000) described a method for the production of native phosphocasein-rich powders using microfiltration and electrodialysis: the latter treatment improved the heat stability of the powder but seriously impaired its rennet coagulation properties. Protein-enriched milk powders may be used as substitutes for SMP and have been shown to have good functional properties for a range of food applications (Mistry and Hassan, 1991). Garem *et al.* (2000) described the manufacture of a milk powder with improved cheesemaking properties by partial removal of whey proteins by a combination of microfiltration and ultrafiltration. The production of micellar casein powders with significant potential application in cheese manufacture was reviewed by Saboya and Maubois (2000).

23.4 Physico-chemical changes that occur during drying

23.4.1 Proteins

On preheating at a temperature above 75°C at pH ~6.5, β-lactoglobulin is denatured, causing significant alterations to its secondary and tertiary structures; at temperatures above 90°C, extensive whey protein denaturation occurs. Denatured β-lactoglobulin and, to a lesser extent, α-lactalbumin, complex *via* sulphydryl-disulphide interchange interactions with κ-casein (Noh and Richardson, 1989; Corredig and Dalgleish, 1999; Fairise *et al.*, 1999). When milk is heated at a pH $< \sim 6.8$, the complexes remain attached to the micelle surface but on heating at more alkaline pH values, the complexes are found in the serum phase (Creamer and Matheson, 1978; Mohammed and Fox, 1987; McKenna *et al.*, 1999). Oldfield (1998) showed that as well as interacting with casein micelles, β-lactoglobulin formed disulphide-linked aggregates with bovine serum albumin (BSA) during the manufacture of milk powder, while α-lactalbumin formed mainly hydrophobic-bonded aggregates.

The rate of the increase in temperature during preheating may affect whey protein interactions Slower indirect heating favours whey protein-whey protein interactions and higher overall denaturation of whey proteins than more rapid direct heating methods, which favour extensive casein-whey protein interactions (Early, 1998). This may be related to the fact that when milk is heated at a relatively low temperature, sulphydryl-sulphydryl interactions predominate, while a higher temperature is required for thermal reduction of disulphide bonds and thus sulphydryl-dispulphide interchange reactions (de Wit, 1981). Therefore, as caseins do not possess free sulphydryl groups (Swaisgood, 1992), low temperature heating will lead to whey protein-whey protein interactions alone.

Holding evaporated concentrate at a temperature above 60°C for an extended period before spray drying can cause aggregation of casein micelles, increasing the viscosity of the concentrate and affecting the reconstitution properties of SMP (Muir, 1980). During the actual processes of evaporation and spray-drying, relatively little denaturation of whey proteins occurs, as the final temperature of milk generally does not exceed 70°C (Singh and Creamer, 1991). Oldfield (1998) reported only slight decreases in the levels of native α-lactalbumin and BSA during evaporation, and no apparent denaturation of β-lactoglobulin and immunoglobulins. However, association of whey proteins with casein micelles can occur during evaporation, probably because decreasing pH conditions reduce protein charge, facilitating association reactions (Oldfield, 1998).

23.4.2 Lactose

In milk, the two isomers of lactose (α- and β-lactose) are present in a temperature-dependent equilibrium. At 20°C, the ratio of α:β isomers is 37:63. The proportion of β-lactose decreases as the temperature is increased. During spray-drying, water is removed rapidly from milk and the lactose present assumes an amorphous glass state which is very hygroscopic and will readily absorb water at moderate or high relative humidities (see Section 23.5.7(*c*)). If amorphous lactose absorbs sufficient water during storage, it acquires molecular mobility and it will crystallise in many forms, with a concomitant release of entrapped water (Roetman, 1979; Saltmarch and Labuza, 1980; Saito, 1985; Joupilla *et al.*, 1997). The release of water causes an increase in caking and plasticization, amongst other reactions. In milk powder, amorphous lactose is the principal constituent and in WMP it forms the continuous matrix in which proteins, fat globules and air vacoules are dispersed. The concentration of lactose in milk powder affects the microstructure of the powder, with low-lactose powders having a porous structure with large vacoules and a high level of free fat (Aguilar and Ziegler, 1994a). As discussed in Section 23.2.4, small lactose crystals may be added to milk concentrate prior to drying to promote crystallisation under relatively mild conditions (i.e., in the liquid rather than powder state) as the added crystals act as nuclei for crystallisation (Sanderson, 1978). In pre-crystallised products, tomahawk-shaped crystals of α-lactose are observed, while in post-crystallised products, needle-shaped β-lactose crystals predominate (Roetman, 1979).

The degradation of lactose during storage has also been reported. Galactose, lactulose and tagatose have been isolated from milk powders and are formed by base-catalysed degradation of lactose and/or the breakdown of the Amadori rearrangement lactose-protein complexes formed during the Maillard reaction (Richards, 1963).

23.4.3 Minerals

During pre-heat treatment at medium or high temperatures, extensive precipitation of calcium phosphate occurs. Evaporation increases the concentrations of lactose and salts and results in a partially reversible transfer of soluble calcium phosphate to the colloidal form, with a resulting decrease in pH (Le Graet and Brule, 1982; Nieuwenhuijse *et al.*, 1988; Oldfield, 1998). The extent of transfer of phosphate to the colloidal phase, which is greater than that of calcium, depends on the temperature of preheating. The concentrations of soluble calcium and phosphorus in reconstituted SMP generally remain lower than those in the original milk, due to irreversible shifts induced during drying (Le Graet and Brule, 1982). Preheating of milk and heating of concentrate reduce calcium ion activity, although this parameter increases slowly on reconstitution and storage of milk powder (Oldfield, 1998).

23.5 Functional properties of milk powders

The behaviour of milk powder is determined by the physical and chemical properties of its primary components, namely proteins, lipids and lactose, both individually and in combination. In the following section, the functional properties of milk powders are described and the effects of process-induced changes during the different stages of manufacture, i.e., milk pre-treatment, drying and storage, on these properties, are discussed. Commonly studied characteristics of milk powder, and the methods used for their measurement, are summarised in Table 23.4.

23.5.1 Bulk density

Bulk density, or packing density, may be defined as the weight of a unit volume of powder (g/mL, g/100mL or g/L; Písecký, 1997). Bulk density depends on interstitial air (trapped between powder particles) and particle density, the latter being determined by the amount of occluded air (air within powder particles, in cavities called vacuoles) and the actual density of powdered material. Bulk density is closely interrelated with many other properties of milk powder, such as instant properties and flowability.

The bulk density of WMP is generally lower than that of SMP, if dried under similar conditions, due to the lower density of milk fat relative to

TABLE 23.4
Methods for analysis of milk powders

Property	*Reference*
Moisture content (oven drying)	IDF No. (1993b)
Protein content (Kjeldahl)	IDF No. (1993a)
Total lactose content	Písecký (1998)
Total fat content (Röse Gottlieb)	IDF No. 123A (1988)
Free fat content	Písecký (1998)
Sulphydryl content	O'Sullivan *et al.* (1999)
Titratable acidity	IDF (1978)
Oxidation in WMP	Ulberth and Roubick (1995) Doka *et al.* (2000)
Insolubility index (mixing and centrifugation)	IDF (1988)
Bulk density (tapping test)	IDF (1986)
Scorched particles	ADMI (1971)
Whey protein nitrogen index	ADMI (1971)
Wettability	IDF (1979)
Dispersibility	IDF (1979) Chen and Lloyd (1994)
Sludge Test	Písecký (1998)
Flowability	Niro (1978), Teunou *et al.* (1999)
Slowly dispersible particles	Písecký (1998)
Particle size distribution	Písecký (1998), Aguilar and Ziegler (1994)
Coffee test	Teehan *et al.* (1997)
Hot water sediment	Písecký (1998)
Hygroscopicity	Písecký (1998)
Degree of caking	Písecký (1998)

protein and lactose. However, this difference is somewhat offset by the fact that the presence of fat in WMP inhibits foaming, which reduces the amount of occluded air (Písecký, 1978). The concentration of lactose in milk affects the bulk density of the powder produced therefrom, with low-lactose powders having a more porous structure, with larger vacoules than normal (Jimenez-Flores and Kosikowski, 1986; Aguilar and Ziegler, 1994a).

Processing steps such as evaporation and pre-heat treatment also influence the bulk density of powders, through determining the extent of denaturation of whey proteins. Denaturation of whey proteins during pre-heat treatment enhances foaming, presumably due to the thermal unfolding of their globular tertiary structure. The bulk density of powder is also markedly affected by processing steps during drying, particularly atomisation. The bulk density of milk powder produced using a nozzle atomiser is higher than that of powder produced with a centrifugal atomiser (Boersen, 1990). Displacement of air by steam during centrifugal atomisation reduces the amount of occluded air (Woodhams and Murray, 1978).

de Vilder *et al.* (1976) reported that increasing T_{inlet} reduced bulk and mean powder particle density, presumably due to an increase in vacuole volume, but also to the formation of a hard surface crust on the powder particles which prevents expansion during the later stages of drying. Increasing T_{outlet} from 70 to 95°C improves the efficiency of drying but further increases in T_{outlet} (95–105°C) cause over-heating, expansion, cracking, high vacuole volume and low powder particle density (Caríc and Kaláb, 1987). In general, the microscopic appearance of milk powder particles is affected directly by the heat treatment applied during manufacture. Increasing T_{inlet} is linked to the formation of wrinkles, or deep folds, on the surface of particles, which are thought to be caused by the presence of casein, as whey powders do not exhibit such wrinkles (Caríc and Kaláb, 1987).

The amount of interstitial air, i.e., air entrapped between powder particles, depends mainly on particle size distribution, shape and degree of agglomeration. The low bulk density of roller-dried milk powder and spray-dried powder produced using a centrifugal atomiser compared to powder produced using a nozzle atomiser is due partly to their irregular shape and narrow size distribution, respectively (Caríc and Kaláb, 1990).

23.5.2 Reconstitution properties of milk powders

Reconstituted milks, made by dissolution of SMP or WMP, may be distinguished from recombined milks, which are prepared by addition of SMP to water, followed by addition of anhydrous milk fat (AMF) and homogenisation, and filled milks, in which the fat added on recombination is a non-milk fat. Ease of recombination of milk powder is affected primarily by its degree of agglomeration, water temperature and heat classification of the powder, wettability, sinkability and solubility (Table 23.5). A key variable is the time required for the hydration of proteins and, in the case of recombined products, molten fat should not be added until hydration is complete. Emulsifiers may be added to facilitate and improve emulsification of added fat, which is achieved using high-shear agitation or homogenisation.

Wettability depends on the ability of the powder to overcome the surface tension between it and water; the failure of a powder to wet sufficiently results in the formation of a scum, adherence of undispersed particles to the walls of the container and the gradual appearance of distinct layers in the reconstituted milk (Litman *et al.*, 1957). The wettability of milk powders is markedly affected by the free fat content of the powder (Jensen, 1975; Woodhams and Murray, 1978), which may be a consequence of excessive pumping, homogenisation after concentration or the formation of lactose crystals which penetrate and damage the membranes surrounding the fat

TABLE 23.5
Parameters involved in the reconstitution of milk powders

Parameter	*Depends on*
Wettability (ability to absorb water on surface)	Hydrophobicity Lecithinisation Agglomeration Particle density
Dispersabilty (disperse without formation of lumps)	Size, agglomeration
Penetrability (ability to penetrate water surface)	Interstitial air Particle size distribution Liquid viscosity
Sinkability (after being moistened)	Particle density
Ease of solubility (rate of dissolving)	Physical and chemical properties

globules. Litman *et al.* (1957) reported that the characteristics of lipids isolated from scum differed markedly from those present in the reconstituted milk; lipids from both sources had similar saponification numbers, but the former had a higher melting point and lower iodine number, indicating that they were less susceptible to oxidation (Litman *et al.*, 1957).

Litman *et al.* (1957) proposed that the failure of milk powder to wet rapidly and sufficiently may be due to the formation of free fat-protein complexes. Problems associated with the presence of free fat may be alleviated or avoided through the addition of a surfactant such as lecithin, as is generally practised during the manufacture of instant WMP (see Section 23.2.4). The wettability of milk powders is also directly affected by the crystallisation of lactose. Lactose present in milk powders which are stored below the glass transition temperature exists in an amorphous state (Jouppila and Roos, 1994b; Teunou *et al.*, 1999). However, if the storage temperature is increased, amorphous lactose crystallises (Jouppila and Roos, 1994a), with subsequent plasticisation and caking of milk powder, which is detrimental to wettability (Mistry and Pulgar, 1996). Caking is known to occur above 44% relative humidity (Teunou *et al.*, 1999). The wettability of milk powders decreases during storage and the extent of the reduction in wettability on storage is reduced if the powders are stored at a low temperature (Litman *et al.*, 1957). The effect of storage at low temperatures is presumably due to the fact that, under these conditions, most of the lipids present in milk have solidified and, consequently, are less mobile and cannot form a film.

The second factor affecting the reconstitutability of milk powder is sinkability. Once the powder has wetted, i.e., the gaseous phase surrounding the particles has been replaced by an aqueous phase, the powder should sink, which in turn facilitates solubilisation. The sinkability of powders is determined largely by particle density (Munns, 1989).

Once the powder particles have been wetted and have sunk, reconstitution depends on solubility, which is a key determinant of overall quality. Standard tests for powder solubility involve mixing powder and water under strictly defined conditions, centrifuging and measuring the volume of insoluble sediment produced, which is expressed as the insolubility index (sometimes referred to as the solubility index). Typical insolubility indices for standard SMP and WMP are <2.0 and <1.5 mL per 100 mL of milk, respectively (Woodhams and Murray, 1978). High protein milk powders have a high insolubility index (Jimenez-Flores and Kosikowski, 1986).

Seasonal variations in the composition of milk strongly influence the insolubility index of milk powder (Newstead *et al.*, 1978; Baldwin and Ackland, 1991), which increases with increasing severity of pre-heat treatment (Baldwin and Ackland, 1991; van Mil and Jans, 1991). The method of atomisation influences powder solubility (Straatsma *et al.*, 1999), as does the method of drying, with the insolubility index of roller-dried milk powders being far higher than that of spray-dried powders (Mistry and Hassan, 1991; Mistry and Pulgar, 1996). A mathematical model to predict the insolubility of milk powder based primarily on heat treatment during drying was developed by Straatsma *et al.* (1999).

Baldwin and Ackland (1991) reported that the insolubility index of WMP increased from 2.5 to 3.2 mL over a 12 month storage period. Similar observations were reported by Jimenez-Flores and Kosikowski (1986), Sharma and Tandon (1986) and Celestino *et al.* (1997a).

The mechanism(s) responsible for the increase in the insolubility index on severe heat treatment or during storage have not been elucidated fully, although there is evidence to suggest that Maillard reaction-derived crosslinks play a role (Baldwin and Ackland, 1991). Under specific conditions, the carbonyl group of lactose and the reactive side chains of proteins (principally the ε-amino group of lysine and, to a lesser extent, the indole group of tryptophan, the imidazole group of histidine, the guanidino group of arginine and the α-amino group of N-terminal amino acids) condense to form a glucosamine which, through an Amadori rearrangement, forms a N-substituted-1-amino-l-deoxy-2-ketose. Such compounds may dehydrate to reductones or enolise to either 1-amino-1,2-endiol or 1-amino-2,3-enediol (Nursten, 1981; Walstra and Jenness, 1984). The former compound may be dehydrated to hydroxymethylfurfural (HMF) or furfural, while the latter is dealdolized to α-carbonyl compounds which are broken down to fission products, which subsequently polymerise to form melanoidins (see Chapter 19). Maillard reaction-derived crosslinking of proteins in milk protein systems has been reported (see Friedman, 1977; Pellegrino *et al.*, 1999). Jones *et al.* (1998) speculated that Maillard reactions occur during spray-drying, although in general these reactions are more predominant in milk powders during storage. In milk powders, Maillard reactions may be favoured by the close proximity of reactive molecules, despite the absence of translational diffusion (Schebor *et al.*, 1999). The activation energy of the Maillard reaction in WMP is 203.2 kJ mol^{-1} (Chong *et al.*, 1996).

It is noteworthy that the concentration of HMF increases with time and temperature of storage of milk powder (Al-Talib, 1984; Jimenez-Flores and Kosikowski, 1986; Baldwin and Ackland, 1991; Schebor *et al.*, 1999) and with the severity of heat treatment, and that insoluble material in reconstituted milk has been found to contain products of lactose-protein interactions (Parris *et al.*, 1990).

A number of alternative mechanisms for the formation of insoluble material in milk powders have been proposed. For instance, it has been suggested that insolubility may be an indirect consequence of lactose crystallisation, as this causes localised increases in calcium ion activity and a concomitant reduction in the stability of casein micelles (Al-Talib, 1984). The possible formation of isopeptide crosslinks (transamination or amidation reactions) or dehydroalanine-derived crosslinks in milk powder and their possible significance for the solubility of milk powders have not been investigated, to the authors' knowledge. Straatsma *et al.* (1999) proposed that insolubility is enhanced by unfolding of β-lactoglobulin and subsequent aggregation with casein.

23.5.3 Heat classification of milk powders

As mentioned in Section 23.2.2, heat classification refers to the extent of heat treatment applied during powder manufacture and generally reflects the severity of treatment during pre-heating before evaporation. The assessment of heat treatments applied to milk and dairy products was comprehensively reviewed by Pellegrino *et al.* (1995). The properties and applications of the major heat classification groups of SMP are outlined in Table 23.6. The vast majority of SMP produced is of the medium-heat class (Early, 1998).

The most common analytical parameter used for heat classification of SMP is the whey protein nitrogen index (WPNI), defined as mg undenatured whey proteins/g powder. Typical WPNI values for low-, medium- and high-heat treated milk powders are >6.0, 1.51–5.99 and <1.50 mg/g powder, respectively (Caríc and Kaláb, 1987). The thermal history of milk powders may also be determined by quantifying the free sulphydryl content (cysteine number) or casein number (Kyle, 1993; Stapelfeldt *et al.*, 1997a). Typical cysteine numbers for low-, medium- and high- heat powders are 32–38, 39–48 and >62%, respectively, while casein number for these powders is of the order <80, 80.1–88 and >88.1%, respectively (see Kyle, 1993).

The validity of heat classification tests may be influenced by variations in the concentration of whey proteins in raw milk, such as caused by seasonal

Table 23.6
Food applications of skim milk powder (SMP) of different heat classes

Heat classification	*Heat treatments typically applied*	*WPNI*	*Functional properties*	*Food applications*
Low heat	70°C for 15 sec	> 6.0	Solubility, lack of cooked flavour	Recombined milk, milk standardisation, cheese making
Medium heat	85°C for 1 min 90°C for 30 sec 105°C for 30 sec	1.5–6.0	Emulsification, foaming, water absorption, viscosity, colour, flavour	Ice cream, chocolate, confectionery
High heat	90°C for 5 min 120°C for 1 min 135°C for 30 sec	<1.5	Heat stability, gelation, water absorption	Recombined evaporated milk
High-high heat	>120°C for >40 min	<1.5	Flavour, water binding, colour	Bakery, recombined evaporated milk

fluctuations in milk composition (Sanderson, 1970); such variations may be monitored by powder manufacturers to allow more accurate control of powder properties. Genetic polymorphism of β-lactoglobulin and season, through their influence on level of whey proteins in milk, affect the WPNI and sulphydryl group content of SMP (Sanderson, 1970; O'Sullivan *et al.*, 1999).

23.5.4 Flowability of powder

The flowability of milk powder is a complex phenomenon, dependent on inter- and intra-particle forces, which can be defined loosely as the cohesion, or resistance of the powder to flow, under its own weight. Powder flowability is affected by the shape and size of powder particles, and the moisture and fat content of the powder. Flowability may be assessed by pouring out a heap of powder under standardized conditions and measuring the angle of repose, α, of the resulting heap. The flowability of milk powders is in the order agglomerated SMP > SMP > agglomerated WMP > WMP (Písecký, 1998; Rennie *et al.*, 1999). The flowability of milk powder decreases with fat content up to ~20% fat but, from 20 to 45% fat, flowability is independent of fat content (Woodhams and Murray, 1978). Resistance to flow decreases with decreasing particle size (Woodhams and Murray, 1978; Rennie *et al.*, 1999) and size distribution and increases with the moisture content of the powder (Rennie *et al.*, 1999). Flow conditioning agents, such as silicates, stearates and phosphates, can increase the free-flowing nature of milk powders (see Onwoluta *et al.*, 1996).

23.5.5 Organoleptic properties

Both the natural taste of milk powders and the development of undesirable flavours during storage are of critical importance. When using milk powder for direct reconstitution, the use of low- or low/medium-heat powder is recommended (Augustin, 1991). However, despite trying to emulate the flavour of fresh milk, reconstituted milk has a distinct flavour and there is a system of IDF-defined standard scales for scoring the flavour of reconstituted milk powders (Hough *et al.*, 1992). It is thought that lactones, even-numbered short-chain fatty acids and furanones contribute to the flavour of milk powders (Shiratsuchi *et al.*, 1995). The volatile flavour compounds in SMP have been identified and shown to originate mainly from breakdown or secondary reactions of residual milk fat, or from feed or forage materials (Shiratsuchi *et al.*, 1994a). It has been suggested that specific off-flavours in SMP described as cowhouse-like, hay-like, sulphuric and quinoline-like may be due to the presence of specific compounds such as tetradecanal, β-ionone and benzothiazole (Shiratsuchi *et al.*, 1994b).

Preheating is likely to play a major role in the development of specific flavours in WMP, such as the intensity of cooked aroma, cooked taste and creaminess (Baldwin *et al.*, 1991). Cooked flavour of milk powder is linked to the production of free sulphydryl groups (Hols and van Mil, 1991) and the Maillard reaction (Shiratsuchi *et al.*, 1995). Baldwin *et al.* (1991) proposed that the cooked flavour in WMP stored at 30°C reaches a maximum after 3 months storage and decreases thereafter and that the intensity of cooked flavour varies with season. Variations in cooked flavour with season may be related to changes in the concentration and profile of whey proteins.

Auldist *et al.* (1996) reported that WMP produced from milk with a high somatic cell count (SCC) did not differ from normal milk powders initially, but that off-flavours developed in the former powders during storage, which was attributed to the high proteolytic and lipolytic activity in milks with a high SCC. The development of off-flavours have been reported in milk powder produced from milk where processing steps included standardisation of the protein content through the addition of cheese whey (Kieseker and Healey, 1996), which may be due to an altered casein-whey protein ratio (and therefore the sulphydryl content per unit weight).

The most common and serious off-flavour encountered in WMP is oxidative rancidity, which is a major determinant of the shelf life of WMP (Driscoll *et al.*, 1985; McGookin and Augustin, 1997). Oxidative rancidity entails the autocatalytic oxidation of fatty acids, particularly unsaturated fatty acids, to peroxides and monocarbonyls, which have distinct off-flavours (Keen *et al.*, 1976). Quantification of free radicals may be used as an early indication of increasing oxidation in WMP (Staplefeldt *et al.*, 1997b). Oxidative rancidity in milk powders increases during storage (Mahran *et al.*, 1984; Liang, 2000) and has been reported to follow zero order reaction kinetics, with an activation energy of approximately 70 kJ/mol (Chong *et al.*, 1996; Liang, 2000).

Lipid oxidation of WMP is greatly affected by storage temperature and heat classification, being highest for low-heat powders at elevated temperatures (Stapelfeldt *et al.*, 1997c). In general, heat treatment increases the oxidative stability of milk powders (Boon, 1976; van Mil and Jans, 1991; Baldwin and Ackland, 1991; McCluskey *et al.*, 1997). Preheat treatment induces the thermal unfolding of the distinct globular structure of β-lactoglobulin, with the concomitant exposure of the buried reducing free sulphydryl group, which has an antioxidant effect (Baldwin and Ackland, 1991). Products of the Maillard reaction between proteins (particularly the caseins) and lactose may also be partially responsible for the increase in oxidative stability of milk powders due to pre-heating (Eriksson, 1982; McGookin and Augustin, 1997).

Many other factors, including season of production, affect the oxidative stability of milk powders (van Mil and Jans, 1991). Celestino *et al.* (1997a) reported that storage of milk for 48 h at 4°C prior to drying increased the susceptibility of the resultant powder to oxidative rancidity and proposed that an increase in lipolysis may be partly responsible. Reducing the amount of free fat in WMP (as discussed in Sections 23.4.2 and 23.5.7(*a*)) also affects the oxidative stability of milk powder.

The quality of WMP is also dependent on water activity, with optimum stability being found in the region 0.11–0.23 (Stapelfeldt *et al.*, 1997c). Storage conditions such as temperature, packaging material and head space in the container also have a marked effect on the oxidative stability of SMP (Driscoll *et al.*, 1985; van Mil and Jans, 1991; McCluskey *et al.*, 1997). It has been reported that SMP stored at 32°C developed an off-flavour after 6 months storage and was unacceptable after 24 months, whereas SMP stored at 21°C was deemed acceptable after 24 months (Driscoll *et al.*, 1985). The shelf life of SMP may also be extended by storing under CO_2 or N_2 (Driscoll *et al.*, 1985) or packaging in tin cans rather than in plastic bags (Driscoll *et al.*, 1985).

McCluskey *et al.* (1997) reported that milk produced from cows fed a diet that had been supplemented with vitamin E (22.5 IU vitamin E/cow/day) had a higher vitamin E content and that the powder produced therefrom was less susceptible to oxidative deterioration. The oxidative stability of WMP may also be increased through the direct addition of phenolic compounds as antioxidants to the milk powders (Radayeva *et al.*, 1974), or through addition of rice bran oil (Nanua *et al.*, 2000). The production of milk rich in phenolic compounds, such as gossypol and genistein, has been advocated, due to their salutary health effects; it is likely that milk powders produced from such milk would also have a high oxidative stability (see Parodi, 1996; O'Connell and Fox, 2001).

During storage, WMP generally becomes darker and more yellow, which is due to lactose crystallisation and the migration of β-carotene and free fat to the surface of powder particles (Nielsen *et al.*, 1997a,b). The Maillard reaction is also thought to cause darkening of milk powders during storage.

A defect sometimes associated with milk powders is the presence of scorched particles, or discoloured specks, resulting from deposits of charred powder, which have been discoloured by Maillard reactions at low a_w and high temperatures within the drying chamber, entering the bulk dry powder (Early, 1998).

23.5.6 Heat stability

The ability of reconstituted milk powder to withstand high temperatures is pertinent in the production of an array of products such as coffee whiteners, dessert products, bakery products and, perhaps most importantly, sterilised reconstituted whole milk. The ability of milk powder to withstand sterilisation may be assessed by the subjective heat stability assay of Davies and White (1966) or by the method of Kieseker and Aitken (1988).

The heat stability of reconstituted milk powder is strongly affected by milk quality (Kelly, 1982). For example, milk powder produced from late lactation milk is inherently unstable (Newstead *et al.*, 1978; Auldist *et al.*, 1996), which may be related to changes in ionic calcium content and protein concentration. Auldist *et al.* (1996) also reported that powder produced from milk with a high SCC exhibited low heat stability and attributed this to high proteolytic activity in mastitic milk.

Pre-heat treatment of milk plays a critical role in enhancing the heat stability of powders produced therefrom. Increasingly severe pre-heat treatments (such as 120–140°C for 1 to 15 min in the production of high heat powders) may produce very heat-stable powders (Sweetsur and Muir, 1981; Fox, 1981; Tan-Kintia and Fox, 1999). In New Zealand, such high-temperature pre-heat treatments are widely used in industry (Newstead and Singh, 1992). How pre-heat treatment of milk for powder manufacture enhances the heat stability of reconstituted powder produced from it has not been elucidated fully, although whey protein denaturation and the precipitation of calcium phosphate under relatively mild conditions and an increase in the stability of caseins to calcium during pre-heat treatment may play a role (Fox, 1981; Tan-Kintia, 1996; O'Connell and Fox, 1999).

The method used to concentrate milk prior to drying also affects the heat stability of the powder produced therefrom. Sweetsur and Muir (1980) and Muir and Sweetsur (1984) reported that the heat stability of concentrated milk prepared by ultrafiltration was markedly higher than that of concentrates prepared by evaporation. The heat stability of concentrates may be increased by altering the protein profile or mineral content by diafiltration or by the use of diaporous rather than conventional membranes, respectively (Muir and Sweetsur, 1984).

Surprisingly, there is little information on the effect of storage of milk powder on the heat stability of milk reconstituted therefrom. Al-Talib (1984) reported a slight decrease in the heat stability of reconstituted milk powder after storage for 90 days, with the reduction being more pronounced when the powder was stored at a higher temperature (37°C). The reduction in the heat stability of reconstituted milk powder on storage may be due to an increase in calcium ion activity, as discussed in Section 23.4.3, as the ethanol stability of reconstituted milk powder also decreases on storage (Al-Talib, 1984). It has also been shown that the age gelation of ultra-high temperature-treated milk prepared from milk powder is affected by the length of time the powder from which the milk was prepared had been stored (Celestino *et al.*, 1997b).

The heat stability of reconstituted WMP is enhanced by the addition of lecithin or buttermilk to milk prior to evaporation and drying (Singh and Tokley, 1990; Singh *et al.*, 1992). The stabilising effect of lecithin is due partly to its surfactant properties (Singh and Tokley, 1990), while the addition of buttermilk may enhance the stability of milk due to the low calcium and β-lactoglobulin content of the former (O'Connell and Fox, 2000).

23.5.7 Other functional properties of milk powders

(a) Free fat content

Fat in WMP is encapsulated within the amorphous lactose matrix, although some fat is also present in capillary pores and cracks and in pools on the surfaces of powder particles (Buma, 1971). Free fat is generally defined as the fat that can be extracted by organic solvents under defined conditions, and generally originates from non-encapsulated fat and from fat globules that are accessible to solvent which penetrates through capillaries.

Free fat is considered to be a desirable attribute of WMP used in chocolate manufacture, and such powder is frequently produced by roller-drying (Dewettinck *et al.*, 1996), although spray-drying processes using simultaneous atomisation of pre-crystallised skim milk powder and high-fat cream have also been developed for this application.

(b) Cheesemaking properties

It is generally regarded that premium quality cheese can not be produced from reconstituted milk powder, which tends to form soft fragile gels on renneting which are difficult to drain and consequently lead to high moisture curds (Lenoir *et al.*, 2000). However, when milk powders are used in the manufacture of cheese it is recommended that low-heat milk powder should be used. The rennetability of milk reconstituted from medium- and high-heat powders is lower than that of normal milk, due to whey protein-κ-casein complex formation leading to the formation of a different type of gel network (Cheftel and Lorient, 1982). The rennetability of reconstituted low-heat SMP decreases slightly during storage while the firmness of renneted curds produced therefrom decreases markedly (Al-Talib, 1984).

(c) Water sorption properties

Milk powders readily absorb water at moderate to high relative humidities, increasing the rate of many reactions. Consequently, the water absorption behaviour of milk powder is of critical importance in determining its swelling, gelling, emulsifying, foaming and organoleptic properties. The water absorption properties of milk powder are conventionally expressed using water sorption curves, which essentially measure hydration as a function of water activity (Saltmarch and Labuza, 1980). Typical absorption isotherms of WMP and SMP are shown in Figure 23.6A. Differences in the absorption isotherm of the two types of powder are partly related to the fact that lipids are relatively non-hydroscopic. As shown, the isotherms of both WMP and SMP have a distinct sigmoidal shape, with a discontinuity at a water activity of ~0.6 (Berlin *et al.*, 1968). It is postulated that this discontinuity is due to lactose crystallisation, which begins at a relative pressure of between 0.33 to 0.50 (Warburton and Pixton, 1978; Saltmarch and Labuza, 1980). Desorption and resorption isotherms do not have a discontinuity (Figure 23.6B), which shows that the water bound within the discontinuous phase in the absorption isotherm is irreversibly or very strongly bound, i.e., water of crystallisation (Berlin *et al.*, 1968). Berlin *et al.* (1968) proposed that the distinct shape of the sorption isotherm of milk powders can be divided into three specific zones. The initial gradual increase is due to water absorption at bulk phase interaction sites (i.e., external surfaces of particles); this is followed by a rapid increase in absorbed water which is due to absorption by the lactose glass, i.e., lactose crystals are arranged into tightly packed lattices which, compared to amorphous lactose, are non-hygroscopic and consequently release water (see Section 23.4.2). The final zone of the sorption isotherms of milk powders (from 0.6 to 0.8 water activity) is due to the diffusion of water into the particles and to protein swelling.

Heldman *et al.* (1965) and Al-Talib (1984) showed that the equilibrium moisture content of milk powder decreases during storage and decreases with the severity of pre-heat treatment.

(d) Other functional properties

There has been little research on the foaming and emulsifying behaviour of milk powders, probably because caseinates are used more widely in applications where such properties are required. In general, milk powders have inferior emulsifying properties to caseinates (Mulvihill and Murphy, 1991;

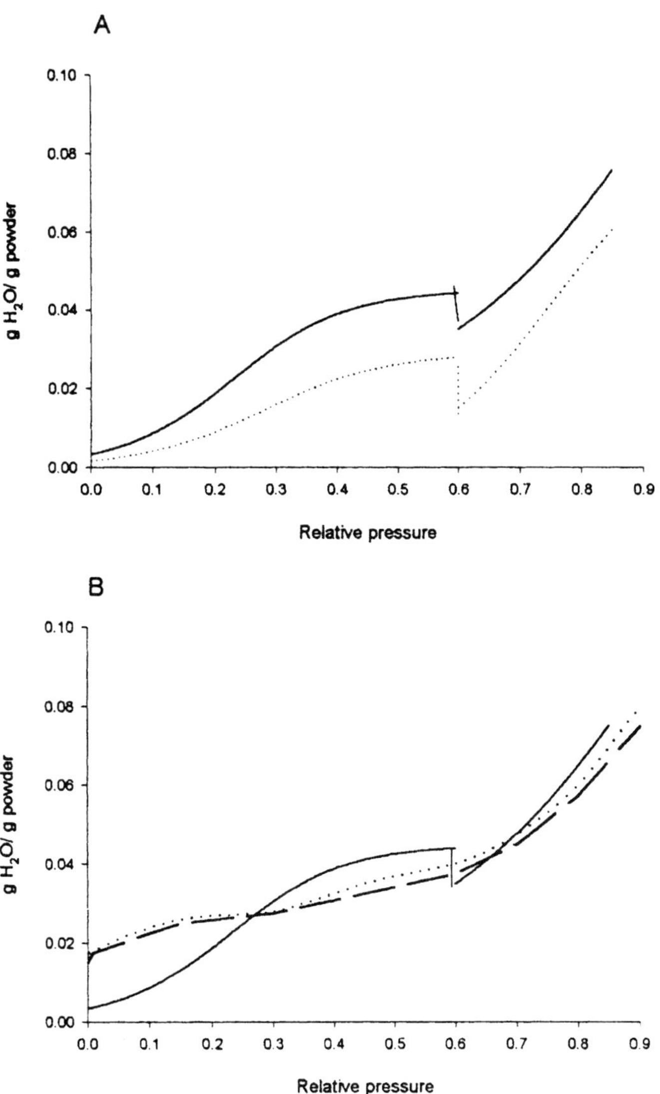

Figure 23.6 (A) Absorption isotherms of whole milk powder (......) and skim milk powder (——) and (B) sorption isotherms of skim milk powder for absorption (——), desorption (......) and reabsorption (- - - -).

Euston and Hirst, 1999). The reason for this is that, in milk powder systems, caseins are present in micellar form and do not have the same molecular mobility and surface active properties as Na caseinate (Mulvihill and Murphy, 1991; Dalgleish, 1996; Euston and Hirst, 1999). However, Euston and Hirst (1999) reported that the stability of emulsions against creaming is greater when milk powders are used, relative to caseinate systems, and proposed that this may be due either to the former system being less susceptible to depletion flocculation or to the formation of a weak gel in milk powder solutions. It has been reported that milk powder has better emulsifying ability than sodium caseinate at pH 5.2 (Singh and Newstead, 1992), which may be related to the dissociation of protein from micelles in this pH region. High-protein milk powders exhibit better foaming properties than standard milk powders (Jimenez-Flores and Kosikowski, 1986; Mistry and Hannan, 1991), which is presumably related to the high protein content of the former. With the emergence of micellar casein powders, the use of milk powders as emulsifying and foaming agents may receive more research attention.

A further important property of milk powder is the ability of powders intended for use as coffee whiteners to disperse and remain soluble when added to hot coffee which has a low pH (coffee stability test). Teehan (1997) reviewed methods for measuring the coffee stability of milk powder. Oldfield *et al.* (2000) reported that increasing severity of preheat treatment of milk before evaporation and increasing total solids level in the concentrate being fed to the atomiser both negatively affected coffee stability of WMP, while lecithinisation improved coffee stability.

23.6 Conclusion

Drying of milk, while initially a relatively straightforward method of preservation, has developed into a process that can produce a diverse range of products, tailored to the requirements of a wide range of food applications. Such diversification has been made possible by increased understanding of the complex physico-chemical changes taking place during evaporation and drying and, in parallel, understanding of the effects of processing variables on these changes, thus enabling modifications to technology to produce desired functional effects. Regarding future developments in this field, the principal areas of development in the area of drying itself are largely concerned with increasing scale of production, as global utilisation of powders continues to grow. For milk powders, a key ongoing research area is the changes that occur during storage of the powder, and thus enhancement of shelf-life. As regards products, it is likely that significant growth and development will occur in the area of milk powders of altered composition, in particular modified protein content and profile, to provide a new generation of highly functional food ingredients.

References

ADMI (1971) Standards for grades of dry milk including analysis, Bulletin 916, American Dry Milk Institute, Inc., Chicago, p. 26.

Aguilar, C.A. and Ziegler, G.R. (1994a) Physical and microscopic characterization of dry whole milk with altered lactose content. 1. Effect of lactose concentration. *J. Dairy Sci.*, **77**, 1189–97.

Aguilar, C.A. and Ziegler, G.R. (1994b) Physical and microscopic characterization of dry whole milk with altered lactose content. 2. Effect of lactose crystallisation. *J. Dairy Sci.*, **77**, 1198–204.

Al-Talib, N.A. (1984) *Effect of Storage Conditions on Some Chemical, Physico-Chemical and Nutritional Properties of Skim Milk Powder*, PhD Thesis, National University of Ireland, Cork.

Augustin, M.A. (1991) Developing non-fat milk powders with specific functional properties. *CSIRO Food Res. Quarterly Reports*, **51**, 16–22.

Auldist, M.J., Coats, S.T., Sutherland, B.J., Clarke, P.T., McDowell, G.H. and Rogers, G.L. (1996) Effect of somatic cell count and stage of lactation on the quality of full cream milk powder. *Aust. J. Dairy Technol.*, **51**, 94–8.

Baldwin, A.J. and Ackland, J.D. (1991) Effect of preheat treatment and storage on the properties of whole milk powder. Changes in physical and chemical properties. *Neth. Milk Dairy J.*, **45**, 169–81.

Baldwin, A.J., Cooper, H.R. and Palmer, K.C. (1991) Effect of preheat treatment and storage on the properties of whole milk powder. Changes in sensory properties. *Neth. Milk Dairy J.*, **45**, 97–116.

Berlin, E., Anderson, B.A. and Pallansch, M.J. (1968) Water vapour sorption properties of various dried milks and wheys. *J. Dairy Sci.*, **51**, 1339–74.

Boersen, A.C. (1990) Spray drying technology – atomisation and straight through agglomeration. *J. Soc. Dairy Technol.*, **43**, 5–7.

Boon, P.M. (1976) The effect of pre-heat treatment on the storage stability of whole milk powder. *N.Z. J. Dairy Sci. Technol.*, **11**, 278–80.

Buma, T.J. (1971) Free fat in spray-dried whole milk. 1. General introduction and brief review of literature. *Neth. Milk. Dairy J.*, **25**, 33–41.

Caríc, M. and Kaláb, M. (1987) Effects of drying techniques on the milk powder quality and microstructure: A review. *Food Microstruct.*, **6**, 171–80.

Celstino, E.L., Iyer, M. and Roginski, H. (1997a). The effects of refrigerated storage of raw milk on the quality of whole milk powder stored for different temperatures. *Int. Dairy J.*, **1**, 119–27.

Celstino, E.L., Iyer, M. and Roginski, H. (1997b) Reconstituted UHT-treated milk: Effects of raw milk, powder quality and storage conditions of UHT milk on its physico-chemical attributes and flavour. *Int. Dairy J.*, **7**, 129–40.

Cheftel, J.-C. and Lorient, D. (1982) Les propriétés fonctionnelles des protéines laitiéres et leur amélioration. *Lait*, **62**, 435–83.

Chen, X.D. (1994) Towards a comprehensive model based control of milk drying process. *Drying Technol.*, **12**, 1105–30.

Chen, X.D. and Lloyd, R.J. (1997) Some aspects of measuring the size and rate of dispersion of milk powder agglomerates using the Malvern Particle Sizer 2600c. *J. Dairy Res.*, **61**, 201–8.

Chong, L.V., Shaw, I.R. and Chen, X.D. (1996) Exothermic reactivities of skim and whole milk powders as measured using a novel procedure. *J. Food Eng.*, **30**, 185–96.

Corredig, M. and Dalgleish, D.G. (1999) The mechanisms of heat-induced interaction of whey proteins with casein micelles in milk. *Int. Dairy J.*, **9**, 233–6.

Creamer, L.K. and Matheson, A.R. (1978) The effect of pH on protein aggregation in heated skim milk. *N.Z. J. Dairy Sci. Technol.*, **13**, 9–15.

Daemen, A.L.H. (1984) The destruction of enzymes and bacteria during the spray-drying of milk and whey. 4. A comparison of theoretical computed results concerning the destruction of phosphatase with those obtained experimentally. *Neth. Milk Dairy J.*, **38**, 55–70.

Dalgleish, D.G. (1979) Proteolysis and aggregation of casein micelles treated with immobilized or soluble chymosin. *J. Dairy Res.*, **46**, 643–61.

Dalgleish, D.G. (1992) The enzymatic coagulation of milk, in *Advanced Dairy Chemistry 1-Proteins*, 2nd edn, Vol. 1, (P.F. Fox ed.) Elsevier Applied Science, London, pp. 579–620.

Dalgleish, D.G. (1996) Food emulsions, in *Emulsions and Emulsion Stability*, (J. Sjöblom ed.) Marcel Dekker, New York, pp. 287–31.

Davies, D.T. and White, J.C.D. (1966) The stability of milk protein to heat. 1. Subjective measurement of heat stability of milk. *J. Dairy Res.*, **33**, 67–81.

de Knegt, R.J. and van den Brink, H. (1998) Improvement of the drying oven method for the determination of the moisture content of milk powder. *Int. Dairy J.*, **8**, 733–8.

Dewettinck, K., de Moor, H. and Huyghebaert, A. (1996) The free fat content of dried milk products and flow properties of milk chocolate. *Milchwissenschaft*, **51**, 25–8.

de Wit, J.N. (1981) Structure and functional behaviour of whey proteins. *Neth. Milk Dairy J.*, **35**, 47–64.

Doka, O., Ajtony, Z., Bicanic, D. and Koehorst, R. (2000) Assessing the extent of degradation in the UV radiation and heat-catalysed oxidised whole milk powder: the UV photoacoustic and diffuse reflectance spectroscopies versus the peroxide value. *Appl. Spect.*, **54**, 1405–8.

Driscoll, N.R., Brennard, C.P. and Hendricks, D.G. (1985) Sensory quality of nonfat dry milk after long-term storage. *J. Dairy Sci.*, **68**, 1931–5.

Early, R. (1990) The use of high-fat and specialized milk powders. *J. Soc. Dairy Technol.*, **43**, 53–6.

Early, R. (1998) Milk concentrates and milk powders, in *The Technology of Dairy Products*, 2nd edn, (R. Early ed.) Blackie Academic and Professional, London, pp. 228–300.

El-Gazzar, F.E. and Marth, E.H. (1991) Ultrafiltration and reverse osmosis in dairy technology: a review. *J. Food Prot.*, **54**, 801–9.

Euston, S.R. and Hirst, R.L. (1999) Comparison of the concentration-dependent emulsifying properties of protein products containing aggregated and non-aggregated milk protein. *Int. Dairy J.*, **10**, 693–701.

Fairise, J.-F., Cayot, P. and Lorient, D. (1999) Characterisation of the protein composition of casein micelles after heating. *Int. Dairy J.*, **9**, 249–54.

Fox, P.F. (1981) Heat-induced changes in milk preceding coagulation. *J. Dairy Sci.*, **64**, 2127–37.

Friedman, M. (1977) *Advances in Experimental Medicine and Biology, Volume 86B: Protein Crosslinking*. Plenum Press, New York.

Garem, A., Schuck, P. and Maubois, J.-L. (2000) Cheesemaking properties of a new dairy-based powder made by a combination of microfiltration and ultrafiltration. *Lait*, **80**, 25–32.

Guo, M.R., Hendricks, G.M., Kinstedt, P.S., Flynn, A. and Fox, P.F. (1996) Nitrogen and mineral distribution in infant formulae. *Int. Dairy J.*, **6**, 963–79.

Hall, C.W. and Hedrick, T.I. (1975) *Drying of Milk and Milk Products*, AVI Publishing Co. Ltd., Westport, CT.

Hansen, R. (1985) *Evaporation, Membrane Filtration and Spray Drying in Milk Powder and Cheese Production*, North European Dairy Journal, Vanløse, Denmark.

Heldman, D.R., Hall, C.H. and Hedrick, T.I. (1965) Vapor equilibrium relationships of dry milk. *J. Dairy Sci.*, **48**, 845–52.

Hols, G. and van Mil, P.J.J.M. (1991) An alternative process for the manufacture of whole milk powder. *J. Soc. Dairy Technol.*, **44**, 49–52.

Holsinger, V.H., McAloon, A.J., Onwulata, C.I. and Smith, P.W. (2000) Cost analysis of encapsulated spray-dried milk fat. *J. Dairy Sci.*, **83**, 2361–5.

Horton, B.S. (1997) What ever happened to the ultrafiltration of milk? *Aust. J. Dairy Technol.*, **52**, 47–9.

Hough, G., Martinez, E. and Barbieri, T. (1992) Sensory thresholds of flavor defects in reconstituted whole milk powder. *J. Dairy Sci.*, **75**, 2370–4.

IDF (1978) Dried milk – Determination of titratable acidity. Standard 86. International Dairy Federation, Brussels.

IDF (1979) Determination of the dispersibility and wettability of instant dried milk. Standard 87. International Dairy Federation, Brussels.

IDF (1986) Dried milk and dried milk products – determination of bulk density. Standard 134. International Dairy Federation, Brussels.

IDF (1988a) Milk based instant foods – determination of fat content – Röse-Gottlieb method. Standard 123 A. International Dairy Federation, Brussels.

IDF (1988b) Dried milk and dried milk products – determination of insolubility index. Standard 129A. International Dairy Federation, Brussels.

IDF (1993a) Milk. Determination of nitrogen content (Kjeldahl method). Standard 20B. International Dairy Federation, Brussels.

IDF (1993b) Dried milk and dried cream – determination of water content. Standard 26B. International Dairy Federation, Brussels.

Jensen, J.D. (1975) Some recent advances in agglomeration, instantising and spray drying. *Food Technol.*, **29(6)**, 60–71.

Jimenez-Flores, R. and Kosikowski, F.W. (1986) Properties of ultrafiltered skim milk powder retentate. *J. Dairy Sci.*, **69**, 329–39.

Jones, A.D., Tier, C.M. and Wilkins, J.P.G. (1998) Analysis of the Maillard reaction products of β-lactoglobulin and lactose in skimmed milk powder by capillary electrophoresis and electrospray mass spectroscopy. *J. Chromatogr. A*, **822**, 147–54.

Jost, R., Maire, J.-C., Maynard, F. and Secretin, M.-C. (1999) Aspects of whey protein usage in infant nutrition, a brief review. *Int. J. Food Sci. Technol.*, **34**, 533–42.

Jouppila, K. and Roos, Y.H. (1994a) Water sorption and time dependent phenomena of milk powders. *J. Dairy Sci.*, **77**, 1798–808.

Jouppila, K. and Roos, Y.H. (1994b) Glass transitions and crystallization in milk powder. *J. Dairy Sci.*, **77**, 2907–15.

Jouppila, K., Kansikas, J. and Roos, Y.H. (1997) Glass transition, water plasticisation, and lactose crystallization in skim milk powder. *J. Dairy Sci.*, **80**, 3152–60.

Keen, A.R., Boon, P.M. and Walker, N.J. (1976) Off-flavours in stored milk powder. 1. Isolation of monocarbonyl classes. *N.Z. J. Dairy Sci. Technol.*, **11**, 180–8.

Kelly, P.M. (1982) The effect of preheat temperature and urea addition on the seasonal variation in the heat stability of skim-milk powder. *J. Dairy Res.*, **49**, 187–96.

Kelly, P.M., Kelly, J., Mehra, R., Oldfield, D.J., Raggett, E. and O'Kennedy, B.T. (2000) Implementation of integrated membrane processes for pilot scale development of fractionated milk components. *Lait*, **80**, 139–53.

Kiesecker, F.G. and Aitken, B. (1988) An objective method for determination of heat stability of milk powders. *Aust. J. Dairy Technol.*, **43**, 26–31.

Kiesecker, F.G. and Aitken, B. (1993) Recombined full-cream milk powder. *Aust J. Dairy Technol.*, **48**, 33–7.

Kieseker, F.G. and Pearce, R.J. (1978) Producing heat stable milk powder. *CSIRO Food Res. Quarterly*, **38**, 35–40.

Kieseker, F.G. and Healy, D. (1996) Protein-adjusted non-fat milk powders. *Aust. J. Dairy Technol.*, **51**, 83–8.

Kyle, W.S.A. (1993) Powdered milk, in *Encyclopedia of Food Science, Food Technology and Nutrition*, (R. McCrae, R.K. Robinson and M.J. Sadler eds.) Academic Press, New York, pp. 3700–13.

Le Great, Y. and Brule, G. (1982) Effect of concentration and drying on mineral equilibria in skim milk and retentates. *Lait*, **62**, 113–25.

Lenoir, J., Remeuf, F. and Schneid, N. (2000) Cheesemaking milk, in *Cheesemaking: From Science to Quality Assurance,* (A. Eck and J.-C. Gillis eds.) Lavoisier Publishing, Paris, pp. 280–96.

Liang, J.-H. (2000) Kinetics of fluorescence formation in whole milk powder during oxidation. *Food Chem.*, **71**, 459–63.

Litman, I.I. and Ashworth, U.S. (1957) Insoluble scum-like materials on reconstituted whole milk powders. *J. Dairy Sci.*, **40**, 403–9.

Mahran, G.A., El-Ghandour, M.A., El-Bagoury, E.H. and Sayed, A.F. (1984) Effect of skim milk powder storage on ice cream quality. *Egyptian J. Dairy Sci.*, **12**, 267–73.

Masters, K. (1991) *Spray Drying Handbook*. Longman Scientific and Technical, Harlow, U.K.

McCluskey, S., Connolly, J.F., Devery, R., O'Brien, B., Kelly, J., Harrington, D. and Stanton, C. (1997) Lipid and cholesterol oxidation in whole milk powder during processing and storage. *J. Food Sci.*, **62**, 331–7.

McGookin, B.J. and Augustin, M.A. (1997) Antioxidant activity of a heated casein-glucose mixture in full cream milk powder. *Aust. J. Dairy Technol.*, **52**, 15–9.

McKenna, A.B. (1997) Examination of whole milk powder by confocal laser scanning microscopy. *J. Dairy Res.*, **64**, 423–32.

McKenna, A.B., Lloyd, R.J., Munro, P.A. and Singh, H. (1999) Microstructure of whole milk powder and of insolubles detected by powder functional testing. *Scanning*, **21**, 305–15.

Mettler, A.E. (1980) Utilization of whey by-products for infant feeding. *J. Soc. Dairy Technol.*, **33**, 67–72.

Mistry, V.V. and Hassan, H.N. (1991) Delactosed high milk protein powder. 2. Physical and functional properties. *J. Dairy Sci.*, **74**, 3716–23.

Mistry, V.V. and Pulgar, J.B. (1996) Physical and storage properties of high milk protein powder. *Int. Dairy J.*, **6**, 195–203.

Mohammed, K.S. and Fox, P.F (1987) Heat-induced microstructural changes in casein micelles before and after coagulation. *N.Z. J. Dairy Sci. Technol.*, **22**, 191–203.

Morrissey, P.A. (1969) The rennet hysteresis of heated milk. *J. Dairy Res.*, **36**, 333–41.

Muir, D.D. (1980) Concentration and milk powder quality, in *Milk and Whey Powders*, Society of Dairy Technology, Middlesex, UK, pp. 73–84.

Muir, D.D. and Sweetsur, A.W.M. (1984) Optimization of the heat stability of protein-rich concentrates prepared by ultrafiltration of skim-milk. *J. Food Technol.*, **19**, 263–71.

Mulvihill, D.M. and Murphy, P.C. (1991) Surface active and emulsifying properties of caseins/caseinates as influenced by state of aggregation. *Int. Dairy J.*, **1**, 13–7.

Munns, R.J. (1989) Optimising milk powders for consumer use. *Food Aust.*, **41**, 938-40.

Munns, R.J. (1991) High fat and full cream powders as food ingredients. *CSIRO Food Res. Quarterly Reports*, **51**, 23–8.

Nanua, J.N., McGregor, J.U. and Godbert, J.S. (2000) Influence of high-oryzanol rice bran oil on the oxidative stability of whole milk powder. *J. Dairy Sci.*, **83**, 2426–31.

Neff, E. and Morris, H.A.L. (1967) Agglomeration of milk powder and its influence on reconstitution properties. *J. Dairy Sci.*, **51**, 330–8.

Newstead, D.F., Baldwin, A.J. and Hughes, I.R. (1978) Factors affecting the viscosity of recombined sweetened condensed milk. *N.Z. J. Dairy Sci. Technol.*, **13**, 65–70.

Nielsen, B., Staplefeldt, H. and Skibsted, L.H. (1997a) Early prediction of the shelf-life of medium-heat whole milk powders using stepwise multiple regression and principal component analysis. *Int. Dairy J.*, **7**, 341–8.

Nielsen, B., Staplefeldt, H. and Skibsted, L.H. (1997b) Differentiation between 15 whole milk powders in relation to oxidative stability during accelerated storage: analysis of variance and canonical variable analysis. *Int. Dairy J.*, **7**, 589–99.

Nieuwenhuijse, J.A., Timmermans, W. and Walstra, P. (1988) Calcium and phosphate paritions during the manufacture of sterilised concentrated milk and their relations to heat stability. *Neth. Milk Dairy J.*, **42**, 387–421.

Niro Atomiser A/S (1978) *Analytical Methods for Dry Milk Products*, 4th edn, Niro Atomiser A/S, Copenhagen.

Noh, B. and Richardson, T. (1989) Incorporation of radiolabelled whey proteins into the casein micelles by heat processing. *J. Dairy Sci.*, **72**, 1724–31.

Nursten, H.E. (1981) Recent developments in studies of the Maillard reaction. *Food Chem.*, **6**, 263–77.

O'Connell, J.E. and Fox, P.F. (1999) Heat-induced changes in the calcium sensitivity of caseins. *Int. Dairy J.*, **9**, 839–47.

O'Connell, J.E. and Fox, P.F. (2000) Heat stability of buttermilk. *J. Dairy Sci.*, **83**, 1728–32.

O'Connell, J.E. and Fox, P.F. (2001) Significance and possible applications of phenolic compounds in the production and quality of milk and dairy products. A review. *Int. Dairy J.*, **11**, 103–20.

O'Donnell, C.P., McKenna, B.M. and Herlihy, N. (1996) Drying of skim milk: opportunities for reduced steam. *Drying Technol.*, **14**, 513–28.

O'Sullivan, E.A., Kelly, P.M., Fitzgerald, R.J., O'Farrell, K., Murphy, M.F. and Harrington, D. (1999) Effect of β-lactoglobulin phenotype on whey protein nitrogen index and sulphydryl content of skim milk powder. *Lait*, **79**, 229–44.

Oldfield, D.J. (1998) *Heat Induced Whey Protein Reactions in Milk: Kinetics of Denaturation and Aggregation as Related to Milk Powder Manufacture*, PhD Thesis, Massey University, Palmerstown North, New Zealand.

Oldfield, D.J., Teehan, C.M. and Kelly, P.M. (2000) The effect of preheat treatment and other process parameters on the coffee stability of instant whole milk powder. *Int. Dairy J.*, **10**, 659–67.

Ontwulata, C.I., Konstance, R.P. and Holsinger, V.H. (1996) Flow properties of encapsulated milkfat powders as affected by flow agents. *J. Food Sci.*, **61**, 1211–5.

Parodi, P.W. (1996) Milk fat components: possible chemopreventative agents for cancer and other diseases. *Aust. J. Dairy Technol.*, **51**, 24–32.

Parris, N., White, A.E. and Farrell, H.M., Jr. (1990) Identification of altered proteins in nonfat dry milk powder prepared from heat-treated skim milk. *J. Agric. Food Chem.*, **38**, 824–9.

Patel, R.S. and Mistry, V.V. (1997) Physicochemical and structural properties of ultrafiltered buffalo milk and milk powder. *J. Dairy Sci.*, **80**, 812–7.

Pellegrino, L., Resimini, L. and Luf, W. (1995) Assessment (indices) of heat treatment of milk, in, *Heat-Induced Changes in Milk*, (P.F. Fox ed.) Special Issue 9501, International Dairy Federation, Brussels, pp. 419–53.

Pellegrino, L., van Boekel, M.A.J.S., Gruppen, H., Resmini, P. and Pagani, M.A. (1999) Heat-induced aggregation and covalent linkages in β-casein model system. *Int. Dairy J.*, **9**, 255–60.

Písecký, J. (1978) Bulk density of milk powders. *Food Technol.*, **43**, 4–7.

Písecký, J. (1997) *Handbook of Milk Powder Manufacture*, Niro A/S, Copenhagen.

Radayeva, I.A., Dmitrieva, L.S. and Bekhova, E.A. (1974) The effect of added antioxidants on keeping quality and on phospholipid fractions of whole milk powder. *Proc. XIX International Dairy Congr.* (New Delhi) E618.

Refstrup, E. (1995) Advances in spray drying of food products. *J. Soc. Dairy Technol.*, **48**, 50–4.

Rennie, P.R., Chen, X.D., Hargreaves, C. and Mackerth, A.R. (1999) A study of cohesion of milk powders. *J. Food Eng.*, **39**, 277–84.

Richards, E.L. (1963) A quantitative study of the changes in dried skim-milk and lactose casein in the 'dry' state during storage. *J. Dairy Res.*, **30**, 223–34.

Roetman, K. (1979) Crystalline lactose and the structure of spray-dried milk products as observed by scanning electron microscopy. *Neth. Milk Dairy J.*, **33**, 1–11.

Saboya, L.V. and Maubois, J.-L. (2000) Current developments of microfiltration technology in the dairy industry. *Lait*, **80**, 541–53.

Saito, Z. (1985) Particle structure in spray-dried whole milk and in instant skim milk powder as related to lactose crystallisation. *Food Microstruct.*, **4**, 333–40.

Saltmarch, M. and Labuza, T.P. (1980) Influence of the physicochemical state of lactose in spray-dried sweet-whey powders. *J. Food Sci.*, **45**, 1231–6.

Sanderson, W.B. (1970) Seasonal variations affecting the determination of the whey protein nitrogen index of skim milk powder. *N.Z. J. Dairy Sci. Technol.*, **5**, 48–52.

Sanderson, W.B. (1978) Instant milk powders: manufacture and keeping quality. *N.Z. J. Dairy Sci. Technol.*, **13**, 137–43.

Schebor, C., Beura, M.P., Karel, M. and Chirife, J. (1999) Color formation due to non-enzymatic browning in amorphous, glassy, anhydrous, model systems. *Food Chem.*, **65**, 427–32.

Sharma, K.S. and Tandon, K.C. (1986) Effect of storage on the nitrogen distribution of skim milk powders. *Indian J. Dairy Sci.*, **39**, 330–2.

Shiratsuchi, H., Shimoda, M., Imayoshi, K., Noda, K. and Osajima, Y. (1994a) Volatile flavour compounds in spray-dried skim milk powder. *J. Agric. Food Chem.*, **42**, 984–8.

Shiratsuchi, H., Shimoda, M., Imayoshi, K., Noda, K. and Osajima, Y. (1994b) Off-flavour compounds in spray-dried skim milk powder. *J. Agric. Food Chem.*, **42**, 1323–7.

Shiratsuchi, H., Yoshimura, Y., Shimoda, M., Noda, K. and Osajima, Y. (1995) Contributors to sweet and milky odour attributes of spray dried skim milk powder. *J. Agric. Food Chem.*, **43**, 2453–7.

Singh, H. and Creamer, L.K. (1991) Denaturation, aggregation and heat stability of milk protein during the manufacture of skim milk powder. *J. Dairy Res.*, **58**, 269–83.

Singh, H. and Newstead, D.F. (1992) Aspects of milk proteins in milk powder manufacture, in *Advanced Dairy Chemistry 1-Proteins*, 2nd edn, (P.F. Fox ed.) Elsevier Applied Science, London, pp. 735–66.

Singh, H. and Tokley, R.P. (1990) Effects of preheat treatments and buttermilk addition on the seasonal variations in the heat stability of recombined evaporated milk and reconstituted concentrated milk. *Aust. J. Dairy Technol.*, **45**, 10–6.

Singh, H., Sharma, R. and Tokley, R.P. (1992) Influence of incorporation of soya lecithin into skim milk powder on the heat stability of recombined evaporated milk. *Aust. J. Dairy Technol.*, **47**, 33–7.

Staplefeldt, H., Berrum, K. and Skibstead, L.H. (1997a) Ellman's reagent for determination of the heat treatment of milk powder. Improved analytical procedure based on a stopped–flow kinetic study. *Milchwissenschaft*, **52**, 146–9.

Staplefeldt, H., Mortensen, G. and Skibstead, L.H. (1997b) Early events in oxidation of whole milk powder detected by electron spin resonance spectrometry. Carry–over effects from butter oil used for instantisation. *Milscwissenschaft*, **52**, 266–9.

Staplefeldt, H., Nielsen, B.R., and Skibstead, L.H. (1997c) Effect of heat treatment, water activity and storage temperature on the oxidative stability of whole milk powder. *Int. Dairy J.*, **7**, 331–9.

Straatsma, J., van Houwelingen, G., Steenbergen, A.E. and de Jong, P. (1999) Spray drying of food products: 2. Prediction of insolubility index. *J. Food Eng.*, **42**, 73–7.

Swaisgood, S. (1992) The chemistry of caseins, in *Advanced Dairy Chemistry 1-Proteins*, 2nd edn, (P.F. Fox ed.) Elsevier Applied Science, London, pp. 63–110.

Sweetsur, A.W.M. and Muir, D.D. (1980) Effect of concentration by ultrafiltration on the heat stability of skim–milk. *J. Dairy Res.*, **47**, 327–35.

Sweetsur, A.W.M. and Muir, D.D. (1981) Role of cyanate ions in the urea-induced stabilisation of the caseinate complex in skim-milk. *J. Dairy Res.*, **48**, 163–6.

Tan-Kintia, R. (1996) *Heat-Induced Changes and the Heat Stability of Milk*, PhD Thesis, National University of Ireland, Cork.

Tan-Kintia, R. and Fox, P.F. (1999) Effect of various preheat treatments on the heat stability of unconcentrated milk. *Int. Dairy J.*, **9**, 219–25.

Teehan, C.C., Kelly, P.M., Devery, R. and O'Toole, A. (1997) Evaluation of test conditions during the measurement of coffee stability of instant whole milk powder. *Int. J. Dairy Technol.*, **50**, 113–21.

Teunnou, E., Fitzpatrick, J.J. and Synnott, E.C. (1999) Characterisation of food powder flowability. *J. Food Eng.*, **39**, 31–7.

Ulberth, F. and Roubicek, D. (1995) Monitoring of oxidative deterioration of milk powder by headspace gas chromatography. *Int. Dairy J.*, **5**, 523–31.

van Mil, P.J.J.M. and Jans, J.A. (1991) Storage stability of whole milk powder: effects of process and storage conditions on product properties. *Neth. Milk Dairy J.*, **45**, 145–67.

van Renterghem, R. and de Block, J. (1996) Furosine in consumption milk and milk powders. *Int. Dairy J.*, **6**, 371–82.

Varnam, A.H. and Sutherland, J.P. (1994) *Milk and Milk Products: Technology, Chemistry and Microbiology*, Chapman and Hall, London.

Walstra, P. and Jenness, R. (1984) *Dairy Chemistry and Physics*, Wiley & Sons, New York.

Walstra, P., Guerts, T.J., Noomen, A., Jellema, A. and van Boekel, M.A.J.S. (1999) *Dairy Technology: Principles of Milk Properties and Processes*, Marcel Dekker, Inc., New York.

Warburton, S. and Pixton, S.W. (1978) The moisture relations of spray dried skimmed. *J. Stored Prod. Res.*, **44**, 143–53.

Warren, R.J. (1980) Packing and marketing of milk and whey powders, in *Milk and Whey Powders*, Society of Dairy Technology, Middlesex, UK, pp. 117–24.

Westergaard, V. (1994) *Milk Powder Technology: Evaporation and Spray Drying*, Niro A/S, Copenhagen.

Woodhams, D.J. and Murray, M.J. (1978) Properties of spray dried milk powders. *N.Z. J. Dairy Sci. Technol.*, **13**, 172–8.

24

ICE CREAM

H.D. Goff

24.1 Introduction

The term "ice cream" is used in the context of this chapter to represent a family of whipped dairy products that are manufactured by freezing and are consumed in the frozen state. These include: ice cream, which contains either dairy or non-dairy fats; premium, higher fat versions; "light", lower fat versions, and ice milk; sherbet; frozen yogurt; and other related products. The structure of ice cream can be described as a complex colloid consisting of three discrete phases, fat globules (some partially coalesced) and their adsorbed interfacial material, air bubbles and their adsorbed interfacial material, and ice crystals, surrounded by a freeze-concentrated aqueous serum or matrix phase that contains the sugars, proteins, polysaccharides and salts (Goff, 1997a). Milk proteins are added as part of the milk solids-not-fat component; the protein content of a mix is usually about 4%. Proteins contribute three very important functional roles to the development of structure in ice cream: emulsification and resulting contribution to partial coalescence and fat structure formation; aeration and foam stability; and solution properties (Goff, 1997a; Walstra and Jonkman, 1998).

Proteins exist in part at the fat interface, together with the added surfactants. Proteins adsorb to the fat globules at the time of homogenization. They are then partially displaced by surfactants during cold aging of the mix emulsion (Goff and Jordan, 1989). The resulting fat globule membrane partly determines the susceptibility of the mix emulsion to partial coalescence, which in turn forms the fat structure that is responsible for many of the structural and textural attributes of ice cream. Proteins also exist at the air interface. Ice cream is a frozen foam, and proteins contribute both to the formation of the foam during aeration and to stabilization of the foam, together with added surfactants and partially-coalesced fat (Pelan

Advanced Dairy Chemistry Volume 1: Proteins, 3rd edn.
Edited by P.F. Fox and P.L.H. McSweeney, Kluwer Academic/Plenum Publishers, 2003.

et al., 1997). Those proteins not at an interface exist in the aqueous, unfrozen phase. The water-holding capacity of proteins leads to enhanced viscosity of the mix (Kinsella *et al.*, 1988), especially due to freeze-concentration, which imparts a beneficial body to the ice cream, increases the melt-down time of ice cream and contributes to reduced icyness (Goff *et al.*, 1989). Milk proteins are also partially incompatible with the added polysaccharides in ice cream, both in the mix but especially during freeze-concentration, and thus may separate into phases (Syrbe *et al.*, 1998). The resulting networks of polysaccharide and aggregated protein may be partially responsible for controlling recrystallization of the ice phase during storage and temperature fluctuations.

As commercially-available isolated milk proteins, blends and fractions become increasingly available, it is important to understand their functional contribution to the structure and texture of ice cream and to ensure that these functional roles are being met. Thus, the objective of this chapter is to overview briefly the ingredients and processes used for ice cream manufacture, focussing on the protein sources, and then to examine in detail the functional contributions of proteins to ice cream.

24.2 Ice cream manufacture

24.2.1 Ingredients

(a) Fat

The ingredients normally found in ice cream formulations are presented in Table 24.1. The fat content is an indicator of the perceived quality and/or value of the ice cream. The fat component of the mix increases the richness of flavour of ice cream, produces a characteristic smooth texture by lubricating the palate, helps to give body, and aids in producing desirable melting properties (Marshall and Arbuckle, 1996; Berger, 1997; Goff,

Table 24.1
Components of typical ice cream mix formulations

Component	*Range (%)*
Milkfat	10–16
Milk solids-not-fat	9–12
Sucrose	9–12
Corn syrup solids	4–6
Stabilizers/Emulsifiers	0–0.5
Total solids	36–45
Water	55–64

1997b). Milkfat, from cream, sweet (unsalted) butter, frozen cream, condensed milk blends or whey cream, is the principal or only fat source for dairy ice cream formulations. Vegetable fats can also be used as fat sources in non-dairy ice cream. Blends of oils are often used in ice cream manufacture, selected to take into account physical characteristics, flavour, availability and cost. Palm kernel oil, coconut oil, palm oil, sunflower oil, peanut oil, fractions thereof, and their hydrogenated counterparts are all used to some extent (Berger, 1997). During freezing of ice cream, the fat emulsion that exists in the mix will partially coalesce or destabilize as a result of emulsifier action, air incorporation, ice crystallization and high shear in the freezer (Goff and Jordan, 1989; Berger, 1997). This partial coalescence of the fat is necessary to set up the structure and texture in ice cream, which is very similar to the structure in whipped cream (Brooker *et al.*, 1986). The process of partial coalescence and the role of the milk protein in this process will be discussed in Section 24.5.1.

(b) Milk solids-not-fat

The milk solids-not-fat (SNF) or serum solids contain the lactose, caseins, whey proteins, minerals (ash), vitamins, acids, enzymes and gases of the milk or milk products from which they were derived. Limitations on their use include off-flavours from certain products and an excess of lactose, which may lead to problems due to excessive freezing point depression or lactose crystallization. Traditionally, the best sources of milk SNF for high quality products have been fresh concentrated skimmed milk or spray dried low-heat skim milk powder. Others include those containing whole milk protein (e.g., condensed or sweetened condensed whole milk, super-heated condensed skimmed milk, dry or condensed buttermilk), those containing casein (e.g., sodium caseinate), or those containing whey proteins (e.g., dried or condensed whey, whey protein concentrate, whey protein isolate) (Marshall and Arbuckle, 1996), as will be discussed in Section 24.3. The specific functionality of proteins in ice cream will be discussed in Section 24.5.

(c) Sweeteners

Sweeteners improve the texture and palatability of ice cream and enhance flavours. Their ability to lower the freezing point of a solution imparts a measure of control over the temperature-hardness relationship (Berger, 1997). The most common sweetening agent is sucrose, alone or in combination with other sugars. Sucrose and lactose are present most commonly in ice cream in the supersaturated or glassy state, with few crystals being present (Caldwell *et al.*, 1992; Berger, 1997). In many ice cream formulations, sweeteners derived from corn syrup are substituted for all or a portion of the sucrose. The use of corn starch hydrolysis products (corn syrups or glucose solids) in ice cream is generally perceived to provide greater smoothness by contributing to a firmer and more chewy body, to provide better melt-down characteristics, to reduce heat shock potential, which improves the shelf-life of the finished product, and to provide an economical source of solids.

(d) Stabilizers
Ice cream stabilizers are a group of hydrocolloid ingredients (usually polysaccharides) used in ice cream formulations to produce smoothness in body and texture, retard or reduce the growth of ice and lactose crystals during storage, especially during periods of temperature fluctuation, known as heat shock, and to provide uniformity to the product and resistance to melting. They also increase the viscosity of the mix, stabilize the mix to prevent serum separation (e.g., carrageenan), aid in suspension of flavouring particles, produce a stable foam with easy cut-off and stiffness at the barrel freezer for packaging, slow down moisture migration from the product to the package or the air, and help to prevent shrinkage of the product volume during storage (Marshall and Arbuckle, 1996). Stabilizers commonly used include: locust bean (carob) gum, guar gum, carboxymethyl cellulose, sodium alginate, xanthan, gelatin and carrageenan. Each stabilizer has its own characteristics and often two or more of these stabilizers are used in combination to lend synergistic properties to each other and improve their overall effectiveness. Carrageenan is a secondary colloid used to prevent serum separation in the mix, which is usually promoted by one of the other stabilizers (Marshall and Arbuckle, 1996; Berger, 1997). This polysaccharide-protein interaction is discussed more fully in Section 24.5.3.

(e) Emulsifiers
Emulsifiers have been used in the manufacture of ice cream mix for many years. They are usually integrated with the stabilizers in proprietary blends but their function and action are very different from that of the stabilizers. They are used to: improve the whipping quality of the mix; produce a drier ice cream to facilitate moulding, fancy extrusion and novelty product manufacture; produce a smoother body and texture in the finished product; and promote superior drawing qualities at the freezer to produce a product with good stand-up properties and melt resistance (Marshall and Arbuckle, 1996; Goff, 1997b). Their mechanism of action can be summarized as follows: they lower the fat/water interfacial tension in the mix, resulting in protein displacement from the fat globule surface, which in turn reduces the stability of the fat globule to partial coalescence that occurs during the whipping and freezing process, leading to the formation of an aggregated fat structure in the frozen product which contributes greatly to texture and melt-down properties (Goff and Jordan, 1989; Krog, 1998; Tharp *et al.*, 1998). The extent of protein displacement from the membrane, and hence the extent of fat destabilization achieved, is a function of the concentration of emulsifier (Berger, 1997; Tomas *et al.*, 1994). Their interaction with proteins and role in structure formation will be described in Section 24.5.1

"Emulsification". Emulsifiers used in ice cream manufacture today are of two main types: mono- and di-glycerides and sorbitan esters. Of the latter, Polysorbate 80 is a very strong promoter of fat destabilization in ice cream (Goff and Jordan, 1989) and is used in many commercial stabilizer/ emulsifier blends.

24.2.2 Processes

(a) Mix manufacture

Ice cream processing operations can be divided into two distinct stages, manufacture of the mix and freezing operations (Figure 24.1). The manufacture of ice cream mix involves the following unit operations: combination and blending of ingredients, batch or continuous pasteurization, homogenization and aging (Marshall and Arbuckle, 1996; Berger, 1997). Pasteurization is designed to kill pathogenic bacteria. In addition, it serves a useful role in reducing the total bacterial load and in solubilization of some of the components (proteins and stabilizers). Both batch ($>\sim$69°C for $\sim$30 min) and continuous (high temperature-short time, HTST, $>\sim$80°C for $\sim$15–25 sec) systems are in common use. Homogenization is

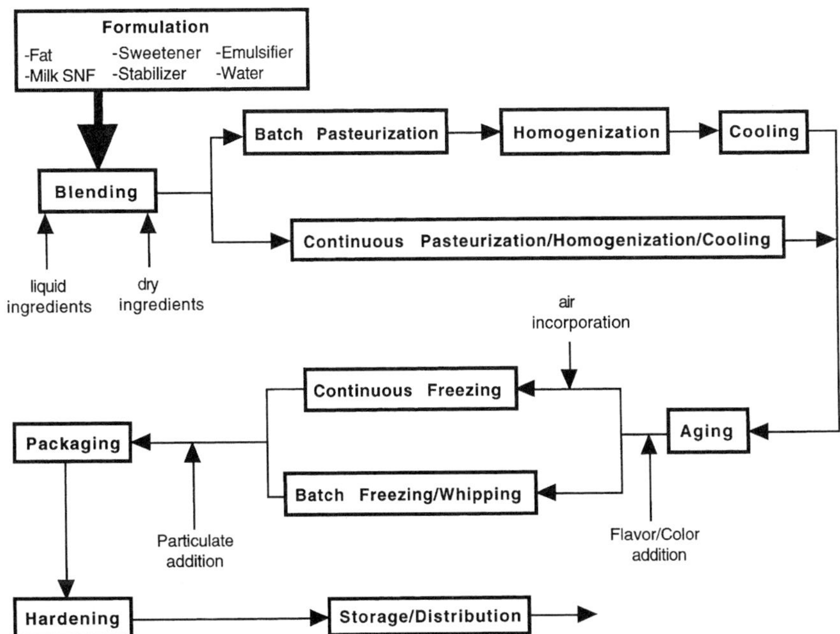

Figure 24.1 Schematic illustration of the processing steps in ice cream manufacture.

responsible for the formation of the fat emulsion by forcing the hot mix through a small orifice under a pressure of 15.5 to 18.9 MPa (2000–3000 psi gauge), depending on the composition of the mix. A large increase in the surface area of the fat globules is responsible in part for the formation of the fat globule membrane, comprised of adsorbed materials that reduces the interfacial free energy of the fat globules. With single-stage homogenizers, fat globules tend to cluster as bare fat surfaces come together or adsorbed molecules are shared. Therefore, a second homogenizing valve is frequently placed immediately after the first with an applied back pressure of 3.4 MPa (500 psi gauge), allowing more time for surface adsorption to occur. An aging time of 4 h or greater is recommended following mix processing prior to freezing to produce a product with a smoother texture and better quality. The temperature of the mix should be maintained as low as possible without freezing (≤4°C). Aging permits hydration of milk proteins and stabilizers (some increase in viscosity occurs during aging), crystallization of the fat and rearrangement of the membranes. The appropriate ratio of solid:liquid fat must be attained at this stage, a function of temperature and the triglyceride composition of the fat used, as a partially crystalline emulsion is needed for partial coalescence during the whipping and freezing steps (van Boekel and Walstra, 1981; Barfod *et al.*, 1991; Boode *et al.*, 1993). Emulsifiers generally displace milk proteins from the fat surface during the aging period (Goff *et al.*, 1987; Goff and Jordan, 1989; Barfod *et al.*, 1991; Dickinson and Tanai, 1992); this is discussed in Section 24.5.1.

(b) Freezing

Freezing of ice cream also occurs in two distinct stages: passing the mix after aeration through a swept-surface heat exchanger, jacketed with a liquid, boiling refrigerant, under high shear conditions to promote extensive ice crystal nucleation and air bubble formation (dynamic freezing), and freezing (hardening) the ice cream quiescently, after the addition of particulate ingredients and packaging, under conditions which promote rapid freezing and the formation of small ice crystals (−30°C or colder, using either forced convection or plate-type conduction freezers) (Marshall and Arbuckle, 1996; Berger, 1997). The dynamic freezing and whipping process is one of the most important unit operations for the development of quality and yield of finished product, due to the incorporation of air creating the foam, the formation of the ice phase, and the destabilization of the fat emulsion due to freeze-concentration of the emulsion, air incorporation and high shear (Goff, 1997b). One objective of ice cream manufacture is to produce ice crystals that are below, or at least not significantly above, the threshold of sensory detection at the time of consumption. This threshold has been suggested by Marshall and Arbuckle (1996) to be 40 to 50 μm. Consequently, the freezing steps of the manufacturing process and the temperature profile throughout the distribution system are critical factors in meeting this objective (Hartel, 1996). Following rapid hardening, the ice cream should be stored at a low, constant temperature, usually −25°C.

24.3 Sources of milk proteins for ice cream

24.3.1 Whole milk protein products

There is a large variety of potential sources of milk proteins for ice cream, either as a mixture of milk SNF which also includes variable levels of lactose and ash components, or in a pure form. Table 24.2 is a list of potential

Table 24.2
References on the suitability of various sources of milk proteins in ice cream formulations

Category and ingredient	*Reference*
Whole milk proteins	
Skim milk powder	Rothhwell (1984); Stampanoni-Koeferli (1996)
Condensed skim or whole milk	Rothwell (1984); Garcia *et al.* (1995)
Freeze-concentrated non-fat milk solids	Garcia *et al.* (1995)
Lactose-hydrolyzed milk solids	Rossi *et al.* (1999)
Whole milk ultrafiltration retentates	Lee and White (1991)
Buttermilk: condensed or dry	Marshall and Arbuckle (1996); Tirumalesha and Jayaprakasha (1998)
Casein proteins	
Sodium caseinate	Parsons *et al.* (1985); Goff *et al.* (1989); Schmidt (1994) Westerbeek (1996); Segall and Goff (1999)
Chymosin-modified casein micelles	Chang *et al.* (1995)
Whey proteins	
Whey solids	Rothwell (1984); Smith *et al.* (1984); Parsons *et al.* (1985)
Delactosed, demineralized whey solids	Addesso and Kleyn (1986); Westerbeek (1996)
Lactose hydrolyzed whey solids	Westerbeek (1996); Morr and Barrantes (1998)
Whey protein concentrate	Huse *et al.* (1984); Lee and White (1991); Hofi *et al.* (1993); Schmidt *et al.* (1993)
Succinylated whey protein concentrate	Thompson *et al.* (1983)
Whey protein isolate	Goff *et al.* (1989); Schmidt (1994); Segall and Goff (1999)
Hydrolyzed whey protein isolate	Segall and Goff (1999)

ingredients and reference citations for literature which includes use of that specific ingredient in ice cream. Whole milk protein blends contain both caseins and whey proteins, and this category includes most of the traditional sources of milk SNF, typically condensed or dry skim milk, as discussed in Section 24.2.2(*a*) "Mix Manufacture." However, most ice cream formulations now use another source or sources of milk SNF or milk protein to replace some or all of skim milk solids, for both functional and economic reasons (Bevan, 1985; Westerbeek, 1996). When assessing replacements for skim milk solids, an important consideration is the levels of protein, lactose and ash in the ingredients being assessed. Lactose is not very sweet and not very soluble, and therefore during the freezing of ice cream, it is freeze-concentrated beyond its solubility and thus, is potentially prone to crystallization. Lactose crystals are very undesirable in ice cream, causing the defect known as sandiness. Lactose, being a disaccharide, also contributes to the freezing point depression in the mix, so its concentration must be controlled closely. The milk salts also affect both the flavour and texture of ice cream. When replacing skim milk solids, sufficient total solids must be added to limit the water content of the mix and meet legal minimum total solids requirements. For these reasons, it is often desirable to find a skim milk replacement with similar concentrations of lactose and protein. Lactose can be reduced through ultrafiltration or modified by limited hydrolysis to its constituent monosaccharides; either change will affect the concentration of the ingredient that can be used and the subsequent protein level achieved in the ice cream. Buttermilk solids have often been cited as a useful substitute for skim milk solids. Buttermilk contains a higher concentration of fat globule membrane phospholipids than skim milk and therefore it can be used for its emulsifying properties to reduce the need for emulsifiers, or in formulations where it is undesirable to add emulsifiers (Marshall and Arbuckle, 1996).

24.3.2 Casein products

It is possible to produce concentrated protein products from the caseins of milk, the most common for use as a food ingredient being sodium caseinate. The use of sodium caseinate in ice cream has been investigated, and a small percentage may be useful in contributing to functional properties, particularly aeration and emulsification (Parsons *et al.*, 1985; Goff *et al.*, 1989). However, the functionality of sodium caseinate is different than that of micellar casein, which needs to be considered when proposing its use. It can contribute positively to aeration, but may lead to an emulsion that is too stable to undergo the required degree of partial coalescence. It is therefore more desirable in the serum phase rather than at the fat interface.

24.3.3 Whey protein products

A great deal of attention has been focussed on the use of whey products in ice cream. Whey contains fat, lactose, whey proteins and water but no casein. While skim milk powder contains ~55% lactose and ■ 36% protein, whey powder contains 72–73% lactose and only about 10–12% protein. Thus, it can aggrevate some of the problems associated with high lactose. However, an increasing number of whey products are available that have a higher protein and a lower lactose content, usually processed by membrane technology. Many of these can give much higher quality products than traditional whey ingredients (Goff *et al.*, 1989). Whey protein concentrates with levels of protein and lactose similar to skim milk can be produced. Protein content can vary from 20–25% to 75% or more. In addition, the level of lactose can be modified by hydrolysis, although the freezing point depression effect of the higher monosaccharide content must be considered. The ash content can be reduced by demineralization. Whey protein isolates, which contain no lactose, are also available for blending with other ingredients to provide the SNF content of ice cream formulations.

24.4 Structure of ice cream

The texture of ice cream is one of its most important quality attributes. It is the sensory manifestation of structure; thus, establishment of the optimum structure is critical to maximal textural quality in ice cream. An understanding of the functional role of proteins in ice cream also depends on conceptualization of structure. The structure of ice cream begins with the mix as a simple emulsion, with a discrete phase of partially crystalline fat globules surrounded by an interfacial layer comprised of proteins and surfactants (Figure 24.2A, B). The continuous, serum phase consists of the unadsorbed casein micelles in suspension in a solution of sugars, unadsorbed whey proteins, salts and high molecular weight polysaccharides. Ice cream is a complex food colloid in that the mix emulsion is subsequently foamed, creating a dispersed phase of air bubbles and is frozen, forming another dispersed phase of ice crystals (Figure 24.3A, B). Air bubbles and ice crystals are usually in the range of 20 to 50 μm (Caldwell *et al.*, 1992). The serum phase is freeze-concentrated. In addition, the partially-crystalline fat phase at refrigerated temperatures undergoes partial coalescence during the concomitant whipping and freezing process, resulting in a network of agglomerated fat, which partially surrounds the air bubbles and gives rise to a solid-like structure (Figure 24.4) (Kalab, 1985; Goff and Jordan, 1989; Boode and Walstra, 1993; Goff, 1997c). Given this context, the functional role of protein can be examined, considering their behaviour at the fat interface, the air interface and in the serum phase.

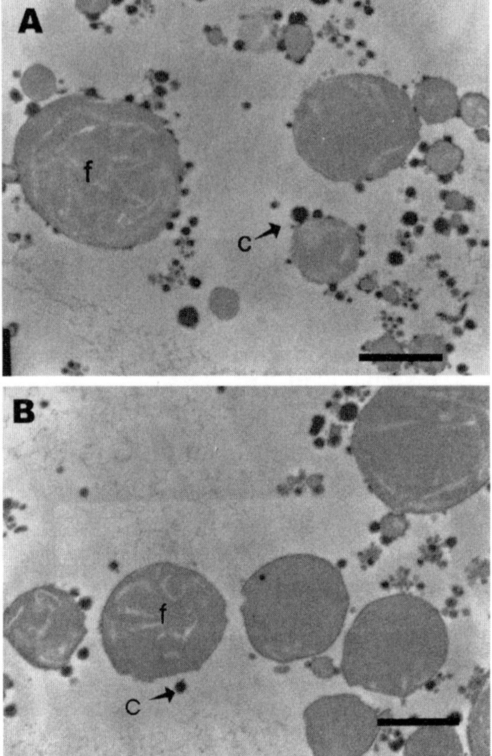

Figure 24.2 Transmission electron micrographs showing the structure of fat and protein in ice cream mix in the absence (A) or presence (B) of emulsifier; f = fat, c = casein micelle, bar = 1 μm. For preparation method, see Goff *et al.* (1987).

24.5 Functional roles of milk proteins in ice cream

24.5.1 Emulsification

The interfacial behaviour of milk proteins in emulsions is well documented, as is the competitive displacement of proteins by small molecule surfactants (Courthadon *et al.*, 1991; Chen and Dickinson, 1993; Dalgleish *et al.*, 1995; Euston *et al.*, 1995, 1996). In ice cream, the emulsion must be stable to withstand mechanical action in the mix state, but must undergo sufficient partial coalescence to establish desirable structural attributes when frozen. These include a dryness at extrusion for fancy molding, slowness of melting, some degree of shape retention during melting, and smoothness during consumption. This implies the use of small molecule surfactants (emulsifiers)

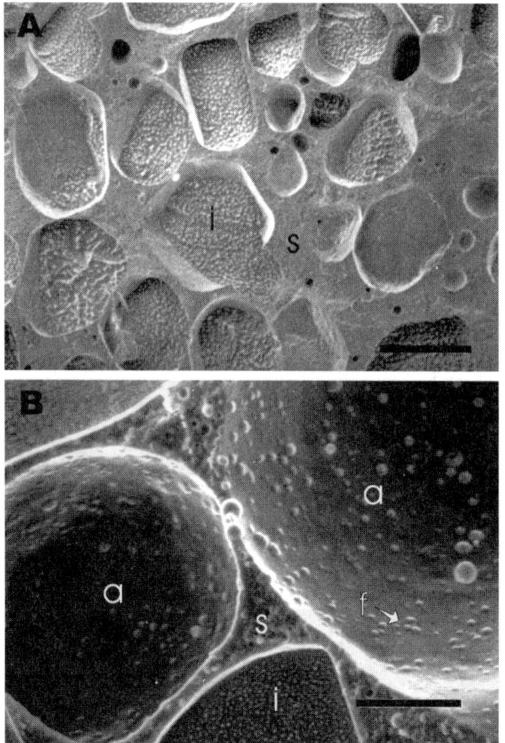

Figure 24.3 Cryo-scanning electron micrographs of ice cream structure. (A) Low magnification showing region dominated by ice crystals. Bar = 30 μm. (B) Higher magnification showing close-up of the interior of air bubble and fat globules. Bar = 10 μm, a = air bubble, f = fat globule, i = ice crystal, s = unfrozen serum phase, For preparation method, see Caldwell *et al.* (1992).

to reduce protein adsorption and produce a weak fat membrane that is sensitive to shear action (Goff *et al.*, 1987, 1989; Goff and Jordan, 1989; Gelin *et al.*, 1994, 1996a,b; Goff, 1997b,c; Campbell and Pelan, 1998). Figure 24.2 demonstrates the action of emulsifiers in ice cream by showing protein adsorption to fat globules in the absence of emulsifier (Figure 24.2A), in which considerable casein micelle adsorption can be seen, or in the presence of emulsifier (Figure 24.2B), which shows little or no adsorption of casein. The reduced steric stability of the globule, which was contributed from the adsorption of micelles accounts for its greater propensity for partial coalescence during shearing. Partial coalescence is responsible for establishing a three-dimensional aggregation of fat globules that provide structural integrity (Figure 24.4). This is especially important if such integrity is needed when the structural contribution from ice is weaker (i.e., before hardening or during melting). Variables that affect the

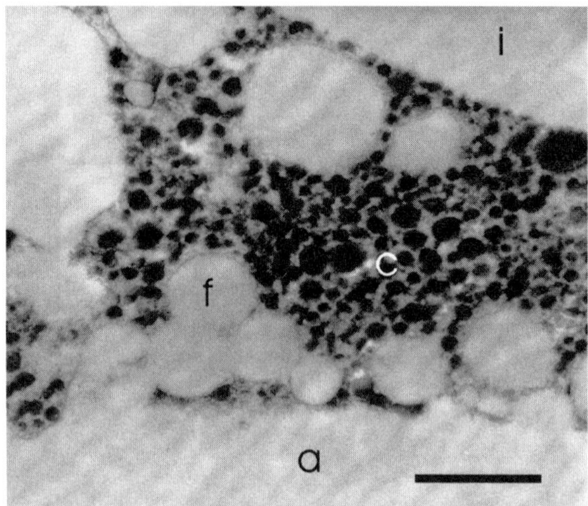

Figure 24.4 A transmission electron micrograph prepared by freeze substitution showing the structure of partially coalesced fat, the freeze-concentration of casein micelles in the unfrozen phase and the interaction of fat at the air interface; a = air bubble, c = casein micelle, f = fat globule, i = ice crystal, bar = 1 μm. For preparation method, see Goff *et al.* (1999a).

destabilization of fat in ice cream have been well studied (Boode and Walstra, 1993; Sakurai *et al.*, 1996; Kokubo *et al.*, 1996, 1998).

With respect to protein contribution to the integrity of fat globules, it is obvious from the studies to date that a weak surface layer is most desirable. Segall and Goff (1999) examined the susceptibility of ice cream emulsions to partial coalescence during shearing when the emulsion was prepared with varying concentration and type of protein, while still retaining sufficient quiescent emulsion stability. The membranes of fat globules stabilized by whey protein isolate were more susceptible than those made from sodium caseinate or casein micelles, while those made from partially hydrolyzed whey proteins did not have sufficient quiescent emulsion stability. However, when casein was added to the whey protein-stabilized emulsion, after homogenization, further adsorption of casein to the whey protein membrane was rapid. Nevertheless, an understanding of protein structures and protein:surfactant interactions at the fat interface may lead to better control over the extent of partial coalescence desirable in the finished product.

24.5.2 Aeration

Milk proteins are well known for their foaming properties (Kinsella *et al.*, 1988) and during the manufacture of ice cream, air is incorporated to about 50% phase volume. Thus, it should not be surprising that milk proteins contribute to stabilizing the air interface in ice cream. Continuous air interfaces in ice cream can always been seen by scanning electron microscopy (Figure 24.3B). During sublimation, these interfaces remain intact, suggesting that a mix of protein, emulsifier and fat forms a continuous layer separating the air bubble from direct contact with ice. This air interface is very important for overall structure and structural stability (Turan *et al.*, 1999). Loss of air can lead to a defect known as shrinkage, the occurrence of which is fairly common and very significant for the loss of quality and acceptability of the product (Dubey and White, 1997). Brooker *et al.* (1986) and Anderson *et al.* (1987) described the process of whipping heavy cream to include the initial adsorption of protein at the air interface and the subsequent adsorption of fat globules and their associated membrane to the existing protein air bubble membrane. Adsorption of globular fat at air interfaces is known to stabilize air bubbles against rapid collapse (Pilhofer *et al.*, 1994; Stanley *et al.*, 1996). Proteins at the fat interface have also been shown to play an important role during the aeration of emulsions (van Camp *et al.*, 1996). However, the actual contribution of protein to the aeration of ice cream and its interaction with both the added emulsifying agents (which are also surface active) and partially coalesced fat at the air interface has been studied less well. Incorporation of air into ice cream is rapid, within seconds, and at the same time, the viscosity of the surrounding matrix is increasing exponentially due to freezing, such that air bubbles after formation become physically entrapped in a semi-solid matrix, making their collapse quite difficult.

Goff *et al.* (1999a) examined air interfaces in ice cream and fat-air interactions using transmission electron microscopy with freeze-substitution. The structures created by increasing levels of fat destabilization in ice cream (achieved through increased emulsifier concentration in the mix and batch versus continuous freezing) were observed as an increasing concentration of discrete fat globules at the air interface (as in Figure 24.3B) and increasing coalescence and clustering of fat globules both at the air interface and within the serum phase (Figure 24.4). Air interfaces at the highest level of fat destabilization were not covered completely by fat globules. It has been suggested that the air interface in ice cream may be covered by a thin layer of non-globular liquid fat (Berger, 1997). However, there was no evidence of a surface layer of free fat in the work of Goff *et al.* (1999a). Further, air interfaces in a fat-free ice cream formulation (Figure 24.5) showed a continuous membrane, very similar to those in a formulation containing fat, offering further evidence that the air bubble membrane itself is comprised of protein, with discrete and partially-coalesced fat globules adsorbed subsequently.

24.5.3 Solution behaviour

Similar to the role of milk protein in aeration, their role in the unfrozen aqueous phase is recognized but less well studied than their role at the fat interface. Milk proteins interact with water and the subsequent hydration is responsible for a variety of functional properties, including rheological behaviour (Kinsella *et al.*, 1988). Thus, freeze-concentration of proteins in ice cream must lead to a sufficient concentration to have a large impact on the viscosity of the unfrozen phase and its subsequent effect on ice crystallization, ice crystal stability and solute mobility (Flores and Goff, 1999). Jonkman *et al.* (1998), who studied the effect of ice cream manufacture on the structure of casein micelles, found that the micelles *per se* were not affected by the process. Although the stability of the micelles was expected to be affected by low temperature, this was offset by an increasing concentration of milk salts in solution during freeze-concentration, such that the micelles remained intact (Figure 24.4) in a similar state to that found in mix.

Polysaccharides are also added to the ice cream mix to enhance its viscosity and to impact on ice crystallization behaviour. Commonly used polysaccharides can be incompatible in solution with milk proteins, leading to a microscopic or macroscopic phase separation (Antipova and Semenova, 1995; Syrbe *et al.*, 1998), a phenomenon that has been studied in milk and ice cream-type systems (Garnier *et al.*, 1995; Bourriot *et al.*, 1999; Schorsch

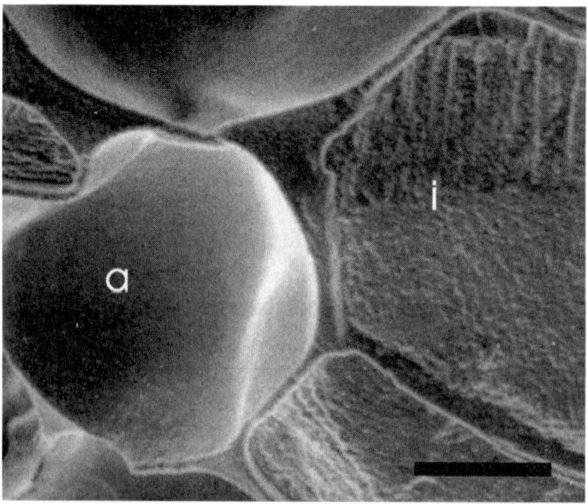

Figure 24.5 A cryo-scanning electron micrograph of the air interface in a fat-free ice cream; a = air bubble, i = ice crystal, bar = 15 µm. For preparation method, see Caldwell *et al.* (1992).

et al., 1999a,b). Goff *et al.* (1999b), who examined the interaction between milk proteins and polysaccharides in frozen systems using labelled polysaccharides and fluorescence microscopy, demonstrated a clear phase separation between the two, leading to discernable networks, of both locust bean gum and milk proteins, created by freezing (Figure 24.6). The same phenomenon can be seen by transmission electron microscopy of ice cream, where it can be seen that when in solution with polysaccharides, the casein aggregates into distinct networks (Figure 24.7). Flores and Goff (1999) demonstrated that milk proteins had a large impact on the size and stability of ice crystals. It thus appears that microscopic phase separation of the milk proteins induced by polysaccharides, and "aggregation" of casein into a weak gel-like network, promoted also by freeze-concentration, may be at least partly responsible for the stability of ice crystals and for the body and texture of the ice cream during consumption.

24.6 Protein analysis in ice cream

Total protein, casein and whey protein can be determined in ice cream with a variety of methods. The traditional measurement of nitrogen by Kjeldahl analysis is the standard technique for total protein in dairy products, with correction for non-protein nitrogen. Rapid instrumental methods based on

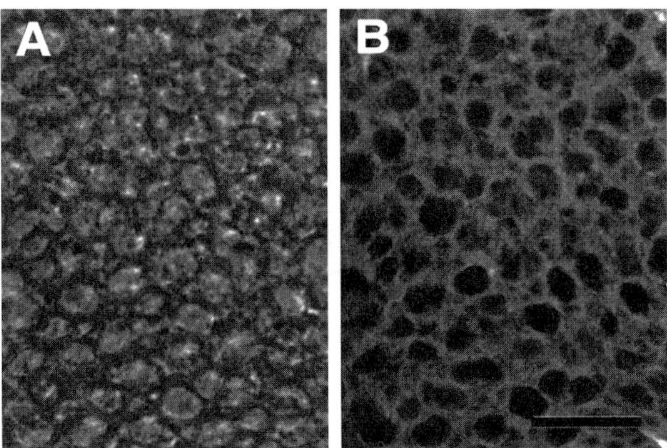

Figure 24.6 Confocal scanning laser micrographs of frozen and melted non-fat ice cream mix, showing phase separation of protein and locust bean gum (tagged with a fluorescent marker). The structure resulted from the formation of ice; however, melting has caused the ice to disappear but the protein and polysaccharide structure to remain. (A) Phase contrast image, showing the structure that results from cryo-aggregated protein. (B) Fluorescent image from the same field as A, showing the structure that results from cryo-gelled, fluorescent-labeled locust bean gum. Bar = 50 μm. For preparation method, see Goff *et al.* (1999b).

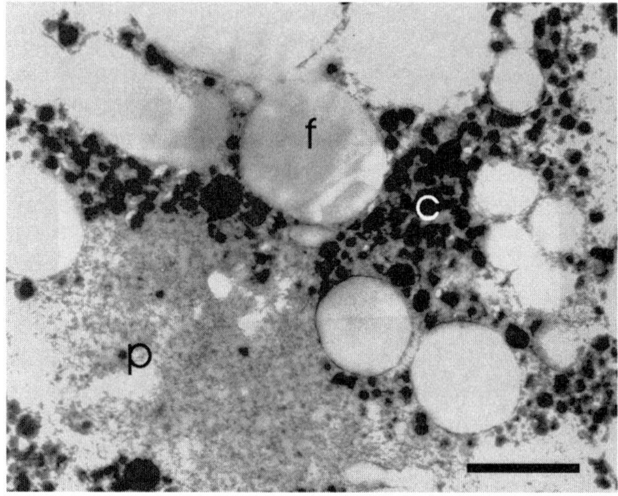

Figure 24.7 A transmission electron micrograph prepared by freeze substitution showing the phase separation and aggregation of casein micelles resulting from the combination of partial coalescence of the fat and addition of polysaccharide; f = fat globule, c = casein micelle, p = polysaccharide network devoid of casein micelles, bar = 1 μm. For preparation method, see Goff *et al.* (1999a).

infra-red analysis are also readily available. In addition to these methods there are several reports in the literature on specific techniques developed for protein analysis and the casein:whey protein ratio in ice cream. These include a method for determining casein and whey protein based on phosphorus:nitrogen ratio and radial immunodiffusion (Douglas *et al.*, 1982), methods for determination of skim milk solids:whey solids ratio by Kjeldahl nitrogen and formol titration (Peeples and Heath, 1979; Arora and Bhatia, 1990) and modified dye binding methods (Kroger *et al.*, 1978; Bruhn *et al.*, 1980).

24.7 Conclusion

Ice cream is a complex food colloid, with at least three discrete phases (ice, air and fat) and a continuous unfrozen aqueous phase. Proteins contribute several important functional roles to ice cream that can be divided into three categories: emulsification, aeration and solution behaviour. They also contribute to the sensory properties (flavour and body) of the product in ways that are exclusive of any of these physical contributions. Several choices are available to manufacturers for their source of milk proteins, including caseins, whey proteins, and fractions and blends of each, with or without lactose and minerals. It should be obvious that the selection of an ingredient to supply milk proteins should be made on the basis of the extent to which that ingredient is able to meet all the expectations (functional roles) placed on it. Future research related to proteins in ice cream will undoubtedly be required either to examine new types of milk protein products that are or have the potential to be available commercially (fractions, modifications, etc.), or to examine new manufacturing techniques which can modify existing protein sources to better deliver the functional demands placed on them, as these functional contributions become more fully elucidated.

References

Addesso, K.M. and Kleyn, D.H. (1986) Development of an acceptable ice cream possessing a reduced sodium content. *J. Food Sci.*, **51**, 1467–70.

Anderson, M., Brooker, B.E. and Needs, E.G. (1987) The role of proteins in the stabilization/destabilization of dairy foams, in *Food Emulsions and Foams*, (E. Dickinson ed.) Royal Society of Chemistry, London, pp. 100–9.

Antipova, A.S. and Semenova, M.G. (1995) Effect of sucrose on the thermodynamic incompatibility of different biopolymers. *Carbohydrate Polymers*, **28**, 359–65.

Arora, K.L. and Bhatia, K.L. (1990) Formol titration method for determining protein content in ice cream. *Ind. J. Animal Sci.*, **60**, 1252–5.

Barfod, N.M., Krog, N., Larsen, G. and Buchheim, W. (1991) Effects of emulsifiers on protein-fat interaction in ice cream mix during ageing. 1: Quantitative analyses. *Fett-Wissenschaft-Technologie*, **93**, 24–35.

Berger, K.G. (1997) Ice cream, in *Food Emulsions*, 3rd edn, (S.E. Friberg and K. Larsson eds.) Marcel Dekker Inc., New York, pp. 413–90.

Bevan, D. (1985) The use of milk products in ice cream manufacture. *Food, Flavourings, Ingredients, Packaging and Processing*, **7(8)**, 31–2.

Boode, K., Walstra, P. and de Groot-Mostert, A.E.A. (1993) Partial coalescence in oil-in-water emulsions. 2. Influence of the properties of the fat. *Colloid Surface A*, **81**, 139–51.

Boode, K. and Walstra, P. (1993) Partial coalescence in oil-in-water emulsions. 1. Nature of the aggregation. *Colloid Surface A*, **81**, 121–37.

Bourriot, S., Garnier, C. and Doublier, J.L. (1999) Phase separation, rheology and microstructure of micellar casein-guar gum mixtures. *Food Hydrocoll.*, **13**, 43–9.

Brooker, B.E., Anderson, M. and Andrews, A.T. (1986) The development of structure in whipped cream. *Food Microstruct.*, **5**, 277–85.

Bruhn, J.C., Pecore, S. and Franke, A.A. (1980) Measuring protein in frozen dairy desserts by dye binding. *J. Food Prot.*, **43**, 753–5.

Caldwell, K.B., Goff, H.D. and Stanley, D.W. (1992) A low-temperature scanning electron microscopy study of ice cream. 1. Techniques and general microstructure. *Food Struct.*, **11**, 1–9.

Campbell, I.J. and Pelan, B.M.C. (1998) The influence of emulsion stability on the properties of ice cream, in *Ice Cream*, (W. Buchheim ed.) Special Issue 9803, International Dairy Federation, Brussels, pp. 25–36.

Chang, J.L., Marshall, R.T. and Heymann, H. (1995) Casein micelles partially hydrolyzed by chymosin to modify the texture of lowfat ice cream. *J. Dairy Sci.*, **78**, 2617–23.

Chen, J. and Dickinson, E. (1993) Time-dependent competitive adsorption of milk proteins and surfactants in oil in water emulsions. *J. Sci. Food Agric.*, **62**, 283–9.

Courthaudon, J.-L., Dickinson, E. and Dalgleish, D.G. (1991) Competitive adsorption of β-casein and nonionic surfactants in oil in water emulsions. *J. Colloid Interface Sci.*, **145**, 390–5.

Dalgleish, D.G., Srinivasan, M. and Singh, H. (1995) Surface properties of oil-in-water emulsion droplets containing casein and Tween 60. *J. Agric. Food Chem.*, **43**, 2351–5.

Dickinson, E. and Tanai, S. (1992) Temperature dependence of the competitive displacement of protein from the emulsion droplet surface by surfactants. *Food Hydrocoll.*, **6**, 163–71.

Douglas, F.W., Jr., Tobias, J., Groves, M.L., Farrell, H.M., Jr. and Edmondson, L.F. (1982) Quantitative determination of total protein, casein, and whey protein of processed dairy products. *J. Dairy Sci.*, **65**, 339–45.

Dubey, U.K. and White, C.H. (1997) Ice cream shrinkage. *J. Dairy Sci.*, **80**, 3439–44.

Euston, S.E., Singh, H., Munro, P.A. and Dalgleish, D.G. (1995) Competitive adsorption between sodium caseinate and oil-soluble and water-soluble surfactants in oil-in-water emulsions. *J. Food Sci.*, **60**, 1151–6.

Euston, S.E., Singh, H., Munro, P.A. and Dalgleish, D.G. (1996) Oil-in-water emulsions stabilized by sodium caseinate or whey protein isolate as influenced by glycerol monostearate. *J. Food Sci.*, **61**, 916–20.

Flores, A.A. and Goff, H.D. (1999) Ice crystal size distributions in dynamically frozen model solutions and ice cream as affected by stabilizers. *J. Dairy Sci.*, **82**, 1399–407.

Garcia, R.S., Marshall, R.T. and Heymann, H. (1995) Low-fat ice creams from freeze-concentrated versus heat-concentrated nonfat milk solids. *J. Dairy Sci.*, **78**, 2345–51.

Garnier, C., Schorsch, C. and Doublier, J.L. (1995) Phase separation in dextran/locust bean gum mixtures. *Carbohydrate Polymers*, **28**, 313–7.

Gelin, J.-L., Poyen, L., Courthadon, J.-L., Le Meste, M. and Lorient, D. (1994) Structural changes in oil-in-water emulsions during the manufacture of ice cream. *Food Hydrocoll.*, **8**, 299–308.

Gelin, J.-L., Poyen, L., Rizzotti, R., Le Meste, M., Courthadon, J.-L. and Lorient, D. (1996a) Interactions between food components in ice cream. Part 1. Unfrozen emulsions. *Food Hydrocoll.*, **10**, 385–93.

Gelin, J.-L., Poyen, L., Rizzotti, R., Dacremont, C., Le Meste, M. and Lorient, D. (1996b) Interactions between food components in ice cream. Part 2. Structure-texture relationships. *J. Text. Stud.*, **27**, 199–215.

Goff, H.D. (1997a) Colloidal aspects of ice cream – a review. *Int. Dairy J.*, **7**, 363–73.

Goff, H.D. (1997b) Ice cream, in *Lipid Technologies and Applications*, (F.D. Gunstone and F.B. Padley eds.) Marcel Dekker Inc., New York, pp. 329–54.

Goff, H.D. (1997c) Instability and partial coalescence in dairy emulsions. *J. Dairy Sci.*, **80**, 2620–30.

Goff, H.D., Liboff, M., Jordan, W.K. and Kinsella, J.E. (1987) The effects of polysorbate 80 on the fat emulsion in ice cream mix: evidence from transmission electron microscopy studies. *Food Microstruct.*, **6**, 193–8.

Goff, H.D. and Jordan, W.K. (1989) Action of emulsifiers in promoting fat destabilization during the manufacture of ice cream *J. Dairy Sci.*, **72**, 18–29.

Goff, H.D., Kinsella, J.E. and Jordan, W.K. (1989) Influence of various milk protein isolates on ice cream emulsion stability. *J. Dairy Sci.*, **72**, 385–97.

Goff, H.D., Verespej, E. and Smith, A.K. (1999a) A study of fat and air structures in ice cream. *Int. Dairy J.*, **9**, 785–97.

Goff, H.D., Ferdinando, D. and Schorsch, C. (1999b) Fluorescence microscopy to study galactomannan structure in frozen sucrose and milk protein solutions. *Food Hydrocoll.*, **13**, 353–64.

Hartel, R.W. (1996) Ice crystallization during the manufacture of ice cream. *Trends Food Sci. Technol.*, **7**, 315–21.

Hofi, M., Fayed, A., El Awamry, Z. and Hofi, A. (1993) Substitution of non fat milk solids in ice cream with ultrafiltration whey protein concentrate. *Egypt J. Food Sci.*, **21**, 139–45.

Huse, P.A., Towler, C. and Harper, W.J. (1984) Substitution of non-fat milk solids in ice cream with whey protein concentrate and hydrolyzed lactose. *N.Z. J. Dairy Sci. Technol.*, **19**, 255–61.

Jonkman, M.J., Walstra, P., van Boekel, M.A.J.S. and Cebula, D.J. (1998) Behaviour of casein micelles under conditions comparable to those in ice cream, in *Ice Cream*, (W. Buchheim ed.) Special Issue 9803, International Dairy Federation, Brussels, p. 179. (1 page)

Kalab, M. (1985) Microstructure of dairy foods. 2. Milk products based on fat. *J. Dairy Sci.*, **68**, 3234–48.

Kinsella, J.E., Whitehead, D.M., Brady, J. and Bringe, N.A. (1988) Milk proteins: possible relationships of structure and function, in *Developments in Dairy Chemistry. 4. Functional Milk Proteins*, (P.F. Fox ed.) Applied Science Publishers, London, pp. 55–95.

Kokubo, S., Sakurai, K., Hakamata, K., Tomita, M. and Yoshida, S. (1996) The effect of manufacturing conditions on the de-emulsification of fat globules in ice cream. *Milchwissenschaft*, **51**, 262–5.

Kokubo, S., Sakurai, K., Iwaki, S., Tomita, M. and Yoshida, S. (1998) Agglomeration of fat globules during the freezing process of ice cream manufacturing. *Milchwissenschaft*, **53**, 206–9.

Krog, N. (1998) The use of emulsifiers in ice cream, in *Ice Cream*, (W. Buchheim ed.) Special Issue 9803, International Dairy Federation, Brussels, pp. 37–44.

Kroger, M., Katz, E.E. and Weaver, J.C. (1978) Determining protein content of ice cream and frozen desserts. *J. Dairy Sci.*, **61**, 274–7.

Lee, F.Y. and White, C.H. (1991) Effect of ultrafiltration retentates and whey protein concentrates on ice cream quality during storage. *J. Dairy Sci.*, **74**, 1170–80.

Marshall, R.T. and Arbuckle, W.S. (1996) *Ice Cream*, 5th edn, Chapman and Hall, New York, NY.

Morr, C.V. and Barrantes, L. (1998) Lactose-hydrolyzed cottage cheese whey nanofiltration retentate in ice cream. *Milchwissenschaft*, **53**, 568–72.

Parsons, J.G., Dybing, S.T., Coder, D.S., Spurgeon, K.R. and Seas, S.W. (1985) Acceptability of ice cream made with processed wheys and sodium caseinate. *J. Dairy Sci.*, **68**, 2880–5.

Peeples, M.L. and Heath, G.L. (1979) Use of protein-formol titration relationships for estimating ratios of skimmilk and whey solids in frozen dairy desserts. *J. Food Sci.*, **44**, 558–9.

Pelan, B.M.C., Watts, K.M., Campbell, I.J. and Lips, A. (1997) The stability of aerated milk protein emulsions in the presence of small molecule surfactants. *J. Dairy Sci.*, **80**, 2631–8.

Pilhofer, G.M., Lee, H.C., McCarthy, M.J., Tong, P.S. and German, J.B. (1994) Functionality of milk fat in foam formation and stability. *J. Dairy Sci.*, **77**, 55–63.

Rossi, M., Casiraghi, E., Alamprese, C. and Pompei, C. (1999) Formulation of lactose-reduced ice cream mix. *Ital. J. Food Sci.*, **11**, 3–18.

Rothwell, J. (1984) Uses for dairy ingredients in ice cream and other frozen desserts. *J. Soc. Dairy Technol.*, **37**, 119–21.

Sakurai, K., Kokubo, S., Hakamata, K., Tomita, M. and Yoshida, S. (1996) Effect of production conditions on ice cream melting resistance and hardness. *Milchwissenschaft*, **51**, 451–4.

Schmidt, K. (1994) Effect of milk proteins and stabilizer on ice milk quality. *J. Food. Quality*, **17**, 9–19.

Schmidt, K., Lundy, A., Reynolds, J. and Yee, L.N. (1993) Carbohydrate or protein based fat mimicker effects on ice milk properties. *J. Food Sci.*, **58**, 761–3, 779.

Schorsch, C., Jones, M. and Norton, I.T. (1999a) Thermodynamic incompatibility and microstructure of milk protein/locust bean gum/sucrose systems. *Food Hydrocoll.*, **13**, 89–99.

Schorsch, C., Clark, A.H., Jones, M. and Norton, I.T. (1999b) Behaviour of milk protein/polysaccharide systems in high sucrose. *Colloid Surface B*, **12**, 317–29.

Segall, K.I. and Goff, H.D. (1999) Influence of adsorbed milk protein type and surface concentration on the quiescent and shear stability of butteroil emulsions. *Int. Dairy J.*, **9**, 683–91.

Smith, D.E., Bakshi, A.S. and Lomauro, C.J. (1984) Changes in freezing point and rheological properties of ice cream mix as a function of sweetener system and whey substitution. *Milchwissenschaft*, **39**, 455–7.

Stampanoni-Koeferli, C.R., Piccinali, P. and Sigrist, S. (1996) The influence of fat, sugar and non-fat milk solids on selected taste, flavor, and texture parameters of a vanilla ice cream. *Food Quality Pref.*, **7**, 69–79.

Stanley, D.W., Goff, H.D. and Smith, A.S. (1996) Texture-structure relationships in foamed dairy emulsions. *Food Res. Int.*, **29**, 1–13.

Syrbe, A., Bauer, W.J. and Klostermeyer, H. (1998) Polymer science concepts in dairy systems - an overview of milk protein and food hydrocolloid interaction. *Int. Dairy J.*, **8**, 179–93.

Tirumalesha, A. and Jayaprakasha, H.M. (1998) Effect of admixture of spray dried whey protein concentrate and butter milk powder on physico-chemical and sensory characteristics of ice cream. *Ind. J. Dairy Sci.*, **51**, 13–9.

Tharp, B.W., Forrest, B., Swan, C., Dunning, L. and Himoe, M. (1998) Basic factors affecting ice cream meltdown, in *Ice Cream*, (W. Buchheim ed.) Special Issue 9803, International Dairy Federation, Brussels, pp. 54–64.

Thompson, L.U., Reniers, D.J., Baker, L.M. and Siu, M. (1983) Succinylated whey protein concentrates in ice cream and instant puddings. *J. Dairy Sci.*, **66**, 1630–7.

Tomas, A., Courthadon, J.L., Paquet, D. and Lorient, D. (1994) Effect of surfactant on some physico-chemical properties of dairy oil-in-water emulsions. *Food Hydrocoll.*, **8**, 543–53.

Turan, S., Kirkland, M., Trusty, P.A. and Campbell, I. (1999) Interaction of fat and air in ice cream. *Dairy Ind. Int.*, **64**, 27–31.

van Boekel, M.A.J.S. and Walstra, P. (1981) Stability of oil-in-water emulsions with crystals in the disperse phase. *Colloid Surface*, **3**, 99–107.

van Camp, J., van Calenberg, S., van Oostveldt, P. and Huyghebaert, A. (1996) Aerating properties of emulsions stabilized by sodium caseinate and whey protein concentrate. *Milchwissenschaft*, **51**, 310–5.

Walstra, P. and Jonkman, M. (1998) The role of milkfat and protein in ice cream, in *Ice Cream*, (W. Buchheim ed.) Special Issue 9803, International Dairy Federation, Brussels, pp. 17–24.

Westerbeek, H. (1996) Milk proteins in ice cream. *Dairy Ind. Int.*, **61(6)**, 21, 23–4.

25

ROLE OF PROTEIN IN CHEESE AND CHEESE PRODUCTS

T.P. Guinee

Abbreviations

ACP	analogue cheese product
AHC	acid/heat-coagulated cheese
CB	cheese base,
CCP	colloidal calcium phosphate
CFLSM	confocal laser scanning microscopy
CFR	curd firming rate
CN	casein
DAM	directly acidified Mozzarella
EU	European union
G'	storage or elastic shear modulus
G''	viscous modulus
HCFUF	high concentration factor ultrafiltration
HHT	high heat treatment
LMMC	low-moisture Mozzarella cheese
MACY	moisture-adjusted cheese yield
MFGM	milk fat globule membrane
MNFS	moisture-in-non-fat substances
PCF	pasteurized processed cheese food
PCP	pasteurized processed cheese product
PDWPC	partially denatured whey protein concentrate
RCT	rennet coagulation time
SCC	somatic cell count
SCT	set-to-cut time
SEM	scanning electron microscopy
TBC	total bacterial count
TEM	transmission electron microscopy
UF	ultrafiltration

Advanced Dairy Chemistry Volume 1: Proteins, 3rd edn.
Edited by P.F. Fox and P.L.H. McSweeney, Kluwer Academic/Plenum Publishers, 2003.

WPD whey protein denaturation
Y_a actual cheese yield
δ phase angle

25.1 INTRODUCTION

Cheese, which accounts for ~30% of total milk usage, is a dairy product of major economic importance. World production of cheese is ~15×10^6 tonnes *per annum*, with an estimated value of US\$ 5.5×10^{10}. Approximately 7% of total production is traded on the global market, the major suppliers being the EU (~50%), New Zealand (~16%) and Australia (~11%) (Sørensen, 1997).

Cheese is an extremely versatile product, which may be consumed directly or indirectly as an ingredient in other foods (Figure 25.1). Cheese is a major ingredient in the catering sector, where it is used in an extensive array of applications, including omelettes, quiches, sauces, chicken *cordon bleu* and pasta dishes. Cheese is also used extensively in the industrial food sector for the preparation of ready-to-use grated/shredded cheeses and cheese blends and for the mass production of cheese-based ingredients such as pasteurized processed cheese products (PCPs), cheese powders and enzyme-modified cheeses (EMCs). These ingredients are, in turn, used by the food service industry (such as burger outlets, pizzerias and restaurants) and by the manufacturers of formulated foods such as soups, sauces and ready-prepared meals.

When used as an ingredient in foods, cheese is required to perform one or more functions (Ginzinger, 1995; Market Tracking International, 1998). In

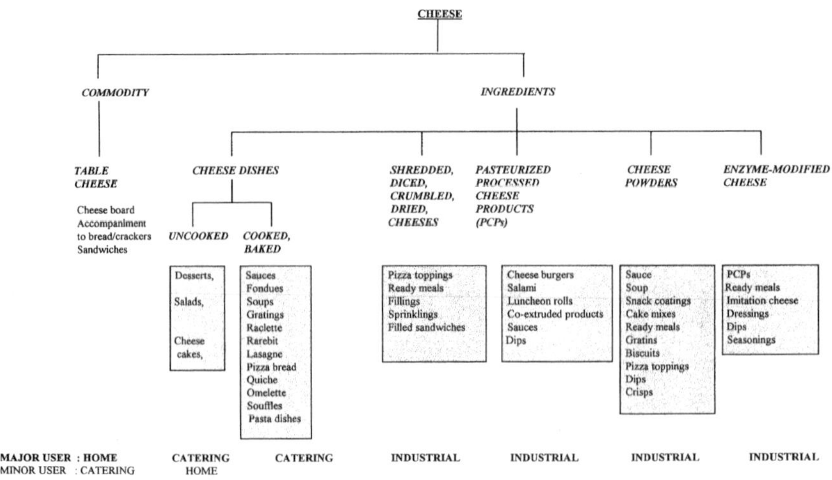

Figure 25.1 Uses of cheese as an ingredient (from Fox *et al.*, 2000).

the unheated state, the cheese may be required to exhibit a number of rheological properties to facilitate its size reduction and use in the preparation of various dishes, e.g., the ability to crumble easily, to slice or to shred cleanly, to bend when in slice form. The rheological properties also determine the textural properties of the cheese during mastication. Cheese is generally required to contribute to the organoleptic characteristics (i.e., taste, aroma and texture) of the food in which it is an ingredient. On grilling or baking, the cheese may be required to melt, flow, brown, oil-off and/or stretch to varying degrees.

Protein, as a major component of most cheese varieties, has a marked influence on the yield of cheese (IDF, 1991) and on its rheological and functional properties (Chen *et al.*, 1979; Kindstedt, 1995; Guinee *et al.*, 2000a). Similarly, the proteins of cheese, along with other added proteins, markedly affect the physico-chemical, rheological, stability and usage appeal characteristics of pasteurized processed cheese products (PCPs) and analogue cheese products (ACPs) (Savello *et al.*, 1989; Abou-El-Nour *et al.*, 1996). The functional properties may be defined as those rheological, physico-chemical, micro-structural and organoleptic (i.e., flavour and texture) characteristics which affect the behaviour of the cheese in food systems during preparation, processing, storage, cooking and/or consumption. Hence, protein has a major influence on cheese quality and on its industrial applications.

In this chapter, some industrial aspects of protein in cheese products, including its contribution to cheese structure, yield, rheology, functionality and to the formation of PCPs and texturized/plasticized curds will be discussed. The gelation of milk protein, in particular casein, by rennet or acid, in the formation of rennet-curd and acid-curd cheeses is discussed in Chapters 18 and 22, respectively; the contribution of protein to cheese flavour has been reviewed (Fox and Wallace, 1997). Protein also contributes to the aeration and foaming characteristics of desserts prepared from fresh cheese and other ingredients. While little information is available on this aspect of cheese, the foaming properties of milk proteins have been reviewed extensively (Stainsby, 1988; Kinsella and Whitehead, 1989; Brooker, 1993; Jana *et al.*, 1994).

25.2 Contribution of protein to cheese structure

Cheese is essentially a concentrated protein gel, which occludes fat and moisture. Gelation is brought about either by:

- acidification (e.g., using starter cultures or food-grade acids and/or acidogens), at a temperature of 20 to 40°C, to the isoelectric pH of casein, i.e., ~4.6,

- sensitisation of the casein to calcium *via* the hydrolysis of the principal micelle-stabilizing casein, κ-CN, by added acid proteinases (i.e., rennets), or
- a combination of acid and heat, e.g., heating milk to ~pH 5.6 at ~90°C (see Chapters 18 and 22).

Para-casein is the principal structural component of the gel in rennet-curd cheeses. At the pasteurization conditions (72°C × 15 s) generally applied to milk for rennet-curd cheese, ≤5% of total whey proteins are denatured and complex with κ-CN (Lau *et al.*, 1990; Fenelon and Guinee, 1999) and are retained in the cheese curd. Casein is the major component of acid-coagulated cheeses. However, a high heat treatment (e.g., 95°C for 2 min) is frequently applied in the manufacture of acid-coagulated cheeses (e.g., Quark and Cream cheese) and acid-heat coagulated cheeses (e.g., Ricotta, Paneer, Mascarpone, some Queso-blanco types) and results in substantial denaturation of whey proteins. The denatured whey proteins interact and complex with the casein micelles and become part of the particulate gel formed on subsequent acidification (van Hooydonk *et al.*, 1987). This interaction has a marked influence on the gel structure and properties of acid- and rennet-curd cheese varieties (Harwalkar and Kalab, 1980, 1988; Guinee *et al.*, 1993; Chapters 18 and 22).

The concentration and type of protein have a major influence on the micro-structure of acid- and rennet-coagulated milk gels, which on dehydration and concentration form the structural fabric of the cheese. The micro-structure of the gel markedly affects its rheological and syneretic properties, the recovery of fat and protein in the cheese and the yield of cheese (Kalab and Harwalkar, 1974; Schafer and Olson, 1975; Harwalkar and Kalab, 1980, 1981; Green *et al.*, 1981, 1983; Marshall, 1986; Banks *et al.*, 1987; Green, 1990a,b; McMahon *et al.*, 1993; Guinee *et al.*, 1995, 1998). The structure of the final cheese influences its rheological, textural and heat-induced functional properties (Emmons *et al.*, 1981; Green *et al.*, 1981, 1990a,b; Korolczuk and Mahaut, 1992; Guinee *et al.*, 1995, 1999, 2000a; McMahon *et al.*, 1996).

25.2.1 Micro-structure of rennet-curd cheeses

The micro-structure of milk gels and cheeses has been studied extensively (Hall and Creamer, 1972; Kalab and Harwalkar, 1974; Kimber *et al.*, 1974; Kalab, 1977; de Jong, 1978a; Green *et al.*, 1981, 1983, 1990; Kiely *et al.*, 1992, 1993; Mistry and Anderson, 1993; Desai and Nolting, 1995; Everett *et al.*, 1995; Bryant *et al.*, 1995; Guinee *et al.*, 1999, 2000a). The physico-chemical properties of the *para*-casein matrix and enclosed components may be deduced from micro-structural observations, compositional analyses and theoretical considerations of the chemistry of the conversion of milk to cheese and partition of components (e.g., milk salts) between the whey and the cheese curd (Walstra and van Vliet, 1986).

Natural rennet-curd cheese is essentially a particulate calcium phosphate *para*-casein matrix, composed of interconnected and overlapped chains of partially fused *para*-casein aggregates (in turn formed from fused *para*-casein micelles). The integrity of the matrix is maintained by various intra- and inter-aggregate hydrophobic and electrostatic attractions. In young cheese, the matrix has an 'internal' structure consisting of a relatively loose network of clearly recognisable particles (*para*-casein micelles and aggregates of *para*-casein micelles) which are in contact with neighbouring particles over part of their surfaces. On-going fusion of *para*-casein particles during maturation leads to a gradual reduction in the extent of internal matrix structure, as reflected by the disappearance of interparticle boundaries and the formation of a more homogeneous mass (Kimber *et al.*, 1974; de Jong, 1978a).

The *para*-casein network is essentially continuous, extending in all directions, although some discontinuities in the matrix at the micro- and macro-structural levels exist. Micro-structural observations made using transmission electron microscopy (TEM) suggest that the hydrolysis of *para*-casein (e.g., by rennet activity) to water-soluble peptides results in parts of the matrix losing contact with the main *para*-casein network, an occurrence which leads to discontinuities or 'breaks' in the *para*-casein matrix at the micro-structural level (de Jong, 1978a). Hence, it is noteworthy that ageing of Mozzarella for 50 d resulted in the degradation of 50% of α_{s1}-CN to α_{s1}-CN (f24–199) and an increase in the porosity of the defatted *para*-casein matrix, as observed using scanning electron microscopy (SEM) (Kiely *et al.*, 1993). Discontinuities at the macro-structural level exist in the form of curd granule junctions or curd chip junctions (in Cheddar and related dry-salted varieties) (Kalab, 1979; Lowrie *et al.*, 1982; Paquet and Kalab, 1988). Curd granule junctions in low-moisture Mozzarella are well defined, ~3–5 μm wide, and appear as veins running along the perimeters of neighbouring curd particles (Kalab, 1977). Unlike the interior of the curd particles, the junctions are comprised mainly of casein, being almost devoid of fat. Factors which contribute to the formation of these junctions include leaching of the fat from the surface of the curd particles and dehydration of surface protein during the cutting, acidification, cooking and pressing stages of curd manufacture. Chip junctions in Cheddar and related dry-salted varieties are clearly discernible on examination of the cheese by light microscopy and, like curd granule junctions, have a higher casein-to-fat ratio than the interior of the chips. The difference in cheese composition at junctions, compared to the interior of the curd particles, probably leads to differences in the molecular attractions between contiguous *para*-casein layers in the interior and exterior of curd particles, and thus to differences in structure-function relationships.

The matrix occludes, within its pores, fat globules (in varying degrees of coalescence), moisture and its dissolved solutes (minerals, lactic acid, peptides and amino acids), microorganisms and enzymes (e.g., residual rennet, proteinases and peptidases from starter and non-starter microorganisms) (Kimber *et al.*, 1974; Laloy *et al.*, 1996; Guinee *et al.*, 2000a). Clumping and coalescence of fat globules occur during manufacture, due to the combined effects of shear stress on the fat globule membrane and the shrinkage of the surrounding *para*-casein matrix which forces the occluded globules into close contact. Concentration and shrinkage of the *para*-casein matrix occurs as a result of its dehydration during the various stages of cheesemaking, e.g., acidification, heating and pressing. In the temperature range used for cheesemaking (~30–55°C) most, or all, of the milk fat is liquid (Norris *et al.*, 1973) and therefore flows on the application of stress. Evidence for fat clumping is provided by scanning electron micrographs which show fissures, or irregularly-shaped openings, in the *para*-casein matrix, which remain after removal of fat during sample preparation (Bryant *et al.*, 1995; Mistry and Anderson, 1993; Figure 25.2). The frequency of these fissures decreases as the fat content is reduced, e.g., from 33.2 to 8.2%, w/w, fat (Mistry and Anderson, 1993).

The cheese matrix also contains starter and non-starter bacteria, which attach themselves, *via* filaments from their cell walls, to the casein matrix (Kimber *et al.*, 1974) and are concentrated near the fat-casein interface (Laloy *et al.*, 1996). Bacterial proteinases and intracellular peptidases, released on autolysis, contribute to the hydrolysis of casein (Fox *et al.*, 1996). In addition, the surface layer of the cheese may be colonised by various microorganisms, including bacteria, moulds and yeasts, which release enzymes (e.g., proteinases, peptidases, lipases) into the outer layer of the cheese matrix at various rates during maturation (Gripon, 1993; Reps, 1993). The enzymes released into the surface layer of the cheese diffuse into the cheese only very slightly owing to their large molecular weight (Lee *et al.*, 1980; Noomen, 1983). However, it is likely that water-soluble products of enzyme hydrolysis (water-soluble fatty acids, low molecular mass peptides and amino acids) and other compounds (e.g., lactic acid and soluble calcium) diffuse through the cheese mass (Lee *et al.*, 1980) and affect zonal variations in pH, *para*-casein hydration, rheology, functionality and flavour (Noomen, 1983; Karahadian and Lindsay, 1987; Gripon, 1993).

Various physico-chemical changes occur in the structural components of the *para*-casein matrix during maturation; these changes are mediated by the residual rennet, microorganisms and their enzymes, and changes in mineral equilibrium between the serum and *para*-casein matrix. The type and level of physico-chemical changes depend on the cheese variety, cheese composition and ripening conditions. These may include:

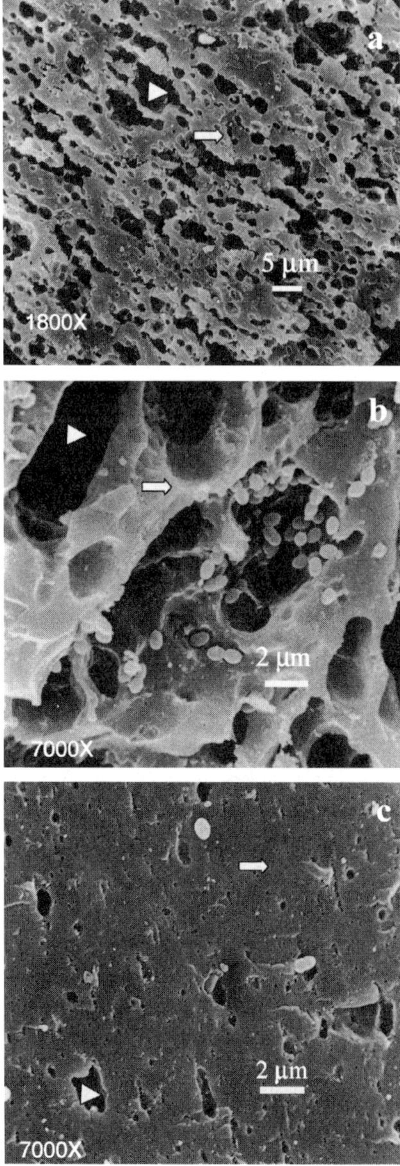

Figure 25.2 Scanning electron micrographs of full- (330 g/kg, a,b) and low- (39 g/kg, c) fat Cheddar cheeses at low (1800×, a) or high (7000×, b,d) magnifications. The arrows correspond to the *para*-casein matrix and the arrowheads to the areas occupied by fat and free serum prior to their removal during sample preparation; bacteria (most likely starter lactococci) are visible in b, being concentrated mainly at the fat-*para*-casein interface (Guinee *et al.*, 1998).

- hydrolysis of the *para*-casein into peptides, of varying molecular mass, and amino acids (FAA); degradation of the FAA into compounds such as amines, aldehydes, alcohols and ammonia (Fox *et al.*, 1996),
- changes in the equilibrium concentrations of calcium and inorganic phosphate between the matrix and the enclosed serum, with the equilibrium being influenced by maturation time, pH and other factors such as the concentration of Na^+ in the moisture phase and soluble Ca (Le Graet *et al.*, 1983; Karahadian and Lindsay, 1987; Guo and Kindstedt, 1995; Guo *et al.*, 1997; Paulson *et al.*, 1998),
- increase in hydration of the *para*-casein, as reflected by the decrease in the level of serum expressed by centrifugation or by hydraulic pressing of the cheese during maturation (Kindstedt, 1995; Guo and Kindstedt, 1995; Guo *et al.*, 1997; Thierry *et al.*, 1998; Boutrou *et al.*, 1999; Guinee *et al.*, 2000a). Hydration is mediated by factors such proteolysis, an increase in pH, and the solubilization of casein-bound calcium (Guo and Kindstedt, 1995),
- physical expansion or swelling of the *para*-casein matrix, at least in Mozzarella cheese, as a result of the increase in casein hydration. The age-related increase in the degree of swelling of the casein matrix of Mozzarella cheese is clearly observed on examination of the cheese by confocal laser scanning microscopy (CLSM) during maturation (Figure 25.3),
- coalescence of fat globules, resulting in the formation of fat pools. This appears to occur in all cheeses, as demonstrated by TEM (Kimber *et al.*, 1974; Laloy *et al.*, 1996). The occurrence of fat coalescence is also supported by the increases in the level of fat that can be expressed from

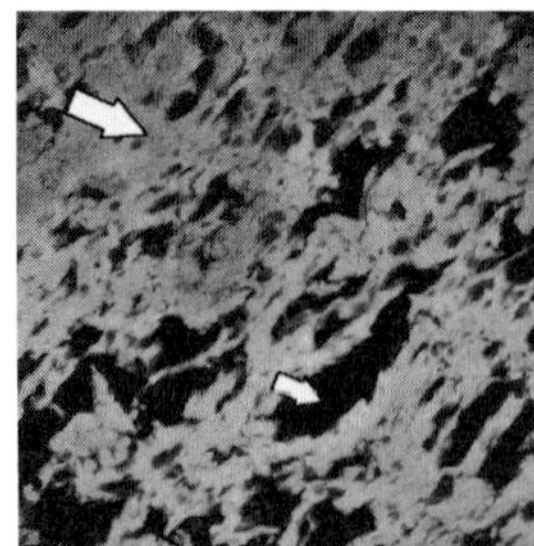
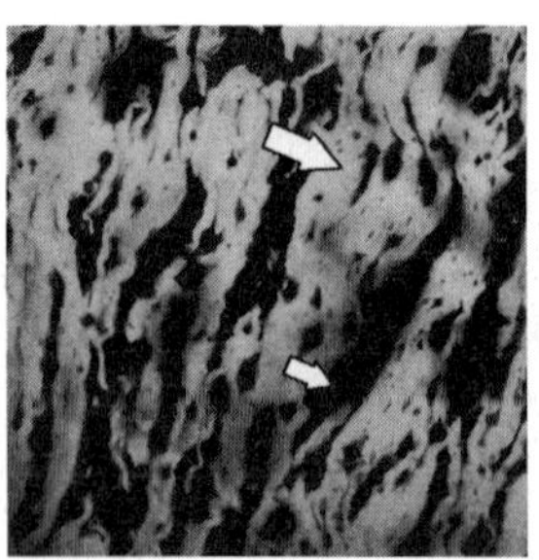
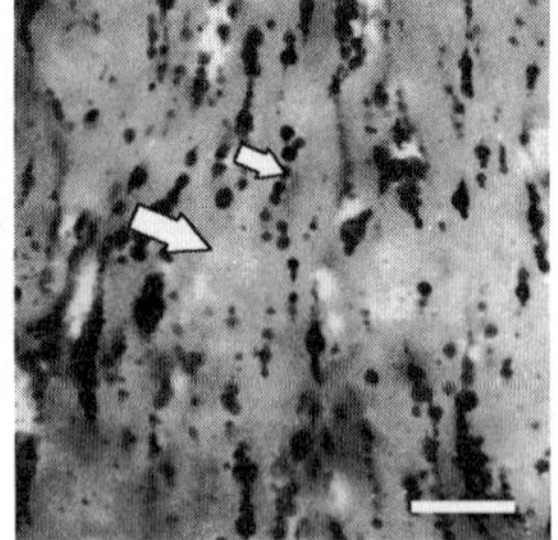

Figure 25.3 Confocal laser scanning micrographs of low-moisture Mozzarella cheese at different stages of production and storage at 4°C: A) Before texturization; salted and milled curd, showing *para*-casein matrix strands (long arrows) and void spaces containing fat globules (short arrows), B) Post texturization (24 h storage at 4°C), stretched curd showing extensive linearization of *para*-casein into fibres (long arrows) and globules/pools of fat (short arrows), C) Post texturization (43 days storage at 4°C); aged cheese showing hydrated swollen *para*-casein fibres forming a continuous protein phase which occludes fat mainly as pools. Bar = 25 μm (reconstructed from Fox *et al.*, 2000).

the cheese when subjected to hydraulic pressure or centrifugation (Figure 25.4), or exudes from the cheese on baking (Kindstedt, 1995; Thierry *et al.*, 1998; Guinee *et al.*, 1997a, 2000a). Fat coalescence may be mediated by an increase in free fat, due to degradation of the fat globule membrane and hydration of the *para*-casein. Hydration results in physical swelling of the protein phase (into space previously occupied by fat), an occurrence which forces the partially denuded fat globules into closer proximity (Kindstedt and Guo, 1995), resulting in coalescence and clumping. It is noteworthy that ~20 to 60% of the total milk fat is liquid (Norris *et al.*, 1973) in the temperature range used for the ripening of different cheese varieties (e.g., 8–22°C) and flows under pressure.

Hence, cheese is a biologically, enzymatically and biochemically dynamic system in which the protein and other structural components undergo physico-chemical and micro-structural changes during ripening, e.g., hydrolysis, hydration and swelling of the *para*-casein matrix and coalescence of non-globular fat. These changes assist in the conversion of fresh 'green' curd to a mature cheese and markedly influence its rheological, textural, functional and flavour characteristics (see Sections 25.5–25.7).

Thus, a storage period is generally required for rennet-coagulated cheeses before they attain the desired attributes (e.g., flavour, aroma, or the ability to melt, flow and stretch on cooking) associated with the particular variety.

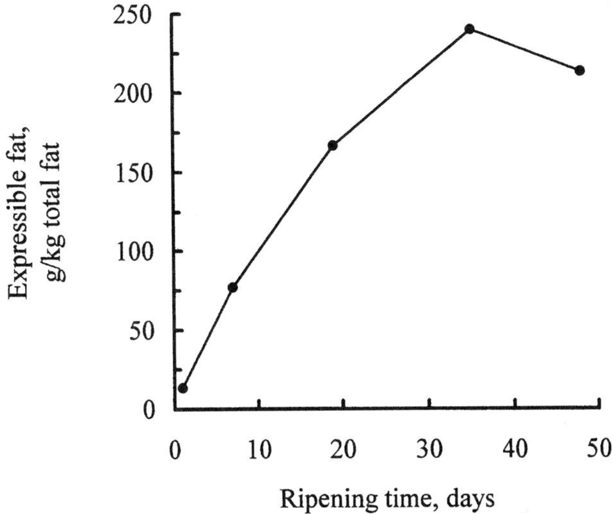

Figure 25.4 Typical changes in level of oil expressed from low-moisture Mozzarella stored at 4°C, on hydraulic pressing (3.2 MPa at 20°C for 3 h). An increase in expressible oil is an index of the increased potential of the cheese to oil-off during baking (redrawn from Guinee *et al.*, 1997a).

25.3 Role of protein in the curd texturization and plasticization processes in the manufacture of Cheddar and *pasta filata* varieties

In the manufacture of Cheddar, Mozzarella and Kashkaval, the curd is subjected to a process of cheddaring whereby drained curd is piled and pressed under its own weight while the pH decreases from ~6.15 to 5.2 as a result of starter growth. In traditional manufacture, piling was achieved by cutting the bed of drained curds into slabs which were piled higher as the pH decreased. In modern cheesemaking, the drained curds are discharged onto continuously moving perforated belts with side plates (e.g., Alfomatic) or into towers (Cheddarmaster) where the respective height and degree of flow of the curd layer or column are controlled by altering belt speed/dimensions or residence time. The cheddaring process creates conditions conducive to curd flow and texturization (i.e., development of a fibrous or chicken breast-like texture). These conditions include:

- holding the curd at ~34–35°C for ~45–60 min
  - to allow continued growth of the starter culture and acid development and hence a decrease in pH from ~6.15 to 5.2,
  - to ensure that the milk fat remains essentially in the liquid state and thereby contributes to the viscous, rather than elastic, character of the curd,
- piling the curd and allowing it to press under its own weight in equipment (e.g., finishing vat, cheddaring tower or cheddaring belts) designed to accommodate a certain degree of curd flow.

Cheddaring promotes a number of physico-chemical conditions which are conducive to the flow and texturization of the curd. These include:

- a decrease in pH and solubilization of micellar calcium which is bound to the casein and acts as a cementing agent between the casein micelles/sub-micelles (van Hooydonk *et al.*, 1986; Swaisgood, 1992),
- an increase in the ratio of soluble Ca to casein-bound Ca; soluble Ca increases from ~5 to 40% of total Ca in the curd as the pH decreases from 6.15 to 5.2 (Guinee *et al.*, 2000b; Figure 25.5),
- an increase in *para*-casein hydration which increases with decreasing pH in the range 6.6 to ~5.15 (Creamer, 1985), and
- an expected increase in the ratio of viscous-to-elastic character of the curd.

The increase in casein hydration with decreasing pH is probably in part a consequence of the increase the ratio of soluble-to-micellar Ca. Studies on model (dilute) casein systems have shown that casein hydration is inversely related to the concentration of casein-bound Ca (Sood *et al.*, 1979). As a consequence of the reduction in casein-bound Ca and the increase in casein

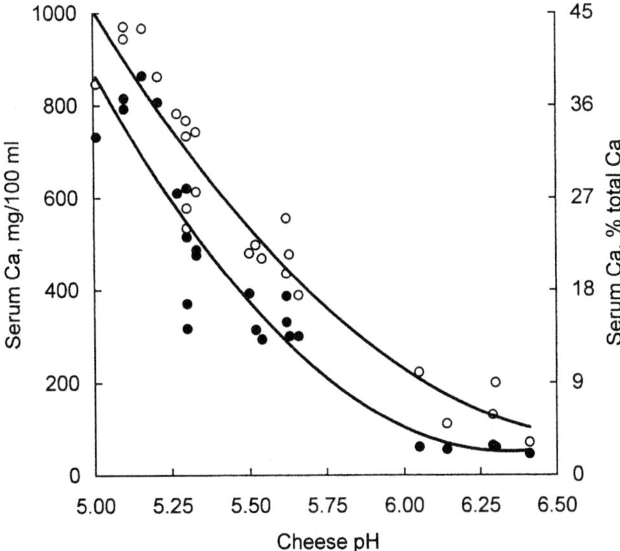

Figure 25.5 Effect of the pH of cheese on the calcium content (○, mg/100 ml; ●, % of total cheese Ca) in cheese serum expressed on hydraulic pressing of various commercial cheeses including Cheddar, Mozzarella, Dubliner, Kashkaval, Provolone and analogue pizza cheese (redrawn from Guinee *et al.*, 2000b).

hydration, the viscoelastic casein matrix undergoes limited flow if physically unrestricted, especially when piled and pressed under its own weight. Curd flow gives the desired planar orientation of the strands of the *para*-casein network (Kalab, 1977, 1979; Lowrie *et al.*, 1982; Figure 25.3A). The physicochemical changes in curd during cheddaring are summarised in Figure 25.6.

The manufacture of Mozzarella and other *pasta filata* cheeses, e.g., Kashkaval and Provolone, is apart from differences in type of starter culture and scald temperature, similar to that of Cheddar up to the completion of the cheddaring process and subsequent milling of the cheddared curd into chips but subsequent treatments of the curd differ markedly. For Cheddar, the curd chips are dry-salted and moulded while the curd chips (pH ~ 5.15) for *pasta filata* cheese are plasticized, cooled and salted. During plasticization, the shredded curds are scalded to ~57°C by kneading and stretching in hot water or dilute brine (e.g., 5% NaCl) at ~78°C. The relatively low curd pH and the high temperature are conducive to limited aggregation of the casein and the formation of *para*-casein fibres of relatively high tensile strength (Taneya *et al.*, 1992; Oberg *et al.*, 1993; Apostolopoulos, 1994; Pagliarini and Beatrice, 1994; Guinee and O'Callaghan, 1997). At a microstructural level, plasticization results in linearization of the *para*-casein matrix into fibres and clumping of fat into droplets which are trapped between, and show the same orientation as, the fibres (Kalab, 1977;

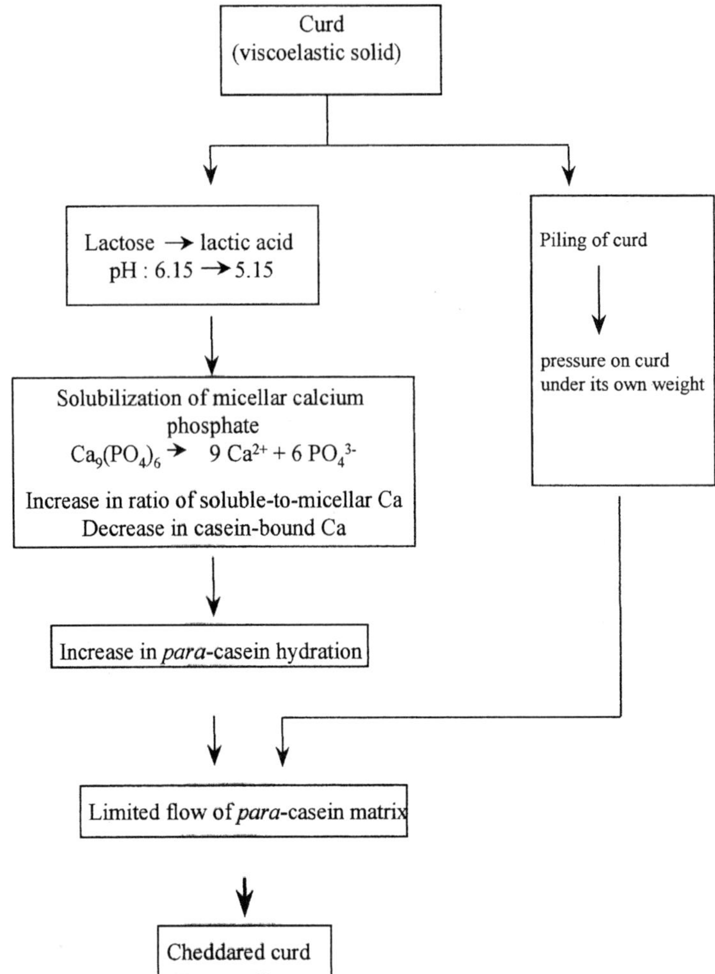

Figure 25.6 Summary of physico-chemical changes in cheese during cheddaring (redrawn from Fox *et al.*, 2000).

Kiely *et al.*, 1992; Taneya *et al.*, 1992; Oberg *et al.*, 1993; Figure 25.3B). When *pasta filata* cheeses are subsequently cooked on pizza pie, the fibres endow the cheese topping with the desired levels of stretch or stringiness and chewiness (Guinee and O'Callaghan, 1997).

The physico-chemical changes responsible for plasticization of the curd have not been elucidated unequivocally but appear to be a consequence of:

- an adequate degree of casein hydration of the cheddared curd, which is controlled by its pH, total calcium concentration and ratio of soluble-to-micellar Ca,

- heat-induced coalescence of free fat which lubricates the flow of the *para*-casein matrix (Paquet and Kalab, 1988; Guinee *et al.*, 1999); extension and shear stresses applied to the curd which assist in the displacement of contiguous planes of the *para*-casein matrix.

The relationship between *para*-casein hydration and pH may be explained by the dominance of two opposing forces over the pH region 6.0 to 5.0, i.e.,

- those that promote charge neutralization, which leads to acute dehydration and aggregation of the *para*-casein which in turn impedes flow at the high temperature, and
- those that promote solubilization of micellar calcium which is conducive to casein hydration and flow.

At pH values in the region 6.0 to 5.2, solubilization of micellar calcium phosphate (MCP) appears to be dominant as decreasing pH results in an increase in the hydration of *para*-casein (Figure 25.7; Creamer, 1985). In contrast, charge neutralization appears to dominate in pH range 5.2–4.6, as reduction in pH leads to a marked decrease in *para*-casein hydration.

The total concentration of calcium in the curd, which is controlled mainly by the pH of the milk at setting and that of the curd at whey drainage (Lawrence *et al.*, 1984), determines the pH at which plasticization of the curd is possible. In the conventional manufacture of Mozzarella, the milk is, typically, set at pH 6.55, the whey is drained off at 6.15 and the

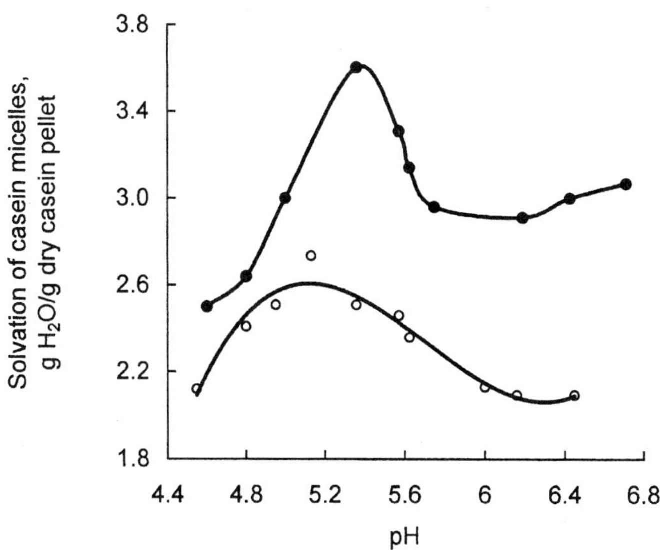

Figure 25.7 Solvation of casein micelles as a function of pH at 20°C for skim milk (●) and rennet-treated skim milk (○) (redrawn from Creamer, 1985).

cheddared curd is, ideally, plasticized at pH ~5.15. At this pH, the concentration of calcium in the curd (~28 mg/g protein) and the ratio of soluble-to-colloidal calcium (~30:70) permit adequate hydration of the *para*-casein for proper plasticization. At higher pH values, the plasticized curd becomes progressively less smooth; this trend reflects the decrease in *para*-casein hydration due to the concomitant reduction in the ratio of soluble-to-colloidal Ca. For similar concentrations of total calcium in cheese, soluble calcium as a percentage of total calcium decreases from ~40% at pH 5.0 to ~15% at pH 5.8 (Guinee *et al.*, 2000b). At a curd pH > 5.4, the curd fails to plasticize properly, as manifested by the non-plastic lumpy cheese mass which has a short consistency and a rough, dull appearance of the surface. Similarly, heating and shearing of cheese during the manufacture of PCPs result in aggregation of the *para*-casein with concomitant exudation of moisture and free fat unless emulsifying salts (e.g., sodium orthophosphates) are added to chelate micellar Ca (see Section 25.8).

However, successful plasticization may be achieved at a high curd pH, e.g., 5.6–5.8, if the total concentration of Ca in the curd is sufficiently low (e.g., <18 mg/g protein), as in the case of directly-acidified Mozzarella (DAM). In this case, acidification is achieved by the addition of food-grade organic acid(s), rather than the conversion of lactose to lactic acid by the starter culture (Shehata *et al.*, 1967; Kindstedt and Guo, 1997a). Typically, the milk pH is adjusted to ~5.6 prior to rennet addition and no further change in pH occurs during curd manufacture which is otherwise similar to that for conventional Mozzarella made using a starter culture. Following whey drainage, the curd, typically with a pH of ~5.6, plasticizes at 57°C. The ability of curd made using direct acidification to plasticize at the relatively high pH can be explained on the basis of the interactive effects of total curd calcium and the ratio of soluble-to-micellar Ca (which changes with pH) on *para*-casein hydration. While soluble Ca decreases from ~40 to 20% of total Ca as the pH is raised from 5.15 to 5.6, the overall concentration of calcium in DAM is markedly lower (e.g., 18 *vs.* 28 mg/g protein). Hence, the estimated concentration of micellar Ca in DAM (~14.5 mg/g protein) is similar to or lower than that (~16.5 mg/g protein) in conventional Mozzarella which is plasticized at pH 5.15. Owing to the inverse relationship between casein-bound Ca and casein hydration (Sood *et al.*, 1979), the degree of *para*-casein hydration in DAM curd at pH 5.6 is, therefore, similar to, or somewhat greater than, that in conventionally-produced Mozzarella curd at 5.15. Thus, the water-binding capacity of DAM curd (pH 5.6) is higher than that of conventionally-produced Mozzarella curd (pH 5.2) during the first three weeks of ageing (Kindstedt and Guo, 1997a).

The above hypothesis on the interactive relationship between pH and Ca level on the plasticizability of Mozzarella curd has been substantiated in a recent study (Guinee, T.P., Feeney, E.P. and Fox, P.F., unpublished results). The following Mozzarella cheeses were compared:

- CL, control, made using a standard process and plasticized at pH 5.15,
- DAM1, directly acidified using lactic acid, set and plasticized at pH 5.6,
- DAM2, directly acidified as for DAM1, except that the milk was set at pH 6.55, and the curd was drained at pH 6.15, acidified with glucono-δ-lactone and plasticized at pH 5.6. Analyses of the cheeses (Table 25.1) indicated that for comparable pH values in the finished cheese (i.e., 5.9), the DAM1 cheese, which plasticized successfully, had a markedly lower Ca content, a higher moisture level, higher stretchability and flow during ripening than the CL or DAM2 cheeses. Moreover, measurement of heat-induced changes in viscoelasticity, using low amplitude strain oscillation, indicated that the phase angle of the molten DAM1 cheese was markedly higher than that of DAM2 and similar to that of CL cheese, over the temperature range 40 to 80°C (Figure 25.8). These results suggest that the *para*-casein matrix of the DAM1 cheese was more hydrated than that of the DAM2 cheese and on heating to >50°C, behaved more like a fluid.

TABLE 25.1
Compositional, functional and viscoelastic analysis of Mozzarella cheese acidified using starter culture (CL) or chemical acidification with lactic acid (DAM1) or with lactic acid and gluconic acid-δ-lactone (DAM2)[1]

	Day	*Cheese code*		
		CL	*DAM1*	*DAM2*
Composition				
Moisture (g/kg)	1	471.9	600.6	490.0
Fat (g/kg)	1	210.6	148.9	179.8
Protein (mg/100 g)	1	273.5	216.4	280.6
Ca (mg/100 g)	1	773	410	857
(mg/g protein)		28.3	18.9	30.5
pH at day 1		5.43	5.96	5.94
Functionality				
Stretchability (cm)	1	19	104	72
	12	101	105	16
Flow (%)	1	28	53	8
	12	36	60	13
Apparent viscosity At 70°C (Pa s)	12	>1000	900	>1000
Viscoelasticity				
G′ at 20°C (kPa)	12	126	54	133
G′ at 80°C (kPa)	12	3.5	1.1	6.8

[1]See text for details of samples.

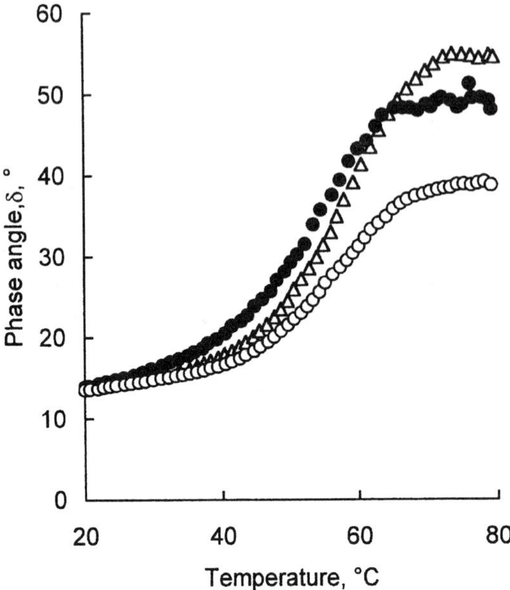

Figure 25.8 Phase angle (δ) as a function of temperature of melting samples of Mozzarella cheese: made using conventional starter culture-induced acidification and curd plasticized at pH 5.15 (Δ); made using direct acidification with lactic acid, milk set at pH 5.6 and curd plasticized at pH 5.6 (●); made using direct acidification with lactic acid and gluconic acid-δ-lactone, milk set at pH 6.55 and curd plasticized at pH 5.6 (○).

25.4 Cheese yield

25.4.1 Relative contributions of casein and fat

Protein contributes directly to cheese yield and indirectly as the structure and rheology of the gel matrix influences the retention of non-protein components such as moisture (and dissolved substances), fat and salts (IDF, 1991; Guinee *et al.*, 1996a). In most rennet-curd varieties, casein is essentially the only protein which contributes to yield. At the pasteurization treatment (~72°C for 15s) used for rennet-curd cheeses, whey protein denaturation is low (i.e., ~5% total) and its contribution to cheese yield is small, e.g., typically 0.07 and 0.12 kg/100 kg cheese milk for full- and half-fat Cheddar, respectively, in milk with a protein content of ~33 g/kg (Lau *et al.*, 1990; Fenelon and Guinee, 1999). Denatured whey proteins complex with the κ-CN (van Hooydonk *et al.*, 1987; Jelen and Rattray, 1995) and partition with the insoluble *para*-casein after rennet treatment. In contrast, native whey proteins are soluble under cheesemaking conditions (mostly, <50°C for <2h) and are largely lost in the cheese whey. Native whey proteins dissolved in the cheese moisture (~4% total) contribute very little to yield, ~0.026 kg/100 kg milk. However, whey protein contributes more to yield when milk is subjected to a high heat treatment (e.g., 90°C × 5 min), as in acid-heat coagulated varieties (e.g., Paneer, Ricotta), or where the cheese is prepared from high concentration factor (e.g., > 6×) milk ultrafiltration retentate, referred to as pre-cheese. The effect of high heat treatment on cheese yield is discussed in Section 25.4.4.

The importance of casein to yield is reflected by the following general equation for the prediction of cheese yield:

$$Y = aF + bC,$$

where, Y is the yield (kg/100 kg), F and C are the concentrations (g/100 g) of milk fat (F) and casein (C), a and b are coefficients, the magnitude of which depend on the contributions of fat and casein to yield. The values of a and b have been found to range from ~1.47 to 1.6 and 1.44 to 1.9, respectively, for Cheddar cheese (IDF, 1991). A recent pilot-scale (500 L) yield study (Guinee and Fenelon, unpublished results) showed that actual (Y_a) and moisture-adjusted (MACY) Cheddar cheese yields were described accurately by the following equation:

$$Y_{a\,\mathrm{Cheddar}} = 1.56F + 1.71C.$$

While this formula is empirical, it indicates that, in general, casein contributes more than its own weight to cheese yield and also contributes more to Cheddar yield than fat. The relatively high contribution of casein is expected as it forms the continuous *para*-casein matrix, which, acting like a sponge, occludes the fat and moisture (serum) phases (Figure 25.3). Occluded moisture contributes directly to cheese yield and indirectly due to the presence of dissolved solids, which include whey proteins, κ-casein glycomacropeptide, lactate and soluble salts. In milk, colloidal Ca and PO_4^{3-}, which are associated with the casein micelles, are present at concentrations of ~21 and 16 mM, respectively; ~90% and ~98% of colloidal Ca and PO_4^{3-}, respectively, are retained during Cheddar manufacture. The combined contributions of dissolved solids and micellar Ca and P to Cheddar cheese yield ranges from ~9% for full-fat (330 g/kg) Cheddar to ~14% for low-fat (60 g/kg) Cheddar (Fenelon and Guinee, 1999).

While fat on its own has little water-holding capacity, its presence in the *para*-casein matrix affects the degree of matrix contraction and hence moisture content and cheese yield. The occluded fat globules physically limit contraction, and hence aggregation, of the surrounding *para*-casein network and therefore reduce the extent of syneresis. Hence, as the fat content of the curd is increased, it becomes more difficult to expel moisture; consequently, the moisture-to-casein ratio generally increases unless the cheesemaking process is modified to enhance casein aggregation, e.g., by increasing the scald temperature (Gilles and Lawrence, 1985; Fenelon and Guinee, 1999).

Owing to its negative effect on syneresis, fat indirectly contributes more than its own weight to cheese yield, i.e., Cheddar cheese yield increases by ~1.16 kg/kg milk fat (Fenelon and Guinee, 1999). This greater than *pro rata* increase is due to the increase in the level of moisture in non-fat substance (MNFS) as the content of fat in the cheese increases. However, if the content of MNFS is maintained constant (e.g., by process modifications), fat contributes less than its own weight to cheese yield (i.e., ~0.9 kg/ kg), due to the fact that ~8–10% of the milk fat is normally lost in the whey.

25.4.2 Effect of casein concentration on yield

Both actual and dry matter Cheddar cheese yields increase linearly with increasing milk protein content in the range 30 to 46 g/kg (Guinee *et al.*, 1996a). However, if cheesemaking conditions are not otherwise modified, the increase in actual yield per unit increase in milk protein concentration is lower than that of moisture-adjusted yield (MACY), an effect attributed to the inverse relationship between cheese moisture and milk protein level (Guinee *et al.*, 1996a; Broome, 1999). The latter trend, which reflects a positive relationship between milk protein level and gel syneresis, is surprising since, all other factors being equal, the porosity of the rennet gel is expected to decrease as the volume fraction of the matrix increases. This trend suggests variations in gel structure (i.e., fineness or coarseness) with milk protein level, which affect whey expression. For unidimensional flow through a porous medium, such as a rennet or acid milk gel, the rate of syneresis (ν) may be expressed by Darcy's law (Walstra *et al.*, 1985):

$$\nu = \frac{B}{\eta} \cdot \frac{\Delta P}{l}$$

where ν is the average linear flow rate (ms^{-1}) of the whey, of viscosity η, through the matrix in the direction of the pressure gradient $\Delta P/l$ and l is the distance through which the serum flows; B is the permeability coefficient of the gel matrix (m^2), which corresponds to the average cross-sectional area of the gel pores. Walstra and van Vliet (1986) reported that the value of B for renneted milk gels decreased as the concentration of milk protein was increased by UF. However, the inverse relationship between milk protein level and cheese moisture indicates that the cumulative syneresis over the course of Cheddar cheesemaking is greater at higher milk protein levels. Moreover, scanning electron microscopical analysis of renneted milk gels at cutting and of curds at the end of cooking indicated that a 2 fold increase in milk protein level, by UF, resulted in a coarser *para*-casein matrix, i.e., less homogeneous distribution with larger pores or channels (Green *et al.*, 1981a). The latter effect, which appeared to become more pronounced as cheesemaking advanced, i.e., from gel cutting to maximum scald, was considered to be the main cause of lower fat retention in curds from milk concentrated more than 2 fold (Green *et al.*, 1981b; Guinee *et al.*, 1994). Indeed, increasing the fineness of the gel network by homogenization or calcium depletion of concentrated milk prior to renneting or cold renneting of ultrafiltered milk resulted in lower syneresis rates and/or improved fat retention (Green *et al.*, 1983; Green, 1985, 1987). Hence, while increasing the concentration of milk protein may reduce the permeability of (unbroken) renneted milk gels (Walstra and van Vliet, 1986), the extent of *para*-casein aggregation and fusion during cutting and the subsequent operations of stirring, heating and acidification is probably greater at the higher protein levels.

25.4.3 Effect of casein hydrolysis on yield

(a) Somatic cell count and stage of lactation

The degree of casein hydrolysis in raw milk is influenced by the level of indigenous proteolytic activity (e.g., plasmin and somatic cell proteinases) and plasminogen activators, which in turn are affected by the stage of lactation, somatic cell count (SCC) and udder infection, e.g., mastitis (Saeman *et al.*, 1988; Politis *et al.*, 1989; Zachos *et al.*, 1992; Burvenich *et al.*, 1995). Increasing SCC in the range 10^5–10^6 cells/ml is paralleled by increases in the concentration of whey proteins and degree of casein hydrolysis and reductions in the concentrations of casein and fat. Elevated SCC results in a marked decrease in β-CN as a proportion of total casein and concomitant increases in γ-CNs, proteose peptones and the ratio of soluble: micellar casein (Anderson and Andrews, 1977; Ali *et al.*, 1980; Schaar *et al.*, 1985; Saeman *et al.*, 1988). These changes ensue from the hydrolysis of β- and α_{s2}-CNs by the elevated activity of plasmin (and probably other proteinases) in the milk; κ-casein is hydrolysed more slowly by plasmin than β- and α_{s2}-CNs.

An increase in the degree of casein hydrolysis in milk, due to high SCC or to the addition of enzymes with proteolytic specificity similar to plasmin (e.g., trypsin), leads to a deterioration in rennet coagulation properties, curd syneretic properties and reduced cheese yield (Donnelly *et al.*, 1984; Okigbo *et al.*, 1985; Mitchell *et al.*, 1986; Politis and Ng-Kwai-Hang, 1988a,b,c; Barbano *et al.*, 1991; Barbano, 1994; Auldist *et al.*, 1996; Klei *et al.*, 1998). An increase in SCC from 10^5 to $>5 \times 10^5$ cells/ml typically results in a reduction of ~3 to 7% in the moisture-adjusted (370 g/kg) yield of Cheddar cheese. However, it is noteworthy that the decrease is relatively large (i.e., ~0.4 kg Cheddar cheese/100 kg milk) on increasing the SCC from 10^5 to 2×10^5 cells/ml, a range which would be considered relatively low for good quality bulk milk. Losses of fat and protein during Cheddar cheese manufacture increased, more or less linearly, by ~0.7 and 2.5%, respectively, with SSC in the range 10^5 to 10^6 cells/ml (Politis and Ng-Kwai-Hang, 1998a). The negative impact of SCC on cheese yield and the recoveries of fat and protein is due largely to the increase in proteolysis α_{s1}-, α_{s2}- and β-CNs to soluble peptides, which are lost in the whey. Moreover, the lower effective concentration of gel-forming casein results in a slower curd-firming rate and a slower rate of aggregation and fusion of *para*-casein micelles during the cheesemaking (Guinee *et al.*, 1997b). A low curd-firming rate is conducive to:

- a greater susceptibility of curd particles to shattering during cutting and the early stages of stirring, resulting in higher losses of curd fines and milk fat (Bynum and Olson, 1982; Mayes and Sutherland, 1989), and
- an impaired syneretic capacity of the curd, with a consequent increase in moisture level (Donnelly *et al.*, 1984).

A high SCC may also inhibit the activity of some strains of lactococci during cheese manufacture, an effect expected to further impair curd-firming rate and reduce firmness at cutting. The negative effect of increased SCC is accentuated when the gel is cut on the basis of a pre-set renneting time, rather than on the basis of firmness, as the slower-than-normal curd-firming results in a lower-than-optimum firmness at cutting. The relatively low recovery of fat from high SCC milks may be due also to an increased permeability of the milk fat globule membrane (MFGM) and a concomitant increase in the level of non-globular fat. The latter effect is thought to result from an increased binding of the indigenous lipoprotein lipase to, and a consequent increase in the hydrolysis of, the MFGM in high SCC milk (Sundheim and Bengtsson-Olivecrona, 1987; Murphy *et al.*, 1989). It is envisaged that any increase in non-globular fat in milk would result in an increase in the level of fat in whey and a reduction in fat recovery.

Marked changes occur in milk composition throughout the year, especially in Ireland, New Zealand and parts of Australia, where milk is produced mainly from spring-calving herds fed predominantly on pasture. Such a seasonal pattern of milk production results in large changes in milk composition, principally due to the physiologically-induced changes in the biosynthetic performance of the mammary gland as influenced by stage of lactation, diet and health. The concentration of total protein, casein and whey protein in Irish bulk herd milk, mainly from spring-calved cows fed mainly on pasture, typically increases gradually from early lactation (e.g., days 15–30) to a maximum in late lactation (days 220–240), and decreases abruptly thereafter, attaining values at very late lactation similar to those in early lactation (White and Davies, 1958; Phelan *et al.*, 1982; Keogh *et al.*, 1982; O'Brien *et al.*, 1999a; Mehra *et al.*, 1999). However, the seasonal profiles of the various constituents are influenced by many factors such as age, diet and health (Rogers and Stewart, 1982; Grandison *et al.*, 1984; O'Brien *et al.*, 1987). In addition, the relative concentrations of the individual caseins (as a percentage of total casein) undergo seasonal changes, especially in factory milk from spring-calving herds (Figure 25.9). These changes probably occur as a result of the general increase in plasmin activity in milk with advancing lactation (Richardson, 1983; Politis *et al.*, 1989).

Seasonal changes in milk protein are most pronounced at the extremes of lactation (Figure 25.9) and result in variations in rennet coagulation properties, cheese composition, recovery of fat and casein, cheese yield and quality (Chapman, 1974; Banks *et al.*, 1981; Grandison *et al.*, 1984; Banks and Muir, 1984; Banks and Tamime, 1987; Guinee *et al.*, 1996a). Undoubtedly, the seasonal and lactational changes in both the concentrations and proportions of different proteins contribute to these variations, especially on consideration of the biochemical characteristics of different caseins (Swaisgood, 1992) and cheese micro-structure (Creamer and Olson, 1982; Creamer *et al.*, 1982). Numerous reports have shown that the rennet coagulation, curd syneretic properties, fat and protein recovery, and hence, dry matter yield efficiencies of late lactation milk are inferior to those of mid-lactation milk (Donnelly *et al.*, 1984; Kefford *et al.*, 1995; Auldist *et al.*, 1996; Sapru *et al.*, 1997). Consequently, cheese made from late lactation milk tends to have a high moisture content and is often of inferior quality (O'Keeffe, 1982). However, late lactation milk should be ideal for cheesemaking owing to its high concentration of casein, which would be expected to enhance rennet coagulability and curd syneresis and give a low moisture content. All other factors being equal, the concentration of casein is positively correlated with curd-firming rate (Guinee *et al.*, 1996b, 1997b)

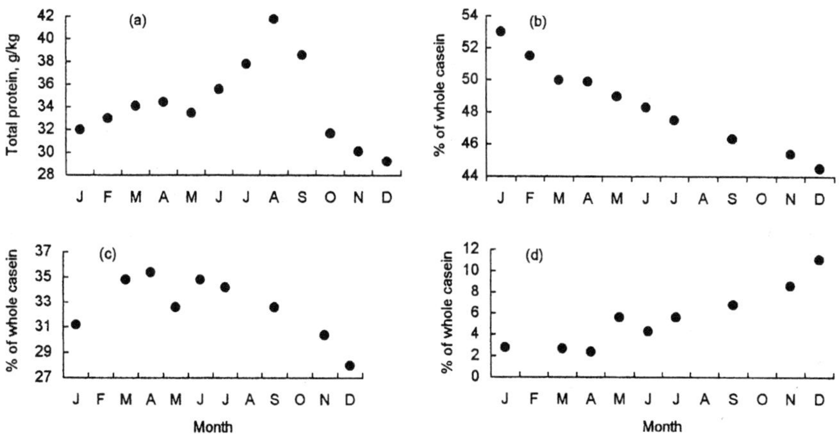

Figure 25.9 Seasonally related changes in the concentration of total protein in Winter/Spring-calving herd milk (a), and of individual caseins in Irish bulk creamery milk, mainly from Winter/Spring-calving herds (b–d). Caseins as a percentage of whole casein: α_s- (b), β- (c) and γ- (d) caseins (redrawn from Donnelly and Barry, 1983).

and inversely related to the moisture content of Cheddar cheese (Guinee *et al.*, 1996a; Broome, 1999). The relatively poor cheesemaking properties of late lactation milk, compared to mid lactation milk, appear to be due to the interactive effects of:

- an increased whey-protein-to-casein ratio,
- increased plasmin activity and a lower effective casein concentration (due to hydrolysis of casein by plasmin),
- a change in the proportions of different caseins (e.g., higher ratio of α_{s1}-CN:β-CN), and
- a higher milk pH and a decrease in the ratios of colloidal Ca-to-casein and ionic Ca-to-casein.

The mean concentration of total Ca in milk from individual cows decreases markedly (from ≈150 to 115 mg/100 ml) during the first 16 d of lactation and increases (from ≈115 to 170 mg/100 ml) in very late lactation i.e., after >300 d (White and Davies, 1958). Over most of the lactation period, the concentration of total Ca fluctuates (i.e., 105 to 130 mg/100 ml) but shows no relationship with stage of lactation (White and Davies, 1958). A similar trend has been observed for the seasonal changes in the concentration of total Ca in manufacturing milk and milk from spring-calving herds (O'Brien *et al.*, 1999b). Similar to total Ca, the concentrations of ionic Ca and soluble Ca in manufacturing milk and spring herd milk fluctuate during lactation but show no relationship with stage of lactation or season (White and Davies, 1958; Keogh *et al.*, 1982). While a similar trend was noted for colloidal Ca, its concentration in creamery and spring herd milk tends to increase slightly (from ≈65 to 70 mg/100 ml) in November and return thereafter to the baseline value (Keogh *et al.*, 1982). The casein content of spring herd milks increases progressively from July (e.g., 2.4%, w/w) to a maximum in mid October (e.g., 3%, w/w) and declines sharply thereafter to a value (e.g., 2.4%, w/w) typical of mid-lactation milk. As the increase in concentration of casein in the mid-to-late lactation period is markedly greater than that of total Ca, the ratios of colloidal Ca-to-casein and ionic Ca-to-casein decreases from 26.6 and 4.8 mg/g casein in mid lactation to ≈23 and 3.9 mg/g casein in late lactation respectively. It is envisaged that decreases in these ratios may impair casein aggregation, rennet coagulation and curd syneresis (Walstra *et al.*, 1985), and thereby reduce dry matter cheese yield. It is noteworthy that in model emulsion systems, the electrokinetic potential of casein-stabilized particles increased as the concentration of added $CaCl_2$ in the continuous aqueous phase was reduced from 0.45 to 0.15 M (Dickinson *et al.*, 1987).

(b) Cold storage of milk and thermization

Modern farm and milk collection practices have resulted in milk being cooled rapidly to <8°C following milking and a relatively low frequency of milk collection from the farm, e.g., every second or third day. Moreover, cold milk is hauled over long distances and is often cold-stored at the cheese plant for 1–3 days, depending on time of year and the manufacturing schedules.

Cold storage and shearing result in a number of physico-chemical changes in the protein and mineral phases which are conducive to alteration of cheesemaking properties, including:

- solubilization of micellar caseins, especially β-CN, and colloidal calcium phosphate (CCP), both of which contribute to an increase in serum casein (Dalgleish and Law, 1988, 1989),
- an increased susceptibility of serum casein to hydrolysis by proteinases from various sources, e.g., psychrotrophic bacteria, somatic cells and/or plasmin, and the concomitant increase in non-protein N (Hicks *et al.*, 1986; Guinot-Thomas, 1995),
- damage to the MFGM and hydrolysis of free fat by lipases from psychrotrophic bacteria and/or milk, resulting in a decrease in the level of milk fat (Hicks *et al.*, 1986).

While there are some differences between studies (Cousin, 1982; Banks *et al.*, 1986), it is generally agreed that cold storage impairs rennet coagulation properties (Fox, 1969; Qvist, 1979), reduces the recovery of protein and fat and reduces cheese yield (Hicks *et al.*, 1982, 1986; Weatherhup *et al.*, 1988). The increase in the rennet coagulation time of manufacturing milk, stored at 4°C for 48 h, ranged from 9 to 60% (Fox, 1969). The large variation in the RCT of cold-aged milk is probably a consequence of differences in composition, microbiological status and somatic cell count. The chemical changes (i.e., increases in serum casein and the ratio of soluble Ca: micellar Ca) and the increase in RCT associated with cold storage are almost complete after 24 h in freshly-drawn milk, preserved with sodium azide to prevent bacterial growth during storage (Qvist, 1979). The chemical changes are largely reversed by pasteurization (72°C × 15 s) or milder heat treatments (e.g., 50°C × 300 s) (Fox, 1969; Ali *et al.*, 1980).

The observed reduction in cheese yield from cold-aged milk may be attributed to the loss of soluble casein in the whey and/or the slower curd-firming rate. The reduced casein level results in slow curd formation and a soft gel at cutting, a situation which is conducive to curd shattering, high losses of fat in the whey, and reduced cheese yield (Mayes and Sutherland, 1989), especially when cutting is performed on a time basis.

In the EU, the permitted TBC in milk for the manufacture of dairy products was at $\leq 4 \times 10^5$ cfu/ml in 1994 and was reduced to $\leq 1 \times 10^5$ cfu/ml in 1998 (Anon, 1992). Improved dairy husbandry practices, combined with the more stringent standards for TBC and SCC, are conducive to reducing the degree of storage-related proteolysis and lipolysis in milk. Thus, in modern dairy companies, the TBC is generally less $<50\ 000$ cfu/ml. The fact that the chemical changes which occur during cold storage are reversed by pasteurization suggests that with modern milk production practices, cold storage of milk for several days probably has little influence on its cheesemaking properties. Moreover, thermization of milk at a sub-pasteurization temperature (e.g., 57–68°C for 10–15 s) on reception at cheese factories is now practised widely and reduces the count of psychrotrophic bacteria in stored milk (Zall, 1980; Zall and Chen, 1981). Studies on Cottage cheese showed significantly higher actual and moisture-adjusted (to 820 g/kg) yields from thermized (74°C × 10 s) milk compared to the control (i.e., 16.85 *vs.* 16.0 kg/100 kg), when milk was cold-stored at 3°C for 7 d (Dzurec and Zall, 1986). Hence, it has been suggested that when is milk is stored on farms for a long period, on-farm thermization may be advantageous for cheese yield.

25.4.4 Effect of protein type on cheese yield: milk pasteurization/incorporation of whey proteins

Theoretically, if all whey proteins were retained in cheese curd, without adversely affecting cheese moisture or quality, a yield increase of ~12% (i.e., 10.7 *vs.* 9.54 kg/100 kg) would be achievable for Cheddar cheese with ~380 g/kg moisture. Moreover, since increasing the severity of milk heat treatment is paralleled by an increase in cheese moisture (Figure 25.10), the increase in actual yield would be even higher (e.g., ~15% following pasteurization at 88°C × 15 s). However, inclusion of a high level of whey protein (i.e., > 35% total), either in denatured or native form, can adversely affect rennet coagulation properties, cheese rheology, heat-induced functionality and overall quality of most rennet-curd cheeses (IDF, 1991; Sections 25.5–25.7). Inclusion of a high content of whey protein in some fresh acid-curd cheeses (e.g., Quark, Cream cheese) alters the textural properties but generally does not impair the eating quality; however, even here, functionality may be altered. In the acid-heat coagulated cheese types (e.g., Ricotta, Paneer, some types of Queso Blanco) the incorporation of a high level of denatured whey protein is a feature of the manufacturing process, as discussed in Section 25.7.

(a) High heat treatment (HHT) of cheese milk

In situ denaturation of whey proteins by HHT is used widely in the commercial manufacture of fresh acid-curd cheeses, e.g., Quark, *Fromage frais* and Cream cheese, with typical heat treatments ranging from 72°C × 15 s to 95°C for 120–300 s (Guinee *et al.*, 1993). The estimated level of whey protein denaturation and yield of Quark (18% moisture) for these treatments are ~3% and 18.6 kg/100 kg, and ~70% and 21.3 kg/100 kg, respectively.

Increasing the level of whey protein in cheese curd from ~0 to ~10% of the total nitrogen by HHT of milk, increases protein recovery and actual and MACY for a range of hard and semi-hard rennet-curd cheeses, including

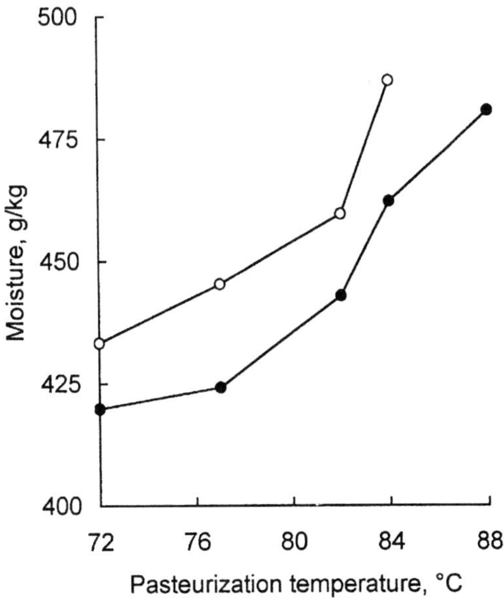

Figure 25.10 Effect of pasteurization temperature (for 15 s) on the moisture content of half-fat Cheddar cheese from curd with a pH at milling of 5.35 (●) or 5.75 (○) (redrawn from Guinee *et al.*, 1998).

Cheddar, Mozzarella and Cheshire (Schafer and Olson, 1975; Marshall, 1986; Banks *et al.*, 1987, 1994a,b; Lau *et al.*, 1990; IDF, 1991). The recovery of fat also increases as the level of whey protein denaturation increases (at least up to ~21% of total) (Schafer and Olson, 1975; Marshall, 1986) provided that the cheesemaking process is altered to restore the gel-forming properties of the HHT milk to those of the unheated milk, e.g., by extending the set-to-cut time when the pH of milk is normal, addition of $CaCl_2$ and/or lowering the set pH.

These effects are attributable mainly to the increase in the concentration of the gel-forming protein and the concomitant reduction in the porosity of the more finely structured gel from HHT milk (see Guinee *et al.*, 1993; Mulvihill and Grufferty, 1995). Consequently, the water-holding capacities of the gel are expected to increase (Section 25.4.2). However, rennet-induced gels from HHT milks with a level of WPD >30% tend to be very fragile and prone to shattering on cutting and stirring, even when the milk has been acidified to pH 6.0 prior to setting. Hence, high losses of fat (e.g., up to 15%) and curd fines (e.g., up to 800 mg/kg) in the cheese whey are possible unless the gel is treated very gently (Guinee *et al.*, 1998). The susceptibility curd particles from HHT milk to fracture persists for a longer time than normal into the stirring period, because of the lower tendency to synerese and the slower rate of firming of the curd particles. Recent studies (Guinee *et al.*, unpublished results) have shown that the contribution of the increase in cheese moisture content to Cheddar cheese yield, as the level of WPD in the cheesemilk is increased from 5 to 40% total, is offset by the reduction in yield as a result of the higher losses of fat in cheese whey.

(b) High concentration factor ultrafiltration (HCFUF)

HCFUF is used widely as an alternative to centrifugation, especially in Europe since the 1980's, to concentrate the gelled milk in the commercial production of fresh acid-curd cheese varieties, such as Quark and Cream cheese. This method gives complete recovery of whey proteins, with higher yields of cheese of very acceptable quality. HCFUF has also been used for the commercial manufacture of rennet-curd cheeses, e.g., cast Feta in Denmark, and Cheddar in Australia using the Siro-Curd process (see Fox *et al.*, 1996). Cheese manufacture involves HCFUF of the milk to produce a high dry matter (e.g., 400 to 500 g/kg) retentate, or liquid pre-cheese, which has a composition close to that of the finished cheese. Rennet and starter culture are added to the pre-cheese, which is then treated in the normal manner, except that little (e.g., Cheddar), or no (cast Feta) expulsion of whey occurs. Since the upper limit of concentration by UF is ~7:1 for whole milk, it is not possible to attain the dry matter level required for hard cheeses such as Cheddar or Gouda. Hence, further whey expulsion from the pre-cheese, after coagulation and cutting, occurs on conventional (e.g., drainage belts, finishing vats) or on specialized (e.g., Siro-Curd) equipment (Garrett, 1987).

The attraction of HCFUF for the manufacture of rennet-curd cheese is the increased yield owing to the high retention of whey proteins and the glycomacropeptides. The exact degree of retention depends on:

- the heat treatment of the milk prior to renneting, which determines the extent of whey protein denaturation and hence whey protein solubility, and
- the level of whey expulsion from the pre-cheese following coagulation and cutting; native whey proteins and the glycomacropeptides are soluble and are lost in the whey.

When complete recovery of the latter components is achievable, as in cast Feta, the estimated saving in skim milk is ~9%, which makes the process economically viable (Lawrence, 1989; IDF, 1991).

(c) Addition of denatured whey proteins to cheesemilk

Several processes have been developed for harvesting proteins from sweet whey by heat denaturation and acidification, and recovery by centrifugation, to yield concentrates containing ~120 g solids/L in the Centriwhey process or ~180 solids g/L in the Lactal process (Banks *et al.*, 1994b). These concentrates are added at the desired level to cheesemilk, which is then treated as in normal cheese manufacture. Denatured whey proteins may also be prepared by concentrating sweet whey by UF and heat-treating and shearing the retentate to give a controlled level of denaturation. The heat-treated retentate may be supplied as a viscous wet slurry or spray dried.

Whey protein concentrates may also be subjected to micro-particulation during preparation (e.g., Simplesse®) whereby the whey protein concentrate is heated and homogenized under acidic conditions (Singer, 1996). The shearing and heating conditions applied during micro-particulation are conducive to limited aggregation of the whey protein molecules and to the formation of discrete spherical particles ranging in size from 0.1 to 2 μm (Desai and Nolting, 1995). The resultant whey protein particles have dimensions similar to native fat globules in milk and mimic their sensory properties. Such particle dimensions are desirable, especially in reduced-fat cheeses, as they impart a creamy perception, similar to that of fat, on ingestion (Singer, 1996). Moreover, owing to the relatively small particle dimensions, the addition of micro-particulated whey protein (Simplesse®) at a level of 10 g/kg does not adversely affect the curd-forming properties of the milk (Lucey and Gorry, 1994). Large whey protein particles (e.g., >3 μm), formed under conditions (e.g., high temperature, low pH and low shear) which promote a high degree of protein aggregation, impart a 'sandy' or 'grainy' mouthfeel (Singer, 1996).

Varying results have been reported with the addition of partially denatured whey protein concentrates (PDWPC) (prepared by the Centri-whey, Lactal or UF processes) to milk for the manufacture of hard or semi-hard cheeses such as Cheddar and Gouda. There is general agreement that the addition of WPC increases moisture content, actual yield and moisture-adjusted yield, with the extent of the increase depending on the level of WPC added and the degree of whey protein denaturation of the added WPC. However, the addition of PDWPC has, generally, been found to give defective body (i.e., greasy, soft) and flavour (i.e., unclean, astringent) characteristics in Gouda and Cheddar cheese. Various PDWPC preparations have been used as fat mimetics to improve the texture of reduced-fat cheeses, which tend to be more firm and elastic than their full-fat equivalents. These preparations include Simplesse® 100 (The NutraSweet Company, IL, USA), Dairy Lo™ (Ault Foods Ltd., Canada; Pfizer Food Science Group, New York, NY, USA), and ALACO PALS™ (New Zealand Milk Products, Inc., Santa Rosa, CA, USA). These materials are added at a level of ~10 to 20 g/kg milk to give an increase in milk protein of ~3.5 g/kg. Their use in half-fat Cheddar cheese increases the moisture content (e.g., 450 *vs.* 427 g/kg) and the actual (e.g., 7.7 *vs.* 7.2 kg/100 kg) and moisture-adjusted cheese yields (Lucey and Gorry, 1994; Fenelon and Guinee, 1997).

25.4.5 Effect of genetic variants of milk protein

Cheesemaking properties and yields are also influenced by genetic variants of milk protein (Ng-Kwai-Hang and Grosclaude, 1992, see Chapter 16). The genetic variants, which have been investigated most thoroughly for their effects on the rennet coagulation and cheesemaking characteristics of milk are those of κ-CN and β-lactoglobulin (β-Lg). Compared to the AA variants, the BB genotypes of β-Lg and κ-CN are generally associated with a higher concentration of casein and superior rennet coagulation properties, as reflected by higher curd firming rates and gel firmness after a given renneting time. The BB-variants of κ-CN and β-Lg have also been associated with superior cheesemaking properties, as reflected by the higher recovery of fat, a lower level of curd fines in cheese whey, and higher actual and moisture-adjusted cheese yields for a range of varieties, including Cheddar, Svecia, Parmigiano-Reggiano, Edam and Gouda, low-moisture Mozzarella and Camembert (Aleandri *et al.*, 1990; van den Berg *et al.*, 1992; Ng-Kwai-Hang and Grosclaudc, 1992; Walsh *et al.*, 1995, 1998). Reported increases in MACY with the κ-CN BB variant range from ~3 to 8%, depending on milk composition and cheese type. The superior rennet coagulation and cheese-yielding characteristics of the κ-CN BB variant compared to the AA variant probably result from its higher level of κ-CN as a percentage of total casein, smaller micelles and lower negative charge. These properties are conducive to a higher degree of casein aggregation and a more compact arrangement of the *para*-casein micelles, which in turn favours more numerous inter-micellar bonds during gel formation. Indeed, using model rennet coagulation studies, it has been shown that for a given casein concentration, the curd firming rate of rennet-treated micelle suspensions is inversely proportional to the cube of the micelle diameter (Horne *et al.*, 1996). The generally higher casein content of milk containing the κ-CN BB variant compared to the AA variant also contributes to its superior rennet coagulation and cheese-yielding properties. Generally, the κ-CN AB variant has been found to exhibit rennet coagulation and cheese-yielding characteristics which are intermediate between those of κ-CN AA and BB.

25.5 Effect of protein on the rheological properties of unheated cheese

The ability of cheese to fulfil its requirements as a food or as an ingredient depends on its functional properties, which may be defined as those physico-chemical, micro-structural, rheological and organoleptic characteristics which affect the behaviour of cheese in food systems during food preparation, processing, cooking and consumption. Protein, a major component in most cheeses, has a major influence on the functional properties of both unheated and heated cheese.

When used as an ingredient in food applications, cheese is required to perform one or more functions. In the unheated state, cheese is generally required to exhibit a number of rheology-based functional properties so as to facilitate its size reduction in the primary stages of the preparation of various dishes, e.g., the ability to crumble easily, to slice or to shred cleanly and to bend when in sliced form (Table 25.2). During size reduction operations, cheese is generally subjected to relatively high stresses (e.g., > 600 kPa) and strains (e.g., $\gg 0.02$) which result in fracture: portioning of cheese into retail sizes, shredding into thin narrow cylindrical pieces (e.g., 2.5 cm long and 0.4 cm diameter), dicing into very small cubes (0.4 cm) and comminution by forcing pre-cut cheese through die plates with narrow

TABLE 25.2
Functional properties of raw cheese which influence its functionality as an ingredient

Property	*Definition*	*Cheeses which generally display the property*	*Positively associated rheological parameters*
Shreddability	The ability of a cheese block to: • Shred into thin strips of uniform dimensions • to resist fracture during shredding • to resist clumping/ balling during shredding	Low moisture Mozzarella, Swiss-type cheese, Medium-aged Cheddar, Gouda, Provolone	Elasticity, springiness, firm, long
Sliceability	The ability to be cut cleanly into this slices without fracturing or crumbling or sticking to cutting implement	Low-moisture Mozzarella, Swiss-type cheese, Provolone, Analogue pizza cheese, PCPs (some)	Elasticity, springiness, firm, long
Gratability	The ability of the cheese to fracture (elastically) into small particles, with a low tendency to stick, on shearing and crushing	Parmesan, Romano	Elastic fracturability
Spreadability	The ability to spread easily when subjected to a shear stress	Mature Camembert, Cream cheese, Mature Blue cheese	Plastic fracturability Soft, adhesiveness, Short
Crumbliness	The ability to breakdown into small irregular shaped pieces when rubbed (at low deformation)	Blue cheese, Cheshire	Elastic fracture at low deformation, Low cohesion

From Fox *et al.* (2000). PCP = Processed cheese product.

apertures. Similarly when eaten, cheese is subjected to a number of strains which reduce it to a paste capable of being swallowed; first, the cheese is bitten (cut by the incisors), compressed (by the molars) on chewing, and sheared (between the palate and the tongue, and between the teeth).

Cheese rheology and the factors that affect it have been reviewed extensively (van Vliet, 1991; Visser, 1991; Rao, 1992; Prentice *et al.*, 1993). The rheology of cheese is a function of various factors, including its composition, micro-structure (i.e., the spatial arrangement of its components and the strength of attractions between the structural elements) and the physico-chemical state of its components (e.g., degree of casein hydrolysis). The concentration and type of protein have a major influence on the latter parameters.

The effects of protein on rheology and texture are confirmed by the positive correlation between the volume fraction of the casein matrix and cheese firmness and fracture stress (de Jong, 1977; Guinee *et al.*, 2000a), and by the effects of gel fineness or coarseness on the rheological characteristics of the gel (Green *et al.*, 1983, 1990a; Guinee *et al.*, 1993).

25.5.1 Effect of protein concentration on rheology

The strength of a gel is dependent on the volume fraction and homogeneity of the gel, which determines the number of stress-bearing strands per unit area of the gel. Considering a gel to which a relatively small stress (i.e., much less than the yield stress) is applied in the direction x, the elastic shear modulus (G′, i.e., ratio of shear stress to shear strain, σ/γ), which is an index of elasticity or strength of the gel, can be related to the number of strands per unit area according to the equation (Walstra and van Vliet, 1986):

$$G' = CNd_2A/dx^2,$$

where N = number of strands per unit area of the gel in a cross section perpendicular to x, bearing the stress; C = is a coefficient related to the characteristic length determining the geometry of the network; dA = change in elastic energy when the aggregates in the strands are moved apart over a distance, dx, on the application of the stress. The number of strands per unit area of a gel are determined by:

- the concentration of gel-forming protein, and
- the fineness or coarseness of the gel, with a fine gel network having a greater number of stress-bearing strands than a coarse gel.

Hence, the G′ of acid casein gels prepared from model milks which were subjected to a standard pasteurization treatment (95°C for 5 min) increased progressively as the total protein content was increased in the range 30 to 55 g/kg (Figure 25.11). The model milks, which were prepared by blending

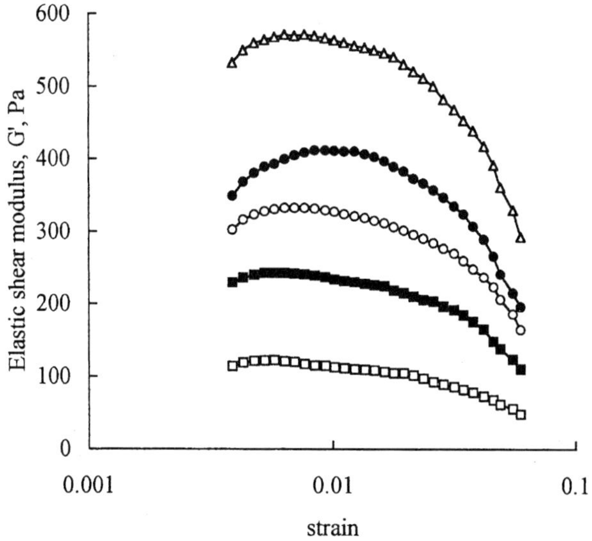

Figure 25.11 Effect of protein on the elastic shear modulus of low-fat (3.7 g/kg) model acid gels from milks prepared by blending micellar casein, whey protein isolate and ultrafiltered milk permeate in fixed proportions in de-ionized water. The model milks were made to different protein levels, adjusted to pH 6.65 and heat-treated (95°C for 5 min). Denatured whey protein in these milks was 38% of total protein. The concentrations (g/kg) of total protein and gel-forming protein (casein plus denatured whey protein) in the milk were 39.2 and 36.3 (□), 44.3 and 41.0 (■), 48.5 and 44.9 (●); 54.2 and 50.1 (○); 59.4 and 54.9 (△) (M.A. Fenelon and T.P. Guinee, unpublished data).

phosphocasein, whey protein isolate and ultrafiltered milk permeate in de-ionized water, had a fixed level of denatured whey protein (38% of total protein) after pasteurization. Similarly, the G′ of acid casein gels prepared from model milks with an equal concentration of total protein (48 g/kg) and subjected to a similar-heat treatment (95°C for 5 min) increased as the level of *in situ* whey protein denaturation was increased (Figure 25.12). The latter trend is expected, as increasing the concentration of denatured whey protein as a proportion of the total gel-forming protein results in a finer gel network (see Guinee *et al.*, 1993; Mulvihill and Grufferty, 1995).

On the application of a stress to a cheese, the volume fraction of the *para*-casein matrix determines the extent of deformation. As the concentration of casein increases, the intra- and inter-strand linkages become more numerous and the matrix becomes more elastic (Ma *et al.*, 1997) and more difficult to deform at room temperature (de Jong, 1976, 1977, 1978a,b; Prentice *et al.*, 1993). At low temperatures (<5°C), milk fat is predominantly solid and adds to the elasticity of the casein matrix (Figure 25.13). The solid fat globules limit the deformation of the casein matrix, as deformation of the

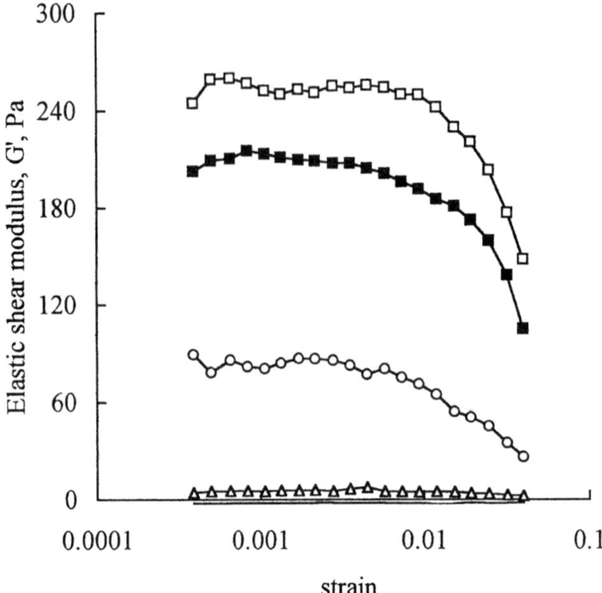

Figure 25.12 Effect of whey protein denaturation on the elastic shear modulus of low-fat (4 g/kg) model acid gels from milks prepared by blending micellar casein, whey protein isolate and ultrafiltered milk permeate in different proportions in deionized water. The model milks were made to ~48 g/kg total protein and the whey protein content was varied from ~11.7 to 40.3% of total protein; the model milks were adjusted to pH 6.55 and heat-treated at 95°C for 5 min. The concentrations of gel-forming protein (i.e., casein and denatured whey protein, g/kg) and denatured whey protein (% of total gel-forming protein) protein were: 43.0 and 34.4 (□); 43.0 and 26.1 (■); 42.9 and 17.3 (○); 42.3 and 4.1 (△) (M.A. Fenelon and T.P. Guinee, unpublished data).

latter would also require deformation of the enmeshed fat globules. However, the contribution of fat to the elasticity of cheese decreases rapidly as the ratio of solid-to-liquid fat decreases with increasing temperature and is very low at 40°C, where all the milk fat is liquid (Figure 25.13). Moreover, G′ for half-fat Cheddar is lower than that of full-fat Cheddar at 4°C, where the fat is solid but similar where fat is liquid, even though the dry matter content of the latter is higher than that of the former (Figure 25.13). A recent study of Cheddar cheeses of varying fat content showed that there were significant positive correlations between the content of intact casein and fracture stress and firmness (Guinee *et al.*, 2000a; Figure 25.14). Hence, reduced-fat Cheddar, which contains a higher volume fraction of *para*-casein matrix than full-fat Cheddar, is firmer and has a higher fracture stress than the latter (Emmons *et al.*, 1980; Bryant *et al.*, 1995; Mackey and Desai, 1995; Drake *et al.*, 1997; Figure 25.15).

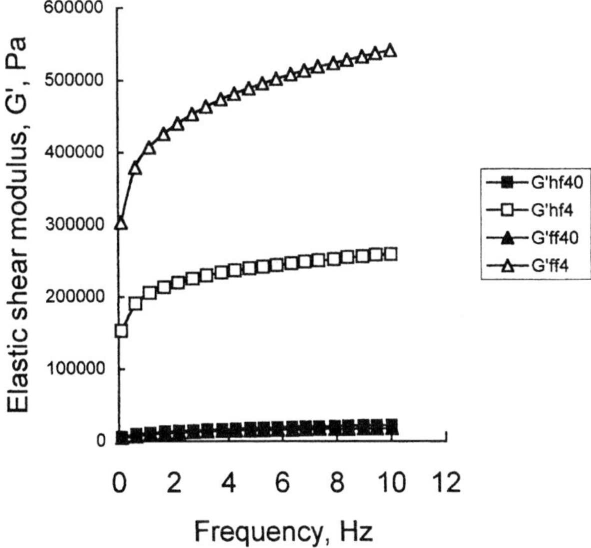

Figure 25.13 Elastic shear modulus as a function of frequency for 120 day-old full-fat (ff, △, ▲) and half-fat (hf, □, ■) Cheddar cheeses at 4 (△, □) or 40 (▲, ■)°C. Samples were tempered at the assay temperature for 20 min and subjected to a low amplitude oscillating strain of 0.06 and frequency of 1 (M.A. Fenelon and T.P. Guinee, unpublished data).

25.5.2 Effect of denatured whey proteins

The structure of acid- or rennet-induced milk gels, i.e., its degree of coarseness or fineness, is influenced markedly by the ratio of casein-to-whey protein, with large effects on the rheological properties of the resultant gel and the cheese formed therefrom (see Guinee *et al.*, 1993; Mulvihill and Grufferty, 1995). The ratio of casein-to-whey protein in cheese can be altered by *in situ* denaturation of whey proteins in the milk, milk ultrafiltration, or by the addition of denatured whey proteins.

(a) High heat treatment of milk

In the manufacture of fresh acid-curd cheeses, such as Quark, high heat treatment of milk (e.g., 95°C × 5 min) prior to culturing gives a smoother and firmer consistency (Figure 25.16). This effect is due to the increase in the level of gel-forming protein and the finer gel structure. High heat treatment causes extensive denaturation of the whey proteins (e.g., >70% of total) and their binding (especially β-lactoglobulin), *via* disulphide interaction, to κ-CN; the denatured whey proteins subsequently become part of the gel, which may be considered as a complex gel. Protein gels formed from two-protein systems have been categorized as filled gels or complex gels

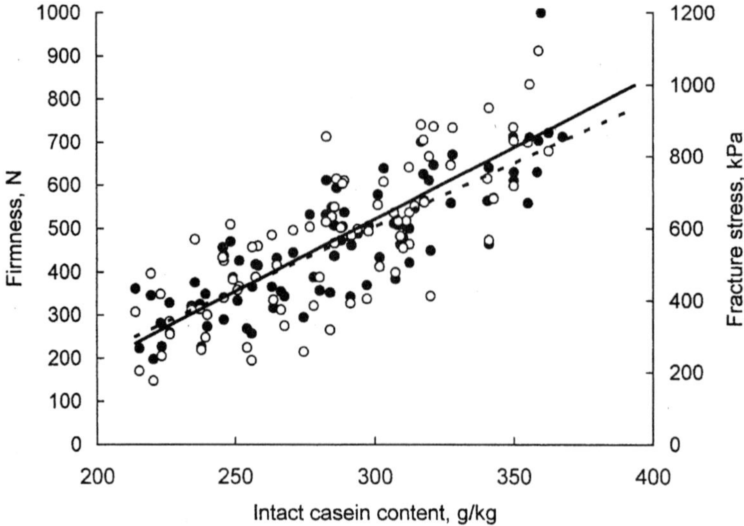

Figure 25.14 Relationship between the content of intact casein and firmness (○) and fracture stress (●) in Cheddar cheeses of varying fat content in the range 60 to 330 g/kg (redrawn from Guinee *et al.*, 2000a).

(Lanier, 1991). In complex gels, the minor protein interacts and complexes with the main gel-forming protein, whereas in filled gels, the minor protein behaves as a non-interactive filler within the gel pores (either in solution or as dispersed particles) unassociated with the main gel-forming protein. Electron microscopic analysis of high heat-treated milk show that these complexes result in the formation of filamentous appendages which protrude from the micelle surface (Harwalkar and Kalab, 1981, 1988; Heertje *et al.*, 1985) and prevent the close approach, and hence extensive fusion, of micelles on subsequent acidification. Hence, the micelles with attached appendages form into thin strands (chains), giving a highly branched, continuous gel network with smaller interstitial spaces or pores. The more finely structured gel has a higher volume fraction, a lower permeability coefficient (and hence lower porosity) and a reduced propensity to spontaneous wheying-off. The relatively low porosity of the gel is conducive to a relatively high water-holding capacity and a low tendency towards syneresis of the resultant fresh cheese, e.g., Quark (Figure 25.16B). Conversely, when micelle fusion is more extensive, the resultant gel has thicker strands, is less continuous, more porous and more susceptible to syneresis on storage (Figure 25.16B). The curd (i.e., Quark) formed after breaking/stirring and concentration of the gel reflects the rheological and syneretic properties of the gel.

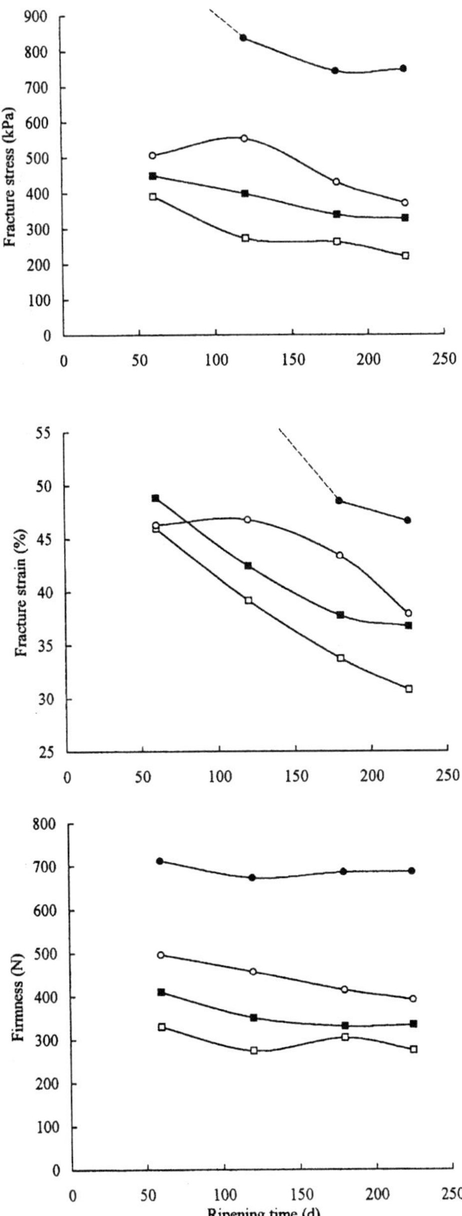

Figure 25.15 Effect of fat level on fracture stress, fracture strain and firmness of Cheddar cheese, compressed at 5 cm/min. Fat level: 6 (●), 17 (○), 22 (■), 33 (□) g/100 g (redrawn from Fenelon and Guinee, 2000).

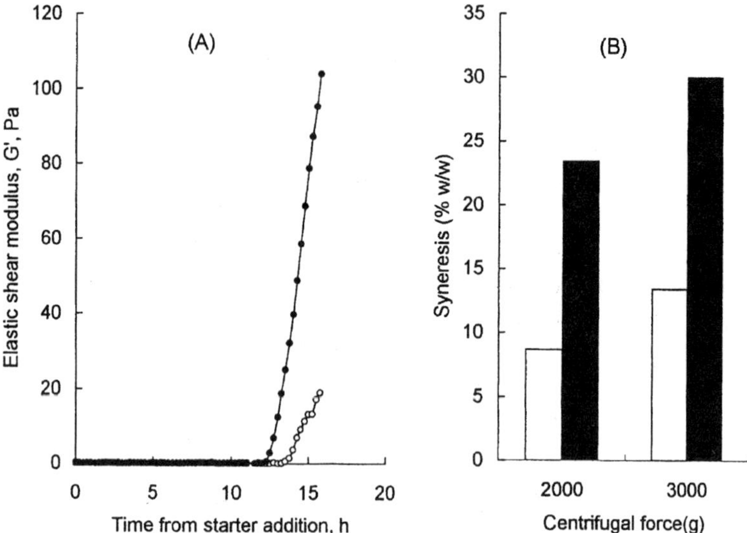

Figure 25.16 Development of elastic shear modulus during fermentation of skim milk pasteurized at 72°C × 15 s (○) or at 90°C × 300 s (●) (A) and the level of syneresis of Quarg cheeses made from the resultant acid-gels: 72°C × 15 s (■) or 90°C × 300 s (□) (B). The skim milk was inoculated with starter culture and incubated at 22°C; following fermentation, the gels (pH 4.6) were stirred gently and samples were weighed in centrifuge tubes and allowed to stand at 8°C for 36–48 h prior to centrifugation. Percentage syneresis was defined as the weight of whey expelled on centrifugation as a percentage of the original sample weight (redrawn from Fox *et al.*, 2000).

Similar to acid milk gels, *in situ* denaturation of whey proteins by high heat treatment of milk limits the degree of aggregation of the *para*-casein micelles in rennet-induced milk gels. However, in contrast to acid-induced gels, the resultant incorporation of whey proteins into the rennet-induced milk gels adversely affects their rheological properties and those of resultant cheese, as reflected by longer gelation times, lower curd firming rates, lower gel strength, lower curd firmness and fracture stress (Singh *et al.*, 1988; Green, 1990b; McMahon *et al.*, 1993; Guinee *et al.*, 1997b, 1998). Indeed, increasing the level of whey protein denaturation to >30% of total, by high heat treatment of milk (e.g., >80°C × 15 s), impairs rennet coagulability to such an extent that the milk is unsuitable for commercial cheese manufacture (Figure 25.17). The negative effect of HHT on the rennet coagulation properties of rennet gels has been attributed to:

- the complexation of denatured whey proteins with κ-CN, leading to the formation of appendages which protrude from the micelle surfaces and render the Phe_{105}-Met_{106} bond of κ-CN less susceptible to hydrolysis by rennet,

- a reduction in the concentration of native micellar calcium phosphate and ionic calcium, and
- the steric hindrance caused by the filamentous appendages (i.e., the denatured whey protein) protruding from the surface of casein micelles (van Hooydonk *et al.*, 1987; Lucey, 1995).

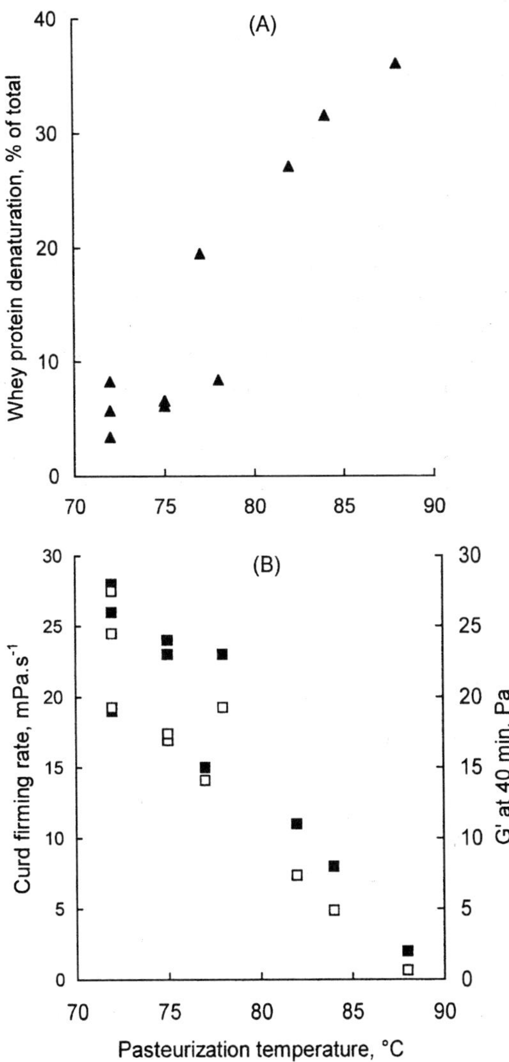

Figure 25.17 Effect of pasteurization temperature (for 15 s) of milk on the level of whey protein denaturation (A) and rennet coagulation properties of milk (B): curd firming rate (■) and elastic shear modulus at 40 min (□) (redrawn from Guinee *et al.*, 1997b).

These complexes impede casein aggregation/fusion during gel formation and post-cutting operations, shrinkage of the gel network and, hence, syneresis (Pearse *et al.*, 1985; Green, 1990a,b; Guinee *et al.*, 1995). The relatively fine structure of the rennet-induced gel from HHT milk increases its water-holding capacity through increased capillary action. Consequently, cheese prepared from HHT milk (e.g., 82°C for 15 s) has a higher moisture content, a lower fracture stress and a lower firmness than cheese made from milk pasteurized at normal temperature (e.g., 72°C for 15 s) (El-Koussey *et al.*, 1977; Marshall, 1986; Guinee *et al.*, 1998; Figure 25.18). Owing to its effect on cheese rheology, high levels of denatured whey proteins in cheese milk may be exploited as a means of improving the texture (reducing the firmness and elasticity) of low-fat cheese which tends to be excessively firm and rubbery (Guinee *et al.*, 1998).

The rennet coagulability of HHT milk can be improved by increasing coagulation temperature, acidification to a lower than normal pH at setting or addition of $CaCl_2$ (Banks *et al.*, 1987; Singh *et al.*, 1988; van Hooydonk and van den Berg, 1988; Lucey, 1995). The beneficial effects of adding $CaCl_2$ include increased $[Ca^{2+}]$ and [CCP] and a concomitant decrease in pH (the addition of $CaCl_2$ to 0.02%, i.e., 1.8 mM Ca, reduces the pH by ~0.05–0.1 units, depending on the concentration of protein). The compositional changes induced by $CaCl_2$ addition contribute to an increase in the extent of aggregation of rennet-treated *para*-casein micelles, an effect thought to be

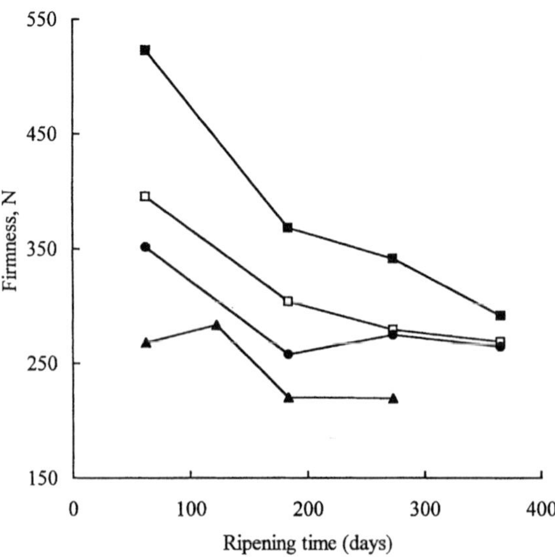

Figure 25.18 Effect of milk pasteurization temperature on the firmness of reduced-fat Cheddar cheese made from milk pasteurized at 72 (■), 77 (□), 82 (●) or 84 (▲)°C for 15 s (redrawn from Guinee *et al.*, 1998).

due to a reduction in micellar negative charge (due to the higher [Ca^{2+}] and lower pH) and inter-micellar repulsion (Horne, 1998). The rennet coagulation properties of severely heat-treated milk are also improved by acidification (e.g., to pH < 6.4), holding for ~2 h at 20°C, and re-neutralization to the natural pH of milk; the level of improvement increases with the degree of acidification in the pH range 6.4 to 5.8 (Singh *et al.*, 1988; Lucey, 1995; Figure 25.19). The improved rennet coagulation properties appear to result from the solubilization of indigenous CCP and the subsequent increase in the [Ca^{2+}]; high heat-treated milk acidified to pH 5.8 and re-neutralized to pH 6.6 has a higher [Ca^{2+}] than non-acidified milk (Singh *et al.*, 1988). However, despite the improvement in rennet coagulation obtained with acidification/re-neutralization, the curds from such milks (e.g., with whey protein denaturation of ≈50% total) are susceptible to shattering on cutting and exhibit impaired syneresis, as reflected by a high level of curd fines, low recovery of fat and high moisture content of the cheese (Guinee, unpublished results; Table 25.3). The impaired syneresis may be associated with the reduced level of indigenous CCP, which is considered to act as a cementing agent between casein sub-micelles in milk

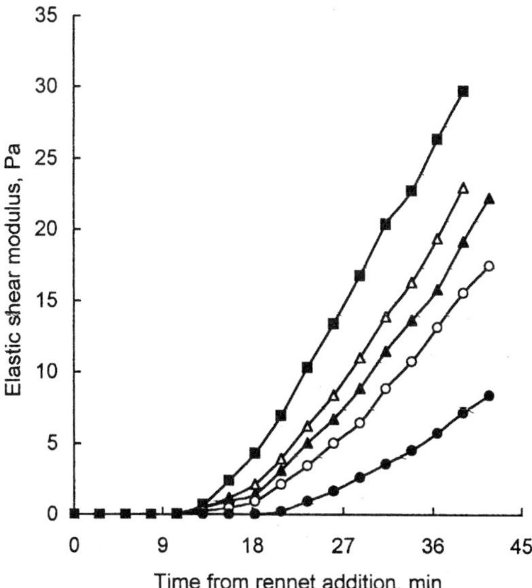

Figure 25.19 Development of elastic shear modulus in rennet-treated milk blends prepared by mixing equal weights of unheated skim milk (protein, 30 g/kg) and low heat skim milk powder reconstituted in de-ionized water (protein, 30 g/kg). The skim milk blends were adjusted to pH 6.6 (○), 6.3 (▲), 6.0 (△) or 5.8 (■) at 21°C, allowed to stand for 2 h at 21°C, and re-neutralized to pH 6.6; G′ for reconstituted low-heat skim milk powder in de-ionized water (protein, 30 g/kg) (●) (T.P. Guinee and B.T. O'Kennedy, unpublished data).

and to contribute to charge neutralization (and attraction between sub-micelles), owing to its binding to negatively charged phosphoserine residues (Horne, 1998). The latter effect may not be as prominent in acidified milk that is not re-neutralized to normal pH since in this case the increased [Ca^{2+}] could contribute charge-neutralization and thereby compensate for the reduction in the concentration of CCP. Hence, good quality Cheddar cheese with an acceptable moisture content has been manufactured on an experimental scale from high heat-treated milk (110°C × 60 s) which had been acidified to pH 5.8 at setting and subjected to an altered cheesemaking process (Banks *et al.*, 1987; Banks, 1990).

(b) Addition of denatured whey proteins
Varying results have been obtained with the addition of partially denatured whey protein concentrates (PDWPC) (prepared by the Centriwhey, Lactal or UF processes) to milk for the manufacture of hard or semi-hard cheese, such as Cheddar and Gouda. There is a general agreement that the addition

Table 25.3
Cheddar cheesemaking characteristics of milks prepared by blending skim milk and reconstituted low-heat skim milk powder (T.P. Guinee & B.T. O'Kennedy, unpublished data)

	Ratio of skim milk to reconstituted skim milk powder in the cheese milk		
	100:00 (Control, T1)	*50:50[1] (T2)*	*50:50[2] (T3)*
Cheese composition[3]			
Moisture (g/kg)	379.4	383.8	396.4
Fat (g/kg)	327.5	322.5	316.9
Protein(g/kg)	254.9	253.7	250.3
Whey composition[3]			
Fat (g/kg)	3.6	4.4	4.8
Protein (g/kg)	9.7	9.5	9.6
Curd fines (mg/kg)	178.0	288.0	279.0
Yield[3]			
MACY (kg/100 g)[4]	9.68	9.53	9.32
Actual (kg/100 g)	9.67	9.59	9.58

[1]Low-heat skim milk powder was reconstituted to a protein level simliar to that in the raw skim milk and then blended with the skim milk at a ratio of 50:50. The blend was then standardized to a protein-to-fat ratio as for normal Cheddar cheese.
[2]Low-heat skim milk powder was blended with raw skim milk as in 1. The standardized milk was then pasteurized, cooled to 30°C, acidified to 6.3, cheddared for 2 h and then neutralized to pH 6.55 prior to rennet addition.
[3]The values shown are the means of quadruplicate trials.
[4] MACY, moisture-adjusted yield; moisture adjusted to 380 g/kg.

of WPC increases the moisture content, actual yield and moisture-adjusted yield, with the extent of the increase being correlated positively with the degree of denaturation of the added WPC (van den Berg, 1979; Brown and Ernstrom, 1982; Banks and Muir, 1985; Baldwin *et al.*, 1986). However, the addition of PDWPC has, generally, been found to cause defective body (greasy, soft) and flavour (unclean, astringent) characteristics in Gouda and Cheddar cheese (van den Berg, 1979), with the intensity of the defects becoming more pronounced with increasing level of the PDWPC added. It has been suggested that these defects may be a consequence of the large size of whey protein particles (aggregates) which do not fit compactly within the pores of, and thereby impede shrinkage of, the *para*-casein matrix (van den Berg, 1979) and therefore give relatively large increases in moisture content.

The addition of whey protein-based fat mimetics, e.g., Simplesse® 100 or Dairy Lo™, to cheese milk improves the texture of reduced-fat Cheddar (as reflected by the reduced fracture stress, fracture strain and firmness) and has little influence on flavour (Lucey and Gorry, 1994; Fenelon and Guinee, 1997). Mackey and Desai (1995), who studied the micro-structure of reduced-fat Cheddar containing Simplesse® 100 using TEM, concluded that the micro-particulated whey protein acts as a non-interacting filler in the cheese matrix, in the formation of a filled gel, and mimics the structural properties of fat. In contrast, Fenelon and Guinee (1997) suggested that denatured whey proteins (Dairy Lo™) interact with casein to form a complex-type gel. This suggestion was supported by the impaired curd-forming properties of the milk with added Dairy Lo™ compared to the control and the similar protein content of the control and experimental cheeses.

25.5.3 Effect of casein hydrolysis on rheology

Factors that promote weakening of the casein matrix reduce the stress required to achieve a given deformation. Hence, the firmness and fracture stress of Cheddar-type cheese decrease as the level of intact *para*-casein decreases (de Jong, 1976, 1977, 1978b; Guinee *et al.*, 2000a). Moreover, for a given cheese variety with a fixed level of casein, the fracture stress and firmness of cheese generally decrease with ripening time due to proteolysis (de Jong, 1976, 1977, 1978b; Creamer and Olson, 1982), and the concomitant increase in casein hydration. It has been inferred from micro-structural observations that proteolysis may result in discontinuities or 'breaks' in the *para*-casein matrix at the micro-structural level (de Jong, 1978a). The early hydrolysis of α_{s1}-CN at the Phe_{23}-Phe_{24} peptide bond by residual chymosin results in a marked weakening of the *para*-casein matrix and reductions in fracture stress and firmness (de Jong, 1976, 1977; Creamer and Olson, 1982).

The sequence of residues 14–24 of α_{s1}-CN is strongly hydrophobic and confers intact α_{s1}-CN with strong self-association and aggregation tendencies in the cheese environment (Creamer *et al.*, 1982). It has been suggested that self-association of α_{s1}-CN in cheese, *via* these hydrophobic patches, leads to extensive cross-linking of *para*-casein molecules and thus contributes to the overall continuity and integrity of the matrix (de Jong 1976, 1978a; Creamer *et al.*, 1982; Lawrence *et al.*, 1987). Hence, de Jong (1978b) reported a linear relationship between the content of intact α_{s1}-CN and the softness in Meshanger cheese (a soft, internal bacterial-ripened Dutch variety), in which proteolysis was varied by altering the quantity of added coagulant. In contrast to the above, the firmness of some cheeses (e.g., brine-salted and/or surface dry-salted varieties that are not packaged for part of their ripening period) may increase initially even though proteolysis occurs during this period. The increase in firmness is a consequence of the loss of moisture and the concomitant increase in the protein concentration (de Jong, 1978b; Visser, 1991). Other factors such as changes in pH and the increase in the salt content in the inner regions (Visser, 1991) as a result of inward diffusion from the surface rind zone, may also contribute to the initial increase in firmness. However, when the composition has stabilized, the softening associated with proteolyis becomes dominant, and the firmness and fracture stress decrease (de Jong, 1978b; Visser, 1991).

Few studies have measured the rheological characteristics of cheeses with similar levels of α_{s1}-CN degradation but with markedly different degrees of β-CN hydrolysis. This is probably because of the fact that β-CN generally undergoes markedly less breakdown than α_{s1}-CN during the ripening of most cheeses such as Cheddar, Gouda and Mozzarella (Visser and de Groot Mostert, 1977; de Jong, 1977; Fox, 1989; Fox and McSweeney, 1996; Yun *et al.*, 1993a; Fox *et al.*, 1996). The slow breakdown appears to be related to:

- intermolecular hydrophobic interactions between the hydrophobic C-terminal region of β-CN which contains the chymosin-sensitive bonds, and/or
- the selective inhibition of chymosin/pepsin hydrolysis of β-CN at the relatively low water activity in cheese, e.g., ~0.91–0.96 for hard/semi-hard varieties (Fox, 1989).

Hence, it is difficult to alter the degree of degradation of β-casein, especially while maintaining the degree of α_{s1}-CN degradation constant. However, differences in the level of β-CN degradation between cheeses have been achieved by the use of either calf rennet or *Cryphonectria parasitica* proteinase as coagulant, with the higher level of degradation being obtained with the latter. A higher degree of β-CN breakdown in Cheddar cheese with a similar level of α_{s1}-CN degradation resulted in a significantly lower fracture strain at all times (270 d), a numerically lower fracture stress at

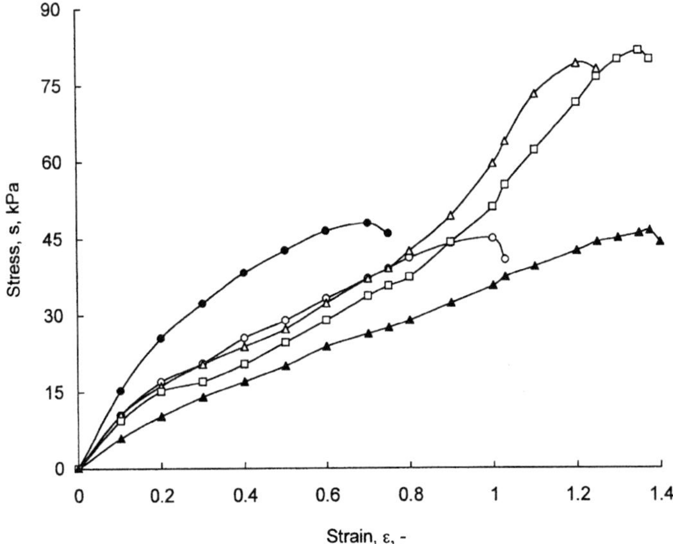

Figure 25.20 Effect of pH on the force compression response of 1 week-old Gouda cheese, the composition of which was, otherwise, similar; cheese pH: 5.02 (●), 5.1 (○), 5.2 (▲), 5.43 (□), 5.58 (△) (redrawn from Visser, 1991).

<100 d, a numerically lower stress at 80% compression and a higher stress at 20% compression at <200 d (Bogenrief and Olson, 1995). Yun *et al.* (1993a,b) reported that higher levels of β-CN degradation (e.g., 30 *vs.* 18% of total) in Mozzarella cheese, with similar levels of α_{s1}-CN degradation, coincided with numerically lower firmness values at $\leqslant$30 d.

25.5.4 Degree of protein hydration/aggregation

The degree of *para*-casein hydration is affected by many parameters, including pH and the ratio of soluble to colloidal calcium. The pH of cheese markedly affects its rheology, both directly due to its influence on *para*-casein hydration (Creamer, 1985), and indirectly due to its influence on the activity of residual rennet and plasmin, and hence, casein hydrolysis (Grufferty and Fox, 1988; Farkye and Fox, 1990; Guinee and Wilkinson, 1992). Hydration of *para*-casein increases as the pH is reduced in the range 6.7 to 5.3–5.4 but decreases as the pH is reduced further from ~5.3 to 4.6 (Creamer, 1985). Owing to the effects of pH on *para*-casein hydration, it is expected that aggregation of the protein increases markedly as the pH is increased from ~5.2 to 5.8 or reduced from ~5.2 to 4.6. Hence, the elastic modulus and fracture stress of cheese decreases as the pH is increased from 4.8 to 5.2 (Creamer and Olson, 1982; Luyten, 1988; Figure 25.20). In contrast, a pH increase in the range 5.2 to 5.6 resulted in a marked increase in fracture stress (to values much higher than those at pH <5.2) and a slight increase in the modulus of elasticity, E. Differences in the hydration of *para*-casein, as affected by pH and colloidal calcium, may help to explain the differences in the textural and rheological characteristics exhibited by different cheese varieties (Lawrence *et al.*, 1987). Hence, for a similar concentration of calcium (29–33 mg Ca/g protein), the fracture stress and fracture strain of hard/semi-hard cheeses with a relatively low pH (e.g., Cheddar) tend to be lower than those of high pH cheeses (e.g., pH 5.35–5.50), such as Emmental and Gouda. The latter cheeses tend to be more elastic and springy and fracture into larger pieces (Prentice *et al.*, 1993). The more extensive hydrolysis of casein by residual coagulant probably also contributes to the lower fracture stress and strain of low pH cheeses; the proteolytic activity of rennet increases as the pH is reduced (Guinee and Wilkinson, 1992).

The age-related change in hydration of *para*-casein, as affected by pH and the ratio of soluble-to-colloidal Ca, is a major factor controlling the rheological changes in surface mould-ripened cheese, e.g., Camembert and Brie, during ripening. In these varieties, softening of the cheese proceeds from the outside to the interior, with the ripened cheese ideally having a decreasing softness gradient from the surface to the mass centre. The ripe cheese also has:

- a decreasing pH gradient from the surface to the centre, as a consequence of oxidation of lactate in the outer region by the surface microflora and the concomitant diffusion of lactic acid from the centre towards to the surface (Lenoir, 1984; Karhadian and Lindsay, 1987), and
- a decreasing Ca gradient in the same direction due to precipitation of soluble calcium phosphate as insoluble $Ca_3(PO_4)_2$ in the high pH surface region and the concomitant diffusion of soluble Ca from the centre towards to the surface (Le Graet *et al.*, 1983; Karahadian and Lindsay, 1987).

Prior to the model studies of Noomen (1983), it was generally considered that the textural changes in these cheeses were due to casein hydrolysis, as mediated by the inward diffusion of proteinases produced by the surface flora (i.e., *P. camemberti*). However, proteinases diffuse very slowly through the aqueous phase of cheese owing to their relatively high molecular mass, e.g., diffusion coefficients of the order of 0.003 cm^2/day compared to 0.2 cm^2/day for NaCl at room temperature (de Jong, 1978b; Lee *et al.*, 1980). Hence, mould proteinases contribute little, *via* direct proteolysis, to softening of the general cheese mass. The results of Noomen (1983) showed that the softening of Camembert-type cheese is due to the interactive effects of:

- hydrolysis of α_{s1}-CN by residual coagulant activity and the concomitant removal of the hydrophobic peptide, α_{s1}-CN (f 1–23); plasmin and

proteinases from the surface mould (e.g., *P. camemberti*) contribute to the hydrolysis of α_{s1}- and β-caseins, especially in the surface regions where the pH is highest (Karahadian and Lindsay, 1987),
- the increase in casein hydration as the pH increases from 4.6 in the fresh cheese to ~6.0 in the centre and ~7.0 at the surface of the ripe cheese at ~30 d.

Moreover, the higher pH at the surface results in the precipitation of Ca and inorganic phosphorus as insoluble colloidal calcium phosphate which accumulates in the rind (Le Graet *et al.*, 1983; Karahadian and Lindsay, 1987). As a consequence, soluble Ca diffuses from the centre towards the surface in an attempt to restore equilibrium. This outward diffusion of soluble Ca is paralleled by the concomitant solubilization of CCP in the low-pH central zones (cf., Section 3, Figure 25.5). The decrease in the concentration of CCP is an additional factor contributing to the hydration of the *para*-casein.

25.6 Effect of protein on the functional properties of unheated cheese

The functional properties of unheated cheese are determined largely by its rheological and flavour/aroma characteristics. While the rheological properties do not directly effect the flavour and aroma, their influence on the rate and extent of breakdown of the cheese mass during mastication may alter the latter characteristics indirectly. Hence, a cheese with low values of fracture stress and strain is expected to deform rapidly and release its fat more quickly for a given residence time in the mouth. Free fat quickly coats parts of the mouth, allowing the aroma and taste sensations of its volatile and non-volatile flavour compounds to be perceived rapidly. The flavour/aroma characteristics of cheese have been reviewed (Fox and Wallace, 1997) and will not be discussed further.

The primary stage of preparation of any food containing cheese requires that the cheese mass be reduced in size to facilitate dispersion, mixing and/or layering onto the food. Size reduction is usually achieved by cutting into relatively large pieces, crumbling, slicing, shredding, dicing, grating and/or shearing; these actions usually involve a combination of compressive and shear stresses. The behaviour of the cheese when subjected to the different size-reduction methods constitutes a group of important functional properties, which are listed in Table 25.2. These properties are related to the rheological characteristics of the cheese, which determine the magnitude of the strain (e.g., change in dimensions, fracture stress) on the application of the stresses applied by the different size-reduction methods. The rheology-related functional properties of the unheated cheese have a major impact on its suitability for a particular application.

Little or no published information is available on the direct effect of protein on the rheology-related functional properties (e.g., sliceability, crumbliness) of unheated cheese. However, the effect of protein on these properties may be deduced from its effect on rheological parameters such as fracture stress, fracture strain and firmness, which were discussed in Section 25.5.

The rheology-related functional characteristics of the unheated cheese determine its suitability for a particular application, as illustrated by the examples below. Mature Camembert or Chaumes, which are soft, "short" and adhesive, are not used in shredded/diced cheese applications, such as pizza pie, because of their tendency to ball and clump. However, the ability of these cheeses to undergo plastic fracture and flow under shear (i.e., spread) makes them ideal for blending with other materials such as butter, milk or flour in the preparation of fondues and sauces. The brittleness and tendency of hard cheeses such as Parmesan and Romano, with low levels of moisture and fat-in-dry matter, to undergo elastic fracturability endows them with excellent gratability (when crushed between rollers) and are suitable for sprinkling onto pasta dishes. However, these properties render the latter cheeses unsuitable for applications that require slices (e.g., sandwiches and cheeseburgers) or shredded cheese. Conversely, other hard cheeses, such as Cheddar and Gouda-type, are unsuitable for grating owing to their lack of brittleness and to their elasticity and relatively high fracture stress and strain, which enables a relatively high degree of recovery to their original shape and dimensions following crushing. Moreover, the relatively high levels of moisture and fat-in-dry matter in the latter cheeses are conducive to a higher degree of flow following fracture, compared to Romano or Parmesan, and hence to the development of tackiness following crushing. Owing to its springiness, elasticity and long body, Swiss-type cheese is ideal for slicing very thinly and therefore is particularly well suited for applications such as sandwiches and stuffed cheese slices. Similarly, the springiness of low-moisture Mozzarella cheese (LMMC) endows it with good shreddability (a low tendency to fracture and form fines or curd dust) and non-stick properties and facilitates uniform distribution on the surface of pizza pies. Owing to their crumbliness, cheeses such as Feta, Cheshire or Caerphilly are particularly well suited for easy inclusion into mixed salads.

25.7 Effect of Protein on the Functional Properties of Heated Cheese

Cheese is used extensively in cooking applications owing to its heat-induced functionality, which is a composite of different attributes, including softening (melting), stretchability, flowability, apparent viscosity and tendency to brown. These attributes have a major impact on the quality of products in which cheese is used, especially grilled cheese sandwiches, pizza pie, cheeseburgers, pasta dishes and sauces. Depending on the application, one or more functional properties may be required.

Various functional properties of heated cheese and their interpretation are given in Table 25.4. Some of the main functional properties of heated cheeses may be defined as heat-induced rheological changes, involving strain displacement on the application of stress. Displacement may occur as flow when the fat globules/pools, which when solid (e.g., at 4°C) contribute to physical support of the matrix, collapse on heating. Displacement may also occur as an extension (stretch) of the molten cheese mass when the melted cheese is subjected to a uniaxial shear stress. In contrast to milk, in which it is considered that heating does not induce fat coalescence (van Boekel and Walstra, 1995), clear evidence for heat-induced coalescence of fat globules in natural cheese is provided by dynamic microscopy of cheese during heating (Paquet and Kaláb, 1988; Auty *et al.*, 1999; Guinee *et al.*, 1999, 2000c; Figure 25.21) and by the release of oil on baking (Rudan and Barbano, 1998). Heat-induced flow involves liquefaction and flow of the fat, the extent of which depends on the thermal stability of the fat globule membrane, the size distribution of fat globules and the degree of fat coalescence (Rayan *et al.*, 1980; van Boekel and Walstra, 1995).

The browning of cheese on heating is also a chemically-related functional property, occurring as a result of Maillard reactions (O'Brien, 1995; Section 25.7.4).

25.7.1 Functionality of different varieties

Studies on the functional properties of different cheese types (natural cheese, processed cheese, analogue cheese) on heating indicate that there are considerable intra- and inter-variety differences in melt time, flowability, stretchability and apparent viscosity (Park *et al.*, 1984; Guinee *et al.*, 2000b; Table 25.5). This trend undoubtedly reflects inter- and intra-variety differences in the conditions of manufacture, composition, degree of maturity and/or formulation (e.g., level and types of added ingredients, and processing conditions) in the case of the PCPs and analogue pizza cheese (Table 25.6). Compared to other varieties, *pasta filata* cheeses (e.g., Mozzarella, Provolone and Kashkaval) are differentiated by their superior stretchability, relatively high apparent viscosity and moderate flowability and melt time (Table 25.5). These functional attributes endow the *pasta filata* cheeses with the characteristics that are typically associated with the melted cheese on pizza pie, i.e., sufficiently rapid melt and desirable levels of stringiness, chewiness and flow. In contrast to the *pasta filata* cheeses, other types of cheese, including analogue pizza cheese and retail Cheddar and Emmental, have relatively low stretchability, low apparent viscosity and high or impaired flowability characteristics. If such cheeses were used on pizza pie, the melted

TABLE 25.4
Functional properties of grilled/baked cheese which influence its functionality as an ingredient

Property	*Definition*	*Cheeses which generally display this property*	*Property related to physico-chemical state*
Metability	The ability of cheese to soften to a molten cohesive mass on heating	Most cheeses after a given storage period, dependent on the variety, PCPs, APCs, cream cheese	Fat liquifaction, fat coalescence
Flowability	The ability of the melted cheese to flow	Most cheeses after a given storage period, dependent on the variety, PCPs, OACs, cream cheese	Fat liquifaction, casein hydration, high degree of fat coalescence, limited oiling-off
Stretchability	The ability of the melted cheese to form cohesive fibres, strings or sheets when extended	Low-moisture Mozzarella, Kash-kaval, young Cheddar (i.e., 15 d)	Moderate degree of casein hydration and casein aggregation, level and type of molecular attraction between *para*-casein molecules
Flow resistance (often referred to as melt resistance)	The resistance to flow of melted cheese	Paneer, PCPs, OACs, natural cheeses from high heat-treated milk	Absence of fat coalescence, heat-induced gelation of a paricular component(s) (e.g. whey protein), thermo-irreversibility of gel system in the uncooked product upon heating
Chewiness (rubbery, tough, elastic)	High resistance to breakdown on mastication	Low-moisture Mozzarella, Kash-kaval, young Cheddar (i.e., 15 d)	As for stretchability
Viscous (soupy)	Low resistance of melted cheese to breakdown on mastication	Mature Cheddar, aged Mozzarella, cream cheese, PCPs, OACs	Relatively high level of casein hydra-tion
Limited cooling-off	Ability of cheese to express a little free-oil on heating, so as to reduce cheese dehydration. Maintain succulence of, and impart surface sheen to, melted cheese	Most natural cheeses (if not very mature or very young) PCPs, APCs	Limited degree of fat coalescence
Blistering			

Key: PCP: processed cheese products; APC: analogue pizza cheese; OAC: other analogue cheese.
From Fox *et al.* (2000).

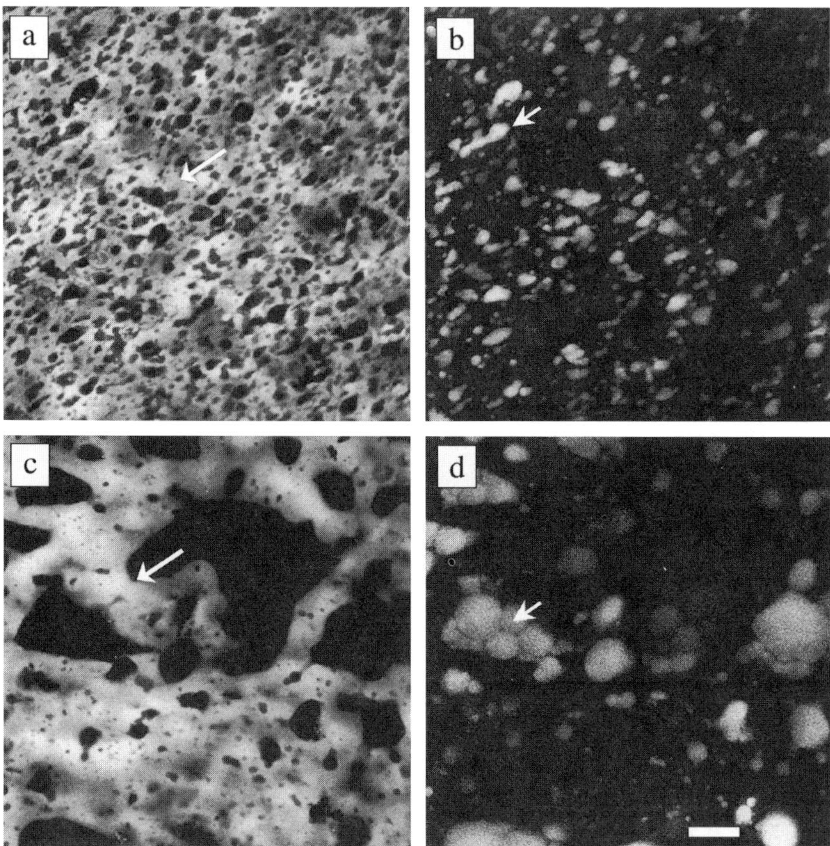

Figure 25.21 Confocal laser scanning micrographs of unheated full-fat Cheddar cheese (a, b) and the same cheese after heating to 90°C at a rate of 5°C min^{-1} and cooling to room temperature (c, d). The micrographs show protein (a, c; long arrows) and fat (b, d; short arrows) as light areas against a dark background. The effect of heating was studied using a warm-stage attachment to the Zeiss LSM 310 confocal scanning laser microscope. Samples were labelled with Nile Blue and images acquired at 488 nm excitation for fat and 633 nm for protein. Bar corresponds to 25 μm (reconstructed from Guinee *et al.*, 2000c).

cheese topping would lack the desired stringiness, would flow excessively and lack the desired chewiness. Conversely, stringiness, which is typical for LMMC and other *pasta filata* cheeses, such as Kashkaval and Provolone, is an undesirable attribute for applications such as sauces, gratins, *cordon-bleu* applications or fondues. Cheeses such as mature Cheddar, Emmental, Raclette and Gouda are more suitable for the latter applications because of their excellent flowability and flavour and absence of stringiness on grilling.

Many parameters contribute to the functional characteristics of heated cheese products (Rayan *et al.*, 1980; Harvey *et al.*, 1982; Rudan and

TABLE 25.5
Functional characteristics of different types of natural cheeses

Cheese type	*Heated/baked cheese*				
	Sample size	*Melt time (s)*	*Flowability (%)*	*Stretchability (cm)*	*Apparent viscosity (Pa.s)*
Pasta filata-type					
–Low-moisture Mozzarella	8	108 (6)	53 (8)	83 (21)	623 (303)
–Kashkaval	2	96 (11)	67 (4)	87 (13)	522 (330)
–Provolone dolce	3	86 (6)	64 (21)	80 (13)	950 (–)
–Provala fumica	1	99	67	77	–
–Provala	1	92	71	76	–
Cheese with eyes					
–Gruyére	1	105	78	67	391
–Jarlsberg	1	82	52	35	371
–Emmental	1	81	74	35	269
Cheddar	8	100 (7)	69 (9)	23 (10)	349 (129)
Analogue pizza cheese	8	105 (13)	42 (19)	27 (8)	668 (307)

†Where the sample size>2, mean values are presented; values in parentheses are standard deviations.
From Fox *et al.* (2000).

Barbano, 1998; Paulson *et al.*, 1998; Rowney *et al.*, 1999). The concentration and type of protein (i.e., casein or whey proteins) and the extent of protein hydrolysis are major determinants of functionality (Arnott *et al.*, 1957; Schulz, 1976; Lazaridis *et al.*, 1981; Mahoney *et al.*, 1982; Lawrence, 1989; Rüegg *et al.*, 1991; Kalab *et al.*, 1991; Yun *et al.*, 1993a,b; Kindstedt, 1993, 1995; Bogenrief and Olson, 1995; Guinee *et al.*, 2000a,b).

25.7.2 Effect of protein on the rheology-related functional parameters of natural cheese on heating

(a) Relationship between protein and the stretchability of pasta filata *cheese: effect of casein concentration, hydrolysis and hydration*

The superior stringiness of *pasta filata* cheese on baking, compared to other natural cheeses, may be attributed to the effects of:

- the plasticization process which contributes to enhanced aggregation of the *para*-casein at the low curd pH and high temperature (57°C),
- the correct values of pH and total calcium, and ratio of soluble Ca-to-total Ca, which interactively provide favourable conditions for stretchability (Section 25.3), and
- the relatively low degree of *para*-casein hydrolysis.

TABLE 25.6
Compositional analyses of different commercial cheese varieties†‡ (T.P. Guinee, unpublished)

Cheese type	*Source*	*Sample size*	*Moisture (g/kg)*	*Protein (g/kg)*	*Fat (g/kg)*	*FDM (g/kg)*	*MNFS (g/kg)*	*S/M (g/kg)*	*Ash*	*Ca*§	P§	*pH 4.6 SN*§	*PTAN*§	*pH*
Pasta filata type														
– Low moisture Mozzarella	Ireland UK Denmark	8	464	260	232	446	604	31	38	27.2	20.6	4.7	0.5	5.53
Kashkaval	Yugoslavia	2	441 (7)	253 (3)	256 (12)	458 (17)	596 (4)	49 (2)	43 (0)	31.0 (5)	22.2 (18)	6.1 (3.9)	0.7 (0.2)	5.37 (0.02)
Provolone dolce	Italy	2	380 (32)	276 (5)	294 (37)	477 (36)	546 (17)	57 (7)	–	–	–	10.4 (9.8)	–	5.54 (0.14)
Provola fumica	Italy	1	434	278	243	428	573	47	–	–	–	7.5	–	5.29
Provola	Italy	1	402	281	285	476	562	44	–	–	–	10.8	–	5.29
Cheese with eyes														
Gruyére	Switzerland	1	341	277	368	558	540	49	–	–	–	10.8	–	5.83
Jarlsberg	Norway	1	404	277	313	524	587	28	–	–	–	8.6	–	5.72
Emmental	Switzerland	1	343	249	380	578	553	13	–	–	–	9.2	–	5.64
Cheddar	Ireland	8	372 (11)	254 (8)	331 (20)	526 (24)	556 (10)	45 (7)	37 (3)	28.0 (1.4)	20.6 (1.3)	20.3 (3.7)	4.6 (2.5)	5.14 (0.12)
Analogue pizza cheese	Ireland	8	489 (37)	184 (19)	250 (19)	490 (34)	651 (44)	35 (4)	42 (3)	344 (21)	281 (44)	2.3 (0.8)	0.2 (0.1)	6.3 (0.1)

†Where sample size > 2, means values are presented; values in parenthesis are standard deviations.
‡FDM, fat-in-dry matter; MNFS, moisture-in-not-fat substances; S/M, salt-in-moisture; pH4.6 SN; nitrogen soluble at pH 4.6; PTAN, nitrogen soluble in 5% phosphotungstic acid.
§Units for Ca (mg/g cheese protein); P (mg/g cheese protein); pH 4.6 SN (g/100 g total N); PTAN (g/100 g total N).
From Fox *et al.* (2000).

Compared to other varieties, e.g., Cheddar, the level of proteolysis in commercial LMMC, as measured by pH 4.6-soluble N as a percentage total N (pH4.6SN%TN), is low. The low level of proteolysis is probably due mainly to the denaturation of rennet during the plasticization process, the short ripening period and the low ripening temperature (typically 3–4°C). A survey of commercial cheeses indicated that the mean concentration of pH4.6SN%TN in Cheddar and LMMC was 20.3 and 4.7, respectively (Guinee *et al.*, 2000b). However, despite the thermal inactivation of residual coagulant during plasticization (Matheson, 1981; Singh and Creamer, 1990), the rate of formation of pH4.6SN in LMMC (500 g/kg moisture) is only slightly lower than that in Cheddar (Figure 25.22). Proteolysis in LMMC is probably mainly as consequence of plasmin activity (Richardson and Pearce, 1981; Grufferty and Fox, 1988; Fox, 1989; Farkye and Fox, 1990, 1991). Undoubtedly, the higher moisture and lower salt content, compared to Cheddar, are more favourable for proteolysis by residual rennet activity in LMMC than in Cheddar (Creamer, 1976). Despite the increase in proteolysis and the concomitant decrease in the level of intact casein,

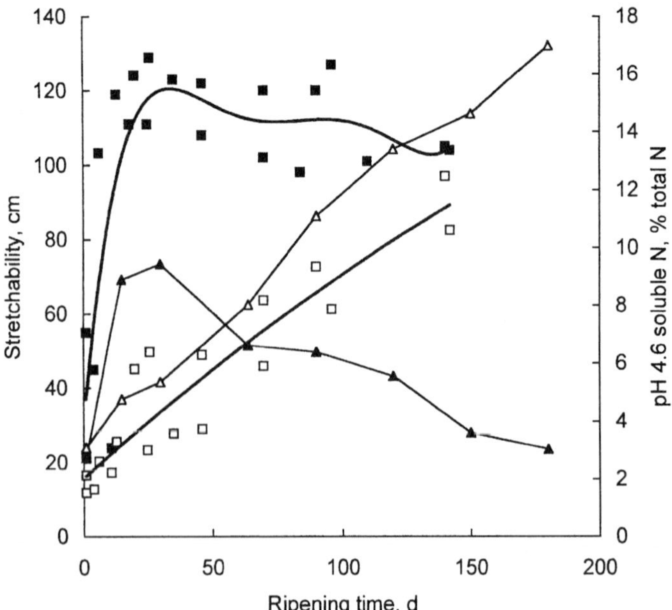

Figure 25.22 Changes in the stretchability (■, ▲) and content of pH 4.6-soluble N (□, △) in low-moisture Mozzarella (■, □) and full-fat Cheddar (▲, △) during ripening at 4°C (Mozzarella) or 8°C (Cheddar). Data presented are the means of three replicate trials for Cheddar, and from two typical Mozzarella cheeses (T.P. Guinee, unpublished data).

the stretchability of LMMC stored at 4°C remains essentially constant to ~140 d when the concentration of pH4.6SN%TN is <12% (Figure 25.22). Indeed, preliminary investigations on the effect of ripening temperature (0 to 15°C) on Mozzarella suggest that the stretchability of LMMC is not significantly impaired until the concentration of pH4.6SN%TN exceeds ~16. In contrast, the stretchability of young Cheddar cheese is acceptable up to ~30 d but then deteriorates rapidly on further ageing as the level of pH4.6SN%TN increases to values ~6 (Figure 25.22). The different stretch-time/pH4.6-soluble N profile of Cheddar and LMMC probably reflects differences in the state of aggregation of *para*-casein (as affected by the occurrence/absence of plasticization process), the ratio of soluble-to-colloidal Ca, the pH and in the type of proteolysis (i.e., hydrolysis of α_{s1}- *vs.* β-CN. Compared to Cheddar, LMMC is plasticized and has a similar Ca content (27.5 mg/g casein), a lower ratio of soluble-to-total Ca (~22 *vs.* 37% total; Figure 25.4), a higher pH (~5.53 *vs.* 5.15), and more extensive degradation of β-CN (Yun *et al.*, 1993b; Fox *et al.*, 1996). A comparative study of 12-week old Cheddar, Gouda and Mozzarella cheeses showed that the latter had the highest level of intact α_{s1}-CN and intermediate levels of β- and γ-CNs (Creamer, 1976).

(b) Effect of casein concentration, hydrolysis and hydration on flowability and apparent viscosity

Of all the functional attributes of melted cheese, flowability has been studied most extensively. Flowability may be defined as displacement of contiguous planes of the *para*-casein matrix as a result of heat-induced stress. This stress appears to be, at least in part, caused by heat-induced liquefaction and coalescence of free fat (formed as a consequence of shearing of the fat globule membrane during plasticization; Paquet and Kalab, 1988; Guinee *et al.*, 1999; Figure 25.3). Once fat coalescence is initiated, the *para*-casein matrix flows to a degree determined by the concentration of casein and the level of casein hydration, which is controlled by the levels of MNFS and total calcium, pH and the ratio of soluble-to-micellar Ca. Moreover, the change in the distribution of the fat phase *per se*, as affected by coalescence, may affect the degree of solvation of the *para*-casein (Hokes *et al.*, 1982). Hence, when heat-induced fat coalescence does not occur, as in full-fat Cheddar cheese prepared from high pressure homogenized milk, the flowability and phase angle of the melted cheese are much lower than those of control full-fat Cheddar prepared from non-homogenized milk in which heat-induced fat coalescence is relatively extensive (Figure 25.23).

Marked inter- and intra-varietal differences occur in the flowability of different cheese types (Park *et al.*, 1984; Rüegg *et al.*, 1991; Kindstedt, 1993; Guinee *et al.*, 2000b), reflecting differences in make procedure, composition and maturity. It is difficult to elucidate the direct effects of altering the concentration of any one compositional component, including protein, on the

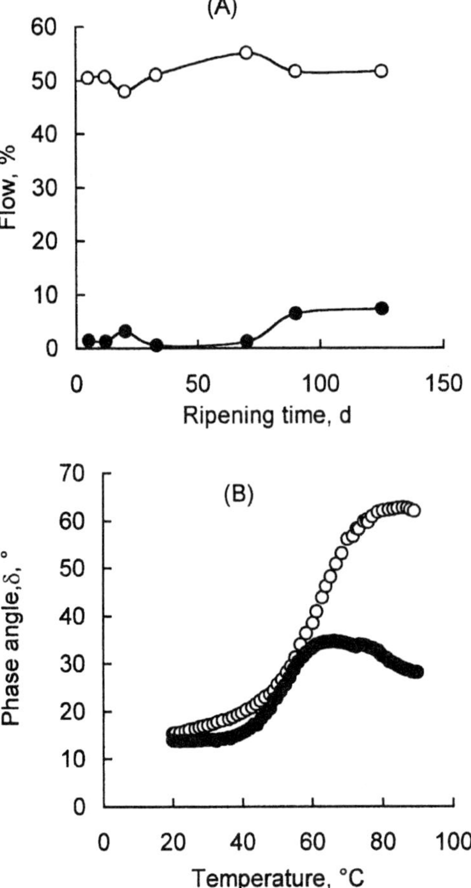

Figure 25.23 Flowability (A) and phase angle (B) of melted full-fat Cheddar cheese prepared from control (not homogenized) (○) or homogenized (20 kPa-stage 1, 5 kPa-stage 2) (●) milk. The samples were subjected to a low amplitude strain of 0.06 and an oscillation frequency of 1 Hz (redrawn from Guinee *et al.*, 2000c).

functionality of melted cheese since the levels of the different components tend to vary simultaneously, e.g., fat reduction is accompanied by increases in the levels of protein and moisture and decreases in the levels of moisture-in-non-fat substance and fat-in-dry matter (Fenelon and Guinee, 2000). The contributions of different compositional components to heat-induced functionality would be best undertaken in model systems the composition of which could be varied more systematically, e.g., model cheeses made from high-concentration milk ultrafiltrates or imitation cheeses.

Guinee *et al.* (2000a) showed that the level of intact casein in Cheddar cheese, with a fat content in the range 70 to 300 g/kg, was positively correlated

($P < 0.05$) with apparent viscosity and negatively with flowability (Figure 25.24). The higher *para*-casein concentration results in a higher volume fraction of the structural matrix (Figure 25.2) and a probable alteration in the number and type of attractions between the *para*-casein molecules (Horne, 1998), factors which are expected to enhance the degree of casein aggregation and matrix elasticity. Indeed, a reduction in the fat content of cheese results in decreases in the level of MNFS and the water binding capacity of the *para*-casein (as reflected by the increase in the level of expressible serum; Guinee *et al.*, 2000a). Moreover, the decrease in the quantity of free oil released on cooking undoubtedly also contributes to the

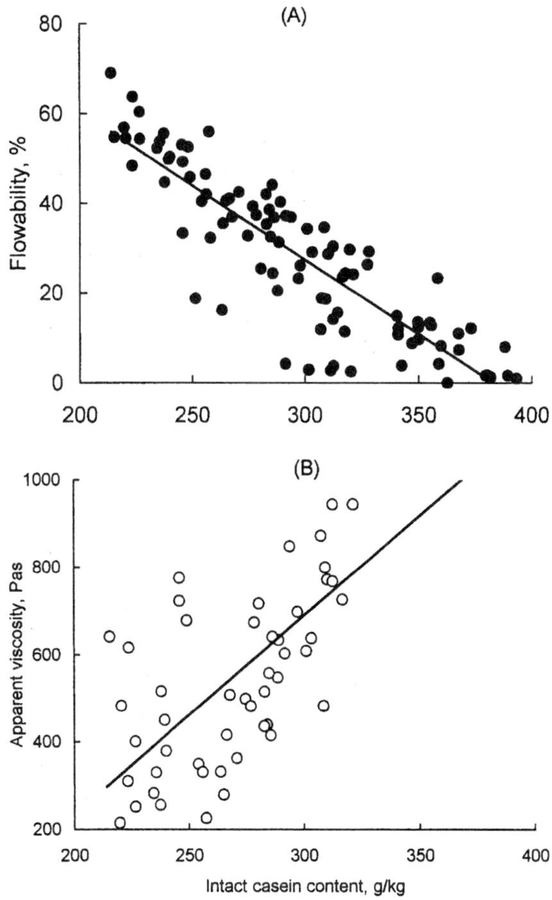

Figure 25.24 Relationship between the level of intact casein and flowability (●) and apparent viscosity (at 70°C) (○) of melted Cheddar cheeses of varying fat content in the range 60 to 330 g/kg (redrawn from Guinee *et al.*, 2000a).

impaired flowability of reduced-fat (high protein) cheese (Rudan and Barbano, 1988).

For a given level of protein, flowability is influenced markedly by the extent of proteolysis. Hence, there is a strong positive correlation between the flowability and concentration of non-protein nitrogen (in the range 0 to 38% total N) in model acid-type cheese, treated with varying levels of a proteinase derived from *Aspergillus oryzae* (Lazaridis *et al.*, 1981). The model cheeses had optimum flowability when the molecular mass of the casein peptides (measured using SDS polyacrylamide gel electrophoresis) was in the range 10–24 kDa (Mahoney *et al.*, 1982); at higher concentrations of peptides of molecular mass <10 kDa, the cheese became too flowable. The flowability of Cheddar cheeses with a similar level of α_{s1}-CN degradation was correlated closely with the extent of β-CN hydrolysis (Bogenrief and Olson, 1995). Yun *et al.* (1993a) reported that Mozzarella cheese made with *C. parasitica* proteinase as coagulant had a higher level of pH4.6SN throughout ripening (8 *vs.* 14 pH4.6SN at 50 d) and a higher level of β-CN degradation at times $\leq$30 d than cheeses made using either chymosin or *Rhizomucor miehei* proteinase. The higher level of primary proteolysis in the *C. parasitica* cheese was associated with higher flowability and free oil release (at $\geq$25 d) and lower apparent viscosity in the melted cheese (Yun *et al.*, 1993b).

(c) Effect of whey protein and casein whey protein interaction on flowability
In some applications of cheese, softening or melt is essential but very limited flow, or a high degree of flow resistance, is required so as to preserve the shape and identity of the cheese. Examples of the latter include fried paneer, grilled or fried burgers containing cheese insets, deep-fried breaded cheese sticks and casseroles in which the identity of cheese pieces following cooking is desirable (Chandan *et al.*, 1979). Most natural cheeses, especially when mature, are unsuitable for these applications owing to their excessive flow and oiling-off on cooking. In the case of cheese insets in deep-fried burgers, the latter attributes result in the cheese permeating the interstices of the coarse meat emulsion, and hence the inset looses its shape and visual effect in the cooked product (Guinee and Corcoran, 1994).

In applications requiring limited flow, PCPs or ACPs, which are specially formulated to endow them with flow resistance, are frequently used in preference to most natural cheeses (Guinee and Corcoran, 1994). Flow resistance may be conferred on cheese products by a number of methods, including:

- reduction in the size of fat globules and increasing their thermal stability to heat-induced coalescence, e.g., in natural cheeses by high pressure (i.e., 25 kPa stage 1, 5 kPa stage 2) homogenization of cheese milk (Tunick and Shieh, 1995; Figure 25.23), or in PCPs and ACPs by the selective use of emulsifying salts, increasing the proportion of intact *para*-casein in the

formulation (Lazaridis *et al.*, 1981; Mahoney *et al.*, 1982), or modification of processing conditions (Rayan *et al.*, 1980; Harvey *et al.*, 1982; Guinee and Corcoran, 1994; Rudan *et al.*, 1998),
- retrogradation and gelation of added starches, especially those with a high ratio of amylose-to-amylopectin, e.g., in analogue cheeses (Neville, 1998; Guinee *et al.*, 1999),
- thermally-induced aggregation and/of gelation of denatured whey proteins or denatured-whey protein/casein complexes on heating/cooking the cheese product or food containing the cheese, e.g., natural, processed and analogue cheeses insets in meat products,
- other micro-structural alterations of the *para*-casein matrix which impede displacement of the contiguous *para*-casein layers on heating, e.g., the addition of Novagel, a blend of micro-crystalline cellulose and guar gum (McMahon *et al.*, 1996).

Flow resistance in natural cheese is generally conferred by the presence of whey proteins, which may be included by several means (Section 25.4). High heat treatment of the milk (e.g., 95°C × 5 min) and the resultant incorporation of a high level of denatured whey protein is a feature of the manufacturing process of acid-heat coagulated cheese types, e.g., Queso Blanco types and Paneer (Chandan *et al.*, 1979; Torres and Chandan, 1981; Kalab *et al.*, 1988) and some fresh cheeses such as Cream cheese (Guinee *et al.*, 1993). The former cheeses exhibit excellent flow resistance and for this reason are used in applications where maintenance of visual identity and shape of the cheese on cooking is desirable, e.g., fried cheese pieces.

The whey protein content of natural hard ripened rennet-curd cheese may be increased by various means (Brown and Ernstrom, 1982; Marshall, 1986; Baldwin *et al.*, 1986; Fernandez and Kosikowski, 1986; Garrett, 1987; Banks *et al.*, 1987, 1994b; Guinee *et al.*, 1995; Lo and Bastian, 1998):

- *in situ* denaturation of the whey proteins by high-heat treatment (HHT) of the cheese milk, e.g., ~65% of total whey proteins are denatured at 100°C × 120 s,
- HCFUF, with or without HHT, of the milk before UF or the rententate after UF, as in the production of pre-cheese, and
- addition of partially denatured whey protein concentrate, prepared by high heat treatment and acidification of whey, to the cheese curd.

Most studies on the effect of high heat treatment have concentrated on aspects of cheese quality other than heat-induced functional attributes, e.g., cheese yield, texture and rheology of the unheated cheese (Section 25.5.2). However, it is generally considered that the inclusion of whey protein, at least by high heat treatment of the cheese milk or by HCFUF of cheese milk, adversely affects the melting behaviour of natural rennet-curd cheeses (Hansen, 1987; Lawrence, 1989). Although there is little direct evidence to indicate that the flowability of natural cheese is adversely affected by the addition of whey proteins, it is expected that flow would be impaired based on the following facts:

- acid-heat coagulated cheeses, which contain denatured whey proteins as a substantial portion (e.g., ~15%) of the total protein, are flow resistant (Chandan *et al.*, 1979),
- flow resistance is conferred on processed cheeses by the addition of a protein (e.g., 3–7%, w/w, whey protein concentrate) on completion of processing, which coagulates on heating to a temperature >70°C (Schulz, 1976),
- in general, the flowability of PCPs and ACPs decreases progressively with the level of added whey protein or cheese base from ultrafiltered milk (Sood and Kosikowski, 1979; Savello *et al.*, 1989),
- in dilute model systems, κ-CN interacts readily with whey proteins (β-lactoglobulin and α-lactalbumin) to form a gel when heated at a temperature >70°C (Doi *et al.*, 1983a,b, 1985),
- the strength of thermally-induced κ-CN/β-lactoglobulin gels, formed in dilute solutions (3 to 11% protein), is increased by chymosin treatment of the mixture (to yield *para*-κ-CN) and by the addition of Ca^{2+} (0–40 mM) (Doi *et al.*, 1983a).

Thus, it is expected that the high heat treatment (typically ~98°C) experienced on baking/grilling is conducive to the formation of thermally-induced *para*-κ-CN/β-lactoglobulin aggregates or gels when whey proteins are present in significant quantities (e.g., 3–7%, w/w) in the cheese. It is likely that on setting, the gels impede the flow of cheese as the fat phase melts and coalesces (Sood and Kosikowski, 1979; Savello *et al.*, 1989). Hence, dynamic viscoelastic analysis indicates that the phase angle of Cheddar cheese made from HHT milk (110°C × 60 s) decreases slightly at temperatures >65°C and attains a value at 80°C which is significantly lower (~29° *vs.* 53°) than that of standard Cheddar from milk pasteurized at 72°C for 16 s (Horne *et al.*, 1994; Figure 25.25). Moreover, the HHT Cheddar underwent less structural degradation as reflected by the higher value of the elastic shear modulus, G′, of the heated cheese at 80°C. The results of Horne *et al.* (1994) suggest that thermal gelation of denatured whey proteins and/or complexes of denatured whey proteins and *para*-κ-CN reduces the flow of the heated cheese when heated to a temperature >65°C. It is noteworthy that the high levels of protein and soluble calcium in cheese (Morris *et al.*, 1988; Guinee *et al.*, 2000b) probably enhance the heat-induced interaction and gelation of *para*-κ-CN/β-lactoglobulin (Doi *et al.*, 1983a; Jelen and Rattray, 1995).

Most evidence suggests that the inclusion of whey proteins in natural rennet-curd cheese adversely affects flowability and other heat-induced functional attributes. However, owing to the fact that an increase in the whey protein content of cheese usually results in an increase in moisture

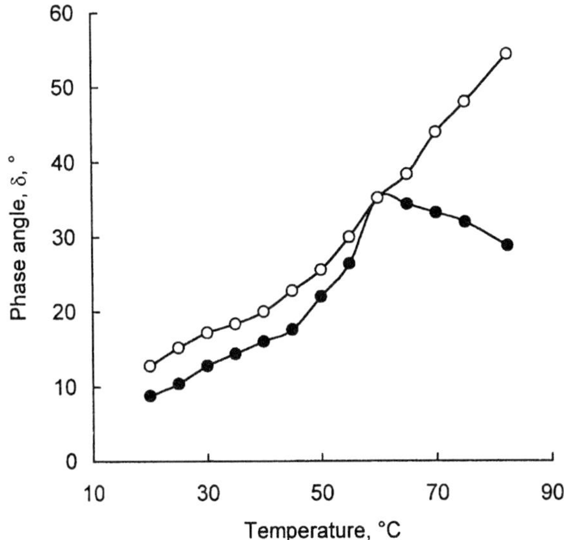

Figure 25.25 Phase angle (δ) as a function of temperature for Cheddar cheese prepared from milk pasteurized at 72°C for 16 s (○) or at 110°C for 60 s (●). The samples were subjected to a low amplitude strain of 2% and an oscillation frequency of 1 Hz (redrawn from Horne *et al.*, 1994).

content and a decrease in the levels of protein and fat, it is difficult to determine the direct effect of whey protein *per se* on heat-induced functionality. The flowability of Mozzarella cheese (age not specified), as measured from the percentage decrease in the height of a cheese disc on heating, decreased by ~40% as the level of whey protein denaturation in cheese milk was increased from ~5 to 35% (Schafer and Olson, 1975). Maubois and Kosikowski (1978) described a method whereby Mozzarella cheese with stretch properties similar to those of a control cheese could be manufactured by HCFUF (to ~ 43% dry matter) and diafiltration at pH 5.8 (to reduce the calcium content); however, little information was given on experimental details or flowability. Covacevich (1981) reported the manufacture of Mozzarella curd, which plasticized satisfactorily, from HCFUF milk retentates (~42%, w/w, dry matter) which had been pre-acidified and salted prior to diafiltration to reduce the Ca content. However, the flowability of the UF Mozzarella at 1 week was less than half of that of commercial Mozzarella (Covacevich, 1981). The UF cheese had a markedly lower pH (~5.15 *vs.* 5.9) and lower levels of moisture (~460 *vs.* 510 g/kg) and MNFS than commercial Mozzarella, changes which are expected to affect flowability adversely (Rüegg *et al.*, 1991; Tunick and Shieh, 1995). Similarly, Madsen and Qvist (1998) reported that the flowability of Mozzarella produced from a pre-cheese (48% dry matter) prepared by HCFUF of milk and containing 7%, w/w, whey protein was significantly lower than that of the control Mozzarella throughout a 5 week ripening period. However, an increase in proteolysis of the *para*-casein as a result of adding a proteinase from *Bacillus licheniformis* or *Bacillus subtilis* (Neut-rase®) to the pre-cheese retentate improved the flowability of the UF cheese; capillary electrophoresis indicated that the whey proteins were not degraded.

McMahon *et al.* (1996) evaluated the effect of different fat replacers, including the whey protein-based fat mimetics, Simplesse® D100 and Dairy-Lo®, on the micro-structure and heat-induced functionality of low-fat (4–5%, w/w) Mozzarella cheese; the estimated level of whey protein in the cheese with these mimetics was 2.3 and 6.0 g/kg, respectively. The fat mimetics resulted in a higher moisture content than the control low-fat cheese (e.g., 553 *vs.* 530 g/kg) but did not significantly affect the flowability or apparent viscosity of the melted cheese over a 28 day ripening period. However, at most times during ripening the flowability of the cheese containing Simplesse® was numerically higher than that of the control while that of the cheese with added Dairy-Lo® was lower than the control (McMahon *et al.*, 1996). The latter trend is probably a consequence of:

- the overall higher level of whey protein in the Dairy-Lo®-containing cheese owing to the higher level of whey protein added in the Dairy-Lo® preparation,
- the relatively high heat treatment of Dairy-Lo®-containing milk compared to the control and Simplesse®-containing milks,
- differences in the size of whey protein particles and their spatial distribution in the cheese matrix (McMahon *et al.*, 1996), and
- the degree of interaction of the whey protein particles in the different preparations with the *para*-casein matrix, as affected by factors such as pH and calcium level (Jelen and Rattray, 1995).

25.7.3 Effect of protein on age-related changes in functionality – extent of protein hydration and hydrolysis

Age-related changes in the functionality of LMMC have been studied extensively (Kindstedt, 1993, 1995; Guinee *et al.*, 1997a) but comparatively little information is available for Cheddar (Bogenrief and Olson, 1995; Olson and Bogenrief, 1995; Guinee *et al.*, 2000a,b), other *pasta filata* cheeses, Parmesan or Emmental. All the above studies have shown that functionality is dynamic, with the various attributes undergoing age-related changes to a degree depending on the composition and functional attribute of the cheese, e.g., whether stretch or flow (Figures 25.22 and 25.26). Changes in the extent of protein hydrolysis and hydration are major factors contributing to the age-related changes in functionality (Figure 25.26), as discussed below.

LMMC made by conventional procedures is generally non-functional on cooking during the first 5–10 days of storage at 4°C after manufacture, with drying-out, crusting and skin formation being common defects in the baked cheese. The lack of desirable heat-induced functionality is due to the low water-binding capacity of the *para*-casein in the unheated cheese and the low propensity to the heated cheese to express free oil on baking. Both factors are conductive to excessive evaporation of moisture from the cheese during baking because of the high cheese temperature (typically 90–100°C in the mass of the melting cheese when heated in a convection oven at 280°C for 4 min; Guinee and O'Callaghan, 1997). The dried-out, crusted cheese lacks succulence and fails to flow or stretch. Moreover, on heating to 70°C, the cheese is extremely tough and chewy, as reflected by the high apparent

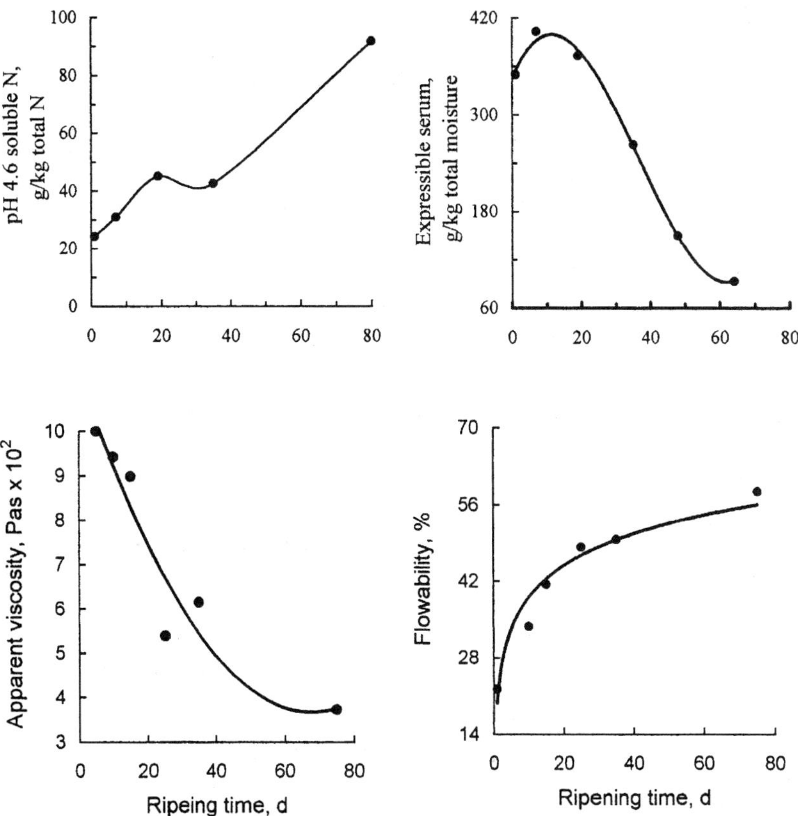

Figure 25.26 Typical changes in the levels of pH 4.6-soluble N, expressible serum (on hydraulic pressing at 3 MPa for 3 h at 21°C), flowability after heating to 280°C for 4 min, and apparent viscosity (at 70°C) of melted low-moisture Mozzarella cheese as functions of storage time at 4°C (redrawn from Guinee *et al*., 1997a).

viscosity (Figure 25.26). Freshly-made Cheddar behaves in a similar manner (Guinee *et al.*, 2000a).

However, the functionality of LMMC and Cheddar improves markedly during the first 2 weeks of ripening at 4°C, as reflected by reductions in melt time and apparent viscosity and increases in flowability and stretchability, and maintains this status until ~40–50 d (Figure 25.26). The improved functionality may be attributed to increases in protein hydration and free fat during ageing of the cheese as reflected by the reduction in the quantity of serum and increase in the quantity of free fat expressed from the cheese when subjected to hydraulic pressing or ultracentrifugation, i.e., for Mozzarella (Guo and Kindstedt, 1995; Kindstedt and Guo, 1997b; Guinee *et al.*, 1997a; Figures 25.4 and 25.26), Cheddar (Guinee *et al.*, 2000a), Emmental (Thierry *et al.*, 1998) or Camembert (Boutrou *et al.*, 1999). The vapour pressure of water bound by the *para*-casein is lower than that of free water and thus has a lower propensity to evaporate during baking (Geurts *et al.*, 1974; Masters, 1976). The exudation of free oil from the shredded cheese during baking also restricts dehydration (Rudan and Barbano, 1988); the free oil forms an apolar surface coat, which impedes the escape of water vapour. The changes in protein hydration appear to be the result of a number of factors, including a small increase in pH (from ~5.15 to ~5.35–5.40 at 5 d) and primary proteolysis and the solubilization of casein-bound calcium (at least in Mozzarella; Guo and Kindstedt, 1995). The physical swelling of the casein associated with its increased hydration probably contributes to the coalescence of non-globular liquid fat.

That hydrolysis of casein contributes to the age-related improvement in functionality of melted cheese is reflected by:

- the parallel increases in primary and secondary proteolysis, e.g., as reflected by nitrogen solubility at pH 4.6, and functional attributes such as flowability and stretchability (or reduction in apparent viscosity) (Figures 25.22 and 25.26),
- the linear relationships between various functional attributes, e.g., flowability, and proteolysis in a particular variety, e.g., Cheddar (Arnott *et al.*, 1957; Guinee *et al.*, 2000a), and
- the increase in flowability with increased level of proteolysis in Mozzarella (Yun *et al.*, 1993a,b) or Cheddar (Bogenrief and Olson, 1995).

Solubilization of casein-bound Ca appears to occur during the ripening of Mozzarella cheeses, as reflected by the age-related increase in the level of soluble Ca in the cheese serum (Guo and Kindstedt, 1995). Calcium in Mozzarella is solubilized when the calcium attached to the casein matrix is partially replaced by sodium (Guo and Kindstedt, 1995; Guo *et al.*, 1997). The *para*-casein in salted Mozzarella binds more water than in unsalted Mozzarella, as reflected by the level of serum expressed from the unsalted cheese on centrifugation, at the various stages during ripening (Guo *et al.*,

1997). Conditions in the cheese which are conducive to this ion exchange effect are the simultaneous occurrence of low concentrations of Na and Ca in the serum phase (i.e., 2.0% NaCl and 0.4% Ca) (Geurts *et al.*, 1972; Guinee and Fox, 1986; Kindstedt and Guo, 1997b). Hence, in a study on the effects of salt on cheese functionality, Paulson *et al.* (1998) found that the flowability of directly-acidified non-fat Mozzarella increased with increasing salt content to ~1% salt-in-moisture (S/M) and remained relatively constant thereafter to ~3.55% S/M. In a similar study, Choi *et al.* (1993) found that the flowability of Mozzarella cheese (age not specified) decreased with increasing S/M level in the range 0.5 to 2.25%.

The increased propensity of the cheese to oil-off on cooking may be associated with age-related degradation of the fat globule membrane and/or the continued coalescence of the partially denuded fat globules (present after texturization). The latter effect may be due to the increased hydration, which physically forces the partially denuded globules into close proximity (Figure 25.3; Kindstedt and Guo, 1997b).

On prolonged storage, e.g., to 75 d, unbaked LMMC generally becomes too soft and sticky while the baked cheese is excessively flowable, has a 'soupy' consistency and lacks the desired chewiness (as reflected by the relatively low apparent viscosity). These changes in functionality are attributable to excessive proteolysis (Arnott *et al.*, 1957; Bogenrief and Olson, 1995; Guinee *et al.*, 1997a). However, the stretchability remains relatively constant even when the product is stored for a prolonged period (up to 4 months at 4°C; Guinee and O'Callagahan, 1997). The constant stretchability of LMMC indicates that the level of primary proteolysis in the cheese at this time (i.e., pH 4.6-soluble N ~ 120 g/kg total N) is insufficient to cause a significant impairment of stretchability (Figure 25.22).

25.7.4 Role of protein in the Maillard reaction and browning of cheese

O'Brien (1995) described the biochemical aspects of the Maillard reaction in detail. Maillard browning essentially involves reactions between an aldehyde group (e.g., of reducing sugars such as lactose or galactose) and a free amino group (e.g., α- and ε-amino groups of amino acids, peptides and proteins) or other reactive N-group (e.g., indolyl group of tryptophan). Browning sometimes occurs during the storage of cheese (e.g., Parmesan) or processed cheese (Piergovanni *et al.*, 1989; Younis *et al.*, 1991; Gopal and Richardson, 1996; Abd-El-Salam *et al.*, 1998). However, browning occurs frequently on heating of cheeses, e.g., Mozzarella and other cheeses made with a thermophilic culture, PCPs or ACPs. While slight browning of cheese may be desirable in some cooked applications (e.g., lasagne, pizza, crustini), intense (dark) browning is unacceptable from aesthetic and nutritional viewpoints. Browning in cheese is thought to be due primarily to the Maillard reaction.

Browning rarely occurs in most natural mature rennet-curd cheeses made with a mesophilic culture (e.g., Cheddar) since these have little or no residual sugars, even after very short ripening time, e.g., at >14 d (Torres *et al.*, 1995). However, these cheeses may be susceptible to browning if residual lactose persists in the cheese as a result of excessive salt, which inhibits starter metabolism (Thomas and Pearce, 1981; Jordan and Cogan, 1993). Moreover, in the absence of a reducing sugar, aldehyde groups, resulting from degradation of amino acids or free fatty acids (Fox *et al.*, 1996; Fox and Wallace, 1997) may predispose the cheese to heat-induced browning, especially if the cheese is mature and has a high concentration of free amino acids.

The degree of browning in rennet-curd cheese made using a thermophilic culture is related to the sugar-fermenting and proteolytic characteristics of the starter culture used (Kindstedt, 1993). Most strains of *Streptococcus thermophilus* and *Lacobacillus delbrueckii* ssp. *bulgaricus*, which are commonly used in the manufacture of LMMC, are unable to metabolize galactose and hence cheese made solely with these cultures is susceptible to browning. However, *Lb. helveticus*, which is frequently used as a component of the starter culture, ferments lactose completely to lactic acid and may also ferment galactose which accumulates on the incomplete metabolization of lactose by the former cultures (Turner and Martley, 1983). Attempts to control the level of browning in pizza cheese include control of the level of residual sugars and/or proteolysis products (Kindstedt, 1993) *via*:

- adjusting the ratio of *Lb. helveticus* (galactose positive) to *Str. thermophilus* in the starter culture,
- the use of galactose-positive strains of *Lb. delbrueckii* ssp. *bulgaricus*,
- galactose-positive, galactose-non-releasing strains of *Str. thermophilus* [galactose positive, galactose-non-releasing strains ferment galactose and do not release galactose, even when lactose is present in the extra-cellular environment] (Mukherjee and Hutkins, 1994),
- the use of proteinase-negative starter strains (to limit the formation of free amino groups),
- the use of curd washing (to physically remove lactose from the cheese curd).

Browning may also occur in PCPs and ACPs, especially if a high concentration of lactose or other reducing sugar (e.g., glucose) is added *via* other ingredients, e.g., maltodextrins, skim milk or whey powder, whey protein concentrates, milk protein or unfermented milk ultrafiltrates (Thomas, 1969). Owing to their generally higher pH, concentration of lactose and buffering capacity (associated with added sodium phosphates),

PCPs are probably more prone to Maillard browning during storage and cooking than natural cheese (O'Brien, 1995).

25.8 CONTRIBUTION OF PROTEIN TO THE FORMATION OF PASTEURIZED PROCESSED CHEESE PRODUCTS (PCPs) AND ANALOGUE CHEESE PRODUCTS (ACPs)

Pasteurised processed cheese products are produced by comminuting and melting into a smooth homogeneous blend, one or more natural cheeses and optional ingredients using heat, mechanical shear and (usually) emulsifying salts. The type and level of optional ingredients permitted are determined by the product type, i.e., whether processed cheese, processed cheese food or processed cheese spread, and include dairy ingredients, vegetables, meats, stabilisers, emulsifying salts, flavours, colours, preservatives and water (see Fox *et al.*, 1996). Cheese, as an ingredient of PCPs, ranges from a minimum of 51% in pasteurized processed cheese spreads and foods to ~95% in pasteurized processed cheese (Code of Federal Regulations, 1986; Fox *et al.*, 1996). In contrast, ACPs generally contain no added cheese, except where a small amount is added to impart a cheese flavour or as required by customer specifications. Similar to PCPs, ACPs contain added stabilisers, emulsifying salts, flavours, colours, preservatives and water. ACPs may be arbitrarily categorized as dairy, partial dairy or non-dairy, depending on whether the fat and/or protein components are from dairy or vegetable sources (Shaw, 1984). Partial dairy analogues, in which the fat is mainly vegetable oil (e.g., soya oil, palm oil, rapeseed and their hydrogenated equivalents) and the protein is dairy-based (usually rennet casein and caseinate), are the most common. Similar to PCPs, the other main ingredients are emulsifying salts and water.

The manufacture of PCPs essentially involves the following major steps:

- shredding of cheese,
- blending with emulsifying salts, water and optional ingredients,
- processing of the blend,
- hot packing and cooling.

Processing refers to the heat treatment of the blend, by direct or indirect steam, with constant agitation. In batch processing, the temperature-time combination varies (70–95°C for 4–15 min) depending on the formulation, extent of agitation, the desired product texture, body and shelf-life characteristics. In continuous cooking, the blend is mixed and heated to 80–90°C in a vacuum mixer from where it is pumped through a battery of tubular heat exchangers and heated to 130–145°C for a few seconds and then flash-cooled to 90°C.

The manufacture of ACP, like that of PCP, involves formulation, processing and hot packing of the hot molten product. While the production method varies somewhat, a typical manufacturing procedure involves the following sequence of steps:

- simultaneous addition of the required amounts of water and dry ingredients (e.g., casein, emulsifying salts),
- addition of oil,
- cooking to ~85°C (using direct steam injection) while continuously shearing until a uniform homogeneous molten mass is obtained (typically 5–8 min),
- addition of flavouring materials (e.g., enzyme-modified cheese, starter distillate) and acid(s) (e.g., citric acid) to the molten mass, followed by blending for a further 1–2 min, and
- hot packing and cooling.

25.8.1 Role of protein in manufacture

The principles of the manufacture of processed cheese have been reviewed extensively (Guinee, 1987; Caríc and Kaláb, 1993; Fox *et al.*, 1996). Application of heat (70 to 90°C) and mechanical shear to natural cheese in the absence of stabilisers usually results in the formation of a heterogeneous, gummy, pudding-like mass which undergoes extensive oiling-off and moisture exudation during manufacture and, especially, on cooling. These defects arise from:

- the coalescence of non-globular fat due to shearing of the fat globule membrane, and
- partial dehydration/aggregation and shrinkage of the *para*-casein matrix induced by the relatively low pH of cheese (for most cheeses, <5.7) and high temperature applied during processing.

The addition of emulsifying salts (10–30 g/kg), such as the sodium salts of citric acid and/or phosphoric acid, during processing promotes emulsification of free fat and rehydration of the *para*-casein and thus contributes to the formation of a smooth, homogeneous, stable product. These salts, referred to as emulsifying salts (e.g., citrates, orthophosphates, pyrophosphates, polyphosphates), generally consist of a monovalent cation (e.g., sodium) and a polyvalent anion (e.g., phosphate). While these salts are not emulsifiers *per se*, they promote, with the aid of heat and shear, a series of concerted physico-chemical changes in the cheese blend which result in rehydration of the aggregated *para*-casein and its conversion into an active emulsifying agent (Caríc and Kaláb, 1993). These changes include calcium sequestration, upward adjustment of the pH and pH stabilisation (buffering), *para*-

casein hydration and dispersal, emulsification and structure formation (Figure 25.27).

(a) Physico-chemical changes during processing
Calcium sequestration involves the exchange of the Ca^{2+} (attached to casein *via* phosphoseryl residues and/or the carboxyl groups of acidic amino acids) of the *para*-casein for the Na^+ of the emulsifying salt. The reduction in the concentration of casein-bound calcium results in partial hydration of the *para*-casein and its partial conversion to a sodium phosphate *para*-caseinate dispersion (sol) (Morr, 1967; Nakajima *et al.*, 1975; Sood *et al.*, 1979; Swaisgood, 1982; Matheis and Whitaker, 1984; Southward, 1985;

A. Overall Reaction

Rennet-curd cheese (calcium *para*-casein network with entrapped moisture and globular fat) + water + emulsifying salts and energy

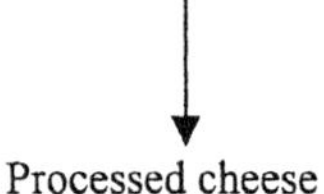

Processed cheese

B. Physico-chemical changes during processing

Change	Main Causative Agent
Ion exchange (Ca sequestration)	Emulsifying salt
pH buffering	Emulsifying salt
Protein dispersion	Emulsifying salt Thermal and mechanical energy
Protein hydration	Emulsifying salt Thermal energy
Emulsification	Dispersed hydrated protein Mechanical energy
Structure formation	Protein-protein interactions Emulsified fat Globules (pseudo-protein particles)

Figure 25.27 Summary of the overall reaction (A) and physico-chemical changes (B) that occur during the manufacture of processed cheese products.

Lee *et al.*, 1986; Mulvihill, 1992; Marchesseau *et al.*, 1997). Attendant on the increased hydration of *para*-casein is a large increase in the level of water-soluble N.

The degree of calcium sequestration and casein hydration (as determined by the proportion of total N that is non-sedimentable on ultracentrifugation of the processed cheese) are dependent on processing conditions and the type and level of emulsifying salts (calcium chelating strength, pH and buffering capacity) (Irani and Callis, 1962; van Wazer, 1971; Cavalier-Salou and Cheftel, 1991; Figure 25.28).

The use of the correct blend of emulsifying salts usually shifts the pH of cheese upwards (typically from ~5.0–5.5 in the natural cheese to 5.6–5.9 in the PCP) and stabilises it by virtue of their high buffering capacity (Caric and Kalab, 1987; Gupta *et al.*, 1984). This change contributes to an enhanced calcium-sequestering ability of the emulsifying salts (Irani and Callis, 1962) and an increased negative charge on the *para*-caseinate. These changes are coincidental with a more open reactive *para*-caseinate conformation with superior water-binding and emulsifying properties (Marchesseau *et al.*, 1997). Hence, the buffering capacity of the emulsifying salts is a critical factor controlling the textural and melting attributes of PCPs and ACPs (Rayan *et al.*, 1980; Thomas *et al.*, 1980; Gupta *et al.*, 1984; Cavalier-Salou and Cheftel, 1991).

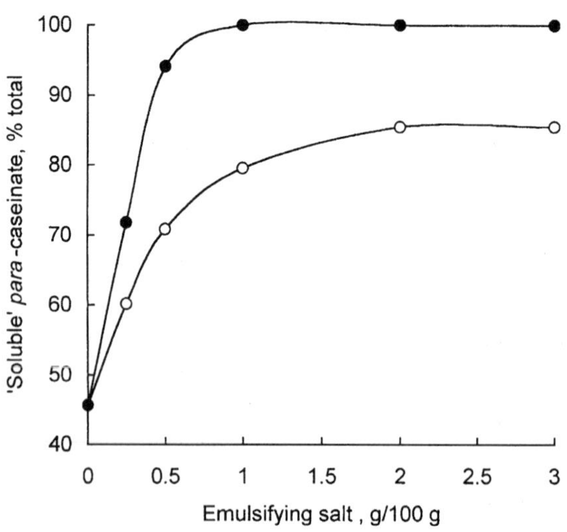

Figure 25.28 Influence of concentration of emulsifying salt on the percentage of total N soluble or non-sedimentable on ultracentrifugation (300 000 g × 1 h at 20°C) of a dispersion prepared by blending a 5 g sample of analogue cheese, distilled water and different levels of trisodium citrate (●) or sodium tripolyphosphate (○) (redrawn from Cavalier-Salou and Cheftel, 1991).

Under the conditions of cheese processing, the dispersed hydrated *para*-caseinate contributes to:

- emulsification – by coating the surfaces of dispersed free fat droplets, and
- emulsion stability by immobilization of a large amount of free water (Philips, 1981; Kinsella, 1982; Dickinson *et al.*, 1987; Dagleish, 1989; Mulvihill, 1992).

(b) Structure formation during cooling

On cooling PCPs, the homogeneous molten viscous mass sets to form a characteristic body, which, depending on blend formulation, processing conditions and cooling rate, may vary from firm and sliceable to soft and spreadable. Little or no information is available on the physico-chemical mechanism(s) responsible for structure formation; factors that probably contribute include: fat crystallization, protein-protein interactions and incorporation of recombined fat globules.

Micro-structural studies on PCPs or ACPs indicate that the structure is an emulsion of discrete, rounded fat droplets of varying size in a continuous protein matrix (Kimura *et al.*, 1979; Taneya *et al.*, 1979; Rayan *et al.*, 1980; Heertje *et al.*, 1981; Lee *et al.*, 1981; Kalab *et al.*, 1987; Savello *et al.*, 1989; Tamime *et al.*, 1990; Guinee *et al.*, 1999). Compared to natural cheese, there is less clumping or coalescence of fat globules and the mean fat globule size is generally smaller than that in natural cheese; the actual size depends on formulation and processing conditions, e.g., emulsifying salts, milk protein additions, processing time and extent of shear. The fat and *para*-casein are distributed more homogeneously and the matrix is less compact and fused than in natural rennet-curd cheeses. High resolution TEM (60 000×) with negative staining of the protein reveals that the matrix consists of strands that are finer than those of natural cheese and appear to be composed of *para*-caseinate particles (20–30 nm diameter) joined end-to-end. It has been suggested that these particles may correspond to casein sub-micelles released from the *para*-casein matrix of the cheese as a result of calcium chelation by the emulsifying salts (Kimura *et al.*, 1979; Taneya *et al.*, 1980; Heertje *et al.*, 1981).

The *para*-caseinate membranes of the emulsified fat globules appear to attach to the matrix strands and anchor the protein matrix, thereby contributing to the continuity of the matrix. The positive correlations between the degree of emulsification and firmness or elasticity, and the inverse relationship between the degree of emulsification and flowability of PCPs support this suggestion (Rayan *et al.*, 1980; Caric *et al.*, 1985; Savello *et al.*, 1989). The incorporation of emulsified *para*-caseinate-coated fat globules, which can be considered as pseudo-protein particles, into the new structural matrix may be considered to increase the effective protein concentration (van Vliet and Dentener-Kikkert, 1982; Marchesseau *et al.*, 1997).

25.8.2 Effect of protein on the rheology of the PCPs and ACPs

(a) Degree of protein hydration/aggregation

In 'hard' PCPs, the strands of the protein matrix are longer and inter-strand connections are more numerous than in 'soft' PCPs (Taneya *et al.*, 1980; Heertje *et al.*, 1981; Caríc *et al.*, 1985; Tamime *et al.*, 1990). Minute electron-dense areas in the strands of the protein phase may correspond to regions of strand overlap and/or reflect areas with a relatively high degree of aggregation and fusion of the *para*-caseinate particles. Hence, the number and area of electron-dense zones in very firm processed cheese food (PCF) that had been cooked to 85°C and held for 5 h were markedly higher than in control PCF that had been cooled after 3 min at 85°C (Kalab *et al.*, 1987). Prolonged holding (e.g., 5 h) at a high temperature is conducive to aggregation and dehydration of *para*-casein. Hence, Csøk (1982) reported that on holding a cooked processed cheese at 95°C, the bound water increased to a maximum (e.g., at ~15 min) and decreased thereafter (Figure 25.29). The initial increase may be attributed to increased solution of the emulsifying salts (not fully solubilized at the end of the heating step) and calcium chelation, while the eventual decrease reflects aggregation of the *para*-caseinate. In commercial practice, it is well known that prolonged

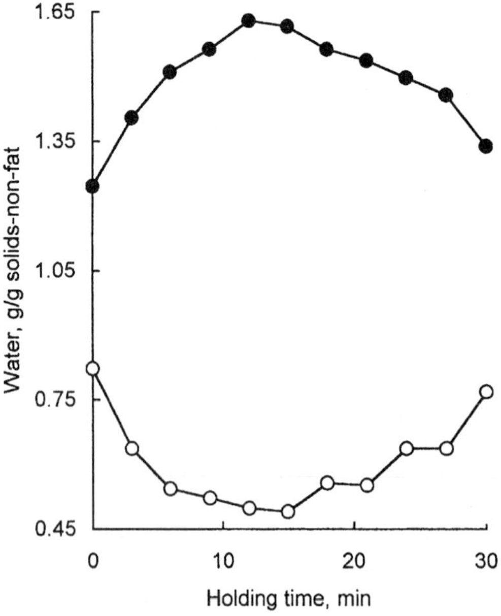

Figure 25.29 Changes in the level free water (○) and total bound water (●) in pasteurized processed cheese as a function of processing time at 95°C (redrawn from Csøk, 1982).

holding of formed PCPs at a high temperature (e.g., due to delay or stoppage of packaging lines) can lead to a process known as 'over-creaming' whereby the product acquires an appearance resembling that of an 'orange-peel' surface, develops an over-firm and heavy pudding-like (coarse) structure which leaks free moisture and exudes beads of free oil (through the 'surface dimples'). The release of moisture and free oil during 'over-creaming' suggests that the process coincides with the onset of protein dehydration, emulsion destabilization and phase inversion; the increased degree of protein aggregation is consistent with the increase in firmness and elasticity. It is noteworthy that at a micro-structural level, clumping and coalescence of fat globules was evident in PCF held for 5 h at 85°C but was absent, or markedly less, in PCF held for 3 min or 1 h (Kalab *et al.*, 1987).

The micro-structure of the protein matrix also appears to change during storage, especially at a high temperature. A study on the age-related changes in processed cheese indicated that firmness and elasticity (i.e., force required to push a wire into a cheese) increased during storage over 3 months, with the extent of the increase being higher as the storage temperature was raised from 10 to 30°C (Tamime *et al.*, 1990). The latter effect may be attributed to continued protein-protein interactions and/or interactions between the cheese proteins and the emulsifying salt, which may affect the degree protein interaction and the degree of emulsification of fat (Tamime *et al.*, 1990). The increase in firmness on raising the storage temperature from 10 to 30°C lends support to the suggestion that protein interactions probably occur during storage; an increase in temperature in this range is conducive to hydrophobic interactions.

The type and level of emulsifying salts have a marked influence on the rheology of PCPs and ACPs (Rayan *et al.*, 1980; Thomas *et al.*, 1980; Gupta *et al.*, 1984; Cavalier-Salou and Cheftel, 1991). Comparative studies on the effect of different emulsifying salts indicate that for a given processing time, both the firmness and elasticity of PCPs or ACPs made with different emulsifying salts decreased in the following general order: tetrasodium pyrophosphate (TSPP) > disodium phosphate (DSP) > trisodium citrate (TSC) > sodium aluminium phosphate (SALP) (Rayan *et al.*, 1980; Cavalier-Salou and Cheftel, 1991). The mean fat globule diameter in the PCPs showed the opposite trend, being largest with SALP and smallest with TSPP. The effects of different salts on firmness undoubtedly reflect differences in pH and calcium chelation that in turn result in differences in casein hydration and emulsification.

(b) Type of protein

Cheese base (CB) is being used increasingly as a cheese substitute in the manufacture of processed cheese, the main advantages being its lower cost and more consistent quality (i.e., intact casein content). Tamime *et al.* (1990) investigated the effect of replacing Cheddar with cheese base produced by coagulating a fermented diafiltered/ultrafiltered milk retentate, cutting the coagulum, cooking the curds and whey to 39°C, whey drainage, and salting and pressing of the curds. Replacement of young Cheddar cheese (which comprised 59%, w/w, of the control processed cheese blend) by a mixture of cheese base and anhydrous milk fat (without altering composition) resulted in a marked increase in firmness. The latter effect was attributed to a number of factors:

- a higher degree of intact casein in the cheese base,
- the presence of whey proteins in the cheese base (~8.7%, w/w) which denature and complex with *para*-κ-CN to form a gel at the high processing temperature (85–90°C for 3 min) (Doi *et al.*, 1983a,b, 1985).

Similarly, substituting a 1:1 blend of mild and mature Cheddar to a level of 33%, w/w, by a freshly made curd, prepared by acid-heat coagulation (pH 5.5, 90°C for 10 min) of milk, resulted in a significant increase in the firmness of the resultant processed cheese (Kaláb *et al.*, 1991). Moreover, increasing the level of substitution of rennet casein by total milk protein, in the range 0 to 50%, resulted in a progressive increase in firmness, and a decrease in flowability, of an ACP (Abou-El-Nour *et al.*, 1996; Figure 25.30). The addition calcium co-precipitate (at levels of 0 to 5%, w/w) also increased the firmness of processed cheese (Thomas, 1970).

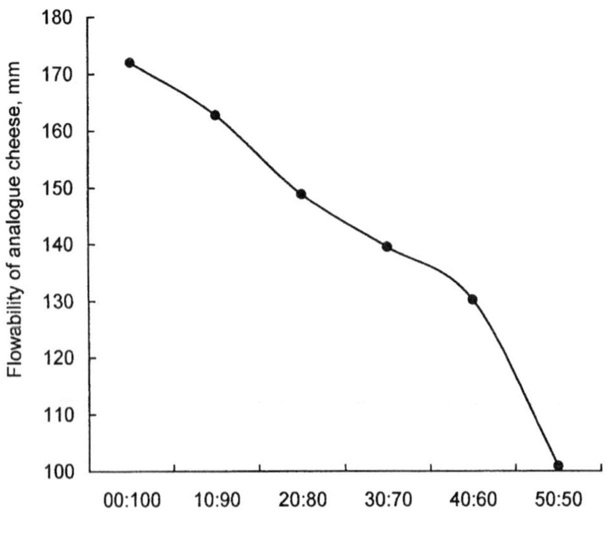

Figure 25.30 Effect of substituting total milk protein for rennet casein on the flowability of melted analogue cheese (redrawn from Abou-El-Nour *et al.*, 1996).

(c) Degree of proteolysis/hydration

Little published information is available on the effect of casein hydration or the concentration of intact *para*-casein in the unheated cheese on the rheological characteristics of the resultant processed cheese. However, it has long been recognized that young hard and semi-hard cheese varieties, such as Cheddar, Gouda and Emmental give firmer, longer-bodied PCPs than mould- or semi-hard and soft smear-ripened varieties, such as Camembert, Blue cheese, Chaumes or Brick cheese (Meyer, 1973). Soft smear-ripened cheeses tend to undergo more extensive proteolysis during ripening than hard and semi-hard varieties and consequently generally have a lower degree of intact casein at the stage considered suitable for eating (Gripon, 1993; Steffen *et al.*, 1993; Reps, 1993). At the commercial level, it is well known that the both the type and degree of maturity of the cheese used has a major influence on the consistency of the product. Hence, block processed cheese with good sliceability and elasticity requires predominantly young cheese (75–90% intact casein) whereas predominantly medium-ripe cheese (60–75% intact casein) is required for spreads. Thus, it has been observed that model processed cheese prepared from acidified curd became excessively soft when a high concentration of peptides of molecular mass <10 kDa is present following treatment of the curd by a fungal proteinase (from *A. oryzae*) prior to processing (Mahoney *et al.*, 1982).

25.8.3 Effect of protein on heat-induced functionality

Whey proteins may be included in PCPs and ACPs by direct addition of whey proteins in the form of whey protein concentrates (WPCs) (Schulz, 1976; Savello *et al.*, 1989; Nishiya *et al.*, 1989), total milk proteinates (Abou El-Nour *et al.*, 1996), co-precipitates (Thomas, 1970), milk ultrafiltates (Sood and Kosikowski, 1979; Anis and Ernstrom, 1984), cheese base (concentrated fermented ultrafiltered milk retentate; Ernstrom *et al.*, 1980; Park *et al.*, 1992) or cheese with a high level of whey protein, e.g., UF cheese or acid-heat coagulated curd (Kalab and Modler, 1985; Collinge and Ernstrom, 1988; Collinge *et al.*, 1988; Kalab *et al.*, 1991). On substituting cheese by UF retentates, the solids of the latter are usually standardized to those of the cheese by using low temperature evaporation of the retentate or by the addition of freeze-dried retentate and/or high-fat cream.

Savello *et al.* (1989) reported that the flowability of ACPs decreased progressively with the addition of whey protein in the range 0 to 3%, w/w (Figure 25.31), which resulted in the formation of fibrous structures that were considered to impede flow on subsequent reheating. The loss of flowability was observed on the addition of whey proteins in either a substantially native (~50% total) or denatured (~75% total) state (Savello *et al.*, 1989). Similarly, increasing the level of substitution of rennet casein in

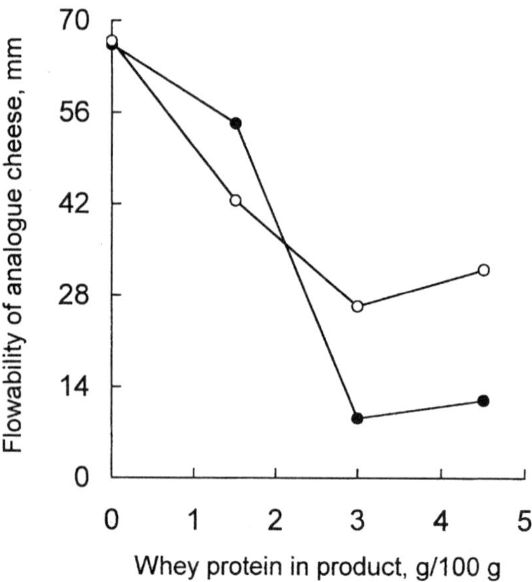

Figure 25.31 Effect of whey protein content, added in a mainly undenatured (●) or denatured (○) form, on the flowability of analogue cheese (redrawn from Savello *et al.*, 1989).

ACPs by total milk proteinate resulted in a progressive increase in firmness and reduction in flow (Figure 25.30); the flow decreased by 25% in products in which 40% of rennet casein was substituted by total milk proteinate (giving 1.9%, w/w, whey protein) (Abou-El-Nour *et al.*, 1996).

The flowability of PCPs is also impaired on partially substituting natural rennet-curd cheese by ultrafiltered milk retentate, with the magnitude of the effect increasing with the level of substitution (Sood and Kosikowski, 1979), processing temperature in the range 66 to 82°C (Collinge and Ernstrom, 1988; Collinge *et al.*, 1988) and rennet treatment of the retentate (Anis and Ernstrom, 1984). Moreover, in processed Cheddar, the extent of flow reduction on adding calcium co-precipitate (to a level of 5%, w/w) appears to depend on the degree of *para*-casein hydrolysis, with the extent decreasing as the cheese becomes older and its proteins become more degraded (Thomas, 1970). However, the level at which flowability became noticeably impaired varied from 25 to 30 g/kg and with the source of the co-precipitate; hence, varying the concentration of co-precipitate provides a convenient means of customizing the degree of flow resistance. In corollary, the flowability of processed cheese with added UF retentate (Sood and Kosikowski, 1979) or natural cheese prepared from HCFUF retentate (Madsen and Qvist, 1998) improved on increasing the extent of casein hydrolysis by treatment of the retentate with various proteinases, e.g., from *Aspergillus oryzae*, *Candida cylindracea*, *Bacillus licheniformis* or *Bacillus subtilis*. The latter results concur with observations indicating a direct correlation between flowability and the concentration of non-protein nitrogen (as % total N) in model chemically-acidified curds treated with different levels of *A. oryzae* proteinase (Lazaridis *et al.*, 1981; Mahoney *et al.*, 1982). The flowability of PCPs made from milk ultrafiltrates may also be improved by reducing the calcium content of the retentate by acidification of the milk prior to ultrafiltration and diafiltration (Anis and Ernstrom, 1984).

In contrast to the above findings, it has been reported that the inclusion of WPC in pasteurized processed Ras-Quark cheese, as a partial replacement for Quark, either had little effect on flowability when included at a level of ~3.6%, w/w, protein (Abd El-Salam *et al.*, 1996) or enhanced flowability when included at a level of ~6%, w/w, protein (Al-Khamy *et al.*, 1997). Similar observations were made by Kalab *et al.* (1991) when a non-flowable acid/heat-coagulated white cheese (AHC) was used at inclusion levels of 0 (control), 8, 16 or 33%, w/w, to replace Cheddar in a pasteurized processed cheese. The flowability of the processed cheese (made using trisodium phosphate as emulsifying salt) increased by ~40% as the level of AHC was increased to 16% w/w (i.e. ~0.64%, w/w, whey protein) and then decreased to a value slightly higher than the control as the level of AHC was increased to 33%, w/w. When trisodium citrate was used as an emulsifying salt, flowability increased by ~20% as the AHC was increased to 8%, w/w, and decreased at higher levels, attaining a value of 80% of the control when 33%, w/w, AHC was added (Kalab *et al.*, 1991).

Discrepancies between the foregoing studies *vis-à-vis* the effect of added whey protein on the flowability of PCPs and ACPs may be due to due differences in:

- pre-treatment of the whey protein and its level of denaturation,
- the method of the whey protein inclusion,
- the overall product formulation (i.e., type of casein, type of emulsifying salts, cheese age), and
- processing conditions.

These factors may determine the degree of aggregation of whey protein, aggregate size and its interaction with the casein/*para*-casein. Mackey and Desai (1995), who investigated the role of Simplesse®100 in reduced-fat Cheddar, suggested that the micro-particulated whey protein acts as a non-interacting filler occluded within the cheese matrix. Hence, the addition of Simplesse®100 to cheese milk has been found to slightly enhance the functionality of Mozzarella, whereas the addition of Dairy Lo, which appears to associate with the casein to form a complex-type gel (Fenelon and Guinee, 1997), impairs functionality (McMahon *et al.*, 1996). Hence, the improvement in functionality observed by Abd El-Salam *et al.* (1996)

and Al-Khamy *et al.* (1997) may be due to the micro-particulation-like effects of the whey pre-treatments on the added WPC, i.e., heating to 60°C and repeated high pressure homogenization at 25 kPa, of ultrafiltered Ras cheese whey. Micro-particulation generally involves heating the whey protein dispersion under acidic conditions (Singer, 1996).

25.9 CONCLUSIONS

Protein is a major component in most cheese varieties, apart from some acid-curd varieties (e.g., Mascarpone; 3%, w/w, protein), brown whey cheeses (e.g., Mysost) and some cheese-like imitation products where it amounts to <20% of dry matter. Controlled destabilization of the proteins, e.g., by selective treatment of the milk with rennet, acid or a combination of acid and heat, leads to limited protein aggregation and the formation of particulate gels. In renneted milk gels, the aggregates are formed from fused *para*-casein micelles whereas in acid/acid-heat coagulated gels, casein or casein-whey protein form the protein phase. The particulate gel, which is comprised of touching protein aggregates and colloidal salts, is a continuous sponge-like structure which encases fat and serum. The structure of the gel determines its permeability and resistance to deformation on cutting and stirring. These parameters, in turn, markedly affect the retention of moisture, fat and protein and therefore cheese yield and yield efficiency.

The interactive effects of gel structure and superimposed cheesemaking operations (acidification, heating, dehydration, salting) determine the biochemical, structural and rheological characteristics of the protein network in the resultant cheese. However, gel structure and its physico-chemical characteristics (e.g., degree of aggregation and hydration) change during storage. Maturation results in several biochemical changes (e.g., proteolysis, increase in pH, protein-protein interactions) and alters the equilibrium concentrations of salts in the serum and protein phases. Consequently, cheese rheology and heat-induced functionality, which are controlled mainly by the structure of the matrix, exhibit dynamic behaviour, undergoing marked changes during ripening. Protein and its degradation products also contribute to cheese flavour (Fox and Wallace, 1997).

Heating of cheese to a high temperature (e.g., 90°C), while shearing, generally results in protein dehydration and the formation of free fat, resulting in contraction of the protein phase, moisture leakage and fat exudation. Re-hydration of the protein by the addition of emulsifying salts assists in a structural transition from a gel to a concentrated *para*-caseinate–stabilised oil-in-water emulsion, which is the basic structure of PCPs and ACPs. The degree of *para*-caseinate aggregation or hydration and size distribution of emulsified oil droplets are major determinants of the rheology and heat-induced functionality of the resultant PCPs. Limited heat-induced destabilisation of cheese is generally a prerequisite for correct functionality in heated cheeses, where some oiling-off and protein aggregation are important for properties such as surface sheen, succulence, stringiness and chewiness of the melted cheese. The addition of whey proteins at concentrations that are conducive to thermal gelation during the formation or subsequent heating of the PCP influence product quality.

Much knowledge is now available on the general contribution of milk proteins to cheese structure, rheology and heat-induced functionality. However, little is known about the direct effects of protein (volume fraction and type) on rheology and heat-induced viscoelastic changes in cheese. For a given variety of natural cheese, it is difficult to quantify the effect of any one particular component (e.g., protein) since the concentrations of different components generally vary simultaneously, e.g., reduction in fat content is paralleled by increases in moisture, protein and Ca. A better understanding of the role of protein could be gleaned from analysis of model systems. However, in such systems it is difficult to evaluate the contribution of protein type and concentration to the nature of molecular interactions within the protein network and therefore to its direct effects on rheology and functionality. The use of model systems, together with non-invasive measuring techniques with resolution at the molecular level (e.g., scanning probe microscopy and ultrasound) will assist in elucidating the contribution of protein, under different conditions, to the physical properties of cheese (Dickinson, 1995).

References

Abd El-Salam, M.H., Al-Khamy, A.F., El-Garawany, G.A., Hamed, A. and Khader, A. (1996) Composition and rheological properties of processed cheese spread as affected by the level of added whey protein concentrates and emulsifying salt. *Egypt. J. Dairy Sci.*, **24**, 309–22.

Abou El-Nour, A., Schurer, G.J., Omar, M.M. and Buchheim, W. (1996) Physicochemical and rheological properties of block-type processed cheese analogue made from rennet casein and total milk protein. *Milchwissenschaft*, **51**, 684–7.

Al-Khamy, A.F., El-Garawany, G.A., Khader, A., Hamed, A. and Abd El-Salam, M.H. (1997) The use of whey protein concentrates in processed cheese spreads. 1-Effect of type of emulsifying salt. *Egypt. J. Dairy Sci.*, **25**, 99–112.

Anderson, M. and Andrews, A.T. (1977) Progressive changes in individual milk protein concentrations associated with high somatic cell counts. *J. Dairy Res.*, **44**, 223–23.

Anis, S.M.K. and Ernstrom, C.A. (1984) Calcium and the meltability of process cheese food made from ultrafiltered retentate. *J. Dairy Sci.*, **67** (Suppl 1), 79 (Abstr.).

Anon. (1992) Council Directive 92/46/EEC. *Official Journal of the European Communities*, No L268/2, 1–31.

Apostolopoulos, C. (1994) Simple empirical and fundamental methods to determine objectively the stretchability of Mozzarella. *J. Dairy Res.*, **61**, 405–13.

Arnott, D.R., Morris, H.A. and Combs, W.B. (1957) Effect of certain chemical factors on the melting quality of process cheese. *J. Dairy Sci.*, **40**, 957–63.

Auldist, M.J. and Hubble, I.B. (1998) Effects of mastitis on raw milk and dairy products. *Aust. J. Dairy Technol.*, **53**, 28–36.

Auldist, M.J., Coates, S., Sutherland, B.J., Mayes, J.J., McDowell, G.H. and Rogers, G.L. (1996) Effects of somatic cell count and stage of lactation on raw milk composition and the yield and quality of Cheddar cheese. *J. Dairy Res.*, **63**, 269–80.

Auty, M.A.E., Fenelon, M.A., Guinee, T.P., Mullins, C. and Mulvihill, D.M. (1999) Dynamic confocal scanning laser microscopy methods for studying milk protein gelation and cheese melting. *Scanning*, **21**, 299–304.

Baldwin, K.A., Baer, R.J., Parsons, J.G., Seas, S.W., Spurgeon, K.R. and Torrey, G.S. (1986) Evaluation of yield and quality of Cheddar cheese manufactured from milk with added whey protein concentrate. *J. Dairy Sci.*, **69**, 2543–50.

Banks, J.M. (1990) Improving cheese yield by the incorporation of whey powder. *Dairy Ind. Int.*, **55** (4), 37–41.

Banks, J.M. and Muir, D.D. (1984) Coagulum strength and cheese yield. *Dairy Ind. Int.*, **49** (9), 17–21, 36.

Banks, J.M. and Muir, D.D. (1985) Effect of incorporation of denatured whey protein on the yield and quality of Cheddar cheese. *J. Soc. Dairy Technol.*, **38**, 27–32.

Banks, J.M. and Tamime, A.Y. (1987) Seasonal trends in the efficiency of recovery of milk fat and casein in cheese manufacture. *J. Soc. Dairy Technol.*, **40**, 64–66.

Banks, J.M., Banks, W., Muir, D.D. and Wilson, A.G. (1980) Cheese yield – composition does matter. *Dairy Ind. Int.*, **46** (4), 15–22.

Banks, J.M., Griffiths, M.W., Phillips, J.D. and Muir, D.D. (1986) The yield and quality of Cheddar cheese produced from thermised milk. *Dairy Ind. Int.*, **51** (7), 31–5.

Banks, J.M., Law, A.J.R., Leaver, J. and Horne, D.S. (1994a) Sensory and functional properties of cheese: incorporation of whey proteins by pH manipulation and heat treatment. *J. Soc. Dairy Technol.*, **47**, 124–31.

Banks, J.M., Law, A.J.R., Leaver, J. and Horne, D.S. (1994b) The inclusion of whey proteins in cheese – an overview, in *Cheese Yield and Factors Affecting its Control*, Special Issue 9402, International Dairy Federation, Brussels, pp. 387–401.

Banks, J.M., Stewart, G., Muir, D.D. and West, I.G. (1987) Increasing the yield of Cheddar cheese by the acidification of milk containing heat-denatured whey protein. *Milchwissenschaft*, **42**, 212–5.

Barbano, D.M. (1994) Overview-influence of mastitis in milk, in *Cheese Yield and Factors Affecting its Control*, Special Issue 9402, International Dairy Federation, Brussels, pp. 387–401.

Bogenrief, D.D. and Olson, N.F. (1995) Hydrolysis of β-casein increases Cheddar cheese meltability. *Milchwissenschaft*, **50**, 678–82.

Boutrou, R., Gaucheron, F., Piot, M., Michel, F., Maubois, J.-L. and Léonil, J. (1999) Changes in the composition of juice expressed from Camembert cheese during ripening. *Lait.*, **79**, 503–13.

Brooker, B.E. (1993) The stabilisation of air in foods containing fat – A review. *Food Struct.*, **12**, 115–22.

Broome, M. (1999) Quality aspects of low concentration ratio membrane processing, in *Quality Quarterly Five Year Compendium*, (R. Richards and R. Chandler eds), Dairy Industry Quality Centre, Weribee, VIC 3030, Australia, pp. 45–7.

Brown, R.J. and Ernstrom, C.A. (1982) Incorporation of ultrafiltration concentrated whey solids into Cheddar cheese for increased yield. *J. Dairy Sci.*, **65**, 2391–5.

Bryant, A., Ustanol, Z. and Steffe, J. (1995) Texture of Cheddar cheese as influenced by fat reduction. *J. Food Sci.*, **60**, 1216–9.

Burvenich, C., Guidry, A.J. and Paape, M.J. (1995) Natural defence mechanisms of the lactating and dry mammary gland, in *Proc. Third IDF International Mastitis Seminar*, (A. Saran and S. Soback eds.) M. Lachmann Printers Ltd., Haifa, Israel, pp. 3–13.

Bynum, D.G. and Olson, N.F. (1982) Influence of curd firmness at cutting on Cheddar cheese yield and recovery of milk constituents. *J. Dairy Sci.*, **65**, 2281–90.

Caric, M. and Kaláb, M. (1993) Processed cheese products, in *Cheese: Chemistry, Physics and Microbiology, Vol. 2. Major Cheese Groups*, 2nd edn, (P.F. Fox ed.) Elsevier Applied Science Publishers, London, pp. 339–83.

Caríc, M., Gantar, M. and Kaláb, M. (1985) Effects of emulsifying salts on the microstructure and other characteristics of process cheese – a review. *Food Microstruct.*, **4**, 297–312.

Cavalier-Salou, C. and Cheftel, J.C. (1991) Emulsifying salts influence on characteristics of cheese analogs from calcium caseinate. *J. Food Sci.*, **56**, 1542–7, 51.

Chandan, R.C., Marin, H., Nakrani, K.R. and Zehner, M.D. (1979) Production and consumer acceptance of Latin American white cheese. *J. Dairy Sci.*, **62**, 691–6.

Chapman, H.R. (1974) The effect the chemical quality of milk has on cheese quality. *Dairy Ind. Int.*, **39**, 329–34.

Chen, A.H., Larkin, J.W., Clark, C.J. and Irwin, W.E. (1979) Texture analysis of cheese. *J. Dairy Sci.*, **62**, 901–7.

Choi, Y.H., Lee, K.I. and Baick, S.C. (1993) Effect of different salting time on the physical properties of Mozzarella cheese. *Kor. J. Dairy Sci.*, **15**, 298–306.

Code of Federal Regulations (1986) Part 133: Cheese and related products, in *Food and Drugs 21. Code of Federal Regulations, Parts 100–69*, US Government Printing Office, Washington, DC, pp. 214–24.

Collinge, S.K. and Ernstrom, C.A. (1988) Relationship between soluble N at pH 4.6 and meltability of pasteurized process cheese food made from UF curd. *J. Dairy Sci.*, **71**, (Suppl. 1), 71 (Abstr.).

Collinge, S.K., Ernstrom, C.A., Yiadom-Farkye, N.A. and Chaistitwanich, R. (1988) Nutritional quality and meltability of cheese from ultrafiltered milk. *J. Dairy Sci.*, **70**, (Suppl. 1), 67 (Abstr.).

Cousin, M.A. (1982) Presence and activity of psychrotrophic microorganisms in milk and dairy products: A review. *J. Food Protect.*, 172–207.

Covacevich, H.R. (1981) Recent experiences in pasta filata cheesemaking by ultrafiltration, in *Proc. Second Biennial Marschall International Cheese Symosium*, Madison, Wisconsin, USA, pp. 237–44.

Covacevich, H.R. and Kosikowski, F.V. (1977) Mozzarella and Cheddar cheese manufacture by ultrafiltration principles. *J. Dairy Sci.*, **61**, 701–9.

Creamer, L.K. (1976) Casein proteolysis in Mozzarella-type cheese. *N.Z. J. Dairy Sci. Technol.*, **11**, 130–1.

Creamer, L.K. (1985) Water absorption by renneted casein micelles. *Milchwissenschaft*, **40**, 589–91.

Creamer, L.K. and Olson, N.F. (1982) Rheological evaluation of maturing Cheddar cheese. *J. Food Sci.*, **47**, 631–6, 46.

Creamer, L.K., Zoerb, H.F., Olson, N.F. and Richardson, T. (1982) Surface hydrophobicity of α_{s1}-I, α_{s1}-casein A and B and its implications in cheese structure. *J. Dairy Sci.*, **65**, 902–6.

Csøk, J. (1982) The effect of holding time on free- and bound-water contents of processed cheese, in *Proc. XXI Int. Dairy Congr., Brief Communications* Volume 1, Book 1, Mir Publishers, Moscow, pp. 475–6.

Dalgleish, D.G. and Law, A.J.R. (1988) pH-induced dissociation of bovine casein micelles. I. Analysis of liberated caseins. *J. Dairy Res.*, **55**, 529–38.

Dalgleish, D.G. and Law, A.J.R. (1989) pH-induced dissociation of bovine casein micelles. II. Mineral solubilization and its relation to casein release. *J. Dairy Res.*, **56**, 727–35.

de Jong, L. (1976) Protein breakdown in soft cheese and its relationship to consistency. I. Proteolysis and consistency of 'Noordhollandase Meshanger' cheese. *Neth. Milk Dairy J.*, **30**, 242–53.

de Jong, L. (1977) Protein breakdown in soft cheese and its relation to consistency. 2. The influence of rennet concentration. *Neth. Milk Dairy J.*, **31**, 314–27.

de Jong, L. (1978a) Protein breakdown in soft cheese and its relation to consistency. 3. The micellar structure of Meshanger cheese. *Neth. Milk Dairy J.*, **32**, 15–25.

de Jong, L. (1978b) The influence of moisture content on the consistency and protein breakdown of cheese. *Neth. Milk Dairy J.*, **32**, 1–14.

Desai, N. and Nolting, J. (1995) Microstructure studies of reduced fat cheeses containing fat substitute, in *Chemistry of Structure-Function Relationships in Cheese*, (E.L. Malin and M.H. Tunick eds.) Plenum Press, New York, pp. 295–302.

Dickinson, E., Whyman, R.H. and Dalgleish, D.G. (1987) Colloidal properties of model oil-in-water food emulsions stabilized separately by α_{s1}-casein, β-casein and and κ-casein, in *Food Emulsions and Foams*, (E. Dickinson ed.) Royal Society of Chemistry, London, pp. 40–51.

Doi, H., Hirmamatus, M., Ibuki, F. and Kanamori, M. (1985) Gelation of the heat-induced complex between κ-casein and α-lactalbumin. *J. Nutr. Sci. Vitaminol.*, **31**, 77–87.

Doi, H., Ideno, S., Huang Kuo, F., Ibuki, F. and Kanamori, M. (1983a) Gelation of the complex between κ-casein and β-lactoglobulin. *J. Nutr. Sci. Vitaminol.*, **29**, 679–89.

Doi, H., Ideno, S., Ibuki, F. and Kanamori, M. (1983b) Participation of the hydrophobic bond in complex formation between κ-casein and β-lactoglobulin. *Agric. Biol. Chem.*, **47**, 407–9.

Donnelly, W.J. and Barry, J.G. (1983) Casein compositional studies. III. Changes in Irish milk composition for manufacturing and role of milk proteinase. *J. Dairy Res.*, **50**, 443–41.

Donnelly, W.J., Barry, J.G. and Buchheim, W. (1984) Casein micelle composition and syneretic properties of late lactation milk. *Irish J. Food Sci. Technol.*, **8**, 121–30.

Drake, M.A. (1997) Rheology of full-fat and low-fat Cheddar cheese as related to type of fat mimetic. *J. Food Sci.*, **62**, 748–52.

Dzurec, D.J. and Zall, R.R. (1986) Why Cottage cheese yields are increased when milk is heated on the farm. *Cult. Dairy Prod. J.*, **21** (3), 8–9.

El-Koussy, L., Amer, S.N. and Ewais, S.M. (1977) Studies on making baby Edam cheese with low fat content, II. Effect of milk heating. *Egypt. J. Dairy Sci.*, **5**, 207–13.

Emmons, D.B., Kalab, M., Larmond, E. and Lowrie, R.J. (1980) Milk gel structure. X. Texture and microstructure in Cheddar cheese made from whole milk and from homogenised low-fat milk. *J. Text. Stud.*, **11**, 15–34.

Ernstrom, C.A., Sutherland, B.J. and Jameson, G.W. (1980) Cheese base for processing. A high yield product from whole milk by ultrafiltration. *J. Dairy Sci.*, **63**, 228–34.

Everett, D.W., Ding, K., Olson, N.F. and Gunasekaran, S. (1995) Applications of confocal microscopy to fat globule structure in cheese, in *Chemistry of Structure-*

Function Relationships in Cheese, (E.L. Malin and M.H. Tunick eds.) Plenum Press, New York, pp. 321–30.

Farkye, N.Y. and Fox, P.F. (1992) Contribution of plasmin to Cheddar cheese ripening: effect of added plasmin. *J. Dairy Res.*, **59**, 209–16.

Farkye, N.Y. and Fox, P.F. (1990) Observations on plasmin activity in cheese. *J. Dairy Res.*, **57**, 413–8.

Fenelon, M.A. and Guinee, T.P. (1997) The compositional, textural and maturation characteristics of reduced-fat Cheddar made from milk containing added Dairy-Lo^{TM}. *Milchwissenschaft*, **52**, 385–9.

Fenelon, M.A. and Guinee, T.P. (1999) The effect of milk fat on Cheddar cheese yield and its prediction, using modifications of the van Slyke cheese yield formula. *J. Dairy Sci.*, **82**, 1–13.

Fenelon, M.A. and Guinee, T.P. (2000) Primary proteolysis and textural changes during ripening in Cheddar cheeses manufactured to different fat contents. *Int. Dairy J.*, **10**, 151–8.

Fernandez, A. and Kosikowski, F.V. (1986) Hot brine stretching and molding of low moisture Mozzarella cheese made from retentate supplemented milks. *J. Dairy Sci.*, **69**, 2551–7.

Fox, P.F. (1969) Effect of cold-ageing on the rennet-coagulation time of milk. *Irish J. Agric. Res.*, **8**, 175–82.

Fox, P.F. (1989) Proteolysis during cheese manufacture and ripening. *J. Dairy Sci.*, **72**, 1379–400.

Fox, P.F. and McSweeney, P.L.H. (1996) Proteolysis in cheese during ripening. *Food Rev. Int.*, **12**, 457–509.

Fox, P.F., O'Connor, T.P., McSweeney, P.L.H., Guinee, T.P. and O'Brien, N.M. (1996) Cheese: physical, biochemical, and nutritional aspects. *Adv. Food Nutr. Res.*, **39**, 163–328.

Fox, P.F. and Wallace, J.M. (1997) Formation of flavour compounds in cheese. *Adv. Appl. Microbiol.*, **45**, 17–85.

Fox, P.F., Guinee, T.P., Cogan, T.M. and McSweeney, P.L.H. (2000) *Fundamentals of Cheese Science*, Aspen Publishers, Inc., Gaithersburg, MD.

Garrett, N.L.T. (1987) The Sirocurd process for cheese manufacture. *J. Soc. Dairy Technol.*, **40**, 68–70.

Geurts, J., Walstra, P. and Mulder, H. (1972) Brine composition and the prevention of the defect 'soft rind' in cheese. *Neth. Milk Dairy J.*, **26**, 168–79.

Geurts, T.J., Walstra, P. and Mulder, H. (1974) Water binding to milk protein, with particular reference to cheese. *Neth. Milk Dairy J.*, **28**, 46–72.

Gilles, J. and Lawrence, R.C. (1985) The yield of cheese. *N.Z. J. Dairy Sci. Technol.*, **20**, 205–14.

Ginzinger, W. (1995) Innovationen bei Käse. *Milchwirschaftliche Berichte aus den Bundesanstalten, Wolfpassing und Rotholz*, No 124, 144–6.

Gopal, P.K. and Richardson, R.K. (1996) A rapid and sensitive method for estimation of galactose in Parmesan cheese. *Int. Dairy J.*, **5**, 399–406.

Grandison, A.S., Ford, G.D., Owen, A.J. and Millard, D. (1984) Chemical composition and coagulating properties of renneted Friesian milk during the transition from winter rations to spring grazing. *J. Dairy Res.*, **51**, 69–78.

Green, M.L., Glover, F.A., Scurlock, E.M.W., Marshall, R.J. and Hatfield, D.S. (1981) Development of structure and texture in Cheddar cheese. *J. Dairy Res.*, **48**, 333–41.

Green, M.L., Turvey, A. and Hobbs, D.G. (1981a) Development of structure and texture in Cheddar cheese. *J. Dairy Res.*, **48**, 343–55.

Green, M.L., Marshall, R.J. and Glover, F.A. (1983) Influence of homogenization of concentrated milks on the structure and properties of rennet curds. *J. Dairy Res.*, **50**, 341–8.

Green, M.L. (1985) Effect of milk pretreatment and making conditions on the properties of Cheddar cheese from milk concentrated by ultrafiltration. *J. Dairy Res.*, **52**, 555–64.

Green, M.L. (1987) Effect of manipulation of milk composition and curd-forming conditions on the formation, structure and properties of milk curd. *J. Dairy Res.*, **54**, 303–13.

Green, M.L. (1990a) Cheddar cheesemaking from whole milk concentrated by ultrafiltration and heated to 90°C. *J. Dairy Res.*, **57**, 559–69.

Green, M.L. (1990b) The cheesemaking potential of milk concentrated up to four-fold by ultrafiltration and heated in the range 90–97°C. *J. Dairy Res.*, **57**, 549–57.

Gripon, J.C. (1993) Mould-ripened cheeses, in *Cheese: Chemistry, Physics and Microbiology, Vol. 2 Major Cheese Groups*, 2nd edn, (P.F. Fox ed.) Chapman & Hall, London, pp. 111–36.

Grufferty, M.B. and Fox, P.F. (1988) Milk alkaline proteinase: a review. *J. Dairy Res.*, **55**, 609–30.

Guinee, T.P. and Fox, P.F. (1986) Transport of sodium chloride and water in Romano-type cheese slices during brining. *Food Chem.*, **19**, 49–64.

Guinee, T.P. (1987) Processed cheese products: physico-chemical aspects, in *Proc. Symposium on Industrial Aspects of Milk Proteins*, Moorepark Research Centre, Fermoy, Cork, Ireland, pp. 111–78.

Guinee, T.P. and Wilkinson, M.G. (1992) Rennet coagulation and coagulants in cheese manufacture. *J. Soc. Dairy Technol.*, **45**, 94–104.

Guinee, T.P., Pudja, P.D. and Farkye, N.Y. (1993) Fresh acid-curd cheese varieties, in *Cheese: Chemistry, Physics and Microbiology, Vol. 2 Major Cheese Groups*, 2nd edn, (P.F. Fox ed.) Chapman and Hall, London, pp. 363–419.

Guinee, T.P., Pudja, P.D. and Mulholland, E.O. (1994) Effect of milk protein standardisation, by ultrafiltration, on the composition and maturation of Cheddar cheese. *J. Dairy Res.*, **61**, 117–31.

Guinee, T.P. and Corcoran, M.O. (1994) Expanded use of cheese in processed meat products. *Farm Food Res.*, **4** (1), 25–8.

Guinee, T.P., Pudja, P.D., Reville, W.J., Harrington, D., Mulholland, E.O., Cotter, M. and Cogan, T.M. (1995) Composition, microstructure and maturation of semi-hard cheeses from high protein ultrafiltered milk retentates with different levels of denatured whey protein. *Int. Dairy J.*, **5**, 543–68.

Guinee, T.P., O'Callaghan, D.J., Mulholland, E.O. and Harrington, D. (1996a) Milk protein standardization by ultrafiltration for Cheddar cheese manufacture. *J. Dairy Res.*, **63**, 281–93.

Guinee, T.P., O'Callaghan, D.J., Pudja, P.D. and O'Brien, N. (1996b) Rennet coagulation properties of retentates obtained by ultrafiltration of skim milks heated to different temperatures. *Int. Dairy J.*, **6**, 581–96.

Guinee, T.P. and O'Callaghan, D.J. (1997) The use of simple empirical method for objective quantification of the strechability of cheese on cooked pizza pies. *J. Food Eng.*, **31,** 147–61.

Guinee, T.P., Mulholland, E.O., Mullins, C. and Corcoran, M.O. (1997a) Functionality of low moisture Mozzarella cheese during ripening. *Proc. 5th Cheese Symposium*, (T.M. Cogan, P.F. Fox. and R.P. Ross eds.) Teagasc, Dublin, pp. 15–23.

Guinee, T.P., Gorry, C.B., O'Callaghan, D.J., O'Kennedy, B.T., O'Brien, N. and Fenelon, M.A. (1997b) The effects of composition and some processing treatments on the rennet coagulation properties of milk. *Int. J. Dairy Technol.*, **50**, 99–106.

Guinee, T.P., Fenelon, M.A., Mulholland, E.G., O'Kennedy, B.T., O'Brien, N. and Reville, W.J. (1998) The influence of milk pasteurization temperature and pH at curd milling on the composition, texture and maturation of reduced fat Cheddar cheese. *Int. J. Dairy Technol.*, **51**, 1–10.

Guinee, T.P., Auty, M.A.E. and Mullins, C. (1999) Observations on the microstructure and heat-induced changes in the viscoelasticity of commercial cheeses. *Aust. J. Dairy Technol.*, **54**, 84–9.

Guinee, T.P., Auty, M.A.E. and Fenelon, M.A. (2000a) The effect of fat on the rheology, microstructure and heat-induced functional characteristics of Cheddar cheese. *Int. Dairy J.*, **10**, 277–88.

Guinee, T.P., Harrington, D., Corcoran, M.O., Mulholland, E.G. and Mullins, C. (2000b) The compositional and functional properties of Mozzarella, Cheddar and Analogue pizza cheeses. *Int. J. Dairy Technol.*, **53**, 51–6.

Guinee, T.P., Auty, M.A.E., Mullins, C., Corcoran, M.O. and Mulholland, E.O. (2000c) Preliminary observations on effects of fat emulsification on the structure-functional relationship of cheddar-type cheese. *J. Text. Stud.*, **31**, 645–63.

Guinot-Thomas, P., Al Ammoury, M., Le Roux, Y. and Laurent F. (1995) Study of proteolysis during storage of raw milk at 4°C: Effect of plasmin and microbial proteinases. *Int. Dairy J.*, **5**, 685–97.

Guo, M.R. and Kindstedt, P.S. (1995) Age-related changes in the water phase of Mozzarella cheese *J. Dairy Sci.*, **78**, 2009–107.

Guo, M.R., Gilmore, J.K.A. and Kindstedt, P.S. (1997) Effect of sodium chloride on the serum phase of Mozzarella cheese. *J. Dairy Sci.*, **80**, 3092–8.

Gupta, S.K., Karahadian, C. and Lindsay, R.C. (1984) Effect of emulsifier salts on textural and flavour properties of processed cheeses. *J. Dairy Sci.*, **67**, 764–78.

Hall, D.M. and Creamer, L.K. (1972) A study of the sub-microscopic structure of Cheddar, Cheshire and Gouda cheese by electron microscopy. *N.Z. J. Dairy Sci. Technol.*, **7**, 95–102.

Hansen, R. (1987) Mozzarella cheese without whey proteins. *Nordeuropaeisk Mejeritdsskrift*, **1**, 21–3.

Harvey, C.D., Morris, H.A. and Jenness, R. (1982) Relation between melting and textural properties of process Cheddar cheese. *J. Dairy Sci.*, **65**, 2291–5.

Harwalkar, V.R. and Kalab, M. (1980) Milk gel structure. XI. Electron microscopy of glucono-δ-lactone-induced skim milk gels. *J. Text. Stud.*, **11**, 35–49.

Harwalkar, V.R. and Kalab, M. (1981) Effect of acidulants and temperature on microstructure, firmness and susceptibility to synerese of skim milk gels. *Scanning Electron Microscopy*, 1981, III, 503–13.

Harwalkar, V.R. and Kalab, M. (1988) The role of β-lactoglobulin in the development of the core-and-lining structure of casein particles in acid-heat-induced milk gels. *Food Microstruct.*, **7**, 173–9.

Heertje, I., Visser, J. and Smits, P. (1985) Structure formation in acid milk gels. *Food Microstruct.*, **4**, 267–77.

Heertje, I., Boskamp, M.J., van Kleef, F. and Gortemaker, F.H. (1981) The microstructure of processed cheese. *Neth. Milk Dairy J.*, **35**, 177–9.

Hicks, C.L., Allauddin, M., Langlois, B.E. and O'Leary, J. (1982) Psychrotrophic bacteria reduce cheese yield. *J. Food Prot.*, **45**, 331–4.

Hicks, C.L., Onuorah, C., O'Leary, J. and Langlois, B.E. (1986) Effect of milk quality and low temperature storage on cheese yield – A summation. *J. Dairy Sci.*, **69**, 649–57.

Hokes, J.C., Mangino, M.E. and Hansen, P.M.T. (1982) A model system for curd formation and melting properties of calcium caseinates. *J. Food Sci.*, **47**, 1235–40, 49.

Horne, D.S. (1998) Casein interactions: casting light on the black boxes, the structure in dairy products. *Int. Dairy J.*, **8**, 171–7.

Horne, D.S., Banks, J.M. and Muir, D.D. (1996) Genetic polymorphism of milk proteins: understanding the technological effects, in *Hannah Research Institute Yearbook*, pp. 70–8.

Horne, D.S., Banks, J.M., Leaver, J. and Law, A.J.R. (1994) Dynamic mechanical 1spectroscopy of Cheddar cheese, in *Cheese Yield and Factors Affecting its Control*, Special Issue No 9402, International Dairy Federation, Brussels, pp. 507–12.

IDF (1991) *Factors Affecting the Yield of Cheese*, Special Issue No 9301, International Fairy Federation, Brussels.

Irani, R.R. and Callis, C.F. (1962) Calcium and magnesium sequestration by sodium and potassium polyphosphates. *J. Am. Oil Chem. Soc.*, **39**, 156–9.

Jana, A.H., Kokane, R.D. and Thakar, P.N. (1994) Lipid-protein interaction in milk products - A review. *Indian J. Dairy Sci.*, **47**, 624–34.

Jelen, P. and Rattray, W. (1995) Thermal denaturation of whey proteins, in *Heat Induced Changes in Milk*, 2nd edn, (P.F. Fox ed.), International Dairy Federation, Brussels, pp. 66–85.

Jordan, K.N. and Cogan, T.M. (1993) Identification and growth of non-starter lactic acid bacteria in Irish Cheddar cheese. *Irish J. Agric. Food Res.*, **32**, 47–55.

Kalab, M. (1977) Milk gel structure. VI. Cheese texture and microstructure, *Milchwissenschaft*, **32**, 449–57.

Kalab, M. (1979) Microstructure of dairy foods. 1. Milk products based on protein. *J. Dairy Sci.*, **62**, 1352–64.

Kalab, M. and Harwalkar, V.R. (1974) Milk gel structure, II. Relation between firmness and ultrastructure of heat-induced skim-milk gels containing 40–60% total solids. *J. Dairy Res.*, **41**, 131–5.

Kalab, M. and Modler, H.W. (1985) Milk gel structure. XV. Electron miscroscopy of whey protein-based cream cheese spread. *Milchwissenschaft*, **40**, 193–6.

Kalab, M., Gupta, S.K., Desai, H.K. and Patil, G.R. (1988) Development of microstructure in raw, fried, and fried and cooked Paneer made from buffalo, cow and mixed milks. *Food Microstruct.*, **7**, 83–91.

Kalab, M., Modler, H.W., Caric, M. and Milanovic, S. (1991) Structure, meltability and firmness of process cheese containing white cheese. *Food Struct.*, **10**, 193–201.

Kalab, M., Yun, J. and King Yiu, S. (1987) Textural properties and microstructure of process cheese food rework. *Food Microstruct.*, **6**, 181–92.

Karahadian, C. and Lindsay, R.C. (1987) Integrated roles of lactate, ammonia, and calcium in texture development of mould surface-ripened cheese. *J. Dairy Sci.*, **70**, 909–18.

Kefford, B., Christian, M.P., Sutherland. B.J., Mayes, J.J. and Grainger, C. (1995) Seasonal influences on Cheddar cheese manufacture: Influence of diet quality and stage of lactation. *J. Dairy Res.*, **62**, 529–37.

Keogh, M.K., Kelly, P.M., O'Keeffe, A.M. and Phelan, J.A. (1982) Studies of milk composition and its relationship to some processing criteria. 2. Seasonal variation in the mineral levels of milk. *Irish J. Food Sci. Technol.*, 13–27.

Kiely, L.J., Kindstedt, P.S., Hendricks, G.M., Levis, J.E., Yun, J.J. and Barbano, D.M. (1992) Effect of pH on the development of curd structure during the manufacture of Mozzarella cheese. *Food Struct.*, **11**, 217–24.

Kiely, L.J., Kindstedt, P.S., Hendricks, G.M., Levis, J.E., Yun, J.J. and Barbano, D.M. (1993) Age related changes in the microstructure of Mozzarella cheese. *Food Struct.*, **12**, 13–20.

Kimber, A.M., Brooker, B.E., Hobbs, D.G. and Prentice, J.H. (1974) Electron microscope studies of the development of structure in Cheddar cheese. *J. Dairy Res.*, **41**, 389–96.

Kimura, T., Taneya, S. and Furuichi, E. (1978) Electron microscopic observation of casein particles in processed cheese, *Proc. 20th Int. Dairy Congr.* (Paris), pp. 239–40.

Kindstedt, P.S. (1993) Mozzarella and pizza cheese. Factors affecting the functional characteristics of unmelted and melted Mozzarella cheese, in *Cheese: Chemistry, Physics and Microbiology, Vol. 2 Major Cheese Groups*, 2nd edn., (P.F. Fox ed.) Chapman and Hall, London, pp. 337–62.

Kindstedt, P.S. (1995) Factors affecting the functional characteristics of unmelted and melted Mozzarella cheese, in *Chemistry of Structure-Function Relationships in Cheese*, (E.L. Malin and M.H. Tunick eds.) Plenum Press, New York, pp. 27–41.

Kindstedt, P.S. and Guo, M.R. (1997a) Chemically-acidified pizza cheese: production and functionality. *Proc. 5th Cheese Symposium*, (T.M. Cogan, P.F. Fox and R.P. Ross eds.) Teagasc, Dublin, pp. 24–30.

Kindstedt, P.S. and Guo, M.R. (1997b) Recent developments in the science and technology of pizza cheese. *Aust. J. Dairy Technol.*, **52**, 41–3.

Kinsella, J.E. (1982) Relationships between structure and functional properties of food proteins, in *Food Proteins*, (P.F. Fox and J.J. Condon eds.) Applied Science Publishers Ltd., London, pp. 51–103.

Kinsella, J.E. and Whitehead, D.M. (1988) Proteins in whey: chemical, physical and functional properties. *Adv. Food Nutr. Res.*, **33**, 343–438.

Klei, L., Yun, J., Sapru, A., Lynch, J., Barbano, D., Sears, P. and Galton, D. (1998) Effects of milk somatic cell count on Cottage cheese yield and quality. *J. Dairy Sci.*, **81**, 1205–13.

Korolczuk and Mahaut (1992) Effect of whey protein addition and heat treatment of milk on the viscosity of UF fresh cheese. *Milchwissenschaft*, **47**, 157–9.

Laloy, E., Vuillemard, J.C., El Soda, M. and Simard, R.E. (1996) Influence of the fat content of Cheddar cheese on retention and localization of starters. *Int. Dairy J.*, **6**, 729–40.

Lanier, T.C. (1991) Interactions of muscle and nonmuscle proteins affecting heat-set gel rheology, in *Interactions of Food Proteins*, (N. Parris and R. Barford eds.) American Chemical Society, Washington, D.C., pp. 268–84.

Lau, K.Y., Barbano, D.M. and Rausmussen, R.R. (1990) Influence of pasteurization on fat and nitrogen recoveries and Cheddar cheese yield. *J. Dairy Sci.*, **73**, 561–70.

Lawrence, R.C. (1989) The use of ultrafiltration technology in cheesemaking. Bulletin 240, International Dairy Federation, Brussels, pp. 2–15.

Lawrence, R.C., Creamer, L.K. and Gilles, J. (1987) Texture development during cheese ripening. *J. Dairy Sci.*, **70**, 1748–60.

Lawrence, R.C., Heap, H.A. and Gilles, J. (1984) A controlled approach to cheese technology. *J. Dairy Sci.*, **67**, 1632–45.

Lazaridis, H.N., Rosenau, J.R. and Mahoney, R.R. (1981) Enzymatic control of meltability in a direct acidified cheese product. *J. Food Sci.*, **46**, 332–5, 9.

Le Graet, Y., Lepienne, A., Brule, G. and Ducruet, P. (1983) Migration du calcium et des phosphates inorganiques dans les fromages à pâte molle de type Camembert au cours de l'affinage. *Lait*, **63**, 317–32.

Lee, B.O., Kilbertus, G. and Alais, C. (1981) Ultrastructural study of processed cheese. Effect of different parameters. *Milchwissenschaft*, **36**, 343–8.

Lee, B.O., Paquet, D. and Alais, C. (1986) Etude biochemique la fonte des fromages. IV. Effet du type de sels de fonte et de la nature de la matière protéique sur la peptisation. Utilisation d'une systeme modele. *Lait*, **66**, 257–67.

Lee, H.J., Olson, N.F. and Lund, D.B. (1980) Diffusion of salt, fatty acids and esterases in Mozzarella cheeses. *J. Dairy Sci.*, **63**, 513–22.

Lenoir, J. (1984) The surface flora and its role in the ripening of cheese, in *Cheesemaking*, Bulletin 171, International Dairy Federation, Brussels, pp. 3–20.

Lo, C.G. and Bastian, E.D. (1998) Incorporation of native and denatured whey protein into cheese curd for manufacture of reduced-fat Havarti-type cheese. *J. Dairy Sci.*, **81**, 16–24.

Lowrie, R.J., Kalab, M. and Nichols, D. (1982) Curd granule and milled curd junction patterns in Cheedar cheese made by traditional and mechanized processes. *J. Dairy Sci.*, **65**, 1122–9.

Lucey, J.A. (1995) Effect of heat on the rennet coagulability of milk, in *Heat Induced Changes in Milk*, 2nd edn, (P.F. Fox ed.) International Dairy Federation, Brussels, pp. 171–87.

Lucey, J.A. and Gorry, C. (1994) Effect of Simplesse® 100 on the manufacture of low fat Cheddar cheese, in *Cheese Yield and Factors Affecting its Control*, Special Issue 9402, International Dairy Federation, Brussels, pp. 439–47.

Luyten, H. (1988) *The Rheological and Fracture Properties of Gouda Cheese*. Ph.D. Thesis, The Agricultural University, Wageningen, The Netherlands.

Ma, L., Drake, M.A., Barbosa-Cánovas, G.V. and Swanson, B.G. (1997) Rheology of full-fat and low-fat Cheddar cheeses as related to type of fat mimetic. *J. Food Sci.*, **62**, 748–52.

Mackey, K.L. and Desai, N. (1995) Rheology of reduced-fat cheese containing fat substitute, in *Chemistry of Structure-Function Relationships in Cheese*, (E.L. Malin and M.H. Tunick eds.) Plenum Press, New York, pp. 21–6.

Madsen, J.S. and Qvist, K.B. (1998) The effect of added proteolytic enzymes on meltability of Mozzarella cheese manufactured by ultrafiltration. *Lait*, **78**, 258–72.

Mahoney, R.R., Lazaridis, H.N. and Rosenau, J.R. (1982) Protein size and meltability in enzyme-treated, direct-acidified cheese products. *J. Food Sci.*, **47**, 670–1.

Marchesseau, S., Gastaldi, E., Lagaude, A. and Cuq, J.-L. (1997) Influence of pH on protein interactions and microstructure of process cheese. *J. Dairy Sci.*, **80**, 1483–9.

Market Tracking International Ltd. (1998) New product and packaging development, in *The International Cheese Market 1999–2003*. Market Tracking International Ltd., London, N19 4RU, pp. 27–61.

Marshall, R. (1986) Increasing cheese yields by high heat treatment of milk. *J. Dairy Res.*, **53**, 313–22.

Masters, K. (1976) *Spray Drying,* George Godwin Limited, London, pp. 51–62.

Matheis, G. and Whitaker, J.R. (1984) Chemical phosphorylation of food proteins: an overview and a prospectus. *J. Agric. Food Chem.*, **32**, 699–705.

Matheson, A.R. (1981) The immunochemical determination of chymosin activity in cheese. *N.Z. J. Dairy Sci. Technol.*, **16**, 33–41.

Maubois, J.-L. and Kosikowski, F.V. (1978) Preparation of Mozzarella cheese by membrane ultrafiltration. *Proc. 20th Int. Dairy Congr.* (Paris), pp. 92–793.

Mayes, J.J. and Sutherland, B.J. (1989) Further notes on coagulum firmness and yield in Cheddar cheese manufacture. *Aust. J. Dairy Technol.*, **44**, 47–8.

McMahon, D.J., Yousif, B.H. and Kalab, M. (1993) Effect of whey protein denaturation on structure of casein micelles and their rennetability after ultra-high temperature processing of milk with or without ultrafiltration. *Int. Dairy J.*, **3**, 239–56.

McMahon, D.J., Yousif, B.H. and Kalab, M. (1993) Effect of whey protein denaturation on structure of casein micelles and their rennetability after ultra-high temperature processing of milk with or without ultrafiltration. *Int. Dairy J.*, **3**, 239–56.

McMahon, D.J., Alleyne, M.C., Fife, R.L. and Oberg, C.J. (1996) Use of fat replacers in low fat Mozzarella cheese. *J. Dairy Sci.*, **79**, 1911–21.

Mehra, R., O'Brien, B., Connolly, J.F. and Harrington, D. (1999) Seasonal variation in the composition of Irish manufacturing and retail milks. 2. Nitrogen fractions. *Irish J. Agric. Food Res.*, **38**, 65–74.

Meyer, A. (1973) *Processed Cheese Manufacture*, Food Trade Press Ltd., London.

Mistry, V.V. and Anderson, D.L. (1993) Composition and microstructure of commercial full-fat and low-fat cheeses. *Food Struct.*, **12**, 259–66.

Mitchell, G.E., Fedrick, I.A. and Rogers, S.A. (1986) The relationship between somatic cell count, composition and manufacturing properties of bulk milk. 2. Cheddar cheese from farm bulk milk. *Aust. J. Dairy Technol.*, **41**, 12–4.

Morr, C.V. (1967) Some effects of pyrophosphate and citrate ions upon the colloidal caseinate-phosphate micelles and ultrafiltrate of raw and heated skim milk. *J. Dairy Sci.*, **50**, 1038–44.

Morris, H.A., Holt, C., Brooker, B.E., Banks, J.M. and Manson, W. (1988) Inorganic constituents of cheese: analysis of juice from a one month-old Cheddar cheese and the use of light and electron microscopy to characterize the crystalline phases. *J. Dairy Res.*, **55**, 255–68.

Mukherjee, K.K. and Hutkins, W. (1994) Isolation of galactose-fermenting thermophilic cultures and their use in the manufacture of low browning Mozzarella cheese. *J. Dairy Sci.*, **77**, 2839–49.

Mulvihill, D.M. (1992) Production, functional properties and utilization protein products, in *Advanced Dairy Chemistry Vol. 1. Proteins*, 2nd edn, (P.F. Fox ed.) Elsevier Applied Science, London, pp. 369–404.

Mulvihill, D.M. and Grufferty, M.B. (1995) Effect of thermal processing on the coagulability of milk by acid, in *Heat Induced Changes in Milk*, 2nd edn, (P.F. Fox ed.) International Dairy Federation, Brussels, pp. 188–205.

Murphy, S.C., Cranker, K., Senyk, G.F., Barbano, D.M., Saeman, A.I. and Galton, D.M. (1989) Influence of bovine mastitis on lipolysis and proteolysis in milk. *J. Dairy Sci.*, **72**, 620–6.

Nakajima, I., Kawaniski, G. and Furuichi, E. (1975) Reaction of melting salts upon casein micelles and their effects on calcium, phosphorus and bound water. *Agric. Biol. Chem.*, **39**, 979–87.

Neville, D.P. (1998) *Studies on the Melting Properties of Cheese Analogues*. M.Sc. Thesis, National University of Ireland, Cork.

Nishiya, T., Tatsumi, K., Ido, K., Tamaki, K. and Hanawa, N. (1989) Functional properties of imitation Mozzarella cheese using rennet casein. *J. Agric. Chem. Soc.*, Japan **63**, 1365–71.

Ng-Kwai-Hang, K.F. and Grosclaude, F. (1992) Genetic polymorphism of milk proteins, in *Advanced Dairy Chemistry-1. Proteins*, 2nd edn, (P.F. Fox ed.) Elsevier, London, pp. 405–55.

Noomen, A. (1983) The role of surface flora in the softening of cheeses with a low initial pH. *Neth. Milk Dairy J.*, **37**, 229–32.

Norris, G.E., Gray, I.K. and Dolby, R.M. (1973) Seasonal variations in the composition and thermal properties of New Zealand milk fat. *J. Dairy Res.*, **40**, 311–21.

O'Brien, B., Mehra, R., Connolly, J.F. and Harrington, D. (1999a) Seasonal variation in the composition of Irish manufacturing and retail milks. 1. Chemical composition and renneting properties. *Irish J. Agric. Food Res.*, **38**, 53–64.

O'Brien, B., Mehra, R., Connolly, J.F. and Harrington, D. (1999b) Seasonal variation in the composition of Irish manufacturing and retail milks. 4. Minerals and trace elements. *Irish J. Agric. Food Res.*, **38**, 87–99.

O'Brien, B., Murphy, J.J., Connolly, J.F., Mehra, R., Guinee, T.P. and Stakelum, G. (1997) Effect of altering the daily herbage allowance in mid lactation on the composition and processing characteristics of bovine milk. *J. Dairy Res.*, **64**, 621–6.

O'Brien, J. (1995) Heat-induced changes in lactose; isomerization, degradation, Maillard browning, in *Heat Induced Changes in Milk*, 2nd edn, (P.F. Fox ed.) International Dairy Federation, Brussels, pp. 134–70.

Oberg, C.J., McManus, W.R. and McMahon, D.J. (1993) Microstructure of Mozzarella cheese during manufacture. *Food Struct.*, **12**, 251–8.

Okigbo, L.M., Richardson, G.H., Brown, R.J. and Ernstrom, C.A. (1985) Coagulation properties of abnormal and normal milk from individual cow quarters. *J. Dairy Sci.*, **68**, 1893–6.

Olson, N.F. and Bogenrief, D.D. (1995) Functionality of Mozzarella and Cheddar cheese, in *Proc. 4th Cheese Symposium*, (T.M. Cogan, P.F. Fox and R.P. Ross eds.) National Dairy Products Research Centre, Teagasc, Moorepark, Fermoy, Cork, Ireland, pp. 81–9.

Pagliarini, E. and Beatrice, N. (1994) Sensory and rheological properties of low-fat filled 'pasta filata' cheese. *J. Dairy Res.*, **61**, 299–304.

Paquet, A. and Kalab. M. (1988) Amino acid composition and structure of cheese baked as a pizza ingredient in conventional and microwave ovens. *Food Microstruct.*, **7**, 93–103.

Park, J., Rosenau, J.R. and Peleg, M. (1984) Comparison of four procedures of cheese meltability evaluation. *J. Food Sci.*, **49**, 1158–62, 70.

Park, J.N., Lee, K.T. and Yu, J.H. (1992) The effect of emulsifying salts on the texture and flavour of block processed cheese made with ultrafiltered cheese base. *Kor. J. Dairy Sci.*, **14**, 234–49.

Paulson, B.M., McMahon, D.J. and Oberg, C.J. (1998) Influence of sodium chloride on appearance, functionality and protein arrangements in non-fat Mozzarella cheese. *J. Dairy Sci.*, **8**, 2053–64.

Pearse, M.J., Linklater, P.M., Hall, R.J. and MacKinlay, A.G. (1985) Effect of heat induced interaction between β-lactoglobulin and κ-casein on syneresis. *J. Dairy Res.*, **52**, 159–65.

Phelan, J.A., O'Keeffe, A.M., Keogh, M.K. and Kelly, P.M. (1982) Studies of milk composition and its relationship to some processing criteria. 1. Seasonal changes in the composition of Irish milk. *Irish J. Food Sci. Technol.*, **6**, 1–11.

Phillips, M.C. (1981) Protein conformation at liquid interfaces and its role in stabilizing emulsions and foams. *Food Technol.*, **35** (1), 50–7.

Piergiovanni, L., de Noni, I., Fava, P. and Schiraldi, A. (1989) Non-enzymatic browning in processed cheeses. Kinetics of the Maillard reaction during processing and storage. *Ital. J. Food Sci.*, **1**, 11–20.

Politis, I. and Ng-Kwai-Hang, K.F. (1988a) Association between somatic cell count of milk and cheese-yielding capacity. *J. Dairy Sci.*, **71**, 1720–7.

Politis, I. and Ng-Kwai-Hang, K.F. (1998b) Effects of somatic cell counts and milk composition on the coagulating properties of milk. *J. Dairy Sci.*, **71**, 1740–6.

Politis, I. and Ng-Kwai-Hang, K.F. (1988c) Effects of somatic cell count and milk composition on cheese composition and cheese making efficiency. *J. Dairy Sci.*, **71**, 1711–9.

Politis, I., Lachance, E., Block, E. and Turner, J.D. (1989) Plasmin and plasminogen in bovine milk: a relationship with involution? *J. Dairy Sci.*, **72**, 900–6.

Prentice, J.H., Langley, K.R. and Marshall, R.J. (1993) Cheese rheology, in *Cheese: Chemistry, Physics and Microbiology. Vol. 1, General Aspects*, 2nd edn, (P.F. Fox ed.) Chapman and Hall, London, pp. 303–40.

Qvist, K.B. (1979) Reestablishment of the original rennetability of milk after cooling. 1. The effect of cooling and LTST pasteurization of milk and renneting. *Milchwissenschaft*, **34**, 467–70.

Rao, M.O. (1992) Classification, description and measurement of viscoelastic properties of solid foods, in *Viscoelastic Properties of Foods*, (M.O. Rao and J.F. Steffe eds.) Elsevier Applied Science, London, pp. 3–47.

Rayan, A.A., Kalab, M. and Ernstrom, C.A. (1980) Microstructure and rheology of process cheese. *Scanning Electron Microscopy* **III**, 635–43.

Reps. A. (1993) Bacterial surface-ripened cheese. in *Cheese: Chemistry, Physics and Microbiology, Vol. 2 Major Cheese Groups*, 2nd edn, (P.F. Fox ed.) Chapman & Hall, London, pp. 137–72.

Richardson, B.C. (1983) Variation of the concentration of plasmin and plasminogen in bovine milk with lactation. *N.Z. J. Dairy Sci. Technol.*, **18**, 247–52.

Richardson, B.C. and Pearce, K.N. (1981) The determination of plasmin in dairy products. *N.Z. J. Dairy Sci. Technol.*, **16**, 209–20.

Rogers, G.L. and Stewart, J.A. (1982) The effects of some nutritional and non-nutritional factors on milk protein concentration and yield. *Aust. J. Dairy Technol.*, **37**, 26–32.

Rowney, M., Roupas, P., Hickey, M.W. and Everett, D.W. (1999) Factors affecting the functionality of Mozzarella cheese. *Aust. J. Dairy Technol.*, **54**, 94–102.

Rudan, M.A. and Barbano, D.M. (1998) A dynamic model for melting and browning of Mozzarella cheese during pizza baking. *Aust. J. Dairy Technol.*, **53**, 95–7.

Rudan, M.A., Barbano, D.M. and Kindstedt, P.S. (1998) Effect of fat replacer (Salatrim) on chemical composition, proteolysis, functionality, appearance and yield of reduced fat Mozzarella cheese. *J. Dairy Sci.*, **81**, 2077–88.

Rüegg, M., Eberhard, P., Popplewell, L.M. and Peleg, M. (1991) Melting properties of cheese, in *Rheological and Fracture Properties of Cheese*, Bulletin 268, International Dairy Federation, Brussels, pp. 36–43.

Saeman, A.I., Verdi, R.J., Galton, D.M. and Barbano, D.M. (1988) Effect of mastitis on proteolytic activity in bovine milk. *J. Dairy Sci.*, **71**, 505–12.

Sapru, A., Barbano, D.M., Yun, J., Klei, L.R., Oltenachu, A. and Blander, D.K. (1997) Cheddar cheese: Influence of milking frequency and stage of lactation on composition and yield. *J. Dairy Sci.*, **80**, 437–46.

Savello, P.A., Ernstrom, C.A. and Kalab, M. (1989) Microstructure and meltability of model process cheese made with rennet and acid casein. *J. Dairy Sci.*, **72**, 1–11.

Schaar, J. (1985) Plasmin activity and proteose-peptone content of individual milks. *J. Dairy Res.*, **52**, 369–78.

Schaar, J., Hannson, B. and Pettersson, H. (1985) Effects of genetic variants of κ-casein and β-lactoglobulin on cheesemaking. *J. Dairy Res.*, **52**, 429–37.

Schafer, H.W. and Olson, N.F. (1975) Characteristics of Mozzarella cheese made by direct acidification from ultra-high temperature processed milk. *J. Dairy Sci.*, **58**, 494–501.

Schulz, M.E. (1976) Preparation of melt-resistant processed cheese. United States Patent, **3**, 962 483.

Shaw, M. (1984) Cheese substitutes: threat or opportunity? *J. Soc. Dairy Technol.*, **37**, 27–31.

Shehata, A.E., Iyer, M., Olson, W.F. and Richardson, T. (1967) Effect of type of acid used in direct acidification procedures on moisture, firmness and calcium levels of cheese. *J. Dairy Sci.*, **50**, 824–6.

Singer, N.S. (1996) Microparticulated proteins as fat mimetics, in *Handbook of Fat Replacers*, (S. Roller and S.A. Jones eds.) CRC Press LLC, Boca Raton, Florida, pp. 175–89.

Singh, H. and Creamer, L.K. (1990) A sensitive quantitative assay for milk coagulants in cheese and whey products. *J. Dairy Sci.*, **73**, 1158–65.

Singh, H., Shalabi, S.I., Fox, P.F., Flynn, A. and Barry, A. (1988) Rennet coagulation of heated milk: influence of pH adjustment before or after heating. *J. Dairy Res.*, **55**, 205–15.

Sood, S.M., Gaind, D.K. and Dewan, R.K. (1979) Correlation between micelle solvation and calcium content. *N.Z. J. Dairy Sci. Technol.*, **14**, 32–4.

Sood, V.K. and Kosikowski, F.V. (1979) Process Cheddar cheese from plain and enzyme treated retentates. *J. Dairy Sci.*, **62**, 1713–8.

Sørensen, H.H. (1997) *The World Market for Cheese*. Bulletin 326, International Dairy Federation, Brussels, pp. 8–17.

Southward, C.R. (1985) Manufacture of edible casein products. 1. Manufacture and properties. *N.Z. J. Dairy Sci. Technol.*, **20**, 79–101.

Stainsby, G. (1989) Dairy foams, in *Advances in Food Emulsions and Foams*, (M. Anderson, B.E. Brooker and E. Dickinson eds.) Elsevier Applied Science Publishers Ltd., London, pp. 221–55.

Steffen, C., Eberhard, P., Bosset, J.O. and Rüegg, M. (1993) Swiss-type varieties, in *Cheese: Chemistry, Physics and Microbiology. Vol. 2. Major Cheese Groups*, 2nd edn, (P.F. Fox ed.) Elsevier Applied Science Publishers, London, pp. 83–110.

Sundheim, G. and Bengtsson-Olivecrona, G. (1987) Isolated milk fat globules as substrate for lipoprotein lipase: Study of factors relevant to spontaneous lipolysis in milk. *J. Dairy Sci.*, **70**, 499–505.

Swaisgood, H.E. (1992) Chemistry of the caseins, in *Advanced Dairy Chemistry Vol. 1. Proteins*, 2nd edn., (P.F. Fox ed.) Elsevier Applied Science, London, pp. 63–110.

Tamime, A.Y., Kalab, M., Davies, G. and Younis, M.F. (1990) Microstructure and firmness of processed cheese manufactured from Cheddar cheese and skim milk powder cheese base. *Food Struct.*, **9**, 23–37.

Taneya, S., Izutsu, T. and Sone, T. (1979) Dynamic viscoelasticity of natural cheese and processed cheese, in *Food Texture and Rheology*, (P. Sherman ed.) Academic Press, London, pp. 369–83.

Taneya, S., Izutsu, T., Kimura, T. and Shioya, T. (1992) Structure and rheology of string cheese. *Food Struct.*, **11**, 61–71.

Taneya, S., Kimura, T., Izutsu, T. and Bucheim, W. (1980) The submicroscopic structure of processed cheese with different melting properties. *Milchwissenschaft*, **35**, 479–81.

Templeton, H.L. and Sommer, H.H. (1936) Studies on the emulsifying salts used in processed cheese. *J. Dairy Sci.*, **19**, 561–672.

Thierry, A., Salvat-Brunaud, D., Madec, M.-M., Michel, F. and Maubois, J.-L. (1998) Affinage de l'emmental: dynamique des populations bactériennes et évolution de la composition de la phase aquese. *Lait*, **78**, 521–42.

Thomas, M.A. (1969) Browning reaction in Cheddar cheese. *Aust. J. Dairy Technol.*, **24**, 185–9.

Thomas, M.A. (1970) Use of calcium co-precipitates in processed cheese. *Aust. J. Dairy Technol.*, **25**, 23–5.

Thomas, M.A., Newell, G., Abad, G.A. and Turner, A.D. (1980) Effect of emulsifying salts on objective and subjective properties of processed cheese. *J. Food Sci.*, **45**, 158–466.

Thomas, T.D. and Pearce, K.N. (1981) Influence of salt on lactose fermentation and proteolysis in Cheddar cheese. *N.Z. J. Dairy Sci. Technol.*, **16**, 253–9.

Torres, J.A., Bouzas, J., Kirby, C., Almonacid-Merin, S.F., Kantt, C.A., Simpson, R. and Banga, J.R. (1995) Time-temperature effects of microbial, chemical and sensory changes during cooling and aging of Cheddar cheese, in *Chemistry of Structure-Function Relationships in Cheese*, (E.L. Malin and M.H. Tunick eds.) Plenum Press, New York, pp. 123–59.

Torres, N. and Chandan, R.C. (1981) Latin American white cheese – A review. *J. Dairy Sci.*, **64**, 552–7.

Tunick, M.H. and Shieh, J.J. (1995) Rheology of reduced-fat Mozzarella, in *Chemistry of Structure-Function Relationships in Cheese*, (E.L. Malin and MH. Tunick eds.) Plenum Press, New York, pp. 7–19.

Turner, K.W. and Martley, F.G. (1983) Galactose fermentation and classification of thermophilic lactobacilli. *Appl. Environ. Microbiol.*, **45**, 1932–4.

van Boekel, M.A.J.S. and Walstra, P. (1995) Effect of heat treatment on chemical and physical changes to milk fat globules, in *Heat Induced Changes in Milk*, 2nd edn, (P.F. Fox ed.) International Dairy Federation, Brussels, pp. 51–65.

van den Berg, G., Escher, J.T.M., de Koning, P.J. and Bovenhuis, H. (1992) Genetic polymorphism of κ-casein and β-lactoglobulin in relation to milk composition and processing properties. *Neth. Milk Dairy J.*, **46**, 145–63.

van den Berg, G. (1979) Increasing cheese yield by inclusion of whey proteins. *Neth. Milk Dairy J.*, **33**, 210–1.

van Hooydonk, A.C.M. and van den Berg, G. (1988) Control and determination of the curd-setting during cheesemaking. Bulletin 225, International Dairy Federation Brussels, pp. 2–10.

van Hooydonk, A.C.M., Hagedoorn, H.G. and Boerrigter, I.J. (1986) pH-induced physico-chemical changes of casein micelles in milk and their on renneting. 1. Effect of acidification on phsico-chemical properties. *Neth. Milk Dairy J.*, **40**, 281–96.

van Hooydonk, A.C.M., Koster P.G. and Boerrigter, I.J. (1987) The renneting properties of heated milk. *Neth. Milk Dairy J.*, **41**, 3–18.

van Vliet, T. and Dentener-Kikkert A. (1982) Influence of the composition of the milk fat globule membrane on the rheological properties of acid milk gels. *Neth. Milk Dairy J.*, **36**, 261–5.

van Vliet, T. (1991) Terminology to be used in cheese rheology, in *Rheological and Fracture Properties of Cheese*, Bulletin 268, International Dairy Federation, Brussels, pp. 5–15.

van Wazer, J.R. (1971) Chemistry of the phosphates and condensed phosphates, in *Phosphates in Food Processing*, (J.M. de Man and P. Melnychyn eds.) AVI Publishing Company Inc., Westport, CT, pp. 1–23.

Visser, J., Minihan, A., Smits, P., Tjan, S.B. and Heertje, I. (1986) Effects of pH and temperature on the milk salt system. *Neth. Milk Dairy J.*, **40**, 351–68.

Visser, J. (1991) Factors affecting the rheological and fracture properties of hard and semi-hard cheese, in *Rheological and Fracture Properties of Cheese*, Bulletin 268, International Dairy Federation, Brussels, pp. 49–61.

Visser, F.M. and de Groot-Mostert, E.A. (1977) Contribution of enzymes from rennet, starter bacteria and milk to proteolysis and flavour development in Gouda cheese. IV. Protein breakdown: a gel electrophoretical study. *Neth. Milk Dairy J.*, **31**, 247–64.

Walsh, C.D., Guinee, T., Harrington, D., Mehra, R., Murphy, J., Connolly, J.F. and FitzGerald, R.J. (1995) Cheddar cheesemaking and rennet coagulation characteristics of bovine milks containing κ-casein AA or BB genetic variants. *Milchwissenschaft*, **50**, 492–5.

Walsh, C.D., Guinee, T.P., Reville, W.D., Harrington, D., Murphy, J.J., O'Kennedy, B.T. and FitzGerald, R.J. (1998) Influence of κ-casein genetic variant on rennet gel microstructure, Cheddar cheesemaking properties and casein micelle size. *Int. Dairy J.*, **8**, 1–8.

Walstra, P. and van Vliet, T. (1986) The physical chemistry of curd making. *Neth. Milk Dairy J.*, **40**, 241–59.

Walstra, P., van Dijk, J.M. and Geurts, T.J. (1985) The syneresis of curd. 1. General considerations and literature review. *Neth. Milk Dairy J.*, **39**, 209–46.

Weatherhup, W., Mullan, M.A. and Kormos, J. (1988) Effect of storing milk at 3 or 7°C on the quality and yield of Cheddar cheese. *Dairy Ind. Int.*, **53** (2), 16–7, 25.

White, J.C. and Davies, D.T. (1958) The relationship between the chemical composition of milk and the stability of the caseinate complex. 1. General introduction, description of samples, methods and chemical composition of samples. *J. Dairy Res.*, **25**, 236–55.

Younis, M.F., Tamime, A.Y., Davies, G., Hunter, E.A. and Abd El-Hady, S.M. (1991) Production of processed cheese using Cheddar cheese and cheese base. 5. Rheological properties. *Milchwissenschaft*, **46**, 701–5.

Yun, J.J., Barbano, D.M. and Kindstedt, P.S. (1993a) Mozzarella cheese: impact of coagulant type on chemical composition and proteolysis. *J. Dairy Sci.*, **76**, 3648–56.

Yun, J.J., Kiely, L.J., Kindstedt, P.S. and Barbano D.M. (1993b) Mozzarella cheese: impact of coagulant type on functional properties. *J. Dairy Sci.*, **76**, 3657–63.

Zachos, T., Politis, I., Gorewit, R.C. and Barbano, D.M. (1992) Effect of mastitis on plasminogen activator activity in milk somatic cells. *J. Dairy Res.*, **59**, 461–7.

Zall, R.R. (1980) Cheesemaking. Can cheesemaking be improved by heat treating milk on a farm? *Dairy Ind. Int.*, **45** (2), 25–30.

Zall, R.R. and Chen, J.H. (1981) Heating and storing milk on dairy farms before pasteurization in milk plants. *J. Dairy Sci.*, **64**, 1540–4.

26

FUNCTIONAL MILK PROTEINS: PRODUCTION AND UTILIZATION

D.M. Mulvihill and M.P. Ennis

26.1 Introduction

Bovine milk contains ~13% solids, which include fat, lactose, protein, and organic and inorganic salts. Normal milk contains ~3.5 g total protein per 100 ml which falls into two main categories based on solubility at pH 4.6 at >~8°C. Under these conditions ~80% of the total nitrogen precipitates and this fraction is referred to as casein, while ~20% remains soluble in the serum or whey, ~15% being whey proteins with the remainder being non-protein nitrogenous components. As discussed in Chapter 1, both the casein and non-casein fractions are heterogeneous; casein includes four principal primary proteins (gene products), α_{s1}-, α_{s2}-, β- and κ-caseins, and several minor proteins, while the non-casein fraction includes β-lactoglobulin, α-lactalbumin, blood serum albumin, immunoglobulins, casein-derived proteose peptones, several minor proteins, including lactotransferrin, and several enzymes. The characteristics of the caseins and whey proteins differ very significantly (Table 26.1).

1. The caseins are insoluble at their isoelectric points (~pH 4.6) at temperatures >~8°C. At very low ionic strength, most, of the whey proteins are also insoluble at their isoelectric points (~pH 5) but they are soluble at this pH in the ionic environment of milk.
2. Coagulation of the caseins can be induced by limited proteolysis using crude proteinase preparations, known as rennets, but the whey proteins remain soluble.
3. Caseins are extremely heat-stable proteins while the whey proteins are heat labile: sodium caseinate dissolved in water does not coagulate in 60 min at 140°C, while the whey proteins are denatured on heating at

Advanced Dairy Chemistry Volume 1: Proteins. 3rd edn.
Edited by P.F. Fox and P.L.H. McSweeney, Kluwer Academic/Plenum Publishers, 2003.

TABLE 26.1
Principal differences between casein and whey proteins

Characteristic	*Casein*	*Whey*
Solubility at pH 4.6	No	Yes
Rennet coagulation	Yes	No
Heat stability	High	Low
Particle size	Large (micelles; MW ~10^8 Da)	Small (molecules; MW ~$1.5–8.0 \times 10^4$ Da)

temperatures >70°C. At its normal pH (~6.7), milk may be heated at 140°C for 20 min before coagulation occurs; however, on heating milk at >~72°C, whey proteins become denatured and interact and complex with the casein micelles.

4. In milk, the caseins occur as large aggregates, micelles, which can be separated from the molecularly dispersed whey proteins by ultracentrifugation (e.g., 100,000 g for 1 h).

Some of these differences between caseins and whey proteins are exploited in industrial methods for casein and whey protein isolation; however, caseins and whey proteins can also be isolated together in various high-protein products, referred to as co-precipitates. Because of the heterogeneity of both protein systems, methods have been developed to isolate individual proteins.

Because of their source and their nutritional value, dehydrated milk protein-enriched products are ‘high esteem’ food ingredients. As well as contributing to the nutritional status of foods, the physico–chemical and functional properties of these protein-enriched products are exploited to modify and enhance the textural and rheological characteristics of foods. The proteins bind and emulsify fat, bind and entrap water and entrap and stabilize air in food products, thus contributing to product stability and sensory appeal.

This chapter provides an overview of methods for the production of milk protein-enriched products and their utilization in foods.

26.2 Production of milk protein products

26.2.1 Production of caseins

Casein has been produced commercially for at least 80 years. Initially used for industrial purposes (e.g., glues, paper glazing and synthetic fibres), it was not until the 1960’s that isolated casein gained importance as a food protein, due mainly to pioneering work in Australia and New Zealand. Today, casein, produced by acid or rennet coagulation, is one of the principal functional food proteins with an annual world production of ~250,000 tonnes. It has some rather unique properties and cannot be replaced by other proteins in certain food applications. Methods for the manufacture of caseins and caseinates have been reviewed by Muller (1971, 1982), Mulvihill (1989, 1992), Fox and Mulvihill (1990) and Mulvihill and Fox (1994).

In the isolation of the casein fraction from milk, fat is first removed by centrifugation to yield a skim milk from which the casein is isolated after destabilizing it and rendering it insoluble. The use of skim milk ensures that the fat content of the casein is sufficiently low to minimize flavour defects arising from deterioration of lipids in the dried casein products. Following destabilization, the insoluble casein is separated from the soluble whey proteins, lactose and salts, washed to remove residual soluble solids and dried (Figure 26.1).

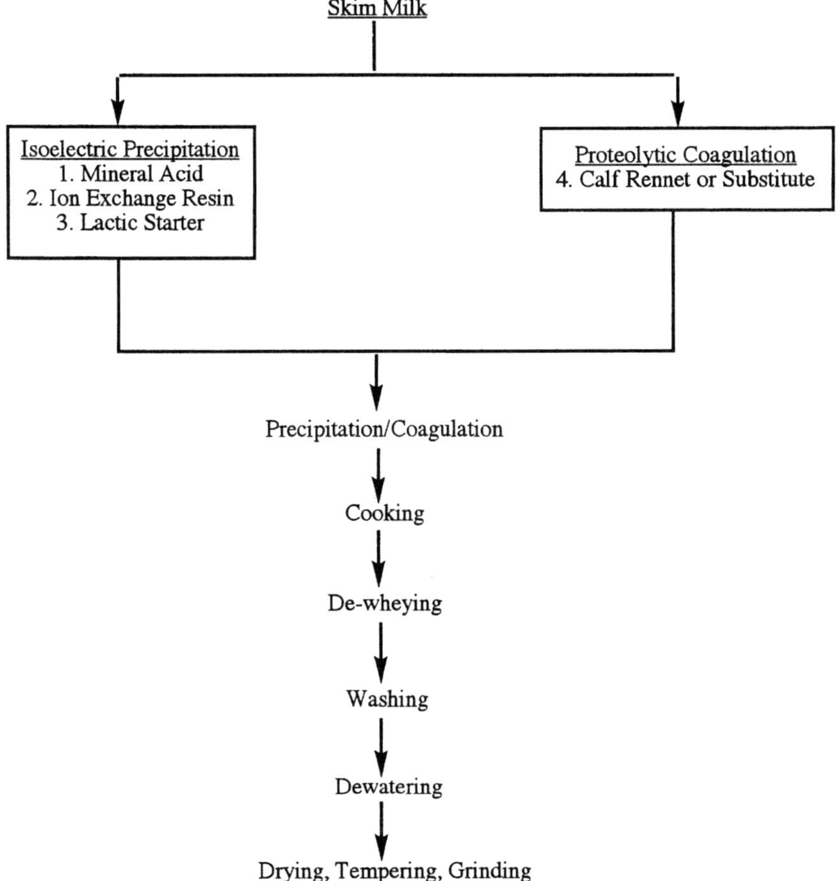

Figure 26.1 Processes for the manufacture of caseins.

(a) Destabilization/precipitation

In the manufacture of mineral acid casein, precipitation is accomplished by sparging dilute (1–2 M) mineral acid, usually HCl, under pressure into milk (preheated to 25–30°C) flowing in the opposite direction to give a pH of ~4.6. Steam is then injected to heat the acidified milk to the required temperature (~50°C). The heated milk is passed through a holding or acidulation tube to ensure complete coagulation and agglomeration of the curd, before separation of the curd and whey (Figure 26.2).

The pH of skim milk can also be reduced to the isoelectric point of casein by mixing skim milk at <10°C with a cation exchange resin in the hydrogen form in a reaction column; this replaces cations in the milk by H^+ to give a pH of ~2.2. The deionized, acidified milk is then mixed with untreated milk to give the final desired pH of ~4.6. The mixture is then heated to ~50°C by direct steam injection (Figure 26.2).

In the manufacture of lactic casein, precipitation is accomplished by inoculating pasteurized skim milk with a mixed-strain or multiple defined-strain starter and incubating at 22 to 26°C. During an incubation period of 14–16 h, the starter slowly ferments some of the lactose to lactic acid and a casein gel network or coagulum, with good water-holding capacity, is formed as the pH of the milk falls slowly under quiescent conditions to the isoelectric pH of the casein. Following coagulation, the coagulum is pumped

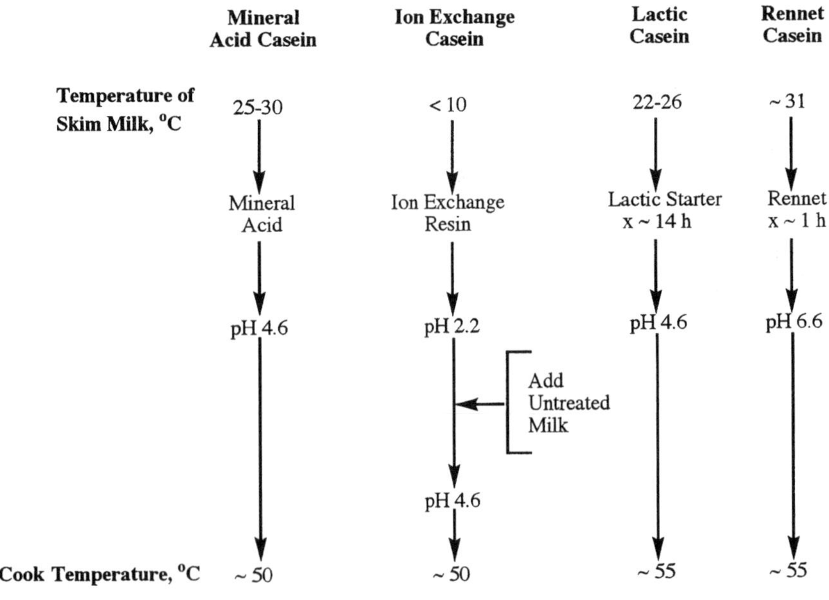

Figure 26.2 Conditions for casein precipitation.

from the coagulation vat and cooked to ~55°C by direct steam injection. To permit the curd particles to agglomerate and to initiate syneresis, a period of contact with whey, termed acidulation, either in a holding pipe or vat, is allowed prior to separation of curd and whey (Figure 26.2).

Numerous proteinases can coagulate milk at its natural pH (~6.7) in a two-stage process. Rennet (crude preparations of gastric proteinase) prepared from calf vells was the proteinase preparation traditionally used, but due to inadequate supply of calf rennet, microbial proteinases are now widely used as rennet substitutes. The primary and secondary phases of rennet coagulation, including the mechanism of gel assembly and the factors that influence gel strength, are now well understood (see Chapter 18). In the first stage, specific hydrolysis of κ-casein yields *para*-κ-casein and casein (glyco)macropeptides, while in the second stage, coagulation of the rennet-altered casein micelles by Ca^{2+} occurs at a temperature above 20°C. A coagulum of this nature, formed from skim milk, can be processed further to yield rennet casein in a manner similar to that used for the manufacture of lactic casein following quiescent acid coagulation. In the traditional method for the manufacture of rennet casein, skim milk at pH ~6.7, is 'set' with rennet (1:4500 rennet:milk) in large jacketed vats at ~31°C, in a manner similar to that practised in cheesemaking. When coagulation has progressed to the desired stage, the gel is pumped from the coagulation vat to a cooking pipe where steam is injected to raise the temperature to 50–55°C for ~45 s before separation of curd and whey (Figure 26.2). Rennet casein has a high ash content, as colloidal calcium phosphate (CCP) is retained in the curd.

(b) De-wheying

Following destabilization of the casein, the curd is separated from the whey prior to washing. The efficiency of the 'de-wheying' step, which is of the utmost importance in determining the volume of whey recovered for further processing, the efficiency of the washing operation and the quality of the final casein produced, depends on the pH and temperature of precipitation and on the equipment used to achieve separation.

(c) Washing and De-watering

During washing, residual whey constituents (lactose, whey proteins, salts) and free acid are removed from the de-wheyed curd particles to a limited extent by washing of the surface of the particles and to a much greater extent by diffusion from within the particles. The rate of diffusion depends on the size and permeability of the curd particles, the difference in the concentration of the constituents between the interior of the particle and the surrounding wash water and on the amount, temperature and movement of the wash water.

Washing systems used include multi-stage counter-current systems and counter-current tower washing systems in which the curd falls through an ascending column of water. A gradient of wash water temperature is normally used during the washing operation; a typical temperature profile for washing acid curd in a four-stage washing system is 55, 65, 75 and 35°C for the first to the fourth stage, respectively.

After washing, casein curd is mechanically de-watered to minimize the quantity of water to be evaporated and thus minimize the energy required during the subsequent thermal drying operation. Ideally, the properties of the casein curd following washing should allow for maximum de-watering under the conditions of operation of the de-watering machine while maintaining the curd in a suitable condition for subsequent drying. Mechanical de-watering devices include roller and screw presses and decanting centrifuges.

(d) Drying, Tempering and Grinding

To produce a stable, long-life casein that meets the internationally recognized compositional standards for edible-grade product, the casein curd is dried to $<12\%$ moisture in any one of a variety of drier types. Traditional driers used are of a semi-fluidized, vibrating type in which casein curd passes along vibrating perforated stainless steel conveyors while warm air is forced up through the perforations, partially fluidizing the curd as it is dried.

Pneumatic ring driers, which consist essentially of a large, stainless steel, ring-shaped duct through which high-velocity, heated air and moist, finely-divided casein curd are circulated continuously, are now also widely used.

Attrition drying, a technique based on the principle of grinding and drying in a single operation, is also widely used in casein plants as it allows the production of a casein product closely resembling spray dried casein. The drier, consisting of a fast-revolving, multi-chambered rotor and a stator with a serrated surface, generates turbulence, vortices and cavitation effects, resulting in highly efficient grinding, which pulverizes the curd into very small particles with a large total surface area. These particles are dried simultaneously in a hot air stream that passes through the drier concurrently with the curd. The dried casein particles have an overall average particle size of $\sim$100 μm and have good wettability and dispersability in water because they are irregularly shaped and many contain cavities due to the rapid evaporative process.

Dried casein is relatively hot as it emerges from the drier and the moisture content of individual particles varies. The dried curd is tempered and blended by pneumatic circulation of the curd between a number of holding bins in order to achieve a cooled final product of uniform moisture content. The casein is then ground in roller or pin-disc mills and the milled material is separated on screens to products of particle-size ranges required by casein users while oversized material is re-cycled for further milling.

26.2.2 Production of caseinates

Acid casein is insoluble in water but will dissolve in alkali under suitable conditions to yield water-soluble caseinates that may be spray or roller dried.

(a) Sodium caseinate
Sodium caseinate, the water-soluble form of casein most commonly used in foods is usually prepared by solubilizing acid casein with NaOH. The steps involved in its manufacture (Figure 26.3) are as follows:

1. Casein curd from a de-watering device (~45% solids) is minced and the finely-divided curd mixed with water at 40°C to give a solids content of ~25%, before passing it through a colloid mill.
2. NaOH (2.5 M) is pumped into the casein slurry, emerging from the mill at <45°C and with the consistency of 'toothpaste', to give a final caseinate pH of 6.6 to 6.8.
3. The mixture is transferred to a vat where solubilization occurs as the mixture is heated and vigorously agitated using a mixer capable of coping with the high viscosity. The slurry is recirculated and/or pumped to a second vat where solubilization is completed as the solution temperature is raised to ~75°C. An in-line pH meter is used to monitor and regulate the addition of the NaOH solution.
4. The caseinate solution is pumped to a balance tank through a heat exchanger in which the temperature is increased to ~95°C. A second in-line pH meter is used to control further addition of NaOH, if necessary, to give a caseinate of the desired pH.
5. The caseinate solution is pumped to the spray drier *via* an in-line viscometer that regulates the addition of hot water to control viscosity and ensure efficient atomization of the solution in the drier.

During the production of caseinate, care must be taken to minimize:

- the time for which the caseinate solution is held at high temperatures, since browning may occur due to reactions between the protein and residual lactose;
- the time for which the casein is exposed to high pH during dissolving, as this may lead to the formation of lysinoalanine and the development of off-flavours.

(b) Other caseinates
Other methods used to produce different caseinate types include (Figure 26.4):

- Production of roller dried sodium caseinate by feeding a mixture of curd (50–65% moisture) and an alkaline sodium salt (Na_2CO_3 or $NaHCO_3$) onto the drying drum of a roller-drier.

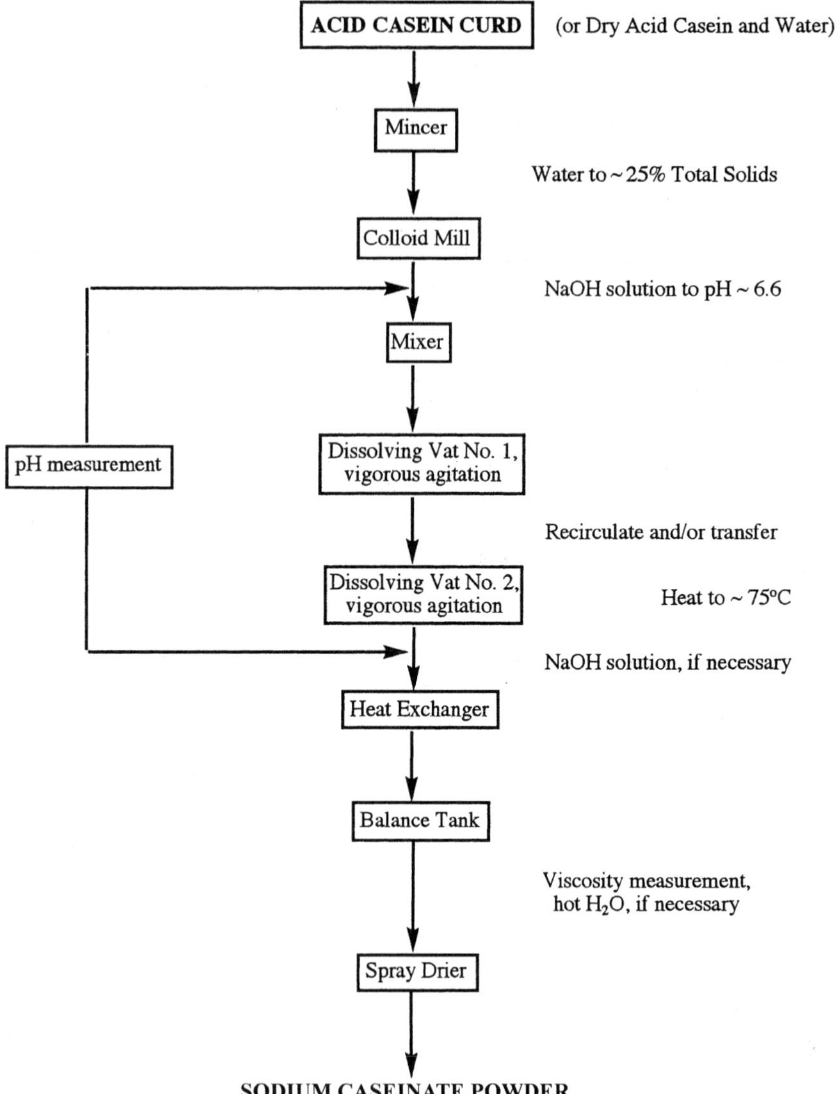

Figure 26.3 Method for the manufacture of sodium caseinate.

- Production of granular sodium caseinate by lowering the moisture content of acid casein curd to <40%, reacting the curd with Na_2CO_3, with agitation, for up to 60 min and drying the resultant caseinate in a pneumatic ring drier or a fluidized bed drier. The resulting caseinate has a higher bulk density and improved dispersability compared to spray and roller-dried products.

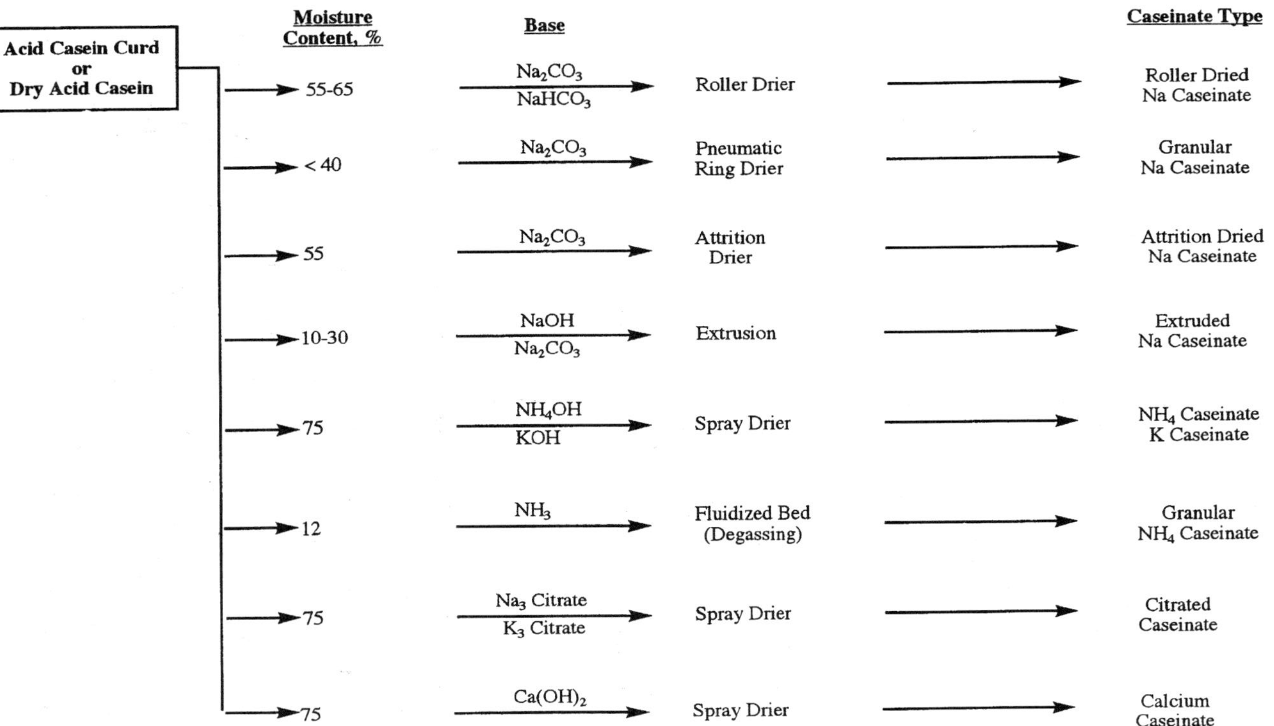

Figure 26.4 Methods for the manufacture of different caseinate types.

- Drying a mixture of acid casein curd (45% dry matter) and Na_2CO_3 in an attrition drier to produce a product that looks like spray dried sodium caseinate but which has a much higher bulk density.
- Conversion of casein to caseinate in the presence of a limited amount of water using extrusion techniques.
- Production of ammonium and potassium caseinates in a manner similar to that used for the production of sodium caseinate by substituting NH_4OH or KOH, respectively, for NaOH.
- Production of granular ammonium caseinate by exposing dry acid casein to gaseous ammonia and removing excess ammonia with a stream of air in a fluidized bed degassing system.
- Production of citrated caseinate by a method similar to that used for the preparation of spray-dried sodium caseinate by using a mixture of trisodium citrate and tripotassium citrate instead of NaOH.
- Production of calcium caseinate by:
  - passing 'soft' casein curd through a mixer to give evenly-sized particles;
  - mixing with water to ~25% total solids;
  - passing the mixture through a colloid mill and adjusting the temperature to give a milled slurry at 35–40°C;
  - mixing the slurry with a metered volume of 10% aqueous $Ca(OH)_2$ slurry to give the desired final pH;
  - agitating and recirculating in a low-temperature conversion tank until conversion is complete (>10 min);
  - heating the dispersion in a tubular heat exchanger to 70°C and pumping directly to a spray drier.

26.2.3 Miscellaneous methods for the preparation of casein and co-precipitates

In addition to the 'traditional' methods described above, several new alternative methods for the preparation of casein or co-precipitates have been reported, some of which may have commercial potential.

One method involves precipitation of milk proteins by addition of ethanol to reduce the dielectric constant of the mixture (Hewedi *et al.*, 1985). Addition of an equal volume of 60% (v/v) ethanol to pasteurized skim milk or to skim milk that had been heated at 90°C × 10 min, both adjusted to pH 6.3, precipitated ~82% and ~90% of total nitrogen, respectively.

Another method uses selective solubilization of lactose from non-fat dry milk using ethanol; the insoluble residue, which may be regarded as similar to a milk protein co-precipitate, containing 2 to 4% lactose, may be obtained by a one-step extraction process under optimum conditions (Hoff *et al.*, 1987).

Ultrafiltration of skim milk to a 4 or 6 volume concentration ratio (initial volume of milk/volume of retentate) to remove lactose, followed by storage of the retentate at −8°C for 1–4 weeks leads to cryo-destabilization of casein and some whey proteins which are sedimentable by centrifugation at 5,000 g for 10 min at 0–5°C (Lonergan, 1983).

A process for the production of 'native' casein involves electrodialysis of skim milk at 10°C against acidified whey to reduce the pH to ~5 under which conditions the casein is in a metastable state and can be separated by centrifugation (Morel, 1991; Noel, 1992).

Molochnikov and Zetteir (1994) mixed an anionic porysaccharide with milk to destabilize the casein which was then recovered as casein concentrate by centrifugation. Tomasula *et al.* (1995) studied the effects of pressure and temperature on the precipitation of casein by CO_2 in a 1000 ml batch reactor. Precipitation occurred at pressures greater than 2700 kPa and temperatures greater than 32°C. Maximum yields were achieved at 2760–3520 kPa and at 38–49°C. The total solids, ash and calcium contents, as well as the physical characteristics of the casein curd, were influenced by the pressure and temperature used to achieve precipitation. A continuous reactor/precipitator has also been developed for the production of casein using high-pressure CO_2 as precipitant (Tomsula *et al.*, 1997).

The use of a cascade of ceramic membranes of differing pore sizes to separate casein and whey protein fractions from skim milk has been reported (Singh *et al.*, 1995). Cross-flow microfiltration of skim milk at 50°C using ceramic membranes of pore size 0.22 μm, followed by batch diafiltration, has been used to produce native phosphocaseinate containing 79% protein (Pouliot *et al.*, 1996), while Maubois (1997) used a membrane of pore size 0.1 μm, to produce a clear microfiltrate close in composition to sweet whey and a retentate enriched in native micellar phosphocaseinate. Although still at the developmental stage, the technique has been used at an industrial level to enrich the casein content of milk for cheese manufacture (Maubois, 1997). Dinkov *et al.* (1997) achieved a level of separation of caseins and whey proteins similar to that obtained during cheese making using an ultrafiltration method in which skim milk at 55°C was ultrafiltered at a pressure drop of 0.4 MPa using a UF-PS (polysulphonic)-100 membrane and a coefficient of volume decrease of 7. Of the total protein in the skim milk, ~86% was in the retentate while ~14% was in the permeate. The retentate had ~99% of the casein and ~21% the whey protein present in the skim milk.

26.2.4 Fractionation of caseins

A number of methods for fractionating casein into β-casein-rich and α_s-/κ-casein-rich fractions on a potentially industrial scale have been developed.

At low temperatures, β-casein exists in solution as monomers (Payens and van Markwijk, 1963), a characteristic exploited by Allen *et al.* (1985) to prepare β-casein by renneting calcium caseinate at 4°C, under which conditions β-casein remains soluble while α_{s1}-, α_{s2}- and *para*-κ-caseins coagulate. In a development of the method, which was successfully scaled up, Ward and Bastian (1996) removed the α_{s1}-, α_{s2}- and *para*-κ-caseins by centrifugation and precipitated the β-casein by heating the supernatant to 30°C. A method for the isolation of β-casein by microfiltration of calcium caseinate at 5°C was reported by Terre *et al.* (1986). The technique was modified to purify β-casein from whole casein at 4°C and pH 4.2–4.6 (Famelart *et al.*, 1989) or from sodium caseinate solution at 4°C, treated with calcium to enhance aggregation of the other caseins (Famelart and Surel, 1994). Murphy and Fox (1991) reported a method for the fractionation of a dilute sodium caseinate solution by ultrafiltration into a β-casein-rich permeate and an α_s-/κ-casein-rich retentate; the β-casein was recovered from the permeate by raising the temperature to 40°C and the aggregated protein recovered by UF (Figure 26.5). In another method, slurried casein feedstock was cooled to −10 to −14°C until the desired amount of β-casein

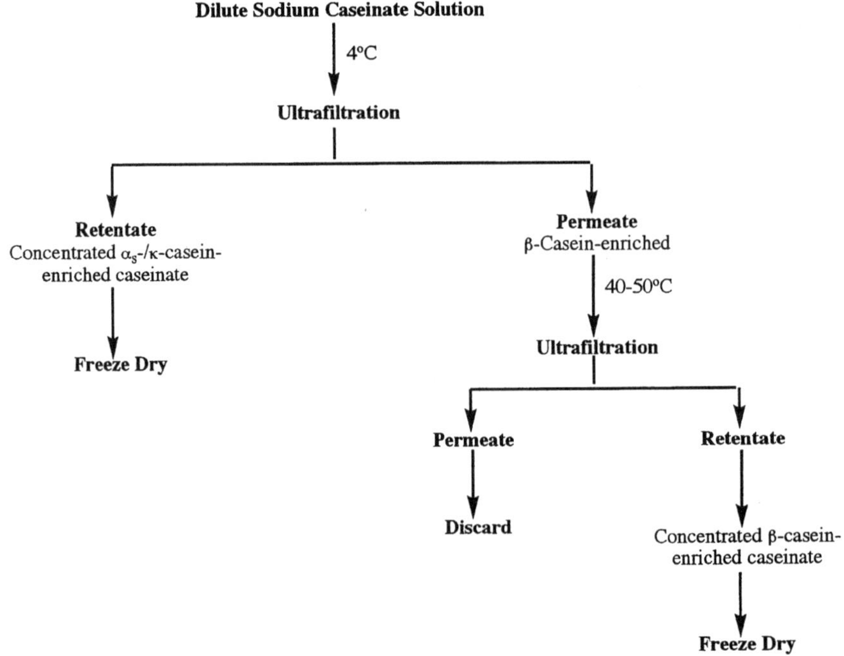

Figure 26.5 Method for preparing α_s-/κ- and β-casein-enriched caseins (Murphy and Fox, 1990).

had dissociated and was then separated from the β-casein-depleted feedstock; a pH range of 5–8 was suitable for rennet casein while a pH range of 3.5 to 8 was suitable for other caseins (Ram *et al.*, 1994). A precipitation method that produces a fraction rich in calcium-sensitive caseins, a fraction containing mainly γ-caseins and α-lactalbumin and a fraction comprising mainly κ-casein and β-lactoglobulin from skim milk has been described (Igarashi, 1995).

A fraction enriched in α_{s1}-casein was prepared by a combination of small-scale batch fractionation of whole casein on DEAE-cellulose and hydrophobic interaction chromatography on a Spherogel TSK-G Phenyl 5 PW column (Sanogo *et al.*, 1989). Hollar *et al.* (1991) used cation-exchange chromatography on a Mono-S column at pH 5.0 in a laboratory-scale ion-exchange method to resolve the four caseins. Leaver and Law (1992) used the cation exchanger, S-Sepharose, on a large scale to prepare β-casein (which remained unbound), κ-casein, α_{s1}-casein and α_{s2}-casein fractions, each >90% pure, from acid casein redissolved in a urea-containing buffer at pH 5.0. Cayot *et al.* (1992) developed a batch anion exchange method using Q-Sepharose Fast Flow gel to purify caseins; the method should be suitable for scale-up. Fractions containing κ- and γ-caseins, β-casein or α_s-caseins, were obtained by eluting with buffers containing differing concentrations of NaCl. Ng-Kwai-Hang and Chin (1994) prepared electrophoretically pure individual caseins by anion exchange HPLC using a urea-containing Tris buffer, pH 7.0, and elution using a NaCl gradient.

According to Christensen and Munksgaard (1989) and Coolbear *et al.* (1996), precipitation techniques are inferior to chromatographic methods such as cation exchange and anion exchange chromatography for preparation of pure caseins; however, chromatographic methods are more appropriate for small scale recovery of caseins while precipitation methods are more suitable for large-scale, industrial fractionation. Separation of caseins and whey proteins by capillary electrophoresis has also been described (Fairise and Cayot, 1998), however, the technique appears unsuitable for scale-up.

26.2.5 Production of whey protein-enriched products

Whey is the liquid remaining after removal of fat and casein from milk during the manufacture of cheese and acid or rennet casein. Sweet whey (minimum pH, 5.6) is obtained from the manufacture of cheese or rennet casein and acid whey (maximum pH, 5.1) from the manufacture of acid casein. The average composition of some types of whey is shown in Table 26.2. Acid whey has a higher mineral/ash content and, if the milk was acidified by the action of starter bacteria, the lactose concentration is reduced. Whey proteins represent only 10% of the total solids of whey and

TABLE 26.2
Average composition and pH of sweet (Rennet casein and Cheddar cheese) and acid (lactic and mineral acid) wheys

Component	*Composition (g/l)*			
	Sweet wheys		*Acid wheys*	
	Rennet casein	*Cheddar cheese*	*Lactic acid casein*	*Mineral acid casein*
Total solids	66.0	67.0	64.0	63.0
Total protein (N × 6.38)	6.6	6.5	6.2	6.1
Non-protein nitrogen (NPN)	0.37	0.27	0.40	0.30
Lactose	52.0	52.0	44.0	47.0
Milk fat	0.20	0.20	0.30	0.30
Minerals (ash)	5.0	5.2	7.5	7.9
Calcium	0.50	0.40	1.6	1.4
Phosphate	1.0	0.50	2.0	2.0
Sodium	0.53	0.50	0.51	0.50
Lactate	–	2.0	6.4	–
pH	6.4	5.9	4.6	4.7

whey powders have a low protein content. However, a number of processes (Figure 26.6) have been developed, and are now being exploited commercially, to recover the whey proteins in more concentrated forms. Processes for the recovery of whey proteins have been reviewed by Marshall (1982), Matthews (1984), International Dairy Federation (1987), Morr (1989), Pearce (1992), Huffman (1996), Mulvihill and Grufferty (1997) and Timmer and van der Horst (1998).

Whey and whey protein-enriched solutions are usually pasteurized using a minimum temperature and holding time and maintained at a low temperature to minimize microbial and physico-chemical deterioration of the proteins and other whey constituents that would adversely alter functional and organoleptic properties of the resulting protein-enriched products.

(a) Whey powders/modified whey powders

Whole whey powders containing less than 15% protein are produced by concentrating whey by evaporation or a combination of reverse osmosis and evaporation, followed by spray drying. Demineralization by 'loose' reverse osmosis, electrodialysis or ion-exchange and/or lactose crystallization are used commercially to reduce the lactose and/or mineral concentration in whey and produce modified whey powders such as demineralized and demineralized-delactosed whey powders which contain ~15–35% protein.

Traditionally, lactose is crystallized from whey concentrated to 58–62% total solids using a multiple-effect falling-film evaporator at a maximum product temperature of ~70°C, followed by controlled cooling of the

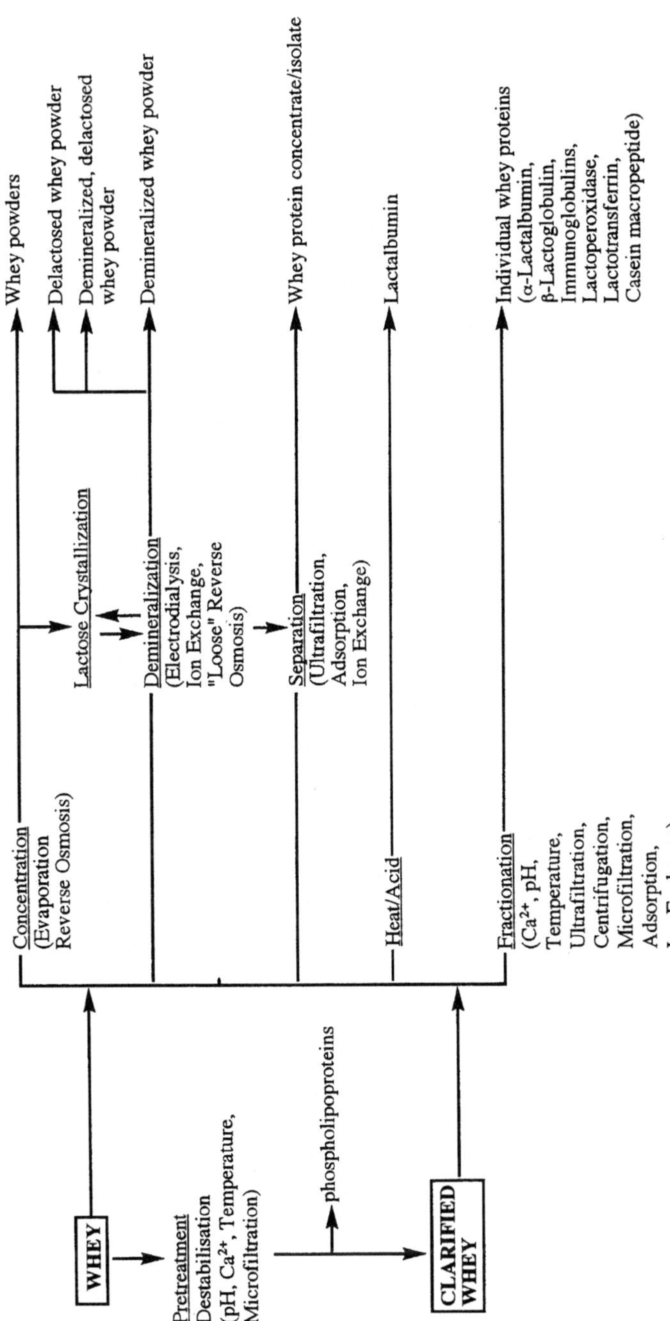

Figure 26.6 Industrial isolation of protein products from whey.

concentrate and seeding to induce nucleation and crystal growth. The lactose crystals are separated from the mother liquor by decanter centrifugation, washed (if desired) and dried. The mother liquor may be concentrated further and spray dried as a whey protein concentrate powder containing ~30% protein.

(b) Production of whey protein concentrates by ultrafiltration-diafiltration Ultrafiltration (UF), a physico–chemical separation technique in which a pressurized solution flows over a porous membrane which allows the passage of only relatively small molecules, facilitates the selective separation of whey proteins from lactose, salts and water under mild conditions of temperature and pH. The retained solution (retentate) flows over the membrane, while, under the influence of pressure, water flows through the membrane, together with the low molecular weight solutes. The protein is retained by the membrane and is concentrated relative to the other solutes in the retentate (Figure 26.7). Fat globules and suspended solids are also retained.

The membranes used in UF are asymmetric microporous structures, the effective layers of which contain pores with diameters ranging from 1 to 20 nm. Commonly-used membrane configurations include tubular, spiral-wound, plate and frame and hollow-fibre, with each configuration offering advantages and disadvantages for particular applications. The membranes, manufactured from synthetic polymers such as polysulphone or polyamide, are resistant to high temperatures (up to 100°C), can withstand a wide pH range (1–13) and can be cleaned with agents normally used in the dairy industry (e.g., HNO_3 and NaOH). Although UF is currently the method of choice for the commercial manufacture of whey protein concentrates (WPC) of varying protein concentration, several major problems limit its operational performance including: high capital and operating costs; membrane fouling and concomitant loss of permeate flux rate; incomplete removal of low molecular weight solutes unless diafiltration (dilution of retentate with water and repeated UF) is used; cleaning, sanitation and related microbial problems; disposal of large volumes of permeate.

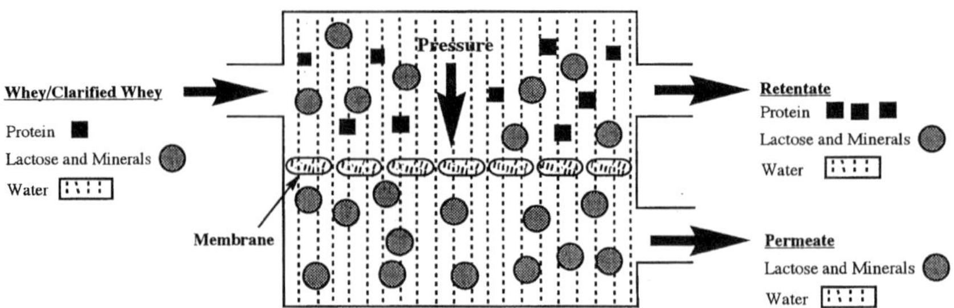

Figure 26.7 Principle of ultrafiltration.

Prior to processing, whey is usually pre-treated by methods involving pH and/or temperature adjustments, addition of calcium or calcium complexing agents and either quiescent standing, centrifugation or microfiltration to dissolve colloidal calcium phosphate and/or to remove insoluble cheese curd or casein fines, milkfat and calcium lipophosphoprotein complexes (Hayes *et al.*, 1974; Breslau *et al.*, 1975; de Wit and de Boer, 1975; Lee and Merson, 1976; de Wit *et al.*, 1978; Matthews *et al.*, 1978; Muller and Harper, 1979; Maubois *et al.*, 1987; Patocka and Jelen, 1987; Kim *et al.*, 1989; Jensen and Larsen, 1994; Karleskind *et al.*, 1995). These pre-treatments increase flux during ultrafiltration, prevent fouling of the membranes, reduce lipid content and modify the properties of the whey protein concentrates. The limit for whey concentration by UF in modern plants is ~24% total solids, with a protein:total solids ratio limit of ~0.72:1. Diafiltration is used to achieve a higher ratio of protein:total solids, ~0.80:1, and a total solid content of ~28%.

Preparation of a whey protein concentrate from raw skim milk using a cascade of ceramic membranes of differing pore sizes has been reported (Liu *et al.*, 1995).

(c) Production of whey protein isolate (WPI)

Whey proteins, at pH values lower than their isoelectric point (~pH 4.6), have a net positive charge and behave as cations that can be adsorbed on cation exchangers. At pH values above their isoelectric point, proteins have a net negative charge and behave as anions that can be adsorbed on anion exchangers. Media with suitable pore sizes and surface characteristics have been developed specifically for the recovery of proteins from dilute solutions, depending upon the pH of the medium. Two major ion exchange fractionation processes have been commercialized for WPI manufacture (Figure 26.8).

The 'Vistec' process uses a cellulose-based exchanger in a stirred tank reactor (Burgess and Kelly, 1979; Palmer, 1982). A series of steps are performed as a fractionation cycle: (1) whey is acidified to pH < 4.6, pumped into a tank reactor and stirred to allow protein adsorption onto the ion exchanger; (2) lactose and other unadsorbed components are filtered off with the water; (3) the resin is resuspended in water and the pH adjusted to >5.5 with alkali, releasing the proteins from the ion exchanger; (4) the aqueous solution of proteins is separated from the resin by filtration in the tank reactor. The protein-rich eluate fraction is concentrated and purified by ultrafiltration, further concentrated by evaporation, and spray dried as WPI containing ~95% protein.

The 'Spherosil' process (Mirabel, 1978; Kaczmarek, 1980) uses either cationic Spherosil S or anionic Spherosil QMA ion exchangers for fractionation in fixed-bed column reactors. Acidified whey at pH < 4.6 is applied to the Spherosil S column reactor to allow protein adsorption. After unadsorbed solutes, including lactose, have been eluted with water, the pH

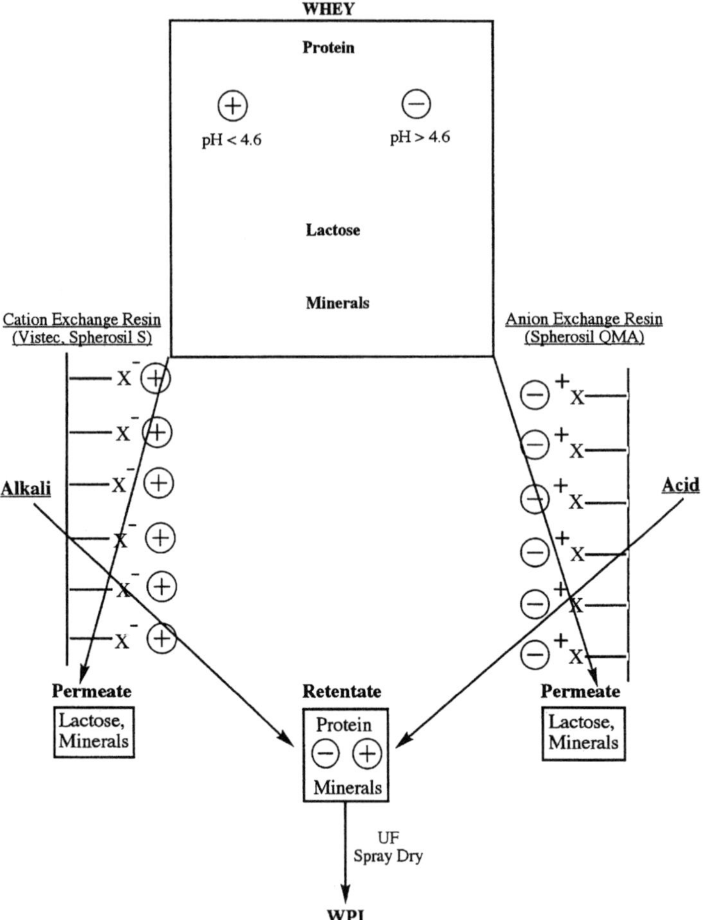

Figure 26.8 Production of whey protein isolate (WPI) by ion exchange adsorption.

is raised by addition of alkali to elute adsorbed proteins from the reactor. The protein-rich eluate is concentrated by UF and evaporation and spray dried to produce WPI. Sweet whey at pH >5.5 is applied to the Spherosil QMA column reactor to permit negatively charged protein molecules to adsorb onto the ion exchanger. After elution of non-protein materials, the proteins are released by lowering the pH with acid and the released proteins concentrated and spray dried as WPI, as for the Spherosil S process. Models for optimization of whey protein extraction processes, using Spherosil QMA in a fluidized bed, have been developed (Carrere *et al.*, 1994, 1996).

Under ideal operating conditions, these adsorption processes recover ~85% of the protein. The recovered concentrates are characterized by high protein and low lactose and lipid concentrations and have good functionality. However, several major problems are associated with these ion exchange processes, including: (1) production of large volumes of rinse water, chemical solutions and deproteinized whey that must be processed or disposed of; (2) the need to concentrate and purify the dilute protein-containing eluate by UF, evaporation and drying; (3) the long time required for each fractionation cycle; (4) microbial contamination of the reactor.

A product described as a whey protein isolate, produced by microfiltration, cross-flow ultrafiltration and spray drying, and which contains 90–93% protein in 99% undenatured form has been reported (Rowan, 1998).

(d) Production of lactalbumin

When whey is heated, the proteins readily denature from their globular conformations to more random structures; sulphydryl and hydrophobic groups are exposed and protein–protein interactions occur. The extent of aggregation and precipitation of the denatured proteins depend on heating temperature and holding time, pH and concentration of calcium. Commercial precipitation conditions (Figure 26.9) depend on whey type and the desired final product characteristics and whey may be pre-concentrated and/or demineralized prior to precipitation (Robinson *et al.*, 1976). The precipitated protein, referred to as lactalbumin, may be recovered by settling and decanting, vacuum filtration, self-desludging centrifuges or horizontal solid-bowl decanters. The precipitate may be washed to reduce the content of minerals and lactose and dried in any one of several types of drier. Depending on precipitation pH and degree of washing, up to 80% of the protein in the whey may be recovered as lactalbumin, containing up to 90% protein on a dry weight basis.

(e) Electrochemical coagulation of whey protein

Electrochemical coagulation processes for the recovery of protein from whey have been described by Pyrgaru (1992) and Janson and Lewis (1994). The latter process used an electrolytic cell in which the anode and cathode compartments were separated by a non ion-selective membrane, which allowed an increase in the acidity of the bulk solution in the anode compartment and an increase in alkalinity of the bulk solution in the cathode compartment during operation. The whey (pH 6.9) moved through the anode compartment in a thin stream (3 mm wide) at a voltage that induced coagulation and was subsequently separated into a coagulum and a protein-depleted whey which was pumped through the cathode compartment to complete the circuit. Under optimum conditions of flow velocity (0.9–1.0 ml min^{-1}), voltage (30 V) and exit pH from the anode compartment (pH 4.6), a 73.5% reduction in protein content between the whey and protein-depleted whey (exit pH 7.43) could be achieved.

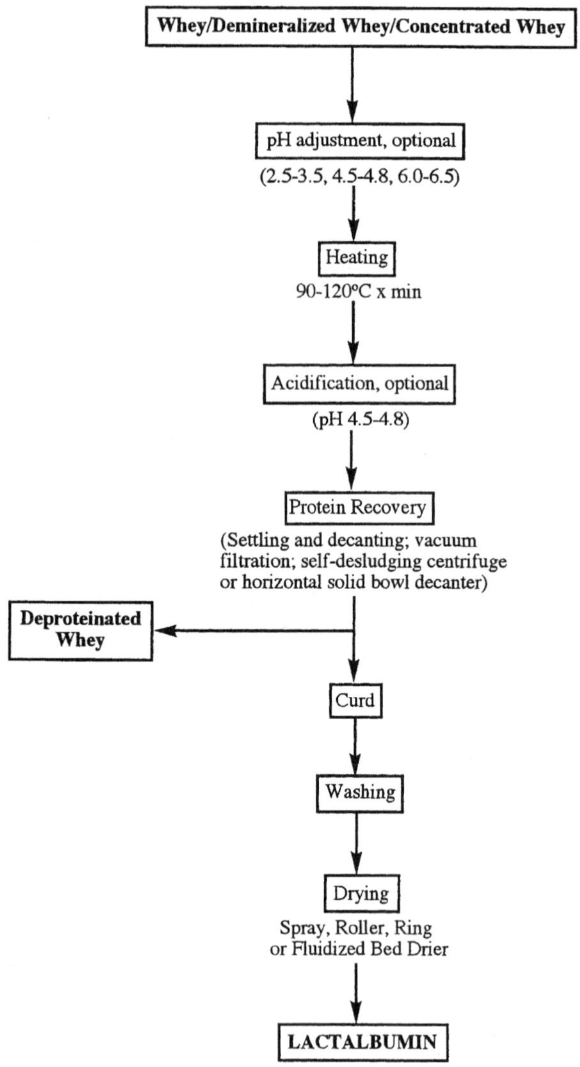

Figure 26.9 Method for manufacture of lactalbumin.

26.2.6 Fractionation of whey proteins

(a) Major whey proteins

Numerous methods that may have commercial scale potential have been described for fractionating whey proteins to produce whey protein concentrates enriched in the major whey proteins, β-lactoglobulin and α-lactalbumin.

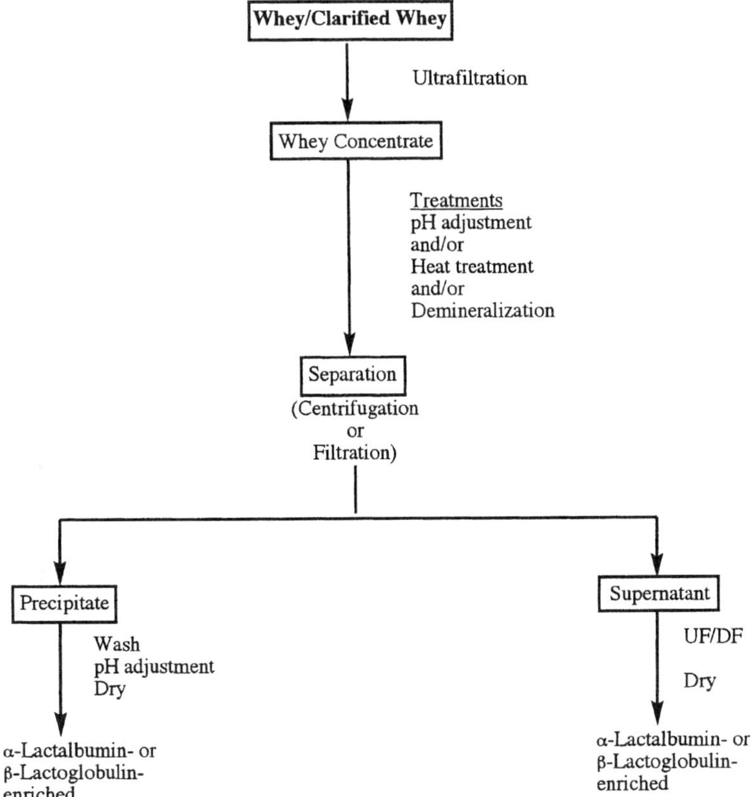

Figure 26.10 Thermal and demineralization methods for fractionation of major whey proteins.

These methods (Figure 26.10) depend on either mild heat treatment of a whey or a whey concentrate or a clarified whey under controlled pH and ionic conditions or demineralization of whey concentrate under controlled conditions to effect selective reversible precipitation of α-lactalbumin or β-lactoglobulin-enriched fractions and the separation of the precipitate from β-lactoglobulin or α-lactalbumin-enriched mother liquors. The precipitate is resolubilized by water addition and pH adjustment and then dried while the soluble protein is further concentrated by ultrafiltration/diafiltration prior to drying.

The methods developed by Amundson *et al.* (1982) and Slack *et al.* (1986) rely on the fact that β-lactoglobulin requires higher ionic strength for solubility in the isoelectric region than does α-lactalbumin. In whey concentrated by UF, acidified to pH 4.65 and demineralised by electrodialysis to an ash content $<0.023\%$, β-lactoglobulin precipitates and can be separated from soluble α-lactalbumin by centrifugation with a yield of $>90\%$.

The method developed by Pearce (1983, 1987) exploits the low heat stability of apo-α-lactalbumin to precipitate it from whey, leaving β-lactoglobulin, blood serum albumin and immunoglobulins in solution. α-Lactalbumin is a calcium-containing metalloprotein which on acidification to ⩽pH 5.0, loses its calcium to form the apo-protein, which denatures and aggregates on heating to ~55°C and can be separated from the remaining soluble whey proteins by centrifugation or filtration. In a method basically similar to that of Pearce (1983), Pierre and Fauquant (1986) prepared β-lactoglobulin, which was 98% pure, and an α-lactalbumin fraction contaminated with serum albumin and other proteins from clarified UF whey concentrate. Maubois *et al.* (1987) proposed a modified method based on the method of Pearce (1983) in which whey was pre-clarified by cooling to 2°C, calcium chloride (1.2 g calcium/kg) was added, pH adjusted to 7.3, heated rapidly to 50°C and held for 8 min to precipitate a lipid fraction which was removed by microfiltration; the clarified whey is then concentrated by UF and fractionated by adjustment to pH 3.8 and heating to 55°C for 30 min to aggregate α-lactalbumin which is separated from soluble β-lactoglobulin by centrifugation. The precipitate is resolubilized by water addition and pH adjustment and then dried while the soluble protein is concentrated further by ultrafiltration/diafiltration prior to drying.

A further modification of the Pearce (1983) process was proposed by de Wit and Bronts (1994). It involved the following steps: incubating a whey protein solution with a calcium-binding ion exchange resin in its acid form to initiate destabilization of α-lactalbumin, separation of the resin, adjustment of the pH to 4.3–4.8 and incubation at 10–50°C to flocculate the α-lactalbumin, fractionation of the proteins by centrifugation or microfiltration, solubilisation of α-lactalbumin by increasing the pH of the flocculated fraction and raising the pH of the β-lactoglobulin-enriched fraction sufficiently to neutralize it before further downstream processing and drying of both fractions. According to Bramaud *et al.* (1997), on treating whey with citrate, α-lactalbumin is converted to a calcium-free apo-form which precipitates at a lower temperature (35°C instead of ~55°C) than the apo-form obtained by pH adjustment alone, resulting in reduced heat denaturation of the protein and diminished co-precipitation of serum albumin during fractionation.

Gesan-Guziou *et al.* (1999) described a further modification of the heat precipitation method in which soluble β-lactoglobulin was separated from the precipitated α-lactalbumin using microfiltration; the precipitated α-lactalbumin was washed, solubilized, and concentrated and purified by ultrafiltration. The purity of the resulting α-lactalbumin and β-lactoglobulin fractions was in the range 52–83 and 85–94%, respectively. The purity of the β-lactoglobulin fraction obtained from acid whey was higher than that from sweet whey, which was attributed to the presence of caseinomacropeptide in the latter. Overall, the recovery of α-lactalbumin and β-lactoglobulin was poor (6 and 51%, respectively) and ways for improving the recoveries were suggested.

Stack *et al.* (1995) described a highly integrated process for the recovery of α-lactalbumin and β-lactoglobulin-enriched fractions from whey (Figure 26.11). The process involves:

- Demineralization of whey or clarified whey by a combination of electrodialysis and ion-exchange to reduce the calcium content to below 120 mg/kg, on a dry weight basis, and the pH to <3.4.
- Heating to 71–98°C for 50–95 sec, followed by rapid cooling to ~10°C.
- Two-stage concentration to 55–63% total solids at <69°C with optional further demineralization between stages.
- Cooling to induce lactose crystallization.
- Separation of lactose crystals from the whey protein-containing liquor.
- Adjusting the whey protein-rich liquor to pH 4.3–4.7 at <10°C and then heating to 35–54°C for 1–3 h to induce flocculation of α-lactalbumin.
- Separating the α-lactalbumin-enriched fraction (flocculant) from the β-lactoglobulin-enriched supernatant by mechanical or membrane separation.
- Washing the α-lactalbumin-enriched flocculant with a solution isoionic with the whey protein liquor and with a pH of 4.3–4.7.
- Neutralization, downstream processing and spray drying of both fractions.

Chemical precipitation methods for separating α-lactalbumin and β-lactoglobulin include the method of Fox *et al.* (1967) who used 3% trichloroacetic acid (TCA) to precipitate all the whey proteins except β-lactoglobulin; however, the acceptability of TCA in food-grade applications seems doubtful. UF retentate may be fractionated with 7% (w/v) NaCl at pH 2.0, under which conditions β-lactoglobulin remains soluble while all other proteins precipitate (Mailliart and Ribadeau-Dumas, 1988). The precipitate can be made free of β-lactoglobulin by washing with 6% NaCl and essentially pure β-lactoglobulin can be recovered from the supernatant by precipitation at 30% NaCl, pH 2.0. The method was modified by Mate and Krochta (1994) to prepare β-lactoglobulin from WPI, with the recovery of β-lactoglobulin from the supernatant by diafiltration instead of salting-out. By adding $FeCl_3$ to whey to a concentration of 4 mM and adjusting the pH to 3.0, α-lactalbumin and immunoglobulins can be precipitated, leaving β-lactoglobulin in solution (Kuwata *et al.*, 1985); Fe^{3+} may be removed subsequently from the α-lactalbumin fraction by ion-exchange or UF. Al-Mashikhi and Nakai (1987a) used sodium hexametaphosphate (1.33 g/1) at pH 4.07, with incubation at 22°C for 1 h, to precipitate 80% of the β-lactoglobulin, leaving α-lactalbumin and immunoglobulins in solution. The method was modified by Cuddigan (1991) for UF whey retentate;

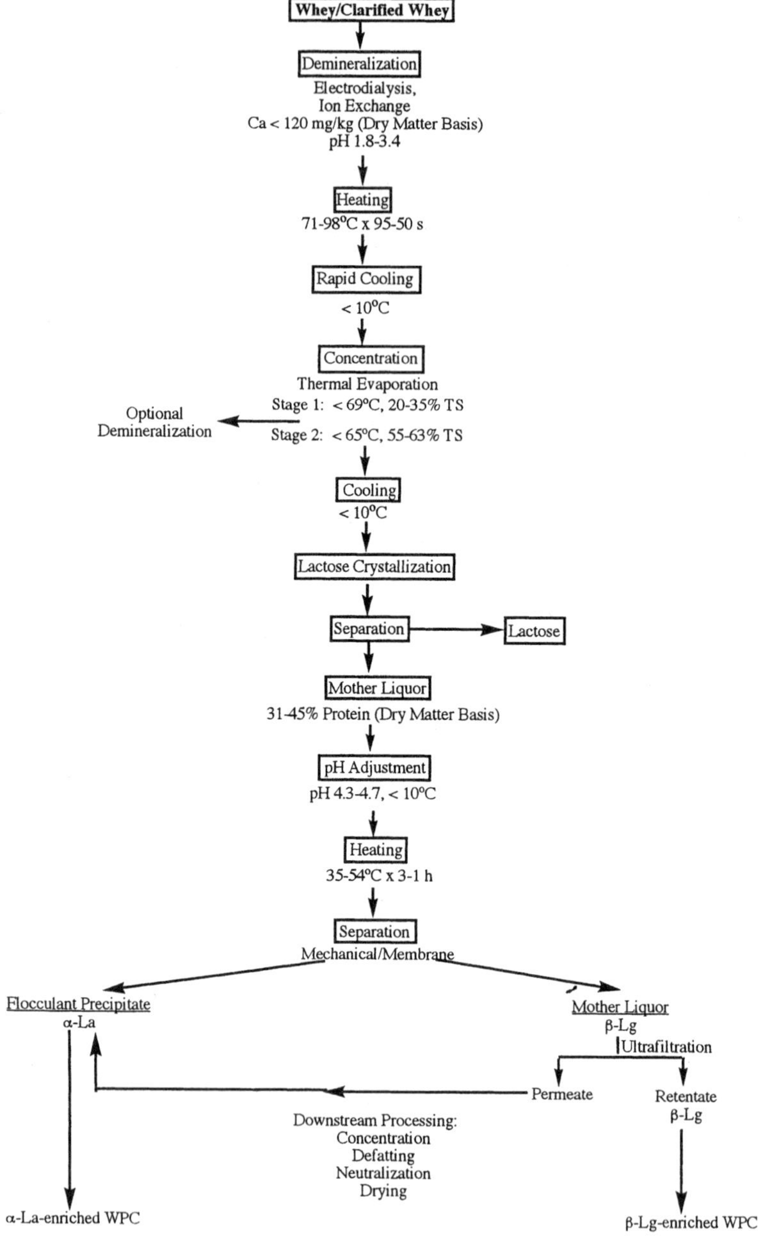

Figure 26.11 Integrated process for recovery of α-lactalbumin (α-la) and β-lactoglobulin (β-lg) enriched fractions from whey (modified from Stack *et al.*, 1995).

unfortunately, different degrees of concentration of the UF retentate were required to obtain the purest β-lactoglobulin in the precipitate and α-lactalbumin in the supernatant.

Mehra and Donnelly (1993) achieved partial fractionation of whey proteins using large pore size, hydrophilic cellulosic membranes. Factors such as whey pH, ionic environment and the presence or absence of milk salts influenced the extent of protein aggregation and thereby influenced the fractionation of the proteins.

Kinekawa and Kitabatake (1996) prepared purified β-lactoglobulin by treating WPI or WPC solutions, pH 2.0, with crude porcine pepsin to hydrolyse casein and most of the whey proteins. The resistance of β-lactoglobulin to proteolysis under the conditions used permitted its recovery in the retentate during UF while the peptides produced on hydrolysis of the other proteins were in the permeate. Methods for preparing β-lactoglobulin-depleted whey protein products involving selective hydrolysis of β-lactoglobulin by papain at pH 8.0 (Schmidt and van Markwijk, 1993), by trypsin, α-chymotrypsin or bacterial proteases at pH 7.0–9.0 (Kaneko *et al.*, 1990) or by thermolysin at a pressure of around 200 MPa (Hayashi *et al.*, 1987) have been described.

Lee and Dungan (1998) performed phase transfer experiments between aqueous solutions of pure α-lactalbumin and β-lactoglobulin and a sodium bis(ethylhexyl) sulfosuccinate (Aerolsol OT; AOT)-in-isooctane reversed micellar phase (also termed a water-in-oil microemulsion) containing nanometer-scale droplets of water suspended in an organic continuum rendered thermodynamically stable by a surfactant. They demonstrated the potential of the process, which may be suited to scaling up for commercial application, for extracting α-lactalbumin and β-lactoglobulin from whey. The two proteins transferred from pure aqueous solution in the feed phase to the aqueous phase of the reversed micelles over a wide range of pH and ionic strength. Depending on the pH, ionic strength and concentration of proteins in the aqueous feed phase, preferential partitioning of α-lactalbumin into the micelles could be achieved, resulting in an α-lactalbumin-enriched reversed micellar phase and a β-lactoglobulin-enriched feed phase.

Methods using strong anion (Skudder, 1985; Sanchez *et al.*, 1985; Thiabault, 1989, 1991) or cation (Kuwata and Ohtomo, 1989; Schutyzer *et al.*, 1987) exchangers have been developed to fractionate whey proteins. All the whey proteins are adsorbed on Spherosil QMA, an ion exchanger used in the production of WPI. However, on continued passage of whey through the exchanger, β-lactoglobulin, which has the highest binding affinity for the resin, displaces adsorbed α-lactalbumin and serum albumin to give an eluate enriched in these proteins; purified β-lactoglobulin can be eluted from the column using 0.1 M HCl (Skudder, 1985; de Wit *et al.*, 1986). Girardet *et al.* (1989), using MonoQ anion exchanger, separated the major whey proteins and even separated genetic variants of β-lactoglobulin by varying the pH, ionic strength and composition of the buffer used. A strongly basic polystyrene anion exchange resin has been used to fractionate α-lactalbumin and β-lactoglobulin from whey (Outinen *et al.*, 1995a, 1996a) and to remove β-lactoglobulin from enzymatic hydrolysates of whey protein (Tossavainen *et al.*, 1996). Extraction of α-lactalbumin and β-lactoglobulin from sweet whey using fluidized ion exchange chromatography has been described (Carrere *et al.*, 1996).

Outinen *et al.* (1996b) compared the heat precipitation methods of Pearse (1983, 1987) and Rialland and Barbier (1988) and two chromatographic methods, that of Skudder (1985) and a modification of it using a polystyrene anion exchange resin, for pilot scale fractionation of whey proteins. These authors found that soluble native β-lactoglobulin could be recovered from heat-treated whey by UF, but recovery of the low density α-lactalbumin precipitate was difficult and significant denaturation of α-lactalbumin occurred when acid whey was used. The two ion exchange processes studied yielded native α-lactalbumin while β-lactoglobulin was denatured extensively during isolation when using Spherosil QMA but less so when the polystyrene Diaion HPA 75 resin was used. The chromatographic processes were considered the more suitable for commercial-scale application.

Affinity chromatographic methods for fractionating whey proteins involving selective binding of β-lactoglobulin to *trans*-retinal immobilised on Celite, a silica-based solid phase, (Wang and Swaisgood, 1993) or to β-lactoglobulin coupled to Sepharose 4B (Chiancone and Gattoni, 1993; MacLeod *et al.*, 1995) have been reported. Gambero *et al.* (1997) used copper ions immobilised on a modified silica support for affinity extraction of α-lactalbumin from whey. Hydrophobic interaction chromatography has been used to separate β-lactoglobulin and α-lactalbumin (de Frutos *et al.*, 1996).

Whey proteins may be separated quantitatively by gel permeation chromatography, including high performance liquid chromatographic (e.g., FPLC) methods (Andrews *et al.*, 1985; Law *et al.*, 1993). Felipe and Law (1997) described a rapid preparative-scale method, using Superdex 75, for the separation of β-lactoglobulin, α-lactalbumin, immunoglobulin and albumin/lactotransferrin fractions. Whey proteins may also be fractionated by capillary electrophoresis; Otte *et al.* (1994) achieved separation, with full recovery, of β-lactoglobulin and α-lactalbumin from whey, the separation being affected by pH and ionic strength, while Fairise and Cayot (1998) also reported separation of the major whey proteins. These electrophoretic techniques, however, would appear to be unsuitable for scale-up.

(b) Minor whey proteins

There is growing commercial interest in the isolation of biologically active proteins such as lactoperoxidase, lactoferrin and immunoglobulins, and biologically or functionally-active peptides from whey. Lactoperoxidase and lactoferrin may have value as antimicrobial agents in milk (Elkstrand, 1989) and other foods (Kussendrager, 1993a); lactoferrin may also have a role in iron transport and metabolism (Spik *et al.*, 1985; see Chapter 10) while immunoglobulins and BSA may be important in the prevention and treatment of cancer and in the treatment of AIDS and other immunocompromising diseases (Horton, 1995).

Lactoperoxidase and lactoferrin. Lactoperoxidase and lactoferrin are positively charged at neutral pH while the other whey proteins are negatively charged. Numerous methods exploit this property to separate these proteins from the remaining whey proteins; the two proteins may be separated subsequently, if desired.

Paul *et al.* (1980) used cation exchange chromatography to isolate lactoperoxidase and lactoferrin from milk or whey; the method was scaled up for industrial application by Prieels and Peiffer (1986). Martin-Hernandez *et al.* (1990), using a modification of the method of Paul *et al.* (1980), separated lactoperoxidase from lactoferrin. Yoshida and Ye-Xiuyun (199la) used CM-Toyopearl or, in an improved method (Yoshida and Ye-Xiuyun, 1991b), Sulphopropyl-Toyopearl, to isolate lactoperoxidase and lactoferrin from sweet or acid whey and resolved the proteins using a NaCl elution gradient. Burling (1989) used microfiltration in combination with cation exchange chromatography on S-Sepharose to extract lactoperoxidase and lactoferrin from milk serum; the former was eluted from the cation exchanger using 0.1–0.4 M NaCl while the latter protein was eluted using 0.5–2.0 M NaCl, at pH 6.5. Processes utilising cation exchange chromatography to prepare the proteins have been patented (Prieels and Peiffer, 1986; Burling, 1989; Kussendrager, 1993b). Law and Reiter (1977) used batchwise addition of CM-Sephadex to whey, followed by recovery and washing of the CM-Sephadex and elution with buffers of differing NaCl concentrations to prepare a fraction containing lactoperoxidase and immunoglobulins and another fraction containing lactoferrin.

Chiu and Etzel (1997) isolated lactoperoxidase and lactoferrin from whey, adjusted to pH 6.5, using a cation exchange membrane and selectively eluted the two proteins from the membrane by varying the concentration of NaCl in the eluting buffer. The method was rapid, showed good recovery and should be suitable for industrial-scale application.

Lactoperoxidase has been isolated from acid whey by size exclusion chromatography on Sephacryl S-200 and hydrophobic chromatography on Butyl Toyopearl 650 M (Yoshida, 1988), while lactoferrin has been prepared from a supernatant of acid whey, made to 1.8 M $(NH_4)_2SO_4$, by hydrophobic interaction chromatography on Butyl Toyopearl 650 M and ion exchange using a DEAE resin (Yoshida, 1989).

Lactoferrin has been isolated from bovine milk on a laboratory scale by ion exchange methods (Gordon *et al.*, 1962; Groves, 1965) and together with immunoglobulins from whey by size exclusion chromatography

(Al-Mashikhi and Nakai, 1987b). Kawakami *et al.* (1987) used a monoclonal antibody immunoaffinity column to isolate lactoferrin of high purity and high iron-binding capacity from either skim milk or cheese whey in a single step. On an industrial scale, the ease of operation and re-usability of the immunoaffinity column may offset the initial high expense of the monoclonal antibodies.

Al-Mashikhi *et al.* (1988) used copper ions immobilised on Sepharose 6B (metal chelate interaction chromatography) to adsorb proteins from whey. Depending on the conditions used, α-lactalbumin and β-lactoglobulin remained unbound; a pH gradient (from pH 8.0 to 2.8) was used to elute the bound proteins as a lactoferrin-rich fraction and an immunoglobulin-rich fraction. The method offered simplicity, high capacity and quantitative adsorption and recovery of activities. Immobilised copper affinity chromatography was also used by de Stefano *et al.* (1994) to partially purify lactoperoxidase from acid whey.

Immobilised ferritin (Pahud and Hilpert, 1976), immobilised heparin (Bläckberg and Hernell, 1980) and immobilised single-stranded DNA (Hutchens *et al.*, 1989) have been used to purify lactoferrin from whey. Chen and Wang (1991) purified lactoferrin from bovine whey by affinity binding of the protein on heparin-Sepharose, microfiltration to remove non-bound proteins and other whey constituents in the permeate, washing of the protein-heparin Sepharose complex by diafiltration, dissociation of active lactoferrin from the heparin-Sepharose by increasing the NaCl concentration in the wash and finally ultrafiltration/diafiltration of the eluate to recover highly pure, highly biologically active lactoferrin in the retentate.

Shimazaki and Nishio (1991) reported that bovine lactoferrin interacted strongly with immobilised Cibacron Blue F3G-A (a textile dye) in column chromatography and that the protein could be eluted using free dye. According to Grasselli and Cascone (1996), different dyes bound whey proteins to varying extents; for example, Red HE-3B had a high affinity for lactoferrin but did not adsorb on lactoperoxidase while Yellow FR had a high capacity for serum albumin but a low capacity for lactoperoxidase and lactoferrin. The selectivity, high protein-binding capacity and low cost of the dyes may make them suitable for large-scale methods for isolating lactoferrin and other whey proteins.

Immunoglobulins. A 'milk immunological concentrate' designed for use in special infant formulae was prepared by UF/diafiltration (DF) of acid whey from colostrum and early lactation milk from immunized cows (Hilpert, 1984). UF, in combination with ion exchange chromatography, has also been used to separate immunoglobulins from lactoferrin (Dubois, 1986; Bottomley, 1989) and for the preparation of immunoglobulins from colostrum or milk from hyperimmunized cows (Taniguchi *et al.*, 1990). Immunoglobulins and lactoferrin have been isolated from whey by size exclusion chromatography (Al-Mashikhi and Nakai, 1987b). Metal chelate interaction chromatography was used by Al-Mashikhi *et al.* (1988) for adsorption of immunoglobulins and lactoferrin from whey, with subsequent elution as separate immunoglobulin-enriched and lactoferrin-enriched fractions.

Chen and Wang (1991) purified immunoglobulin G by affinity binding using Protein G as ligand in a method similar to that already described by the same authors for the purification of lactoferrin. Baick and Yu (1995) used size exclusion chromatography, ion exchange chromatography and affinity chromatography on Protein A Sepharose and on Hi-Trap Protein G to isolate immunoglobulin G from a crude immunoglobulin fraction which was precipitated from acid whey by 33% saturation with ammonium sulphate. An immobilised monoclonal antibody system for the recovery of immunoglobulins from milk has been described (Gani *et al.*, 1982).

Thomas *et al.* (1992) used a 'formed-in-place' membrane system to selectively enrich immunoglobulin G from cheese whey. At pH 5.0–5.5, immunoglobulin G is positively charged which, in combination with its large molecular size, greatly reduces permeation of the protein through the membrane, resulting in concentration of the immunoglobulin G in the retentate, whereas other whey proteins were largely in the permeate.

Macropeptide. Numerous methods for isolating the κ-casein macropeptide [κ-CN(f106–169); CMP] split off from κ-casein during enzymatic coagulation and present in cheese and rennet casein wheys at 1.2–1.5 g/l have been described. Eustache (1977) recovered CMP-enriched protein by UF of the supernatant remaining following heat-induced coagulation of proteins from whey. Berrocal and Neeser (1991, 1993) recovered CMP by selective precipitation with ethanol from the supernatant remaining following heat-induced flocculation of protein in delactosed whey or whey protein concentrate. In a method with potential for scale-up, Saito *et al.* (1991) used ethanol precipitation and ion exchange chromatography to isolate CMP from the supernatant of heat-coagulated whey proteins; the CMP was fractioned further into sialo-CMP and asialo-CMP by affinity chromatography using peanut lectin Sepharose 4B. Other ion exchange methods involve adsorption of CMP onto an anion exchanger and recovery by desorption with dilute acid or salt solutions (Burton and Skudder, 1987; Marshall, 1991; Tanimoto *et al.*, 1992; Outinen *et al.*, 1995b). A simple method for the preparation of CMP involves passing whey through an ion exchange column under conditions where the CMP does not bind but emerges in a desalted, concentrated form (Kawasaki and Dosako, 1992). Using a method suitable for scale-up, Kawasaki *et al.* (1993) isolated CMP from a WPC solution, acidified to pH 3.5, by removing whey proteins by UF, the CMP being present in the permeate. On adjusting the pH of this permeate to 7.0, the CMP associated and was separated from low molecular weight whey components by UF/DF, the associated CMP being in the retentate.

26.2.7 Co-precipitate production

Following precipitation of caseins from milk by acidification or renneting, the whey proteins remain soluble in the whey. However, these can be co-precipitated with the casein by first heating milk, at its natural pH, to a temperature which denatures the whey proteins and induces their complexation with casein, followed by precipitation of the milk protein complex by acidification to pH 4.6 or by a combination of added $CaCl_2$ and acidification (Buchanan *et al.*, 1965; Muller *et al.*, 1967). Products obtained in this manner are referred to as casein-whey protein co-precipitates (Figure 26.12). Yields of 92–95% of total milk protein are obtained, compared to <80% for acid or rennet caseins. Processes for the manufacture of similar products with improved solubility have been described (Connolly, 1983; Lankveldt, 1984;

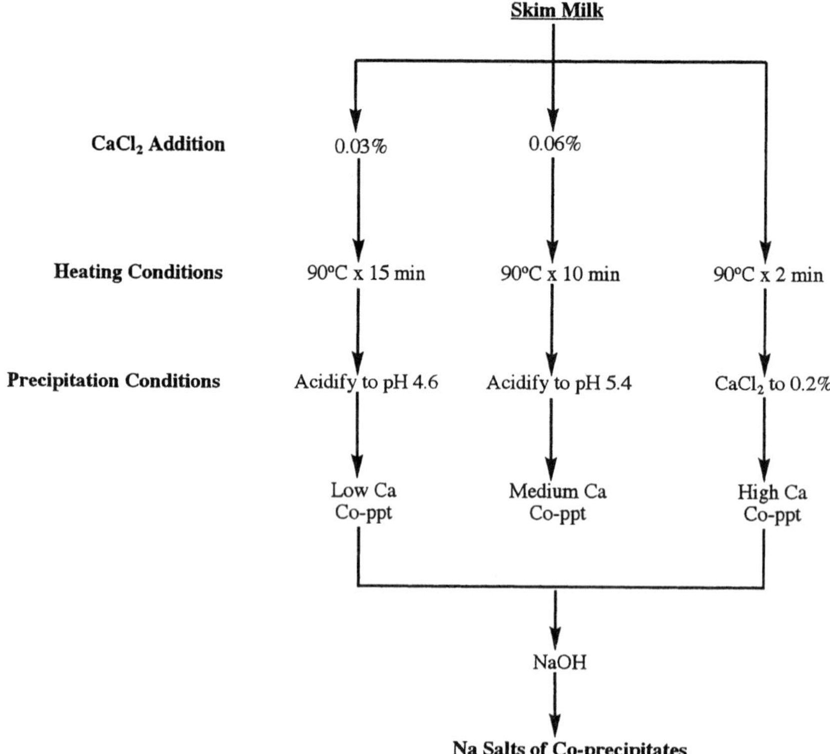

Figure 26.12 Methods for the manufacture of conventional co-precipitates.

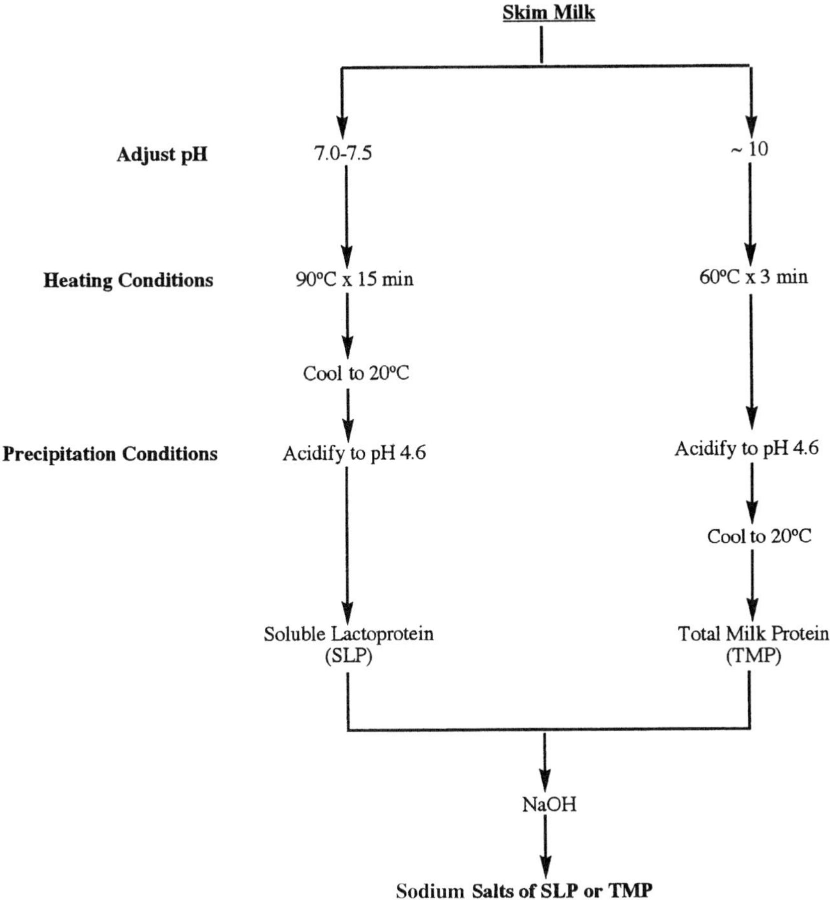

Figure 26.13 Methods for the manufacture of soluble lactoprotein (SLP) and total milk protein (TMP).

Grufferty and Mulvihill, 1987). These processes involve preadjusting milk to pH 7.0–7.5, heating at 90°C × 15 min or preadjusting milk to pH 10 and heating at 60°C × 3 min prior to isoelectric precipitation (Figure 26.13). The isoelectric or pH/$CaCl_2$ induced precipitates are processed in a manner similar to that used for caseins.

26.2.8 Production of milk protein concentrate

Directly processing skim milk by UF/DF yields milk protein concentrate (MPC) containing up to ~80% protein, in which the casein is in a similar, micellar, form to that found in milk and the whey proteins are also in their native form (Puhan, 1990; Novak, 1996). The products have a relatively high ash content, since protein-bound minerals are retained. Murashov and Murashova (1996) prepared MPC by mixing skim milk or buttermilk with cheese whey, adjusting the pH to 5.5 to 7.0, heating to 65–90°C and holding, followed by cooling and reverse osmosis to concentrate the proteins.

26.2.9 Physically modified milk protein products

Processes for the thermo-mechanical coagulation of protein solutions into water-dispersible protein particles of 1–20 μm diameter, referred to as microparticulated protein, constitute the basis of several patents for producing protein-based replacers for fats and oil-in-water emulsions (see Cheftel and Dumas, 1993). Singer *et al.* (1988) described one such process for the microparticulation of whey proteins under acidic conditions. WPC solution prepared by ultrafiltration and vacuum evaporation of sweet whey and containing 40–50% (w/w) total solids (of which 45–55% are undenatured proteins) is adjusted to a pH value in the range 3.7 to 4.2 with food-grade acid such as hydrochloric and/or citric acids and heated to a temperature in the range from 80 to 120°C for 15 min to 3 s while being sheared at a rate from 5,000 to 600,000 per min, respectively. Denaturation of the whey proteins occurs during heating but aggregation is prevented by the continued shearing action and approximately spherical particles of diameter 0.1 to 3.0 μm are formed that are suitable as a fat replacer. In a further process (Singer and Dunn, 1990), a mixture of egg white protein and skim milk protein is heated and simultaneously sheared at neutral pH. During the heating and shearing, the casein micelles in the skim milk act as nuclei around which the egg white proteins coagulate, thus ensuring control of particle size of the final microparticulated product within narrow limits, consistent with optimum sensory quality of the final product as a fat replacer. The thermo-mechanical coagulation of solutions of WPI and calcium caseinate (33% total solids and 20% protein) either at acidic (3.5–3.9) or neutral (6.5–6.7) pH using a long-barrel twin-screw extruder operating at 85–100°C and a screw speed of 100–200 rpm was described by Cheftel and Dumay (1993). Over 50% of the particles in the resulting microparticulated products had a diameter within the range 6–11 μm, the products were easily dispersed at a level of 5–10%, they developed desirable "creaminess" in low-fat and fat-free foods and were capable of withstanding the heating to which the foods were subjected during processing. Hoffmeister (1993) described how sweet whey is used for the manufacture of speciality processed WPC suitable for use in no-fat or low-fat dairy products. The whey, with a high content of undenatured whey protein, is ultrafiltered and the retentate is heat-treated to induce controlled protein denaturation. The final dry product (96% total solids) contains approximately 35% protein, 52.5% lactose, 5% fat and 9% ash and has a pH of 6.0–6.6 (10% solution).

26.2.10 Milk protein hydrolysates and milk protein-derived bioactive peptide products

Milk proteins may be hydrolysed by a wide variety of proteolytic enzymes to hydrolysates containing small peptides and amino acids. Many peptides derived from caseins have biological functions while the functional and nutritional properties of whey proteins are also modified by proteolytic digestion (Mulvihill and Fox, 1994). Methods for the preparation of milk protein hydrolysates and milk protein-derived bioactive peptide products are discussed in Chapter 14.

26.3 Food uses of milk protein products

The utilization of milk protein products as food ingredients depends on their physico–chemical and functional properties. These properties are described in Chapters 27–29 and have been reviewed extensively by Fox and Mulvihill (1983), Kinsella (1984), de Wit (1989a) and Mulvihill and Fox (1989), and therefore are not discussed further here.

Details of many of the food uses of milk protein products are proprietary to food processors and are not reported in the literature. However, reviews on uses of milk proteins in foods include Southward and Goldman (1978), International Dairy Federation (1982), Southward and Walker (1982), Zadow (1986), Southward (1986, 1989), Hugunin (1987), de Wit (1989b, 1998) and Chandan (1997). The following are brief outlines of some reported food applications of dairy protein products.

26.3.1 Dairy products

Milk protein products are used widely to supplement the protein content and, therefore, enhance the sensory characteristics of conventionally-processed dairy products and are also used in the production of a range of imitation and low-fat and non-fat dairy products (Exhibit 26.1). Imitation cheeses (cheese analogues) are made from vegetable fat, caseins, salts and water and are used in pizza, lasagne and sauces and on burgers, grilled sandwiches, macaroni and other dishes at a significant cost-saving compared to the use of natural cheese. The functional properties of casein which favour their use in imitation cheese include fat and water binding, texture enhancing, melting properties, stringiness and enhancement of shreddability. While caseins (both acid and rennet) and caseinates have been used most commonly for cheese analogues, co-precipitates also have potential in this area.

Sodium caseinate is used to produce powdered coffee creamers which are cheaper, have a longer shelf life and are more convenient to use (e.g., they require no refrigeration) than fresh coffee creams. In these products, which

Exhibit 26.1
Applicatons of milk protein products in dairy-type products

Caseins/Caseinates/Co-precipitates

Used in:	Imitation cheeses (vegetable oil, caseins/caseinates, salts and water).
Effect:	Fat and water binding, texture enhancing, melting properties, stringiness and shredding properties.
Used in:	Coffee creamers (vegetable fat, carbohydrate, sodium caseinate, stabilizers and emulsifiers).
Effect:	Emulsifier, whitener, gives body and texture, promotes resistance to feathering, sensory properties.
Used in:	Cultured milk products, e.g., yoghurt.
Effect:	Increase gel firmness, reduces syneresis.
Used in:	Milk beverages, imitation milk, liquid milk fortification, milk shakes.
Effect:	Nutritional, emulsifier, foaming properties.
Used in:	High fat powders, shortening, whipped toppings and butter-like spread.
Effect:	Emulsifier, texture enhancing, sensory properties.

Whey Proteins

Used in:	Yoghurt, Quarg, Ricotta cheese.
Effect:	Yield, nutritional, consistency, curd cohesiveness.
Used in:	Cream cheeses, cream cheese spreads, sliceable/squeezable cheeses, cheese fillings and dips.
Effect:	Emulsifier, gelling, sensory properties.

Microparticulated Milk Proteins

Used in:	Low-fat and/or non-fat milks, yoghurts, dairy spreads, natural and processed cheeses, cheese dips and spreads.
Effect:	Fat replacement, sensory properties

also contain vegetable fat, a carbohydrate source and added emulsifier and stabilizers, sodium caseinate acts as an emulsifier/fat encapsulator and whitener, imparts body and flavour and promotes resistance to feathering (i.e., coagulation of cream in hot coffee solutions).

Sodium caseinate is used to increase gel firmness and reduce syneresis in yoghurts, and is added to milk shakes for its emulsifying and foaming properties. In the manufacture of imitation milks, the principal ingredients are caseins/caseinates, vegetable fat and carbohydrate, e.g., corn syrup. The main advantages associated with imitation milk products are low cost and the absence of lactose to which some people are intolerant. There is also interest in fortifying liquid milk with casein products such as sodium, potassium or calcium caseinates or co-precipitates. Sodium caseinate is used as an emulsifying and fat-encapsulating agent in the manufacture of high-fat powders for use as shortenings in baking or cooking. Dry whipping fats or whipping creams contain casein products. A number of butter-like dairy spreads are manufactured using milk and/or vegetable fat and various casein products. In these applications, casein acts mainly as an emulsifier and in the case of dairy spreads, it also enhances texture and flavour.

Whey protein products are used widely in yoghurts and various cheeses to improve the yield, nutritional value and consistency. Some of the casein in Quarg can be replaced by thermally-modified WPC, resulting in an increase in yield and nutritional value. The use of sweet UF WPC in the manufacture of Ricotta cheese increases the cohesiveness of the curd. Emulsions prepared using heat-denatured whey proteins and fat are used as a protein base for formulated cream cheeses and cream cheese spreads.

The viscosity and stability of yoghurts are improved by replacing some skim milk solids by WPC. Sliceable and squeezeable cheese-type products, based on the emulsifying and gelling properties of whey proteins, may be produced by heat treatment of pre-cheese mixes containing skim milk and WPC solids dispersed in an emulsion of milk fat in WPC. Whey protein concentrates are also used in cheese fillings and dips as they tend to complement cheese flavour and produce a soft end product. Microparticulated milk protein products are used as fat replacement ingredients in low-fat and non-fat milks, yoghurts, dairy spreads, natural and processed cheeses, cheese dips and spreads.

26.3.2 Beverages

Casein products are used as stabilizers or for their whipping and foaming properties in drinking chocolate, fizzy drinks and fruit beverages (Exhibit 26.2). There is also a large market for sodium caseinate as an emulsifier in cream liqueurs and to a lesser extent in wine aperitifs. Cream liqueurs typically contain 16% (w/w) milk fat, 3.3% sodium caseinate, 19% added sugar and 14% ethanol. Trisodium citrate is also added to prevent calcium-induced age gelation. Casein products have also been used in the wine and beer industries as fining agents, to reduce colour and astringency and to aid in clarification.

Fruit juices, soft drinks or milk-based beverages supplemented with whey protein concentrates are highly nutritious products. For use in soft drinks, defatted WPC with a low ash content, good solubility at pH 3.0 and a bland flavour are required. The WPC must also be resistant to physical deterioration or flavour changes on storage of the product and it must not mask the typical flavour of the soft drink *via* protein-flavour component interactions. WPCs are added to milk-like flavoured drinks to impart viscosity, body and colloidal stability and they have been included as protein supplements in powdered orange beverages and in frozen orange juice concentrates.

Exhibit 26.2
Applications of milk protein products in beverages

Caseins/Caseinates/Co-precipitates	
Used in:	Drinking chocolate, carbonated drinks and fruit beverages.
Effect:	Stabilizer, whipping and foaming properties.
Used in:	Cream liqueurs, wine aperitifs.
Effect:	Emulsifier.
Used in:	Wine and beer industry.
Effect:	Fines removal, clarification, reduce colour and astringency.
Whey Proteins	
Used in:	Soft drinks, fruit juices, powdered or frozen orange beverages.
Effect:	Nutritional.
Used in:	Milk-based flavoured beverages.
Effect:	Viscosity, colloidal stability.

26.3.3 Dessert-type products

Sodium caseinate is used in ice cream substitutes and frozen desserts to improve whipping properties, body and texture and to act as a stabilizer (Exhibit 26.3). It is also used extensively in mousses, instant puddings and whipped toppings for similar reasons and also because it acts as an

Exhibit 26.3
Applications of milk protein products in dessert-type products

Caseins/Caseinates/Co-precipitates	
Used in:	Ice cream, frozen desserts.
Effect:	Whipping properties, body and texture.
Used in:	Mousses, instant pudding, whipped topping.
Effect:	Whipping properties, film former, emulsifier, imparts body and flavour.
Whey Proteins	
Used in:	Ice cream, frozen juice bars, frozen dessert coatings.
Effect:	Skim milk solid replacement, whipping properties, emulsifying, body/texture.
Microparticulated Milk Proteins	
Used in:	Low-fat and non-fat puddings, frozen desserts and ice creams.
Effect:	Fat replacement, sensory properties.

emulsifier and film-former. The basic ingredients of whipped toppings are vegetable fat, sugar, protein (sodium caseinate), emulsifier, stabilizers and water. After blending the ingredients at 38–46°C, the mixture is pasteurized and homogenized and then either cooled rapidly to below freezing point or spray dried.

In the manufacture of ice cream, skim milk solids can be replaced by whey powder. A higher level of replacement (up to 25%) may be possible by using delactosed, demineralized whey powder or UF-WPC with no adverse effect on flavour, texture or appearance. WPC has also been used in frozen juice bars and in compound coatings, especially chocolate coatings, for frozen desserts. Microparticulated milk protein products are used as fat replacement ingredients in low-fat and non-fat puddings, frozen dairy desserts and ice-creams.

26.3.4 Bakery products

Milk proteins do not have properties close enough to those of wheat gluten to enable them to replace the latter protein to any great extent in bakery products; however, their use as a nutritional supplement and for functional effects in cereal-based products has considerable potential (Exhibit 26.4). In bakery applications, the functionality of a milk protein product is related to the composition of the product and the production conditions, the nature of the interactions with the wheat flour, which depend on the fractionation method used to produce the milk protein product, and baking parameters such as mixing time and water absorption (Erdogdu-Arnoczky *et al.*, 1996). The limiting amino acid in most cereal proteins is lysine and since caseins are particularly rich in lysine, they make excellent supplements for cereals. Only about 4% casein in a casein-wheat flour mixture is required to increase the lysine content by ~60%. The protein efficiency ratio (PER) of white wheat flour is only 1.1 compared with 2.5 for casein and, on blending casein and wheat flour to give a mixture containing 75% wheat protein and 25% casein, the PER is increased to ~1.8. Another important functional characteristic of milk protein products in bakery applications is water binding, which affects dough consistency.

Casein/caseinates are added to breakfast cereals, milk biscuits, protein-enriched bread and biscuits, high-protein bread and cookies as a nutritional supplement and to frozen baked cakes and cookies as an emulsifier and to improve texture. The type of casein/caseinate must be chosen carefully to be compatible with the particular bakery application. Co-precipitates are used in pastry glaze to improve colour; in milk biscuits, cake mixes for diabetics, high-protein biscuits and cookies as a nutritional supplement and in fortified bread to improve dough consistency, sensoric properties and to increase volume and yield.

Exhibit 26.4
Applications of milk protein products in bakery, confectionary and pasta products

Bakery Products

Caseins/Caseinates/Co-precipitates

Used in:	Bread, biscuits/cookies, breakfast cereals, cake mixes, pastries, frozen cakes and pastries, pastry glaze.
Effect:	Nutritional, sensory, emulsifier, dough consistency, texture, volume/yield.

Whey Proteins

Used in:	Bread, cakes, muffins, croissants.
Effect:	Nutritional, emulsifier, egg replacer.

Confectionery

Caseins/Caseinates/Co-precipitates

Used in:	Toffee, caramel, fudges.
Effect:	Confers firm, resilient, chewy texture; water binding, emulsifier.
Used in:	Marshmallow and nougat.
Effect:	Foaming, high temperature stability, improves flavour and brown colour.

Whey Proteins

Used in:	Aerated candy mixes, meringues, sponge cakes.
Effect:	Whipping properties, emulsifier.

Pasta Products

Used in:	Macaroni, pasta, imitation pasta.
Effect:	Nutritional, texture, freeze-thaw stability, microwaveable.

Whole whey protein products generally depress loaf volume, an effect which has been associated with proteose peptone components. When whey is concentrated by ultrafiltration, the depressant appears to be removed since dough fortified with UF-WPC suffers only slight depression of loaf volume. The concentration of whey lipids during UF also contributes to good baking characteristics.

Replacement of eggs by whey protein in cake manufacture would have economic and nutritional advantages. However, simply replacing whole eggs by WPC in Madeira-type cakes results in poor quality cakes but much better results are obtained when the fat and WPC are pre-emulsified.

Various types of WPC have been used in convenience breakfast bakery products, like muffins and croissants, to increase their nutritional value.

26.3.5 Confectionery

Caseins are used in toffee, caramel, fudge and other confections as they form a firm, resilient, chewy matrix on heating and they contribute water binding and aid emulsification (Exhibit 26.4). WPCs have limited use in these products as they result in a softer coagulum and the high lactose content tends to cause crystallization during storage. However, whey proteins are very suitable for use in aerated candy mixtures and are incorporated as a frappé, which is a highly aerated sugar syrup containing the whipping protein. Casein hydrolyzates are used as foaming agents in place of egg albumen in marshmallow and nougat as they confer stability to high cooking temperatures and good flavour and browning properties. Replacement of egg white by WPC in the manufacture of meringues results in acceptable products only when defatted WPCs are used, while the manufacture of acceptable sponge cakes requires fat-containing rather than defatted WPCs.

26.3.6 Pasta products

Milk protein products are often incorporated into the flour base for pasta manufacture to enhance nutritional quality and to improve texture (Exhibit 26.4). Products fortified by addition of sodium or calcium caseinate, low calcium co-precipitate or WPC prior to extrusion include macaroni and pasta.

Enrichment of pasta flours with undenatured whey protein products produces firmer cooked noodles which are also more freeze-thaw stable and suitable for microwave cooking. 'Imitation' or 'synthetic' pasta-type products containing a substantial proportion of milk protein have been manufactured also.

26.3.7 Meat products

In the meat industry, milk proteins are used mainly in comminuted meat products rather than in prime cuts; however, they are also used in injection brines for non-comminuted products like cooked hams (Exhibit 26.5).

Exhibit 26.5
Applications of milk protein products in meat products

Caseins/Caseinates/Co-precipitates	
Used in:	Comminuted meat products.
Effect:	Emulsifier, water binding, improves consistency, releases meat proteins for gel formation and water binding.
Whey Proteins	
Used in:	Frankfurters, luncheon rolls.
Effect:	Pre-emulsion, gelation.
Used in:	Injection brine for fortification of whole meat products.
Effect:	Gelation, yield.

Caseins in comminuted meat products contribute to fat emulsification, water binding and improved consistency as they release meat proteins for gel formation and water binding. While sodium caseinate is the preferred additive in meat applications, various types of co-precipitates have also been used.

In frankfurters and luncheon rolls, meat protein may be replaced by whey proteins. In these systems, whey proteins are used to prepare pre-emulsions of part of the fat and to support network formation, *via* gelation, during subsequent cooking.

Soluble, low viscosity whey protein concentrates are suitable for use in injection brines for fortification of whole meat products. Injection of fresh and cured meats with whey protein solution may increase the yield substantially.

26.3.8 Dietary, pharmaceutical and medical applications

Since milk protein products are of high nutritional quality, they are used extensively in dietary preparations (Exhibit 26.6) for people who are ill or convalescing, for malnourished children in developing countries on a therapeutic diet and for people on weight-reducing diets. Caseins are used in special preparations to enhance athletic performance and have been incorporated into formula diets for astronauts while in space.

While casein products are not used generally in infant formulae, they are used extensively in specialized preparations for infants with specific nutritional problems. Caseinates, co-precipitates and milk protein concentrates are used in low-lactose formulae for lactose-intolerant infants while various types of caseinate have been used in infant foods with a specific mineral balance, e.g., low-sodium infant formulae for children with specific renal problems. Casein hydrolyzates are used in specialized foods for premature infants, in formulae for infants suffering from diarrhoea, gastroenteritis, galactosaemia and malabsorption. Special casein hydrolyzates, low in phenylalanine, may be prepared for use in formulae for feeding infants with phenylketonuria. Casein products are also added to various foods for children and infants and to drinks as a nutritional supplement.

Modified low-mineral whey powders are used to produce 'humanized' infant formulae which have a whey protein-to-casein ratio close to that of human milk. Hypoallergenic, peptide-based formulae have been developed based on whey protein hydrolyzates. Selected individual caseins and whey proteins have been proposed as possible ingredients for the 'next generation' of 'more humanized' infant formulae.

Milk protein hydrolyzates are used for intravenous nutrition for patients suffering from intestinal disorders, protein metabolism disorders and for post-operative patients. Special casein preparations are used as a protein source in foods for patients suffering from cancer, pancreatic disorders or anaemia.

EXHIBIT 26.6
Applications of milk protein products in dietary pharmaceutical and healthcare products

Used in:	Special Dietary Preparations • Ill or convalescent patients. • Dieting patients/people. • Athletes. • Astronauts
Used in:	Infant Foods • Nutritional fortification. • "Humanized" infant formulae. • Low-lactose infant formulae. • Specific mineral balance infant foods. • Casein hydrolyzates: used for infants suffering from diarrhoea, gastroenteritis, galactosemia, malabsorption, phenylketonuria. • Whey protein hydrolyzates used in hypoallergenic formulae preparations. • Nutritional fortification.
Used in:	Intravenous Feeds • Patients suffering from metabolic disorders and intestinal disorders or for post-operative patients.
Used in:	Special Food Preparations • Patients suffering from cancer, pancreatic disorders or anemia.
Used in:	Healthcare Preparations • β-Casomorphins used in sleep or hunger regulation or insulin secretion. • Immuno-enhancing diets for HIV patients. • Anti-tumor diets for cancer patients.
Used in:	Miscellaneous Products • Toothpastes. • Cosmetics. • Therapeutic creams.

Peptides with various types of biological activity have been identified in enzymatic digests of caseins and whey proteins (see Mulvihill and Fox, 1994) and this has generated interest in the exploitation of milk protein hydrolysates tailored for use in specific healthcare and pharmaceutical applications. Peptides with opioid activity, called exorphins, immunomodulating peptides, platelet-modifying peptides, bacteriocidal peptides, calcium binding phosphopeptides and peptide inhibitors of angiotensin converting enzyme (ACE) have been found in enzymatic digests of milk proteins. Enzymatic digests of β-casein contain exorphins referred to as β-caseinomorphins, tetra-to-heptapeptides that can regulate sleep, hunger or insulin secretion. Exorphins have also been isolated from hydrolysates of α_{s1}-casein, κ-casein, α-lactalbumin, β-lactoglobulin and lactoferrin.

Dietary whey proteins are reported to be immuno-enhancing when administered to HIV patients, and to have anti-tumor effects in the dietary treatment of head and neck cancer patients and to improve wound healing and help fight post-operative infections (see International Dairy Federation, 1998). The use of milk protein products in such products as toothpaste, cosmetics and in a variety of therapeutic creams has been advocate.

26.3.9 Convenience Foods

Milk protein products are used widely in convenience foods, i.e., foods that require a minimum of preparation by the consumer (Exhibit 26.7). Either skim milk powder or whey powder-caseinate blends are used as whitening agents in gravy mixes. Whey solids are used in dehydrated soup mixes and sauces to impart a milky or dairy flavour, as flavour enhancers and to provide emulsifying and stabilizing effects. Caseinates are used as emulsifying agents and viscosity controllers in canned cream soups and sauces and for the preparation of dry emulsions for use in dehydrated cream soups and sauces. In some convenience foods, caseinate-whey protein blends are used as cheap replacements for skim milk powders. Whey protein products are used as replacements for egg yolk in salad dressing and microparticulated milk protein products are used to replace lipids in a variety of convenience foods. Milk protein products are also used as texture, stability and flavour enhancers in microwaveable foods.

EXHIBIT **26.7**

Applications of milk protein products in convenience foods and textured products

Convenience Foods	
Used in:	Gravy mixes, soup mixes, sauces, canned cream soups and sauces, dehydrated cream soups and sauces, salad dressings, microwavable foods, low-fat convenience foods.
Effect:	Whitening agents, dairy flavour, flavour enhancer, emulsifier, stabilizer, viscosity controller, freeze-thaw stability, egg yolk replacement, fat replacement
Textured Products	
Used in:	Puffed snack foods, protein-enriched snack-type products, meat extenders, surimi, restructured fish.
Effect:	Structuring, texturing, nutritional.

26.3.10 Textured products

Textured milk protein-based foods, in the form of cheese, have been manufactured from milk for thousands of years. However, milk protein-enriched products are now also used in the production of recently-developed textured foods (Exhibit 26.7). Rewetted acid caseins, acidified rennet casein or co-precipitate can be mixed with carbonates or bicarbonates of alkali metals or alkali earth metals and extruded to produce puffed snack foods while caseinates can be co-extruded with wheat flour to produce protein-enriched snack-type food products. Texturing agents for use in *fromage frais* and yoghurts can be prepared by heat and shear processing whey protein- and casein-containing solutions.

Fibrous meat-like structures formed from caseins by fibre spinning techniques are produced for use as extenders in comminuted meats. Whey proteins may be co-spun with the casein to produce stronger fibres than those containing casein alone. Meat-like structure can also be formed from casein or co-precipitates by renneting followed by thermoplastic extension which involves a combination of heat treatment and extrusion or working. Microwave heating of whey protein solutions results in simultaneous expansion and gelation to give textured products with possible applications in comminuted meats. WPC and WPI are used as gelling agents in surimi and other restructured seafood products.

26.3.11 Edible films and coatings

Edible films and coatings are generally considered as barriers to unwanted transfer of moisture, gases, aromatics or other small molecules. Edible milk protein films are generally flavourless, tasteless, flexible and vary in opacity from translucent to transparent, depending on the protein source, formulation and composition (Exhibit 26.8). The water solubility of the films can be reduced by treating the films with buffers at the pI of the proteins or by enzymatically cross-linking the proteins using transglutaminase.

EXHIBIT 26.8
Applications of milk protein products in packaging and microencapsulation

Used in:	Edible films, coatings.
Effect:	Control tensile strength, stretchability and opacity, moisture and gas barrier, sensory properties.
Used in:	Microencapsulation.
Effect:	Confer handling and dispersion properties, reduce environmental reactivity, confer oxidative stability, prevent volatile loss.

Incorporation of lipids significantly reduces the water vapour permeability of the films while inclusion of plasticizers to increase film flexibility tends to increase the water vapour permeability. Edible films based on caseinates, micellar casein or rennet casein can significantly reduce moisture loss, while whey protein films show excellent gas barrier properties due to the highly polar nature of the proteins. Milk protein films show good tensile strength and low stretchability when compared to, for example, polyethylene films and may be expected to be good lipid barriers due to the high content of hydrophilic groups in the proteins. Potential applications of milk proteins in edible films have been reviewed by Chen (1995).

Commercially-available WPC appears to be an attractive material for the manufacture of edible films with good functional properties (Banerjee and Chen, 1995). Coating of dry-roasted peanuts with WPI solution delayed the development of nut rancidity, the WPI coating acting as an effective oxygen barrier (Mate *et al.*, 1996). Edible WPI and β-lactoglobulin films are reported to exhibit similar oxygen and water vapour permeabilities (Mate and Krochta, 1996).

26.3.12 Microencapsulation

Food ingredients are microencapsulated to:

- reduce reactivity with the outside environment,
- improve handling properties,
- provide more uniform dispersion in the final product,
- mask unwanted flavours,
- control release in the final product and control water evaporation (Dziezak, 1988; Shahidi and Han, 1993).

Milk proteins possess many of the properties required of a good wall material for microencapsulation (Exhibit 26.8). Rosenberg and Young (1993) prepared microencapsulated anhydrous milk-fat (AMF) with good microstructural characteristics by emulsifying the fat in solutions of WPC/WPI and spray-drying the resulting emulsion. Using WPI or WPI/lactose as the encapsulating material conferred good oxidative stability to AMF by preventing oxygen uptake (Moreau and Rosenberg, 1996) and resulted in good retention of volatile esters in the core (Rosenberg and Sheu, 1996). Sodium caseinate and WPI effectively emulsified orange oil in the first stage of microencapsulation (Kim *et al.*, 1996). WPI was less effective than sodium caseinate at retaining the oil on spray drying but sodium caseinate-encapsulated oil showed poor stability to oxidation; however, both proteins showed microencapsulation properties superior to those of gum arabic (Kim and Morr, 1996). Whey-based microparticulated protein conforms to the definition of a WPC, a GRAS (generally regarded as safe) substance, and may be used in many applications as a fat replacer (Giese, 1996). Casein-dextran conjugates have been used as secondary emulsifiers in the encapsulation of numerous biologically-active proteins (Shimizu and Nakane, 1995).

26.4 Future developments

As outlined above, numerous protein products are recovered from milk at present. It is likely that the range will be extended in the future as commercial methods developed to effect separation of individual caseins and whey proteins become more widely used. It is likely that enzymatic and physical methods developed to modify the nutritional and functional properties of milk proteins will be used more widely to confer new physico–chemical and functional properties to milk protein products. While the main area of application of milk protein products will continue to be as functional ingredients in formulated foods, it is likely that there will be increased use of milk protein-based products in dietary, pharmaceutical and medical products.

References

Allen, I.M., McAuliffe, A.G. and Donnelly, W.J. (1985) Simplified approaches to casein fractionation. *Ir. J. Food Sci. Technol.*, **9**, 85 (Abstract).

Al-Mashikhi, S.A., Li-Chan, E. and Nakai, S. (1988) Separation of immunoglobulins and lactoferrin from cheese whey by chelating chromatography. *J. Dairy Sci.*, **71**, 1747–55.

Al-Mashikhi, S.A. and Nakai, S. (1987a) Reduction of β-lactoglobulin content of cheese whey by polyphosphate precipitation. *J. Food Sci.*, **52**, 1237–42, 44.

Al-Mashikhi, S.A. and Nakai, S. (1987b) Isolation of bovine immunoglobulins and lactoferrin from whey proteins. *J. Dairy Sci.*, **70**, 2486–92.

Amundson, C.H., Watanawanichakorn, S. and Hill, C.G. (1982) Production of enriched protein fractions of β-lactoglobulin and α-lactalbumin from cheese whey. *J. Food Process. Preserv.*, **6**, 55–71.

Andrews, A.T., Taylor, M.D. and Owen, A.J. (1985) Rapid analysis of bovine milk proteins by fast protein liquid chromatography. *J. Chromatogr.*, **348**, 177–85.

Baick, S.C. and Yu, J.H. (1995) Separation of immunoglobulin from Holstein colostrum and its immunological response. *Food Biotechnol.*, **4**, 117–21.

Banerjee, R. and Chen, H. (1995) Functional properties of edible films using whey protein concentrate. *J. Dairy Sci.*, **78**, 1673–83.

Berrocal, R. and Neeser, J.R. (1991) Process for production of a κ-casein glycomacropeptide. *European Patent Application 0 453 782 A1.*

Berrocal, R. and Neeser, J.R. (1993) Process for preparation of κ-casein glycomacropeptide. *Swiss Patent CH 681 543 A5.*

Bläckberg, L. and Hernell, O. (1980) Isolation of lactoferrin from human whey by a single chromatographic step. *FEBS Lett.*, **109**, 180–4.

Bottomley, R.C. (1989) Isolation of an immunoglobulin rich fraction from whey. *European Patent Application 0 320152 A2.*

Bramaud, C., Aimar, P. and Daufin, G. (1997) Whey protein fractionation: isoelectric precipitation of α-lactalbumin under gentle heat treatment. *Biotechnol. Bioeng.*, **56**, 391–7.

Breslau, B.R., Cross, R.A. and Goulet, J. (1975) Production of a crystal clear, bland tasting, protein solution from cheese whey. *J. Dairy Sci.*, **58**, 782 (abstract).

Buchanan, R.A., Snow, N.S. and Hayes, J.F. (1965) The manufacture of 'calcium co-precipitate'. *Aust. J. Dairy Technol.*, **20**, 139–42.

Burgess, K.J. and Kelly, J. (1979) Technical note: selected functional properties of a whey protein isolate. *J. Food Technol.*, **14**, 325–9.

Burling, H. (1989) Process for extracting pure fractions of lactoperoxidase and lactoferrin from milk serum. *PCT International Patent Application WO 89/04608 A1.*

Burton, J. and Skudder, P. (1987) Whey protein. *United Kingdom Patent Application GB 2 188 526 A.*

Carrere, H., Sochard, S., Bascoul, A., Wilhelm, A.M. and Delmas, H. (1994) Whey protein extraction: fluidised ion exchange chromatography isotherms determination and process modelling. *Food Bioprod. Process.*, **72**, 216–26.

Carrere, H., Bascoul, A., Floquet, P., Wilhelm, A.M. and Delmas, H. (1996) Whey proteins extraction by fluidized ion exchange chromatography: simplified modelling and economical optimization. *Chem. Eng. J.* **64**, 307–17.

Cayot, P., Courthaudon, J.-L. and Lorient, D. (1992) Purification of α_s-, β- and κ-caseins by batchwise ion-exchange separation. *J. Dairy Res.*, **59**, 551–6.

Chandan, R. (1997) *Dairy-based Ingredients*. Eagan Press, St. Paul, Minnesota.

Cheftel, J.C. and Dumay, E. (1993) Microcoagulation of proteins for development of "creaminess". *Food Rev. Int.*, **9**, 473–502.

Chen, H. (1995) Functional properties and applications of edible films made of milk proteins. *J. Dairy Sci.*, **78**, 2563–83.

Chen, J.-P. and Wang, C.-H. (1991) Microfiltration affinity purification of lactoferrin and immunoglobulin G from cheese whey. *J. Food Sci.*, **56**, 701–6, 713.

Chiancone, E. and Gattoni, M. (1993) Selective removal of β-lactoglobulin directly from cows milk and preparation of hypoallergenic formulas: a bioaffinity method. *Biotechnol. Appl. Biochem.*, **18**, 1–8.

Chiu, C.K. and Etzel, M.R. (1997) Fractionation of lactoperoxidase and lactoferrin from bovine whey using a cation exchange membrane. *J. Food Sci.*, **62**, 996–1000.

Christensen, T.M.I.E. and Munksgaard, L. (1989) Quantitative fractionation of casein by precipitation or ion exchange chromatography. *Milchwissenschaft*, **44**, 480–4.

Connolly, P.B. (1983) Method of producing milk protein isolates and milk protein/vegetable protein isolates and compositions of same. *United States Patent US 4 376 072.*

Coolbear, K.P., Elgar, D.F., Coolbear, T. and Ayers, J.S. (1996) Comparative study of methods for the isolation and purification of bovine κ-casein and its hydrolysis by chymosin. *J. Dairy Res.*, **63**, 61–71.

Cuddigan, N.M. (1991) *Fractionation of Whey Proteins by Sodium Hexametaphosphate and Functional Properties of the Resulting Fractions*, M.Sc. Thesis, National University of Ireland, Cork.

de Frutos, M., Cifuentes, A. and Diez-Masa, J.C. (1996) Behaviour of whey proteins in hydrophobic interaction chromatography. *J. High Resol. Chromatog.*, **19**, 521–6.

de Stefano, G., Pizzuto, P. and Sciancalepore, V. (1994) Recupero della lattoperossidasi dal siero acido con la cromatografia di affinita al rame immobilizzato. *Ind. Aliment.*, **33**, 387–9.

de Wit, J.N. (1989a) Functional properties of whey proteins, in *Developments in Dairy Chemistry-4-Functional Proteins*, (P.F. Fox ed.) Elsevier Applied Science Publishers, London, pp. 285–322.

de Wit, J.N. (1989b) The use of whey protein products, in *Developments in Dairy Chemistry-4-Functional Proteins*, (P.F. Fox ed.) Elsevier Applied Science Publishers, London, pp. 323–46.

de Wit, J.N. (1998) Nutritional and functional characteristics of whey proteins in food products. *J. Dairy Sci.*, **81**, 597–608.

de Wit, J.N. and Bronts, H. (1994) Process for the recovery of alpha-lactalbumin and beta-lactoglobulin from a whey product. *European Patent Application 0 604 864.*

de Wit, J.N. and de Boer, R. (1975) Ultrafiltration of cheese whey and some functional properties of the resulting whey protein concentrate. *Neth. Milk Dairy J.*, **29**, 198–211.

de Wit, J.N., Klarenbeek, G. and de Boer, R. (1978) A simple method for the clarification of whey. *Proc. 20th Int. Dairy Congr.* (Paris), E919–20.

de Wit, J.N., Klarenbeek, G. and Adamse, M. (1986) Evaluation of functional properties of whey protein concentrates and whey protein isolates. Effects of processing history and composition. *Neth. Milk Dairy J.*, **40**, 41–56.

Dinkov, K., Andreev, A. and Panaoitov, P. (1997) Separation of casein and whey proteins of cow milk using two types of ultrafiltration. *Milchwissenschaft*, **52**, 127–30.

Dubois, E. (1986) Precédé de séparation de certaines protéines du lactosérum ou du lait. *French Patent 2 605 322.*

Dziezak, J.D. (1988) Microencapsulation and encapsulated ingredients. *Food Technol.*, **42**, 136–51.

Elkstrand, B. (1989) Antimicrobial factors in milk – a review. *Food Biotechnol.*, **3**, 105–26.

Erdogdu-Arnoczky, N., Czuchajowska, Z. and Pomeranz, Y. (1996) Functionality of whey and casein in fermentation and in breadbaking by fixed and optimized procedures. *Cereal Chem.*, **73**, 309–16.

Eustache, J.-M. (1977) Extraction of glycoproteins and sialic acid from whey. *United States Patent 4 042 576.*

Fairise, J.-F. and Cayot, P. (1998) New ultrarapid method for the separation of milk proteins by capillary electrophoresis. *J. Agric. Food Chem.*, **46**, 2628–33.

Famelart, M.H., Hardy, C. and Brule, G. (1989) Optimisation of the preparation of β-casein-enriched solution. *Lait*, **69**, 47–57.

Famelart, M.H. and Surel, O. (1994) Caseinate at low temperatures – calcium use in β-casein extraction by microfiltration. *J. Food Sci.*, **59**, 548–53, 587.

Felipe, X. and Law, A.J.R. (1997) Preparative-scale fractionation of bovine, caprine, and ovine whey proteins by gel permeation chromatography. *J. Dairy Res.*, **64**, 459–64.

Fox, K.K., Holsinger, V.H., Posati, L.P. and Pallansch, M.J. (1967) Separation of β-lactoglobulin from other milk serum proteins by trichloroacetic acid. *J. Dairy Sci.*, **50**, 1363–7.

Fox, P.F. and Mulvihill, D.M. (1983) Functional properties of caseins, caseinates and co-precipitates, in *Physico–chemical Aspects of Dehydrated Protein-rich Milk Products*, Proceedings of International Dairy Federation Symposium, Helsingor, Denmark, International Dairy Federation, Brussels, pp. 188–259.

Fox, P.F. and Mulvihill, D.M. (1990) Casein, in *Food Gels*, (P. Harris ed.) Elsevier Applied Science Publishers, London, pp. 121–74.

Gambero, A., Kubota, L.T., Gushikem, Y., Airoldi, C., Granjeiro, J.M., Taga, E.M. and Alcantara, E.F.C. (1997) Use of chemically modified silica with β-diketoamine groups for separation of α-lactalbumin from bovine milk whey by affinity chromatography. *J. Colloid Interface Sci.*, **185**, 313–6.

Gani, M.M., May, K. and Porter, K. (1982) A process and apparatus for the recovery of immunoglobulins. *European Patent 0 059 598 A1.*

Gésan-Guiziou, G., Daufin, G., Timmer, M., Allersma, D. and van der Horst, C. (1999) Process steps for the preparation of purified fractions of α-lactalbumin and β-lactoglobulin from whey protein concentrate. *J. Dairy Res.*, **66**, 225–36.

Giese, J. (1996) Fats, oils, and fat replacers. *Food Technol.*, **50(4)**, 78–83.

Girardet, J.M., Paquet, D. and Linden, G. (1989) Effects of chromatographic parameters on the fractionation of whey proteins by anion exchange FPLC. *Milchwissenschaft*, **44**, 692–6.

Gordon, W.G., Ziegler, J. and Basch, J.J. (1962) Isolation of an iron-binding protein from cow's milk. *Biochim. Biophys. Acta*, **60**, 410–11.

Grasselli, M. and Cascone, O. (1996) Separation of lactoferrin from bovine whey by dye affinity chromatography. *Neth. Milk Dairy J.*, **50**, 551–61.

Groves, M.L. (1965) Preparation of some iron-binding proteins and α-lactalbumin from bovine milk. *Biochim. Biophys. Acta*, **100**, 154–62.

Grufferty, M.B. and Mulvihill, D.M. (1987) Proteins recovered from milks heated at alkaline pH values. *J. Soc. Dairy Technol.*, **40**, 82–5.

Hayashi, R., Kawamura, Y. and Kunugi, S. (1987) Introduction of high pressure to food processing: Preferential proteolysis of β-lactoglobulin in milk whey. *J. Food Sci.*, **52**, 1107–8.

Hayes, J.F., Dunkerley, J.A., Muller, L.L. and Griffin, A.T. (1974) Studies on whey processing by ultrafiltration. II. Improving permeation rates by preventing fouling. *Aust. J. Dairy Technol.*, **29**, 132–40.

Hewedi, M.M., Mulvihill, D.M. and Fox, P.F. (1985) Recovery of milk protein by ethanol precipitation. *Ir. J. Food Sci. Technol.*, **9**, 11–23.

Hilpert, H. (1984) Preparation of a milk immunoglobulin concentration from cow's milk, in *Human Milk Banking*, (A.F. Williams and J.F. Baum eds.) Raven Press, New York, pp. 17–28.

Hoff, J.E., Nielsen, S.S., Peng, I.C. and Chambers, J.V. (1987) Ethanol extraction of lactose from non-fat dry milk: production of protein raffinate. *J. Dairy Sci.*, **70**, 1785–96.

Hoffmeister, A. (1993) A new milk protein concentrate for low-fat foods. *Food Market. Technol.*, **7(3)**, 4–6, 8.

Hollar, C.M., Law, A.J.R., Dalgleish, D.G. and Brown, R.J. (1991) Separation of major casein fractions using cation-exchange fast protein liquid chromatography. *J. Dairy Sci.*, **74**, 2403–9.

Horton, B.S. (1995) Commercial utilization of minor milk components in the health and food industries. *J. Dairy Sci.*, **78**, 2584–9.

Huffman, L.M. (1996) Processing whey protein for use as a food ingredient. *Food Technol.*, **50**(2), 49–52.

Hugunin, A.G. (1987) Applications of UF whey protein: developing new markets, Bulletin 212, International Dairy Federation, Brussels, pp. 135–44.

Hutchens, T.W., Magnuson, J.S. and Yip, T.-T. (1989) Rapid purification of porcine colostral whey lactoferrin by affinity chromatography on single-stranded DNA agarose. Characterisation, amino acid composition and N-terminal amino acid sequence. *Biochim. Biophys. Acta*, **999**, 323–9.

Igarashi, Y. (1995) An improved procedure for the preliminary fractionation of milk proteins. *Int. Dairy J.* **5**, 305–10.

International Dairy Federation (1982) *Dairy Ingredients in Food Products*, Bulletin 147, International Dairy Federation, Brussels.

International Dairy Federation (1987) *Trends in Whey Utilization*, Bulletin 212, International Dairy Federation, Brussels.

International Dairy Federation (1998) *Whey. Proceedings of Second International Whey Conference*, Chicago, IL, International Dairy Federation, Brussels.

Janson, H.V and Lewis, M.J. (1994) Electrochemical coagulation of whey protein. *J. Soc. Dairy Technol.*, **47**, 87–90.

Jensen, J. and Larsen, P.H. (1994) Process for recovery of high value protein from whey. *Swedish Patent Application SE 470 375 B.*

Kaczmarek, J. (1980) Whey protein separation and processing, in *Proceedings of 1980 Whey Production Conference, Chicago, IL*, Whey Products Institute, USDA, Philadelphia, PA, pp. 68–80.

Kaneko, T., Kojima, T. Kuwata, T. and Yamamoto, Y. (1990) Selective enzymatic degradation of beta-lactoglobulin contained in cows milk serum protein. *European Patent Application 0 355 399 A1.*

Karleskind, D., Laye, I., Mei, F.-I. and Morr, C.V. (1995) Chemical pretreatment and microfiltration for making delipidized whey protein concentrate. *J. Food Sci.*, **60**, 221–6.

Kawakami, H., Shinmoto, H., Dosako, S.I. and Sogo, Y. (1987) One-step isolation of lactoferrin using immobilized monoclonal antibodies. *J. Dairy Sci.*, **70**, 752–9.

Kawasaki, Y. and Dosako, S. (1992) Process for producing κ-casein glycomacropeptides. *European Patent Application 0 488 589 A1.*

Kawasaki, Y., Kawakami, H., Tanimoto, M., Dosako, S., Tomizawa, A., Kotake, M. and Nakajima, I. (1993) pH-Dependent molecular weight changes of κ-casein glycomacropeptide and its preparation by ultrafiltration. *Milchwissenschaft*, **48**, 191–5.

Kim, Y.D. and Morr, C.V. (1996) Microencapsulation properties of gum arabic and several food proteins: spray-dried orange oil emulsion particles. *J. Agric. Food Chem.*, **44**, 1313–20.

Kim, Y.D., Morr, C.V. and Schenz, T.W. (1996) Microencapsulation properties of gum arabic and several food proteins: liquid orange oil emulsion particles. *J. Agric. Food Chem.*, **44**, 1308–13.

Kim, S.H., Morr, C.V. and Surak, J.G. (1989) Effect of whey pretreatment on composition and functional properties of whey protein concentrates. *J. Food Sci.*, **54**, 25–9.

Kinekawa, Y.-I. and Kitabatake, N. (1996) Purification of β-lactoglobulin from whey protein concentrate by pepsin treatment. *J. Dairy Sci.*, **79**, 350–6.

Kinsella, J.E. (1984) Milk proteins: physico–chemical and functional properties. *CRC Crit. Rev. Food Sci. Nutr.*, **21**, 197–262.

Kussendrager, K. (1993a) Lactoferrin and lactoperoxidase. Bio-active milk proteins. *Int. Food Ing.*, **6**, 17–21.

Kussendrager, K.D. (1993b) Process for isolating lactoferrin and lactoperoxidase from milk and milk products, and products obtained by such process. *PCT International Patent Application WD 93/13676 A1.*

Kuwata, T and Ohtomo, H. (1989) Method for removing β-lactoglobulin from bovine milk whey. *European Patent 0 321 605.*

Kuwata, T., Pham, A.M., Ma, C.Y. and Nakai, S. (1985) Elimination of β-lactoglobulin from whey to simulate human milk protein. *J. Food Sci.*, **50**, 605–9.

Lankveldt, J.M.G. (1984) Texturizing milk proteins, in *Milk Proteins, '84. Proceedings of the International Congress on Milk Proteins, Luxembourg*, (T.E. Galesloot and B.J. Tindbergen eds.) Pudoc Publishers, Wageningen, pp. 129–35.

Law, A.J.R., Leaver, J., Banks, J.M. and Horne, D.S. (1993) Quantitative fractionation of whey proteins by gel permeation FPLC. *Milchwissenschaft*, **48**, 663–6.

Law, B.A. and Reiter, B. (1977) The isolation and bacteriostatic properties of lactoferrin from bovine milk whey. *J. Dairy Res.*, **44**, 595–9.

Leaver, J. and Law, A.J.R. (1992) Preparative-scale purification of bovine caseins on a cation-exchange resin. *J. Dairy Res.*, **59**, 557–61.

Lee, D.N. and Merson, R.L. (1976) Chemical treatments of cottage cheese whey to reduce fouling of ultrafiltration membranes. *J. Food Sci.*, **41**, 778–86.

Lee, H.-Y. and Dungan, S.R. (1998) Selective solubilization of α-lactalbumin and β-lactoglobulin into reversed micelles from their mixtures. *J. Food Sci.* **63**, 601–5.

Liu, Y.S., Viotto, L.A. and Rizvi, S.S.H. (1995) Fractionation of skim milk by ceramic membranes: III. Separation of whey proteins and evaluation of their functionality. *J. Dairy Sci.*, **78**(Supplement 1), 148 (Abstr.)

Lonergan, D.A. (1983) Isolation of casein by ultrafiltration and cryodestabilization. *J. Food Sci.*, **48**, 1817–21.

MacLeod, A., Fedio, W.M. and Ozimek, L. (1995) Separation of beta-lactoglobulin variant A and variant B from cheese whey by biospecific subunit exchange affinity chromatography. *Milchwissenschaft*, **50**, 440–4.

Mailliart, P. and Ribadeau-Dumas, B. (1988) Preparation of β-lactoglobulin and β-lactoglobulin-free proteins from whey retentate by NaCl salting out at low pH. *J. Food Sci.*, **53**, 743–5, 752.

Marshall, K.R. (1982) Industrial isolation of milk proteins: whey protein, in *Developments in Dairy Chemistry-1-Proteins*, (P.F. Fox ed.) Applied Science Publishers, London, pp. 339–73.

Marshall, S.C. (1991) Casein macropeptide from whey, A new product opportunity. *Food Res. Quart.*, **51**, 86–91.

Martin-Hernandez, M.C., van Markwijk, B.W. and Vreeman, H.J. (1990) Isolation and properties of lactoperoxidase from milk. *Neth. Milk Dairy J.* **44**, 213–31.

Mate, J.I., Frankel, E.N. and Krochta, J.M. (1996) Whey protein isolate edible coatings: effect on the rancidity process of dry roasted peanuts. *J. Agric. Food Chem.*, **44**, 1736–40.

Mate, J.I. and Krochta, J.M. (1994) β-Lactoglobulin separation from whey protein isolate on a large scale. *J. Food Sci.* **59**, 1111–4.

Mate, J.I. and Krochta, J.M. (1996) Comparison of oxygen and water vapor permeabilities of whey protein isolate and β-lactoglobulin edible films. *J. Agric. Food Chem.*, **44**, 3001–4.

Matthews, M.E. (1984) Whey protein recovery processes and products. *J. Dairy Sci.*, **67**, 2680–92.

Matthews, M.E., Doughty, R.K. and Short, J.L. (1978) Pretreatment of acid casein whey to improve processing rates in ultrafiltration. *N.Z. J. Dairy Sci. Technol.*, **13**, 216–20.

Maubois, J.-L. (1997) *Current Uses and Future Perspectives of MF Technology in the Dairy Industry,* Bulletin 320, International Dairy Federation, Brussels, pp. 37–40.

Maubois, J.L., Pierre, A., Fauquant, J. and Piat, M. (1987) *Industrial Fractionation of Main Whey Proteins*, Bulletin 212, International Dairy Federation, Brussels, pp. 154–9.

Mehra, R.J. and Donnelly, W.J. (1993) Fractionation of whey protein components through a large pore size, hydrophylic, cellulosic membrane. *J. Dairy Res.*, **60**, 89–97.

Mirabel, B. (1978) Nouveau precédé d'extraction des protéins du lactosérum. *Annal. Nutr. Aliment.*, **23**, 243–53.

Molochnikov, V. and Zetteir, K.H. (1994) Process for separation of natural milk proteins. *German Federal Republic Patent Application DE 42 27 666 A1.*

Moreau, D.L. and Rosenberg, M. (1996) Oxidative stability of anhydrous milkfat microencapsulated in whey proteins. *J. Food Sci.*, **61**, 39–43.

Morel, F. (1991) Production de caseine "native" pour la première fois à l'echelle industrielle. SAFIR ourvre la voie. *Process* **1058**, 53–6.

Morr, C.V. (1989) Whey proteins: manufacture, in *Developments in Dairy Chemistry-4-Functional Proteins*, (P.F. Fox ed.) Elsevier Applied Science Publishers, London, pp. 245–84.

Muller, L.L. (1971) Manufacture and uses of casein and co-precipitate. *Dairy Sci. Abstr.*, **33**, 659–74.

Muller, L.L. (1982) Manufacture of caseins, caseinates and co-precipitates, in *Developments in Dairy Chemistry-1-Proteins*, (P.F. Fox ed.) Elsevier Applied Science Publishers, London, pp. 315–37.

Muller, L.L. and Harper, W.J. (1979) Effects on membrane processing of pretreatments of whey. *J. Agric. Food Chem.*, **27**, 662–4.

Muller, L.L., Hayes, J.F. and Snow, N.S. (1967) Studies on co-precipitates of milk proteins. Part 1. Manufacture with varying calcium contents. *Aust. J. Dairy Technol.* **22**, 12–8.

Mulvihill, D.M. (1989) Caseins and caseinates: manufacture, in *Developments in Dairy Chemistry – 4 – Functional Proteins*, (P.F. Fox ed.) Elsevier Applied Science Publishers, London, pp. 97–130.

Mulvihill, D.M. (1992) Production, functional properties and utilization of milk protein products, in *Advanced Dairy Chemistry, Volume 1-Proteins*, (P.F. Fox ed.) Elsevier Applied Science Publishers, London, pp. 369–404.

Mulvihill, D.M. and Fox, P.F. (1989) Physico–chemical and functional properties of milk proteins, in *Developments in Dairy Chemistry-4-Functional Proteins*, (P.F. Fox ed.) Elsevier Applied Science Publishers, London, pp. 131–72.

Mulvihill, D.M and Fox, P.F. (1994) Developments in the production of milk proteins, in, *New and Developing Sources of Food Proteins*, (B.J.F. Hudson ed.) Chapman and Hall, London, pp. 1–30.

Mulvihill, D.M and Grufferty, M.B. (1997) Production of whey-protein-enriched products, in *Food Proteins and Lipids*, (S. Damodaran ed.) Plenum Press, New York, pp. 77–93.

Murashov, V.U. and Murashova, L.S. (1996) Manufacture of Milk Protein Concentrate. *USSR Patent SU 1 824 700 A1.*

Murphy, J.M. and Fox, P.F. (1991) Fractionation of sodium caseinate by ultrafiltration. *Food Chem.*, **39**, 27–38.

Ng-Kwai-Hang, K.F. and Chin, D. (1994) Semipreparative isolation of bovine casein components by HPLC chromatography. *Int. Dairy J.*, **4**, 99–100.

Noel, R. (1992) Process for separation of calcium phosphocaseinate and whey from skim milk, and more generally separation of a protein from a biological liquid. *French Patent Application FR 2 671 945 A1.*

Novak, A. (1996) *Application of Membrane Filtration in the Production of Milk Protein Concentrates*, Bulletin 311, International Dairy Federation, Brussels, pp. 26–7.

Otte, J.A.H.J., Kristiansen, K.R., Zakora, M. and Qvist, K.B. (1994) Separation of individual whey proteins and measurement of α-lactalbumin and β-lactoglobulin by capillary zone electrophoresis. *Neth Milk Dairy J.*, **48**, 81–97.

Outinen, M., Harju, M., Tossavainen, O. and Antila, P. (1995a) Process for fractionating whey proteins and components so obtained. *PCT International Patent Application WO 95/19714 A1.*

Outinen, M., Tossavainen, O., Syvaoja, E.-L. and Korhonen, H. (1995b) Chromatographic isolation of kappa-casein macropeptide from cheese whey with a strong basic anion exchange resin. *Milchwissenschaft*, **50**, 570–4.

Outinen, M., Tossavainen, O. and Syvaoja, E.-L. (1996a) Chromatographic fractionation of α-lactalbumin and β-lactoglobulin with polystyrenic strongly basic anion exchange resins. *Lebensm. Wissens. Technol.*, **29**, 340–3.

Outinen, M., Tossavainen, O., Tupasela, T., Koskela, P., Koskinen, H., Rantamaki, P., Syvaoja, E.-L., Antila, P. and Kankare, V. (1996b) Fractionation of proteins from whey with different pilot scale processes. *Lebensm. Wissens. Technol.*, **29**, 411–7.

Pahud, J.J. and Hilpert, H. (1976) Affinity chromatography of lactoferrin on immobilized ferritin. *Protides Biol. Fluids*, **23**, 571–4.

Palmer, D.E. (1982) Recovery of proteins from food factory waste by ion exchange, in, *Food Proteins*, (P.F. Fox and J.J. Condon eds.) Applied Science Publishers, London, pp. 341–52.

Patocka, J. and Jelen, J. (1987) Calcium chelation and other pretreatments for flux improvement in ultrafiltration. *J. Food Sci.*, **52**, 1241–4.

Paul, K.G., Ohlsson, P.I. and Hendriksson, A. (1980) The isolation and some liganding properties of lactoperoxidase. *FEBS Letters*, **110**, 200–4.

Payens, T.A.J. and van Markwijk, B.W. (1963) Some features of the association of β-casein. *Biochim. Biophys. Acta*, **71**, 517–30.

Pearce, R.J. (1983) Thermal separation of β-lactoglobulin and α-lactalbumin in bovine Cheddar cheese whey. *Aust. J. Dairy Technol.*, **38**, 144–9.

Pearce, R.J. (1987) Fractionation of Whey Proteins, Bulletin 212, International Dairy Federation, Brussels, pp. 150–3.

Pearce, R.J. (1992) Whey protein recovery and whey protein fractionation, in, *Whey and Lactose Processing*, (J.G. Zadow ed.) Elsevier Applied Science, London, pp. 271–316.

Pierre, A. and Fauquant, J. (1986) Industrial process for production of purified proteins from whey. *Lait*, **66**, 405–19.

Pouliot, M., Pouliot, Y. and Britten, M. (1996) On the conventional cross-flow microfiltration of skim milk for the production of native phosphocaseinate. *Int. Dairy J.*, **6**, 105–11.

Prieels, J.P. and Peiffer, R. (1986) Process for the purification of proteins from a liquid such as milk. *UK Patent Application GB2, 171, 102, A1.*

Puhan, Z. (1990) Manufacture and use of milk protein concentrates, in *Proceedings of Dairy Products Technical Conference*, American Dairy Products Institute, Chicago, IL, pp. 33–41.

Pyrgaru, Y.M. (1992) Separation of proteins from whey. *USSR Patent SU 1 722 383.*

Ram, S., Loh, D.W., Love, D.C. and Elston, P.D. (1994) A process for producing beta-casein enriched products. *PCT International Patent Application WO 94/06306 A1.*

Rialland, J.-P. and Barbier, J.-P. (1988) Process for selectively separating the α-lactalbumin from the proteins of whey. *United States Patent 4 782 138.*

Robinson, B.P., Short, J.L. and Marshall, K.R. (1976) Traditional lactalbumin manufacture, properties and uses. *N.Z. J. Dairy Sci. Technol.*, **11**, 114–26.

Rosenberg, M. and Young, S.L. (1993) Whey proteins as microencapsulating agents. Microencapsulation of anhydrous milkfat – structure evaluation. *Food Struc.*, **12**, 31–41.

Rosenberg, M. and Sheu, T.-Y. (1996) Microencapsulation of volatiles by spray-drying in whey protein-based wall systems. *Int. Dairy J.*, **6**, 273–84.

Rowan, C. (1998) The whey ahead. *Food Manuf.*, **73**, 19–20.

Saito, T., Yamaji, A. and Itoh, T. (1991) A new isolation method of caseino-glycopeptide from sweet cheese whey. *J. Dairy Sci.*, **74**, 2831–7.

Sanchez, V., Biscans-Milcan, B., Couderc, J.-P. and Barre, P. (1985) Precédé et dispositif pour la separation chromatographique des macromolécules biologiques. *European Patent 0 189 611.*

Sanogo, T., Paquet, D., Aubert, F. and Linden, G. (1989) Purification of α_{s1}-casein by fast protein liquid chromatography. *J. Dairy Sci.*, **72**, 2242–6.

Schmidt, D.G. and van Markwijk, B.W. (1993) Enzymatic hydrolysis of whey proteins. Influence of heat treatment of α-lactalbumin and β-lactoglobulin on their proteolysis by pepsin and papain. *Neth. Milk Dairy J.*, **47**, 15–22.

Schutyser, J.A.J., Buser, T.J.W., van Olden, D. and Overeem, T. (1987) The isolation of proteins from whey with a new strongly acidic silica-based ion exchanger. *J. Liquid Chromatogr.* **10**, 2151–75.

Shahidi, F. and Han, X-Q. (1993) Encapsulation of food ingredients. *CRC Crit. Rev. Food Sci. Nutr.*, **33**, 501–47.

Shimazaki, K. and Nishio, N. (1991) Interacting properties of bovine lactoferrin with immobilised Cibacron Blue F3G-A in column chromatography. *J. Dairy Sci.*, **74**, 404–8.

Shimizu, M. and Nakane, Y. (1995) Encapsulation of biologically active proteins in a multiple emulsion. *Biosci. Biotech. Biochem.*, **59**, 492–6.

Singer, N.S. and Dunn, J.M. (1990) Protein microparticulation: the principle and the process. *J. Amer. Coll. Nutr.*, **9**, 388–97.

Singer, N.S., Yamamoto, S. and Latella, J. (1988) Protein product base. *United States Patent 4 734 287.*

Singh, S., Viotto, L.A. and Rizvi, S.S.H. (1995) Fractionation of skim milk by ceramic membranes: II. Separation of caseins and evaluation of their functional properties. *J. Dairy Sci.*, **78**(Supplement 1), 147.

Skudder, P.J. (1985) Evaluation of a porous silica-based ion-exchange medium for the production of protein fractions from rennet and acid whey. *J. Dairy Res.*, **52**, 167–81.

Slack, A.W., Amundson, C.H. and Hill, C.G. (1986) Nitrogen solubilities of β-lactoglobulin and α-lactalbumin enriched fractions derived from ultrafiltered cheese whey retentate. *J. Food Proc. Preserv.*, **10**, 19–29.

Southward, C.R. (1986) Utilization of milk components: casein, in *Modern Dairy Technology, Volume 1. Advances in Milk Processing*, (R.K. Robinson ed.) Elsevier Applied Science Publishers, London, pp. 317–68.

Southward, C.R. (1989) Use of casein and caseinates, in *Developments in Dairy Chemistry-4-Functional Proteins*, (P.F. Fox ed.) Elsevier Applied Science Publishers, London, pp. 173–244.

Southward, C.R. and Goldman, A. (1978) Co-precipitates and their application in food products. II. Some properties and applications. *N.Z. J. Dairy Sci. Technol.*, **13**, 97–105.

Southward, C.R. and Walker, N.J. (1982) Casein, caseinates and milk protein co-precipitates, in *CRC Handbook of Processing and Utilization in Agriculture, Volume 1. Animal Products*, (I.A. Wolf ed.) CRC Press Inc., Boca Raton, Florida, pp. 445–552.

Spik, G., Montreuil, J., Crichton, R.R. and Mazurier J. (1985) *Proteins of Iron Storage and Transport*, Elsevier Science Publishers, Amsterdam.

Stack, F.M., Hennessy, M., Mulvihill, D.M. and O'Kennedy, B.T. (1995) Process for the fractionation of whey constituents. *European Patent 0765 125 B1.*

Taniguchi, H., Goto, M., Okamoto, T., Sakuchi, I., Ano, T., Kirihara, O. and Ando, K. (1990) Process for preparing a therapeutic agent for rotavirus infection. *European Patent Application 0 391 416 A1.*

Tanimoto, M., Kawasaki, Y., Dosako, S., Ahiko, K. and Nakajima, I. (1992) Large-scale preparation of κ-casein glycomacropeptide from rennet casein whey. *Biosci. Biotech. Biochem.*, **56**, 140–1.

Terre, E., Maubois, J.L., Brulé, G. and Pierre, A. (1986) Precede d'obtention d'une matiere enrichie en caseine beta, appareillage pour la raise en oeuvre de ce precede, et application des produits obtenus par ce precede comme aliments, complements alimentaires ou additifs en industrie alimentaire et pharmaceutique ou dans la preparation de peptides a activite physiologique. *French Patent 2 592 769.*

Thibault, P.A. (1989) Precédé d'élimination sélective et qualitative des lactoglobulines a partir d'une matiere de départ contenant des protéines de lactosérum. *European Patent 0 398 802.*

Thibault, P.A. (1991) Process for the selective and quantitative elimination of lactoglobulins from a starting material containing whey proteins. *United States Patent 5 077 067.*

Thomas, R.L., Cordle, C.T., Criswell, L.G., Westfell, P.H. and Barefoot, S.F. (1992) Selective enrichment of proteins using formed-in-place membranes. *J. Food Sci.*, **57**, 1002–5.

Timmer, J.M.K. and van der Horst, H.C. (1998) Whey processing and separation technology: state-of-the-art and new developments, in *Whey: Proceedings of Second International Whey Conference*, Chicago, IL, International Dairy Federation, Brussels, pp. 40–65.

Tomasula, P.M., Craig Jr, J.C. and Boswell, R.T. (1997) A continuous process for casein production using high-pressure carbon dioxide. *J. Food Eng.*, **33**, 405–19.

Tomasula, P.M., Craig Jr, J.C., Boswell, R.T., Cook, R.D., Kurantz, M.J. and Maxwell, M. (1995) Preparation of casein using carbon dioxide. *J. Dairy Sci.*, **78**, 506–14.

Tossavainen, O., Outinen, M., Harju, M. and Makinen-Kiljunen, S. (1996) Removal of β-lactoglobulin residues from an enzymatic whey protein hydrolysate. *Milchwissenschaft*, **51**, 628–32.

Wang, Q. and Swaisgood, H.E. (1993) Characteristics of β-lactoglobulin binding to the all-*trans*-retinal moiety covalently immobilised on Celite. *J. Dairy Sci.*, **76**, 1895–901.

Ward, L.S. and Bastian, E.D. (1996) A method for isolating β-casein. *J. Dairy Sci.*, **79**, 1332–9.

Yoshida, S. (1988) Isolation of lactoperoxidase of 89,000 Daltons and a globulin of 81,000 Daltons from milk whey. *J. Dairy Sci.*, **71**, 2021–7.

Yoshida, S. (1989) Preparation of lactoferrin by hydrophobic interaction chromatography from milk acid whey. *J. Dairy Sci.*, **72**, 1446–50.

Yoshida, S and Ye-Xiuyun. (1991a) Isolation of lactoperoxidase and lactoferrin from bovine milk acid whey by carboxymethyl cation exchange chromatography. *J. Dairy Sci.*, **74**, 1439–44.

Yoshida, S and Ye-Xiuyun. (1991b) Isolation of lactoperoxidase and lactoferrin from bovine milk rennet whey and acid whey by sulphopropyl cation-exchange chromatography. *Neth. Milk Dairy J.*, **45**, 273–80.

Zadow, J.G. (1986) Utilization of milk components: whey, in, *Modern Dairy Technology, Volume 1. Advances in Milk Processing*, (R.K. Robinson ed.) Elsevier Applied Science Publishers, pp. 273–316.

27

INTERFACIAL, EMULSIFYING AND FOAMING PROPERTIES OF MILK PROTEINS

E. DICKINSON

27.1 INTRODUCTION

The emulsifying and foaming properties of milk proteins are amongst their most important functional properties (Morr, 1982; Mitchell, 1986; Doxastakis, 1989; Mulvihill and Fox, 1989; Phillips *et al.*, 1994; Damodaran, 1997) and the characteristics of most dairy products depend sensitively on how the milk protein fraction is organized at fat–serum and air–serum interfaces (Mulder and Walstra, 1974). Like all soluble food proteins, the caseins and the whey proteins exhibit high surface activity. The main thermodynamic driving force for protein adsorption at hydrophobic surfaces is the removal of non-polar side-chains from the unfavourable environment of the bulk aqueous phase (Dickinson *et al.*, 1988; Dickinson, 1989a; Dickinson and McClements, 1996). Both the caseins and the whey proteins have an especially strong tendency to adsorb at oil–water and air–water interfaces and to protect emulsions and foams against various physico-chemical mechanisms of instability. The relative effectiveness of different samples can vary substantially, however, depending on the exact protein composition, the molecular structures of the main individual protein monomers, the local solution environmental conditions, and the state of aggregation. In the case of milk proteins used as commercial ingredients, the state of aggregation is particularly strongly influenced by the prior processing conditions.

This chapter will review recent advances in understanding the structure and properties of adsorbed milk protein layers and the relationship of the molecular attributes to the physico-chemical mechanisms of emulsion and foam stabilization. The fundamental principles of colloid stabilization by milk proteins are set out in detail elsewhere (Dickinson and Stainsby, 1982;

Advanced Dairy Chemistry Volume 1: Proteins, 3rd edn.
Edited by P.F. Fox and P.L.H. McSweeney, Kluwer Academic/Plenum Publishers, 2003.

Dickinson, 1992a; McClements, 1999) and will not be restated here. Rather, the aim will be to highlight new experimental findings or concepts that build upon standard textbook material or present challenges to current thinking.

Understanding the functional properties of milk proteins in real dairy products is complicated by the considerable compositional and structural complexity of milk itself, and of the variability of commercial ingredients like sodium caseinate, whey protein concentrate, or skim milk powder. Approaching the problem from a fundamental molecular viewpoint, therefore, requires a consideration first of the properties of the isolated individual protein components. In discussing adsorbed layer structure and dynamic interfacial behaviour, it has proved convenient to distinguish between the properties of the disordered flexible caseins and those of the compact globular whey proteins (Dickinson, 1989b, 1992b, 1997a, 1998a; Dalgleish, 1997, 1999). To a first approximation, the caseins may be regarded as charged multiblock copolymers that adsorb to give mobile monolayers with some polymer regions in direct contact with the interface and others protruding into the aqueous phase (Dickinson, 1997a, 1998a, 1999a; Dalgleish, 1999). In contrast, the globular protein layer is probably much better approximated as a dense two-dimensional assembly of highly interacting solid particles with the characteristic mechanical and spectroscopic properties of a thin layer of heat-set globular protein gel (Ball and Jones, 1995; Boerboom *et al.*, 1996; Dickinson, 1999b). We will start with a summary of recent work on β-casein, probably the most widely studied model food protein.

27.2 ADSORBED LAYER STRUCTURE OF PURE PROTEINS

Bovine β-casein has all the key characteristics of an excellent surface-active agent and polymeric stabilizer. It is a flexible linear polyelectrolyte of not too high molecular mass (209 amino acid residues) carrying a moderate net charge ($-15\ e$) at neutral pH. Unlike many other proteins, β-casein has no internal covalent crosslinks and no tendency to polymerize through intermolecular disulfide bonds. Its highly non-uniform distribution of hydrophilic and hydrophobic residues produces a distinctly amphiphilic molecular structure that resembles a simple water-soluble surfactant or a block copolymer. β-Casein self-assembles into roughly spherical surfactant-like micelles above a certain 'critical' concentration (in the range 10^{-2}–10^{-1} % (w/w)) (Schmidt, 1982; Leclerc and Calmettes, 1997). This reversible self-association in water is driven mainly by attractive hydrophobic interactions, and the extent of aggregation increases with temperature and ionic strength (Schmidt and Payens, 1972).

Based on various experimental studies at both hydrophobic solid surfaces (Dalgleish, 1990; Mackie *et al.*, 1991; Brooksbank *et al.*, 1993; Fragneto *et al.*, 1995; Kull *et al.*, 1997) and liquid interfaces (Dalgleish and Leaver,

1991; Dickinson *et al.*, 1993a; Atkinson *et al.*, 1995, 1996; Dalgleish, 1996; Puff *et al.*, 1998), it can be reliably inferred that the β-casein molecule adsorbs with an extensive hydrophobic region anchored directly at the surface and a hydrophilic region (40–50 residues at the N-terminus) protruding extensively into the aqueous phase. This representation has been supported by statistical mechanical modelling of β-casein adsorption at a hydrophobic plane surface using self-consistent field (SCF) theory (Leermakers *et al.*, 1996; Atkinson *et al.*, 1996, 1997; Dickinson *et al.*, 1997a; Dickinson, 1999). In reasonably good agreement with neutron reflectivity experiments on β-casein at air–water and oil–water interfaces (Dickinson *et al.*, 1993a; Atkinson *et al.*, 1995, 1996), the segment density profile predicted by the SCF theory at neutral pH and low ionic strength has a dense inner layer (<2 nm) and an extended outer region with a gradually decreasing segment density reaching *ca.* 1% (w/v) at a distance 10 nm from the surface. Recent X-ray reflectometry experiments at the surface of cow's milk (Holt and White, 1999) are also consistent with a similar interfacial structure dominated by preferentially adsorbed β-casein. On average, the region of the β-casein molecule lying furthest from the surface is the sequence of hydrophilic residues at the N-terminus. Its molecular extremity lies at an average distance of *ca.* 5 nm from the surface at pH 7, and this distance is reduced to *ca.* 3.5 nm at pH 5.5. For a hypothetical variant of β-casein having the 5 phosphoserines replaced by polar (uncharged) residues, the SCF calculation predicts a thinner stabilizing layer (Leermakers *et al.*, 1996). The reduction in thickness of the adsorbed β-casein layer following dephosphorylation has also been confirmed experimentally by neutron reflectivity at the air–water interface (Atkinson *et al.*, 1997) and other interfacial techniques (Husband *et al.*, 1997). These results demonstrate the crucial role of the cluster of charged phosphoserine residues in β-casein in maintaining the steric stabilizing layer, whilst also preventing interfacial precipitation (multilayers). The phosphoserine-rich 'blob' can be considered to be repelled from the surface by electrostatic forces, but restrained from moving into the bulk solution by a stretched 'spring' section anchored strongly to the surface (Brooksbank *et al.*, 1993; Horne and Leaver, 1995). The mobility of the 'blob' has been demonstrated in ^{31}P NMR experiments on β-casein-stabilized emulsions (ter Beek *et al.*, 1996). When the effective electrical charge on the 'blob' is reduced (dephosphorylation) or screened (salt addition), the 'spring' relaxes back to give a thinner adsorbed layer (Horne and Leaver, 1995). Specific binding of calcium ions by the phosphoserines also causes the 'blob' to move towards the surface, again implying a thinning of the layer.

As with its adsorption behaviour, the reversible micelle formation of β-casein in aqueous solution is sensitive to the balance of hydrophobic and electrostatic interactions within the amphiliphilic polymer molecule. Removing just the end 20 residues from the hydrophobic C-terminus destroys the ability of the protein to self-associate in aqueous solution (Berry and Creamer, 1975). The aggregation behaviour of β-casein in solution at concentrations above the critical micelle concentration (CMC) is correlated with adsorbed amounts above monolayer coverage. It is well established (Graham and Phillips, 1979; Nylander and Wahlgren, 1994; Nylander, 1998) that this additional protein is much more weakly bound than the monolayer formed by adsorption below the CMC.

The saturated monolayer coverage of β-casein (and α_{s1}-casein) at various kinds of hydrophobic surfaces— solid–water, oil–water and air–water — typically lies in the range 2–3 mg m^{-2}. [Sodium caseinate also gives a similar saturation coverage on emulsion droplets (Dickinson, 1997a, 1998; Dalgleish, 1997, 1999), but the protein loads for micellar caseins are considerably higher (Haque *et al.*, 1988; Singh *et al.*, 1993; Courthaudon *et al.*, 1999).] The similarity of the spatial dimensions at the different surfaces suggests that in emulsions and foams, the casein molecule does not significantly penetrate the non-aqueous phase (Dalgleish, 1998). Although the β-casein molecule is highly disordered in solution, it appears that some ordered secondary structure (α-helix) is induced on adsorption at the teflon/water interface (Casssens *et al.*, 1999a).

Effective steric stabilization implies that there is a strong repulsive interaction when two protein-coated surfaces come close together (Dickinson and Stainsby, 1982). Direct experimental evidence for long-range repulsive forces between adsorbed β-casein layers has been demonstrated for hydrophobized mica surfaces using an interferometric surface force apparatus (Nylander and Wahlgren, 1997) and for latex particle surfaces using a particle scattering technique (Murray *et al.*, 1998a). This long-range interaction is predicted by SCF theory (Dickinson *et al.*, 1997b) and is consistent also with a scaling theory approach in which β-casein is modelled as a multiblock polymer (Aguié-Béghin *et al.*, 1999).

The polypeptide chain of α_{s1}-casein (199 residues) is marginally shorter than that of β-casein, and at neutral pH it carries a higher net charge ($-22e$). The α_{s1}-casein molecule also has a distinctly amphiphilic character, but the separation of hydrophilic and hydrophobic residues is statistically more random than in β-casein (Dickinson and Matsumura, 1994). The occurrence of hydrophobic residues at both ends of the α_{s1}-casein molecule has certain definite consequences for its adsorption and self-assembly behaviour (Dickinson, 1998a, 1999a; Horne, 1998). At neutral pH and low ionic strength (0.01 M), the protein is nearly monomeric (Schmidt, 1982). Due to charge screening, dimers are formed at ionic strength 0.05 M, octamers at >0.2 M, and the protein is salted out of solution at $\geqslant 0.5$ M. This self-association is driven by a combination of specific electrostatic and hydrophobic interactions (Alaimo *et al.*, 1999). In contrast to the roughly spherical shape of the small β-casein micelles, α_{s1}-casein forms long chain-like aggregates through a series of consecutive association steps (Schmidt, 1982).

As with β-casein, there is strong adsorption of α_{s1}-casein at hydrophobic solid and liquid surfaces to form an effective steric stabilizing monolayer. However, in contrast to the dangling tail predicted for adsorbed β-casein, theory predicts a loop-like conformation for adsorbed α_{s1}-casein (Dickinson *et al.*, 1997a). The absence of the dangling tail with α_{s1}-casein is consistent with the inaccessibility of peptide groups at positions 21–25 to attack by proteolytic enzymes (Shimizu *et al.*, 1986) and with an experimentally determined hydrodynamic layer thickness that is only about a half that of β-casein (Dalgleish, 1993). The predicted maximum average extension from the surface of the loop-like region at around residue number 60 is <3 nm at neutral pH (and even less at pH values closer to the isoelectric point). Electrostatic interaction appears to play a considerable role in relation to the properties of the adsorbed layer of α_{s1}-casein. Dispersed particle surfaces stabilized by α_{s1}-casein are more highly charged than those coated with β-casein (Dalgleish *et al.*, 1985; Dickinson *et al.*, 1987, 1998a). Recent SCF calculations predict (Dickinson *et al.*, 1997b) that the pair interaction potential between α_{s1}-casein-coated surfaces is substantially attractive at ionic strength $\geqslant 0.1$ M, and this appears to be consistent with the much greater sensitivity of α_{s1}-casein-stabilized emulsions to flocculation by salt (Dickinson *et al.*, 1987, 1998a; Casanova and Dickinson, 1998a).

Compared with the caseins, the major whey protein in bovine milk, β-lactoglobulin, gives an adsorbed monolayer that is thinner and of lower protein surface coverage (2–3 nm and <2 mg m^{-2} at neutral pH). Another important difference, as revealed by neutron reflectivity measurements at the air–water interface (Atkinson *et al.*, 1995), is a gradually increasing layer thickness and surface coverage with elapsed time (25% over 4–5 h). These features are attributable to the soft globular character of the native β-lactoglobulin molecule and its unfolding and slow reorganization following adsorption, as detected by Fourier transform infrared spectroscopy (FTIR) (Fang and Dalgleish, 1997) and differential scanning calorimetry (DSC) (Corredig and Dalgleish, 1995). The unfolding and aggregation of β-lactoglobulin after adsorption has some analogy with its behaviour on heating. In a DSC study of β-lactoglobulin adsorbed on emulsion droplets, the protein was found (Corredig and Dalgleish, 1995) to have no distinct thermal transition, suggesting the appropriateness of the term 'surface denaturation' for this whey protein (Dalgleish, 1999). Immediately following adsorption, the β-lactoglobulin monolayer can be regarded as a closely packed monolayer of deformable particles (de Feijter and Benjamins, 1982). This then converts into a two-dimensional gel-like layer (Dickinson, 1999b; Wijmans and Dickinson, 1999b) following strengthening of physical interactions (electrostatic, hydrophobic, hydrogen bonding) and slow covalent crosslinking *via* sulfhydryl-disulfide interchange (Dickinson and Matsumura, 1991).

27.3 Surface rheology of milk proteins

The essential stabilizing function of milk proteins in foams and emulsions is that they enable the fluid interface to resist tangential stresses from the adjoining flowing liquids. Two main types of surface rheology may be distinguished according to whether the protein layer is subjected to a shear deformation or a dilatational deformation. Elastic and mechanical characteristics of adsorbed and spread films at air–water and oil–water interfaces are amongst the most important and widely studied surface properties of proteins (MacRitchie, 1986; Dickinson *et al.*, 1988; Lucassen-Reynders, 1993; Murray and Dickinson, 1996; Murray, 1998; Benjamins and Lucassen-Reynders, 1998). Similar general trends are typically obtained with both adsorbed and spread monolayers (Krägel *et al.*, 1998; Murray *et al.*, 1999).

Milk proteins exhibit a wide range of surface rheological behaviour. Large differences in intermolecular interactions in adsorbed layers of the disordered caseins and the globular whey proteins are especially sensitively reflected in the surface shear rheology (Castle *et al.*, 1987; Dickinson *et al.*, 1990; Benjamins and Voorst Vader, 1992). For instance, at neutral pH, the apparent surface shear viscosity at the hydrocarbon–water interface is some 10^3–10^4 times larger for β-lactoglobulin than for β-casein (Murray and Dickinson, 1996). The highly viscoelastic character of the β-lactoglobulin layer is attributed to the high packing density and strong protein–protein interactions, as compared with the loose packing and weak protein–protein interactions in the β-casein monolayer. The surface viscosity of α_{s1}-casein (or sodium caseinate) is rather higher than that of β-casein, but still much lower than that of α-lactalbumin or β-lactoglobulin (Murray and Dickinson, 1996). The surface viscosity of adsorbed casein is enhanced by conditions which favour casein aggregation in the bulk phase, e.g., by calcium ions (Hunt *et al.*, 1993). Large increases in the surface viscosity of the casein layer can also be induced by introducing intermolecular covalent bonds using the cross-linking enzyme, transglutaminase (Færgemand *et al.*, 1997, 1999; Dickinson, 1997b). The surface shear viscosity of globular protein layers generally falls substantially over the range 0–60°C (Castle *et al.*, 1987), but at higher temperatures it may increase irreversibly due to the formation of new intermolecular covalent crosslinks (Dickinson and Hong, 1994).

The wide range of surface shear viscosity values of the different individual milk proteins can be exploited in investigations of interfacial interactions and competitive adsorption in mixed protein systems (Dickinson *et al.*, 1990). For the α_{s1}-casein + β-casein system, there is a rapid approach to steady-state interfacial composition and surface rheological behaviour (Dickinson *et al.*, 1988; Anand and Damodaran, 1996). For the casein(ate) + gelatin system, the combined surface tension and interfacial rheology data are consistent with a model of a mixed protein film consisting of a primary layer of casein and a secondary layer of gelatin (Dickinson *et al.*, 1985; Galazka and Dickinson, 1995). The exchangeability of adsorbed proteins with the bulk aqueous phase is enhanced by macromolecular mobility and flexibility at the interface (Dickinson and Matsumura, 1994). As there is little reversible exchange between the bulk phase and the closely-packed viscoelastic monolayer formed from globular proteins, competitive adsorption involving adsorbed whey proteins is rather limited (Dickinson *et al.*, 1989, 1990; Cao and Damodaran, 1993). In addition to detecting interfacial interactions in binary protein systems, surface shear viscosity measurement can also be used as a sensitive indicator of interfacial complexation between milk proteins and polysaccharides (Dickinson and Euston, 1991; Dickinson and Galazka, 1992; Dickinson *et al.*, 1998b).

The essential difference between shear deformation and dilatational deformation is that, whereas in the former case the surface concentration remains constant, in the latter case, the surface area is enlarged (or diminished) with the consequence that the surface concentration inevitably changes (Lucassen, 1981). The dilatational rheology of adsorbed protein layers at oil–water and air–water interfaces is relevant to the large changes in surface area that occur during the formation of protein-stabilized emulsions and foams (Walstra, 1983, 1989; Walstra and Smulders, 1997). While β-casein generally gives lower elastic moduli than globular proteins, in comparison with the wide range of shear viscosity values for proteins, the dilatational film properties appear less sensitive to protein structure (Murray and Dickinson, 1996). Also, in contrast to the surface shear properties, the dilatational moduli often appear relatively independent of film ageing.

In a recent study at both air–water and sunflower oil–water interfaces (Lucassen-Reynders and Benjamins, 1999), it was found that both the disordered β-casein and the globular BSA show purely elastic behaviour of the surface dilatational modulus over a considerable range of surface pressures. It was inferred that, at low and moderate surface concentrations, relaxation processes do not play any part over time-scales from 1 to 100 s. This implies that diffusional interchange with the aqueous solution does not occur, and that relaxation processes in the layer take place over shorter time-scales. But relaxation processes do certainly occur over long time-scales (up to 10^5 s) in closely-packed viscoelastic layer interfaces (Lucassen-Reynders and Benjamins, 1999). While the behaviour is qualitatively similar at different fluid interfaces, it has been suggested (Murray *et al.*, 1998b, 1999) that the slightly higher modulus at the hydrocarbon oil–water interface is due to a greater degree of unfolding as a result of better solvency of the oil for the non-polar side-chains of the protein. Lower values of dilatational moduli at the triglyceride oil–water interface have been attributed to less severe non-ideality in the surface equation state (Benjamins and Lucassen-Reynders, 1998).

For closely-packed monolayers of α-lactalbumin or β-lactoglobulin, a substantial dilatational deformation of the interface will not result in a uniform change in monolayer structure. Rather, a much more patchy localized accumulation or depletion of protein will occur, with the thin protein gel layer exhibiting brittle fracture, and distinct tears and holes appearing (Murray and Dickinson, 1996). This type of behaviour has been reproduced recently in a computer simulation of the surface rheology of a globular protein monolayer modelled as a bonded network of adsorbed spherical particles (Wijmans and Dickinson, 1998, 1999a). Formation and breaking of bonds during the compression-expansion cycle of the simulated dilatational deformation leads to a strong element of irreversibility.

27.4 Competitive adsorption and protein–surfactant interactions

In addition to milk proteins, dairy colloids contain small-molecule surfactants such as polysorbates (Tweens), monoglycerides and phospholipids (Euston, 1997). In products such as whipped toppings or ice-cream, the primary role of the small-molecule surfactant (emulsifier) is to enhance the amount of fat destabilization during the whipping of the dairy topping or during the freezing/aeration stage of ice-cream making (Goff and Jordan, 1989; Barfod *et al.*, 1991; Gelin *et al.*, 1994; Goff, 1997a; Pelan *et al.*, 1997). The mechanism of destabilization is related to the displacement of milk proteins from the fat droplet surface, which induces adhesion and partial coalescence (clumping) of fat globules at the surface of incorporated air cells (Goff and Jordan, 1989; Barfod *et al.*, 1991). Hence, the competitive adsorption with surfactants is an important aspect of the interfacial functionality of milk proteins.

Due to their higher surface activity, milk proteins saturate air–water and oil–water interfaces at much lower concentrations than do small-molecule surfactants (Dickinson and Woskett, 1989; Dickinson, 1993b). Hence, protein predominates at the interface when the surfactant concentration is low. However, the more efficient packing in the saturated monolayer leads to a lower interfacial tension for surfactants than for proteins at high concentrations, and, as a result, the protein is invariably competitively displaced from the interface (Dickinson, 1993b; Bos *et al.*, 1997). Generally speaking, water-soluble surfactants are more effective at displacing milk proteins from the oil–water interface than are oil-soluble surfactants (Dickinson, 1995). Interpretation of the behaviour is complicated by various kinds of protein–surfactant interactions, both weak and strong, and specific and non-specific (Bos *et al.*, 1997).

Competitive adsorption of milk proteins and surfactant at macroscopic interfaces has been studied by a range of direct and indirect experimental techniques. Confocal laser scanning microscopy has been used to observe the displacement of sodium caseinate from the oil–water interface by monoglyceride or lecithin (Heertje *et al.*, 1990, 1996). Phase separation and competitive adsorption in β-casein + monoolein layers at the air–water interface has been probed by Brewster angle microscopy (Rodriguez Patino *et al.*, 1999). On the mesoscopic scale, the developing structure of films of β-casein and β-lactoglobulin in the presence of increasing amounts of polyoxyethylene sorbitan monolaurate (Tween 20) has been investigated by atomic force microscopy (Mackie *et al.*, 1999), producing structures that can be modelled successfully by Brownian dynamics simulations (Wijmans and Dickinson, 1999b). Changes in adsorbed amount and layer thickness during displacement of β-casein or β-lactoglobulin from the air–water interface by hexaoxyethylene *n*-dodecyl ether ($C_{12}E_6$) have been studied by specular neutron reflectance (Dickinson *et al.*, 1993b; Horne *et al.*, 1998). Displacement of β-lactoglobulin from the oil–water interface by a water-soluble surfactant leads to a reduction in surface shear viscosity by several orders of magnitude (Courmaudon *et al.*, 199la; Chen and Dickinson, 1995a; Wüstneck *et al.*, 1996; Krägel *et al.*, 1999), as well as a more modest reduction in the dilatational rheological parameters (Murray *et al.*, 1998b). Using the fluorescence recovery after photobleaching (FRAP) technique, large increases have been detected in the mobility of globular protein adsorbed layers in thin films in the presence of water-soluble surfactants (Clark *et al.*, 1990; Clark, 1995).

The presence of small-molecule surfactants affects the emulsifying and stabilizing properties of milk proteins in a number of ways. The more rapid reduction in interfacial tension than with protein alone assists in the formation of smaller droplets during homogenization (Dickinson *et al.*, 1989). Once formed, the stabilizing layer around the droplets may be stronger or weaker depending on the interfacial composition and the nature of the protein–surfactant interactions. For a wide range of pure and commercial milk protein emulsifying agents and different types of surfactants, the competitive adsorption of protein with surfactant during or after emulsification has been found to reduce the protein load on the droplets (de Feijter *et al.*, 1987; Courthaudon *et al.*, 1991b,c,d; Dickinson and Tanai, 1992a,b; Dickinson and Gelin, 1992; Dickinson *et al.*, 1993c; Chen and Dickinson, 1993, 1995c; Dickinson and Hong, 1994; Tomas *et al.*, 1994; Euston *et al.*, 1995, 1996; Dalgleish *et al.*, 1995; Fang and Dalgleish, 1996a,b; Cornec *et al.*, 1996). Typically, the protein surface coverage is inferred from the concentration of unadsorbed protein remaining in the serum phase following centrifugation of the emulsion. In the case of water-soluble non-ionic surfactant, whether the addition occurs before or after emulsification has relatively little influence on the competitive adsorption behaviour (Courthaudon *et al.*, 1991c). For an anionic surfactant like sodium dodecyl sulfate (SDS), which forms interfacial complexes with the adsorbed protein, a much higher concentration of surfactant is required for complete protein displacement (de Feijter *et al.*, 1987; Dickinson and Woskett, 1989; Chen and Dickinson, 1995c). Depending on the pH, the complexation between SDS and milk proteins may lead to an increase or decrease in emulsion stability (Demetriades and McClements, 2000). Generally speaking, the protein surface coverage is lower with triglyceride oil droplets than with hydrocarbon oil droplets, and the amount of surfactant required for complete protein displacement is also lower (Courthaudon *et al.*, 1991d). The concentration of water-soluble surfactant required for partial or complete displacement of protein is reduced by the presence of monoglyceride in the triglyceride oil phase (Dickinson and Tanai, 1992a), but it increases with ageing or heating of the adsorbed layer for the case of β-lactoglobulin-stabilized emulsions (Chen and Dickinson, 1993; Dickinson and Hong, 1994).

Any specific binding of surfactant to protein has an effect on the surface activity of the protein and its competitive adsorption. The β-lactoglobulin + Tween 20 system has received extensive study, with the existence of a 1:1 complex inferred from surface tension data (Coke *et al.*, 1990), diffusion measurements (Clark *et al.*, 1991; Wilde and Clark, 1993) and binding isotherms (Bos *et al.*, 1997). Detailed studies of foam films have indicated an increase in film thickness as the surfactant/protein molar ratio increases from $R = 0.2$ to 0.9 (Clark, 1995). These changes occur prior to the onset of enhanced interfacial mobility at $R = 0.9$, as determined by the FRAP technique (Coke *et al.*, 1990). It is proposed (Bos *et al.*, 1997) that the trapping of Tween 20/β-lactoglobulin complex in adsorbed multilayers can account for this thicknening of the film. Further evidence of direct adsorption of such a complex comes from dynamic surface tension measurements with the overflowing cylinder apparatus (Clark *et al.*, 1993).

The behaviour of milk protein systems containing lecithin is highly complicated. This is due to the sensitivity of the phospholipid interactions and liquid crystalline phase behaviour to small changes in molecular features (head group size and charge, fatty acid length and degree of unsaturation) (Bergenståhl and Claesson, 1990; Bos *et al.*, 1997). In general, it appears that the zwitterionic phospholipids tend to prefer to coexist with milk proteins at surfaces, and are therefore less effective than simple nonionic or anionic surfactants in displacing proteins from the oil–water interface (Courthaudon *et al.*, 1991d; Dickinson and Iveson, 1993; Fang and Dalgleish, 1996a,b). In thin films containing a mixture of caseinate and lecithin, it has been reported (Husband and Wilde, 1998) that the lecithin does not dominate the interface, but simply reduces the strength of protein–protein interactions, thereby inducing surface mobility and allowing stepwise thinning of the caseinate film (Koczo *et al.*, 1996).

The bulk rheological properties of heat-set whey protein emulsion gels are substantially affected by the protein–surfactant interactions and the competitive adsorption (Jost *et al.*, 1989; Xiong and Kinsella, 1991; McClements *et al.*, 1993a; Dickinson and Hong, 1995, 1996, 1997; Dickinson *et al.*, 1996; Dickinson and Yamamoto, 1996a,b; Chen and Dickinson, 1998a,b, 1999a,b; Dickinson and Chen, 1999). Protein-coated droplets function as active filler particles, thereby increasing the overall elastic modulus, whereas surfactant-coated droplets function as inactive fillers which reduce the modulus. The relative magnitude of the effect of the filler depends on various other factors such as oil volume fraction, average particle size, surfactant type and strength of the protein gel matrix. The different effects on emulsion gel rheology of oil-soluble surfactants (e.g., monoglycerides) and water-soluble surfactants (e.g., Tweens) are consistent with their differing competitive adsorption behaviour (Chen and Dickinson, 1993a). The bulk rheological behaviour obtained with lecithin-containing systems is very sensitive to the source and purity of the lecithin sample and whether it has been introduced before or after emulsification (Jost *et al.*, 1989; Dickinson and Yamamoto, 1996a,b).

27.5 TESTING THE EMULSIFYING AND FOAMING PROPERTIES OF PROTEINS

A substantial part of the literature on milk protein functionality involves comparisons of emulsifying and foaming properties of samples that differ in composition and purity, which may also have been subjected to a variety of physical, chemical and biochemical treatments (Haertlé and Chobert, 1999). Among the various kinds of experimental methods used to characterize foams (Wilde and Clark, 1996) and, especially, emulsions (Hill, 1996), it is convenient to distinguish between the fundamental physico-chemical measurements and the simple empirical tests. The former give insight into specific mechanisms of (in)stability with some aspirations for broad absolute validity. In contrast, the main purpose of the empirical test is to provide a useful relative yardstick for comparing different samples or conditions within the same local study.

As described in detail elsewhere (Dickinson, 1992a; McClements, 1999), most fundamental physico-chemical measurements on emulsions involve determinations of the droplet-size distribution immediately after emulsification, and then as a function of time following a well-defined period of storage, shearing, processing, etc. Furthermore, measurements are typically made of the rate of creaming (or serum separation), the degree of flocculation and various rheological properties. The basic approach to studying foams tends to be more indirect because of the great difficulty of measuring bubble-size distributions *in situ*. Hence, fundamental studies of foam stability involve determinations of foam drainage kinetics, the thinning and rupture of isolated liquid films, and the coalescence and disproportionation of individual bubbles. As important as the method used to characterize the emulsion or foam is the equipment employed to produce it. In fundamental studies, oil-in-water emulsions are normally made by some kind of high-pressure recirculating valve device (Haque and Kinsella, 1989a) or single-pass jet homogenizer (Burgaud *et al.*, 1990), with the most recent research particularly directed towards new Microfluidization™ technology operating at ultra-high pressures (Dalgleish *et al.*, 1996, 1997; Leaver *et al.*, 1999; Paquin, 1999). In fundamental laboratory studies on foams, bubbles are generated by sparging through a sinter or a small number of fixed-size orifices. Controlling the average droplet size and degree of polydispersity in fine emulsions is easier to achieve than controlling bubble size and polydispersity in coarse foams.

Many empirical studies involve preparing protein-stabilized emulsions and foams by high-speed mixing. The simplest emulsification test involves determining the emulsifying capacity (EC) as the point of emulsion breaking/inversion on gradual addition of oil to a vigorously mixed protein solution. However, the most widely adopted test for comparing the emulsifying properties of food proteins is determination of the so-called emulsifying activity index (EAI) from a turbidity measurement following emulsion dilution in aqueous SDS solution (to dissociate flocculated droplets) (Pearce and Kinsella, 1978). In refinements of the EAI test, turbidimetry can be compared with or replaced by light-scattering determination of the average droplet size (Haque and Kinsella, 1989b; Relkin *et al.*, 1993). In the case of foam whipping tests, reproducibility between different laboratories is especially difficult to achieve (Phillips *et al.*, 1990b). Foamability (or overrun) is most often expressed simply in terms of the foam volume generated from a solution of known concentration during a defined period of foam generation, and then foam stability in terms of the change in the foam volume with time (Wilde and Clark, 1996). Also popular as an empirical measure of foam stability is the time-dependent decay of the electrical conductivity (Kato *et al.*, 1983; Wright and Hemmant, 1987).

Two of the most significant factors affecting the foaming properties of milk proteins are pH and ionic strength (Mulvihill and Fox, 1989; Phillips *et al.*, 1994). The chemical nature of any added salts is also important (Phillips *et al.*, 1991). Foam stability is especially enhanced near the isoelectric point, e.g., at pH $\sim$ 5 for whey protein isolate (Phillips *et al.*, 1990a). This correlates with increased film strength for the more aggregated protein state, as does moderate heat treatment, but heat-induced denaturation/aggregation of whey protein gives a lower overrun by reducing the amount of protein available for new film formation during whipping. This same general principle applies to the emulsifying properties of whey proteins subjected to prior heat treatment (Das and Kinsella, 1990) or high-pressure treatment (Galazka *et al.*, 1995, 1996; Dickinson and James, 1998). Generally speaking, the more aggregated the protein is, the more inefficient it is in forming a coherent stabilizing film at the oil–water interface, and hence the poorer its emulsifying efficiency. In foams containing a mixture of monomeric and polymeric whey proteins, it has been demonstrated (Zhu and Damodaran, 1994) that the monomeric protein contributes positively to foamability whereas the polymeric species contribute to foam stability. Conditions that favour protein aggregation also tend to favour emulsion flocculation, which in turn implies reduced stability with respect to creaming (Dickinson, 1988) and also to coalescence in shear flow (Dickinson and Williams, 1994).

Comparison of the properties of peptide fragments with those of native milk proteins can provide useful insight into the relationship between molecular structure and interfacial function. Emulsifying and foaming properties have been reported recently for the well-defined β-casein peptides produced by plasmin hydrolysis (Caessens *et al.*, 1997, 1999b,c; Smulders *et al.*, 1999). The emulsion-forming properties of amphiphilic and hydrophobic peptides were found to be comparable to those of β-casein. However, the differences in emulsion stability are much greater, with emulsions containing hydrophobic peptides being very susceptible to bridging flocculation and those containing amphiphilic peptides exhibiting extensive coalescence (Smulders *et al.*, 1999). The results are largely consistent with the generally accepted structure of adsorbed β-casein, with the hydrophilic N-terminus being especially important for the stability-related properties (Dickinson, 1997a, 1998).

The emulsion forming and stabilizing properties of hydrolyzates made by treating milk proteins with trypsin have been studied (Agboola and Dalgleish, 1996a,b; Huang *et al.*, 1996; Swaisgood *et al.*, 1996). Sodium caseinate has the same rate of hydrolysis at the surface or in solution, but β-lactoglobulin is hydrolyzed much faster in the adsorbed state. Based on particle size analysis, emulsions made with hydrolyzed caseinate and β-lactoglobulin were found to be less stable on storage than those formed with unmodified proteins, with or without subsequent hydrolysis (Agboola and Dalgleish, 1996a). On the other hand, proteolysis leads to an improvement in the emulsifying properties of milk fat globule membrane (Corredig and Dalgleish, 1997). The enzyme treatment renders caseinate-stabilized emulsions less stable to reduction in pH or alcohol addition, but more stable to calcium ions (Agboola and Dalgleish, 1996b). There is an optimum degree of proteolysis which is dependent on the protein structure and the particular physico-chemical property being determined (Lieske and Konrad, 1996). With commercial whey protein hydrolyzates of varying degree of hydrolysis, it has been found that maximum emulsifying capacity occurs in the range 10–20% hydrolysis, with more extensive hydrolysis producing peptides too short to act as effective emulsifiers (Singh and Dalgleish, 1998).

Polymerization with the crosslinking enzyme, transglutaminase, has a considerable influence on the emulsion-stabilizing properties of milk proteins (Dickinson and Yamamoto, 1996c; Yildirim *et al.*, 1996; Dickinson, 1997a, 1999c; Chanyongvorakul *et al.*, 1997; Færgemand *et al.*, 1998;

Dickinson *et al.*, 1999b). In experiments with crosslinked β-casein polymers of increasing degree of polymerization, it was found that there was a reduction in the emulsifying activity index (EAI) but an improvement in the storage stability (Liu and Damodaran, 1999). Treating a concentrated liquid-like protein-stabilized emulsion with transglutaminase converts it into a solid-like protein-stabilized emulsion gel (Dickinson and Yamamoto, 1996c). With a β-lactoglobulin-stabilized system, the permanence of the 'chemical' crosslinks formed between the adsorbed protein molecules on different droplets gives this emulsion gel a very different rheological character from the equivalent heat-set emulsion gel. Another way in which the stability and rheology of milk protein-based emulsions and foams can be affected by covalent protein–protein crosslinking is *via* the formation of disulfide bonds (Okumura *et al.*, 1990; Monahan *et al.*, 1993, 1996; McClements *et al.*, 1993; Damodaran and Anand, 1997). [Conversely, the *breaking* of internal disulfide bonds increases the unfolding ability of the whey protein molecules, with consequent improvement in foaming and emulsifying properties (Kella *et al.*, 1989)]. Conjugation with polysaccharides can also lead to substantial improvements in the stabilizing properties of milk proteins (Dickinson and Galazka, 1991a; Kato *et al.*, 1992; Dickinson and Izgi, 1996; Nagasawa *et al.*, 1996).

A large variety of commercial milk protein preparations are available (Jost, 1993), but the functional properties of different samples of commercial whey protein concentrates are very variable (Morr, 1982). Probably the two most important factors in relation to emulsifying and foaming properties are the solubility (i.e., aggregation state) and the lipid content (Karlskind *et al.*, 1996). While commercial samples of sodium caseinate tend to be less variable in emulsifying and foaming properties than commercial whey proteins, differences in product quality can still be directly traced to variations in the sources of commercial caseinates (Lynch and Mulvihill, 1997). Caseinate fractions enriched with β-casein exhibit improved emulsifying capacity but reduced film stability, whereas the opposite is the case with caseinate fractions enriched with α_s-/κ-caseins (Murphy and Fox, 1991). Calcium-enriched caseinates give higher protein loads due to the more aggregated state of the protein (Mulvihill and Murphy, 1991). Functional properties of sodium caseinate made by extruding skim milk powder appear relatively unaffected by the processing conditions (Britten *et al.*, 1993). However, extensive heating of sodium caseinate solutions at $\geqslant$120°C does cause changes in the emulsifying and foaming properties (Guo *et al.*, 1996).

27.6 Flocculation, creaming and rheology of emulsions

Milk protein-stabilized emulsions are susceptible to various types of flocculation, with important implications for creaming and rheology.

Flocculation is the aggregation of droplets without disruption of the protective stabilizing layer at the interface. For milk fat globules, this definition includes clustering but not clumping (Dickinson and Stainsby, 1982). Flocculation may be reversible or irreversible, depending on the strength of the interactions between the droplets. The non-Newtonian rheological behaviour of protein-stabilized emulsions is invariably attributable to flocculation (Dickinson, 1998b), except at very high volume fractions (Hemar and Horne, 1999). There are several different mechanisms of flocculation (Dickinson, 1993a, 1997a); these will be discussed in turn.

Destabilization by flocculation is commonly induced by lowering the pH towards the isoelectric point, p*I*, or by addition of calcium ions. Lowering the net charge on the adsorbed protein leads to a reduction in the surface charge density and hence loss of electrostatic stabilization, possibly accompanied by loss of steric stabilization also as a consequence of the collapse of charged stabilizing tails. When the pH is lowered through the p*I* to pH 3.5, whey protein-stabilized emulsions become stable again, but caseinate emulsions remain flocculated even though pH 3.5 is well below p*I* (Hunt and Dalgleish, 1995; Agboola and Dalgleish, 1996c). Addition of calcium ions can lead to strong crosslinking between adsorbed β-casein layers (Velev *et al.*, 1998). Dispersed systems stabilized by α_{s1}-casein, β-casein or sodium caseinate are highly susceptible to flocculation by calcium ions (Dickinson *et al.*, 1987; Agboola and Dalgleish, 1995; Dickinson and Davies, 1999). This is not unexpected given the well-known calcium-sensitivity of these caseins in solution. Rather less obviously predictable, however, is the observation of high calcium ion sensitivity for β-lactoglobulin-stabilized emulsions (Agboola and Dalgleish, 1996d), since the native globular whey protein (unlike α_{s1}- or β-casein) is not precipitated by ionic calcium in aqueous solution (Dalgleish, 1997). Electrostatic stabilization is destroyed by addition of excess electrolyte which screens out the double-layer repulsion between the charged protein-coated emulsion droplets (Dickinson and Stainsby, 1982). Such salt-induced flocculation occurs, for instance, when NaCl is added to *ca.* 0.2 M to an α_{s1}-casein-stabilized emulsion (Dickinson *et al.*, 1987, 1998a; Casanova and Dickinson, 1998). The presence of electrolytes also affects the amount adsorbed and the competitive adsorption in milk protein-stabilized emulsions (Hunt and Dalgleish, 1996; Srinivasan *et al.*, 1999).

Effective steric protection of polymer-coated droplets requires a good quality solvent for the polymeric stabilizer. Under poor quality solvent conditions, the entropic stabilizing repulsion of the adsorbed protein layer becomes a destabilizing steric *attraction* (Dickinson and Stainsby, 1982). Lowering of solvent quality occurs, for instance, when ethanol is present in the aqueous phase (Agboola and Dalgleish, 1996c). Another condition for effective steric stabilization of emulsion droplets is the presence of sufficient protein emulsifier to cover fully the oil–water interface during emulsification. Where this condition is not satisfied, flocculation may occur due to protein bridging between droplets (clustering) (Dickinson *et al.*, 1997c; Rientjes and Walstra, 1993). This effect may be enhanced by competitive adsorption of protein during emulsification (Dickinson *et al.*, 1989; Dickinson and Galazka, 1991b). Bridging flocculation can also arise *after* emulsion formation when adsorbed protein interacts with surfactant (Chen and Dickinson, 1995b) or polysaccharide (Dickinson and Pawlowsky, 1997), or as a result of slow disulfide bond formation between whey protein molecules adsorbed on different droplets (McClements *et al.*, 1993b).

In addition to all these surface-related phenomena, there is also the kind of emulsion flocculation behaviour that is induced by non-adsorbed species such as micelles, small aggregates and macromolecules. This weak and reversible aggregation, called depletion flocculation (Dickinson, 1992a), has its entropic origin in the exclusion of non-adsorbed species from the narrow gap between closely approaching droplet surfaces. Depletion flocculation of milk protein-stabilized emulsions is caused by the presence of various kinds of excess soluble material in the aqueous continuous phase—caseinate sub-micelles (Dickinson and Golding, 1997a), surfactant micelles (Dickinson *et al.*, 1999a) or polysaccharides (Cao *et al.*, 1991; Tuinier and de Kruif, 1999).

Flocculation, by whatever mechanism, typically leads to reduced creaming stability, except in concentrated emulsions where the formation of a coherent gel-like network structure can actually have a stabilizing influence (Dickinson, 1998a). Whereas rapid creaming of coarse or dilute emulsions can often be detected visually, the quantitative evaluation of the slow creaming of fine or concentrated emulsions requires the use of a non-invasive technique such as nuclear magnetic resonance imaging (Kauten *et al.*, 1991; Newling *et al.*, 1997; McDonald *et al.*, 1999) or ultrasonic velocity scanning (Howe *et al.*, 1986; Cao *et al.*, 1991; Gouldby *et al.*, 1991; Fillery-Travis *et al.*, 1993; Dickinson *et al.*, 1997c). In addition to measuring the rate of creaming or serum separation, these instrumental methods can also give supplementary structural information on the developing cream layer, such as the height-dependent oil volume fraction and local droplet-size distributions.

Studies of fine caseinate-stabilized oil-in-water emulsions of constant average droplet size prepared at neutral pH with different amounts of protein have shown creaming stability that is strongly dependent on the concentration of unadsorbed protein (Dickinson *et al.*, 1997c; Dickinson and Golding, 1997a). In moderately dilute aqueous solution, sodium caseinate can be considered to exist as small spherical sub-micellar particles (10–20 nm) in equilibrium with free casein molecules (Creamer and Berry, 1975). [Recent experiments suggest the formation of rod-like caseinate aggregates at high concentrations (Farrer and Lips, 1999)]. The existence of these putative spherical caseinate sub-micelles has been used to interpret the formation of layered structures in thin aqueous films (Koczo *et al.*, 1996). It can be estimated that these very small particles have the potential to generate an appreciable depletion force between sub-micron emulsion droplets at a concentration of a few per cent by volume (Dickinson, 1997c). In support of this prediction, it has been observed that a 35% (w/v) oil-in-water emulsion made with 2% (w/w) caseinate (<1% (w/w) free protein) has good creaming stability over a period of several weeks, whereas the equivalent emulsion made with 4% (w/w) caseinate (2–3%, w/w, free protein) exhibits extensive serum separation within 24 h (Dickinson and Golding, 1997a). Further evidence for flocculation is provided by light microscopy (Dickinson *et al.*, 1997c) and rheological measurement (Dickinson and Golding, 1997b). This depletion effect is completely reversible towards dilution of the aqueous phase and is not accompanied by droplet coalescence. The state of flocculation is strongly influenced by the concentrations of other species such as ethanol, calcium ions and surfactants (Dickinson and Golding, 1998a,b; Dickinson *et al.*, 1999a). Evidence for depletion forces between β-casein-coated emulsion droplets has also been inferred using a new magnetic chaining technique (Dimitrova and Leal-Calderon, 1999).

The rheology of a concentrated emulsion (45% (w/v) oil) made from a binary mixture of α_{s1}-casein + β-casein containing excess unadsorbed protein depends in a sensitive way on temperature and protein composition (Casanova and Dickinson, 1998b; Dickinson *et al.*, 1999c). This behaviour is the result of the complex, and still poorly understood, self-assembly properties of mixed caseins (Horne, 1998). Particularly noteworthy is the observation that replacement of just 5% of the α_{s1}-casein by β-casein in a pure α_{s1}-casein-stabilized emulsion leads to an increase in the low-stress apparent viscosity by two orders of magnitude (Dickinson *et al.*, 1999c). A correlation has been identified between the casein composition corresponding to the maximum viscosity and that corresponding to maxima in surface coverage and mean droplet size. All three maxima are indicative of some sort of α_{s1}-casein–β-casein complexation, as has previously been independently invoked (Cayot *et al.*, 1991) to explain the reduced EAI of binary mixtures of α_{s1}-casein + β-casein as compared with the individual proteins. By tuning the ionic calcium content appropriately, these mixed α_{s1}/β-casein emulsions can be made to exhibit strongly temperature-dependent rheology over the range 30–40°C. By adjusting the calcium ion content in the range 15–30 mM (depending on the pH), thermoreversible gelation can also be made to occur in the pH range 5.5–6.8 over this same narrow temperature interval using commercial sodium casein caseinate as emulsifier (Dickinson and Casanova, 1999).

So long as sufficient protein is present to cover fully the droplet surface, fine milk protein-stabilized emulsions normally retain good stability towards coalescence during extended quiescent storage, even in the presence of competitively adsorbing surfactants. On the other hand, the coalescence of droplets (clumping of fat globules) can be induced readily by subjecting the same emulsions to moderate or intense shearing. The shear stability is sensitively dependent on the type and concentration of the adsorbed milk protein (Segall and Goff, 1999), and the destabilization mechanism probably often involves shear-induced flocculation as an intermediate stage in the process (Dickinson and Williams, 1994). Oil droplets containing fat crystals are especially susceptible to (partial) coalescence when the emulsion is agitated or stirred (Boode and Walstra, 1993a,b; Boode *et al.*, 1993). This aggregation process is important in the phase inversion of cream (Campbell *et al.*, 1996) and in the manufacture of ice-cream (Gelin *et al.*, 1994; Goff, 1997a,b; Pelan *et al.*, 1997). In the absence of fat crystals, the destabilization rate of oil droplets is controlled by the properties of the protective stabilizing layer around the droplets. The situation is complicated by the fact that aggregates induced at low shear rates may be disrupted at high shear rates. For sodium caseinate-based emulsions, the orthokinetic stability is especially sensitive to calcium ions, with the critical Ca^{2+} concentration depending on whether the calcium salt has been introduced into the system before or after homogenization (Chen *et al.*, 1993; Agboola and Dalgleish, 1996d; Schokker and Dalgleish, 1998; Dickinson and Davies, 1999).

References

Agboola, S.O. and Dalgleish, D.G. (1995) Calcium-induced destabilization of oil-in-water emulsions stabilized by caseinate or by β-lactoglobulin. *J. Food Sci.*, **60**, 399–404.

Agboola, S.O. and Dalgleish, D.G. (1996a) Enzymatic hydrolysis of milk proteins used for emulsion formation. 1. Kinetics of protein breakdown and storage stability of the emulsions. *J. Agric. Food Chem.*, **44**, 3631–36.

Agboola, S.O. and Dalgleish, D.G. (1996b) Enzymatic hydrolysis of milk proteins used for emulsion formation. 2. Effects of calcium, pH, and ethanol on the stability of the emulsions. *J. Agric. Food Chem.*, **44**, 3637–42.

Agboola, S.O. and Dalgleish, D.G. (1996c) Effects of pH and ethanol on the kinetics of destabilization of oil-in-water emulsions containing milk proteins. *J. Sci. Food Agric.*, **72**, 448–54.

Agboola, S.O. and Dalgleish, D.G. (1996d) Kinetics of the calcium-induced instability of oil-in-water emulsions: studies under shearing and quiescent conditions. *Lebensm. Wiss. Technol.*, **29**, 425–32.

Aguié-Béghin, V., Leclerc, E., Daoud, M. and Douillard, R. (1999) Asymmetric multiblock copolymers at the gas–liquid interface: phase diagram and surface pressure. *J. Colloid Interface Sci.*, **214**, 143–55.

Alaimo, M.H., Wickham, E.D. and Farrell, H.M., Jr. (1999) Effect of self-association of α_{s1}-casein and its cleavage fractions α_{s1}-casein (136–196) and α_{s1}-casein (1–197) on aromatic circular dichroic spectra: comparison with predicted models. *Biochim. Biophys. Acta*, **1431**, 395–409.

Anand, K. and Damodaran, S. (1996) Dynamics of exchange between α_{s1}-casein and β-casein during adsorption at air–water interface. *J. Agric. Food Chem.*, **44**, 1022–8.

Atkinson, P.J., Dickinson, E., Horne, D.S. and Richardson, R.M. (1995) Neutron reflectivity of adsorbed β-casein and β-lactoglobulin at the air–water interface. *J. Chem. Soc. Faraday Trans.*, **91**, 2847–54.

Atkinson, P.J., Dickinson, E., Horne, D.S., Leermakers, F.A.M. and Richardson, R.M. (1996) Theoretical and experimental investigations of adsorbed protein structure at a fluid interface. *Ber. Bunsen-Ges. Phys. Chem.*, **100**, 994–8.

Atkinson, P.J., Dickinson, E., Horne, D.S., Leaver, J., Leermakers, F.A.M. and Richardson, R.M. (1997) β-Casein adsorbed layer structures predicted by self-consistent-field modelling: comparison with experiment. In *Food Colloids: Proteins, Lipids and Polysaccharides*, (E. Dickinson and B. Bergenståhl eds.) Royal Society of Chemistry, Cambridge, pp. 217–28.

Ball, A. and Jones, R.A.L. (1995) Conformational changes in adsorbed proteins. *Langmuir*, **11**, 3542–8.

Barfod, N.M., Krog, N., Larsen, G. and Buchheim, W. (1991) Effects of emulsifiers on protein/fat interaction in ice-cream mix during ageing. 1. Quantitative analyses. *Fat. Sci. Technol.*, **93**, 24–9.

Benjamins, J. and van Voorst Vader, F. (1992) The determination of the surface shear properties of adsorbed protein layers. *Colloids Surf.*, **65**, 161–74.

Benjamins, J. and Lucassen-Reynders, E.H. (1998) Surface dilational rheology of proteins adsorbed at air–water and oil–water interfaces. In *Proteins at Liquid Interfaces*, (D. Möbius and R. Miller eds.) Elsevier, Amsterdam, pp. 341–84.

Bergenståhl, B.A. and Claesson, P.M. (1990) Surface forces in emulsions. In *Food Emulsions*, 2nd edn, (K. Larsson and S.E. Friberg eds.) Marcel Dekker, New York, pp. 41–96.

Berry, G.P. and Creamer, L.K. (1975) The association of bovine β-casein. *Biochemistry*, **14**, 3542–5.

Boerboom, F.J.G., de Groot-Mostert, A.E.A., Prins, A. and van Vliet, T. (1996) Bulk and surface rheological behaviour of aqueous milk protein solutions: a comparison. *Neth. Milk Dairy J.*, **50**, 183–98.

Boode, K. and Walstra, P. (1993a) Partial coalescence in oil-in-water emulsions. 1. Nature of the aggregation. *Colloids Surf. A*, **81**, 121–37.

Boode, K. and Walstra, P. (1993b) Kinetics of partial coalescence in oil in-water emulsions. In *Food Colloids and Polymers: Stability and Mechanical Properties*, (E. Dickinson and P. Walstra eds.) Royal Society of Chemistry, Cambridge, pp. 23–30.

Boode, K., Walstra, P. and de Groot-Mostert, A.E.A. (1993) Partial coalescence in oil-in-water emulsions. 2. Influence of the properties of the fat. *Colloids Surf. A*, **81**, 139–51.

Bos, M., Nylander, T., Arnebrant, T. and Clark, D.C. (1997) Protein/emulsifier interactions. In *Food Emulsifiers and their Applications*, (L. Hasenhuettl and R.W. Hartel eds.) Chapman and Hall, New York, pp. 173–210.

Britten, M., Bastrash, S., Fortin, J. and Fichtali, J. (1993) Emulsifying and foaming properties of sodium caseinate obtained from extrusion process. *Milchwissenschaft*, **48**, 303–6.

Brooksbank, D.V., Davidson, C.M., Horne, D.S. and Leaver, J. (1993) Influence of electrostatic interactions on β-casein layers adsorbed on polystyrene latices. *J. Chem. Soc. Faraday Trans.*, **89**, 3419–25.

Burgaud, I., Dickinson, E. and Nelson, P.V. (1990) An improved high-pressure homogenizer for making fine emulsions on a small scale. *Int. J. Food Sci. Technol.*, **25**, 39–46.

Caessens, P.W.J.R., Gruppen, H., Visser, S., van Aken, G.A. and Voragen, A.G.J. (1997) Plasmin hydrolysis of β-casein: foam and emulsion properties of the fractionated hydrolysate. *J. Agric. Food Chem.*, **45**, 2935–41.

Caessens, P.W.J.R., de Jongh, H.H.J., Norde. W. and Gruppen, H. (1999a) The adsorption-induced secondary structure of β-casein and of distinct parts of its sequence in relation to foam and emulsion properties. *Biochim. Biophys. Acta*, **1430**, 73–83.

Caessens, P.W.J.R., Gruppen, H., Slangen, C.J., Visser, S. and Voragen, A.G.J. (1999b) Functionality of β-casein peptides: importance of amphipathicity for emulsion stabilizing properties. *J. Agric. Food Chem.*, **47**, 1856–62.

Caessens, P.W.J.R., Visser, S., Gruppen, H., van Aken, G.A. and Voragen, A.G.J. (1999c) Emulsion and foam properties of plasmin derived β-casein peptides. *Int. Dairy J.*, **9**, 347–51.

Campbell, I., Norton, I. and Morley, W. (1996) Factors controlling the phase inversion of oil-in-water emulsions. *Neth. Milk Dairy J.*, **50**, 167–82.

Cao, Y. and Damodaran, S. (1995) Co-adsorption of β-casein and bovine serum albumin at the air/water interface from a binary mixture. *J. Agric. Food Chem.*, **43**, 2567–73.

Cao, Y., Dickinson, E. and Wedlock, D.J. (1991) Influence of polysaccharides on the creaming of casein-stabilized emulsions. *Food Hydrocoll.*, **5**, 443–54.

Casanova, H. and Dickinson, E. (1998a) Influence of protein interfacial composition on salt stability of mixed casein emulsions. *J. Agric. Food Chem.*, **46**, 72–6.

Casanova, H. and Dickinson, E. (1998b) Rheology and flocculation of oil-in-water emulsions made with mixtures of α_{s1}-casein + β-casein. *J. Colloid Interface Sci.*, **206**, 82–9.

Castle, J., Dickinson, E., Murray, B.S. and Stainsby, G. (1987) Mixed protein films adsorbed at the oil–water interface. *ACS Symp. Ser.*, **343**, 118–34.

Cayot, P., Courthaudon, J.-L. and Lorient, D. (1991) Emulsifying properties of pure and mixed α_{s1}-casein and β-casein fractions—effects of chemical glycosylation. *J. Agric Food Chem.*, **39**, 1369–73.

Chanyongvorakul, Y., Matsumura, Y., Sawa, A., Nio, N. and Mori, T. (1997) Polymerization of β-lactoglobulin and bovine serum albumin at oil–water interfaces in emulsions by transglutaminase. *Food Hydrocoll.*, **11**, 449–55.

Chen, J. and Dickinson, E. (1993) Time-dependent competitive adsorption of milk proteins and surfactants in oil-in-water emulsions. *J. Sci. Food Agric.*, **62**, 283–9.

Chen, J. and Dickinson E. (1995a) Surface shear viscosity and protein–surfactant interactions in mixed protein films adsorbed at the oil–water interface. *Food Hydrocoll.*, **9**, 35–42.

Chen, J. and Dickinson E. (1995b) Protein/surfactant interfacial interactions. 1. Flocculation of emulsions containing mixed protein + surfactant. *Colloids Surf. A*, **100**, 255–65.

Chen, J. and Dickinson E. (1995c) Protein/surfactant interfacial interactions. 3. Competitive adsorption of protein + surfactant in emulsions. *Colloids Surf. A*, **101**, 77–85.

Chen, J. and Dickinson E. (1998a) Viscoelastic properties of protein-stabilized emulsions: effect of protein–surfactant interactions. *J. Agric. Food Chem.*, **46**, 91–7.

Chen, J. and Dickinson E. (1998b) Protein/emulsifier interactions and the viscoelasticity of whey protein emulsion gels. In *Emulsifiers: Functionality and Applications*, (K. Berger and R.J. Hamilton eds.) Society of the Chemical Industry, London, pp. 19–24.

Chen, J. and Dickinson E. (1999a) Effect of surface character of filler particles on rheology of heat-set whey protein emulsion gels. *Colloids Surf. B*, **12**, 373–81.

Chen, J. and Dickinson E. (1999b) Effect of monoglycerides and diglycerol esters on viscoelasticity of heat-set whey protein emulsion gels. *Int. J. Food Sci. Technol.*, **34**, 493–501.

Chen, J., Dickinson, E. and Iveson, G. (1993) Interfacial interactions, competitive adsorption and emulsion stability. *Food Struct.*, **12**, 135–46.

Clark, D.C. (1995) Application of state-of-the-art fluorescence and interferometric techniques to study coalescence in food dispersions. In *Characterization of Food: Emerging Methods*, (A. Gaonkar ed.) Elsevier, Amsterdam, pp. 23–57.

Clark, D.C., Coke, M., Mackie, A.R., Finder, A.C. and Wilson, D.R. (1990) Molecular diffusion and thickness measurements of protein-stabilized thin liquid films. *J. Colloid Interface Sci.*, **138**, 207–18.

Clark, D.C., Coke, M., Wilde, P.J. and Wilson, D.R. (1991) Molecular diffusion at interfaces and its relation to disperse phase stability. In *Food Polymers, Gels and Colloids*, (E. Dickinson ed.) Royal Society of Chemistry, London, pp. 272–8.

Clark, D.C., Wilde, P.J., Bergink-Martens, J.M., Kokelaar, A.J.J. and Prins, A. (1993) Surface dilational behaviour of a mixture of aqueous β-lactoglobulin and Tween 20 solutions. In *Food Colloids and Polymers: Stability and Mechanical Properties*, (E. Dickinson and P. Walstra eds.) Royal Society of Chemistry, Cambridge, pp. 354–64.

Coke, M., Wilde, P.J., Russell, E.J. and Clark, D.C. (1990) The influence of surface composition and molecular diffusion on the stability of foams formed from protein/detergent mixtures. *J. Colloid Interface Sci.*, **138**, 489–504.

Cornec, M., Mackie, A.R., Wilde, P.J. and Clark, D.C. (1996) Competitive adsorption of β-lactoglobulin and β-casein with Span 80 at the oil–water interface and the effects on emulsion behaviour. *Colloids Surf. A*, **114**, 237–44.

Corredig, M. and Dalgleish, D.G. (1995) A differential microcalorimetric study of whey proteins and their behaviour in oil-in-water emulsions. *Colloids Surf. B*, **4**, 411–22.

Corredig, M. and Dalgleish, D.G. (1997) Studies on the susceptibility of membrane-derived proteins to proteolysis as related to changes in their emulsifying properties. *Food Res. Int.*, **30**, 689–97.

Courthaudon, J.-L., Dickinson, E., Matsumura, Y. and Clark, B.C. (1991a) Competitive adsorption of β-lactoglobulin + Tween 20 at the oil–water interface. *Colloids Surf.*, **56**, 293–300.

Courthaudon, J.-L., Dickinson, E. and Dalgleish, D.G. (1991b) Competitive adsorption of β-casein and non-ionic surfactants in oil-in-water emulsions. *J. Colloid Interface Sci.*, **145**, 390–5.

Courthaudon, J.-L., Dickinson, E., Matsumura, Y. and Williams, A. (1991c) Influence of emulsifiers on the competitive adsorption of whey proteins in emulsions. *Food Struct.*, **10**, 109–15.

Courthaudon, J.-L., Dickinson, E. and Christie, W.W. (1991d) Competitive adsorption of lecithin and β-casein in oil-in-water emulsions. *J. Agric. Food Chem.*, **39**, 1365–8.

Courthaudon, J.-L., Girardet, J.-M., Campagne, S., Rouhier, L.-M., Campagna, S., Linden, G. and Lorient, D. (1999) Surface-active and emulsifying properties of casein micelles compared to those of sodium caseinate. *Int. Dairy J.*, **9**, 411–2.

Creamer, L.K. and Berry, G.P. (1975) A study of the properties of dissociated bovine casein micelles. *J. Dairy Res.*, **42**, 169–83.

Dalgleish, D.G. (1990) The conformations of proteins on solid/water interfaces: caseins and phosvitin on polystyrene latices. *Colloids Surf.*, **46**, 141–55.

Dalgleish, D.G. (1993) The sizes and conformations of the proteins in adsorbed layers of individual caseins on latices and in oil-in-water emulsions. *Colloids Surf. B*, **1**, 1–8.

Dalgleish, D.G. (1996) Conformations and structures of milk proteins adsorbed to oil–water interfaces. *Food Res. Int.*, **29**, 541–7.

Dalgleish, D.G. (1997) Adsorption of protein and the stability of emulsions. *Trends Food Sci. Technol.*, **8**, 1–6.

Dalgleish, D.G. (1998) Casein micelles as colloids: surface structures and stabilities. *J. Dairy Sci.*, **81**, 3013–8.

Dalgleish, D.G. (1999) Interfacial structures and colloidal interactions in protein-stabilized emulsions. In *Food Emulsions and Foams: Interfaces, Interaction and Stability*, (E. Dickinson and J.M. Rodriguez Patino eds.) Royal Society of Chemistry, Cambridge, pp. 1–16.

Dalgleish, D.G. and Leaver, J. (1991) The possible conformations of milk proteins adsorbed on oil/water interfaces. *J. Colloid Interface Sci.*, **141**, 288–94.

Dalgleish, D.G., Dickinson, E. and Whyman, R.H. (1985) Ionic strength effects on the electrophoretic mobility of casein-coated polystyrene latex particles. *J. Colloid Interface Sci.*, **108**, 174–9.

Dalgleish, D.G., Srinivasan, M. and Singh, H. (1995) Surface properties of oil-in-water emulsion droplets containing casein and Tween 60. *J. Agric Food Chem.*, **43**, 2351–5.

Dalgleish, D.G., Tosh, S.M. and West, S.J. (1996) Beyond homogenization: the formation of very small emulsion droplets during the processing of milk by a Microfluidizer. *Neth. Milk Dairy J.*, **50**, 135–48.

Dalgleish, D.G., West, S.J. and Hallett, F.R. (1997) The characterization of small emulsion droplets made from milk proteins and triglyceride oil. *Colloids Surf. A*, **123**, 145–53.

Damodaran, S. (1997) Protein-stabilized foams and emulsions. In *Food Proteins and their Applications*, (S. Damodaran and A. Paraf eds.) Marcel Dekker, New York, pp. 57–110.

Damodaran, S. and Anand, K. (1997) Sulfhydryl–disulfide interchange-induced interparticle protein polymerization in whey protein-stabilized emulsions and its relation to emulsion stability. *J. Agric Food Chem.*, **45**, 3813–20.

Das, K.P. and Kinsella, J.E. (1990) Effect of heat denaturation on the adsorption of β-lactoglobulin at the oil–water interface and on coalescence stability of emulsions. *J. Colloid Interface Sci.*, **139**, 551–60.

de Feijter, J.A. and Benjamins, J. (1982) Soft-particle model of compact macromolecules at interfaces. *J. Colloid Interface Sci.*, **90**, 289–92.

de Feijter, J.A., Benjamins. J. and Tamboer, M. (1987) Adsorption displacement of proteins by surfactants in oil-in-water emulsions. *Colloids Surf.*, **27**, 243–66.

Demetriades, K. and McClements, D.J. (2000) Influence of sodium dodecyl sulfate on the physicochemical properties of whey protein-stabilized emulsions. *Colloids Surf. A*, **161**, 391–400.

Dickinson, E. (1988) The structure and stability of emulsions. In *Food Structure—Its Creation and Evaluation*, (J.M.V. Blanshard and J.R. Mitchell eds.) Butterworths, London, pp. 41–57.

Dickinson, E. (1989a) Protein adsorption at liquid interfaces and the relationship to foam stability. In *Foams: Physics, Chemistry and Structure*, (A.J. Wilson ed.) Springer-Verlag, Berlin, pp. 39–53.

Dickinson, E. (1989b) Surface and emulsifying properties of caseins. *J. Dairy Res.*, **56**, 471–7.

Dickinson, E. (1992a) *An Introduction to Food Colloids*, Oxford University Press, Oxford.

Dickinson, E. (1992b) Structure and composition of adsorbed protein layers and the relationship to emulsion stability. *J. Chem. Soc. Faraday Trans.*, **88**, 2973–83.

Dickinson, E. (1993a) Emulsion stability. In *Food Hydrocolloids: Structures, Properties and Functions*, (K. Nishinari and E. Doi eds.) Plenum, New York, pp. 387–98.

Dickinson, E. (1993b) Proteins in solution and at interfaces. In *Interactions of Surfactants with Polymers and Proteins*, (E.D. Goddard and K.P. Ananthapadmanabhan eds.) CRC Press, Boca Raton, FL, pp. 295–317.

Dickinson, E. (1995) Recent trends in food colloids research In *Food Macromolecules and Colloids*, (E. Dickinson and D. Lorient eds.) Royal Society of Chemistry, Cambridge, pp. 1–19.

Dickinson, E. (1997a) Properties of emulsions stabilized with milk proteins. *J. Dairy Sci.*, **80**, 2607–19.

Dickinson, E. (1997b) Enzymic crosslinking as a tool for food colloid rheology control and interfacial stabilization. *Trends Food Sci. Technol.*, **8**, 334–9.

Dickinson, E. (1997c) Aggregation processes, particle interactions and colloidal structure. In *Food Colloids: Proteins, Lipids and Polysaccharides*, (E. Dickinson and B. Bergenståhl eds.) Royal Society of Chemistry, Cambridge, pp. 107–26.

Dickinson, E. (1998a) Proteins at interfaces and in emulsions: stability, rheology and interactions. *J. Chem. Soc. Faraday Trans.*, **94**, 1657–69.

Dickinson, E. (1998b) Rheology of emulsions—the relationship to structure and stability. In *Modern Aspects of Emulsion Science*, (B.P. Binks ed.) Royal Society of Chemistry, Cambridge, pp. 145–74.

Dickinson, E. (1999a) Caseins in emulsions: interfacial properties and interactions. *Int. Dairy J.*, **9**, 305–12.

Dickinson, E. (1999b) Adsorbed protein layers at fluid interfaces: interactions, structure and surface rheology. *Colloids Surf. B*, **15**, 161–76.

Dickinson, E. (1999c) Enzymatic crosslinking of milk proteins. *Proceedings of 25th International Dairy Congress* (Aarhus), **2**, 209–15.

Dickinson, E. and Casanova, H. (1999) A thermoreversible emulsion gel based on sodium caseinate. *Food Hydrocoll.*, **13**, 285–9.

Dickinson, E. and Chen, J. (1999) Heat-set whey protein emulsion gels: role of active and inactive filler particles. *J. Dispersion Sci. Technol.*, **20**, 197–213.

Dickinson, E. and Davies, E. (1999) Influence of ionic calcium on stability of sodium caseinate emulsions. *Colloids Surf. B*, **12**, 203–12.

Dickinson, E. and Euston, S.R. (1991) Stability of food emulsions containing both protein and polysaccharide. In *Food Polymers, Gels and Colloids*, (E. Dickinson ed.) Royal Society of Chemistry, Cambridge, pp. 132–46.

Dickinson, E. and Galazka, V.B. (1991a) Emulsion stabilization by ionic and covalent complexes of β-lactoglobulin with polysaccharides. *Food Hydrocoll.*, **5**, 281–96.

Dickinson, E. and Galazka, V.B. (1991b) Bridging flocculation induced by competitive adsorption: implications for emulsion stability. *J. Chem. Soc. Faraday Trans.*, **87**, 963–9.

Dickinson, E. and Galazka, V.B. (1992) Emulsion stabilization by protein–polysaccharide complexes. In *Gums and Stabilisers for the Food Industry*, Vol. 6, (G.O. Phillips, D.J. Wedlock and P.A. Williams eds.) IRL Press, Oxford, pp. 351–62.

Dickinson, E. and Gelin, J.-L. (1992) Influence of emulsifier on competitive adsorption of α_s-casein and β-lactoglobulin in oil-in-water emulsions. *Colloids Surf.*, **63**, 329–35.

Dickinson, E. and Golding, M. (1997a) Depletion flocculation of emulsions containing unadsorbed sodium caseinate. *Food Hydrocoll.*, **11**, 13–8.

Dickinson, E. and Golding, M. (1997b) Rheology of sodium caseinate stabilized oil-in-water emulsions. *J. Colloid Interface Sci.*, **191**, 166–76.

Dickinson, E. and Golding, M. (1998a) Influence of alcohol on stability of oil-in-water emulsions containing sodium caseinate. *J. Colloid Interface Sci.*, **197**, 133–41.

Dickinson, E. and Golding, M. (1998b) Influence of calcium ions on creaming and rheology of emulsions containing sodium caseinate. *Colloids Surf. A*, **144**, 167–77.

Dickinson, E. and Hong, S.-T. (1994) Surface coverage of β-lactoglobulin at the oil–water interface: influence of protein heat treatment and various emulsifiers. *J. Agric. Food Chem.*, **42**, 1602–6.

Dickinson, E. and Hong, S.-T. (1995) Influence of water-soluble nonionic emulsifier on the rheology of heat-set protein-stabilized emulsion gels. *J. Agric. Food Chem.*, **43**, 2560–6.

Dickinson, E. and Hong, S.-T. (1996) Rheology of heat-set protein-stabilized emulsion gels: influence of emulsifier–protein interactions. In *Gums and Stabilisers for the Food Industry*, Vol. 8, (G.O. Phillips, D.J. Wedlock and P.A. Williams eds.) Oxford University Press, Oxford, pp. 319–28.

Dickinson, E. and Hong, S.-T. (1997) Influence of an anionic surfactant on the rheology of heat-set β-lactoglobulin-stabilized emulsion gels. *Colloids Surf. A*, **127**, 1–10.

Dickinson, E. and Iveson, G. (1993) Adsorbed films of β-lactoglobulin + lecithin at the hydrocarbon–water and triglyceride–water interfaces. *Food Hydrocoll.*, **6**, 533–41.

Dickinson, E. and Izgi, E. (1996) Foam stabilization by protein–polysaccharide complexes. *Colloids Surf.*, **113**, 191–201.

Dickinson, E. and James, J.D. (1998) Rheology and flocculation of high-pressure-treated β-lactoglobulin-stabilized emulsions. *J. Agric. Food Chem.*, **46**, 2565–71.

Dickinson, E. and Matsumura, Y. (1991) Time-dependent polymerization of β-lactoglobulin through disulfide bonds at the oil–water interface in emulsions. *Int. J. Biol. Macromol.*, **13**, 26–30.

Dickinson, E. and Matsumura, Y. (1994) Proteins at liquid interfaces: role of the molten globule state. *Colloids Surf. B*, **3**, 1–17.

Dickinson, E. and McClements, D.J. (1996) *Advances in Food Colloids*, Blackie, Glasgow.

Dickinson, E. and Pawlowsky, K. (1997) Effect of ι-carrageenan on flocculation, creaming and rheology of a protein-stabilized emulsion. *J. Agric. Food Chem.*, **45**, 3799–806.

Dickinson, E. and Stainsby, G. (1982) *Colloids in Foods*, Applied Science Publishers, London.

Dickinson, E. and Tanai, S. (1992a) Protein displacement from the emulsion droplet surface by oil-soluble and water-soluble surfactants. *J. Agric. Food Chem.*, **40**, 179–83.

Dickinson, E. and Tanai, S. (1992b) Temperature dependence of the competitive displacement of protein from the emulsion droplet surface by surfactants. *Food Hydrocoll.*, **6**, 163–71.

Dickinson, E. and Williams, A. (1994) Orthokinetic coalescence of protein-stabilized emulsions. *Colloids Surf. A*, **88**, 481–97.

Dickinson, E. and Woskett, C.M. (1989) Competitive adsorption between proteins and small-molecule surfactants in food emulsions. In *Food Colloids*, (R.D. Bee, P. Richmond and J. Mingins eds.) Royal Society of Chemistry, London, pp. 74–96.

Dickinson, E. and Yamamoto, Y. (1996a) Effect of lecithin on the viscoelastic properties of β-lactoglobulin-stabilized emulsion gels. *Food Hydrocoll.*, **10**, 301–7.

Dickinson, E. and Yamamoto, Y. (1996b) Viscoelastic properties of heat-set whey protein-stabilized emulsion gels with added lecithin. *J. Food Sci.*, **61**, 811–6.

Dickinson, E. and Yamamoto, Y. (1996c) Rheology of milk protein gels and protein-stabilized emulsion gels cross-linked with transglutaminase. *J. Agri. Food Chem.*, **44**, 1371–7.

Dickinson, E., Flint, F.O. and Hunt, J.A. (1989) Bridging flocculation in binary protein-stabilized emulsions. *Food Hydrocoll.*, **3**, 389–97.

Dickinson, E., Golding, M. and Casanova, H. (1999c) Creaming, flocculation and rheology of casein-stabilized emulsions. In *Food Emulsions and Foams: Interfaces, Interactions and Stability*, (E. Dickinson and J.M. Rodriguez Patino eds.) Royal Society of Chemistry, Cambridge, pp. 327–41.

Dickinson, E., Golding, M. and Povey, M.J.W. (1997c) Creaming and flocculation of oil-in-water emulsions containing sodium caseinate. *J. Colloid Interface Sci.*, **185**, 515–29.

Dickinson, E., Hong, S.-T. and Yamamoto, Y. (1996) Rheology of heat-set emulsion gels containing β-lactoglobulin and small-molecule surfactants. *Neth. Milk Dairy J.*, **50**, 199–207.

Dickinson, E., Horne, D.S., Phipps, J.S. and Richardson, R.M. (1993a) A neutron reflectivity study of the adsorption of β-casein at fluid interfaces. *Langmuir*, **9**, 242–8.

Dickinson, E., Horne, D.S., Pinfield, V.J. and Leermakers, F.A.M. (1997a) Self-consistent-field modelling of casein adsorption: comparison of results for α_{s1}-casein and β-casein. *J. Chem. Soc. Faraday Trans.*, **93**, 425–32.

Dickinson, E., Horne, D.S. and Richardson, R.M. (1993b) Neutron reflectivity study of the competitive adsorption of β-casein and water-soluble surfactant at the planar air–water interface. *Food Hydrocoll.*, **7**, 497–505.

Dickinson, E., Mauffret, A., Rolfe, S.E. and Woskett, C.M. (1989) Adsorption at interfaces in dairy systems. *J. Soc. Dairy Technol.*, **42**, 18–22.

Dickinson, E., Murray, B.S. and Stainsby, G. (1985) Time-dependent surface viscosity of adsorbed films of casein + gelatin at the oil–water interface. *J. Colloid Interface Sci.*, **106**, 259–62.

Dickinson, E., Murray, B.S. and Stainsby, G. (1988) Protein adsorption at air–water and oil–water interfaces. In *Advances in Food Emulsions and Foams*, (E. Dickinson and G. Stainsby eds.) Elsevier Applied Science, London, pp. 123–62.

Dickinson, E., Owusu, R.K., Tan, S. and Williams, A. (1993c) Oil-soluble surfactants have little effect on competitive adsorption of α-lactalbumin and β-lactoglobulin in emulsions. *J. Food Sci.*, **58**, 295–8.

Dickinson, E., Pinfield, V.J., Horne, D.S. and Leermakers, F.A.M. (1997b) Self-consistent-field modelling of adsorbed casein: interaction between two protein-coated surfaces. *J. Chem. Soc. Faraday Trans.*, **93**, 1785–90.

Dickinson, E., Ritzoulis, C. and Povey, M.J.W. (1999a) Stability of emulsions containing both sodium casemate and Tween 20. *J. Colloid Interface Sci.*, **212**, 466–73.

Dickinson, E., Ritzoulis, C., Yamamoto, Y. and Logan, H. (1999b) Ostwald ripening of protein-stabilized emulsions: effect of transglutaminase crosslinking. *Colloids Surf. B*, **12**, 139–46.

Dickinson, E., Rolfe, S.E. and Dalgleish, D.G. (1988) Competitive adsorption of α_{s1}-casein and β-casein in oil-in-water emulsions. *Food Hydrocoll.*, **2**, 397–405.

Dickinson, E., Rolfe, S.E. and Dalgleish, D.G. (1989) Competitive adsorption in oil-in-water emulsions containing α-lactalbumin and β-lactoglobulin. *Food Hydrocoll.*, **3**, 193–203.

Dickinson, E., Rolfe, S.E. and Dalgleish, D.G. (1990) Surface shear viscometry as a probe of protein–protein interactions in mixed protein films adsorbed at the oil–water interface. *Int. J. Biol. Macromol.*, **12**, 189–94.

Dickinson, E., Semenova, M.G. and Antipova, A.S. (1998a) Salt stability of casein emulsions. *Food Hydrocoll.*, **12**, 227–35.

Dickinson, E., Semenova, M.G., Antipova, A.S. and Pelan E.G. (1998b) Effect of high-methoxy pectin on properties of casein-stabilized emulsions. *Food Hydrocoll.*, **12**, 425–32.

Dickinson, E., Whyman, R.H. and Dalgleish, D.G. (1987) Colloidal properties of model oil-in-water food emulsions stabilized separately by α_{s1}-casein, β-casein and κ-casein. In *Food Emulsions and Foams*, (E. Dickinson ed.) Royal Society of Chemistry, London, pp. 40–51.

Dimitrova, T.D. and Leal-Calderon, F. (1999) Forces between emulsion droplets stabilized with Tween 20 and proteins. *Langmuir*, **15**, 8813–21.

Doxastakis, G. (1989) Milk proteins. In *Food Emulsifiers*, (G. Charalambous and G. Doxastakis eds.) Elsevier, Amsterdam, pp. 9–62.

Euston, S.R. (1997) Emulsifiers in dairy products and dairy substitutes. In *Food Emulsifiers and their Applications*, (L. Hasenhuett and R.W. Hartel eds.) Chapman and Hall, New York, pp. 173–210.

Euston, S.E., Singh, H., Munro, P.A. and Dalgleish, D.G. (1995) Competitive adsorption between sodium caseinate and oil-soluble and water-soluble surfactants in oil-in-water emulsions. *J. Food Sci.*, **60**, 1124–31.

Euston, S.E., Singh, H., Munro, P.A. and Dalgleish, D.G. (1996) Oil-in-water emulsions stabilized by sodium caseinate or whey protein isolate as influenced by glycerol monostearate. *J. Food Sci.*, **61**, 916–20.

Færgemand, M., Murray, B.S. and Dickinson, E. (1997) Crosslinking of milk proteins with transglutaminase at the oil–water interface. *J. Agric. Food Chem.*, **45**, 2514–9.

Færgemand, M., Murray, B.S., Dickinson, E. and Qvist, K.B. (1999) Cross-linking of adsorbed casein films with transglutaminase. *Int. Dairy J.*, **9**, 305–12.

Færgemand, M., Otte, J. and Qvist, K.B. (1998) Emulsifying properties of milk proteins crosslinked with microbial transglutaminase. *Int. Dairy J.*, **8**, 715–23.

Fang, Y. and Dalgleish, D.G. (1996a) Competitive adsorption between dioleoylphosphatidylcholine and sodium caseinate on oil–water interfaces. *J. Agric. Food Chem.*, **44**, 59–64.

Fang Y. and Dalgleish D.G. (1996b) Comparison of the effects of three different phosphatidylcholines on casein-stabilized oil-in-water emulsions. *J. Am. Oil Chem. Soc.*, **73**, 437–42.

Fang, Y. and Dalgleish, D.G. (1997) Conformation of β-lactoglobulin studied by FTIR: effect of pH, temperature, and adsorption to the oil–water interface. *J. Colloid Interface Sci.*, **196**, 292–8.

Farrer, D. and Lips, A. (1999) On the self-assembly of sodium caseinate. *Int. Dairy J.*, **9**, 281–6.

Fillery-Travis, A.J., Gunning, P.A., Hibberd, D.J. and Robins, M.M. (1993) Coexistent phases in concentrated polydisperse emulsions flocculated by non-adsorbing polymer. *J. Colloid Interface Sci.*, **159**, 189–97.

Fragneto, G., Thomas, R.K., Rennie, A.R. and Penfold, J. (1995) Neutron reflection study of bovine β-casein adsorbed on OTS self-assembled monolayers. *Science*, **267**, 657–60.

Galazka, V.B. and Dickinson, E. (1995) Surface properties of protein layers adsorbed from mixtures of gelatin with various caseins. *J. Texture Stud.*, **26**, 401–9.

Galazka, V.B., Dickinson, E. and Ledward, D.A. (1996) Effect of high pressure on the emulsifying behaviour of β-lactoglobulin. *Food Hydrocoll.*, **10**, 213–9.

Galazka, V.B., Ledward, D.A., Dickinson, E. and Langley, K.R. (1995) High pressure effects on the emulsifying behaviour of whey protein concentrate. *J. Food Sci.*, **60**, 1341–3.

Gelin, J.-L., Poyen, L., Courthaudon, J.-L., Le Meste, M. and Lorient, D. (1994) Structural changes in oil-in-water emulsions during the manufacture of ice-cream. *Food Hydrocoll.*, **8**, 299–308.

Goff, H.D. (1997a) Instability and partial coalescence in whippable dairy emulsions. *J. Dairy Sci.*, **80**, 2620–30.

Goff, H.D. (1997b) Colloidal aspects of ice-cream—a review. *Int. Dairy J.*, **7**, 363–73.

Goff, H.D. and Jordan, W.K. (1989) Action of emulsifiers in promoting fat destabilization during the manufacture of ice-cream. *J. Dairy Sci.*, **72**, 18–29.

Gouldby, S.J., Gunning, P.A., Hibberd, D.J. and Robins, M.M. (1991) Creaming in flocculated oil-in-water emulsions. In *Food Polymers, Gels and Colloids*, (E. Dickinson ed.) Royal Society of Chemistry, Cambridge, pp. 244–61.

Graham, D.E. and Phillips, M.C. (1979) Proteins at liquid interfaces. *J. Colloid Interface Sci.*, **70**, 415–26.

Guo, M.R., Fox, P.P., Flynn, A. and Kindstedt, P.S. (1996) Heat-induced modifications of the functional properties of sodium caseinate. *Int. Dairy J.*, **6**, 473–83.

Haertlé, T. and Chobert, J.-M. (1999) Recent progress in processing of dairy proteins: a review. *J. Food Biochem.*, **23**, 367–407.

Haque, Z. and Kinsella, J.E. (1989a) Emulsifying properties of food proteins—development of a standard emulsification method. *J. Food Sci.*, **54**, 39–44.

Haque, Z. and Kinsella, J.E. (1989b) Relative emulsifying activity of bovine serum albumin and casein as assessed by three different methods. *J. Food Sci.*, **54**, 1341–4.

Haque, Z., Leman, J. and Kinsella, J.E. (1988) Emulsifying properties of food proteins—bovine micellar casein. *J. Food Sci.*, **53**, 1107–10.

Heertje, I., Nederlof, J., Hendrickx, A.C.M. and Lucassen-Reynders, E.H. (1990) The observation of the displacement of emulsifiers by confocal scanning laser microscopy. *Food Struct.*, **9**, 305–16.

Heertje, I., van Aalst, H., Blonk, J.C.G., Don, A., Nederlof, J. and Lucassen-Reynders, E.H. (1996) Observations on emulsifiers at the interface between oil and water by confocal scanning light microscopy. *Lebensm. Wiss. Technol.*, **29**, 217–26.

Hemar, Y. and Horne, D.S. (1999) Rheology of highly concentrated protein-stabilized emulsions. In *Food Emulsions and Foams: Interfaces, Interactions and Stability*, (E. Dickinson and J.M. Rodriguez Patino eds.) Royal Society of Chemistry, Cambridge, pp. 318–26.

Hill, S.E. (1996) Emulsions. In *Methods of Testing Protein Functionality*, (G.M. Hall ed.) Chapman & Hall, London, pp. 153–85.

Holt, S.A. and White, J.W. (1999) The molecular structure of the surface of commercial cow's milk. *Phys. Chem. Chem. Phys.*, **1**, 5139–45.

Horne, D.S. (1998) Casein interactions: casting light on the black boxes, the structure in dairy products. *Int. Dairy J.*, **8**, 171–7.

Horne, D.S. and Leaver, J. (1995) Milk proteins on surfaces. *Food Hydrocoll.*, **9**, 91–5.

Horne, D.S., Atkinson, P.J., Dickinson, E., Pinfield, V.J. and Richardson, R.M. (1998) Neutron reflectivity study of competitive adsorption of β-lactoglobulin and nonionic surfactant at the air–water interface. *Int. Dairy J.*, **8**, 73–7.

Howe, A.M., Mackie, A.R. and Robins, M.M. (1986) Technique to measure emulsion creaming by velocity of ultrasound. *J. Dispersion Sci. Technol.*, **7**, 231–43.

Huang, X.L.L., Catignani, G.L. and Swaisgood, H.E. (1996) Improved emulsifying properties of β-barrel domain peptides obtained by membrane fractionation of a limited tryptic hydrolysate of β-lactoglobulin. *J. Agric. Food Chem.*, **44**, 3437–43.

Hunt, J.A. and Dalgleish, D.G. (1995) Heat stability of oil-in-water emulsions containing milk proteins: effect of ionic strength and pH. *J. Food Sci.*, **60**, 1120–3.
Hunt, J.A. and Dalgleish, D.G. (1996) The effect of the presence of KCl on the adsorption behaviour of whey protein and caseinate in oil-in-water emulsions. *Food Hydrocoll.*, **10**, 159–65.
Hunt, J.A., Dickinson, E. and Horne, D.S. (1993) Competitive displacement of proteins in oil-in-water emulsions containing calcium ions. *Colloids Surf. A*, **71**, 197–203.
Husband, F.A. and Wilde, P.J. (1998) The effects of caseinate sub-micelles and lecithin on the thin film drainage and behaviour of commercial caseinate. *J. Colloid Interface Sci.*, **205**, 316–22.
Husband, F.A., Wilde, P.J., Mackie, A.R. and Garrood, M.J. (1997) A comparison of the functional properties of β-casein and dephosphorylated β-casein. *J. Colloid Interface Sci.*, **195**, 77–85.
Jost, R. (1993) Functional characteristics of dairy proteins. *Trends Food Sci. Technol.*, **4**, 283–8.
Jost, R., Dannenberg, F. and Rosset, J. (1989) Heat-set gels based on oil/water emulsions: an application of whey protein functionality. *Food Microstruct.*, **8**, 23–8.
Karlskind, D., Laye, I., Morr, C.V. and Schenz, T.W. (1996) Emulsifying properties of lipid-reduced and calcium-reduced protein concentrates. *J. Food Sci.*, **61**, 54–8.
Kato, A., Mifuru, R., Matsudomi, N. and Kobayashi, K. (1992) Functional casein–polysaccharide conjugates prepared by controlled dry heating. *Biosci. Biotechnol. Biochem.*, **56**, 567–71.
Kato, A., Takhashi, A., Matsudomi, N. and Kobayashi, K. (1983) Determination of foaming properties by conductivity measurements. *J. Food Sci.*, **48**, 62–5.
Kauten, R.J., Maneval, J.E. and McCarthy, M.J. (1991) Fast determination of spatially localized volume fractions in emulsions. *J. Food Sci.*, **56**, 799–801, 847.
Kella, N.K.D., Yang, S.T. and Kinsella, J.E. (1989) Effect of disulfide bond cleavage on structural and interfacial properties of whey proteins. *J. Agric. Food Chem.*, **37**, 1203–10.
Koczo, K., Nikolov, A.D., Wasan, D.T., Borwankar, R.P. and Gonsalves, A. (1996) Layering of sodium caseinate sub-micelles in thin liquid films—a new stability mechanism for food dispersions. *J. Colloid Interface Sci.*, **178**, 694–702.
Krägel, J., Bree, M., Wüstneck, R., Makievski, A.V., Grigoriev, D.O., Senkel, O., Miller, R. and Fainerman, V.B. (1998) Dynamics and thermodynamics of spread and adsorbed food protein layers at the water/air interface. *Nahrung-Food*, **42**, 229–31.
Krägel, J., Wüstneck, R., Husband, F., Wilde, P.J., Makievski, A.V., Grigoriev, D.O. and Li, J.B. (1999) Properties of mixed protein/surfactant adsorption layers. *Colloids Surf. B*, **12**, 399–407.
Kull, T., Nylander, T., Tiberg, F. and Wahlgren, N.M. (1997) Effect of surface properties and added electrolyte on the structure of β-casein layers adsorbed at the solid/aqueous interface. *Langmuir*, **13**, 5141–7.
Leaver, J., Law, A.J.R. and Horne, D.S. (1999) Caseinate-stabilized emulsions: influence of ageing, pH and oil phase on the behaviour of individual protein components. In *Food Emulsions and Foams: Interfaces, Interactions and Stability*, (E. Dickinson and J.M. Rodriguez Patino eds.) Royal Society of Chemistry, Cambridge, pp. 258–68.
Leclerc, E. and Calmettes, P. (1997) Interactions in micellar solutions of β-casein. *Phys. Rev. Lett.*, **78**, 150–3.

Leermakers, F.A.M., Atkinson, P.J., Dickinson, E. and Horne, D.S. (1996) Self-consistent-field modelling of adsorbed β-casein: effects of pH and ionic strength on surface coverage and density profile. *J. Colloid Interface Sci.*, **178**, 681–93.

Lieske, B. and Konrad, G. (1996) Physicochemical and functional properties of whey protein as affected by limited papain proteolysis and selective ultrafiltration. *Int. Dairy. J.*, **6**, 13–31.

Liu, M. and Damodaran, S. (1999) Effect of transglutaminase-catalyzed polymerization of β-casein on its emulsifying properties. *J. Agric. Food Chem.*, **47**, 1514–9.

Lucassen, J. (1981) Dynamic properties of free liquid films and foams. In *Anionic Surfactants: Physical Chemistry of Surfactant Action*, (E.H. Lucassen-Reynders ed.) Marcel Dekker, New York, pp. 217–65.

Lucassen-Reynders, E.H. (1993) Interfacial elasticity in emulsions and foams. *Food Struct.*, **12**, 1–12.

Lucassen-Reynders, E.H. and Benjamins, J. (1999) Dilational rheology of proteins adsorbed at fluid interfaces. In *Food Emulsions and Foams: Interfaces, Interactions and Stability*, (E. Dickinson and J.M. Rodriguez Patino eds.) Royal Society of Chemistry, Cambridge, pp. 195–206.

Lynch, A.G. and Mulvihill, D.M. (1997) Effect of sodium caseinate on the stability of cream liqueurs. *Int. J. Dairy Technol.*, **50**, 1–7.

Mackie, A.R., Gunning, A.P., Wilde, P.J. and Morris, V.J. (1999) Orogenic displacement of protein from the air/water interface by competitive adsorption. *J. Colloid Interface Sci.*, **210**, 157–66.

Mackie, A.R., Mingins, J. and North, A.N. (1991) Characterization of adsorbed layers of a disordered coil protein on polystyrene latex. *J. Chem. Soc. Faraday Trans.*, **87**, 3043–9.

MacRitchie, F. (1986) Spread monolayers of proteins. *Adv. Colloid Interface Sci.*, **25**, 341–85.

McClements, D.J. (1999) *Food Emulsions: Principles, Practice and Techniques*, CRC Press, Boca Raton, FL.

McClements, D.J., Monahan, F.J. and Kinsella, J.E. (1993a) Effect of emulsion droplets on the rheology of whey protein isolate gels. *J. Texture Stud.*, **24**, 411–22.

McClements, D.J., Monahan, F.J. and Kinsella, J.E. (1993b) Disulfide bond formation affects stability of whey protein isolate emulsions. *J. Food Sci.*, **58**, 1036–9.

McDonald, P.J., Ciampi, E., Keddie, J.L., Heidenreich, M. and Kimmich, R. (1999) Magnetic resonance determination of the spatial dependence of the droplet size distribution in the cream layer of oil-in-water emulsions: evidence for the effects of depletion flocculation. *Phys. Rev. E.*, **59**, 874–84.

Mitchell, J.R. (1986) Foaming and emulsifying properties of proteins. In *Developments in Food Proteins*, Vol. 4, (B.F.J. Hudson ed.) Elsevier Applied Science, London, pp. 291–338.

Monahan, F.J., McClements, D.J. and Kinsella, J.E. (1993) Polymerization of whey proteins in whey protein-stabilized emulsions. *J. Agric. Food Chem.*, **41**, 1826–9.

Monahan, F.J., McClements, D.J. and German, J.B. (1996) Disulfide-mediated polymerization reactions and physical properties of heated WPI-stabilized emulsions. *J. Food Sci.*, **61**, 504–9.

Morr, C.V. (1982) Functional properties of milk proteins and their use as functional ingredients. In *Developments in Dairy Chemistry*, (P.F. Fox ed.) Applied Science, London, pp. 375–99.

Mulder, H. and Walstra, P. (1974) *The Milk Fat Globule*, Pudoc, Wageningen, Netherlands.

Mulvihill, D.M. and Fox, P.F. (1989) Physico-chemical and functional properties of milk proteins. In *Developments in Dairy Chemistry*, Vol. 4, (P.F. Fox ed.) Elsevier Applied Science, London, pp. 131–72.

Mulvihill, D.M. and Murphy, P.C. (1991) Surface-active and emulsifying properties of caseins/caseinates as influenced by state of aggregation. *Int. Dairy J.*, **1**, 13–37.

Murphy, J.M. and Fox, P.F. (1991) Functional properties of α_s-/κ- or β-rich casein fractions. *Food Chem.*, **39**, 211–28.

Murray, B.S. (1998) Interfacial rheology of mixed food protein and surfactant adsorption layers with respect to emulsion and foam stability. In *Proteins at Liquid Interfaces*, (D. Möbius and R. Miller eds.) Elsevier, Amsterdam, pp. 179–220.

Murray, B.S. and Dickinson, E. (1996) Interfacial rheology and the dynamic properties of adsorbed films of food proteins and surfactants. *Food Sci. Technol. Int.*, (Japan) **2**, 131–45.

Murray, B.S., Dickinson, E., McCarney, J.M., Nelson, P.V. and Whittle, M. (1998a) Observation of the dynamic colloidal interaction forces between casein-coated latex particles. *Langmuir*, **14**, 3466–9.

Murray, B.S., Færgemand, M., Trotereau, M. and Ventura, A. (1999) Comparison of the dynamic behaviour of protein films at air–water and oil–water interfaces. In *Food Emulsions and Foams: Interfaces, Interactions and Stability*, (E. Dickinson and J.M. Rodriguez Patino eds.) Royal Society of Chemistry, Cambridge, pp. 223–35.

Murray, B.S., Ventura, A. and Lallemant, C. (1998b) Dilatational rheology of protein + non-ionic surfactant films at the air–water and oil–water interfaces. *Colloids Surf. A*, **143**, 211–9.

Nagasawa, K., Takahashi, K. and Hattori, M. (1996) Improved emulsifying properties of β-lactoglobulin by conjugating with carboxymethyl dextran. *Food Hydrocoll.*, **10**, 63–7.

Newling, B., Glover, P.M., Keddie, J.L., Lane, D.M. and McDonald, P.J. (1997) Concentration profiles in creaming oil-in-water emulsion layers determined with stray field magnetic resonance imaging. *Langmuir*, **13**, 3621–6.

Nylander, T. (1998) Protein adsorption in relation to solution association and aggregation. In *Biopolymers at Interfaces*, (M. Malmsten ed.) Marcel Dekker, New York, pp. 409–52.

Nylander, T. and Wahlgren, N.M. (1994) Competitive and sequential adsorption of β-casein and β-lactoglobulin on hydrophobic surfaces and the interfacial structure of β-casein. *J. Colloid Interface Sci.*, **162**, 151–62.

Nylander, T. and Wahlgren, N.M. (1997) Forces between adsorbed layers of β-casein. *Langmuir*, **13**, 6219–25.

Okumura, K., Miyake, Y., Taguchi, H. and Shimabayashi, Y. (1990) Formation of stable protein foam by intermolecular disulfide cross-linkages in thiolated α_{s1}-casein as a model. *J. Agric. Food Chem.*, **38**, 1303–6.

Paquin, P. (1999) Technological properties of high pressure homogenizers: the effect of fat globules, milk proteins and polysaccharides. *Int. Dairy J.*, **9**, 329–35.

Pearce, K.N. and Kinsella, J.E. (1978) Emulsifying properties of proteins: evaluation of a turbidimetric technique. *J. Agric. Food Chem.*, **26**, 716–23.

Pelan, B.M.C., Watts, K.M., Campbell, I.J. and Lips, A. (1997) The stability of aerated milk protein emulsions in the presence of small molecule surfactants. *J. Dairy Sci.*, **80**, 2631–8.

Phillips, L.G., German, J.B., O'Neill, T.E., Foegeding, E.A., Harwalkar, V.R., Kilara, A., Lewis, B.A., Mangino, M.E., Morr, C.V., Regenstein, J.M., Smith, D.M. and Kinsella, J.E. (1990b) Standardized procedure for measuring foaming properties of three proteins: a collaborative study. *J. Food Sci.*, **55**, 1441–4, 1453.

Phillips, L.G., Schulman, W. and Kinsella, J.E. (1990a) pH and heat treatment effects on foaming of whey protein isolate. *J. Food Sci.*, **55**, 1116–9.

Phillips, L.G., Whitehead, D.M. and Kinsella, J. (1994) *Structure–Function Properties of Food Proteins*, Academic Press, San Diego.

Phillips, L.G., Yang, S.T. and Kinsella, J.E. (1991) Neutral salt effects on stability of whey protein isolate foams. *J. Food Sci.*, **56**, 588–9.

Puff, N., Cagna, A., Aguié-Béghin, V. and Douillard, R. (1998) Effect of ethanol on the structure and properties of β-casein adsorption layers at the air/buffer interface. *J. Colloid Interface Sci.*, **208**, 405–14.

Relkin, R., Liu, T. and Launay, B. (1993) Effects of pH on gelation and emulsification of a β-lactoglobulin concentrate. *J. Dispersion Sci. Technol.*, **14**, 335–54.

Rientjes, G.J. and Walstra, P. (1993) Factors affecting the stability of whey-based emulsions. *Milchwissenschaft*, **48**, 63–7.

Rodriguez Patino, J.M., Carrera Sanchez, C. and Rodriguez Nino, J.M. (1999) Is Brewster angle microscopy a useful technique to distinguish between isotropic domains in β-casein–monoolein mixed monolayers at the air–water interface? *Langmuir*, **15**, 4777–88.

Schmidt, D.G. (1982) Association of caseins and casein micelle structure. In *Developments in Dairy Chemistry*, (P.F. Fox ed.) Applied Science, London, pp. 61–86.

Schokker, E.P. and Dalgleish, D.G. (1998) The shear induced destabilization of oil-in-water emulsions using caseinate as emulsifier. *Colloids Surf. A*, **145**, 61–9.

Segall, K.I. and Goff, H.D. (1999) Influence of adsorbed milk protein type and surface concentration on the quiescent and shear stability of butteroil emulsions. *Int. Dairy J.*, **9**, 683–91.

Shimizu, M., Ametani, A., Kaminogawa, S. and Yamauchi, K. (1986) The topography of α_{s1}-casein adsorbed to an oil/water interface: an analytical approach using proteolysis. *Biochim. Biophys. Acta*, **869**, 259–64.

Singh, A.M. and Dalgleish, D.G. (1998) The emulsifying properties of hydrolyzates of whey proteins. *J. Dairy Sci.*, **81**, 918–24.

Singh, H., Fox, P.F. and Cuddigan, M. (1993) Emulsifying properties of protein fractions prepared from heated milk. *Food Chem.*, **47**, 1–6.

Smulders, P.E.A., Caessens, P.W.J.R. and Walstra, P. (1999) Emulsifying properties of β-casein and its hydrolysates in relation to their molecular properties. In *Food Emulsions and Foams: Interfaces, Interactions and Stability*, (E. Dickinson and J.M. Rodriguez Patino eds.) Royal Society of Chemistry, Cambridge, pp. 61–9.

Srinivasan, M., Singh, H. and Munro, P.A. (1999) Adsorption behaviour of sodium and calcium caseinates in oil-in-water emulsions. *Int. Dairy J.*, **9**, 337–41.

Swaisgood, H.E., Huang, X.L.L. and Catignani, G.L. (1996) Characteristics of the products of limited proteolysis of β-lactoglobulin. *ACS Symp. Ser.*, **650**, 166–77.

ter Beek, L.C., Ketelaars, M., McCain, D.C., Smulders, P.E.A., Walstra, P. and Hemminga, M.A. (1996) Nuclear magnetic resonance study of the conformation and dynamics of β-casein at the oil/water interface in emulsions. *Biophys. J.*, **70**, 2396–402.

Tuinier, R. and de Kruif, C.G. (1999) Phase separation, creaming, and network formation of oil-in-water emulsions induced by an exocellular polysaccharide. *J. Colloid Interface Sci.*, **218**, 201–10.

Tomas, A., Courthaudon, J.-L., Paquet, D. and Lorient, D. (1994) Effect of surfactant on some physicochemical properties of dairy oil-in-water emulsions. *Food Hydrocoll.*, **8**, 543–53.

Velev, O.D., Campbell, B.E. and Borwankar, R.P. (1998) Effect of calcium ions and environmental conditions on the properties of β-casein stabilized films and emulsions. *Langmuir*, **14**, 4122–30.

Walstra, P. (1983) Formation of emulsions. In *Encyclopedia of Emulsion Technology*, Vol. 1, (P. Becher ed.) Marcel Dekker, New York, pp. 57–127.

Walstra, P. (1989) Principles of foam formation and stability. In *Foams: Physics, Chemistry and Structure*, (A.J. Wilson ed.) Springer-Verlag, Berlin, pp. 1–15.

Walstra, P. and Smulders, I. (1997) Making emulsions and foams: an overview. In *Food Colloids: Proteins, Lipids and Polysaccharides*, (E. Dickinson and B. Bergenståhl eds.) Royal Society of Chemistry, Cambridge, pp. 367–81.

Wijmans, C.M. and Dickinson, E. (1998) Simulation of interfacial shear and dilatational rheology of an adsorbed protein monolayer modelled as a network of spherical particles. *Langmuir*, **14**, 7278–86.

Wijmans, C.M. and Dickinson, E. (1999a) Brownian dynamics simulation of a bonded network of reversibly adsorbed particles: towards a model of protein adsorbed layers. *Phys. Chem. Chem. Phys.*, **1**, 2141–7.

Wijmans, C.M. and Dickinson, E. (1999b) Brownian dynamics of the simulation of the displacement of a protein monolayer by competitive adsorption. *Langmuir*, **15**, 8344–8.

Wilde, P.J. and Clark, D.C. (1993) The competitive displacement of β-lactoglobulin by Tween 20 at the oil–water interface. *J. Colloid Interface Sci.*, **155**, 48–54.

Wilde, P.J. and Clark, D.C. (1996) Foam formation and stability. In *Methods of Testing Protein Functionality*, (G.M. Hall ed.) Chapman & Hall, London, pp. 110–52.

Wright, D.J. and Hemmant, J. (1987) Foaming properties of protein solutions: comparison of large-scale whipping and conductimetric methods. *J. Sci. Food Agric.*, **41**, 361–71.

Wüstneck, R., Krägel, J., Miller, R., Wilde, P. and Clark, D.C. (1996) The adsorption of surface-active complexes between β-casein, β-lactoglobulin and ionic surfactants and their shear rheological behaviour. *Colloids Surf. A*, **114**, 255–65.

Xiong, Y.L. and Kinsella, J.E. (1991) Influence of fat globule membrane composition and fat type on the rheological properties of milk-based composite gels. *Milchwissenschaft*, **46**, 207–12.

Yildirim, M., Hettiarachchy, N.S. and Kalapathy, U. (1996) Properties of biopolymers from crosslinking whey protein isolate and soybean 11S globulin. *J. Food Sci.*, **61**, 1129.

Zhu, H.M. and Damodaran, S. (1994) Heat-induced conformational changes in whey protein isolate and its relation to foaming properties. *J. Agric. Food Chem.*, **42**, 846–55.

28

THERMAL DENATURATION, AGGREGATION AND GELATION OF WHEY PROTEINS

H. SINGH AND P. HAVEA

28.1 INTRODUCTION

Whey is the soluble fraction of milk, rich in proteins, minerals and lactose, that is separated from the casein during the manufacture of rennet-coagulated cheese, rennet casein or acid casein. The whey produced from rennet-coagulated casein or cheese is referred to as sweet whey whereas that from mineral or lactic acid-coagulated casein is called acid whey. For decades, the dairy industry considered whey as a waste product, which was usually disposed of or used as animal feed. In recent years, whey has been recognised as a major source of nutritional and functional ingredients for the food industry worldwide (de Wit, 1990). A number of whey protein products are now manufactured from different whey types and they vary with respect to the concentration of protein, minerals, lipids and lactose. The manufacture of these products involves combinations of several processes such as ultrafiltration, diafiltration, reverse osmosis, ion exchange, evaporation and drying. Commercial whey protein products include various whey powders, whey protein concentrates (WPCs), whey protein isolates (WPIs) and fractionated proteins, such as α-lactalbumin and β-lactoglobulin (Mulvihill, 1992). The most important commercial whey protein products are WPC (up to about 85% protein) and WPI (approximately 95% protein). The principal whey proteins are β-lactoglobulin, α-lactalbumin, bovine serum albumin (BSA) and immunoglobulins (Ig). Whey protein products also contain various other components, including lipids, lactose and minerals. Typical values for the composition of industrial WPCs are shown in Table 28.1. These products have applications in meat products, beverages, baked products and infant foods. The important functional properties of whey protein products include water binding, emulsification, foaming and

Advanced Dairy Chemistry Volume 1: Proteins, 3rd edn.
Edited by P.F. Fox and P.L.H. McSweeney, Kluwer Academic/Plenum Publishers, 2003.

TABLE 28.1
Typical composition of industrial whey protein concentrates

Component	*Range (%)*
Moisture	3.30 – 4.50
Protein	73.0 – 86.0
Lactose	0.80 – 7.60
Total lipids	3.30 – 7.40
Phospholipids	0.80 – 1.55
Ash	2.50 – 5.30

whipping, and gelation. The functional properties of whey protein products depend not only on their composition but also on various processing treatments during manufacture.

Heat treatment is one of the major processing operations and causes the denaturation of whey proteins, resulting in considerable alterations in their functional properties. Consequently, whey protein products with similar composition can exhibit very different functional properties, due to differences in the extent of denaturation. The denaturation of whey proteins is often assumed to be a process involving two steps: modification of the native state of the protein to an activated state or unfolding and a subsequent (irreversible) aggregation of unfolded molecules. Recently, it has become clear that there may be several states of protein that may be present between the native and denatured states. Little is known about these intermediate states of the whey proteins and their relevance to functionality.

One of the most important functional properties of the whey proteins is their ability to form heat-induced gels, capable of holding large amounts of water and other food components. The heat-induced gelation of whey proteins involves a series of stages, starting with the unfolding and aggregation of protein molecules in aqueous solution. The gelation properties of whey protein products are affected by the composition, protein concentration, pH, heating temperature and time, ionic strength and presence of other ingredients. The gelation behaviour of commercial whey protein products tends to be rather unpredictable, because of inconsistency of product quality and composition, and lack of understanding of the nature of the proteins in industrial whey protein products and how they behave under different heating conditions. A better understanding of this behaviour can lead to better control of the industrial processing of the whey protein products, and development of new processes or products with predictable or tailor-made functional properties. Current knowledge on the heat-induced denaturation and gelation of whey proteins is discussed in this review, with specific emphasis on the interactions of whey proteins which occur during these processes.

28.2 Heat-induced unfolding and aggregation of whey proteins

In aqueous solutions, whey proteins exist in compact, organised, globular conformations. The major forces responsible for maintaining the globular structure of whey proteins are hydrophobic interactions, hydrogen bonding, ion-pair interactions and van der Waals' interactions. Changes in temperature and solution conditions cause proteins to denature and lose their native structure and the compact molecule begins to unfold into a disordered, random structure. The transformation from the initial native molecule to the unfolded state is co-operative and reversible. This unfolding of globular proteins is accompanied by an endothermal heat effect, which can be observed by differential scanning calorimetry (DSC). Using this technique, the temperature of unfolding can be measured either as the onset temperature of an endotherm or, more commonly, as the endotherm peak temperature. DSC has been used widely to study the kinetics of the unfolding of β-lactoglobulin and other whey proteins (Ruegg *et al.*, 1975; de Wit and Klarenbeek, 1984; Park and Lund, 1984). Whey proteins have a thermal transition temperature range of 62–78°C (Table 28.2). Recent work has shown that there may be several intermediate conformations between the native state and the final unfolded state (Freire *et al.*, 1992; Griko *et al.*, 1994). These intermediate states have a compact structure with a native-like secondary structure, but their tertiary structure is calorimetrically similar to that of the unfolded state. The "molten globule" state has been postulated to be a unique intermediate state with a partially folded conformation, having some similarity to, yet being distinct from, both the native and fully unfolded forms (Ptitsyn, 1995a,b). It has been suggested (Hirose, 1993) that this state is involved in various functional properties of whey proteins, particularly heat-induced gel formation.

When the whey proteins have been heated to a temperature at which they unfold at least partially, the hydrophobic amino acid residues buried deep

Table 28.2
Thermal denaturation temperature and enthalpy of whey proteins [adapted from de Wit (1984) and Kinsella and Whitehead (1989)]

Protein	*Td (°C)*	*Ttr (°C)*	*ΔH (kJ/mol)*
β-Lactoglobulin	78	83	311
α-Lactalbumin	62	68	253
BSA	64	70	803
Ig	72	89	500

Td is the initial denaturation temperature; Ttr is the temperature at the DSC peak maximum; △H is the enthalpy of denaturation.

within the native protein structure are exposed, resulting in an increase in the reactivity of such groups. There is also an increased reactivity of sulphydryl groups which can undergo oxidation to disulphide (S—S) or cysteic acid (—SO_3H) groups or sulphydryl-disulphide interchange reactions (de Wit, 1984; Mulvihill and Donovan, 1987a; Kella and Kinsella, 1988). Through sulphydryl-disulphide interchange and hydrophobic interactions, the unfolded protein molecules may associate with each other to form aggregates. Depending on the environmental conditions, e.g., pH, salt concentration and protein concentration, this aggregation may be followed by precipitation, coagulation and/or gelation (Donovan and Mulvihill, 1987a,b; de Wit, 1989). At a low concentration of salt and a high pH value, the unfolded proteins may remain dissociated due to the electrostatic repulsions between the charged residues. As the concentration of salt is increased or the pH is reduced, the electrostatic repulsion between charged residues decreases, allowing aggregation to occur more readily. At a high net charge, linear aggregates are formed, whereas at a low net charge, random aggregates develop.

The rate of aggregation of whey proteins may be observed by size exclusion chromatography, electrophoresis or solubility at the isoelectric point. These techniques measure the concentration of native protein before or after heat treatment. Differences in these concentrations, measured as a function of time at various temperatures, can indicate the kinetics of aggregation. Dynamic and static light scattering techniques have also been used to follow directly the kinetics of aggregation in heated whey protein solutions. Work in our laboratory using one-dimensional (1D) and two-dimensional (2D) polyacrylamide gel electrophoresis (PAGE) has examined the interactions of whey proteins during the heat-induced gelation of solutions of β-lactoglobulin and commercial WPCs (McSwiney *et al.*, 1994a,b; Gezimati *et al.*, 1997; Havea *et al.*, 1998). 1D-PAGE under native or dissociating (in the presence of sodium dodecyl sulphate, SDS) conditions provides information on the rate of loss of different proteins, and permits estimation of the proportions of protein that were linked by hydrophobic or disulphide interactions and an indication of the intermediates formed during heat treatment. A typical profile for the loss of β-lactoglobulin, based on the results of native-PAGE and SDS-PAGE, during the heat treatment of 12% WPC solution at 75°C is shown in Figure 28.1. The loss of native β-lactoglobulin, as determined by quantitative native-PAGE, is faster than the loss of protein as determined by SDS-PAGE, indicating that a proportion of protein molecules are linked by non-covalent interactions, which are destroyed by SDS. 2D-PAGE can be used to determine the composition of various intermediate protein aggregate species formed during heating and to differentiate between the aggregates that were disulphide linked and those that were non-covalently linked.

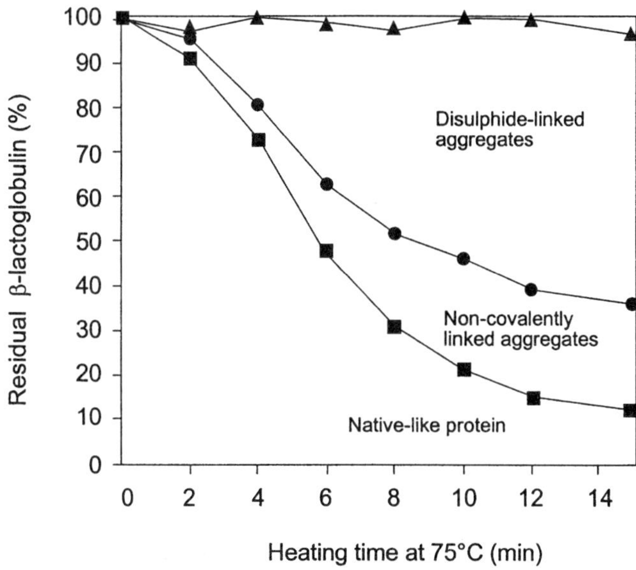

Figure 28.1 Loss of native-like (■), "SDS-monomeric" (●) and "total-reducible" (▲) β-lactoglobulin on heating WPC solutions (120 g/kg, pH 6.9) at 75°C for various times. Obtained using native-PAGE, SDS-PAGE and SDS-reducing PAGE (from Singh *et al.*, 2000, reproduced with permission).

Because of the heterogeneity of the whey protein system and because the individual proteins exhibit different responses towards heat, the thermal denaturation and aggregation of total whey protein reflects the collective response of the component proteins (de Wit and Klarenbeek, 1984). The denaturation of whey proteins has been studied extensively in skim milk (Lyster, 1970; Hillier and Lyster, 1979; Manji and Kakuda, 1986; Dannenberg and Kessler, 1988; Dalgleish, 1990; Oldfield *et al.*, 1998), cheese whey (Hillier and Lyster, 1979), simulated milk ultrafiltrate (Park and Lund, 1984), buffered solutions (Gough and Jenness, 1962; Harwalkar, 1980a,b) and distilled water (de Wit and Swinkels, 1980; Relkin and Launay, 1990). The thermal behaviour of individual proteins is different in different systems and is influenced by heating conditions, pH, protein concentration, ionic strength, concentrations of calcium and lactose, and the presence of chelating agents such as citrate and phosphate. For example, the resistance of individual pure whey protein solutions (8–10% concentration, 0.7 M phosphate buffer, pH 6.0) to thermal denaturation follows the order: β-lactoglobulin > Ig > BSA > α-lactalbumin (de Wit and Klarenbeek, 1984). However, the resistance of the same proteins to thermal denaturation in milk (Larson and Rolleri, 1955; Dannenberg and Kessler, 1988; Singh and Creamer, 1991) and rennet whey (Donovan and Mulvihill, 1987a,b) follows the order: α-lactalbumin > β-lactoglobulin > BSA > Ig.

To understand the heat-induced aggregation of the whey proteins in whey protein products such as WPC and WPI, individual whey proteins and mixtures of purified proteins have been studied separately by many workers under a variety of conditions (e.g., Griffin *et al.*, 1993; Hines and Foegeding, 1993; Matsuura and Manning, 1994; Hollar *et al.*, 1995; Qi *et al.*, 1995). Because β-lactoglobulin is the most abundant whey protein, it dominates the overall behaviour of whey protein products. Therefore, most of the research work on the aggregation behaviour of whey proteins has been carried out on β-lactoglobulin in model systems.

28.2.1 Heat-induced aggregation of β-lactoglobulin

The thermal denaturation and aggregation of β-lactoglobulin has been studied extensively using a range of techniques, such as DSC, nuclear magnetic resonance, spectroscopic methods (circular dichroism, infrared, Raman, fluorescence), and light, neutron and X-ray scattering.

At neutral pH, β-lactoglobulin exists primarily as a dimer, with each monomer consisting of an eight-stranded β-barrel and a three-turn helix that lies parallel to three of the β-strands (Papiz *et al.*, 1986). The single free sulphydryl group belongs to Cys_{121} which is located on the outside of the β-barrel and largely buried beneath the α-helix. Upon heating to approximately 70°C, the dimers dissociate to monomers (McKenzie, 1971; Hambling *et al.*, 1992), followed by some critical change in the conformation of the β-lactoglobulin monomer (e.g., Iametti *et al.*, 1996; Qi *et al.*, 1997). These changes lead to the exposure of hydrophobic groups (Iametti *et al.*, 1996; Relkin, 1998) and the free sulphydryl group (Cys_{121}) (Iametti *et al.*, 1996; Hoffmann and van Mil, 1997; Prabakaran and Damodaran, 1997), resulting in a reactive monomer which can propagate an aggregation reaction, leading to the formation of non-native dimers, trimers, tetramers and larger aggregates by polymerisation. This aggregation is irreversible and involves a combination of thiol-catalysed disulphide bond interchange and hydrophobic interactions. The disulphide linkage involved in the inter-molecular sulphydryl-disulphide interchange reaction is probably Cys_{66}-Cys_{160}, which is found in one of the external loops of β-lactoglobulin. The other disulphide bond is buried in the inner parts of the protein and is less available for this reaction (Papiz *et al.*, 1986).The extent of aggregation and the relative contribution of hydrophobic and sulphydryl-disulphide interactions to heat-induced aggregation depend on the time and temperature of heating, pH, protein concentration and ionic strength.

There is no clear agreement on the reaction order for β-lactoglobulin denaturation and aggregation. Pseudo first-order (Harwalkar, 1980b), first-order (de Wit and Swinkels, 1980) and second-order (Relkin and Launay, 1990) reactions have been reported. Recent reports based on the results of light scattering and electrophoresis suggest a 1.5 reaction order for the thermal aggregation of β-lactoglobulin. Based on the light scattering results, Roefs and de Kruif (1994) proposed that the aggregation of β-lactoglobulin can be described by analogy with a radical addition polymerisation reaction. According to this model, the overall reaction order in the temperature range 60–75°C and at near neutral pH is 1.5. This kinetic model recognises an initiation, a propagation and a termination step. The free sulphydryl group plays the role of the radical and is capable of building oligomers by sulphydryl-disulphide mechanisms. However, some recent studies have shown that β-lactoglobulin unfolds and aggregates through a series of parallel and consecutive steps, some of which involve thiol-disulphide interchange and others which involve hydrophobically-driven associations. Electrophoretic studies (McSwiney *et al.*, 1994a) have shown that the monomers and dimers of β-lactoglobulin can be dissociated from the larger aggregates formed on heating β-lactoglobulin in SDS solution, which is inconsistent with the simple thiol-catalysed polymerisation reaction. Several groups (Cairoli *et al.*, 1994; Huang *et al.*, 1994a,b; McSwiney *et al.*, 1994b) have suggested that during heating, β-lactoglobulin probably forms intermediates which bear some similarity to the "molten globule" state (Hirose, 1993; Ptitsyn, 1995a,b).

Recent work (Hoffmann *et al.*, 1996; Prabakaran and Damodaran, 1997) has shown that the reactive monomers initially form dimers and trimers *via* a thiol-disulphide exchange reaction and that the conversion of dimer to polymer is the rate-limiting step in aggregation (Prabakaran and Damodaran, 1997). The dimers may be important intermediates in the further aggregation of β-lactoglobulin (Manderson *et al.*, 1998). The formation of higher molecular weight aggregates (trimers, tetramers and other polymers) probably involves interactions between the exposed thiol group and disulphide bonds of two dimers or between those of a dimer and another reactive monomer (Havea, 1998; Manderson *et al.*, 1998). Presumably, the size of the aggregates increases mainly *via* addition of reactive monomers, dimers and small aggregates to larger aggregates (Schokker *et al.*, 1999).

Schokker *et al.* (1999) studied the early stages of the heat-induced aggregation of β-lactoglobulin at neutral pH using size exclusion chromatography combined with multi-angle laser light scattering (SEC-MALLS) and electrophoretic techniques. Upon heating, the amount of native β-lactoglobulin, as indicated by the size of peak 5, decreased (Figure 28.2), resulting in the formation of non-native monomers (Laligant *et al.*, 1991) and various small aggregates with estimated molecular weights of 31.5 ± 0.5, 53.0 ± 0.7 and 76.9 ± 1.0 kDa, respectively. These molecular weight values corresponded to β-lactoglobulin dimer, trimer and tetramer, respectively. Proteins larger than approximately 100 kDa (Figure 28.2; peak 1) eluted in the void volume of the column.

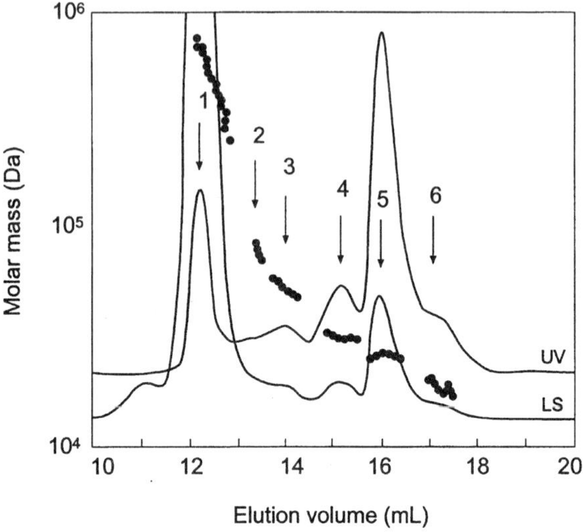

Figure 28.2 Size exclusion chromatography combined with multi-angle laser light scattering (Superdex 75): estimated molar mass profile of β-lactoglobulin aggregates after heat treatment of 5 min at 78.5°C (●). UV absorption and light scattering at 90° elution profiles are also shown. Arrows indicate (1) large aggregates, (2) tetramers, (3) trimers, (4) dimers, (5) native β-lactoglobulin and (6) non-native monomers (from Schokker *et al.*, 1999, reproduced with permission).

The aggregation behaviour of β-lactoglobulin A is different from that of the B variant, although the results are not always conclusive because of the different experimental conditions used. At concentrations lower than 5%, the aggregation rate of β-lactoglobulin B is greater than that of the A variant, but at concentrations greater than 5%, the A variant appears to be more sensitive to thermal aggregation (Nielsen *et al.*, 1996). Variant A forms relatively many intermediate-sized aggregates, whereas the B variant forms relatively more very large aggregates (Manderson *et al.*, 1998). Heat-induced aggregates formed from either variant at near neutral pH are held together by disulphide and non-covalent bonds, but the contribution of non-covalent bonds is greater for the A variant (Manderson *et al.*, 1998). Heat-induced gels of β-lactoglobulin A are stronger than gels made from β-lactoglobulin B (e.g. Huang *et al.*, 1994b; McSwiney *et al.*, 1994b). Differences in the aggregation behaviour may be explained in terms of charge, i.e., variant A is more negatively charged than variant B at neutral pH (Basch and Timasheff, 1967), thermal stability, i.e., variant A has a lower denaturation temperature than β-lactoglobulin B (e.g., Imafidon *et al.*, 1991; Huang *et al.*, 1994a), and the reactivity of certain groups, e.g., possibly the Cys_{66}-Cys_{160} disulphide bond in the A variant is less reactive due to the proximity of the negatively charged Asp_{64} (Manderson *et al.*, 1998).

28.2.2 Heat-induced aggregation of other whey proteins

When pure α-lactalbumin is heated under mild conditions (about 80°C, pH approximately 6.7), it does not form aggregates (Paulsson *et al.*, 1986; Matsudomi *et al.*, 1992, 1993; Calvo *et al.*, 1993; Hines and Foegeding, 1993). The lack of aggregate formation is explained largely by the lack of a free thiol group in α-lactalbumin (Eigel *et al.*, 1984). However, it has been shown by DSC studies (Rüegg *et al.*, 1977) that α-lactalbumin undergoes a reversible transition at 64°C. When α-lactalbumin is heated under more severe conditions (100°C for 10–30 min), disulphide-linked polymers, as well as modified monomers, are formed (Chaplin and Lyster, 1986). The latter are probably in the "molten globule" state (Kuwajima, 1989; Hirose, 1993). Schnack and Klostermeyer (1980) showed that significant loss of disulphide bonds occurred when α-lactalbumin was heated under the same conditions as those used by Chaplin and Lyster (1986). It was suggested that, as the disulphide bond is lost, a group similar to a thiol is formed simultaneously, which is probably responsible for the intermolecular disulphide interactions of α-lactalbumin. There is no report in the literature on non-covalent interactions of α-lactalbumin.

Thermal interactions of BSA have received little attention from food chemists. BSA is one of the most heat sensitive among the whey proteins, under a range of heating conditions at near neutral pH (de Wit and Klarenbeek, 1984). A concentration of 4% (w/v) BSA was required for the formation of a self-supporting gel following heating at 80°C for 30 min in 100 mM sodium phosphate buffer, pH 6.8 (Matsudomi *et al.*, 1993). Both β-lactoglobulin and α-lactalbumin cannot form self-supporting gels under these conditions of concentration and pH. Gezimati *et al.* (1996a,b) suggested that BSA undergoes thermal interactions in a similar fashion to that previously reported for β-lactoglobulin (McSwiney *et al.*, 1994a,b). Studies using native- and SDS-PAGE showed that BSA formed aggregates that were held together by hydrophobic interactions in addition to the well-documented sulphydryl-disulphide bond interchange reaction, except that BSA began to aggregate and gel at a lower temperature. The similarity should be expected on the basis that both BSA and β-lactoglobulin contain a free thiol group, Cys_{34} for BSA (Carter and Ho, 1994) and Cys_{121} for β-lactoglobulin (Hambling *et al.*, 1992).

Little information is available on the aggregation behaviour of Ig and other minor whey proteins.

28.2.3 Interactions and aggregations in whey protein mixtures

Because the commercial whey protein products are mixtures of different proteins, there has been much interest in investigating the nature of protein interactions which occur in protein mixtures during heating, and their effect on protein aggregation and gel formation. Analysis of mixtures of β-lactoglobulin, α-lactalbumin and BSA by DSC suggested no interactions, as indicated by their denaturation temperatures, although the reversibility of the denaturation of α-lactalbumin disappeared when heated in a mixture with β-lactoglobulin (Paulsson and Dejmek, 1992). Using size exclusion chromatography and electrophoresis, it was shown that β-lactoglobulin could form heat-induced complexes with α-lactalbumin (Matsudomi *et al.*, 1992; Calvo *et al.*, 1993; Hines and Foegeding, 1993; Dalgleish *et al.*, 1997; Gezimati *et al.*, 1997; Havea, 1998). The faster loss of native α-lactalbumin when heated in the presence of β-lactoglobulin, compared with when it was heated alone, was demonstrated using size exclusion chromatography (Elfagm and Wheelock, 1978; Hines and Foegeding, 1993; Calvo *et al.*, 1993), DSC (Paulsson and Dejmek, 1990) or gel filtration by high performance liquid chromatography (HPLC) (Matsudomi *et al.*, 1992). In contrast, loss of β-lactoglobulin was not affected by the presence of α-lactalbumin (Hines and Foegeding, 1993). Elfagm and Wheelock (1978) reported similar results at concentrations of 0.2% (w/v) β-lactoglobulin and 0.1% (w/v) α-lactalbumin. However, when the ratio of β-lactoglobulin to α-lactalbumin was higher (3:1), α-lactalbumin had a marked effect in reducing the amount of aggregated β-lactoglobulin. Heated mixtures of α-lactalbumin and β-lactoglobulin contained aggregates that were held together mainly by disulphide bonds and to a lesser extent by non-covalent bonding (Matsudomi *et al.*, 1992; Dalgleish *et al.*, 1997). During the early stages of heating, aggregates were formed that contained relatively more β-lactoglobulin than α-lactalbumin, whereas in later stages, they contained equal amounts of both proteins (Dalgleish *et al.*, 1997).

Recently, Schokker *et al.* (2000) examined the heat-induced aggregation of mixtures of α-lactalbumin and β-lactoglobulin A and B at 75°C using size exclusion chromatography (SE-HPLC) in combination with MALLS and electrophoretic techniques. In the absence of α-lactalbumin, β-lactoglobulin A was converted into a series of disulphide-bonded and hydrophobically associated aggregates, but with a greater proportion of hydrophobically associated aggregates than for β-lactoglobulin B. The presence of α-lactalbumin diminished the proportion of smaller aggregates and increased the number of very large aggregates in both variant protein mixtures, but did not significantly alter the rate of aggregation.

A few studies have shown that α-lactalbumin enhances the rigidity of β-lactoglobulin gels (Matsudomi *et al.*, 1992, 1993; Hines and Foegeding, 1993; Gezimati *et al.*, 1997; Rojas *et al.*, 1997). These studies did not permit a clear explanation of the likely protein interactions that are involved in the enhancement of gel rigidity. Some of the suggestions (Gezimati *et al.*, 1996a, 1997) include the possibility that the aggregates formed by the interactions between β-lactoglobulin and α-lactalbumin are more extensively cross-linked by disulphide bonds than the aggregates formed by β-lactoglobulin-β-lactoglobulin interactions. In addition, there may be a greater number of disulphide bridges bonding the aggregates together in β-lactoglobulin and α-lactalbumin mixtures.

Hines and Foegeding (1993) reported that the rate of loss of native β-lactoglobulin from solution at 80°C was increased by the presence of BSA, indicating some synergistic effect. Matsudomi *et al.*, (1994) found that when mixtures of BSA and β-lactoglobulin were heated at 80°C and examined at room temperature, the gels formed from the mixtures were stronger, again suggesting a synergistic effect. It is expected that when such a mixture is heated at a lower temperature (e.g., $\leqslant$70°C), the number of β-lactoglobulin molecules that undergo the thermal transition will be less than the number of BSA molecules; hence, BSA will be the dominant species in the aggregates and the gel formed will be composed largely of BSA molecules. In contrast, when heating at a high temperature (> 75°C), each protein will form aggregates and the rates of aggregation of the two proteins will be more comparable. The interactions between the two proteins are, therefore, dependent on the rate and temperature of heating (Gezimati *et al.*, 1996a,b).

The heat-induced interactions between α-lactalbumin and BSA were studied by Matsudomi *et al.* (1993) and Havea *et al.* (2000). Matsudomi *et al.*, (1993) showed that addition of $\geqslant$3% α-lactalbumin to 6% BSA caused a significant increase in gel hardness following heating at 80°C for 30 min in 100 mM sodium phosphate buffer, pH 6.8. BSA and α-lactalbumin interacted to form soluble aggregates through thiol-disulphide interchange reactions during gel formation. These heterogeneous aggregates had lower molecular weights than those formed by BSA alone. The enhancing effect of α-lactalbumin on the hardness of BSA gels was attributed to the formation of a finer, more uniform gel matrix. It was also reported that during gel formation, BSA was involved in the formation of hydrophobically-associated aggregates, which were dissociated under SDS-PAGE conditions, whereas there was no evidence of α-lactalbumin being involved (Matsudomi *et al.*, 1993). However, Havea *et al.* (2000) showed that when 10% solutions of a 1:1 mixture of α-lactalbumin and BSA were heated at 75°C, large BSA disulphide-bonded aggregates were formed which were able to catalyse the formation of modified α-lactalbumin monomers, α-lactalbumin dimers and α-lactalbumin trimers and adducts of α-lactalbumin with BSA.

There are no published reports on thermally induced interactions in a mixture of β-lactoglobulin, α-lactalbumin and BSA. However, it is expected that the interactions of these proteins in the different mixtures discussed above occur when the three proteins are heated together. Because of the different thermal transition temperatures of these proteins, it is likely that BSA will undergo aggregation to form the initial aggregates and that β-lactoglobulin and α-lactalbumin will form aggregates at some later stage. All kinds of homogeneous as well as heterogeneous aggregates between these proteins are expected during heat treatment. It is not clear if there are specific preferential interactions among some of these proteins; this area needs further investigation.

28.2.4 Heat-induced interactions and aggregation of whey proteins in WPC solutions

The heat-induced interactions of whey proteins in commercial WPC solutions have not been explored in detail as yet. The purpose of studies on model protein systems is to elucidate the likely mechanisms that may govern the interactions of whey proteins during heat treatment. However, it is difficult to relate these findings directly to those taking place in WPC systems. This is mainly because of the presence of many other components, such as immunoglobulins, casein fines, lactose, fat and minerals, which could have significant effects on the interactions of these proteins. Havea *et al.* (1998) studied the possible protein interactions leading to aggregation and gelation during the heating of WPC solutions, using 1D- and 2D-PAGE. They showed that when the WPC solutions were heated at 75°C, aggregates consisting of β-lactoglobloulin, α-lactalbumin, BSA, caseins and minor whey proteins were formed by both hydrophobic and disulphide interactions. Several intermediates, such as β-lactoglobulin dimers and trimers, 1:1 β-lactoglobulin and α-lactalbumin complexes, and α-lactalbumin dimers, were identified by 2D-PAGE. Figure 28.3 shows 2D-PAGE patterns of a WPC solution (120 g/kg, pH 6.9) that was heated at 75°C for 6 min. The series of spots labelled "A" (Figure 28.3a) are proteins that were dissociated from the aggregate bands in the native gel sample strip under SDS conditions. These proteins were linked to the aggregates by non-covalent associations. The series of spots labelled "B" (Figure 28.3b) are proteins that dissociated from aggregate bands in the SDS-PAGE sample strip and could be dissociated only under reduced conditions. These proteins were linked to the aggregates by disulphide bonds.

The extent of aggregation, the formation of intermediate molecular weight products and the nature of the bonds involved in the formation of aggregates were dependent on the protein concentration at heating. Higher proportions of low molecular weight intermediate species (dimers, trimers, etc.) formed at low WPC concentrations whereas large aggregates formed at high WPC concentrations. At low WPC concentrations, the aggregates formed were linked predominantly *via* disulphide bonds, whereas at high WPC concentrations, a considerable proportion of the aggregates also involved non-covalent linkages (ionic, hydrophobic association). Using size exclusion chromatography and dynamic light scattering, Ju and Kilara (1998) investigated the effects of heating WPI solutions on the formation of aggregates. When various concentrations (1–9%) of WPI solutions were

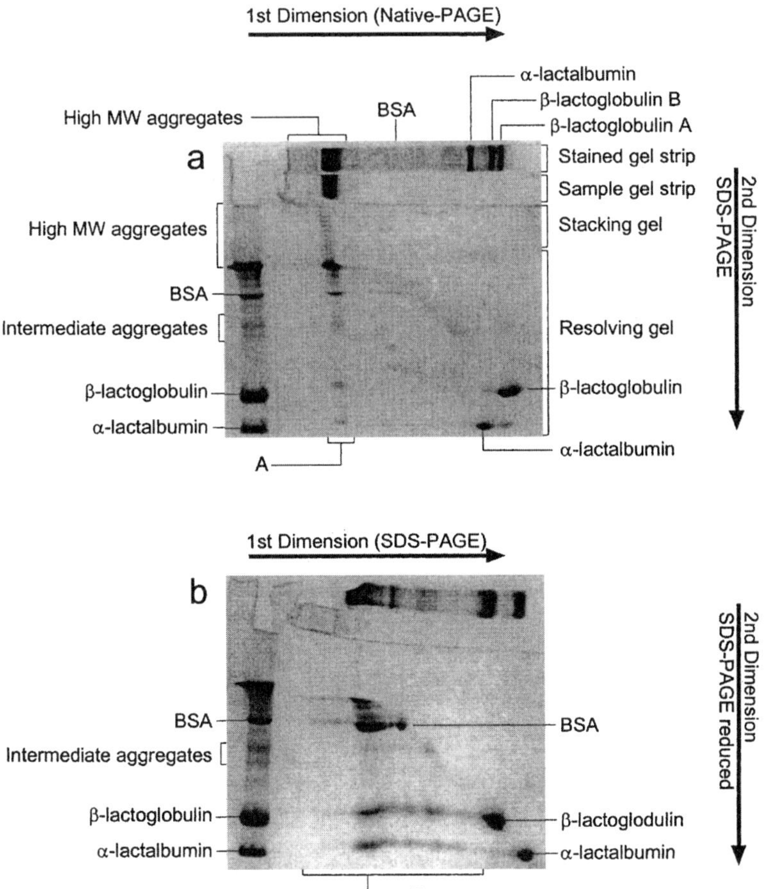

Figure 28.3 2D-PAGE pattern of a WPC solution (120 g/kg, pH 6.9) heated at 75°C. (a) Native-SDS 2D PAGE, (b) SDS-SDS reduced 2D PAGE. At the top of each gel there is a stained gel strip showing the protein bands from the first dimension separation running from left to right. Immediately below this strip is a second strip, which had contained the same protein bands used as samples for the second dimension separation. On the left-hand side, a portion of the corresponding sample was run to help identify the protein bands.

heated at 80°C for 30 min, soluble aggregates were formed, the extent and the size of which increased with an increase in WPI concentration.

28.3 Heat-induced gelation of whey proteins

As mentioned earlier, the formation of heat-induced whey protein gels involves protein unfolding and aggregation as the primary processes. A gel is formed when the extent of aggregation exceeds some critical level and a three-dimensional, self-supporting network is formed which traps the solvent in the system. When the extent of aggregation is below some critical minimum, soluble aggregates or a precipitate will form. Therefore, gel formation and the properties of the gel depend on the type and number of protein-protein interactions which in turn are affected by variables such as type of protein, protein concentration, temperature, pH, ionic strength and the presence of other ingredients such as lactose (Kinsella, 1976; Modler, 1985). The structure of protein gels can vary widely depending on these conditions, and has an impact on gel properties, such as rheological properties, sensory qualities and water-holding capacity.

Rheological measurements are used commonly to determine physical changes in a protein solution undergoing gel formation. As gels usually exhibit pronounced viscoelastic behaviour, the gelation process can be characterised by small amplitude dynamic testing, normally involving the application of a sinusoidal oscillatory stress or strain under constant frequency. The measured response is usually resolved into an in-phase elastic component, the storage modulus (G') and an out-of-phase viscous component, the loss modulus (G''). The phase angle (tan δ) between the applied strain and the resultant stress is given by the ratio of loss modulus to storage modulus. Because of the fragile nature of the network, deformation should be small and within the elastic region. The gelling system starts as a low viscosity solution ($G'' \gg G'$) with large tan δ values and converts to a viscoelastic solid ($G' > G''$) with very low tan δ values. G' increases with the cross-link density until the aggregation reactions reach completion. At long times, both G' and G'' are expected to reach final plateau values, although an equilibrium value is rarely achieved within practical times. Typical changes in G', G'' and tan δ during heat-induced gel formation in whey protein solutions are shown in Figure. 28.4. The information obtained on the kinetics of gelation may be used to derive fundamental information on gelation mechanisms and to study the effects of process variables on gel properties. Different information regarding the structure of gels may be obtained from large deformation testing; the response to large deformation and failure properties may be correlated better with sensory evaluation and the texture of gels.

28.3.1 Mechanisms of gel formation

The mechanism of gel formation is not understood fully. Early investigators proposed a two-stage process for the gelation of globular proteins: the unfolding of globular protein molecules and the association of the unfolded (denatured) protein molecules to form a three-dimensional network (Ferry, 1948; Hermansson, 1979):

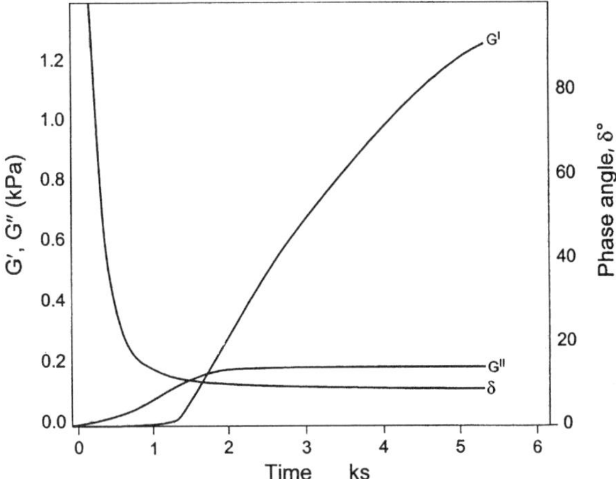

Figure 28.4 Typical changes in storage modulus (G'), loss modulus (G'') and tan δ during heat-induced gel formation in whey protein solutions.

$$xP_N \rightarrow xP_D \rightarrow x(P_D)_x$$

where x = number of protein molecules, P_N = native protein molecules and P_D = denatured protein molecules. The second, aggregation, step proceeds more slowly than the first, allowing the unfolded protein molecules to orient themselves to a certain degree. Thus, the rates of both unfolding and aggregation are important in the formation of protein gels. Since this model was proposed, many aspects of protein structure and folding have been elucidated. Nevertheless, the idea that the protein attains a different conformation, however transitory, prior to irreversible aggregation and gelation remains valid (Clark, 1992; Doi, 1993).

Based on recent studies, a four-step sequence of events has been proposed in the heat-induced gelation of proteins (Schmidt, 1981; Aguilera, 1995):

- denaturation (unfolding) of native proteins N → D(U)
- aggregation of unfolded molecules D(U) → A
- strand formation of aggregates A → S
- association of strands and network formation S → G

Globular proteins adopting a single conformation in their native state (N) are transformed to the unfolded (U) or denatured (D) state under certain conditions (e.g., heating). Some degree of conformational change in protein structure is a prerequisite to expose parts of the molecule that facilitate intermolecular interactions. These could be alterations in tertiary structures, such as the "molten globule" state (Chrunyk and Wetzel, 1993; Li *et al.*, 1994). The unfolding could also proceed to a point at which secondary and tertiary structures change to a significant extent (Dill *et al.*, 1989). Complete unfolding of globular proteins does not occur except on heating at high temperatures under alkaline conditions.

Interactions of unfolded protein molecules lead to the formation of spherical or bead-like "soluble" aggregates (A) (Beveridge *et al.*, 1984; McSwiney *et al.*, 1994a,b). These aggregates are the basic building blocks leading to strand(s) or pro-gels, and ultimately to gels (G). Although the molecular mechanism(s) responsible for the formation of aggregates during and after protein unfolding are not well known, aggregation of unfolded protein molecules is considered to proceed through a sequence of bimolecular association steps to yield higher order polymers. It is generally assumed that the driving forces for aggregation are non-specific interactions between the hydrophobic regions of unfolded polypeptide chains, but sulphydryl-disulphide interchange reactions and ionic interactions are likely to participate as well (Clark and Ross-Murphy, 1987). The work of McSwiney *et al.* (1994a,b) showed that the development of rheologically significant structures did not begin until most of the protein (β-lactoglobulin) had aggregated, measured as the loss of native protein after heating by PAGE analysis (Figure 28.5). Both hydrophobic and sulphydryl-disulphide interactions were involved in the formation of "soluble" aggregates because they could be completely dispersed into monomeric protein with SDS and mercaptoethanol but not with SDS alone.

The most outstanding microstructural feature of aggregate gels formed from whey proteins is the presence of a homogeneous network of connected

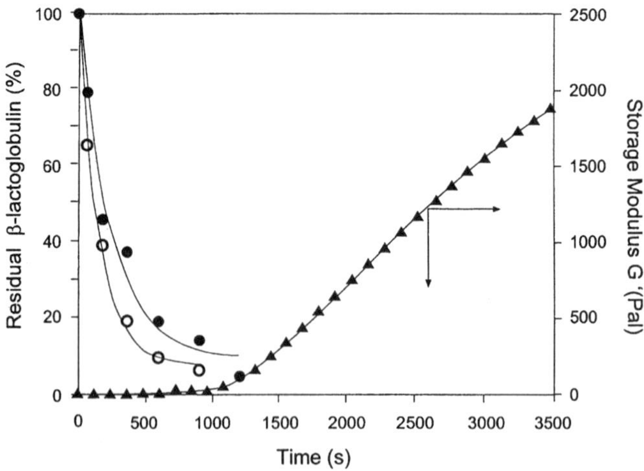

Figure 28.5 The relationships between loss of β-lactoglobulin from native-PAGE (○) or SDS-PAGE (●) and changes in storage modulus G' (▲) during heating at 75°C at pH 7.0 (from McSwiney *et al.*, 1994a, reproduced with permission).

protein particles or aggregates forming a three-dimensional matrix and interstices filled by a liquid or aqueous solution (Aguilera, 1995). Tombs (1974) postulated two models for the formation of globular protein gels from aggregates: random clustering, usually referred to as "particulated" gels, and string of beads, usually referred to as "fine-stranded" gels. β-Lactoglobulin, BSA and whey protein products are known to form either type of gel, depending on pH and ionic strength (Stading and Hermansson, 1991). The network of particulated gels is composed of spherical particles linked together, forming the strands of the network (Stading *et al.*, 1993). β-Lactoglobulin gels formed at a fast heating rate (5–12°C/min) showed strands of spherical particles of uniform size linked as a "string of beads" in a chain of single particles. Those formed at slower heating rates resulted in thicker strands of large particles, with a broad particle size distribution, fused together at many points (Stading *et al.*, 1993). Figure 28.6 illustrates

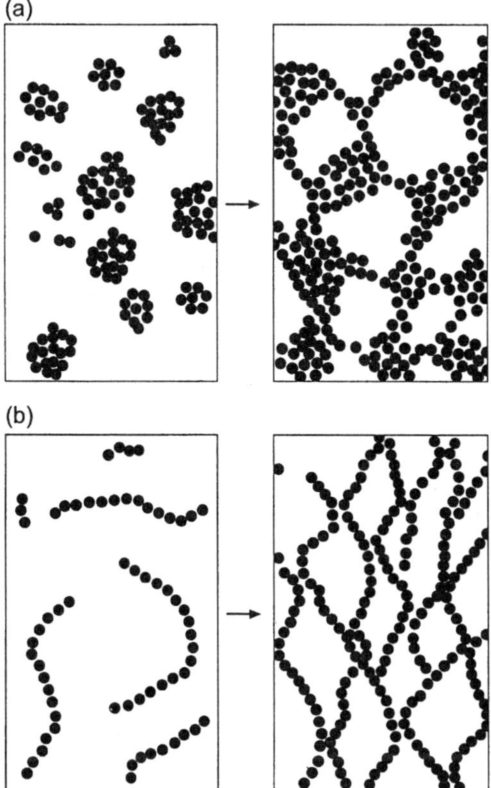

Figure 28.6 Formation of "particulated" (a) and "fine-stranded" (b) gel structures (adapted from Doi, 1993).

diagrammatically the formation of "particulated" (a) and "fine-stranded" (b) gel structures (Doi, 1993).

Several stabilising forces that affect gelation at the molecular level have been suggested: electrostatic forces, covalent bonds and hydrophobic interactions (Mangino, 1992a). Although it is not clear whether they stabilise the aggregates themselves or the strand of aggregates, it appears that the primary overall stabilising forces are hydrophobic interactions (Clark *et al.*, 1981) and possibly ionic interactions (Kohnhorst and Mangino, 1985).

28.3.2 Properties of whey protein gels

The appearance and properties of whey protein gels are affected by a number of factors, in particular protein concentration, time and temperature of heating, and ionic strength. These factors in turn influence the types of forces that hold the components together (Mangino, 1992b). If the gel network is too weak, the viscosity of the system will increase, but fluid flow will be possible and a gel will not form. However, if the protein-protein interactions are too strong, the network will collapse and water will be expelled from the structure. A critical balance between the attractive forces necessary to form a network and the repulsive forces necessary to prevent its collapse is required for gel formation (Mangino, 1992a,b).

Generally, heating a protein solution above the minimum denaturation temperature of the constituent proteins is required for gel formation, although heating at temperatures below the denaturation temperature may also result in gelation but requires longer heating times before any significant structure begins to develop. The effects of heating temperature and time are also dependent on protein type and concentration. When other factors are maintained, gel hardness increases with increasing heating temperature and time. Heating rate also affects the gelation process. Rapid heating does not allow enough time for the proteins to unfold and aggregate in a sequential manner even if the temperature is above the denaturation temperature, whereas slow heating allows the proteins enough time for unfolding and aggregation resulting in a much stronger gel (Stading and Hermansson, 1990).

The strength of whey protein gels is affected by the concentration and purity of the protein. There are different observations on the lowest protein concentration required to form a gel. Schmidt *et al.* (1978) found that a protein concentration of 7.5% or greater was needed in order to form a strong gel (WPC, pH 7.0, heated for 10 min at 100°C). For WPI in water, 14% protein concentration was needed at pH 7.0 after heating for 15 min at 90°C (Zirbel and Kinsella, 1988). Pure solutions of β-lactoglobulin and BSA could form gels at 5 and 4% protein, respectively, at pH 8.0 after heating for 15 min at 90°C (Matsudomi *et al.*, 1991). Paulsson *et al.* (1986) reported that the minimum protein concentration required for gelation at pH 6.6 in the presence of 1% NaCl was 1 and 2%, respectively, in the temperature range 60–90°C. As the protein concentration is increased, the number of potential interactions between molecules is enhanced, resulting in an increased gel strength, a reduced gelling time and a finer gel network (Paulsson *et al.*, 1986; Matsudomi *et al.*, 1991).

As pH affects molecular conformation and intermolecular interactions, it is not surprising that it influences gel network structure and rheological characteristics. Several investigators have found that opaque gels are formed in the pH range 4–6, whereas the gels are translucent above and below this pH range (Zirbel and Kinsella, 1988; Stading and Hermannson, 1991; Langton and Hermansson, 1992). Opaque gels have been described as soft and creamy and they tend to lose water during compression. Transparent gels formed at pH below 4 are weak (low values for fracture stress) and brittle (low values for fracture strain) whereas those formed at pH above 6.0 are strong and rubbery with high fracture stress and strain values. Shimada and Cheftel (1988) reported a decrease in the firmness of WPC gels and an increase in elasticity from pH 6.5 to 9.5. Gels formed at pH values from 7 to 9 were more elastic and less coagulated than gels formed at low pH (<6.0). The differences in gel properties with pH are attributed to variations in electrostatic interactions and disulphide bonding. At pH values close to the isoelectric point, excessive protein-protein interactions result in the formation of dense aggregates, resulting in low gel strength. When the pH is too far away from the isoelectric point, electrostatic repulsions inhibit protein-protein interactions, causing low gel strength. There is an optimum balance between protein-protein and protein-solvent interactions, resulting in maximum gel strength at 1 or 2 pH units from the isoelectric point. pH also affects the reactivity of thiol groups; there is an increase in intermolecular disulphide bond formation with an increase in pH (Hillier *et al.*, 1980; Xiong and Kinsella, 1990). Between pH 6.0 and pH 10.0, the strength of whey protein gels is inversely related to the pH (Hillier *et al.*, 1980; Zirbel and Kinsella, 1988; Xiong and Kinsella, 1990; Xiong, 1992).

Tang *et al.* (1993, 1995) reported that, in moving from extreme pH values towards the isoelectric point, heat-induced WPC gels exhibited maximum G' values (Figure 28.7). Egelandsdal (1980) and Stading and Hermansson (1991) reported similar results. At the pH values where the maxima occurred, a balance between protein-protein interactions and the gel matrix structure was achieved. Excessive protein-protein interactions led to a geometric structure that was weak in some mechanical or rheological properties, including gel stiffness, G'. However, weak protein-protein interactions also resulted in low gel stiffness. The maximum in G' was thus governed by the balance between these two factors and would occur somewhere between weak and excessive protein-protein interactions. Therefore, between the pH values where the maxima occurred, excessive

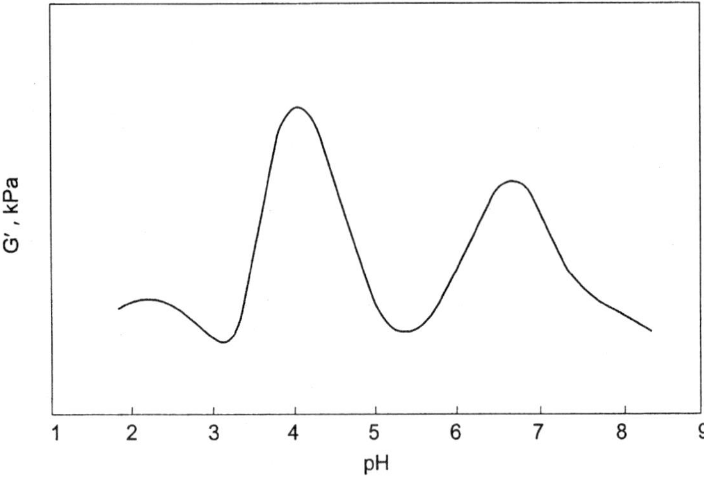

Figure 28.7 Changes in storage modulus (G') as a function of pH during heat treatment of WPC solutions (adapted from Tang *et al.*, 1995).

protein-protein interactions led to lower gel stiffness; outside this pH region, weak protein-protein interactions also led to lower stiffness. At the pH values where the maxima occurred, an optimum balance between the extent of protein-protein interactions and the gel geometric structure led to a maximum in gel stiffness, see Figure 28.7 (Tang *et al.*, 1993, 1995).

It is well established that salts have a major effect on the properties of whey protein gels, especially at pH values far from the isoelectric point where the proteins carry a large net charge. Addition of NaCl or $CaCl_2$ to a dialysed solution of WPC or WPI results in an increase in gel strength until maximum values are reached, and then the gel strength decreases at higher salt concentrations (Schmidt *et al.*, 1978, 1979; Mulvihill and Kinsella, 1988; Kuhn and Foegeding, 1991a,b). Tang *et al.* (1995) also observed a similar effect of KCl on the storage modulus, G', of heated WPC solutions. However, there is some disagreement on the optimum concentrations of salts required to achieve maximum gel strength. Schmidt *et al.* (1979) reported maximum gel hardness at 11.1 mM $CaCl_2$ for 10% WPC gels at pH 7.0. This is in agreement with the results of Mulvihill and Kinsella (1988) who reported maximum gel hardness for 10% β-lactoglobulin with 10 mM $CaCl_2$ or 200 mM NaCl. These results differ somewhat from those of Kuhn and Foegeding (1991a) who found that 20 mM $CaCl_2$ or 50–70 mM NaCl was required for maximum gel hardness for 10% WPI gels at pH 7.5. The mineral composition of the whey protein product used and whether or not the sample was dialysed may explain these differences. A maximum in a gel property has generally been attributed to an optimum balance between protein-protein and protein-solvent interactions at a particular salt concentration. Calcium ions influence protein-protein interactions by shielding electrostatic repulsion and also by forming calcium bridges between protein molecules. Addition of calcium was shown to increase the aggregation rate of WPI (Xiong, 1992); however, above a certain optimum concentration, coagulation occurred as a result of excessive protein-protein interactions. In general, divalent calcium ions have a much greater effect on gel properties than monovalent sodium or potassium ions (Tang *et al.*, 1993, 1995). This is due to the higher binding affinity of calcium to specific sites, specific conformational changes and intermolecular bridging (Foegeding *et al.*, 1986). WPI gels formed in the presence of sodium or calcium differed in appearance and deformability measured as shear strain at failure (Kuhn and Foegeding, 1991a,b). WPI (10%, pH 7.5) gels containing NaCl had high shear strains and were opaque, even though the shear stress (gel strength) for both gels were very similar. Increasing the NaCl concentration reduced the shear strain to a minimum at 148 mM NaCl, followed by a gradual increase at higher levels. In contrast, increasing the $CaCl_2$ concentration increased the shear strain to a maximum at 100 mM, followed by a gradual decrease at higher levels of addition. Kuhn and Foegeding (199la,b) also demonstrated that a range of divalent cations (calcium, magnesium and barium) caused similar increases in shear stress and shear strain at failure of WPI gels and that the increase in shear stress at failure was much larger than that caused by a range of monovalent cations (sodium, lithium, potassium, rubidium and cesium).

28.4 Concluding remarks

Commercial WPCs are used in a wide range of applications in the food industry. However, the utilisation of WPCs is limited because of significant variations in product quality and composition, which also vary in their functional properties. The variations in these products are due largely to variations in the composition of the milk as well as the processing history. The utilisation of these products can be enhanced greatly if the variations in product quality and composition are eliminated. Extensive research has been carried out on the factors that influence the functional properties of WPCs. A considerable amount of information has been established on the effects of factors such as pH, protein concentration, salts and ionic strength on the thermal denaturation, aggregation and gelation of whey proteins. However, there is relatively little understanding of how the various whey proteins interact during thermal treatment.

To understand the mechanisms of how the whey proteins interact during heat treatment, individual proteins have been studied extensively in single protein solutions or in different mixtures. Because of the wide range of solvent media and heating conditions under which these studies have been carried out, the results vary considerably. However, the results generally show that the interactions between the whey proteins are largely explained by the formation of disulphide bonds with some contributions by hydrophobic associations between denatured protein molecules. Understanding the mechanisms of how the whey proteins interact during thermal treatments can lead to a better control of whey product manufacture and the production of whey protein products with predictable functional properties.

References

Aguilera, J.M. (1995) Gelation of whey proteins. *Food Technol.*, **49**, 83–9.

Basch, J.J. and Timasheff, S.N. (1967) Hydrogen ion equilibrium of the genetic variants of bovine β-lactoglobulin. *Arch. Biochem. Biophys.*, **118**, 37–47.

Beveridge, T., Jones, L. and Tung, M.A. (1984) Progel and gel formation and reversibility of gelation of whey, soybean, and albumin protein gels. *J. Agric. Food. Chem.*, **32**, 307–13.

Cairoli, S., Iametti, S. and Bonomi, F. (1994) Reversible and irreversible modifications of β-lactoglobulin upon exposure to heat. *J. Protein Chem.*, **13**, 347–54.

Calvo, M.M., Leaver, J. and Banks, J.M. (1993) Influence of other whey proteins on the heat-induced aggregation of α-lactalbumin. *Int. Dairy J.*, **3**, 719–27.

Carter, D.C. and Ho, J.X. (1994) Structure of serum albumin. *Adv. Protein Chem.*, **45**, 153–203.

Chaplin, L.C. and Lyster, R.L.J. (1986) Irreversible heat denaturation of bovine α-lactalbumin. *J. Dairy Res.*, **53**, 249–58.

Chrunyk, B.A. and Wetzel, R. (1993) Breakdown in the relationship between thermal and thermodynamic stability in an interleukin-1-beta point mutant modified in a surface loop. *Protein Eng.*, **6**, 733–8.

Clark, A.H. (1992) Gels and gelling. In *Physical Chemistry of Foods*, (H.G. Schwartberg and R.W. Hartel eds.) New York, Marcel Dekker, pp. 263–305.

Clark, A.H. and Ross-Murphy, S.B. (1987) Structural and mechanical properties of bio-polymer gels. *Adv. Polym. Sci.*, **83**, 60–192.

Clark, A.H., Saunderson, D.H.P and Suggett, A. (1981) Infrared and laser Raman spectroscopic studies of thermally-induced globular protein gels. *Int. J. Pept. Protein Res.*, **17**, 353.

Dalgleish, D.G. (1990) Denaturation and aggregation of serum proteins and caseins in heated milk. *J. Agric. Food Chem.*, **38**, 1995–9.

Dalgleish, D.G., Senaratne, V. and Francois, S. (1997) Interactions between α-lactalbumin and β-lactoglobulin in the early stages of heat denaturation. *J. Agric. Food Chem.*, **45**, *3459–63*.

Dannenberg, F. and Kessler, H.-G. (1988) Reaction kinetics of the denaturation of whey proteins in milk. *J. Food Sci.*, **53**, 258–63.

de Wit, J.N. (1984) Functional properties of whey protein in food systems. *Neth. Milk Dairy J.*, **38**, 71–89.

de Wit, J.N. (1990) Thermal stability and functionality of whey proteins. *J. Dairy Sci.*, **73**, 3602–12.

de Wit, J.N. (1989) Functional properties of whey proteins. In *Developments in Food Proteins 4*, (P.F. Fox ed.) Elsevier Science Publishers Ltd., London, pp. 285–321.

de Wit, J.N. and Klarenbeek, G. (1984) Effects of various heat treatments on structure and solubility of whey proteins. *J. Dairy Sci.*, **67**, 2701–10.

de Wit, J.N. and Swinkels, G.A.M. (1980) A differential scanning calorimetric study of the thermal denaturation of bovine β-lactoglobulin: thermal behaviour at temperatures up to 100°C. *Biochim. Biophys. Acta*, **624**, 40–8.

Dill, K.A., Alonso, D.O.V. and Hutchinson, K. (1989) Thermal stabilities of globular proteins. *Biochemistry*, **28**, 5439–49.

Doi, E. (1993) Gels and gelling of globular proteins. *Trends Food Sci. Technol.*, **4**, 1–5.

Donovan, M. and Mulvihill, D.M. (1987a) Thermal denaturation and aggregation of whey proteins. *Ir. J. Food Sci. Technol.*, **11**, 87–100.

Donovan, M. and Mulvihill, D.M. (1987b) Effects of calcium modification and sodium dodecyl sulphate binding on the thermostability of whey proteins. *Ir. J. Food Sci. Technol.*, **11**, 77–85.

Egelandsdal, B. (1980) Heat-induced gelling in solutions of ovalbumin. *J. Food Sci.*, **45**, 570–3, 581.

Eigel, W.N., Butler, J.E., Ernstrom, C.A., Farrell, H.M., Jr., Harwalkar, V.R., Jenness, R. and Whitney, R.Mc.L. (1984) Nomenclature of proteins of cow's milk: 5th revision. *J. Dairy Sci.*, **67**, 1599–631.

Elfagm, A.A. and Wheelock, J.V. (1978) Interactions of bovine α-lactalbumin during heating. *J. Dairy Sci.*, **61**, 28–32.

Ferry, J.D. (1948) Protein gels. *Adv. Protein Chem.*, **4**, 1–5.

Foegeding, E.A., Alien, C.E. and Dayton, W.R. (1986) Effect of heating rate on thermally formed myosin fibrinogen and albumin gels. *J. Food Sci.*, **51**, 104–8, 112.

Freire, E., Murphy, K.P., Sanchez-ruiz, J.M., Galisteo, M.L. and Privalov, P.L. (1992) The molecular basis of cooperativity in protein folding: thermodynamic dissection of interdomain interactions in phosphoglycerate kinase. *Biochemistry*, **31**, 250–6.

Gezimati, J., Singh, H. and Creamer, L.K. (1996a) Aggregation and gelation of bovine β-lactoglobulin, α-lactalbumin and serum albumin. In *Macromolecular Interactions in Food Technology*, (N. Parris, A. Kato, L.K. Creamer and J. Pearce eds.) American Chemical Society (ACS Series no. 650), Washington DC, pp. 113–23.

Gezimati, J., Singh, H. and Creamer, L.K. (1996b) Heat-induced interactions and gelation of mixtures of β-lactoglobulin and serum albumin. *J. Agric. Food Chem.*, **44**, 804–10.

Gezimati, J., Creamer, L.K. and Singh, H. (1997) Heat-induced interactions and gelation of mixtures of β-lactoglobulin and α-lactalbumin. *J. Agric. Food Chem.*, **45**, 1130–6.

Gough, P. and Jenness, R. (1962) Heat denaturation of β-lactoglobulin A and B. *J. Dairy Sci.*, **45**, 1033–9.

Griffin, W.G., Griffin, M.C.A., Martin, S.R. and Price, J. (1993) Molecular basis of thermal aggregation of bovine β-lactoglobulin A. *J. Chem. Soc. Faraday Trans.*, **89**, 3395–406.

Griko, Y.V., Freire, E. and Privalov, P.L. (1994) Energetics of the α-lactalbumin states: a calorimetric and statistical thermodynamic study. *Biochemistry*, **33**, 1889–99.

Hambling, S.G., McAlpine, A.S. and Sawyer, L. (1992) β-Lactoglobulin. In *Advances in Dairy Chemistry*, *Vol. 1*, (P.F. Fox ed.) Elsevier Applied Science Publishers, London, pp. 141–90.

Harwalkar, V.R. (1980a) Measurement of thermal denaturation of β-lactoglobulin at pH 2.5. *J. Dairy Sci.*, **63**, 1043–51.

Harwalkar, V.R. (1980b) Kinetics of thermal denaturation of β-lactoglobulin at pH 2.5. *J. Dairy Sci.*, **63**, 1052–7.

Havea, P. (1998) *Studies on Heat-induced Interactions and Gelation of Whey Protein*, PhD Thesis, Massey University, Palmerston North, New Zealand.

Havea, P., Singh, H., Creamer, L.K. and Campanella, O.H. (1998) Electrophoretic characterization of the protein products formed during heat treatment of whey protein concentrate solutions. *J. Dairy Res.*, **65**, 79–91.

Havea, P., Singh, H. and Creamer, L.K. (2000) Formation of new protein structures in heated mixtures of BSA and α-lactalbumin. *J. Agric. Food Chem.*, **48**, 1548–56.

Hermansson, A.M. (1979) Aggregation and denaturation of protein involved in gel formation. In *Functionality and Protein Structure*, (A. Pour-El. ed.) ACS Symposium Series no. 92. American Chemical Society, Washington, DC, pp. 81–103.

Hillier, R.M. and Lyster, R.L.J. (1979) Whey protein denaturation in heated milk and cheese whey. *J. Dairy Res.*, **46**, 95–102.

Hillier, R.M., Lyster, R.L. and Cheeseman, G.C. (1980) Gelation of reconstituted whey powder by heat. *J. Sci. Food Agric.*, **31**, 1152–7.

Hines, M.E. and Foegeding, E.A. (1993) Interactions of α-lactalbumin and bovine serum albumin with β-lactoglobulin in thermally induced gelation. *J. Agric. Food Chem.*, **41**, 341–6.

Hirose, M. (1993) Molten globule state of food proteins. *Trends Food Sci. Technol.*, **4**, 48–51.

Hoffmann, M.A.M., Roefs, S.P.F.M., Verheul, M., van Mil, P.J.J.M. and de Kruif, K.G. (1996) Aggregation of β-lactoglobulin studied by in situ light scattering. *J. Phys. Chem.*, **101**, 6988–94.

Hoffmann, M.A.M. and van Mil, P.J.J.M. (1997) Heat-induced aggregation of β-lactoglobulin: role of the free thiol group and disulphide bonds. *J. Agric. Food. Chem.*, **45**, 2942–8.

Hollar, C.M., Parris, N., Hsieh, A. and Cockley, K.D. (1995) Factors affecting the denaturation and aggregation of whey proteins in heated whey protein concentrate mixtures. *J. Dairy Sci.*, **78**, 260–7.

Huang, X.-L., Catignani, G.L., Foegeding, E.A. and Swaisgood, H.E. (1994a) Comparison of the gelation properties of β-lactoglobulin genetic variants A and B. *J. Agric. Food Chem.*, **42**, 1064–7.

Huang, X.-L., Catignani, G.L. and Swaisgood, H.E. (1994b) Relative structural stabilities of β-lactoglobulins A and B as determined by proteolytic susceptibility and differential scanning calorimetry. *J. Agric. Food Chem.*, **42**, 1276–80.

Iametti, S., Cairoli, S., De Oregon, B. and Bonomi, F. (1996) Modifications of high-order structures upon heating of β-lactoglobulin: dependence on the protein concentration. *J. Agric. Food Chem.*, **43**, 53–8.

Imafidon, G.I., Ng-Kwai-Hang, K.F., Harwalkar, V.R. and Ma, C.-Y. (1991) Effect of polymorphism on the thermal stability of β-lactoglobulin. *J. Dairy Sci.*, **74**, 2416–22.

Ju, Z.Y. and Kilara, A. (1998) Textural properties of cold-set gels induced from heat-denatured whey protein isolates. *J. Food Sci.*, **63**, 288–92.

Kella, N.K.D. and Kinsella, J.E. (1988) Structural stability of β-lactoglobulin in the presence of kosmotropic salts, a kinetic and a thermodynamic study. *Int. J. Pept. Protein Res.*, **32**, 396–405.

Kinsella, J.E. (1976) Functional properties of proteins in foods: a survey. *CRC Crit. Rev. Food Sci. Nutr.*, **7**, 219–80.

Kinsella, J.E. and Whitehead, D.M. (1989) Proteins in whey: chemical, physical, and functional properties. *Adv. Food Nutr. Res.*, **33**, 343–438.

Kohnhorst, A.L. and Mangino, M.E. (1985) Prediction of the strength of whey protein gels based on compositional properties. *J. Food Sci.*, **50**, 1403–5.

Kuhn, P.R. and Foegeding, E.A. (199la) Mineral salt effects on whey protein gelation. *J. Agric. Food Chem.*, **39**, 1013–6.
Kuhn, P.R. and Foegeding, E.A. (1991b) Factors influencing whey protein rheology: dialysis and calcium chelation. *J. Food Sci.*, **56**, 789–91.
Kuwajima, K. (1989) The molten globule state as a clue for understanding the folding and cooperativity of globular-protein structure. Protein: *Struct. Funct. Genet.*, **6**, 87–103.
Laligant, A., Dummay, E., Valencia, C.C., Cuq, J.L. and Cheftel, J.C. (1991) Surface hydrophobicity and aggregation of β-lactoglobulin heated near neutral pH. *J. Agric. Food Chem.* **39**, 2147–55.
Langton, M. and Hermansson, A.M. (1992) Fine-stranded and particulate gels of β-lactoglobulin and whey protein at varying pH. *Food Hydrocolloids*, **5**, 523–39.
Larson, B.L. and Rolleri, C.D. (1955) Heat denaturation of specific serum proteins in milk. *J. Dairy Sci.*, **38**, 351–60.
Li, H., Hardin, C.C. and Foegeding, E.A. (1994) Thermal denaturation and cation-mediated aggregation of β-lactoglobulin. *J. Agric. Food Chem.*, **42**, 2411–20.
Lyster, R.L.J. (1970) The denaturation of α-lactalbumin and β-lactoglobulin in heated milk. *J. Dairy Res.*, **37**, 233–43.
Manderson, G.A., Hardman, M.J. and Creamer, L.K. (1998) Effect of heat treatment on the conformation and aggregation of β-lactoglobulin A, B, and C. *J. Agric. Food Chem.*, **46**, 5052–61.
Mangino, M.E. (1992a) Properties of whey protein concentrates. In *Whey and Lactose Processing*, (J.G. Zadow ed.) Elsevier Applied Science, London, pp. 231–70.
Mangino, M.E. (1992b) Gelation of whey protein concentrates. *Food Technol.*, **1**, 114–7.
Manji, B. and Kakuda, Y. (1986) Thermal denaturation of whey proteins in skim milk. *Can. Inst. Food Sci. Technol. J.*, **19**, 161–6.
Matsudomi, N., Oshita, T., Sasaki, E. and Kobayashi, K. (1992) Enhanced heated-induced gelation of β-lactoglobulin by α-lactalbumin. *Biosci. Biotechnol. Biochem.*, **56**, 1697–700.
Matsudomi, N., Oshita, T., Kobayashi, K. and Kinsella, J. (1993) α-Lactalbumin enhances the gelation properties of bovine serum albumin. *J. Agric. Food Chem.*, **41**, 1053–7.
Matsudomi, N., Oshita, T. and Kobayashi, K. (1994) Synergistic interaction between β-lactoglobulin and bovine serum albumin in heat-induced gelation. *J. Dairy Sci.*, **77**, 1487–93.
Matsudomi, N., Rector, D. and Kinsella, J.E. (1991) Gelation of bovine serum albumin and β-lactoglobulin: effects of pH, salts and thiol reagents. *Food Chem.*, **40**, 55–70.
Matsuura, J.E. and Manning, M.C. (1994) Heat-induced gel formation of β-lactoglobulin: a study on the secondary and tertiary structure as followed by circular dichroism spectroscopy. *J. Agric. Food Chem.*, **42**, 1650–56.
McKenzie, H.A. (1971) β-Lactoglobulins. In *Milk Proteins. Chemistry and Molecular Biology*, (H.A. Mckenzie ed.) Academic Press, New York, Vol. II, pp. 257–330.
McSwiney, M., Singh, H. and Campanella, O.H. (1994a) Thermal aggregation and gelation of bovine β-lactoglobulin. *Food Hydrocolloids*, **8**, 441–53.
McSwiney, M., Singh, H., Campanella, O.H. and Creamer L.K. (1994b) Thermal gelation and denaturation of bovine β-lactoglobulins A and B. *J. Dairy Res.*, **61**, 221–32.
Modler, H.W. (1985) Functional properties of non-fat dairy ingredients - a review. Modification of lactose and products containing whey proteins. *J. Dairy Sci.*, **68**, 2206–18.

Mulvihill, D.M. (1992) Production, functional properties and utilization of milk protein products. In *Advanced Dairy Chemistry, Proteins, Vol. 1*, (P.F. Fox ed.) Elsevier Science Publishers Ltd., London, pp. 369–404.

Mulvihill, D.M. and Donovan, M. (1987) Whey proteins and their thermal denaturation - a review. *Irish J. Food Sci. Technol.*, **11**, 43–75.

Mulvihill, D.M. and Kinsella, J.E. (1988) Gelation of β-lactoglobulin: effects of sodium chloride and calcium chloride on the rheological and structural properties of gels. *J. Dairy Sci.*, **53**, 231–6.

Nielsen, B.T., Singh, H. and Latham, J.M. (1996) Aggregation of bovine β-lactoglobulin A and B on heating at 75°C. *Int. Dairy J.*, **6**, 519–27.

Oldfield, D.J., Singh, H., Taylor, M.W. and Pearce, K.N. (1998) Kinetics of denaturation and aggregation of whey proteins in skim milk heated in an ultra-high temperature (UHT) pilot plant. *Int. Dairy J.*, **8**, 311–8.

Papiz, M.Z., Sawyer, L., Eliopoulos, E.E., North, A.C.T., Findlay, J.B.C., Sivarasadarao, R., Jones, T.A., Newcomer, M.E. and Kraulis, P.J. (1986) The structure of β-lactoglobulin and its similarity to plasma retinol-binding protein. *Nature*, **324**, 383–5.

Park, K.H. and Lund, D.B. (1984) Calorimetric study of thermal denaturation of β-lactoglobulin. *J. Dairy Sci.*, **67**, 1699–706.

Paulsson, M. and Dejmek, P. (1990) Thermal denaturation of whey proteins in mixtures with caseins studied by differential scanning calorimetry. *J. Dairy Sci.*, **73**, 590–600.

Paulsson, M. and Dejmek, P. (1992) Surface film pressure of β-lactoglobulin, α-lactalbumin and bovine serum albumin at the air-water interface studied by Wilhelmy plate and drop volume. *J. Colloid Interface Sci.*, **150**, 394–403.

Paulsson, M., Hegg, P.-Q. and Castberg, H.B. (1986) Heat-induced gelation of individual whey proteins. A dynamic rheological study. *J. Food Sci.*, **51**, 87–90.

Prabakaran, S. and Damodaran, S. (1997) Thermal unfolding of β-lactoglobulin: characterization of initial unfolding events responsible for heat-induced aggregation. *J. Agric. Food Chem.*, **45**, 4303–8.

Ptitsyn, O.B. (1995a) Molten globulin and protein folding. *Adv. Protein Chem.*, **47**, 83–229.

Ptitsyn, O.B. (1995b) Structures of folding intermediates. *Curr. Opinion Struct. Biol.*, **5**, 74–8.

Qi, X.L., Brownlow, S., Holt, C. and Sellers, P. (1995) Thermal denaturation of β-lactoglobulin: effect of protein concentration at pH 6.75 and 8.05. *Biochim. Biophys. Acta*, **1248**, 43–9.

Qi, X.L., Holt, C., McNulty, D., Clarke, D.T., Brownlow, S. and Jones, G.R. (1997) Effect of temperature on the secondary structure of β-lactoglobulin at pH 6.7, as determined by CD and IR spectroscopy: a test of the molten globule hypothesis. *Biochem. J.*, **324**, 341–6.

Relkin, P. (1998) Reversibility of heat-induced conformational changes and surface exposed hydrophobic clusters of β-lactoglobulin: Their role in heat-induced sol-gel state transition. *Int. J. Biol. Macromol.*, **22**, 59–66.

Relkin, P. and Launay, B. (1990) Concentration effects on the kinetics of β-lactoglobulin heat denaturation: a differential scanning calorimetric study. *Food Hydrocolloids*, **4**, 19–32.

Roefs, S.P.F.M. and de Kruif, K.G. (1994) A model for the denaturation and aggregation of β-lactoglobulin. *Eur. J. Biochem.*, **226**, 883–9.

Rojas, S.A., Douglas, G.H., Senaratne, V., Dalgleish, D.G. and Flores, A. (1997) Gelation of commercial fractions of β-lactoglobulin and α-lactalbumin. *Int. Dairy. J.*, **7**, 79–85.

Ruegg, M., Moor, U. and Blanc, B. (1975) Hydration and thermal denaturation of β-lactoglobulin. A calorimetric study. *Biochim. Biophys. Acta*, **400**, 334–42.

Ruegg, M., Moor, U. and Blanc, B. (1977) A calorimetric study of the thermal denaturation of whey proteins in simulated milk ultrafiltrate. *J. Dairy Res.*, **44**, 509–20.

Schmidt, R.H. (1981) Gelation and coagulation. In *Protein Functionality in Foods*, (J.P. Cherry ed.) ACS Symposium Series no. 147. American Chemical Society, Washington, DC, pp. 131–47.

Schmidt, R.H., Illingworth, L.B., Ahmed, E.M. and Richter, R.L. (1978) The effect of dialysis on heat-induced gelation of whey protein concentrate. *J. Food Prot.* Pres., **2**, 111–21.

Schmidt, R.H., Illingworth, L.B., Deng, J.C. and Cornell, A.J. (1979) Multiple regression and response surface analysis of calcium chloride and cystine on heat-induced whey protein gelation. *J. Agric. Food Chem.*, **27**, 529–32.

Schnack, U. and Klostermeyer, H. (1980) Thermal decomposition of α-lactalbumin. 1. Destruction of cystine residues. *Milchwissenchaft*, **35**, 206–11.

Schokker, E.P., Singh, H., Finder, D.N., Norris, G.E. and Creamer L.K. (1999) Characterization of intermediates formed during heat-induced aggregation of β-lactoglobulin A,B at neutral pH. *Int. Dairy J.*, **9**, 791–800.

Schokker, E.P., Singh, H. and Creamer, L.K. (2000) Heat-induced aggregation of β-lactoglobulin A and B with α-lactalbumin. *Int. Dairy. J.*, **10**, 843–53.

Shimada, K. and Cheftel, J.C. (1988) Texture characteristics, protein solubility, and sulphydryl group/disulfide bond contents of heat-induced gels of whey protein. *J. Agric. Food Chem.*, **36**, 1018–25.

Singh, H. and Creamer, L.K. (1991) Denaturation, aggregation and heat stability of milk protein during the manufacture of skim milk powder. *J. Dairy Res.*, **58**, 269–83.

Singh, H., Ye, A. and Havea, P. (2000) Milk protein interactions and functionality of dairy ingredients. *Aust. J. Dairy Technol.*, **55**, 29–35.

Stading, M. and Hermansson, A.M. (1990) Viscoelastic behavior β-lactoglobulin gel structures. *Food Hydrocolloids*, **4**, 121–35.

Stading, M. and Hermansson, A.M. (1991) Large deformation properties of β-lactoglobulin gel structures. *Food Hydrocolloids*, **5**, 339–52.

Stading, M., Langton, M. and Hermansson, A.-M. (1993) Microstructure and rheological behaviour of particulated β-lactoglobulin gels. *Food Hydrocolloids*, **7**, 195–212.

Tang, Q., McCarthy, O.J. and Munro, P.A. (1993) Oscillatory rheological study of the gelation mechanism of whey protein concentrate solutions: effects of physicochemical variables on gel formation. *J. Dairy Res.*, **60**, 543–55.

Tang, Q., McCarthy, O.J. and Munro, P.A. (1995) Oscillatory rheological study of the effects of pH and salts on gel development in heated whey protein concentrate solutions. *J. Dairy Res.*, **62**, 469–77.

Tombs, M.P. (1974) Gelation of globular proteins. *Faraday Discuss. Chem. Soc.*, **57**, 158–64.

Xiong, Y.L. (1992) Influences of pH and ionic environment on thermal aggregation of whey proteins. *J. Agric. Food Chem.*, **40**, 380–4.

Xiong, Y.L. and Kinsella, J.E. (1990) Mechanism of urea-induced whey protein gelation. *J. Agric. Food Chem.*, **38**, 1887–93.

Zirbel, F. and Kinsella, J.E. (1988) Factors affecting the rheological properties of gels made from whey protein isolate. *Milchwissenschaft*, **43**, 689–756.

29

PROTEIN HYDRATION AND VISCOSITY OF DAIRY FLUIDS

A.J. Carr, C.R. Southward and L.K. Creamer

29.1 Introduction

The rheological properties of milk, dairy products, dairy process streams and food ingredients derived from milk are important for the efficient processing of dairy products, as well as yielding products with desirable properties. For protein-rich dairy powders, it is generally more economic to process concentrated streams. Consequently, knowledge of the factors that affect their viscosity and performance as ingredients is important. The viscosity of concentrated solutions can be very sensitive to relatively minor alterations to any protein interactions, such as the degree of aggregation or altering the ratio of whey protein to casein, within the protein phase of the mixture. Many of the observed effects can be traced back to the fundamental properties of the particular proteins involved and follow well-known physical principles. In each section of this chapter, the fundamental aspects of the behaviour of the relevant milk proteins are outlined, followed by a brief consideration of the interactions (including processing-induced interactions) and concluding with a more detailed review of the known characteristics of a range of milk protein product solutions.

29.1.1 Basics of protein structure

Most proteins are globular and have a well-ordered, three-dimensional structure which enables them to interact in quite specific ways with large or small molecules in order to perform biological functions. Another group is the structural proteins, which are important in well-ordered structures such as skin, hair and feathers. Most of the whey proteins are globular, whereas the caseins are neither globular proteins nor structural proteins but have a

Advanced Dairy Chemistry Volume 1: Proteins, 3rd edn.
Edited by P.F. Fox and P.L.H. McSweeney, Kluwer Academic/Plenum Publishers, 2003.

nutritional role for neonates. This role involves the transport of mineral nutrients and the ability to supply a steady flow of amino acids to the growing infant. Unlike the usual globular proteins, the casein molecules do not have a well-defined hydrophilic surface and a hydrophobic interior. In the native casein micelle, however, particular segments of the various casein molecules are oriented so that the micelle as a whole has a hydrophilic surface and an interior that has regions of high hydrophobicity interspersed with hydrophilic clusters which include much of the mineral matter of the micelle.

29.1.2 Changes in viscosity characteristics with shear rate

In general, the rheological data reported in the literature, particularly those from biochemical laboratories, were obtained at low protein concentrations and at low shear rates. Shear rates used in industrial processing, with the exception of sedimentation (10^{-6} to 10^{-4} s^{-1}), can be much higher – typical shear rates are: stirring, 10^1 to 10^3 s^{-1}; pumping, 10^2 to 10^3 s^{-1}; spraying, 10^3 to 10^4 s^{-1}, and rubbing, 10^4 to 10^5 s^{-1} (Bylund, 1995).

Assessing the viscosity of fluids for sensory purposes is somewhat more complex. For example, Shama and Sherman (1973) suggested that the shear rate used in the oral evaluation of viscosity was probably inversely proportional to the viscosity of the fluid. Reported values range from 36.7 s^{-1} (Zamora, 1995) to 500 s^{-1} (Costell *et al.*, 1994); a comprehensive review on determining sensory viscosity is given by Costell and Duran (2000).

Nevertheless, Newtonian behaviour is unlikely at higher protein concentrations, and, consequently, caution should be used when extrapolating data obtained under low shear conditions to high shear conditions.

29.2 WHEY PROTEIN PRODUCTS

29.2.1 Proteins in whey

The proteins in whey are β-lactoglobulin (approximately 50% of whey protein), α-lactalbumin (approximately 25%), bovine serum albumin (BSA), immunoglobulins (IgG), other minor components, mostly derived from blood serum, and the proteose-peptones and, in the case of sweet (cheese) whey, various caseinomacropeptides (CMP). All except the proteose-peptones and CMP can be irreversibly aggregated *via* sulphydryl-disulphide interchange.

29.2.2 Hydration of whey proteins

Whey proteins that are in their native globular state have a hydration of about 0.2 g water/g protein, which is comparable with that of other globular proteins. However, heat-denatured whey proteins, although retaining most of their secondary structure, are linked together and, depending on the environment during denaturation and during measurement, can have a perceived hydration of over 10 g water/g protein. A non-syneresing gelled protein solution could be taken as an extreme example.

29.2.3 Viscosity of solutions of pure native whey proteins

The rheological properties of solutions of native globular whey proteins are much like those of any other globular protein and, in dilute solution, are a function of molecular mass, molecular shape (usually expressed as axial ratio) and protein–protein interaction, e.g., dimerisation (Tanford, 1961).

Under varying shear rate, Pradipasena and Rha (1977) found that the apparent viscosity of β-lactoglobulin solutions was Newtonian up to 5% (w/w) and pseudoplastic at higher concentrations. At a constant shear rate, the apparent viscosity of 10–30% (w/w) β-lactoglobulin solutions increased with shearing time. However, at a concentration of 40% (w/w), β-lactoglobulin exhibited time-dependent shear-thinning behaviour. No doubt, these types of concentration-dependent behaviour are related to the partial denaturation of the protein under these shear conditions, which is consistent with the reversible denaturation observed by Andrews (1991) for pure β-lactoglobulin.

29.2.4 Whey protein concentrates (WPCs) and isolates (WPIs)

These products are made by separating the more-or-less native whey proteins from the lower molecular weight molecules, primarily lactose, to give products containing 35–90 + % protein (on a dry matter basis). Despite efforts to avoid denaturation, some non-native protein material is usually present as a consequence of heat treatments, shearing and pH changes.

The rheology of solutions of WPC has been well characterised and the viscosity characteristics of such solutions are not markedly different from those expected on the basis of studies on β-lactoglobulin (Pradipasena and Rha, 1977).

Hermansson (1975) reported that WPC solutions in the intermediate concentration range exhibit measurable yield stresses, although later studies (Tang *et al.*, 1993) do not support this observation.

(a) Effects of concentration

As the concentration of WPC solutions is increased, three types of behaviour are observed. At low concentrations (<12%, w/w), WPC solutions are Newtonian (Hermansson, 1975) but become increasingly pseudoplastic at intermediate concentrations (Herbert, 1972; Hermansson, 1975). At higher concentrations (>20%, w/w, Hermansson, 1975;

>35%, w/w, Tang *et al.*, 1993), WPC solutions exhibit time-dependent shear thinning, although the structure broken down at high shear rates appears to re-form after resting for 15 min (Tang *et al.*, 1993).

For a given shear rate, the logarithm of apparent viscosity for whey protein systems increases linearly with concentration (Tang *et al.*, 1993), as expected from Eilers' relation.

$$\eta = \eta_{ref}\left(1 + \frac{1.25\phi}{1 - \phi/\phi_{max}}\right)^2$$

where η = viscosity of the suspension, η_{ref} = viscosity of the medium, ϕ = volume fraction of the dispersed particles, and ϕ_{max} = maximum attainable volume fraction. The value of ϕ_{max} for a system containing spheres of the same size (monodisperse systems) is 0.64 (Hemar and Horne, 1998). For systems with spheres of different sizes (polydisperse systems) it may be higher. Snoeren *et al.* (1982) calculated a ϕ_{max} of 0.79 for skim milk concentrates.

(b) Effects of salts

In general, the addition of NaCl has little effect on solutions of WPC (Hermansson, 1975; Tang *et al.*, 1993). However, addition of $CaCl_2$ (to >0.6 M) at pH 7.0 results in a marked increase in viscosity (Tang *et al.*, 1993).

(c) Effects of pH

Increasing the pH (in the range 6–10) of WPC dispersions results in a slight increase in apparent viscosity (Hermansson, 1975), in contrast to caseinate systems (Section 29.3.4(*a*)).

(d) Effects of temperature

The effect of temperature, in the range from 5 to 60°C, on the apparent viscosity of WPC solutions is described by the Arrhenius equation (Tang *et al.*, 1993), as found for concentrated milk (Section 29.4.2(*a*)), pure casein fractions (Section 29.3.2(*a*)) and caseinate (Section 29.3.4(*c*)).

29.2.5 Factors that affect the functionality of WPC

If the whey proteins have aggregated, for example, as a consequence of heat treatment, then the rheological properties of solutions of whey proteins are altered because the space occupied by the protein phase will be greater, i.e., the protein could be thought of as being more hydrated. Extensive aggregation results in gel formation, i.e., the protein and its associated water occupy the whole of the space (see Chapter 28). One very important factor is whether the whey proteins have been partially denatured. This often results in β-lactoglobulin changing to a different conformation, with the disulphide bonds in non-native arrangements and with a cysteine becoming available

(Manderson *et al.*, 1998, 1999; Havea *et al.*, 1998). Products containing such species have a tendency towards aggregation at neutral pH and at temperatures close to 60°C. Quite apart from heat-induced protein rearrangements, shear-induced, alkaline pH-induced and acid-induced denaturation are also common. β-Lactoglobulin and possibly other whey proteins have the ability to bind fatty acids and other hydrophobic molecules which increase their stability against denaturation. To confuse the picture, acid treatments can remove bound ligands from the proteins and divalent cations appear to enhance denaturability. All the above factors mean that processing operations must be well controlled to obtain reproducible results. There are also factors outside the control of experimenters (processors) which relate to the changing composition of the whey, e.g., the levels of IgG and BSA are high early in lactation and, if milk is stored, the level of proteose-peptones will increase as a consequence of enzyme action.

29.3 CASEIN-BASED PRODUCTS

The caseins, which make up nearly 80% of the proteins in milk, could not be studied readily until methods were developed to overcome their very strong associations with one another. The primary structures of the caseins are now well known and their behaviour in dilute solution has been studied. These proteins have regions within the sequence of high, medium or low hydrophobicity, and of high negative charge, high positive charge or low net charge (Figure 29.1). However, their conformation is still not well characterised although it is known that casein conformation is environment-dependent and that other casein molecules form part of that environment.

29.3.1 Properties of the individual caseins

The self-association and binary and ternary association properties of α_{s1}-, α_{s2}-, β- and κ-caseins as a function of pH, temperature and concentrations of monovalent salts (ionic strength) and calcium ions have been studied (see Chapters 1, 3, 4 and 5).

(a) α_s-Caseins

α_{s1}-Casein and α_{s2}-casein behave very similarly, despite the marked differences in their sequences (see Chapter 3) and some differences in the distribution of charged and hydrophobic residues along the amino acid sequence (see Chapter 3 and Figure 29.1). Unlike the globular whey proteins, they do not have a stable, uniquely folded, tertiary structure with a distinct inside and outside. In dilute solution with little salt (low ionic strength) and high pH, α_{s1}-casein exists as a monomer (Ho and Chen, 1967) but at neutral pH, the monomers aggregate *via* hydrophobic interactions (Rollema, 1992) to form loosely-bound complexes. Based on studies on

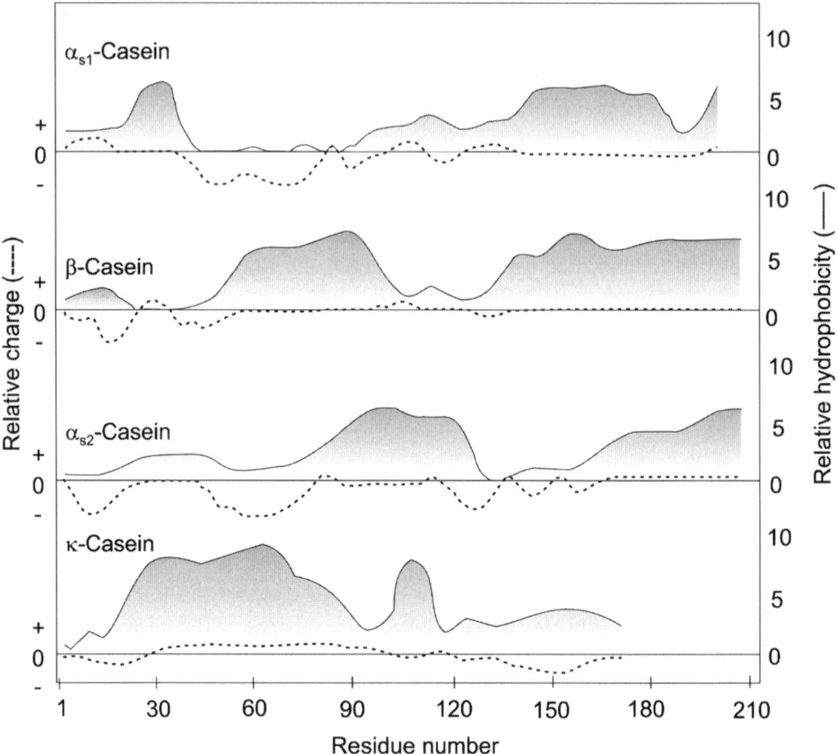

Figure 29.1 Diagrammatic representation of the distribution of net charge and hydrophobicity along the casein sequences. The traces shown for κ-casein are for the aglyco B variant. Glycosylation will effectively reduce the hydrophobicity near residue 130 by shielding the hydrophobic amino acid side chains from the solvent.

polymorphs and on peptides produced by selective cleavage, it is very likely that one of the two important regions of the α_{s1}-casein molecule for self-association (Figure 29.2A) is the segment between residues 10 and 30 (Creamer *et al.*, 1982).

(b) β-Casein

This protein is probably very similar to the archetypal casein, and, like the α_s-caseins, has regions of high and low hydrophobicity, and regions of high negative charge (Figure 29.1) and self-associates strongly. However, β-casein forms an equilibrium mixture of protein monomers and polymer aggregates at neutral pH, i.e., undergoes micellar aggregation (Figure 29.2B). The equilibrium constant and the size of the aggregates are determined by the pH, ionic strength and temperature of the solution (Rollema, 1992). In contrast to the α_s-caseins, most of the hydrophobic residues are in the

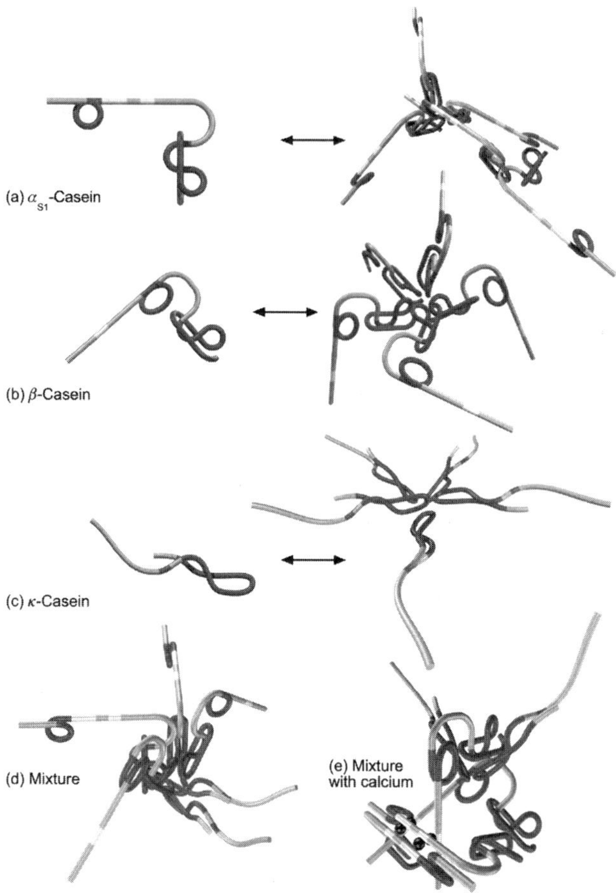

Figure 29.2 Models of the association of the caseins in solutions free of divalent cations: (A) α_{s1}-casein, (B) β-casein, (C) κ-casein, (D) inter-casein aggregation; and (E) the effect of calcium ions on the inter-casein aggregation. In all cases, the aggregates are large and only a small proportion of the individual molecular aggregates is shown. Within each casein molecule the hydrophobic regions are darkest, the acidic "phosphate cluster" regions are white, and calcium ions are dark circles.

C-terminal 150 residues of β-casein (Figure 29.1) and the important region of the molecule for this self-association reaction is close to the C-terminus (Berry and Creamer, 1975).

(c) κ-Casein

This protein also has quite distinct regions; the N-terminal 15 amino acid sequence is polar, as is the C-terminal 50 amino acid sequence with a large number of threonine residues (Figures 29.1 and 29.2C). This casein probably has more β-sheet structure than the other caseins and this appears to be in the *para*-κ-casein region of the molecule (Creamer *et al.*, 1998). Bovine κ-casein has two cysteine residues which are oxidised to give the disulphide-bonded κ-casein that exists in milk. This polymeric protein is located mainly on the external surface of the natural casein micelle (see Chapters 1 and 5). Even after reduction of the disulphide bonds of κ-casein aggregates, bovine κ-casein still self-associates strongly and preferentially adsorbs on to the outer surface of the micelles.

(d) Viscosity of solutions of individual caseins
Measurement of the intrinsic viscosity of the major casein molecules was difficult because of the very strong self-association of the purified caseins. For α_{s1}-casein, a low ionic strength and a high pH were necessary (Ho and Chen, 1967); for β-casein, a low ionic strength at a low temperature sufficed (Noelken and Reibstein, 1968). The measurements were made in order to estimate the size and shape of the casein molecules (Tanford, 1961). However, these studies were done before it was realised that these caseins were not at all comparable with globular proteins.

29.3.2 Interactions in mixtures of caseins

Binary mixtures of the three most important caseins (α_{s1}-, β- and κ-caseins) show that inter-casein association is stronger than self-association and, in ternary mixtures, α_{s1}-:κ-casein complexes are most favoured (Figure 29.2D), followed by α_{s1}-:β-casein complexes (Creamer and Berry, 1975). In these mixtures of caseins, the effects of pH, temperature, ionic strength, proportion of each casein type, presence of divalent cations and phosphate (when added to the casein:divalent cation mixture) are important (Figure 29.2E) and affect the behaviour of the mixtures (Creamer and Yamashita, 1976; Yamashita *et al.*, 1976; Creamer *et al.*, 1977).

The characteristics of casein–casein molecular interactions in solution are quite different from those of whey protein interactions and it is the balance between the intermolecular and the intramolecular strengths of association between the casein molecules that determines the state of aggregation and consequently the apparent viscosity of the mixture.

(a) Effects of temperature
The self-association of β-casein is strongly temperature-dependent (Pearce, 1975), that of κ-casein is less so and that of α_{s1}-casein is essentially independent of temperature. The association of either β- or κ-casein with the other caseins is, similarly, temperature-dependent.

(b) Effects of calcium
Addition of low concentrations of calcium (or certain other divalent cations, such as manganese or cobalt but not magnesium) to α_{s1}- or α_{s2}-casein

solutions buffered at neutral pH (and at a physiological ionic strength) has no measurable effect. This is because the calcium is tightly bound to the highly anionic regions of the protein (Ho and Waugh, 1965a), although the net negative charge of each aggregate diminishes markedly and the aggregates increase in size. Addition of more calcium initiates the coalescence of the moderately-sized particles into substantially larger particles, which make the solution turbid, and that then precipitate. Solutions of β-casein become turbid, and precipitate only at higher temperatures (>25°C) and κ-casein solutions do not become turbid on addition of calcium.

(c) Effects of phosphate

Addition of phosphate to casein solutions at pH 7 causes no obvious change and the phosphate does not appear to bind to casein despite the strongly cationic regions of the various casein molecules (Figure 29.1). Addition of low concentrations of ionic calcium to the casein-phosphate mixture has little effect, but further additions cause the formation of colloidal suspensions. In comparison with phosphate-free systems, the quantity of associated water is reduced, indicating that the particles are less hydrated than those in calcium caseinate (Creamer and Yamashita, 1976) and suggesting that the proteins are in closer proximity to one another.

With all casein systems, there is essentially no evidence that the individual casein molecules behave as distinct entities within the various aggregates that form under the particular experimental conditions. On going from sodium caseinate to calcium caseinate to calcium phospho-caseinate, the ability of a casein molecule to dissociate from the complex is reduced. Even though β-casein dissociates readily from these complexes at low temperature, the extent of dissociation is in the order: sodium caseinate > calcium caseinate > calcium phospho-caseinate.

Whereas the casein–calcium interactions that lead to varying hydration can be considered as a balance between repulsion of like charges on the protein segments and the attractive forces arising from the hydrophobic effect, the changes induced by including phosphate in the mixture arise from some very specific interactions (see Horne, 1998). This concept imparts some understanding of the very different effects caused by citrate, a calcium chelator, and various phosphates on casein-based products.

(d) Effects of proteolytic enzymes

Coagulants. Chymosin and pepsin cleave the Phe_{105}-Met_{106} bond of κ-casein very rapidly. This reaction is used to coagulate milk for cheesemaking and for making rennet casein. This enzymatic reaction is equally rapid in sodium caseinate, but no coagulation occurs, and, although *para*-κ-casein itself is insoluble, it remains an integral part of soluble aggregates of the other caseins (Berry and Creamer, 1976). However, in the presence of calcium salts, and especially when calcium and phosphate are present, the protein–mineral complexes in which a high proportion of κ-casein is hydrolyzed, are very insoluble.

Plasmin. This enzyme, which occurs naturally in milk (see Chapter 11.3), is sometimes present in casein and caseinate products (Richardson, 1983a,b). It hydrolyses the casein molecules at specific sites and reduces the viscosity of concentrated sodium or ammonium caseinate mixtures (Richardson, 1982). The specific cleavage of β-casein to γ_1-, γ_2- and γ_3-caseins and some of the proteose-peptones is of particular interest because the γ-caseins have a low solubility in neutral-pH buffers at higher temperatures but are readily incorporated within the soluble complexes of the other caseins. This behaviour is reminiscent of that of *para*-κ-casein.

29.3.3 Relationship between protein hydration and solution viscosity

Protein hydration is generally considered to be the water that is more-or-less immobilised by a protein (Damodaran, 1997). Casein micelles are hydrated (about 2.5 g water/g protein). Centrifugation does not remove the water held within the micelle and hydrodynamic methods, such as viscosity measurements, indicate that the micelles occupy more space than that taken up by the protein alone. Wide-band nuclear magnetic resonance (NMR) showed (Lelievre and Creamer, 1978) that most of this water was freely exchangeable with the solvent water that surrounds the micelles. Nevertheless, protein hydration, and particularly micelle hydration, is a very useful tool for studying casein-casein interactions, e.g., cheese texture (Creamer, 1985) or in heat-treated milks (Creamer and Matheson, 1980).

Korolczuk (1981a,b) measured the apparent viscosity of acid casein or sodium caseinate dissolved in 2.5% (w/v) borax solution as a function of protein concentration and tried to fit the results to equations loosely based on the Einstein equation which describes the viscous behaviour of suspensions of small spherical particles that do not interact with one another. He assumed that, because suspensions of casein micelles can be treated in this way and used to determine micelle size and hence the hydration of the micelle, the same rationale could be applied to caseinate solutions. Korolczuk (1982a) determined that a single equation could relate apparent viscosity to caseinate concentration and protein hydration in a simple fashion. This work was expanded (Korolczuk, 1982b,c) to cover caseins and milk protein concentrates over a range of pH and temperature values.

29.3.4 Acid casein and caseinates

The industrial processes for isolating casein from milk rely on destabilising the casein micelles by addition of acid to the milk (acid or whole casein) or by addition of a specific proteinase, such as chymosin or pepsin (rennet casein), followed by curd washing, drying and grinding (Bylund, 1995). Sodium caseinate, the major soluble derivative of acid casein, is used in numerous food products. A second commercial derivative is calcium caseinate which is not soluble but can be dispersed in aqueous solutions. Rennet casein, which is very insoluble in water at neutral pH, can be dispersed/dissolved with the aid of calcium-chelating salts such as sodium citrate or any of a range of monovalent alkaline phosphate salts. The flow properties of solutions, dispersions or slurries of these commercially-important products have been determined under different conditions of pH, temperature and ionic strength.

(a) Simple casein solutions

Acid casein can be dissolved by the use of alkalis or acids. These soluble derivatives can be dried, stored and then redissolved (Korolczuk, 1982a).

Casein concentration effects. Commercial sodium caseinate dissolves readily in water to give a translucent solution. This solution contains no ionic material other than the protein itself and Na^+. The latter tend to be localised close to the anionic regions of the protein (Ho and Waugh, 1965b) and the repulsive forces between the protein molecules dominate over the hydrophobic forces which tend to draw the proteins into aggregates. Thus, pH, which affects the number of negative charges along the protein chain, and salt content, which affects the shielding of these ionic charges, can have substantial overall effects on solution viscosity, which is governed by the space that the particles or aggregates appear to occupy.

Hermansson (1975) compared the solution characteristics of some commercially available sodium caseinates, WPCs and soy protein isolates. She found that, in contrast to solutions of whey proteins, caseinate solutions do not show a yield stress (Hermansson, 1975); below 12% (w/w), caseinate solutions are almost Newtonian and have low viscosity. Above 12%, they are slightly pseudoplastic and the consistency index increases greatly with increasing concentration (the apparent viscosity of the solutions increased from 0.02 to 0.55 Pa.s and from 0.55 to 1434 Pa.s over concentration ranges from 4 to 12% and 12 to 20%, respectively).

Towler (1974), who examined 7.5 to 15% (w/w) solutions of sodium caseinate, found that they exhibited shear thinning and that the 15% sodium caseinate solutions were pseudoplastic when high shear rates were used. Roeper and Winter (1982) reported that the viscosity of sodium caseinate solutions deviated significantly from Newtonian behaviour at a concentration of approximately 22% (w/v).

Fichtali *et al.* (1993) noted that the activation energy for flow tends to increase with the concentration of sodium caseinate solutions, with values of 3.49, 7.32 and 9.09 kcal/mol for 10, 13 and 16% (w/w) solids concentration,

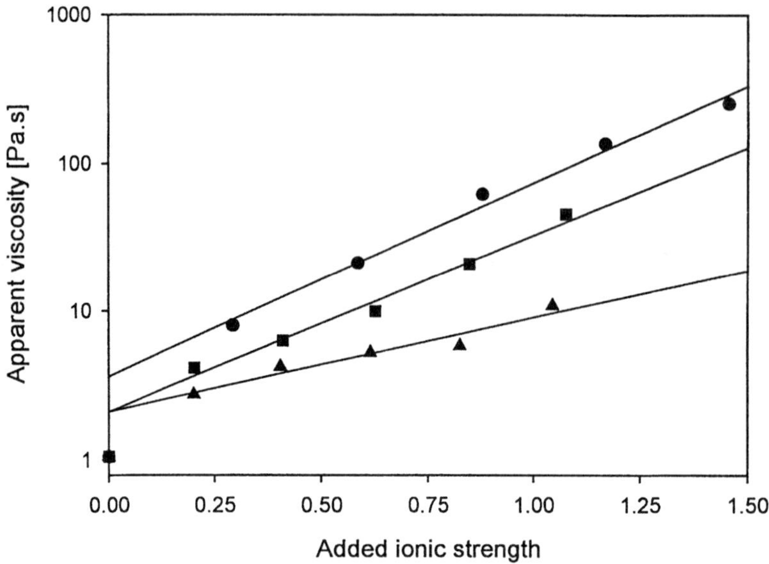

Figure 29.3 Effect of salt addition [NaCl, (●); KCl, (■); NH_4Cl (▲)] on the apparent viscosity of sodium caseinate (25°C, 14%, w/w) (Reprinted, with permission from Elsevier, Carr *et al.*, 2002).

respectively. This implies, as expected, that more energy is required to shear more concentrated solutions of sodium caseinate (Fichtali *et al.*, 1993).

Salt effects. Hermansson (1975) observed that with sodium caseinate solutions (>10%, w/w), addition of salt (0.2, 0.5 or 1.0 mol L^{-1} NaCl) caused a considerable increase in viscosity, but no change in flow character; swelling and solubility were reduced markedly.

Carr *et al.* (2002) observed that the apparent viscosity of caseinate solutions increased with increases in salt (NaCl, KCl or NH_4Cl) concentration over an added ionic strength range of 0 to 1.1 mol L^{-1} (Figures 29.3 and 29.4). At 25°C, addition of salt to an added ionic strength of 0.2 M caused a proportionately larger increase in apparent viscosity than subsequent salt additions (Figure 29.3). The proportionately larger increases in apparent viscosity upon the initial addition of salt were not observed in experiments conducted at 50°C (Figure 29.4). However Carr *et al.* (2002) noted that at high shear rates and high ionic strength, sodium caseinate solutions exhibit the Weisenberg effect of 'rod-climbing'.

pH effects. Casein is soluble at pH values below pH 3.5 or above pH 5.4. Dolby (1961a) reported that the apparent viscosity of 15% casein solutions showed a minimum between pH 6.0 and pH 8.0 and Hayes and Muller (1961) noted that comparable casein solutions showed a similar minimum,

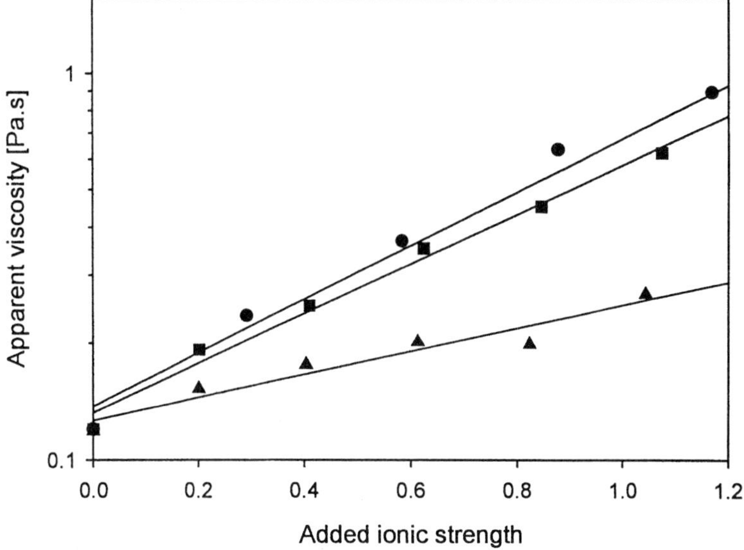

Figure 29.4 Effect of salt addition [NaCl, (●); KCl, (■); NH_4Cl (▲) on the apparent viscosity of sodium caseinate (50°C, 14%, w/w) (Reprinted, with permission from Elsevier, from Carr *et al.*, 2002).

the exact position depending on the casein sample (Dolby, 1961a) and the alkali used to dissolve the casein (Hayes and Muller, 1961). A maximum or a plateau was also seen between pH 8.0 and pH 10.0. Salzberg and Georgevits (1956), Puri *et al.* (1972) and Hermansson (1975) also noted these general trends. It is possible that these trends are based on changes in the inter- and intramolecular interactions within the solutions. The casein molecules carry more net negative charge at higher pH and above pH 10 they will tend to behave as separate entities. Below pH 8, there are likely to be strong, localised inter-molecular, but not intra-aggregate, attractive interactions. Between pH 8 and pH 10 there will be a shift from relatively short-range interactions to longer-range interactions as the aggregates present between pH 6 and 8 swell and start to interact with one another.

Korolczuk (1982b,c), who measured casein viscosity at pH 2.4–2.9 as a function of temperature and concentration, found that the viscosity increased faster with increasing concentration than for the same material dissolved in 2.5% borax solution (Korolczuk, 1982b). Increasing temperature initially reduced the viscosity but above 60°C the viscosity increased.

(b) Mixed cation caseinates

In efforts to produce commercial casein with a consistently low viscosity, Hayes and Muller (1961) examined the effect of adding calcium to caseinate solutions and found that, between pH 5.75 and pH 7.0, the viscosity increased and then decreased to less than that of the original caseinate. This result has been confirmed by Hayes *et al.* (1969), Towler (1974) and Carr *et al.* (2002) (Figure 29.5A). Sergeeva and D'Yachenko (1973) observed that the caseinates of the alkaline earth metals formed less viscous solutions than the caseinates of alkali metals.

A. Khaleque and C.R. Southward (unpublished results) dried a range of calcium/sodium caseinates prepared by varying the ratio of $Ca(OH)_2$ to NaOH. Reconstituted caseinate solutions showed very similar trends of viscosity versus Ca:Na ratio (Figure 29.5B) as described in Figure 29.5A. Thus, the drying process did not affect the viscosity characteristics of the calcium/sodium caseinate solutions. It also indicates that there is at least some equilibration of the cations within the protein dispersion.

The reason for the low viscosity of the calcium caseinate "solutions" compared with sodium caseinate solutions of the same concentration and pH lies in the differences between the structure of the calcium caseinate particles in the calcium caseinate dispersions and the sodium caseinate aggregates in the sodium caseinate solutions (Section 29.3.4(*a*)). The calcium caseinate particles contain all the casein types and also contain calcium ions that are essentially held immobilised in the regions of high negative charge that contain phosphoserine residues, e.g., residues 11–21 and 35–44 in β-casein, and 41–51 and 61–70 in α_{s1}-casein (Figures 29.1 and 29.2).

Because the manufacture of calcium caseinate involves the neutralisation of an acidic material, well-washed acid casein curd, with an alkali, $Ca(OH)_2$, and nearly all the calcium is tightly bound to the strong anionic sites of the protein, there are very few free anions or cations present within the calcium caseinate dispersions. This means that the negative charges on and within the particles are not electrostatically shielded, and this results in rearrangements of the casein segments within the particles to minimise intermolecular repulsions. Consequently, the calcium caseinate particles are poorly hydrated, compact, opaque and have a strong inter-particle repulsion. In the final calcium caseinate particle, κ-casein is most likely to be close to the outer surface of the particles, and the principal binding forces within the particles are inter- and intra-casein molecule hydrophobic association and what is often loosely called "calcium bridging".

In contrast, the casein–casein interactions in sodium caseinate solutions are dominated by electrostatic repulsions between the strands of the casein molecules, because the cation–polyanion interactions for the monovalent cations are much weaker (Ho and Waugh, 1965b) than the cation-polyanion interactions for the divalent cations (Ho and Waugh, 1965a). This repulsion overcomes the hydrophobic association energy. In admixture, the partial neutralisation of the phosphate clusters by the divalent cations allows longer range interactions to occur between the particles, which becomes apparent as a swelling of the calcium caseinate particles.

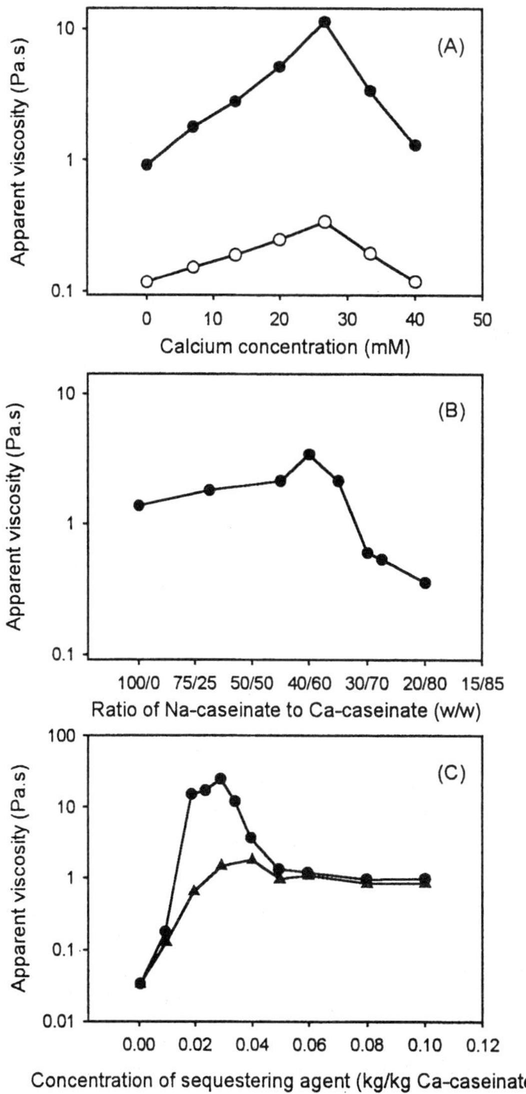

Figure 29.5 (A) Effect of calcium chloride concentration on the apparent viscosity of a sodium caseinate solution (14%, w/w) at 25°C (●) or 50°C (○) (Reprinted, with permission from Elsevier, from Carr *et al.*, 2002). (B) Effect of variation of sodium and calcium contents on the apparent viscosity of caseinates (A. Khaleque and C.R. Southward, 1972, personal communication, 2000). (C) Apparent viscosity (100 s^{-1}) of a calcium caseinate solution as a function of the calcium sequestering agents sodium tripolyphosphate (●) or trisodium citrate (▲) (Redrawn from Towler, 1974).

Replacement of a proportion of the sodium by calcium (and *vice versa*) in these low-ion systems might be expected to give a range of products with intermediate viscosity properties. However, as seen in Figure 29.5B, this is not the case and there is a peak in the viscosity-mineral composition plot. The increase in the proportion of calcium in the system gradually introduces strong crosslinks between the phosphorylated regions of the casein molecules. The sodium caseinate aggregates, although aggregated quite loosely, have a concentration of negative charge on the surface. This would be the likely target for calcium chelation, thus enlarging aggregate size. This enlargement would be reflected in a higher solution viscosity. Further increases in the quantity of calcium in the system continue to enlarge the effective size of the aggregates although there are small centres of strong association, mainly caused by the interaction between calcium and the phosphorylated segments. At the point of maximum viscosity, there is also a marked change in the appearance of the solutions from translucent to opaque as the particles become more discrete, less hydrated and there is greater interparticle electrostatic repulsion.

Addition of sequestrants. Hayes and Muller (1961) showed that addition of ethylenediaminetetraacetic acid (EDTA) to a borax solution of whole casein that had been precipitated at a high pH (5.05) altered the viscosity as a function of pH to that of a low-pH (4.45) casein. Various phosphate salts have been used to modify the properties of dairy products for nearly a century and deMan and co-workers (e.g., Vujicic and deMan, 1968) explored some of these interactions. Addition of either of the calcium-sequestering salts, trisodium citrate and sodium tripolyphosphate (TPP), to calcium caseinate at a ratio of 0.1:1.0 (w/w) increased the viscosity of the solution by comparable amounts (Towler, 1974). However, at a ratio of about 0.030:1.0, the apparent viscosity of the TPP mixture was nearly 20-fold greater than that of the trisodium citrate mixture (Figure 29.5C). The result for citrate addition is comparable to that for addition of EDTA (Hayes and Muller, 1961) but the effect of TPP indicates an even greater interaction effect between the casein molecules, induced by the complex phosphate ions, at certain critical ratios of calcium:TPP:casein.

Solubilisation of rennet casein. Towler (1974) examined the differences in flow properties between solutions of sodium caseinate and rennet casein dispersed with TPP. The apparent viscosity of all solutions showed a shear-thinning (pseudoplastic) effect, which was greater for the rennet casein solutions and was much greater for 15% (w/w) than for 10% solutions, which, in turn, was much greater than that for 7.5% solutions. The viscosity of the 15% rennet casein solution increased by nearly 80-fold as the shear rate changed from 10^4 s^{-1} to about 20 s^{-1}. Between 25 and 60°C at a shear rate of 100 s^{-1}, there was a linear relationship between the logarithm of the apparent viscosity and the inverse of the temperature (in K). It was also noted that at casein concentrations of 7.5 to 15%, both rennet casein and sodium caseinate solutions were pseudoplastic and the degree of pseudoplasticity increased as the concentration of casein was increased. Towler (1974) also found that 15% rennet casein solutions were thixotropic only when subjected to a high shear rate and that sodium caseinate solutions under similar conditions showed only pseudoplastic behaviour.

All the above results for rennet casein were obtained with a TPP to rennet casein ratio of 0.1:1.0. At the lowest ratio of TPP to casein (0.035:1.0) for full solubility, the apparent viscosity was approximately 5 Pa.s which decreased to about 700 mPa.s at a ratio of 0.1:1.0. Further increases in TPP content did not reduce the apparent viscosity significantly.

Konstance and Strange (1991) prepared 15% (w/w) mixtures of commercial samples of rennet casein, sodium caseinate and calcium caseinate in 0.05 to 0.55 M NaCl, $CaCl_2$ or NaH_2PO_4. After high-speed blending, samples of these dispersions were adjusted to pH values in the range 3.0 to 6.0 and diluted to 10% (w/v); sub-samples were held overnight at a temperature between 20 and 60°C. The protein content and the apparent viscosity of the non-sedimentable material were determined and the results fitted to a multiparameter mathematical model. In general, increasing the concentration of NaCl or $CaCl_2$ reduced solubility and, therefore, viscosity, whereas increasing the concentration of NaH_2PO_4 increased both viscosity and protein solubility and the type of the protein was relatively less important.

(c) Temperature effects

The correlation between the viscosity and the temperature of casein solutions is well established (Sergeeva and D'Yachenko, 1973; Towler, 1974; Korolczuk, 1982b; Roeper and Winter, 1982; Mohanty *et al.*, 1988; Fichtali *et al.*, 1993) and follows the Arrhenius relationship. Over the limited range investigated (25 to 60°C), Towler (1974) observed a linear relationship between the logarithm of the apparent viscosity and the reciprocal of the absolute temperature for solutions of either rennet casein or sodium caseinate. This relationship appears to hold true for all types of casein. However, some casein solutions deviate at high temperatures. Sergeeva and D'Yachenko (1973) observed that, although the viscosity of calcium caseinate solutions generally decreased as the temperature was increased, the solutions gelled reversibly at around 88°C. Korolczuk (1982b) observed that, between 25 and 80°C, the viscosity of neutral caseinate solutions decreased as the temperature was increased. However, solutions of acidified casein deviate from this relationship at high temperatures. Acidic casein solutions at pH 2.4–2.9 were characterised by a higher viscosity than neutral pH solutions of the same protein concentration and the solution viscosity fell only in the temperature range 25–60°C; at higher temperatures, the viscosity increased. Mohanty *et al.* (1988) reported that the viscosity of acidic casein solutions decreased logarithmically with increasing temperature up to 70°C. The increase in the viscosity of acidic casein solutions during heating at a high temperature can be interpreted as resulting from hydrophobic interactions (Korolczuk, 1982b).

29.3.5 External factors that affect casein viscosity

(a) Seasonal and lactational factors

In early studies, it was found that there was a seasonal (time of year) effect on the viscosity of casein mixtures although this could well have been a consequence of the time-of-lactation effect (Dolby, 1964). The seasonal effect was also found to depend on the genetic variant of α_{s1}-casein (Southward and Dolby, 1968) and on the proportion of α_{s1}-casein to the other caseins (i.e., β- and/or κ-casein) (Ram *et al.*, 1994). However, the extent of disulphide bonding of κ-casein was important also (Towler *et al.*, 1981).

Both α_{s1}- and β-caseins contribute to the viscosity of casein solutions through hydrophobic and hydrogen-bonded interactions (Towler *et al.*, 1981). However, κ-casein interacts strongly with the other casein components through non-covalently-bonded interactions and most of the κ-casein molecules are linked to one another by disulphide bridges, thus giving rise to relatively large aggregates in solution (Towler *et al.*, 1981).

(b) Adventitious enzyme activity

It was observed (Richardson, 1982) that the viscosity of some casein solutions decreased rapidly to appear "water thin" after a period of 3 weeks. This change in viscosity is attributed to proteolytic activity. Milk contains indigenous proteinases such as plasmin (alkaline milk proteinase) and lysosomal proteinases and may contain proteinases produced by bacteria. Plasmin is responsible for the hydrolysis of α_{s1}-, α_{s2}- and β-caseins, and Grufferty and Fox (1988) suggested that the lactational/seasonal decrease in the viscosity of casein, described by Dolby (1964) and Southward and Dolby (1968, 1971), could be due to the activity of plasmin which increases in late-lactation milk (e.g., Richardson, 1983b). Plasmin activity may therefore be an important factor in limiting the storage life of certain casein products.

(c) Processing factors

pH of precipitation. The effects of some processing conditions on the viscosity of caseinate solutions were explored by Dolby (1961a,b) and Hayes and Muller (1961). The pH of precipitation was very important because too low a pH allowed anions, such as sulphate, to precipitate with the casein whereas at too high a pH, calcium ions were bound (Hayes and Muller, 1961). A precipitation pH of 4.7, instead of 4.5, nearly doubled the viscosity of a 15% (w/v) casein solution in 2.4% (w/w) borax solution (Dolby, 1961b). See also Section 29.3.4.

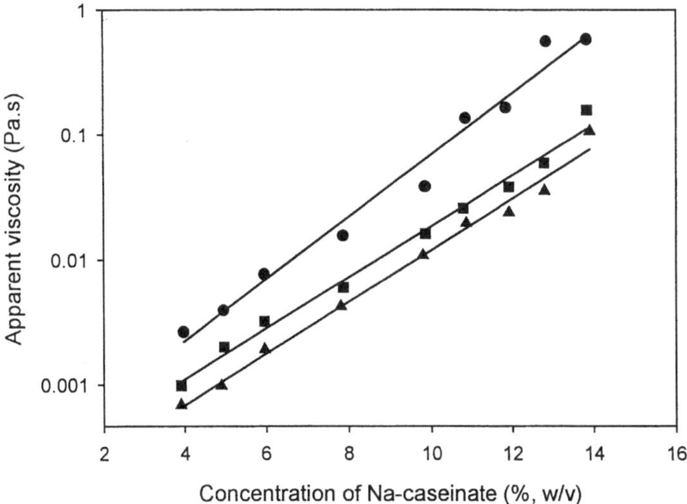

Figure 29.6 Apparent viscosity of sodium caseinate (●) and sodium caseinate heated at 120°C (■) or 132°C (▲) for 60 min as a function of concentration (redrawn from the data of Guo *et al.*, 1996).

Heat treatments. Processing temperature, both for washing and for drying the casein curd, should be restricted; otherwise, the viscosity of the dissolved casein will be too high (Dolby, 1961b).

Increasing the severity of preheat treatments for sodium caseinate prior to drying reduces the apparent viscosity of the reconstituted powder (Guo *et al.*, 1996) (Figure 29.6). The decrease in viscosity is correlated with a reduction in the size of the casein molecules due to heat-induced degradation. For sodium caseinate, viscosity is proportional to molecular weight according to the Mark-Houwink equation:

$$[\eta] = KM^a$$

where $[\eta]$ is intrinsic viscosity, M is molecular mass and K and a are constants (Shaw, 1980; Rha and Pradipasena, 1986). Increasing the severity of preheat treatment for sodium caseinate prior to drying reduces the apparent viscosity of the reconstituted powder (Guo *et al.*, 1996) (Figure 29.6). The decrease in viscocity is correlated with a reduction in the size of the casein molecules due to heat-induced degradation. For globular proteins, such as whey proteins, a can be regarded as zero.

29.3.6 Commercial uses of casein

(a) Industrial uses

One of the earliest uses for caseinate (i.e., acid casein that has been dissolved in alkali) was because of its adhesive function, i.e., as a glue in interior timber construction for laminated beams, plywood and furniture, and as a binder of pigments to paper in paper coating for high-quality printing [the major use of casein during the mid-20th century (Southward and Walker, 1982)]. Other paper-based adhesive applications of caseinate have included aluminium foil laminating, cap liners, match-striker strips, cigarette side seaming and bottle and can labelling (Salzberg *et al.*, 1974). Its adhesive properties have also been exploited in other technical applications, e.g., in water-based paints, [wallboard] joint cements, leather tanning and finishing, and horticultural spreaders (Southward and Walker, 1982).

The application of casein adhesives has always been dependent on the flow properties of the solution, i.e., its consistency or viscosity. This relates to the application of wood glues to timber by brushing and the application of urea-based casein adhesives to bottle labels in wiping stations at speeds of up to 90,000 bottles/h. Obviously, the least amount of water in the adhesive to produce the desired flow properties means that either less drying time or less heat energy for the evaporation of free water is needed, with a concomitant reduction in costs. Consequently, the viscosity of these casein-based solutions has a significant influence on the operations described. Urea is used in many casein-based formulations for bottle labels because of its solvation properties (Salzberg and Marino, 1975). In that role, it is effectively replacing water, thus increasing the solids in the adhesive formulation without increasing the solution viscosity.

Substantial quantities of casein were used in paper coating up to the 1970s because casein exhibited a number of desirable characteristics for this application. These included its easy dispersion and preservation, good pigment-dispersing power, strong film-forming characteristics, high adhesive strength for pigments and good ink and varnish holdout (Salzberg and Marino, 1975). However, casein also has some disadvantages for paper coating, amongst which is the high viscosity of concentrated solutions of casein, which limits the solids content of the coating mix and hence the speed of the coating processes (Salzberg and Georgevits, 1956).

Muller and Hayes (1963) and Salzberg and Simonds (1965) addressed the issue of high viscosity in casein-based paper-coating mixtures by treating the casein (either in milk/alkaline solution or in acid solution) with an enzyme, such as pangestin (an alkaline proteinase preparation) or pepsin, to cause slight proteolysis of the casein. The purpose of this treatment was to reduce the viscosity of the casein solution (and hence the paper-coating mixture), with the aim of producing a coating mix of higher solids content (with less water to evaporate) and with a viscosity equivalent to that of a coating mix containing a standard casein. Close control of the proteolytic treatment was necessary in order to prevent a reduction in the adhesive strength of the modified casein (Muller and Hayes, 1963), the viscosity of which could be reduced to about one-fifth of the normal value without deleterious effects. Other mechanisms to reduce the viscosity of casein solutions included the addition of urea or dicyandiamide, effectively increasing solvation (Salzberg and Marino, 1975), and the incorporation of disulphide-bond reducing agents such as sodium mercaptoacetate, cysteine or sodium bisulphite (Ronai and Weisberg, 1954). Later studies on both viscosity reduction and the adhesive strength of casein in paper coating were carried out to evaluate combined treatments of casein with dicyandiamide (solvation) and ammonium mercaptoacetate or 2-mercaptoethanol (disulphide bond reduction) (Towler *et al.*, 1981). In this last-named study, it was found that viscosity reduction was more significant at low shear (up to 500 s^{-1}), which relates to the mechanical forces involved in stirred tanks and low-shear pumping. At higher shear rates (up to 18,000 s^{-1}), however, the reduction in viscosity by the addition of these chemicals was much less significant. The limiting processing step in the paper-coating process was considered to be the viscosity of the coating solution at the coating nip, where the coating solution is actually applied to the paper. At this point, shear rates of the order of 10^6 s^{-1} are considered to apply (Salzberg *et al.*, 1961). These are probably of the same order as for the application of adhesives to bottle labels, as mentioned above. Consequently, there is some doubt as to whether any of the treatments mentioned above would show a significant reduction in the viscosity of the coating solution at that stage of the process. Nevertheless, as the addition of urea or dicyandiamide (promoting solvation) increases the solids content of the casein coating solution, it concomitantly reduces the amount of water to be evaporated during drying, and therefore may have an economic benefit in this application.

When the world price of casein increased substantially in the early 1970s, there was a significant move to cheaper adhesives, such as starch and cellulose derivatives and synthetic polymers (generally termed latexes) for paper coating. As a result of these factors, it is believed that most of the treatments referred to above are not currently used. Estimates of world casein consumption suggest that the proportion of casein used in non-food technical applications (compared with edible uses) declined from approximately 80% in the mid-1960s to about 20% by the late 1980s (Southward, 1994). That proportion may be of a similar magnitude at present.

(b) Food uses

In foods, the nutritional quality of casein products is obviously an important property. However, caseinates are in particular demand as functional food ingredients, in which their surface properties predominate. These include film formation, e.g., in edible films (Gennadios *et al.*, 1994) and encapsulation of pharmaceutical preparations; fat and oil emulsification in meats, desserts, cream products, whipped toppings, beverages, processed cheese and spreads; and aeration and foam stability in frozen desserts, whipped toppings, beverages and confectionery. The viscosity of caseinates has some importance when they are used to control the consistency in food products such as soups, spreads and processed cheese-like foods (Southward, 1989). In some uses, such as comminuted meats, a high viscosity caseinate may be desired (for "water binding") (e.g., Rund, 1973). On the other hand, in the processing of sodium caseinate, the viscosity of its solutions limits the concentration that can be pumped to the spray drier. That observation prompted a study, similar to the work described above relating to paper coating, in which solutions of sodium caseinate were hydrolysed using proteolytic enzymes such as papain and pepsin to reduce the solution viscosity. Trends similar to those described by Muller and Hayes (1963) were obtained, but the viscosity reduction was too low to be commercially useful (Hooker *et al.*, 1982). However, viscosity is probably not as significant in commerce as the other functional properties listed above.

29.4 Skim milk and whole milk protein systems

The following sections deal with products which contain all the milk proteins. Sections 29.4.1 to 29.4.4 are concerned primarily with the viscous properties of the process streams prior to milk powder manufacture and Sections 29.4.5 and 29.4.6 describe the properties of dissolved or dispersed powders.

29.4.1 Skim milk

(a) Milk

Milk is a dilute aqueous solution with a low viscosity, about twice that of water. The higher viscosity of milk is caused by the dissolved substances and the dispersed particles. For practical purposes, milk displays essentially Newtonian properties. Some workers have reported the occurrence of a very low yield stress for milk, particularly at low temperatures (Wayne and Shoemaker, 1988; Kristensen *et al.*, 1997). However, it is difficult to determine whether the apparent yield stress is a true yield point or an artefact of measurement.

In general, the viscosity of dilute solutions is governed by the volume fraction of the dispersed particles and the viscosity of the medium. The interdependence of these factors is described by Einstein's equation which assumes that the suspended particles are hard spheres and do not interact with one another:

$$\eta = \eta_{ref}(1 + 2.5\phi)$$

where η = viscosity of the suspension, η_{ref} = viscosity of the medium and ϕ = the sum of the hydrodynamic volumes of all the dispersed particles significantly larger than the solvent molecules. The hydrodynamic volumes of the dispersed particles found in milk products are governed by the size and shape of the molecules and their association with the solvent.

(b) Casein micelles

Skim milk contains five components, casein micelles, whey proteins, lactose, salts and water, which are important in determining the rheological characteristics of milk as a consequence of processing. The casein micelles contain 2–4 g water/g protein and this quantity varies with the ratio of κ-casein to the other caseins, mineral content, pH and previous heat treatment. A relatively small proportion of this water is bound to the milk proteins in the same way that water is bound to other proteins (Fennema, 1985; Rupley and Careri, 1991). The remainder, which is in free exchange with the bulk water (Lelievre and Creamer, 1978), is held between the protein strands within the micelle, with the highest proportion towards the outer regions of the natural micelle.

(c) Heat and environmental effects

The effects of heat on milk are wide-ranging and include: whey protein aggregation, protein-sugar complex formation, interaction of the whey proteins with κ-casein, dissociation of κ-casein from the micelle surface and changed mineralisation of the micelle. Many of these reactions alter micelle hydration and hence the particle size and the rheological properties of the milk. Clearly, greater micelle hydration will make the milk more viscous. Also, an increase in the extent of whey protein denaturation, which will change globular particles into non-globular particles, will increase the viscosity (all other things being equal).

Micelle hydration increases with pH and decreases with heat treatment (Creamer and Matheson, 1980), and the loss of κ-casein from the micelle increases with decreasing pH (Singh and Creamer, 1992). The proportion of β-lactoglobulin and α-lactalbumin associated with the casein micelles does not exceed 55% over a range of temperature (Oldfield *et al.*, 1998).

29.4.2 Concentrated milks

The viscosity of milk increases exponentially with increasing total solids, in the range 44 to 50% total solids (Bloore and Boag, 1981; Hayashi and Kudo, 1989; Velez-Ruiz and Barbosa-Canovas, 1998). The relationship between the concentration of either whole or skim milk and viscosity depends on the volume fraction of the proteinaceous material and the fat present in milk and on the viscosity of the medium (Snoeren *et al.*, 1983).

(a) Temperature effect

The viscosity of milk at any particular concentration decreases with increases in temperature. After fitting power law and Herschel-Bulkley models to viscosity data for milk at concentrations up to 48.6% (w/w), Velez-Ruiz and Barbosa-Canovas (1998) found that the effect of temperature on the consistency index followed the Arrhenius equation. Horne

(1998) suggested that as the temperature is increased, the strength of the hydrophobic interactions causes the micelles to tighten up and become more compact, allowing the suspension to flow more freely and so contribute to the decrease in apparent viscosity.

(b) Prior heat treatment
Heat treatments (often known as preheating), which are used to control some characteristics of the product functional properties in its end-use, tend to increase the apparent viscosity of milk. Morr (1969) found that heat treatment increases the viscosity of milk products primarily by increasing the quantity of imbibed water retained in the heat-induced microstructure. de Jong and van der Linden (1998) proposed a multi-component polymerisation model to predict the rheological properties of heat-treated milk. This model relates heat-induced protein aggregation to viscosity changes in milk. The model recognises the denaturation/aggregation kinetics of β-lactoglobulin and its interaction with casein micelles. Jeurnink and de Kruif (1995), however, concluded that the observed increases in viscosity due to heating skim milk could not be explained by increases in micelle dimensions alone but were caused by temporary clustering of the micelles.

29.4.3 Age thickening of concentrates

On holding at any temperature >50°C, the viscosity of concentrates has been found to decrease to a minimum and then to increase (Beeby, 1966). Further, at any particular concentration, the increase in viscosity begins earlier the higher the temperature of holding. Beeby (1966) concluded that the initial decrease in the viscosity of concentrated milk on holding at a particular temperature is due to relatively weak intermolecular forces that are broken by thermal agitation. The subsequent increase in viscosity may be explained by the aggregation of protein. As the concentration of milk solids is increased, the solubility product of many of the salts in milk is exceeded and these salts probably become associated with the casein micelles, thereby decreasing the net negative charge or the surface properties of the micelles and thus destabilising the system, leading to aggregation during heat treatments. In its early stages, aggregation is accompanied by an increase in viscosity and, as the process continues, gelation eventually occurs (Muir, 1980). In their review on changes in milk components during evaporation, Singh and Newstead (1992) noted that it is likely that this type of aggregation and gelation is caused by the deposition of calcium and phosphate on the surface of the casein micelles. Snoeren *et al.* (1984a) suggested that age thickening is due to the loosening of casein micelles. As a consequence of concentration, the pH decreases and the ionic strength increases which results in an increase in the voluminosity of the casein micelles and favours the solubility of β-casein. Snoeren *et al.* (1983)

suggested that the mechanism controlling the thickening of whole milk concentrates is also linked to the denaturation of whey proteins. According to Snoeren *et al.* (1982), heat-induced changes in the viscosity of skim milk concentrates can be traced back to denaturation of the whey proteins, and consequently to a highly increased hydration and bulkiness, in comparison to undenatured whey protein.

There is some debate over the relationship between age thickening of the concentrate and the quality of the spray-dried powder. Snoeren *et al.* (1981) showed that the properties of spray-dried milk were related only to the initial basic viscosity (defined as the viscosity at an infinite shear rate) of the concentrate and not to the structural viscosity (which is deemed to be the contribution to the apparent viscosity arising from the hydrodynamic interactions between the particles) caused by holding (at 50°C) of the concentrate ("age thickening"). The basic viscosity was affected by the preheat treatment of the milk and by increasing the dry matter content of the concentrate. Therefore, Snoeren *et al.* (1981) concluded that the structural viscosity, observed on holding, must be completely or almost completely disrupted by the atomising disk in the spray dryer.

However, it is not known whether the suspected decrease in structural viscosity due to the high shear rates encountered during spray drying is reversible on reducing the shear rate, i.e., does a concentrate that has undergone age thickening have the same apparent viscosity at low shear rates as it had prior to being subjected to high shear rates or the same apparent viscosity as it did prior to holding? The question of what minimum shear rate is required during spray drying to enable the breakdown of the structural viscosity is also unanswered. It is clear from this discussion that it is important to understand the pseudoplastic behaviour of milk concentrates with regard to the degree of age thickening. It is also clear that the viscosity of the concentrate should, ideally, be determined at a rate of shear that is equal to that exerted by the disk in the spray dryer, but this is difficult because of high atomisation shear rates. Baldwin *et al.* (1980) reported that the thickening of the concentrate during holding can cause problems during pumping. The shear rates resulting from pumping are probably not sufficient to disrupt the structural viscosity resulting from holding. If the shear rate resulting in the presumed disruption of the structural viscosity during spray drying could be simulated, then problems associated with age thickening may be reduced.

Reddy and Datta (1994) fitted power law models to viscosity data for concentrated reconstituted whole milk containing 40 to 65% solids and at a temperature of 40 to 65°C. The consistency coefficient (the power-law parameter which is related to the viscosity) was found to be dependent on both temperature and concentration. The flow behaviour index (which indicates the deviation from Newtonian flow) was found to be independent of temperature.

In conclusion, the apparent viscosity of milk concentrate entering the spray dryer appears to be dependent on the degree of concentration, the holding temperature, especially at temperatures above about 40°C (Bloore and Boag, 1981), the period of holding and variations in the salt balance.

(a) Concentrated UHT milk

In the manufacture of UHT milk concentrates, most of the interesting changes in rheological properties occur after production is completed. The post-production changes in the rheology of products showing age-thickening or gelation, occur over a long time. The rheological changes involved are not believed to be the result of a diffusion process (Prentice, 1984). Although the mean free path of any particle is necessarily small because of the close packing, movement is not so restricted that it would account for changes on a time-scale of weeks or months. What appears to happen is that slow changes occur in the protein, whereby it gradually unfolds somewhat and complexes of fat and protein appear, while the soluble whey proteins form a network-like structure which tends to enmesh the casein. The end product of the reactions is for the network-like structure to pervade throughout the milk, giving rise to gelation (Prentice, 1984). Age thickening on storage of concentrated milks has also been associated with the crystallisation of lactose in the concentrate (Baucke and Sanderson, 1970).

29.4.4 Whole milk concentrates

The production of whole milk powder and concentrates involves an additional step, i.e., homogenisation. It is well documented that when whole milk concentrate is homogenised, its viscosity is increased (Reuter and Randhahn, 1978; de Vilder and Moermans, 1983; Snoeren *et al.*, 1984b). The increase in the viscosity of homogenised whole milk concentrate may be explained by the increase in the apparent volume fraction (Snoeren *et al.*, 1984b). It appears that the serum in the space between the casein micelles adhering to the fat globules becomes part of the fat-protein complex.

de Vilder *et al.* (1979) carried out a comprehensive investigation of the effect of homogenisation on whole milk concentrates. The viscosity of the non-homogenised concentrate increased from 3.1 to 23.0 Pa.s with increasing dry matter content (43.2 to 54.7%). Single-stage homogenisation increased the viscosity from 6.9 to 780.0 Pa.s but two-stage homogenisation increased it from 6.2 to 440.0 Pa.s. Mulder and Walstra (1974) attributed the increase in viscosity following single-stage homogenisation of milk to the formation of fat clusters. During two-stage homogenisation, these clusters were largely broken in the second stage, so that a lower viscosity was observed although the level of homogenisation was virtually unchanged.

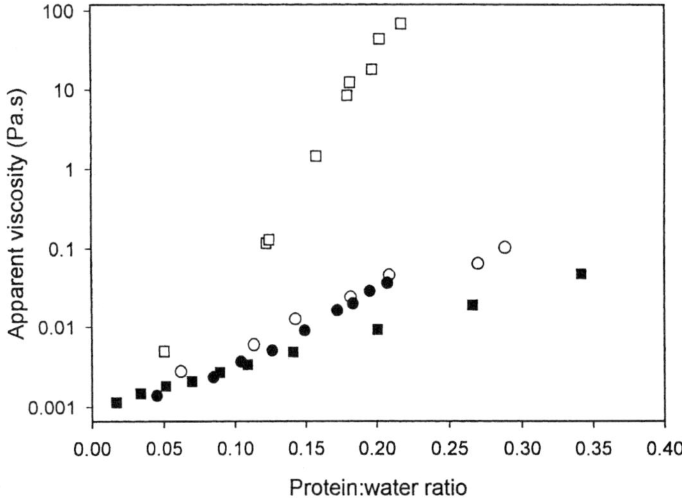

Figure 29.7 Apparent viscosity of reconstituted MPC85 (●) and reconstituted low-heat skim milk powder (○) at 15°C, pH 7.0, 291 s^{-1} (redrawn from Carr, 1999); sodium caseinate (□), 25°C, pH 6.7 (redrawn from Carr, 1994); WPC (■), 22°C, pH 7.0, 291 s^{-1} (redrawn from Tang, 1993) as a function of protein to water ratio.

The viscosity of whole milk concentrate has been observed to increase with increasing homogenisation pressure (14 to 20 MPa) (de Vilder *et al.*, 1979).

29.4.5 Milk protein concentrates (MPCs)

These products are manufactured by membrane filtration of skim milk to obtain a stream with a high protein to lactose ratio which is then concentrated and dried in a similar manner to that for other skim milk powders. The apparent viscosity of MPC containing 85% protein, MPC85, on a total solids basis is higher than that of typical skim milk (both reconstituted from powders). If the apparent viscosity is compared on a protein concentration basis, the apparent viscosity of MPC85 appears to be lower than that of low-heat skim milk. However, when the comparison is on a protein to water ratio (Carr, 1999), the apparent viscosity is the same as that of low-heat skim milk (Figure 29.7). This discrepancy is due to the inclusion in the calculation of protein concentration and total solids (w/w) of components that do not directly affect apparent viscosity, such as salts and lactose.

Increasing the severity of heat treatments prior to the ultrafiltration/diafiltration operations during the manufacture of MPC85 has been shown to increase the apparent viscosity of the reconstituted powder (Carr, 1999).

The increase in the viscosity of MPC85 with heat treatment was correlated with an increase in the apparent diameter of the casein micelle.

29.4.6 Coprecipitates

These products are manufactured from skim milk by a moderately severe heat treatment, followed by precipitation between pH 4.4 (with acid) and pH 6.5 (with added calcium salts). The viscosity of aqueous solutions of soluble coprecipitates (which contain both casein and heat-denatured whey proteins) at different concentrations, temperatures and shear rates was measured by Hayes *et al.* (1969), who found that low-calcium coprecipitate had a viscosity profile similar to that of acid casein. The viscosity of solutions of medium- and high-calcium coprecipitate was found to be relatively high at pH values above 7. Southward and Goldman (1975) observed that the viscosity-shear rate characteristics of the coprecipitates show patterns similar to those of various caseins that were measured by Towler (1974). Thus, the viscosity-shear rate plots for high-calcium coprecipitate and rennet casein, both of which have a relatively high calcium content (near 3%), are very similar. Furthermore, the low-calcium analogues (low-calcium coprecipitate and sodium caseinate) have similar but lower viscosity profiles, as observed by Hayes *et al.* (1969). Medium-calcium coprecipitate has intermediate viscosity characteristics. Thus, it appears that the "occluded" calcium in the coprecipitates and caseins is the prime factor causing the variations observed in the viscosity of these milk protein products at similar temperatures and concentrations, whereas the coprecipitated whey proteins have little or no effect.

29.5 Conclusions

It is apparent from the review of the literature described in this chapter that a comparatively large number of studies have been made on the viscosity of solutions of casein. No doubt, this is the result of the relative ease of separation of relatively pure "whole" casein from the other milk components and of its commercial importance throughout the 20th Century. The techniques used to isolate the individual caseins, which have been developed and significantly refined during the last 30–40 years, have also led to a better understanding of the properties of each of the casein fractions in terms of their effects on the viscosity characteristics of solutions of whole casein.

The role of casein and milk protein viscosity, especially in concentrated milk systems, became more important internationally during the latter half of the 20th Century as trade in milk powders developed. Heat-induced interactions between κ-casein and the whey proteins, produced during the manufacture of milk powders, have been of special importance because of the subsequent use of the powders in canned, evaporated milks. The isolation of whey proteins has assumed commercial significance only since the 1970s, as a result of the development of large-scale economically viable membrane equipment. Perhaps this is the reason for the relatively small number of studies on the viscosity of whey proteins that have been reported.

One application of milk protein products which appears to be growing at present is fortification of yoghurt and other acidified milk products. The effect of milk protein ingredients on the viscosity of such acidified foods is obviously important and needs further study to provide the basis for full exploitation of this application. The viscosity of a solution of a "pure" milk protein does not, of course, always translate directly to solutions or dispersions of the final product which may contain the [added] milk protein. Thus, the presence or addition of minerals, carbohydrates and fats may influence the properties of the milk protein ingredient. Specific processing steps, such as heating or homogenisation, may also influence the properties of the added protein or the product. These aspects are the reason for the extensive and continuing interest in the so-called "functional" properties of milk and other proteins, e.g., whipping, foaming and emulsification, that have been reported in the literature during the past 30 years. In most cases, it is not possible to duplicate exactly what happens in the final user's production plant. However, some means of determining trends and predicting changes under scenarios of "what ifs" may be achieved by a combination of the more fundamental studies of the viscosity of milk proteins and their testing in so-called simple model systems.

Variations in breed of cow, lactation, climate, diet and plane of nutrition affect the protein composition of the milk. In addition, the storage of the milk prior to pasteurisation can influence the degree of both specific and non-specific proteolysis as well as the pH and consequently the mineral balance between micellar and serum phases. All these factors influence protein–protein interactions in the process streams as well as the final products.

Changes in milk protein manufacturing techniques, especially those used in membrane processing for producing whey proteins and MPCs, which are in relatively rapid development, may also need to be reviewed continually in order to assess their effects, real and potential, on the rheological properties of the resultant protein products.

Acknowledgements

The authors gratefully acknowledge the careful editorial work by Claire Woodhall and the financial support from the New Zealand Dairy Board and the Foundation for Research, Science and Technology (Contracts DRI 801 and DRIX0001).

References

Andrews, A.T. (1991) Partial denaturation and renaturation of β-lactoglobulin at air–water interfaces. *Biochem. Soc. Trans.*, **19**, 272S.

Baldwin, A.J., Baucke, A.G. and Sanderson, W.B. (1980) The effect of concentrate viscosity on the properties of spray dried skim milk powder. *N.Z. J. Dairy Sci. Technol.*, **15**, 289–97.

Baucke, A.G. and Sanderson, W.B. (1970) Viscosity increase in concentrated skim milk, in *New Zealand Dairy Research Institute Annual Report*, New Zealand Dairy Research Institute, Palmerston North, p. 44.

Beeby, R. (1966) Heat-induced changes in the viscosity of concentrated skim milk. *Proc. XVII Int. Dairy Congr.* (Munich), **E/F**, 115–21.

Berry, G.P. and Creamer, L.K. (1975) The association of β-casein. The importance of the C-terminal region. *Biochemistry*, **14**, 3542–5.

Berry, G.P. and Creamer, L.K. (1976) A comparison between the protein–protein association in caseinate and rennet-treated caseinate solutions. *N.Z. J. Dairy Sci. Technol.*, **11**, 127–9.

Bloore C.G. and Boag, I.F. (1981) Some factors affecting the viscosity of concentrated skim milk. *N.Z. J. Dairy Sci. Technol.*, **16**, 143–54.

Bylund, G. (1995) *Dairy Processing Handbook*, Tetra Pak Processing Systems AB, Lund.

Carr, A.J. (1994) *Rheology of Sodium Caseinate Solutions*, M.Tech Thesis, Massey University, Palmerston North.

Carr, A.J. (1999) *The Functional Properties of Milk Protein Concentrates*, Ph.D Thesis, Massey University, Palmerston North.

Carr, A.J., Munro, P.A. and Campanella, O.H. (2002) Effect of added monovalent or divalent cations on the rheology of sodium caseinate solutions. *Int. Diary J.*, 12, in press.

Costell, E. and Durán, L. (2000) Sensory and instrumental measures of viscosity, in *Trends in Food Engineering*, (J.E. Lozano, C. Anon, E. Parada-Arias and G.V. Barbosa-Canovas eds.) M.M. Gongora-Nieto, associate ed., Technomic Publishing Company Inc., Lancaster, Pennsylvania, pp. 53–64.

Costell, E., Pastor, M.V. and Durán, L. (1994) Rheological parameters as stimuli of perceived sensory viscosity in peach nectars, in *Progress and Trends in Rheology*, Volume IV, (C. Gallegos ed.) Steinkopf Verlag GmbH & Co. KG, Darmstadt, pp. 218–20.

Creamer, L.K. (1985) Water absorption by renneted casein micelles. *Milchwissenschaft*, **40**, 589–91.

Creamer, L.K. and Berry, G.P. (1975) A study of the properties of dissociated bovine casein micelles. *J. Dairy Res.*, **42**, 169–83.

Creamer, L.K. and Matheson, A.R. (1980) Effect of heat treatment on the proteins of pasteurized skim milk. *N.Z. J. Dairy Sci. Technol.*, **15**, 37–49.

Creamer, L.K. and Matheson, A.R. (1981) Separation of bovine caseins using hydrophobic interaction chromatography. *J. Chromatogr.*, **210**, 105–11.

Creamer, L.K. and Yamashita, S. (1976) The role of phosphate in casein micelle structure. I. The effect of inorganic phosphate on calcium-caseinate aggregation. *N.Z. J. Dairy Sci. Technol.*, **11**, 257–62.

Creamer, L.K., Berry, G.P. and Mills, O.E. (1977) A study of the dissociation of β-casein from the bovine casein micelle at low temperature. *N.Z. J. Dairy Sci. Technol.*, **12**, 58–66.

Creamer, L.K., Zoerb, H.F., Olson, N.F. and Richardson, T. (1982) Surface hydrophobicity of α_{s1}-I, α_{s1}-casein A and B and its implications in cheese structure. *J. Dairy Sci.*, **65**, 902–6.

Creamer, L.K., Plowman, J.E., Liddell, M.J., Smith, M.H. and Hill J.P. (1998) Micelle stability: κ-casein structure and function. *J. Dairy Sci.*, **81**, 3004–12.

Damodaran, S. (1997) Food proteins; an overview, in *Food Proteins and Their Applications*, (S. Damodaran and A. Paraf eds.) Marcel Dekker, Inc., New York, pp. 1–24.

de Jong, P. and van der Linden, H.J.L.J. (1998) Polymerization model for prediction of heat-induced protein denaturation and viscosity changes in milk. *J. Agric. Food Chem.*, **46**, 2136–42.

de Vilder, J. and Moermans, R. (1983) The continuous measurement of the viscosity of the concentrate during the production of milk powder. *Milchwissenschaft*, **38**, 449–52.

de Vilder, J., Martens, R. and Naudts, M. (1979) The influence of the dry matter content, the homogenization and the heating of concentrate on physical characteristics of whole milk powder. *Milchwissenschaft*, **34**, 78–84.

Dolby, R.M. (1961a) The viscosity index of lactic casein. 1. Methods of determination. *Aust. J. Dairy Technol.*, **16**, 256–60.

Dolby, R.M. (1961b) The viscosity index of lactic casein. 2. The effect of manufacturing conditions. *Aust. J. Dairy Technol.*, **16**, 261–4.

Dolby, R.M. (1964) Seasonal variation in the viscosity index of New Zealand commercial casein. *Aust. J. Dairy Technol.*, **19**, 7–8.

Fennema, O.R. (1985) Water and ice, in *Food Chemistry*, 2nd edn, (O.R. Fennema ed.) Marcel Dekker Inc., New York, pp. 23–67.

Fichtali, J., van de Voort, F.R. and Doyon, G.J. (1993) A rheological model for sodium caseinate. *J. Food Eng.*, **19**, 203–11.

Frisch, H.L. and Simha, R. (1956) The viscosity of colloidal suspensions and macromolecular solutions, in *Rheology Theory and Applications*, Vol. 1, (F.R. Eirich ed.) Academic Press, New York, pp. 525–613.

Gennadios, A., McHugh, T.H., Weller, C.L. and Krochta, J.M. (1994) Edible coatings and films based on proteins, in *Edible Coatings and Films to Improve Food Quality*, (J.M. Krochta, E.A. Baldwin and M.O. Nisperos-Carriedo eds.) Technomic Publishing Company, Inc., Lancaster, Pennsylvania, pp. 231–77.

Grufferty, M.B. and Fox, P.F. (1988) Functional properties of casein hydrolysed by alkaline milk proteinase. *N.Z. J. Dairy Sci. Technol.*, **23**, 95–108.

Guo, M.R., Fox, P.F., Flynn, A. and Kindstedt, P.S. (1996) Heat-induced modifications of the functional properties of sodium caseinate. *Int. Dairy J.*, **6**, 473–83.

Havea, P., Singh, H., Creamer, L.K. and Campanella, O.H. (1998) Electrophoretic characterization of the protein products formed during heat treatment of whey protein concentrate solutions. *J. Dairy Res.*, **65**, 79–91.

Hayashi, H. and Kudo, N. (1989) Effect of viscosity on spray drying of milk, in *Reports of Research Laboratory – Technical Research Institute, Snow Brand Milk Products Co.*, No. 88. Snow Brand Milk Products Co., Tokyo, pp. 53–9.

Hayes, J.F. and Muller, L.L. (1961) Factors affecting the viscosity of solutions of acid-precipitated caseins. *Aust. J. Dairy Technol.*, **16**, 265–9.

Hayes, J.F., Muller, L.L. and Fraser, P. (1969) Studies on co-precipitates of milk proteins. Part 5 – Investigations on viscosity of co-precipitates in dispersions of high concentration. *Aust. J. Dairy Technol.*, **24**, 75–8.

Hemer, Y., Horne, D.S. (1998) Electrostatic interactions in adsorbed protien layers probed by a sedimentation technique. *J. Coll. Interface Sci.*, **206**, 138–45.

Herbert, R.S. (1972) *Whey Protein Isolate Preparation and Functional Properties Evaluation*, Ph.D. Thesis, University of Wisconsin, Madison, Wisconsin.

Hermansson, A.-M. (1975) Functional properties of proteins for foods—flow properties. *J. Texture Stud.*, **5**, 425–39.

Ho, C. and Chen, A.H. (1967) The polymerization of bovine α_s-casein B. *J. Biol. Chem.*, **242**, 551–3.

Ho, C. and Waugh, D.F. (1965a) Interactions of bovine caseins with divalent cations. *J. Am. Chem. Soc.*, **87**, 889–92.

Ho, C. and Waugh, D.F. (1965b) Interactions of bovine α_s-casein with small ions. *J. Am. Chem. Soc.*, **87**, 110–7.

Hooker, P.H., Munro, P.A. and O'Meara, G.M. (1982) Reduction of the viscosity of sodium caseinate solutions by enzymic hydrolysis. *N.Z. J. Dairy Sci. Technol.*, **17**, 35–40.

Horne, D.S. (1998) Casein interactions: casting light on the *black boxes*, the structure in dairy products. *Int. Dairy J.*, **8**, 171–7.

Jeurnink, T.J.M. and de Kruif, K.G. (1995) Calcium concentration in milk in relation to heat stability and fouling. *Neth. Milk Dairy J.*, **49**, 151–65.

Konstance, R.P. and Strange, E.D. (1991) Solubility and viscous properties of casein and caseinates. *J. Food Sci.*, **56**, 556–9.

Korolczuk, J. (1981a) Voluminosity and viscosity of casein solution. I. The correlation between the voluminosity, protein concentration and viscosity. *Milchwissenschaft*, **36**, 414–6.

Korolczuk, J. (1981b) Voluminosity and viscosity of casein solutions. II. The correlation between the voluminosity and the empirical constants of regression equations. *Milchwissenschaft*, **36**, 467–9.

Korolczuk, J. (1982a) Hydration and viscosity of casein solutions. *Milchwissenschaft*, **37**, 274–6.

Korolczuk, J. (1982b) Viscosity and hydration of neutral and acidic milk protein concentrates and caseins. *N.Z. J. Dairy Sci. Technol.*, **17**, 135–40.

Korolczuk, J. (1982c) Effect of temperature on viscosity and hydration of casein in neutral and acidic solutions. *N.Z. J. Dairy Sci. Technol.*, **17**, 273–6.

Kristensen, D., Jensen, P.Y., Madsen, F. and Birdi, K.S. (1997) Rheology and surface tension of selected processed dairy fluids: influence of temperature. *J. Dairy Sci.*, **80**, 2282–90.

Lelievre, J. and Creamer, L.K. (1978) An NMR study of the formation and syneresis of renneted milk gels. *Milchwissenschaft*, **33**, 73–6.

Manderson, G.A., Hardman, M.J. and Creamer, L.K. (1998) Effect of heat treatment on the conformation and aggregation of β-lactoglobulin A, B, and C. *J. Agric. Food Chem.*, **46**, 5052–61.

Manderson, G.A., Hardman, M.J. and Creamer, L.K. (1999) Effect of heat treatment on bovine β-lactoglobulin A, B, and C explored using thiol availability and fluorescence. *J. Agric. Food Chem.*, **47**, 3617–27.

Mohanty, B., Mulvihill, D.M. and Fox, P.F. (1988) Hydration-related properties of casein at pH 2.0–3.0. *Food Chem.*, **27**, 225–36.

Morr C.V. (1969) Protein aggregation in conventional and ultra high-temperature heated skimmilk. *J. Dairy Sci.*, **52**, 1174–80.

Muir, D.D. (1980) Concentration and milk powder quality, in *Milk and Whey Powders*, Society of Dairy Technology, Wembley, Middlesex, pp. 73–84.

Mulder, H. and Walstra, P. (1974) *The Milk Fat Globule: Emulsion Science as Applied to Milk Products and Comparable Foods*, PUDOC, Wageningen, Commonwealth Agricultural Bureaux, Farnham Royal, UK.

Muller, L.L. and Hayes, J.F. (1963) The manufacture of low-viscosity casein. *Aust. J. Dairy Technol.*, **18**, 184–8.

Noelken, M. and Reibstein, M. (1968) Conformation of β-casein B. *Arch. Biochem. Biophys.*, **123**, 397–402.

Oldfield, D.J., Singh, H. and Taylor, M.W. (1998) Association of β-lactoglobulin and α-lactalbumin with the casein micelles in skim milk heated in an ultra-high temperature plant. *Int. Dairy J.*, **8**, 765–70.

Pearce, K.N. (1975) A fluorescence study of the temperature-dependent polymerization of bovine β-casein A^1. *Eur. J. Biochem.*, **58**, 23–9.

Pradipasena, P. and Rha, C.K. (1977) Pseudoplastic and rheopectic properties of a globular protein (β-lactoglobulin) solution. *J. Texture Stud.*, **8**, 311–25.

Prentice, J.H. (1984) *Measurements in the Rheology of Foodstuffs*, Elsevier Applied Science Publishers, London.

Puri, B.R., Mohindroo, U. and Malik, R.C. (1972) Studies on physico–chemical properties of caseins. III. Viscosities of casein solutions in different alkalies. *J. Indian Chem. Soc.*, **49**, 855–63.

Ram, S.P., Roeper, J., Thompson, C.J., Wilson, S.C., Lewis, D.S. and Elston, P.D. (1994) Physico–chemical and functional properties of the fractionated casein products. *Proc. 24th Int Dairy Congr.* (Melbourne), Brief Communications, 457.

Reddy, C.S. and Datta, A.K. (1994) Thermophysical properties of concentrated reconstituted milk during processing. *J. Food Eng.*, **21**, 31–40.

Reuter, H. and Randhahn, H. (1978) Relation between fat globule size distribution and viscosity of raw milk. *Proc. 20th Int. Dairy Congr.* (Paris), Brief Communications, 281–2.

Rha, C.K. and Pradipasena, P. (1986) Viscosity of proteins, in *Functional Properties of Food Macromolecules*, (J.R. Mitchell and D.A. Ledward eds.) Elsevier Applied Science Publishers, London, pp. 79–120.

Richardson, B.C. (1982) The effect of storage on the viscosity of some casein solutions. *N.Z. J. Dairy Sci. Technol.*, **17**, 277–82.

Richardson, B.C. (1983a) The proteinases of bovine milk and the effect of pasteurization on their activity. *N.Z. J. Dairy Sci. Technol.*, **18**, 233–45.

Richardson, B.C. (1983b) Variation of the concentration of plasmin and plasminogen in bovine milk with lactation. *N.Z. J. Dairy Sci. Technol.*, **18**, 247–52.

Roeper, J. and Winter, G.J. (1982) Viscosity of sodium caseinate solutions at high concentrations. *Proc. XXI Int. Dairy Congr.* (Moscow), Vol. **1**, Book 2, 97–8.

Rollema, H.S. (1992) Casein association and micelle formation, in *Advanced Dairy Chemistry, Volume 1-Proteins*, 2nd edn, (P.F. Fox ed.) Elsevier Applied Science, London, pp. 111–40.

Ronai, K.S. and Weisberg, S.M. (1954) Modified proteins for stabilizing latex paints. *Ind. Eng. Chem.*, **46**, 774–7.

Rund, K.F. (1973) Viscous hydrolysed milk protein in meat processing. *Fleischerei*, **24** (11), 17–9.

Rupley, J.A. and Careri, G. (1991) Protein hydration and function. *Adv. Prot. Chem.*, **41**, 37–172.

Salzberg, H.K. and Georgevits, L.E. (1956) Enhancement of casein for paper coating. *Tappi*, **39**, 656–62.

Salzberg, H.K. and Marino, W.L. (1975) Casein in coating paper and paperboard, in *Protein Binders in Paper and Paperboard Coating*, TAPPI Monograph Series –No. 36, (R Strauss ed.) Technical Association of the Pulp and Paper Industry, Atlanta, Georgia, pp. 2–74.

Salzberg, H.K. and Simonds, M.R. (1965) Low viscosity casein. *United States Patent* 3186918.

Salzberg, H.K., Georgevits, L.E. and Cobb, R.M.K. (1961) Casein in paper coating, in *Synthetic and Protein Adhesives in Paper Coating*, TAPPI Monograph Series – No. 22, (L.H. Silvernail and W.M. Bain eds.) Technical Association of the Pulp and Paper Industry, New York, pp. 103–66.

Salzberg, H.K., Britton R.K. and Bye, C.N. (1974) Casein adhesives, in *Testing of Adhesives*, TAPPI Monograph Series – No. 35, (R.G. Meese ed.) Technical Association of the Pulp and Paper Industry, Atlanta, Georgia, pp. 30–51.

Sergeeva, V.F. and D'Yachenko, P.F. (1973) The physico–chemical properties of the caseinates of the alkali and alkaline earth metals. *Tr. Vses. Nauchno-Issled. Inst. Molochn. Promsti.*, **30**, 61–8.

Shama, F. and Sherman, P. (1973) Identification of stimuli controlling the sensory evaluation of viscosity: oral methods. *J. Texture Stud.*, **4**, 111–8.

Shaw, D.J. (1980) Introduction to colloid and surface chemistry, 3rd edn, Butterworth and Co (Publishers) Ltd, London, p 221–2.

Singh, H. and Creamer, L.K. (1992) Heat stability of milk, in *Advanced Dairy Chemistry, Volume 1 – Proteins*, 2nd edn, (P.F. Fox ed.) Elsevier Applied Science, London, pp. 621–56.

Singh, H. and Newstead, D.F. (1992) Aspects of proteins in milk powder manufacture, in *Advanced Dairy Chemistry, Volume 1 – Proteins*, 2nd edn, (P.F. Fox ed.) Elsevier Applied Science, London, pp. 735–65.

Snoeren, T.H.M., Damman, A.J. and Klok, H.J. (1981) Effect of concentrate viscosity on the properties of skim milk powder. *Zuivelzicht*, **73**, 1004–7.

Snoeren, T.H.M., Damman, A.J. and Klok, H.J. (1982) The viscosity of skim-milk concentrates. *Neth. Milk Dairy J.*, **36**, 305–16.

Snoeren, T.H.M., Damman, A.J. and Klok, H.J. (1983) The viscosity of whole milk concentrate and its effect on the properties of whole milk powder. *Zuivelzicht*, **75**, 847–9.

Snoeren, T.H.M., Brinkhuis, J.A., Damman, A.J. and Klok, H.J. (1984a) Viscosity and age-thickening of skim-milk concentrate. *Neth. Milk Dairy J.*, **38**, 43–53.

Snoeren, T.H.M., Damman, A.J. and Klok, H.J. (1984b) Effect of homogenisation on the viscosity of whole milk concentrate and properties of whole milk powder. *Zuivelzicht*, **76**, 64–6.

Southward, C.R. (1989). Use of casein and caseinates, in *Developments in Dairy Chemistry – 4, Functional Milk Proteins*, (P.F. Fox ed.) Elsevier Applied Science, London, pp. 173–244.

Southward, C.R. (1994) Utilisation of milk components: casein, in *Modern Dairy Technology, Volume 1, Advances in Milk Processing*, 2nd edn, (R.K. Robinson ed.) Chapman & Hall, London, pp. 375–432.

Southward, C.R. and Dolby, R.M. (1968) Seasonal variation in the viscosity index and adhesive strength of casein from the milk of individual cows. *J. Dairy Res.*, **35**, 25–30.

Southward, C.R. and Dolby, R.M. (1971) Seasonal variation in the viscosity index and adhesive strength of casein from the milk of individual cows. II. *J. Dairy Res.*, **38**, 343–51.

Southward, C.R. and Goldman, A. (1975) Co-precipitates – a review. *N.Z. J. Dairy Sci. Technol.*, **10**, 101–12.

Southward, C.R. and Walker, N.J. (1982) Casein, caseinates, and milk protein coprecipitates, in, *CRC Handbook of Processing and Utilization in Agriculture*. (I.A. Wolff ed.) CRC Press, Inc., Boca Raton, Florida, pp. 445–552.

Tanford, C. (1961) *Physical Chemistry of Macromolecules*, John Wiley and Sons Inc., New York.

Tang, Q. (1993) *Rheology of Whey Protein Solutions and Gels*, PhD Thesis, Massey University, Palmerston North, New Zealand.

Tang, Q., Munro, P.A. and McCarthy, O.J. (1993) Rheology of whey protein concentrate solutions as a function of concentration, temperature, pH and salt concentration. *J. Dairy Res.*, **60**, 349–61.

Towler, C. (1974) Rheology of casein solutions. *N.Z. J. Dairy Sci. Technol.*, **9**, 155–60.

Towler, C., Creamer, L.K. and Southward, C.R. (1981) The effect of disulphide-bond reducing agents on the viscosity of casein solutions. *N.Z J. Dairy Sci. Technol.*, **16**, 155–65.

Velez-Ruiz, J.F. and Barbosa-Canovas, G.V. (1998) Rheological properties of concentrated milk as a function of concentration, temperature and storage time. *J. Food Eng.*, **35**, 177–90.

Vujicic, I. and deMan, J.M. (1968) Binding of polyphosphates to casein. *Can. Inst. Food Technol. J.*, **1**, 171–2.

Wayne, J.E.B. and Shoemaker, C.F. (1988) Rheological characterization of commercially processed fluid milks. *J. Texture Stud.*, **19**, 143–52.

Yamashita, S., Creamer, L.K. and Berry, G.P. (1976) The aggregation of bovine caseins in dilute calcium chloride solutions. *N.Z. J. Dairy Sci. Technol.*, **11**, 169–75.

Zamora, M.C. (1995) Relationships between sensory viscosity and apparent viscosity of corn starch pastes *J. Texture Stud.*, **26**, 217–30.

INDEX